Carnivore Behavior, Ecology, and Evolution

VOLUME 2

Carnivore Behavior, Ecology, and Evolution

VOLUME 2

John L. Gittleman, EDITOR

Department of Ecology and Evolutionary Biology
The University of Tennessee, Knoxville

Comstock Publishing Associates, a division of

Cornell University Press | Ithaca and London

First published 1996 by Cornell University Press.

Printed in the United States of America

⊚ The paper in this book meets the minimum requirements of the American National Standard for Information Sciences—Permanence of Paper for Printed Library Materials, ANSI Z39.48-1984.

Library of Congress Cataloging-in-Publication Data

Carnivore behavior, ecology, and evolution.
(Revised for volume 2)
    Bibliography: p.
    Includes index.
    1. Carnivora.   I. Gittleman, John L.
QL737.C2C33   599.74′045   88-47725
ISBN 0-8014-2190-X (alk. paper) (v. 1)
ISBN 0-8014-9525-3 (pbk. : alk. paper) (v. 1)

Contents

Preface

Since publication of the first volume of *Carnivore Behavior, Ecology, and Evolution* in 1989, carnivores have become even more critical as a locus for raising and solving important biological problems. Studies of the African lion and the dwarf mongoose illustrate the power of new genetic techniques of DNA fingerprinting for understanding the evolution of social behavior; differential rates of mortality in the giant panda and other endangered carnivores are now known to influence dispersal and life history patterns that in turn are basic to these species' survival; and reintroduction efforts on behalf of the black-footed ferret and the red wolf are establishing essential guidelines for the preservation and management of endangered species. These are but a few of the examples that could be mentioned to convey how well carnivores can provide the test cases that allow us to ask the right questions in relatively new areas of organismal biology.

Carnivore Behavior, Ecology, and Evolution was the first volume to bring together critical, up-to-date reviews of major areas pertaining to carnivores as a whole. My motivation for editing a second volume arises from the need to evaluate a fresh set of conceptual ideas and empirical data. Although once again ordering its contributions in three main sections—on behavior, ecology, and evolution—this book includes a completely original assemblage of chapters. As before, emphasis is placed on presenting critical reviews of rapidly expanding and developing areas of carnivore biology. The contributions are again devoted largely to terrestrial carnivores (Fissipedia), despite mounting evidence of their close phylogenetic linkage with the aquatic pinnipeds. The rationale for maintaining this taxonomic separation is that comparative (adaptive) trends appear to be more meaningful within the terrestrial lineage, and that other books are available on the Pinnipedia.

What are the important changes in the study of carnivores since the last volume? Without question, the overarching theme is "conservation." As expressed in the general introduction by George Schaller, emphasis is placed on combining theory with empirical data, for this approach provides a solid foundation for setting forth integrated conservation goals and avoiding the solving of problems in piecemeal fashion. Simply put, conservation without good biology is bound to fail. With this in mind, the chapters throughout this book address the need to combine different levels of explanation (both proximal and functional); to use information from a variety of sources (e.g., field, zoo, and molecular biology); to select the most modern and most suitable methods for specific problems and species of study; and to appreciate the political and biological difficulties inherent in studying carnivores.

In essence, the second volume of *Carnivore Behavior, Ecology, and Evolution* is a new, independent book that is intended to complement the original one, not in any way replace it. Carnivores continue to provide crucial test cases in all areas of biology. Most of the chapters presented here are either directly or indirectly tied together by conservation questions against a backdrop of rigorous integrative biology. At a time when tens of thousands of animal and plant species are going extinct every month, it is little wonder that carnivores —so many of them in such grave danger themselves—remain "keystones" for providing solutions to these problems.

JOHN L. GITTLEMAN

Knoxville, Tennessee

Acknowledgments

I am very grateful to the contributors to this volume not only for offering new and exciting ideas in carnivore biology but also for thinking hard about critical issues that mark the future of carnivore work. I think it is safe to say that all of us feel very lucky to be studying such magnificent animals. In return, we hope that some of the ideas presented in this book will help ensure the continued existence of carnivores (and other organisms) throughout the world.

Consistent with the editing of the first volume, all of the chapters were evaluated for content, style, and accuracy by me and at least one outside reviewer. I am most appreciative to the following, who served as external reviewers: Marc Bekoff, Annalisa Berta, Gordon Burghardt, Tim Caro, John Endler, John Flynn, Laurence Frank, Todd Fuller, Robert Hunt, James Malcolm, Stephen O'Brien, Craig Packer, Ed Price, Daniel Promislow, Jeff Thomason, Blaire Van Valkenburgh, Peter Waser, Bob Wayne, and Lars Werdelin. Many other reviewers offered comments and suggestions in response to direct solicitations from the authors; as Editor I much appreciate their assistance as well. I also wish to thank John Eisenberg, David Macdonald, and George Schaller for early encouragement, and for ideas in the selection of topics and contributors.

I am grateful to the University of Tennessee and particularly the (now deceased) Department of Zoology for creating a supportive and vibrant environment to work in; Sandy Echternacht must be singled out in this respect, for without his administrative skills little time would be free to work on projects like this.

Cornell University Press again helped to smooth out the editorial process and assisted at each stage of publication. In particular, I am deeply grateful to Helene Maddux, who insisted the book maintain quality but at the same time meet its deadlines, and to Bill Carver, who has not only my awe for his thorough, rigorous, and creative copyediting but the admiration of all the contributors to this book. I especially thank Robb Reavill, who, beginning with the first volume and throughout the process of assembling the present book, created a constructive and helpful working relationship.

Last, as with all of my work, I dedicate this book to my wife, Karen E. Holt, for the unfailing fascination, love, and sheer fun she has shown in thinking about the ideas expressed here.

J. L. G.

Contributors

MARC BEKOFF, Department of Environmental, Population, and Organismic Biology, University of Colorado, Boulder, Colorado 80309-0334, USA

AUDRONE R. BIKNEVICIUS, Department of Biological Sciences, Ohio University, Athens, Ohio 45701-2979, USA

T. M. CARO, Department of Wildlife, Fish, and Conservation Biology, University of California, Davis, California 95616, USA

TIM W. CLARK, School of Forestry and Environmental Studies, Yale University, New Haven, Connecticut 06511, USA

JULIET CLUTTON-BROCK, Department of Zoology, The Natural History Museum, Cromwell Road, London SW7 5BD, United Kingdom

SCOTT CREEL, Field Research Center for Ecology and Ethology, Rockefeller University, Millbrook, New York 12545, USA

TAMAR DAYAN, Zoology Department, Tel Aviv University, Ramat Aviv, 69978, Tel Aviv, Israel

JOHN J. FLYNN, Department of Geology, Field Museum of Natural History, Chicago, Illinois 60605, USA

LAURENCE G. FRANK, Department of Psychology, University of California, Berkeley, California 94720, USA

WILLIAM L. FRANKLIN, Department of Animal Ecology, 124 Sciences II, Iowa State University, Ames, Iowa 50011, USA

TODD K. FULLER, Department of Forestry and Wildlife Management, University of Massachusetts, Amherst, Massachusetts 01003, USA

JOHN L. GITTLEMAN, Department of Ecology and Evolutionary Biology, The University of Tennessee, Knoxville, Tennessee 37996-1610, USA

MATTHEW E. GOMPPER, Program in Ecology, Evolution, and Conservation Biology, The University of Nevada, 1000 Valley Road, Reno, Nevada 89512-0013, USA

ROBERT M. HUNT, JR., University of Nebraska State Museum, W436 Nebraska Hall, Lincoln, Nebraska 68588-0549, USA

DALE JAMIESON, Department of Philosophy, University of Colorado, Boulder, Colorado 80309, USA

KENNETH G. JOHNSON, Department of Forestry, Wildlife, and Fisheries, The University of Tennessee, Knoxville, Tennessee 37901-1071, and Florida Division of Wildlife, Bureau of Research, 566 Commercial Blvd., Naples, Florida 33942, USA

WARREN E. JOHNSON, Laboratory of Viral Carcinogenesis, National Cancer Institute, Frederick, Maryland 21702-1201, USA

KLAUS-PETER KOEPFLI, Department of Biology, University of California, Los Angeles, California 90024-1606, USA

M. G. L. MILLS, National Parks Board, Kruger National Park, Private Bag X402, Skukuza 1350, South Africa

ALESSIA ORTOLANI, Department of Wildlife, Fish, and Conservation Biology, University of California, Davis, California 95616, USA

RICHARD P. READING, School of Forestry and Environmental Studies, Yale University, New Haven, Connecticut 06511, USA

GEORGE B. SCHALLER, The Wildlife Conservation Society, The International Wildlife Conservation Park, Bronx, New York 10460, USA

ZHONGMIN SHEN, Energy, Environment, and Resources Center and Department of Economics, The University of Tennessee, Knoxville, Tennessee 37996-0550, USA, and Center of Environmental Sciences, Beijing University, Beijing, People's Republic of China

DANIEL SIMBERLOFF, Department of Biological Science, Florida State University, Tallahassee, Florida 32306-2043, USA

BLAIRE VAN VALKENBURGH, Department of Biology, University of California, Los Angeles, California 90024-1606, USA

PETER M. WASER, Department of Biological Sciences, Purdue University, West Lafayette, Indiana 47907-1392, USA

ROBERT K. WAYNE, Department of Biology, University of California, Los Angeles, California 90024-1606, USA

LARS WERDELIN, Department of Palaeozoology, Swedish Museum of Natural History, Box 50007, S-104 05 Stockholm, Sweden

SENSHAN YANG, Department of Plant and Soil Science and Graduate Program in Ecology, The University of Tennessee, Knoxville, Tennessee 37901-1071, USA, and Agronomy Department, Nanjing Agricultural University, Nanjing, Jiangsu, People's Republic of China

YIN YAO, Department of Botany and Graduate Program in Ecology, The University of Tennessee, Knoxville, Tennessee 37996-1610, USA, and Department of Biology, Lanzhou University, Lanzhou, Gansu, People's Republic of China

CHENGXIA YOU, Graduate Program in Ecology, The University of Tennessee, Knoxville, Tennessee, 37996-1610, USA, and Academia Sinica, Kunming Institute of Ecology, Kunming, Yunnan, People's Republic of China

Introduction: Carnivores and Conservation Biology

George B. Schaller

The topics in this book are highly diverse, but whether they involve molecular biology, taxonomy, or various aspects of behavior or ecology, most touch on conservation. This underlying theme is both timely and important. At the close of this century, one that has seen far more environmental destruction than any other in historic times, the future of some of our most magnificent carnivores is in doubt. Humankind has always been fascinated by carnivores, has always felt an emotional response to them, a response of exaltation or fear, delight or loathing. We keep ungulates mainly as servants and sustenance but have taken the dog and cat into our homes. We have made natural icons of the wolf and tiger, symbols of wildness and wilderness whose survival has become an act of conscience for us. On reading the titles of the chapters in this book, I was reminded of many conservation issues concerning ethics and aesthetics and resource management. I comment here on only a few. These and others every biologist needs to ponder if he or she is to retain a moral footing in a world of conservation conflicts that will become ever more frequent and strident.

When I sit by the coals of a campfire, listening to the whoop of a spotted hyena, or lie in my sleeping bag during the cold hours before dawn, knowing that somewhere nearby a snow leopard glides among the crags, my mind sometimes drifts back to thoughts of the Pleistocene and its extinct carnivores. I wonder how the lithe hunting hyena (*Euryboas lunensis*), the lion-sized giant cheetah (*Acinonyx pardinensis*), the cave bear (*Ursus spelaeus*), the spectacular array of smilodont or sabertoothed cats, and many others, all now vanished, once fit into the ecological scene. How electrifying it would be to step back in time to study their social systems, their predatory habits, their interactions with those carnivores and other animals that still persist. Today's assemblage of terrestrial carnivores is greatly impoverished when compared with that of the Pleistocene. Yet in spite of all our studies, we still know little about most of the existing species.

1

As the millennium ends we are fortunate that somehow most carnivores have survived the recent centuries of persecution and habitat destruction; they have been remarkably adaptable and resilient. The geographic ranges of many have contracted and some subspecies have been exterminated, but most have endured. The sea mink (*Mustela marcodon*) and Falkland Island fox (*Dusicyon australis*) are among the few whose place on earth is now vacant. What of the next millennium?

Wozencraft (1989) recognized 236 carnivore species, of which only a modest number have been studied in detail. There are 70 mongoose and civet species, most of them small, solitary, nocturnal, and devoted to tropical thickets, all characteristics that make research difficult (but see Rabinowitz 1991); by contrast, the social and diurnal mongooses—suricate (*Suricata suricatta*), banded mongoose (*Mungos mungo*), dwarf mongoose (*Helogale parvula*)— have been the focus of detailed research. Among the 63 mustelids, the north-temperate ones such as the sea otter (*Enhydra lutris*), Eurasian badger (*Meles meles*), American pine marten (*Martes americana*), and American mink (*Mustela vison*) have a voluminous literature, but, as with the civets and mongooses, the tropical mustelids have been largely ignored. Similarly, there has as yet been little work on the 18 procyonid species, with the notable exception of raccoons (*Procyon lotor*) and coatis (*Nasua nasua*).

The cats, comprising 37 species, include the lion (*Panthera leo*), tiger (*Panthera tigris*), snow leopard (*Panthera (Uncia) uncia*), puma (*Felis concolor*), and other large animals whose beautiful and powerful images stir our emotions. They offer a combination of aesthetic and intellectual experience, surely one reason for their allure to researchers. With the exception of the clouded leopard (*Neofelis nebulosa*), all of the large species have been the focus of at least one major project, with some projects, such as those on lions in Tanzania's Serengeti National Park and tigers in Nepal's Chitwan National Park, persisting for decades. The North American lynx (*Lynx canadensis*) and bobcat (*Lynx rufus*) have also received thorough studies, but most of the other small cats have found no one to chronicle their lives. Yet work on ocelots (*Felis (Leopardus) pardalis*: Emmons 1988) has shown that with persistence even small, solitary cats can yield valuable data. The three hyena species (spotted hyena, *Crocuta crocuta*; brown hyena, *Hyaena brunnea*; striped hyena, *Hyaena hyaena*) and the related aardwolf (*Proteles cristatus*) are relatively well known, research on the spotted and brown hyenas having been particularly intensive.

Among the 34 canid species, such north-temperate ones as the red fox (*Vulpes vulpes*) and gray wolf (*Canis lupus*) have their dedicated scientific devotees, as can be expected, but species in other parts of the world—golden jackal (*Canis aureus*), Ethiopian wolf (*Canis simensis*), African wild dog (*Lycaon pictus*), dhole (*Cuon alpinus*), maned wolf (*Chrysocyon brachyurus*)— have also received study; still, most foxes of the genera *Dusicyon* and *Vulpes* (18 species) have not (see Ginsberg and Macdonald 1990). Of the seven bear species, the American black bear (*Ursus americanus*), brown bear (*Ursus arctos*), and polar bear (*Ursus* (*Thalarctos*) *maritimus*) have had the benefit of much detailed research, three species have had only modest attention, and the habits of one, the Malayan sun bear (*Helarctos malayanus*), remain almost unknown. If popular attention were a measure of scientific knowledge, the giant panda (*Ailuropoda melanoleuca*) would be among the best-known mammals, but only some basic information has so far been collected on this animal, whose striking black-and-white image is widely recognized as a symbol of the world's conservation effort. Field studies on red pandas (*Ailurus fulgens*) are also in their initial stages (see Reid et al. 1991).

This brief overview shows, not surprisingly, that much research has focused on large, often charismatic species, especially those that occupy fairly open habitats. Indeed, the diversity of carnivores in steppes, savannas, and woodlands may be as great as, or greater than, that in tropical rainforest (see Redford et al. 1990). Animals with valuable pelts, or those in demand by sports hunters, have also been the focus of many studies in North America, Europe, and Russia, parts of the world where the biomass of biologists and funds for research have also been greatest. If a species is known to be endangered—because of persecution, fragmentation of habitat, restricted distribution, or other cause—it may have the good fortune of attracting attention and receiving study, as happened with the last 500 Ethiopian wolves (Gottelli and Sillero-Zubiri 1992) and 1000 giant pandas (Schaller et al. 1985).

Yet fewer than 15% of the terrestrial carnivore species have received so much as one intensive study. Even the status of most species remains obscure. Long-term research on particular populations, well exemplified by the work on Isle Royale wolves (Allen 1993), have been all too rare. A species may adapt its social system to different conditions of prey abundance and distribution (Macdonald 1979; Moehlman 1987; Crawshaw and Quigley 1991), and to determine the limits of such plasticity requires projects in various localities, especially if the animal occupies different habitats over a wide geographic range, as, for instance, do puma, brown bear, and tiger. The fates of the large carnivores will ultimately depend on our ability to manage them effectively on a long-term basis in increasingly restricted areas. The genetic problems created by inbreeding within small populations are being documented. But there has been little work on the confrontation and conflict that inevitably arise when predators such as tigers come in contact with humans and their livestock. We know how to protect large carnivores, but not how to achieve a measure of peaceful coexistence with them.

No terrestrial carnivore species is currently threatened with imminent extinction, but even seemingly secure species can be affected suddenly by unforeseen events, such as the fashion craze for spotted cat skins in the 1960s. Between 1962 and 1967 the Peruvian Amazon alone exported at least 5345 jaguar pelts and 16,499 ocelot pelts (Robinson and Redford 1991). In 1989, the *Exxon Valdez* oil spill in Alaska's Prince William Sound killed between

3500 and 5500 sea otters, nearly half the population; some 250 oiled sea otters were saved, at a cost of about $90,000 apiece (Strohmeyer 1993). To cite an example of a disaster to a pinniped carnivore, in 1988 a viral disease related to canine distemper killed at least 18,000 harbor seals (*Phoca vitulina*) in the North Sea. The virus attacked animals whose immune system had apparently been suppressed by PCBs and other toxic chemicals in the water.

The status of African wild dogs has within a few years changed from tenuous security to endangered, again partly as a result of disease. Several packs vanished from the Serengeti ecosystem, and there was a suspicious correlation between packs that had been anesthetized, vaccinated, and radio-collared and those that disappeared (Burrows et al. 1994). Burrows (1992) hypothesized that the animals were so stressed by being handled that they produced an adrenocorticosteroid (cortisol) that suppressed the immune system, which in turn activated latent rabies viruses. Other factors may be involved as well, given that population's history of disease-induced declines even when they had not been handled (Creel 1992). Whatever the reason, these declines not only point to a lack of knowledge concerning the role of metabolic stress and disease in regulating carnivore populations, but also raise a basic issue in research: do invasive field techniques, including vaccination, sometimes cause such stress or other problems that the animal's health is affected?

The ethical ramifications of field research on the welfare of animals are receiving belated attention (Farnsworth and Rosovsky 1993). Science is not value-free, not without moral obligations, not to be used merely to amass information. According to Bekoff et al. (1992:474), "Not only must individual scientists develop a coherent and principled ethic concerning animal use, but journals and professional societies must strive to articulate an operational and practical consensus ethic."

What stance, then, should a biologist adopt toward hunting for pleasure, a multibillion-dollar industry that in North America indirectly funds much wildlife research? Indigenous peoples once performed rituals to foster successful hunts and afterward appease the animal's disturbed spirit. The modern hunter's agenda is often seen as beer, blood, and brag. But a more benign "nonconsumptive" method has in some areas been adopted for the puma: hunters track and chase the cat with dogs until it seeks the safety of a tree, but it is not killed. Yet, as has been suggested for African wild dogs, the stress of being chased affects cortisol levels. After repeated hunts, the puma almost loses its ability to produce this steroid, and in extreme cases its metabolic balance can be so seriously upset as to cause death (Gates 1993).

The term "sustainable use" has become a hopeful mantra of the conservation community, and it has been enthusiastically adopted by most governments, which often view the concept as synonymous with exploitation. In certain limited circumstances, as with the northern fur seals (*Callorhinus ursinus*) on the Pribilof Islands, the concept has been successfully applied, but, as Robinson (1993) noted, there is a fundamental contradiction between resource potential and human needs and desires: sustainable use, as currently applied, will inevitably lead to loss of biodiversity.

Mongolia briefly permitted snow leopard hunts at $13,000 per cat, and Venezuela is considering jaguar hunts for at least $10,000 per cat, with a focus on those that prey on livestock. Both species are in Appendix I of CITES (Convention on International Trade in Endangered Species). "Well managed, the jaguar [*Panthera onca*] can be turned into a highly valuable resource by the government and by the people at the local level," wrote two American wildlife managers (quoted in Lewis 1992:64). Jaguar population size and production in Venezuela remain unknown; all jaguars, not just the livestock killers, will become vulnerable; and local subsistence hunting will have to be curtailed to protect both the jaguar and its prey. Besides, few of the financial benefits derived from the indulgence of foreign hunters tend to filter down to local people. Conservation and management must first of all be based on solid research, and then on careful long-term monitoring, if sustainable use is to be truly sustainable.

There is, of course, no imperative to kill jaguar for pleasure. Carnivores such as the jaguar may be more valuable alive than dead. Resource economics provides a means of calculating the potential value of a species. For instance, the restoration of gray wolves to the Greater Yellowstone ecosystem could bring a net profit of $18 million to the local economy the first year, even if the wolves kill some livestock, mainly through increased tourism (Brandt 1993): the howl of a wolf impinges uniquely on our spirit. The release of 14 wolves into Yellowstone in March 1995 has become a test of how much we truly value the species. Venezuela would not reap such benefits from protecting jaguars, and there may even be some economic loss. But to me it seems condescending to suggest to other countries that the only value of a species is the one on its price tag. National pride, concern for natural and cultural heritage, ethics, and aesthetics all are part of a country's value system too.

One winter during the early 1950s, while I was at the University of Alaska, government predator-control agents killed about 250 wolves on Alaska's North Slope, for no apparent reason. The public took little notice. Forty years later, wolf management remained just as primitive. In November 1992, Alaska's Department of Fish and Game authorized the killing by aerial shooting of 300–400 wolves the first year and 100–300 each year thereafter in an area of about 48,500 km². The reason: to increase moose (*Alces alces*) and caribou (*Rangifer tarandus*) populations for the benefit of hunters—even though ungulates were at high densities and the impact of predation by black bears was as great as or greater than predation by wolves (Steinhart 1993). This time, the proposal created a national uproar. Walter Hickel, then Alaska's governor, stated his credo on national television: "You can't just let nature run wild." The wolf hunt was delayed, but only temporarily, and in one guise or another, a wolf pogrom continues in Alaska as well as in parts of Canada. Of concern to the wildlife profession, in matters like these, are those biologists who jeopardize our scientific credibility by misapplying or ignoring their data for the sake of political and economic expediency. Peter De Fazio, Congressman from Oregon, termed it "voodoo biology." Renée Askins said at a wolf conference (Williams 1993:50): "I'm angry that my own profession of wildlife biology has reached the point that it sees itself beyond ethical reproach, as though science is exempt from moral scrutiny or accountability." The distinction between science and advocacy has been too rigid. We must all proclaim a personal commitment to the carnivores, must all become visionaries who will not casually yield beliefs. As Dave Foreman said (Letto 1992:28), "It is not our job to make the final compromise; our job is to be advocates for the natural world—we speak for wolf."

It is clear that the basic problems in conservation biology are economic, social, and political. This being so, every biologist has an obligation and responsibility to help protect and preserve the species he or she studies by going beyond the collecting of data to finding solutions to problems at the local level, as well as nationally and internationally at the policy level. Fortunately, more and more biologists have entered the fray to alter the ethos of greed and indifference that threatens even species that were considered to be reasonably safe.

Take the case of the tiger. There are probably fewer than 7700 remaining in the world; three of eight subspecies have vanished since the 1940s, and three others are seriously endangered (Jackson 1993). Many Asian peoples believe that tiger parts have medicinal value: eyeballs help cure epilepsy and malaria, blood builds up the constitution, flesh is a tonic for the stomach, bones are good for ulcers, rheumatism, typhoid, and other maladies. An upsurge in the tiger-bone trade, especially to China, is decimating the species from India to Russia. Since bones from other large cats are also acceptable in medicinal products, the trade has potential impact on all such species. China established a tiger-breeding facility whose main purpose was to sell tiger parts on the international market. It was argued that legal exploitation of tiger products would decrease the illegal market (when in fact the converse would most likely happen), and that sustainable use of this endangered species would ultimately assist conservation. With bones selling for $100–$300/kg, tigers had become a lucrative commodity. Using the Pelly Amendment, a part of the Fishermen's Protective Act of 1967 that allows the United States to invoke trade sanctions against any country that undermines an international effort to save endangered species, the United States declared its intention to impose such sanctions on China and Taiwan, hoping to reduce the trade in tiger parts (and rhinoceros horn). Both China and Taiwan then announced a ban on the use of these products; nevertheless, an import embargo on wildlife products was imposed on Taiwan by the United States in July 1994 and maintained until 1995. Is farming the endangered tiger for its bones an acceptable form of sustainable use? Does it promote conservation of the species?

A banquet with sliced bear paw as the ultimate delicacy may cost $1000 in

Tokyo. But the real value of bears in Asia lies in their gall bladders. Bile is used to cure ulcers, fevers, and other illnesses. A gram sells for about 18 times the price of gold. Bears are killed throughout Asia and now even in North America for this valuable commodity. As Mills and Servheen (1990:ix) noted, in Asia "a bear in the wild, lacking any intrinsic ideological value, is worth nothing." More than an estimated 8000 Asiatic black bears (*Ursus thibetanus*) and brown bears are kept in captivity in China for bile production. With a tube inserted surgically in its gall bladder, a bear is confined in a tight cage while its bile is drawn off, drop by drop. Private interests in North America are currently trying to gain control of wildlife from government agencies, to convert public property into private gain. They want to establish "game ranches" to sell animals for meat, medicine, and sport; hunters want the right to sell parts or all of the animals they have killed, such as the gall bladders of bears; landowners want complete control over wildlife on their property. Geist (1988) has argued strongly that wildlife management in North America was able to develop as a fairly effective system only after the government took over control of wildlife early this century and curtailed traffic in it. Should wildlife depend on market forces for survival?

Carnivores have, of course, long been economically valuable. Millions of pelts from fur farms and such wild caught animals as arctic (*Alopex* (*Canis*) *lagopus*), red, and *Dusicyon* foxes reach the market annually. Killed as pests or for products, the species have so far survived. But when the value of an individual rises to excessive heights because of rarity or demand, the future of the species becomes bleak. Sea otter pelts had a value of $2000–$2500 each during the fur boom of the 1920s, and the species barely survived. Skins of giant pandas may sell for $10,000 or more, a huge windfall for both poacher and middleman. Between 1985 and 1991, Chinese courts handled 123 cases of panda poaching and pelt smuggling. Three offenders were executed and 275 others went to prison. But the killing continues, fueled by the wealth of Hong Kong, Taiwan, and Japan, where panda skins are status symbols. For nearly a decade, starting in the mid-1980s, China rented pandas for short-term exhibit to the world's zoos. Millions were earned both by the zoos and by China, but little improvement was forthcoming in either captive breeding or protection of pandas in the wild. This embarrassing interlude of mismanagement and greed could not have persisted as long as it did without the active apathy of many who merely observed this casual waste of animals. What the interlude exhibited was a failure of collective vision.

Captive-breeding programs of seriously endangered species are necessary hedges against disasters. They provide basic biological knowledge and produce animals for restoration to their native habitat, if such survives. The black-footed ferret (*Mustela nigripes*) and red wolf (*Canis rufus*) would be extinct without such human intervention. Zoos—perhaps better called "wildlife conservation parks" to reflect the role they must now assume—currently face dilemmas that also are of concern to field biologists. During the next century, human population pressure and the increasing rate of habitat destruction will see a flood of animals seeking the safety of a zoo ark. At present, zoos contribute significantly to the conservation of only about 20 full species (Rahbek 1993). Given the constraints of space and money (Conway 1989), zoos will have to make tough choices and cooperate more fully in decisions about what to maintain and what to discard and lose.

There are currently over 650 Siberian tigers (*Panthera tigris altaica*) in the world's zoos. Is it justified to keep so many of that one subspecies when the Chinese subspecies is in urgent need of help? The acquisition of animals to diversify a captive gene pool inevitably becomes an issue when the last wild survivors are removed from their natural habitat. Should we worry about preserving the genetic sanctity of subspecies? It can be argued that conservation strategies ought to focus on full species, with the aim of preserving the whole spectrum of genetic variability, as determined by molecular analyses of populations and subspecies. White tigers, a rare natural anomaly that is popular with the public, have been so inbred by zoos that genetic problems have erupted. Should the guiding criterion in maintaining such animals be zoo-gate receipts or an animal's welfare? And how do you dispose of genetic defectives, unwanted hybrids, and other surplus animals?

Captive breeding may become a troublesome distraction, if money and energy are devoted to that rather than to the basic goal, that of protecting the species in nature. The giant panda has suffered from such a misplaced focus. Zoos are expensive to build and maintain—the total operating and capital-improvement budgets of the 159 accredited zoos in North America were $1.23 billion in 1991 (Davis 1993)—and the funds for captive breeding might in some instances be better spent on habitat protection. It costs 50 times as much to keep black rhinoceroses (*Diceros bicornis*) and African elephants (*Loxodonta africana*) in zoos as it does to protect the same numbers of these animals in the wild (Leader-Williams 1990). Sadly, however, funds designated for zoos by municipalities are not available for conservation.

Captive breeding can help a few species, but for many it is not a panacea. Captive breeding for what? Perhaps as a remembrance of a better past, and for ethical, scientific, and educational reasons. Above all, it is a way of buying time until, one hopes, a species can be reintroduced back into the wild. The limited success achieved with such species as the swift fox (*Vulpes velox*), Eurasian lynx (*Lynx lynx*), and black-footed ferret, however, has obscured the fact that perhaps only 10% of attempted reintroductions succeed, and that all are extremely expensive. (Restoration by translocation of wild animals has been a more successful option.) Some European farmers have objected to the restoration of lynx, claiming that the cats kill livestock; Asian farmers are even less likely to welcome tigers back. If a country wants free-living large carnivores, its first priority must be to keep those it still has.

As wilderness vanishes, it is difficult to be sanguine about saving the snow leopard and brown bear and maned wolf. Our responsibility is not just to gather knowledge but also to serve a larger cause. We must fight for what remains, and try to restore what has been squandered. We must be mediators between our science and our culture, conveying with eloquence, wisdom, passion, clarity, and fact that to save carnivores and their environment is as important to our future as it is to theirs. The words of Lopez (1978:226) about wolves apply to many other carnivores as well: "Throughout history, man has externalized his bestial nature, finding a scapegoat upon which he could heap his sins and whose sacrificial death would be atonement. He has put his sins of greed, lust, and deception on the wolf and put the wolf to death—in literature, in folklore, and in real life." We cannot ease the burdens of the past, but we can atone by assuring the carnivores a future.

References

Allen, D. 1993. *Wolves of Minong*. Ann Arbor: Univ. Michigan Press.

Bekoff, M., Gruen, L., Townsend, S., and Rollin, B. 1992. Animals in science: Some areas revisited. *Anim. Behav.* 44:473–484.

Brandt, E. 1993. How much is a gray wolf worth? *Natl. Wildlife* 31(4):4–12.

Burrows, R. 1992. Rabies in wild dogs. *Nature* 359:277.

Burrows, R., Hofer, H., and East, M. 1994. Demography, extinction and intervention in a small population: The case of the Serengeti wild dogs. *Proc. Roy. Soc. London B.* 256:281–292.

Conway, W. 1989. The prospects for sustaining species and their evolution. In: D. Western and M. Pearl, eds. *Conservation for the Twenty-First Century*, pp. 199–209. Oxford: Oxford Univ. Press.

Crawshaw, P., and Quigley, H. 1991. Jaguar spacing, activity and habitat use in a seasonally flooded environment in Brazil. *J. Zool. (Lond.)* 223:357–370.

Creel, S. 1992. Cause of wild dog deaths. *Nature* 360:633.

Davis, M. 1993. Do zoos compete with habitat? *Wild Earth* 3(1):48–51.

Emmons, L. 1988. A field study of ocelots (*Felis pardalis*) in Peru. *Rev. Ecol.* 43:133–157.

Farnsworth, E., and Rosovsky, J. 1993. The ethics of ecological field experimentation. *Conserv. Biol.* 7:463–472.

Gates, P. 1993. Stressful findings. *BBC Wildlife* 11(1):33.

Geist, V. 1988. How markets in wildlife meat and parts, and the sale of hunting privileges, jeopardize wildlife conservation. *Conserv. Biol.* 2(1):15–26.

Ginsberg, J., and Macdonald, D. 1990. *Foxes, Wolves, Jackals, and Dogs: An Action Plan for the Conservation of Canids*. IUCN/SSC canid specialist group. Gland, Switzerland: IUCN.

Gottelli, D., and Sillero-Zubiri, C. 1992. The Ethiopian wolf—an endangered endemic canid. *Oryx* 26:205–215.

Jackson, P., ed. 1993. *Cat News* 19. IUCN/SSC cat specialist group. Gland, Switzerland: IUCN.

Leader-Williams, N. 1990. Black rhinos and African elephants: Lessons for conservation funding. *Oryx* 24:23–29.

Letto, J. 1992. One hundred years of compromise. *Buzzworm* 4(2):26–32.

Lewis, D. 1992. A job for the jaguar. *BBC Wildlife* 10(11):64–65.

Lopez, B. 1978. *Of Wolves and Men*. New York: Charles Scribner's.

Macdonald, D. 1979. The flexible social system of the golden jackal, *Canis aureus*. *Behav. Ecol. Sociobiol.* 5:17–38.

Mills, J., and Servheen, C. 1990. *The Asian Trade in Bears and Bear Parts*. Washington, D.C.: World Wildlife Fund and Traffic USA.

10   *George B. Schaller*

Moehlman, P. 1987. Social organization in jackals. *Amer. Scientist* 75:366–375.

Rabinowitz, A. 1991. Behaviour and movements of sympatric civet species in Huai Kha Khaeng Wildlife Sanctuary, Thailand. *J. Zool. (Lond.)* 223:281–298.

Rahbek, C. 1993. Captive breeding—a useful tool in the preservation of biodiversity? *Biodiversity and Conservation* 2:426–437.

Redford, K., Taber, A., and Simonetti, J. 1990. There is more to biodiversity than the tropical rain forest. *Conserv. Biol.* 4:328–330.

Reid, D., Hu, J., and Huang, Y. 1991. Ecology of the red panda *Ailurus fulgens* in the Wolong Reserve, China. *J. Zool. (Lond.)* 225:347–364.

Robinson, J. 1993. The limits to caring: Sustainable living and the loss of biodiversity. *Conserv. Biol.* 7:20–28.

Robinson, J., and Redford, K. 1991. *Neotropical Wildlife Use and Conservation.* Chicago: Univ. Chicago Press.

Schaller, G., Hu, J., Pan, W., and Zhu, J. 1985. *The Giant Pandas of Wolong.* Chicago: Univ. Chicago Press.

Steinhart, P. 1993. A new aerial wolf war? *Defenders* 68(1):30–36.

Strohmeyer, J. 1993. *Extreme Conditions.* New York: Simon and Schuster.

Williams, T. 1993. Alaska's war on the wolves. *Audubon* 95:44–50.

Wozencraft, W. 1989. Appendix: Classification of the Recent carnivora. In: J. L. Gittleman, ed. *Carnivore Behavior, Ecology, and Evolution,* pp. 569–593. Ithaca, N.Y.: Cornell Univ. Press.

PART I

BEHAVIOR

INTRODUCTION

Textbooks in animal behavior, behavioral ecology, and ethology (e.g., Krebs and Davies 1993; Slater and Halliday 1994) show three recent developments in behavioral study that relate centrally to carnivores: ethical considerations, as prompted by changes in an animal's behavior, that every investigator should adopt prior to carrying out research; proximal mechanisms that influence population and species differences in behavior patterns; and new methods (e.g., endocrine analyses and statistical comparative methods) for getting at previously intractable problems. The chapters in this section well represent how each of these issues necessarily involves carnivores.

Just as carnivores continue to play "keystone" roles in ecology, the ethics of studying animals are now being shaped by issues confronted in the study of carnivores. Generally, this is unsurprising, because even though carnivores are charismatic in their complex social structure, arresting coloration, or menacing morphology, their dangerous and often cryptic nature makes observing them very difficult. This difficulty, unfortunately, may encourage more aggressive, almost cavalier, approaches to working on carnivores. After introducing a general perspective that ethical considerations are inherent in *any* animal study, Bekoff and Jamieson outline a series of questions related to, among other things, why carnivores should be studied, the motivations for developing various approaches to data collection and analysis, the severity of the problems encountered in manipulating individual animals, and whether the information gained in a study is "worth" the ethical and perhaps physical dangers that may be imposed on the animals. Some of these issues may seem too philosophical or vague, but that view clearly reflects more on the naive stage we have reached in understanding such issues than it does on the importance of the subject. Ideally, the remaining chapters of the book should be viewed in terms of many of these ethical questions.

The chapters by Creel and Frank underscore the importance of proximal mechanisms for reaching a more complete understanding of the evolution of carnivore social structures. Creel, using detailed study of the dwarf mongoose (*Helogale parvula*) as a test case, shows that communal breeding in carnivores is directly related to the physiology of estrus cycling in dominant and subordinate females. The technology of analyzing endocrines in the urine or feces opens up a new line of inquiry into the factors significantly affecting reproductive suppression in communal carnivores, and also the reproductive mechanisms that may be involved in breeding captive endangered-carnivore species. The theme of proximal (endocrine) factors extends to the chapter by Frank, which critically reviews various aspects of the long-term field study of the spotted hyena (*Crocuta crocuta*). Specifically, he shows that the female "masculinization" of an erect phallus, which is critical to establishing the female-based social system of many carnivores, is influenced by hormonal androgens. As Frank points out, Harrison Matthews (1939) long ago suggested that the spotted hyena is endocrinologically an unusual species, and that hormonal effects on such features as aggression, dominance, ontogeny, neonatal mortality, and sex ratio could be profitably studied in this species. The chapters by both Creel and Frank constitute instructive guidelines for synthesizing field and laboratory approaches in carnivore research.

The remaining chapters in this section, though dealing with completely different problems, are pulled together by a common thread: coloration in carnivores and sympatry in canids are both characteristics that everyone seems to know something about and to offer explanatory hypotheses for, but few empirical tests have actually been carried out. Dating back to classic studies in evolutionary biology (e.g., Wallace 1889; Poulton 1890), variation in color patterns across carnivores, such as the black-and-white markings of the striped skunk (*Mephitis mephitis*) or giant panda (*Ailuropoda melanoleuca*), were accepted as clear examples of warning coloration, and similar functional patterns are given in modern treatments of the adaptive nature of animal coloration. Nevertheless, as Ortolani and Caro effectively argue, little systematic comparative analysis has tested any such hypothesis. Using an original comparative data base combined with modern phylogenetic analysis, the authors identify various behavioral and ecological correlates of color differences. For example, pale coloration is associated with living in desert habitats, spotted coats correlate with arboreal living, and tail tips are found in grassland dwellers. This comparative analysis is a major advancement in testing explanations for carnivore color patterns against a solid empirical approach and, more important, showing what functional trends do not seem to be supported.

Just as with coloration, considerable speculation surrounds the ecology and population structure of sympatric canids (Rosenzweig 1966). Johnson, Fuller, and Franklin review various hypotheses on dietary competition and morphological separation, specifically citing well-studied guilds of canids living in selected geographical areas. Included in the analysis is a highly useful classification scheme that should be used in general biogeographical studies as well as in specific hypothesis testing of character displacement. The later chapter by Dayan and Simberloff is also instructive in this regard, and a larger point conveyed in this chapter is that, even though canids are the most intensively studied and best-known group of carnivores, some of the fundamental characteristics they present severely impede our understanding of their basic biology.

Embedded in the chapters of this part is another general theme, one that goes back to the overarching problem of "conservation." In order to gather useful data, in either a laboratory or a field setting, it is necessary to collect longitudinal information. With carnivores, this requirement presents a serious constraint. The limited budgets of the ever-shrinking funding agencies are making it more difficult to raise the funds needed to gather the kind of data necessary for carefully examining biological problems in carnivores. And, even more worrisome, funding sources are increasingly reluctant to support projects on species that are nocturnal, difficult to find and watch, impossible to observe in relatively large sample sizes, and so on. As many of the chapters convincingly show, merging different perspectives—proximate and ultimate, laboratory and field, single species and comparative—is a useful means for compensating some of the difficulties encountered in behavioral studies of carnivores.

John L. Gittleman

References

Krebs, J. R., and Davies, N. B. 1993. *An Introduction to Behavioural Ecology.* Oxford: Blackwells.

Matthews, H. 1939. Reproduction of the spotted hyaena (*Crocuta crocuta* Erxleben). *Phil. Trans. Soc. B.* 230:1–78.

Poulton, E. B. 1890. *The Colours of Animals.* London: International Science Series, 68.

Rosenzweig, M. L. 1966. Community structure in sympatric Carnivora. *J. Mamm.* 47:607–612.

Slater, P. J. B., and Halliday, T. R. eds. 1994. *Behaviour and Evolution.* Cambridge: Cambridge Univ. Press.

Wallace, A. R. 1889. *Darwinism.* London: Macmillan.

CHAPTER 1

Ethics and the Study of Carnivores: Doing Science While Respecting Animals

MARC BEKOFF AND DALE JAMIESON

The human relationship to nature is a deeply ambiguous one. Human animals are both a part of nature and distinct from it. They are part of nature in the sense that, like other forms of life, they were brought into existence by natural processes, and, like other forms of life, they are dependent on their environment for survival and success. Yet humans are also reflective animals with sophisticated cultural systems. Because of their immense power and their ability to wield it intentionally, humans have duties and responsibilities that other animals do not (Bekoff and Jamieson 1991).

One striking feature of humans is that they are curious animals, and this curiosity has produced a wealth of knowledge about humans, nonhuman animals (hereafter "animals"), other forms of life, and the abiotic environment. Along with the acquisition of knowledge *about* nature come numerous intrusions *into* nature, even from activities such as photographing wildlife that seem to be harmless (Duffus and Wipond 1992). As Cuthill (1991:1008) has observed, "We have to tamper with nature to understand it." Though it sometimes appears that an examination of some part of nature is not harmful, often what seemed initially to be a minor intrusion turns out to have serious consequences for what has been affected (Caine 1992). Sometimes human intrusions even make it difficult to gain the knowledge that we seek, or to attain well-meaning goals (as may be the case with wild pandas: Bertram 1993; Schaller 1993).

In this chapter we focus primarily on ethical questions that arise in studies of behavior and behavioral ecology in wild carnivores under field conditions (see also Cuthill 1991; Kirkwood 1992). There are many important questions that we will not address, including those involved in the physiological analyses of carnivore behavior under field conditions (e.g., hibernation). Furthermore, we say little about domestic dogs and cats, even though they are carnivores and are used in many research projects, including some in behavioral ecology (for references, see Beck 1973; Daniels and Bekoff 1989a, b, c; Bradshaw 1992;

Thorne 1992; Orlans 1993, chap. 13; Clutton-Brock, this volume). A 1988 report from the U.S. Department of Agriculture states that 140,471 dogs and 42,271 cats were used in various sorts of experimentation in that year alone (cited in Singer 1990:37), and Kew (1991:160) reports data indicating that about 10,000 dogs and 5,000 cats are killed annually in scientific research in the United Kingdom. Domestic dogs and cats are also used for various nonscientific purposes that involve breeding for human-desired traits, many of which are injurious to the animals themselves (Daniels and Bekoff 1990).

A great many problems, both methodological and ethical, are unique to field studies of particular carnivores or other species. Many of the issues we are concerned with, however, are not restricted to the study of wild carnivores, and our essay should also inform discussions that center on other taxa, perhaps even insects (Wigglesworth 1980; Eisemann et al. 1984; Fiorito 1986; Lockwood 1987), and on various human activities that involve using captive animals in zoos and wildlife parks (Jamieson 1985a, 1995; Peterson 1989; Chiszar et al. 1990; Kiley-Worthington 1990; Bostock 1993) and in research laboratories.

Divergent Views on the Role of Ethics

Although a consideration of ethical questions that centers on how humans use animals in science could be viewed as anti-science, it is rather in the best traditions of science. For it involves applying to science itself the scientific spirit of skepticism, rationality, and a demand for evidence. Indeed, consideration of the ethical issues meriting serious attention in field studies of carnivore behavior may make for better science; to this end, Roger Ewbank (1993), editor-in-chief of the journal *Animal Welfare,* has recently called for papers that specifically address these problems.

There is a strong trend for more, rather than less, concern for the ethical issues that arise from animal use in a wide variety of contexts (Huntingford 1984; Rowan 1984, 1988; Dickinson 1988; Rolston 1988; Hettinger 1989; Goodall 1990; Rachels 1990; Bekoff and Jamieson 1990, 1991; Cuthill 1991; Elwood 1991; Bekoff et al. 1992; Hargrove 1992; Bekoff and Gruen 1993; Benton 1993; Broom and Johnson 1993; Cavalieri and Singer 1993; Farnsworth and Rosovsky 1993; Guillermo 1993; Orlans 1993; Peterson and Goodall 1993; Preece and Chamberlain 1993; Quinn 1993; Singer 1993a; Zimmerman et al. 1993; Wilkie 1993; Bekoff 1994a; for historical accounts, see Carson 1972 and Ryder 1989). And the trend is seen among people with diverse backgrounds and interests (Plous 1991; Galvin and Herzog 1992; McAdam 1992). Those interested in animal welfare comprise a group that is as heterogeneous as many professional societies (Bekoff et al. 1992). Scientists such as Richard Dawkins (1993) have called for legal rights for chimpanzees, gorillas, and orang-utans; even Desmond Morris (1990) has become a spokesperson for animal welfare, and presents what he calls a "Bill of Rights for animals" (pp. 168–169). Top-level administrators in the United States such as President's Clinton's science advisor, John Gibbons, have indicated the need for ethical reflection on the use of animals in research. Nowadays many professional societies have guidelines, which are constantly updated, to which authors must adhere if they are to publish in the journals sponsored by these societies (American Society of Mammalogists 1987; Anonymous 1992; Stamp Dawkins and Gosling 1992; see also Rollin 1989; Bekoff et al. 1992; and Bekoff 1993a, for lists). Indeed, the Association for the Study of Animal Behaviour (United Kingdom) and the Animal Behavior Society (United States) jointly devoted a special issue of the journal *Animal Behaviour* to ethics in research on animal behavior.

Despite the great interest in animal welfare, there are divergent views concerning what is justifiable with respect to the treatment of animals in research and in other activities. Moreover, it is clear that people have diverse opinions about the moral acceptability of various of these activities. For example, in a recent debate about experimentally induced infanticide in birds (Bekoff 1993a; Emlen 1993), the authors disagree about what should have been done to prevent further harm to the chicks who were being killed by intruding females. Not only is this question important to consider, but so is the question concerning the value of the knowledge that accumulates from studies of this sort. Suffice it to say, people differ in terms of the benefits that they perceive to be coming from different types of research. Given that there is growing concern about animal welfare, it is important to note that Cuthill (1991) concluded that in his opinion there *were* no ethical problems with the field studies that he reviewed. Although Elwood (1991:847) concluded from his review of studies on rodents that "it is clear that many of the experiments conducted in the field of infanticide and maternal aggression could have been improved from the ethical point of view," he also notes that "in recent years there has been a decreased use of the more questionable experimental paradigms." Nevertheless, even in reviews of experimental field studies in which individuals are removed or added to groups there often is no mention of ethical considerations, for either the target animals or the nontarget animals (e.g., Sih et al.'s 1985 review of field experiments used to study competition; see also Pimm 1991, chap. 12).

Moreover, there are still those who claim that concern for issues of animal welfare is the privilege of "those who have independent sources of wealth, no family obligations, and a lamented shortage of concrete worries" (Hardy 1990:11). Some even claim that "animal usage is not a moral or ethical issue" (White 1990:43; for a reply, see Bekoff 1992). Jamieson (1985b) has called people who do not see issues concerning the regulation and use of animals as moral issues "moral privatists." These people do indeed take an ethical position, but do not acknowledge it. For the view that it is nobody else's business what one does with animals is itself an ethical position.

In this essay we will raise many questions but not always provide clear-cut answers. We hope to open the door for informed, nonpolarized debate. This approach does not, however, mean that these questions are unanswerable, or that we are neutral about how they are answered. Indeed, many of the questions we ask are being answered every day in animal care and use committees, and in the choices that we make about experimental design. Our general point of view is that we should have respect for the animals that we study, and for the diverse habitats in which they live, and that respect for individual animals cannot be separated from embracing a sound, thoroughgoing environmental ethic (Ryder 1992). Many people agree with these general sentiments, but a great deal of work remains to be done in translating them into clear, specific directions for action.

Philosophical Background

This chapter is not intended to be a primer on moral philosophy, but any serious consideration of the ethics of carnivore research must begin with a discussion of the philosophical background. Public attitudes about animals are changing, in part owing to a shift in the philosophical climate. Beginning with Peter Singer's landmark *Animal Liberation* (1975, 1990), there has been an explosion of writing and activism on behalf of animals (see Magel 1989; Bekoff and Jamieson 1991; and Jamieson 1993a for bibliographies; and Finsen and Finsen 1994 for a discussion of the animal-rights movement). Concern about animals is only likely to grow in the future. Behavioral research, because it does not seem to contribute directly to human health or welfare, may be especially vulnerable to criticism, and if the research community does not effectively regulate itself in ways that are consistent with prevailing public values and attitudes, then it is likely to be increasingly regulated by government agencies. For these reasons it is important for carnivore researchers to be aware of some of the major currents of thought regarding the human and humane treatment of animals.

In recent years philosophers have devoted intense attention to questions about the moral status of animals. Though the debate has been vigorous, few moral philosophers today would defend our present treatment of animals in scientific contexts. As DeGrazia (1991:49) notes, "There is no well-developed theory explicitly addressing the moral status of animals that supports such current practices as factory farming, animal research, and hunting." Although there is a great deal of disagreement about the theoretical bases for these results, and many different views about the details of what morality requires, most contemporary moral philosophers would support significantly more restrictive policies regarding our treatment of animals than those currently in place.

Arguments for animal protection grow largely out of traditional moral theory. The basic structure of such arguments typically involves embracing some theory or principle, and then arguing that partisans of this theory or principle either have overlooked its applications to animals or have held beliefs about animals that we now know to be false.

Singer, for example, is a utilitarian in the tradition of Bentham and Mill. He believes that right actions are those that maximize utility summed over all those who are affected by the actions. Much of what we do to animals only appears to be justified, from a utilitarian point of view, because we ignore the consequences of these actions on the animals themselves. Once we take the interests of animals seriously it becomes clear that the misery that we cause them in factory farming and in some areas of research outweighs any benefits that we may gain from exploiting them. Tom Regan (1983) argues from the perspective of rights theory. One traditional view in moral philosophy is that humans have rights, and therefore it is wrong to kill or torture them even if the overall consequences of doing so would be good. Regan then goes on to show that many of the reasons that have been given for supposing that humans have rights also imply that many animals have rights. If we are to be consistent rights theorists, we must therefore recognize the rights of many animals. Thus, Regan believes that just as it is wrong to kill or torture a human even if the benefits of doing so would be very great, so it is wrong to treat animals in these ways even if doing so benefits humans. Animal-protectionist conclusions have also been reached from a neo-Aristotelian basis (Rollin 1981), from common-sense morality (Sapontzis 1987), and from concerns about character (Midgley 1983). What this suggests is that some of the most profound conclusions regarding our treatment of animals are relatively insensitive to initial theoretical assumptions. That most major theories appear to converge from different directions on the conclusion that we ought to alter our behavior with respect to animals makes the case for animal protection even more compelling in many people's minds.

The widespread criticism within the philosophical community of our practices with respect to animals may come as a surprise to many nonphilosophers. Efforts have been made to portray philosophers such as Singer and Regan as fruitcakes or radicals, and to promote the idea that critics of animal-liberationist philosophy such as Cohen (1986), Frey (1980, 1983), and Carruthers (1989, 1992) represent the true philosophical consensus. Although we cannot discuss their work in detail here, the writings of Cohen, Frey, and Carruthers have been subjected to criticism that many find devastating (see for example McGinn 1980; Singer 1990; E. Johnson 1991; Bekoff and Jamieson 1991; Jamieson and Bekoff 1992; Jamieson 1993a; Pluhar 1993; Rothschild 1993). Carruthers himself (1992:194) has changed his mind about the force of his previous arguments.

Although there is widespread agreement within the philosophical community that our practices with respect to animals need to be reformed, there are differences regarding the extent of the needed reforms. Regan is an "abolitionist"; he believes that all invasive animal research is immoral. Other philosophers, such as Singer, hold that at least in principle some animal research can be defended. There are many subtleties and complications in this rigorous, well-reasoned body of literature that cannot be explored here, but any serious rethinking of the ethics of carnivore research should take these challenges seriously.

Why Study Animals?

Perhaps the most fundamental question regarding field research is why do it at all. Even the least-invasive research can be disruptive, and costs time and money. In recent years anthropology has been going through a disciplinary soul-searching, and it is probably time for behavioral biology to go through one as well. Many people study animals for deeply personal reasons—they like being outdoors, they like animals, they don't know what else they would do with their lives—but these sentiments hardly amount to a justification. Several other reasons for doing this sort of research are also frequently given: that animal research benefits humans, that it benefits animals, and that it benefits the environment.

Animal research that benefits humans falls into two categories. One category subsumes research that contributes to human health; the other subsumes research that provides economic benefits. Little field research on carnivores can be defended on the grounds that it contributes to human health. Animal models for human diseases and disorders are better constructed under laboratory conditions, and even then many of them are quite controversial on both scientific and moral grounds. Animal research that contributes economic benefits often concerns the control of predators. Much of this research employs morally questionable methods, and also raises questions about where science ends and industry begins. Predator management may be informed by science, but in itself it is not science; and if producing direct economic benefits were the only justification for studying animals, then very little behavioral research would be justified.

The idea that behavioral research benefits animals and the environment is an appealing one. The thought is that only by studying carnivores will we know how to preserve them, and that only by preserving carnivores can we protect the natural environment. As noble as these sentiments are, they are rife with dangers. For this attitude can lead very quickly to transforming science into wildlife management; and wildlife management itself poses important moral challenges.

Humans face an environmental crisis in part because of their attempts to control, dominate, and manage nature. These attempts have led to the destruction of important aspects of nature, and even to serious threats to human well-being. In attitude and intention, much wildlife management is more of the same. A new generation seems to think that though in the past we were incompetent managers, we now know how to do it right. Our track record, however, does not inspire confidence (Mighetto 1991; for discussion of various projects and wildlife research in general, see Gray 1993). Ludwig et al. (1993) call into question the idea that we can manage animal populations in a sustainable way. They argue that science is probably incapable of predicting sustainable levels of exploitation of an animal population, and that even if it were possible to make such predictions, human shortsightedness and greed would prevent us from acting on them.

Some people argue that human intervention has already so disrupted natural processes that we have no choice but to manage populations and habitats—that the decision *not* to manage is just a form of unreflective, irrational management. Moreover, some would say that management approaches are part of a strategy to convince humans that much more must be done to preserve habitat and to save the animals that we have. No doubt there is something to these arguments. Still, we should be suspicious of them because they are so self-serving. Whether people are pro- or anti-environment, they seem to agree that humans should manage and manipulate nature. But if we are convinced that we must manage the natural world, we should remember that the best of intentions are no substitute for minimizing harm.

The deepest problem with wildlife management may be the tendency to confuse scientific ideas with philosophical ones. When managers argue that a population should be "culled" or otherwise managed and controlled, as for example has been suggested for Alaskan wolves (e.g., Grooms 1993, chap. 12), their view is often seen as a scientific recommendation. Appeals are made to what are presented as irrefutable scientific data concerning interactions between wolves and their prey. But the models that are used are often simplistic (important variables are omitted, including the effects of weather and food availability), and estimates of the actual numbers of supposed problem predators can be inaccurate, as can be estimates of the number of individuals hunted and killed annually by sports hunters or by managers. It is also possible that only a small percentage of people living in a given area actually want to control predators such as wolves (Grooms 1993), and the arguments put forth for doing so can therefore be self-serving. Indeed, Alaskans as a whole have been identified "as one of the two demographic groups with the greatest affection for wolves" (Grooms 1993:168). Economic factors (e.g., tourism, fees for hunting licenses) frequently provide the primary motivation for controlling predator populations, and their compelling nature can result in failures to look for other ways to deal with predator/prey interactions. But most important, recommendations about population control presuppose various philosophical positions, each with its ethical dimensions and assumptions. And when these presuppositions are made explicit, it is clear that not everyone agrees with them. Scientific data, even those that are judged to be accurate, are not the only sorts of information that are relevant to establishing management practices. Issues concerned with population control bring with them commitments to various views about human authority, the value of species versus individuals, and the importance of human versus nonhuman interests (Shapiro 1989; Jamieson 1995).

The purest motivation for studying carnivores may be simply the desire to understand them. But even if this is our motivation, we should proceed cautiously and reflectively. For in quenching our thirst for knowledge we impose costs on these animals. In many cases they would be better off if we were willing to accept our ignorance, secure in the knowledge that they are leading their own lives in their own ways (for a similar point about cetaceans, see Jamieson and Regan 1985). If, however, we do make the decision to study animals, we should recognize that we are doing it primarily for ourselves and not for them, and we should proceed respectfully and harm them as little as possible.

The Context of Research

Field research is conducted in contexts that influence how data are collected and used, and the results that one obtains. When scientists "go into the field" they go to a particular place, often a foreign country, with a human population, with its own government, culture, way of life, and attitudes toward both foreigners and the animals the scientists are studying. In the past, many researchers marched into foreign lands (getting permission from government officials, if needed), studied the animals in whom they were interested, and departed, often with little thought about the legacy they were leaving behind. Not only must foreign researchers be sensitive to local customs concerning animal use (e.g., Brenes 1988; Croft 1991; Rabinowitz 1986, 1991; Wenzel 1991; Freeman et al. 1992; Lynge 1992; Bekoff and Gruen 1993; Schaller 1993), but they should also do what they can to help develop the scientific capability of the host country. They should appreciate that it is a privilege to be permitted to work where they want to, and they should recognize that there may be a good deal of indigenous knowledge about the animals they are interested in studying. Furthermore, foreign researchers should not assume that they are doing their host country a favor by coming there.

There are many reasons to respect local customs and abilities. First, better science will be done in the long run if different kinds of knowledge are exploited and expertise is democratized. Second, the study of animal behavior, like anthropology, is historically rooted in Western dominance, imperialism, and colonialism (Haraway 1989). Making a contribution to the society in which researchers work is part of showing respect for the human population and making small amends for the exploitation of the past. Finally, the enlistment of local researchers may help foreigners come to terms with local attitudes and customs.

But the use of local workers may also present problems, because of difficulties in communicating about what needs to be done (e.g., Rabinowitz 1991;

Schaller 1993; see also Terborgh 1993) and what data should be collected, and because of problems that might develop concerning the ownership of data and how they are to be disseminated and used. Rabinowitz (1986:155), an American scientist who has worked in various countries, also makes the point that it is often difficult to work with officials from his home country, who too often interfere with his research abroad. He recalls an incident when he was studying jaguars in Belize in which an official for a U.S. agency was willing to break Belize law to allow a visitor who "puts a lot of U.S. dollars into this country" to hunt jaguars. Such incidents create conflicts between researchers, who are concerned to protect the animals they study, and the sometimes differing values of people whose decisions affect the future of the work. Conflicts in values also inevitably arise between those who are trained in the traditions of western science and those in host countries who may have very different attitudes toward knowledge, animals, and nature.

Responsibility for Data Collection, Analysis, and Dissemination

In recent years there has been increasing concern about scientific integrity and even fraud in science (for discussion and references, see Goodstein 1991; LaFollette 1992; Altmann 1993; Bulger et al. 1993; Dresser 1993). Pushed along by such requirements as that of the National Institutes of Health that all training grants must have an ethics component, these issues are beginning to attract the attention of practicing scientists. Indeed, a recent book in which articles on animal behavior are reprinted begins with a section concerned with integrity in science (Sherman and Alcock 1993). Furthermore, an entire issue of the journal *Ethics and Behavior* (1993:3[3]) is devoted to the general themes of whistleblowing and misconduct in science.

Trust is an important part of virtually all scientific endeavors (Woolf 1988; Hardwig 1985, 1991; Kitcher 1993; Webb 1993). Science proceeds by amplifying, theorizing, or rejecting one's own findings and those of others. In many areas of science, including field studies, replication is difficult and in some cases even rare. Against this background the very act of citing a report can be seen as an act of trust. Using other scholars' reports of recent work, such as information contained in book reviews, also involves trust. Though there is usually no reason to believe that someone is knowingly altering his or her view of a colleague's work, it happens: the senior author recently learned of a case in which a book was given a more favorable review than it would otherwise have received, because of the ill health of the author of the book being considered. Trust is also placed in the hands of journal editors, but this, too, is sometimes misguided: there are instances, for example, of collusion among editors so as to influence when certain types of results appear (Collins and Pinch 1979:258).

Trust also comes into question when principal investigators or project leaders who play a major role in writing a research proposal do not do much of the field work that is later reported in published articles. Thus, there is often a gap between the design of an experiment and the actual procedures by which data are collected. Individuals disappear into the field and emerge with data, but the type of guidance that is given during the process of data collection is often inconsistent or even nonexistent, and is rarely open to public scrutiny. These problems can be acute in field studies because of the difficulty involved in asserting control over many of the variables that are encountered in field projects, and also because decisions often have to be made on the spot, lest data be lost; wild animals typically do not wait around while scientists make decisions.

Because field research is often a highly collaborative activity, questions of authorship arise (for general discussion, see part V of Bulger et al. 1993). Principal investigators generally are listed as authors even if they have had little to do with data collection. The fact that the principal investigator raised the money for the research and took part in data analysis is often regarded as sufficient for his or her being the lead author. In recent years in literary studies there have been interesting discussions of authorship (e.g., Foucault 1984), and the discussion is spreading into ethics. Professional research assistants, graduate students, and postdoctoral fellows are often in a very weak position to assert claims to authorship. They may have done much or all of the data collection and made important contributions to the conceptual framework, yet they may be listed as junior authors, mentioned in the acknowledgments, or excluded from recognition altogether. There are also risks for principal investigators whose names appear on every publication from their laboratory or research project, quite apart from growing suspicions about actual participation in authorship. For if such a coauthor has played little role in the collection of the data on which the report is based, then he or she cannot be sure that everything that was claimed to have been done was done; a great deal of trust is invested in all of the participants in a field project.

Though these problems are not unique to field endeavors, they can become magnified in research efforts of this type. All those involved in such projects need to be aware of the trust that is invested in all co-workers, and aware as well that each individual shares responsibility for the integrity of the project as a whole and for the integrity of the results.

Observation and Manipulation

Field projects bring along a whole host of problems concerning observation and manipulation that are usually not of much concern to laboratory endeavors (Cuthill 1991), although both sorts of research can involve intrusions into the privacy of animals' lives. The problems most apparent in field research are those that center on the lack of control over the behavior of the animals being studied, over variables that influence the behavior of the animals being studied, and, as mentioned above, the potential lack of control over the individuals who do the actual research. Furthermore, because animals living under field conditions are generally more difficult to observe than those living under more confined conditions, various manipulations are often used to make them more accessible to study (see Mills, this volume). These include such activities as handling, trapping (and retrapping) using various sorts of mechanical devices (which might include luring using live animals as bait), and marking (and remarking) individuals, none of which is unique to field studies, but all of which can have important and diverse effects on wild animals who may not be accustomed to being handled by humans or even to their presence.

Just "being there" and visiting individuals, groups, nests, dens, or ranging areas can also have a significant influence on the behavior of the animals (for discussion and bibliographies, see Michener 1989; Bekoff and Jamieson 1991; Cuthill 1991; Bekoff et al. 1992; Caine 1992; Shackley 1992; Isbell and Young 1993; Mainini et al. 1993). Kenney and Knight (1992) found that magpies who are not habituated to human presence spend so much time avoiding the humans that the avoidance effort itself takes time away from such essential activities as feeding. Burley et al. (1982) showed that mate choice in zebra finches is influenced by the color of the leg band used to mark individuals, and there may be all sorts of other human-engendered influences that have *not* been documented. Similar data for most carnivores are lacking (but see Laurenson and Caro 1993). Furthermore, the filming of animals so as to establish permanent records can also have a negative influence on the animals being filmed; reflections from camera bodies, the noise of motor-driven cameras and other sorts of video devices, and the heat and brightness of spotlights (A. Pusey, pers. comm.) can all be disruptive.

Although many problems are encountered both in laboratory and field research, the consequences for wild animals may be different from and greater than those experienced by captive animals, whose lives are already changed by the conditions under which they live. This can be so whether experiments entail handling, trapping, and marking individuals or not. Consider experimental procedures that include (1) visiting the home ranges, territories, or dens of animals, (2) manipulating food supply and other resources, (3) changing the size and composition of groups (age, sex ratio, kin relationships) by removing or adding individuals, (4) playing back vocalizations, (5) depositing scents, (6) distorting phenotypes, (7) using dummies, and (8) manipulating the gene pool. All of these manipulations can change the behavior of individuals, including their movement patterns, how they make use of space, the amount of time they devote to various activities (including hunting and anti-predatory behavior), and the various types of social interactions they engage in (including care-giving, social play, and dominance interactions). These changes can in turn influence the behavior of groups as a whole, including group hunting or foraging patterns, care-giving behavior, and dominance relationships, and can also influence nontarget individuals (Cuthill 1991; see also Hofer et al. 1993 for discussion of how snaring migratory herbivores can influence the population dynamics and demography of spotted hyenas). Though we often cannot know about various aspects of the behavior of animals before we arrive in the field, our presence does seem to influence what animals do, at least when we initially enter into their worlds with some degree of regularity. For example, the coyotes that were observed in one long-term study (Bekoff and Wells 1986) initially spent a great deal of time staring and barking at observers who remained at least 1–2 km distant, but eventually settled down and seemed to go about their daily activities with little concern; their behavior reverted to what we assume was typical of what they did before we arrived. Furthermore, changes in the behavior of individuals and groups, including those animals who are (re)introduced to a specific area, can have far-ranging effects on local and distant ecosystems. Consider, for example, the reintroduction of red wolves into areas where coyotes already live (Wayne and Gittleman 1995).

The point here is that what appear to be relatively small changes at the individual level can have drastic and wide-ranging effects in both the short and the long term. Certain activities that were designed outside of the field situation may be impossible to perform, others, though possible to implement, may have unanticipated negative effects on the animals, and still others that seemed impossible to perform may turn out to be possible and less intrusive than those manipulations that were previously planned. On-the-spot decisions often need to be made, and expectations about what these changes will mean to the lives of the animals who are involved, and how ethical considerations might inform these decisions, need to be given serious attention. We should take as our guiding principle that when we are unsure about how our activities will influence the lives of the animals being studied, we should err on the side of the animals and not engage in these practices until we know their consequences. Recently, it has been claimed that we should take this stance even regarding insects (Wigglesworth 1980; Eisemann et al. 1986). Eisemann et al. (1984:167) conclude that this attitude "helps to preserve in the experimenter an appropriately respectful attitude towards living organisms whose physiology, though different, and perhaps simpler than our own, is as yet far from completely understood."

Questions without Answers: A Baker's Dozen

Field workers engage in many different activities, and we cannot address them all here. The following list of questions is meant, rather, to provoke thought and open discussion: it is certainly not exhaustive, and there is no order to the questions being posed, in terms of their relative importance, although some questions might be viewed as being more important than others. The fact that there may be little consensus about the answers to these questions at this time does not mean that there are not better and worse answers. As a general principle we should err on the side of the animals, and never forget that respect for the animals is of utmost importance. But real progress in the future will involve developing ever more precise guidelines about what is permissible.

Here we pose 13 questions, among which there is some unavoidable overlap in the areas considered. The first four questions are concerned with general issues that center on ethical responsibilities to wild and domestic animals, captive-breeding programs, concerns about individuals and groups, and the validity of data stemming from studies of captive animals and individuals who are studied and manipulated in the field. Questions 4, 5, and 6 are concerned with methods of study and the relationship between researchers and the animals they use, and questions 7, 8, and 9 consider whether or not researchers should interfere in natural predation or employ staged encounters either between predators and prey or in studies of infanticide or dominance. The tenth question deals with relationships between animal cognition and animal welfare, the eleventh asks what principles we should use as ethical guides, and questions 12 and 13 are concerned with scientists' responsibility for how their results are used and how to deal with ethical misconduct.

1. Do wild animals have different moral status than domestic animals? This is an important question, because field studies are performed on both domestic and wild carnivores, often in the same habitat. Callicott (1980/1989:30) writes that "domestic animals are creations of man. They are living artifacts, but artifacts nonetheless, and they constitute yet another mode of the extension of the works of man into the ecosystem" (see also Katz 1993). According to Howard (1993:234–235), "Domestic species are genetically programmed to depend upon humans for their safe existence and, fortunately, they always die relatively humanely rather than suffering one of nature's brutal deaths" (for a reply to Howard, see Bekoff and Hettinger 1994). Since these animals may live longer, have a higher quality of life, and die less painfully than do wild animals, Howard concludes that animal research actually "produces an improvement of life for some individuals of these species" (p. 235; for similar views made without any empirical support, see Gallup and Suarez 1987; Lansdell 1988; Greenough 1992; Grandin 1992; and Raynor 1992; for rebuttal, see Singer 1992a). Greenough (1992:9) goes as far as to claim, "In fact, it is very rare for research animals to be subjected to significant amounts of pain. For most animals, life in the laboratory is considerably more comfortable than [that] for their counterparts in the wild." Colwell (1989:33) maintains that "our moral responsibility for the appropriate care of *individual* organisms in agriculture, zoos, or gardens does not depend on whether they are wild or domesticated in origin." He also writes, "I contend, however, that the role of domesticated *species* as coevolved members of our ancestral component community . . . places them in a biologically and ethically distinct class from 'wild' species" (emphasis original).

2. Are we ever justified, and if so under what conditions, in bringing wild animals into captivity? The most frequently cited reason for bringing animals

into captivity is to preserve endangered species by allowing individuals to live in a protected environment that will facilitate breeding, thus maintaining the species' gene pool. It is sometimes said that the goal of these programs is the eventual return of these animals to the wild, but there are serious philosophical questions involved here. For example, do animals have a right to liberty? Do species have interests? Can the welfare of individuals be sacrificed in the interests of species (for discussion, see Regan 1983; Jamieson 1985a, 1995; Rachels 1989; Shapiro 1989; Singer 1990, 1993a; DeBlieu 1991, Varner and Monroe 1991; Bekoff and Gruen 1993; Norton et al. 1995). It is noteworthy, in fact, that some of the most virulent critics of captive-breeding programs are themselves among the scientists who have been devoting their lives to these efforts and who sincerely want them to succeed.

To quote Schaller (1993:233–234), "The realization that the panda has so suffered and declined in numbers while we chronicled its life burdens me painfully. Enthusiasm and goodwill count for little when the enemy is a vast bureaucracy of local officials who myopically use obstruction, evasion, outdated concepts, activity without insight, and other tragic efforts to avoid central-government guidelines and create ecological mismanagement on a dismaying scale." And, with respect to what he calls the "rent-a-panda" program, Schaller writes (p. 249), "The politics and greed, coupled with the shameful indifference to the panda's welfare that has characterized much of the rental business, will not vanish." Peterson (1989) also is skeptical of captive-breeding programs, and focuses on the Species Survival Plan for Siberian tigers. He notes that at the time of his writing extant groups of captive Siberian tigers were "poorly distributed in terms of sex ratio and age structure" (p. 301), that only a few individuals could actually be allowed to breed, and that others might have to "be removed—probably killed (or, to use the preferred expression, 'euthanized')" (p. 301). Peterson also stresses that the ultimate goal of most captive-breeding programs, the return of endangered animals to the wild, will probably never be attained. Rabinowitz, too, is uncertain about many captive-breeding programs because "they provide no comprehensive management of captive populations and no follow-up programs to reintroduce the young to the wild" (1991:165). Furthermore, he points out, "the proper techniques of reintroduction are rarely used." Schaller (1993) is critical of attempts that entail "rescuing" pandas for purposes of protecting them and developing breeding stock. He notes the deplorable conditions at one research center and explains that "the panda rescue work, a legacy of the 1983 bamboo die-off, continued well into 1987, long after there was any justification for it" (p. 223). He asserts that "if most of those that were rescued after the bamboo die-off were given their liberty they would perhaps replenish the forests" (p. 224). Here there is some tension between the seemingly defensible act of rescuing animals who might otherwise have starved to death because of the lack of food, and the motivation of helping them along so that they might be used to replenish wild populations.

No one, however, should deny the paramount importance of the goals of captive-breeding programs. Despite the logistical and financial difficulties entailed in implementing a captive-breeding and reintroduction program, Rabinowitz (1991:166) concludes, "no price can be put on saving even a single species that might otherwise have been lost. However, a halfhearted or haphazard and incorrect approach is both a waste of resources and a source of potential harm to the animals involved." Given the fact that many experts are extremely skeptical of attaining the goals of captive breeding, specifically for establishing healthy and self-sustaining animal populations that can be successfully reintroduced to the wild, we need to reassess what we are doing and why we are doing it (see also Schaller, this volume).

3. What is the relationship between good science and animal welfare? This question arises in particular with respect to research on captive animals. Much of what we learn about animals living in zoos (for example) does not generalize beyond that sort of environment, because of the way in which individuals are housed, fed, and otherwise maintained. Thus, we need to be careful when designing a field study based solely on data from captive individuals, for the nature of the questions being asked will greatly influence the usefulness of the data that are collected. For example, information about scentmarking by captive felids (e.g., Asa 1993, Mellon 1993) appears to be useful for furthering our understanding of what happens in the wild, and detailed studies of behavioral development can also inform field research on carnivores (Bekoff 1989).

Various field techniques can affect behavior and welfare, as well, and perhaps influence the validity of the data that are collected. Field researchers should study the behavioral effects of the techniques that they use in studying wild animals (e.g., trapping animals, marking individuals, fitting individuals with radio collars) to determine whether there are consequent behavioral changes that might influence the validity of the data that are collected, and this information might also inform welfare decisions. For example, Laurenson and Caro (1994) studied the effects of nontrivial handling on free-living cheetahs, including making them wear a radio collar, radio tracking them by air, and examining their lairs, and found no detectable influence on the cheetahs' behavior. But they stress that although their subjective assessments proved to be supported, researchers must not depend solely on their intuitions. Laurenson and Caro also note that little work that focuses on the influence of researchers' presence and their experimental manipulations on the behavior of the animals being studied has been done on other mammals (e.g., Ramsay and Stirling 1986; Orford et al. 1989). But in a long-term study of coyotes (Bekoff and Wells 1986), it was found, for example, that shiny cameras and spotting scopes made the animals uneasy, and camera bodies and spotting scopes were accordingly painted dull black so that they would reflect little light. Furthermore, when visiting dens we always wore the same clothes, so that we would be presenting the coyotes a roughly consistent odor and a consistent visual image. Many studies concerned with human intrusion have been done on birds (e.g., Pietz et al. 1993), and those who study carnivores can learn many lessons from this research (for references, see Bekoff and Jamieson 1991; Laurenson and Caro 1993).

Those who study behavior and behavioral ecology in the field are in a good position to make important contributions to animal welfare, but unfortunately they often play only a minor role in informing legislation on matters of animal welfare (Cuthill 1991). Field workers can help to provide guidelines concerning dietary requirements, space needs, the type of captive habitat that would be the most conducive to maintaining the natural activity budgets of the animals being held captive, information on social needs in terms of group size and age and sex composition, and information about the nature of the bonds that tend to be formed between animals and human researchers (Rollin 1981; Davis and Balfour 1992).

4. Should we subject individuals to harmful or painful or disturbing experiences so that we can learn how animals deal with these types of situations and how their behavior is influenced? Various carnivore studies involve trapping individuals either in leg-hold or live traps (Mills, this volume). Is this practice justified in efforts to learn more about how animals can be trapped and how trap-lines can be monitored in ways that make the experience of being trapped less harmful or painful? It seems highly unlikely that anyone who has ever worked with trapped animals could claim that being trapped is not harmful or painful—both physically and psychologically—for the individuals involved, and more research is needed on developing alternatives to leg-hold traps and other devices that restrict an animal's movements. Researchers should have to provide persuasive evidence that there are no alternatives to the trapping methods used, and they should be able to demonstrate that they are using the most humane methods (trapping, visiting trapped animals) available. Finally, it is incumbent on researchers to show that the stress of being trapped does not influence the behavior patterns of the animals being studied, patterns that might be precisely those of interest in the study at hand.

5. How can we minimize the number of animals who are used? Elwood (1991) points out that in certain sorts of studies, it is imperative to pay strict attention to the experimental protocol, and that stress must be minimized in the animals being studied. If studies produce results whose validity can be legitimately questioned, because, for example, the data come from stressed animals, then attempts to repeat the studies, in one form or another, will result in yet more animals being used, perhaps to no further research benefit. In cases where animals have to be followed or located repeatedly, it is worth asking whether it might not be sufficient to mark or radio-collar a single individual if all others are individually identifiable on the basis of reliable behavioral or other markers? Not only would this entail less handling of individuals, but minimal labeling of the animals might also lead to less disruption of ongoing behavior.

In recent years there has been a movement in laboratory research to reduce the number of animals used by refining experimental techniques. Many of these refinements concern the use of different instruments or analgesics, but in some cases changes in experimental designs and statistical methods can also result in the use of fewer animals. Field and laboratory researchers should cooperate in designing projects that use the smallest numbers of animals possible.

6. What is the proper relationship between researchers and the animals they study? Because some form of bonding between the animals who are being studied and the researchers is probably inevitable (Davis and Balfour 1992; Bekoff 1994b, c), these bonds should be exploited in such a way as to benefit the animals. As L. E. Johnson (1991:122) notes, "Certainly it seems like a dirty double-cross to enter into a relationship of trust and affection with any creature that can enter into such a relationship, and then to be a party to its premeditated and premature destruction."

Bonding can also help to produce "better" science. "Better" means different things to different people, but the notion seems to involve practices that help to expedite data collection, contribute to a more complete understanding of the animals, reduce the number of animals to be studied, and provide less contrived explanations of the behavior patterns observed (Bekoff 1993b). Whatever one thinks about the ultimate scientific respectability of folk-psychological attributions to animals, there is little question that such attributions are helpful in hypothesis generation (Dennett 1987), and that these attributions will be better motivated on the part of those who have significant relationships with the animals in question than on the part of those who do not.

We can also ask if humans should name the animals whom they use, and whether doing so might have an effect on ongoing research (Montgomery 1991; Bekoff and Allen 1996). We believe that naming is a good idea, and in primate research, in particular, it has been the norm. Furthermore, we advocate using the words "he" or "she" or "who" rather than "it" or "which." Whether or not named animals behave according to the names they are given (e.g., Brutus, Samson, Dolly, Wimpy, Dumpy), and whether or not researchers are biased by names to see things that are not there, remain open questions. Naming animals certainly can influence the treatment to which individuals are subjected (Bekoff 1993b), because named animals are typically treated more like subjects than objects, individuals with personalities and unique characteristics (Bekoff and Gruen 1993). Having strong relationships with named individuals invites the use of anecdotes, and when these are used judiciously they can play an important role in describing and interpreting the behavior of nonhumans, and informing and motivating empirical research in areas such as animal cognition (Burghardt 1991; Bekoff 1995; Bekoff and Allen 1996).

7. Should humans interfere with natural predation? It should be noted first that most carnivores have only limited to moderate success in their predatory efforts, and in many cases in which these efforts occur (e.g., at night, in inaccessible habitat, opportunistically) humans could not intervene even if they so desired. Still, it is interesting to note that in a case that was reported in a popular magazine devoted to birds (Bonta 1991), the vast majority of letters

from nonscientists (see *Bird Watcher's Digest* September/October 1991, November/December 1991, March/April 1992) maintained that a family that watched a rat snake eat some young birds should have intervened. Sapontzis (1984:36) maintains that although not all predation is avoidable, "where we can prevent predation without occasioning as much or more suffering than we would prevent, we are obligated by the principle that we are obligated to alleviate unavoidable animal suffering" (see also Sapontzis 1987, chap. 13). Sapontzis stresses that issues of practicality must figure into ethical deliberations, and that it is not immediately obvious how "unnecessary predation" can be avoided. In light of these views it is interesting that not a single field biologist who has studied predation by wild animals and who responded to a request for an answer to this question believed that natural predation should be interfered with. Though it appears that most scientists would reject the view of Sapontzis and the readers of *Bird Watcher's Digest,* it is interesting to speculate on why this should be so. For in general, students of behavior do not take a "hands-off" approach to nature. If they did, they would not be in the field in the first place.

Kirkwood (1992) presents a thoughtful essay on the welfare of wild animals and considers questions such as whether we should intervene on behalf of free-living wild animals, and, if so, to what extent and how it should be done. Although he acknowledges that there are many different views on the matter, he claims (p. 143) that "most would probably agree that when wild animals are harmed by man's very recent (in evolutionary terms) changes to the environment (such as oil spills, power lines, roads, and environmental contamination) there is a reasonable case, on welfare grounds, to intervene." (For relevant papers, see Ferrer and Hiraldo 1992 and McOrist and Lenghaus 1992.) Kirkwood also writes about veterinary intervention to treat injured or sick wild animals. He calls for "an international code on intervention for wildlife welfare to provide guidance on ethics, methods and standards" (p. 151). Though we cannot explore these issues in depth here, we do believe that there are circumstances in which humans may have to intervene for the welfare of wild animals, including some of those listed above (see also Schaller 1993). One also needs to give serious consideration to the idea that if any experimental manipulation— including the mere presence of researchers—leads to harm for either the target animal or (indirectly) any other individual, then we are obligated to intervene on the animals' behalf.

8. Should studies be allowed that employ staged encounters between predators and prey, or between animals among whom the formation and maintenance of social-dominance relationships could potentially bring harm to one or more individuals (Huntingford 1984)? The general form of this question is to ask whether researchers should intentionally stimulate predatory, agonistic, or other types of encounters that would not otherwise have happened, the result of which could be harm to the participants. This is not a simple question that submits to easy answers, and people can change their minds. To wit, in the 1970s the senior author performed such studies on the development of preda-

tory behavior in captive coyotes, but on reflection found it impossible to justify them and decided that he would no longer do this sort of research (Bekoff and Jamieson 1991:26, note 20). My decision centered on the psychological pain and suffering to which the prey (mice and young chickens) were subjected by being placed in a small arena in which there were no possibilities for escape, as well as the physical consequences of being stalked, chased, caught, maimed, and killed. One result of my decision was that detailed information about the development and predatory behavior could not be obtained, because similar data cannot be collected under nonstaged field conditions. Staged-encounter studies are also performed in the field. For example, in a recent study, Small and Keith (1992) released radio-collared arctic and snowshoe hares to learn how arctic foxes preyed on them.

9. Should humans interfere with infanticide, or use staged encounters to study it? An exchange between Bekoff (1993a) and Emlen (1993) concerning research reported in Emlen et al. (1989) highlights some of the important issues concerning the experimental study of infanticide (see also Elwood 1991). Bekoff questioned whether the research had to be done in the way in which it was, namely allowing chicks to be maimed or killed by intruding females after the chicks' mothers had been killed by the researchers. One interesting aspect of Emlen's response is that he and his colleagues did indeed stop their experiments when they became bothered by the maiming and killing of wattled jacana chicks by females other than their mothers. It would have been helpful to see this displeasure in the original paper. Another issue that developed in this debate concerned whether or not the experimental study of infanticide was justifiable in terms of the "costs" to the animals involved and the "benefits" of the knowledge obtained. Of the people polled by one of us (MB) concerning the value of this sort of study of infanticide, there was an almost 50-50 split concerning whether or not the study should have been undertaken in the first place. The basis of most of the no votes was that we do not know enough about infanticide to justify field experiments such as those that Emlen et al. performed (see for example Dagg 1982; Bartlett et al. 1993).

10. How can the knowledge that we gain from studies of wild carnivores inform us of their cognitive abilities and of our ethical obligations to them? Some have gone as far as to claim that if gaining knowledge of the cognitive skills of wild animals does nothing more than inform the debate about animal welfare, then these efforts are worthwhile (e.g., Byrne 1991; see also Bekoff and Jamieson 1990, 1991; Duncan and Petherick 1991; Duncan 1993; Bekoff 1995; Bekoff and Allen 1996). But students of behavior have had a very difficult time in conceptualizing cognitive ethology. Many seem to suppose that behavior is what is observed, and that cognitive states can only be inferred; and for that reason cognition is beyond the reach of empirical science. Although we believe that this view is incorrect, building a theoretical foundation for cognitive ethology poses difficult problems (Jamieson and Bekoff 1993). Even if such a foundation *can* be built, the relationships between cognition and moral standing

are complicated. It is logically consistent to hold, on the one hand, that we *may* owe moral respect to creatures who are not cognitive and, on the other hand, that we may *not* owe such respect to creatures who *are* cognitive.

Questions about the relation between what animals are like and how they should be treated are often in the background of discussions about the complexity of behavior and the "inner" lives of animals. These questions need to move from the background to the foreground, and be openly discussed. Sometimes in animal research we seem to be confronted by a paradox: by conducting research that involves treating animals as if they were not morally important we sometimes discover that they have rich, complex cognitive and emotional lives. Indeed, what we may discover in some cases may convince us that it is wrong to treat them in the ways that made the discoveries possible.

11. What principles should we use as ethical guides? In particular, if human-caused pain in animals is less than or equal to what the animal would experience in the wild, is it then permissible to inflict the pain (Rolston 1988; Hettinger 1994)? For many animals it is difficult to know whether this condition is satisfied, for we do not know how most individual animals in nature experience pain. For this reason we must be careful that this principle not become just a rationalization for researchers doing what they really want to do on other grounds. Many other principles have been proposed that perhaps should guide us in our treatment of animals: utilitarian ones, rights-based ones, and so forth. Furthermore, scientists often operate on the basis of implicit, unstated principles and guidelines. All such principles need to be brought into the open and explicitly discussed.

12. Are scientists responsible for how their results are used? For example, if we learn about how wolves live, are we responsible for making sure that this information is not used to harm them? This is not a purely "academic" question, since a great deal of research on carnivores is funded by agencies that want to reduce their populations or control their behavior. Information about the behavior of tigers or wolves may be useful to those who simply want to make rugs out of them. Those who study marine mammals have been struggling with these issues for at least a decade. Purely scientific information about populations, migration routes, and behavior can be used by those who are involved in the commercial exploitation of the animals who are being studied. Even where hunting bans and restrictions are put in place, they may be only temporary.

One idea worth considering is that a scientist who studies a particular animal may be morally required to be an advocate for that animal in the way that physicians are supposed to be advocates for their patients. On this view, the welfare of the animals whom a scientist studies should come first, perhaps even in preference to the goal of obtaining peer-reviewed scientific results. Some scientists, such as Jane Goodall and Dian Fossey, have exemplified this ethic, but they have had many critics from within the scientific community.

13. What responsibility does the research community bear in preventing ethical misconduct, and how should this responsibility be exercised? In recent

years various communities of researchers have taken steps to deal with problems of misconduct by adopting codes of professional ethics and refusing to publish papers that violate the ethical guidelines. At the same time, researchers have too often behaved like physicians in being reluctant to take steps against their own. In cases of conflict there has been a tendency for many scientists to side with the more powerful members of the community against the less powerful (e.g., Lang 1993). Unfortunately, there is still little agreement about what the collective ethic should be with regard to many of the questions that we ask, and little sense that the research community should be obligated to encourage high ethical standards within its own community (for discussion of some of these issues, see *Ethics and Behavior* 1993:3(1), which is devoted to whistle-blowing and misconduct in science). In many circles there is even a sense of complacency about research misconduct. Yet, in the most exhaustive empirical study to date on the reported incidence of misconduct, Swazey et al. (1993) found that reports of fraud, falsification, and plagiarism occur at a surprisingly high rate. They conclude that this is a serious problem, one that needs immediate attention.

Concluding Remarks

There is a continuing need to develop and improve general guidelines for research on free-living (and captive) carnivores (and other animals). These guidelines should be aspirational as well as regulatory. As Quinn (1993:130) notes, "We're not destroying the world because we're clumsy. We're destroying the world because we are, in a very literal and deliberate way, at war with it." We should not be satisfied that things are better than they were in the bad old days, and we should work for a future in which even these enlightened times will be viewed as the bad old days. Progress has already been made in the development of guidelines, and the challenge is to make them more binding, effective, persuasive, and specific. If possible, we should also work for consistency among countries that share common attitudes toward animals; research in some countries (e.g., the United States) is less regulated than research in other countries (Gavaghan 1992). In this evolving process, interdisciplinary input from both field workers and philosophers is necessary; no single discipline can do the necessary work alone (Farnsworth and Rosovsky 1993; Shrader-Frechette and McCoy 1993). Researchers who are exposed to the pertinent issues, and who think about them and engage in open and serious debate, can then carry these lessons into their research projects and impart this knowledge to colleagues and students. Not knowing all of the subtleties of philosophical arguments—on the details of which even professional ethicists disagree—should be neither a stumbling block nor an insurmountable barrier to learning; finding oneself on the horns of the dilemma and then doing nothing is not the way to go. Few field workers understand how binoculars or radio transmitters work, but this does not prevent them from using them. The same should be true of the details of philosophical tradition and argument.

Flip, simplistic, and unargued dismissals from those who are deeply concerned with animal welfare will do nothing but divide those whose inputs are needed if we are to improve our behavior with respect to animals. Nor should we dismiss animal-welfare concerns on the basis of facile anthropocentric arguments, or because responding to them will cost time or money. We should use the fewest animals we can, the least invasive techniques available. We should share data when possible, and carefully survey available literature so that unnecessary duplication does not occur. Great concern should be shown for the frequency with which animals are visited, the types of traps that are used, the types, color, weight, and shape of tags that are used to label individuals, where tags are placed, the weight of radio transmitters, the behavior of the researchers, the consistency of the behavior and dress of the researchers, the color and noise of cameras and other equipment, and any number of similar matters.

How we refer to some of our research practices also needs reconsideration. We should use words that accurately describe our actions, not euphemisms that may be intended to mask them. When animals are killed it should be explicitly stated that they are killed; words like euthanize, cull, sacrifice, and collect often are used to deflect attention from the act of killing and to give a false sense of acceptability. We need to eliminate research projects that should not be undertaken at all. Many proposed projects are not carried out because they lack scientific merit, but surely a lack of ethical merit is every bit as important. Of course, decisions like these can put one on a slippery slope, and it is not easy to know where to draw the line. For example, Hauser (1993) considers the interesting and difficult question of whether or not vervet monkey infants cry wolf to elicit parental care. Field data are ambiguous, and Hauser suggests that "one fruitful approach to resolving these issues would be to manipulate the physical condition of mother and offspring" (p. 1243). Hauser does not tell us, however, what these manipulations would consist of. Obviously, some experiments would be more ethical than others, and sorting among them would be very difficult when balancing the knowledge that would be gained against the type of manipulations that would have to be done. But just because these sorts of decisions are difficult does not mean that there are not better and worse answers.

Although it is important for debate to be as constructive and nonpolarized as possible, we should not be afraid of honest disagreement (e.g., Bekoff, 1993a; Emlen 1993). We will come to a consensus about the ethics of specific practices only if we expose our differences to the light of day, and frankly discuss the issues that are involved.

Recently, Slater (1993:188) claimed that "the assumptions one brings to bear in designing experiments are not necessarily the same as those one should adopt in deciding how to treat animals." Slater was responding to the views of Kennedy (1992), who, Slater believes, approaches problems of animal welfare "too much as a scientist." Basically, Kennedy's views on animal welfare are informed and motivated by his fear of the ill effects of false assumptions; thus, he is unhappy with erring on the side of animals in the absence of hard data that would lead one to this conclusion (see also Crockett 1993). Slater's claim needs further fleshing out by and for those interested in animal welfare. We need to know more about the assumptions that one does or should use in deciding how to treat animals, and how they connect, or even if they *can* connect, with the assumptions that influence experimental research that is supposedly more objective. It is highly likely that those who demand hard data bearing on animal welfare will not be convinced by Slater's view, and highly likely that they will shut the door to further discourse because Slater's position is merely stated, not argued. Still, Slater's challenging statement is one that should motivate considerable discussion among those interested in animal welfare.

One obvious and important affinity between assumptions underlying scientific research and views on animal welfare is that neither is completely objective. Each of us comes to science with biases, just as each of us brings biases to our views of nonhuman animals. Slater (as have many others) notes that studying animal welfare from an objective scientific viewpoint poses difficult problems. As Fumento (1993:366) observed, scientists "don't like to see a 'blend[ing of] the natural sciences, values, and social sciences,' because inevitably this leads to the subjugation of scientific truth." Perhaps the problems associated with studying animal welfare from an objective scientific viewpoint are insoluble, for among the reasons that an objective view cannot be attained is that it may be impossible to be objective about the use of animals by humans. If this is the case, what are we to do about it? Certainly we cannot let the animals suffer simply because of our inability to accept the fact that an objective study of animal welfare is impossible (Bekoff and Jamieson 1991). We should use common sense in our interactions with animals (Bekoff 1992, 1993c). Who could doubt that the pain and suffering of carnivores and many other animals are real, and that this realization must inform our treatment of them? Neither dismissing the notion that humans have obligations to the animals whom they use in research nor ignoring the issues attendant upon that will make the problems disappear.

Finally, we should teach our children and our students well (see, for example, Dickinson 1988). They will live and work in a world in which, increasingly, science will not be seen as a self-justifying activity, but rather as another human institution whose claims on the public treasury must be justified (Jamieson 1993b). More than ever before it is important for students to realize that to question science is not to be anti-science or anti-intellectual, and that to question how humans interact with animals is not in itself to demand that humans never use animals. Questioning science will make for better, more responsible science, and questioning the ways in which humans use animals will make for more informed decisions about animal use. By making such decisions in an informed and responsible way, we can help to ensure that in the future we will not repeat the mistakes of the past, and that we will move toward a world in which humans and other animals will be able to share peaceably the resources of a finite planet.

Acknowledgments

We thank Susan Townsend, Lori Gruen, Tom Daniels, Hal Herzog, Dan Blumstein, Gordon Burghardt, and especially Ned Hettinger for discussing many of these issues with us, and Ann Pusey, David Scheel, Hal Herzog, and Tom Daniels for taking time to respond in detail to our requests for information concerning natural predation. Michael Mooring kindly shared his experiences as a foreign researcher in Zimbabwe. Ned Hettinger, Elizabeth J. Farnsworth, Susan Townsend, Colin Allen, Gordon Burghardt, Lewis Petrinovich, and John Gittleman provided helpful comments on an ancestral version of this chapter.

References

Altmann, S. 1993. Professional ethics: The ABS code of ethics. *Anim. Behav. Soc. Newsletter* Fall:3–5.

American Society of Mammalogists. 1987. Acceptable field methods in mammalogy: Preliminary guidelines approved by the American Society of Mammalogists. *J. Mammalogy* 68 (4, Suppl.):1–18.

Anonymous. 1992. Guidelines for the protection of bat roosts. *J. Mammal.* 73:707–710.

Asa, C. S. 1993. Relative contributions of urine and anal-sac secretions in scent marks of large felids. *Amer. Zool.* 33:167–172.

Bartlett, T. Q., Sussman, R. W., and Cheverud, J. M. 1993. Infant killing in primates: A review of observed cases with specific reference to the sexual selection hypothesis. *Amer. Anthropol.* 95:958–990.

Beck, A. M. 1973. *The Ecology of Stray Dogs: A Study of Free-Ranging Urban Animals.* Baltimore: York Press.

Bekoff, M. 1989. Behavioral development of terrestrial carnivores. In: J. L. Gittleman, ed. *Carnivore Behavior, Ecology, and Evolution,* pp. 89–124. Ithaca, N.Y.: Cornell Univ. Press.

Bekoff, M. 1992. Scientific ideology, animal consciousness, and animal protection: A principled plea for unabashed common sense. *New Dir. Psychol.* 10:79–94.

Bekoff, M. 1993a. Experimentally induced infanticide: The removal of birds and its ramifications. *Auk* 110:404–406.

Bekoff, M. 1993b. Should scientists bond with the animals whom they use? Why not? *Psycholoquy* 4(37):1–8.

Bekoff, M. 1993c. Common sense, cognitive ethology and evolution. In: P. Cavalieri and P. Singer, eds. *The Great Ape Project: Equality beyond Humanity,* pp. 102–108. London: Fourth Estate.

Bekoff, M. 1994a. Cognitive ethology and the treatment of non-human animals: How matters of mind inform matters of welfare. *Anim. Welfare* 3:75–96.

Bekoff, M. 1994b. Scientists, animals, and their social bonds. *New Sci.* 21 May:44–45.

Bekoff, M. 1994c. Should scientists bond with the animals whom they use? Why not? *Intl. J. Comp. Psychol.* 7:78–86.

Bekoff, M. 1995. Cognitive ethology and the explanation of nonhuman animal behavior. In: J.-A. Meyer and H. L. Roitblat, eds. *Comparative Approaches to Cognitive Science,* pp. 119–149. Cambridge, Mass.: MIT Press.

Bekoff, M., and Allen, C. 1996. Cognitive ethology: Slayers, skeptics, and proponents. In: R. W. Mitchell, N. Thompson, and L. Miles, eds. *Anthropomorphism, Anecdotes, and Animals: The Emperor's New Clothes?* Albany, N.Y.: SUNY Press.

Bekoff, M., and Gruen, L. 1993. Animal welfare and individual characteristics: A conversation against speciesism. *Ethics and Behavior* 3:163–175.

Bekoff, M., Gruen, L., Townsend, S. E., and Rollin, B. E. 1992. Animals in science: Some areas revisited. *Anim. Behav.* 44:473–484.

Bekoff, M., and Hettinger, N. 1994. Animals, nature, and ethics. *J. Mammal.* 75:219–223.

Bekoff, M., and Jamieson, D. 1990. Cognitive ethology and applied philosophy: The significance of an evolutionary biology of mind. *Trends Ecol. Evol.* 5:156–159.

Bekoff, M., and Jamieson, D. 1991. Reflective ethology, applied philosophy, and the moral status of animals. *Persp. in Ethology* 9:1–47.

Bekoff, M., and Wells, M. C. 1986. Social ecology and behavior of coyotes. *Adv. Study Behav.* 16:251–338.

Benton, T. 1993. *Natural Relations: Ecology, Animal Rights, and Justice.* New York: Verso.

Bertram, B. 1993. Prisoners of fate. *Nature* 363:219.

Bonta, M., 1991. Snake attack. *Bird Watcher's Digest* July/August:96.

Bostock, S. St. C. 1993. *Zoos and Animal Rights.* New York: Routledge.

Bradshaw, J. W. S., ed. 1992. *The Behavior of the Domestic Cat.* Wallingford, Conn.: CAB International.

Brenes, A., ed. 1988. *The Comparative Psychology of Natural Resource Management.* Cosenza, Italy: University of Calabria.

Broom, D. M., and Johnson, K. G. 1993. *Stress and Animal Welfare.* London: Chapman and Hall.

Bulger, R. E., Heitman, E., and Reiser, S. J., eds. 1993. *The Ethical Dimensions of the Biological Sciences.* New York: Cambridge Univ. Press.

Burghardt, G. M. 1991. Cognitive ethology and critical anthropomorphism: A snake with two heads and hognose snakes that play dead. In: C. Ristau, ed. *Cognitive Ethology: The Minds of Other Animals,* pp. 53–90. Hillsdale, N.J.: Lawrence Erlbaum Associates.

Burley, N., Krantzberg, G., and Radman, P. 1982. Influence of color-banding on the conspecific preferences of zebra finches. *Anim. Behav.* 30:444–455.

Byrne, R. W. 1991. Brute intellect. *Sciences* July:42–47.

Caine, N. G. 1992. Humans as predators: Observational studies and the risk of pseudoreplication. In: H. Davis and D. Balfour, eds. *The Inevitable Bond: Examining Scientist-Animal Interactions,* pp. 357–364. New York: Cambridge Univ. Press.

Callicott, J. B. 1980/1989. Animal liberation: A triangular affair. In: J. B. Callicott, ed. *In Defense of the Land Ethic: Essays in Environmental Philosophy,* pp. 15–38. Albany, N.Y.: SUNY Press.

Carruthers, P. 1989. Brute experience. *J. Philos.* 86:258–269.

Carruthers, P. 1992. *The Animals Issue: Moral Theory in Practice.* New York: Cambridge Univ. Press.

Carson, G. 1972. *Men, Beasts, and Gods: A History of Cruelty and Kindness to Animals.* New York: Scribners.

Cavalieri, P., and Singer, P., eds. 1993. *The Great Ape Project: Equality beyond Humanity.* London: Fourth Estate.

Chiszar, D., Murphy, J. B., and Illiff, W. 1990. For zoos. *Psychol. Rec.* 40:3–13.

Cohen, C. 1986. The case for the use of animals in research. *New England J. Med.* 315:865–870.

Collins, H. M., and Pinch, T. F. 1979. The construction of the paranormal: Nothing unscientific is happening. In: R. Wallis, ed. *On the Margins of Science: The Social Construction of Rejected Knowledge,* pp. 237–272. Keele, U.K.: Keele Univ. Press.

Colwell, R. K. 1989. Natural and unnatural history: Biological diversity and genetic engineering. In: W. R. Shea and B. Sitter, eds. *Scientists and Their Responsibilities*, pp. 1–40. Canton, Mass.: Watson Pub. International.

Crockett, C. M. 1993. Rigid rules for promoting psychological well-being are premature. *Amer. J. Primat.* 30:177–179.

Croft, D. B., ed. 1991. *Australian People and Animals in Today's Dreamtime: The Role of Comparative Psychology in the Management of Natural Resources*. New York: Praeger.

Cuthill, I. 1991. Field experiments in animal behaviour: Methods and ethics. *Anim. Behav.* 42:1007–1014.

Dagg, I. 1982. Harems and other horrors: Sexual bias in behavioral biology. Waterloo, Ont.: Otter Press.

Daniels, T. J., and Bekoff, M. 1989a. Spatial and temporal resource use by feral and abandoned dogs. *Ethology* 81:300–312.

Daniels, T. J., and Bekoff, M. 1989b. Population and social biology of free-ranging dogs, *Canis familiaris*. *J. Mammal.* 70:754–762.

Daniels, T. J., and Bekoff, M. 1989c. Feralization: The making of wild domestic animals. *Behav. Processes* 19:79–94.

Daniels, T. J., and Bekoff, M. 1990. Domestication, exploitation, and rights. In: M. Bekoff and D. Jamieson, eds. *Interpretation and Explanation in the Study of Animal Behavior, Vol. II: Explanation, Evolution, and Adaptation*, pp. 345–377. Boulder, Colo.: Westview Press.

Davis, H., and Balfour, D., eds. 1992. *The Inevitable Bond: Examining Scientist-Animal Interactions*. New York: Cambridge Univ. Press.

Dawkins, R. 1993. Gaps in the mind. In: P. Cavalieri and P. Singer, eds. *The Great Ape Project: Equality beyond Humanity*, pp. 80–87. London: Fourth Estate.

DeBlieu, J. 1991. *Meant to Be Wild: The Struggle to Save Endangered Species through Captive Breeding*. Golden, Colo.: Fulcrum Pub.

DeGrazia, D. 1991. The moral status of animals and their use in research: A philosophical review. *Kennedy Inst. of Ethics J.* March:48–70.

Dennett, D. 1987. *The Intentional Stance*. Cambridge, Mass.: MIT Press.

Dickinson, P. 1988. *Eva*. New York: Dell Publishing.

Dresser, R. 1993. Defining scientific misconduct: The relevance of mental state. *J. Amer. Med. Assoc.* 269:895–897.

Duffus, D. A., and Wipond, K. J. 1992. A review of the institutionalization of wildlife viewing in British Columbia, Canada. *Northwest Env. J.* 8:325–345.

Duncan, I. J. H. 1993. Welfare is to do with what animals feel. *J. Agric. Env. Ethics* 6 (Spec. Suppl. 2):8–14.

Duncan, I. J. H., and Petherick, J. C. 1991. The implications of cognitive processes for animal welfare. Unpublished paper presented at symposium on "Cognition and Awareness in Animals: Do Farm Animals' Perceptions Affect Their Production or Well-Being?" Lexington, Kentucky.

Eisemann, C. H., Jorgensen, W. K., Merritt, D. J., Rice, M. J., Cribb, B. W., Webb, P. D., and Zalucki, M. P. 1984. Do insects feel pain? A biological review. *Experientia* 40:164–167.

Elwood, R. W. 1991. Ethical implications of studies on infanticide and maternal aggression in rodents. *Anim. Behav.* 42:841–849.

Emlen, S. T. 1993. Ethics and experimentation: Hard choices for the field ornithologist. *Auk* 110:406–409.

Emlen, S. T., Demong, N. J., and Emlen, D. J. 1989. Experimental induction of infanticide in female wattled jacanas. *Auk* 106:1–7.

Ewbank, R. 1993. Editorial. *Anim. Welfare* 2:2.

Farnsworth, E. J., and Rosovsky, J. 1993. The ethics of ecological field experimentation. *Conserv. Biol.* 7:463–472.

Ferrer, M., and Hiraldo, F. 1992. Man-induced sex-biased mortality in the Spanish imperial eagle. *Biol. Conserv.* 60:57–60.

Finsen, L., and Finsen, S. 1994. *The Animal Rights Movement.* New York: Twayne.

Fiorito, G. 1986. Is there "pain" in invertebrates? *Behav. Processes* 12:383–388.

Foucault, M. 1984. *The Foucault Reader* (P. Robinow, ed.). New York: Pantheon.

Freeman, M. M. R., Wein, E. E., and Keither, D. E. 1992. *Recovering Rights: Bowhead Whales and Inuvialuit Subsistence in the Western Arctic.* Edmonton, Alta.: Canadian Circumpolar Institute.

Frey, R. G. 1980. *Interests and Rights: The Case against Animals.* New York: Oxford Univ. Press.

Frey, R. G. 1983. *Rights, Killing, and Suffering.* Oxford: Blackwell.

Fumento, M. 1993. *Science under Siege: Balancing Technology and the Environment.* New York: Morrow.

Gallup, G. G., and Suarez, S. D. 1987. Antivivisection: Questions of logic, consistency, and conceptualization. *Theoret. Philos. Psychol.* 7:81–94.

Galvin, S. L., and Herzog, H. A., Jr. 1992. Ethical ideology, animal rights activism, and attitudes toward the treatment of animals. *Ethics and Behavior* 2:141–149.

Gavaghan, H. 1992. Animal experiments the American way. *New Sci.* 16 May:32–36.

Goodall, J. 1990. *Through a Window.* Boston: Houghton-Mifflin.

Goodstein, D. 1991. Scientific fraud. *Amer. Scholar* 60:505–515.

Grandin, T. 1992. Author's response to M. Bekoff's "Some thoughts about cattle restraint." *Anim. Welfare* 1:230–231.

Gray, G. G. 1993. *Wildlife and People: The Human Dimensions of Wildlife Ecology.* Urbana: Univ. Illinois Press.

Greenough, W. T. 1992. More on monkeys. *Discover* June:9.

Grooms, S. 1993. *The Return of the Wolf.* Minocqua, Wisconsin: NorthWord Press.

Guillermo, K. S. 1993. *Monkey Business.* Washington, D.C.: National Press Books.

Haraway, D. 1989. *Primate Visions: Gender, Race, and Nature in the World of Modern Science.* New York: Routledge.

Hardwig, J. 1985. Epistemic dependence. *J. Philos.* 82:335–349.

Hardwig, J. 1991. The role of trust in knowledge. *J. Philos.* 88:693–708.

Hardy, D. T. 1990. *America's New Extremists: What You Need to Know about the Animal Rights Movement.* Washington, D.C.: Washington Legal Foundation.

Hargrove, E., ed. 1992. *The Animal Rights/Environmental Ethics Debate.* Albany, N.Y.: SUNY Press.

Hauser, M. D. 1993. Do vervet monkey infants cry wolf? *Anim. Behav.* 45:1242–1244.

Hettinger, N. 1989. The responsible use of animals in biomedical research. *Between the Species* 5:123–131.

Hettinger, N. 1994. Valuing predation in Rolston's environmental ethics: Bambi lovers versus tree huggers. *Env. Ethics* 16:3–20.

Hofer, H., East, M. L., and Campbell, K. L. I. 1993. Snares, commuting hyaenas and migratory herbivores: Humans as predators in the Serengeti. In: N. Dunstone and M. L. Gorman, eds. *Mammals as Predators,* pp. 347–366. Oxford: Clarendon Press.

Howard, W. E. 1993. Animal research is defensible. *J. Mammal.* 74:234–235.

Huntingford, F. A. 1984. Some ethical issues raised by studies of predation and aggression. *Anim. Behav.* 32:210–215.

Isbell, L. A., and Young, T. P. 1993. Human presence reduced predation in a free-ranging vervet monkey population in Kenya. *Anim. Behav.* 45:1233–1235.

Jamieson, D. 1985a. Against zoos. In: P. Singer, ed., *In Defense of Animals,* pp. 108–117. New York: Harper and Row.

Jamieson, D. 1985b. Experimenting on animals: A reconsideration. *Between the Species* 1:4–11.

Jamieson, D. 1993a. Ethics and animals: A brief review. *J. Agric. Env. Ethics* 6 (Spec. Suppl. 1):15–20.

Jamieson, D. 1993b. What will society expect of the future research community. Unpublished paper presented at National Research Council's Commission on Physical Sciences, Mathematics, and Applications, 13–15 August.

Jamieson, D. 1995. Zoos revisited. In: B. G. Norton, M. Hutchins, E. F. Stevens, and T. L. Maple, eds. *Ethics of the Ark: Zoos, Animal Welfare, and Conservation,* pp. 52–66. Washington, D.C.: Smithsonian Institution Press.

Jamieson, D., and Bekoff, M. 1992. Carruthers on nonconscious experience. *Analysis* 52:23–28.

Jamieson, D., and Bekoff, M. 1993. On aims and methods of cognitive ethology. *Philos. Sci. Assoc.* 2:110–124.

Jamieson, D., and Regan, T. 1985. Whales are not cetacean resources. In: M. W. Fox and L. Mackley, eds. *Advances in Animal Welfare Science, 1984,* pp. 101–111. The Hague: Martinus Nijhoff.

Johnson, E. 1991. Carruthers on conscious and moral status. *Between the Species* 7:190–192.

Johnson, L. E. 1991. *A Morally Deep World: An Essay on Moral Significance and Environmental Ethics.* New York: Cambridge Univ. Press.

Katz, E. 1993. Artefacts and functions: A note on the value of nature. *Env. Values* 2:223–232.

Kennedy, J. S. 1992. *The New Anthropomorphism.* New York: Cambridge Univ. Press.

Kennedy, S. P., and Knight, R. L. 1992. Flight distances of black-billed magpies in different regimes of human density and persecution. *The Condor* 94:545–547.

Kew, B. 1991. *The Pocketbook of Animal Facts and Figures.* London: Green Print.

Kiley-Worthington, M. 1990. *Animals in Circuses and Zoos: Chiron's World?* Essex: Little Eco-Farms Pub.

Kirkwood, J. 1992. Wild animal welfare. In: R. D. Ryder and P. Singer, eds. *Animal Welfare and the Environment,* pp. 139–154. London: Duckworth.

Kitcher, P. 1993. *The Advancement of Science: Science without Legend, Objectivity without Illusions.* New York: Oxford Univ. Press.

LaFollette, M. C. 1992. *Stealing into Print: Fraud, Plagiarism, and Misconduct in Scientific Publishing.* Berkeley: Univ. California Press.

Lang, S. 1993. Questions of scientific responsibility: The Baltimore case. *Ethics and Behavior* 3:3–72.

Lansdell, H. 1988. Laboratory animals need only humane treatment: Animal "rights" may debase human rights. *Internat. J. Neurosci.* 42:169–178.

Laurenson, M. K., and Caro, T. M. 1994. Monitoring the effects of non-trivial handling in free-living cheetahs. *Anim. Behav.* 47:547–557.

Lockwood, J. A. 1987. The moral status of insects and the ethics of extinction. *Florida Entomol.* 70:70–89.

Ludwig, D., Hilborn, R., and Walters, C. 1993. Uncertainty, resource exploitation, and conservation: Lessons from history. *Science* 260:17.

Lynge, F. 1992. *Arctic Waters: Animal Rights, Endangered Peoples.* Hanover, N.H.: Univ. Press of New England.

Magel, C. 1989. *Keyguide to Information Sources in Animal Rights.* Jefferson, N.C.: McFarland.

Mainini, B., Neuhaus, P., and Ingold, P. 1993. Behaviour of marmots *Marmota marmota* under the influence of different hiking activities. *Biol. Conserv.* 64:161–164.

McAdam, D. 1992. Radicals and others. *Science* 255:1448–1449.

McGinn, C. 1980. Book review of Frey 1980. *Times Lit. Suppl.* 1 August:865.

McOrist, S., and Lenghaus, C. 1992. Mortalities of little penguins (*Eudyptula minor*) following exposure to crude oil. *Vet. Record* 22 February:161–162.

Mellon, J. D. 1993. A comparative analysis of scent-marking, social and reproductive behavior in 20 species of small cats (*Felis*). *Amer. Zool.* 33:151–166.

Michener, G. R. 1989. Ethical issues in the use of wild animals in behavioral and ecological research. In: J. W. Driscoll, ed. *Animal Care and Use in Behavioral Research: Regulations, Issues, and Applications*. Beltsville, Md.: Information Center.

Midgley, M. 1983. *Animals and Why They Matter*. Athens, Ga.: Univ. Georgia Press.

Mighetto, L. 1991. *Wild Animals and American Environmental Ethics*. Tucson: Univ. Arizona Press.

Montgomery, S. 1991. *Walking with the Great Apes: Jane Goodall, Dian Fossey, Birute Galdikas*. Boston: Houghton-Mifflin.

Morris, D. 1990. *The Animal Contract*. New York: Warner Books.

Norton, B. G., Hutchins, M., Stevens, E. F., and Maple, T. L., eds. 1995. *Ethics of the Ark: Zoos, Animal Welfare, and Conservation*. Washington, D.C.: Smithsonian Institution Press.

Orford, H. J. L., Perrin, M. R., and Berry, H. H. 1989. Contraception, reproduction, and demography of free-ranging Etosha lions (*Panthera leo*). *J. Zool. (Lond.)* 216:717–734.

Orlans, F. B. 1993. *In the Name of Science: Issues in Responsible Animal Experimentation*. New York: Oxford Univ. Press.

Peterson, D. 1989. *The Deluge and the Ark: A Journey in Primate Worlds*. Boston: Houghton Mifflin.

Peterson, D., and Goodall, J. 1993. *Visions of Calaban: On Chimpanzees and People*. Boston: Houghton Mifflin.

Pietz, P. J., Krapu, G. L., Greenwood, R. J., and Lokemoen, J. T. 1993. Effects of harness transmitters on behavior and reproduction of wild mallards. *J. Wildl. Mgmt.* 57:696–703.

Pimm, S. L. 1991. *The Balance of Nature? Ecological Issues in the Conservation of Species and Communities*. Chicago: Univ. Chicago Press.

Plous, S. 1991. An attitude survey of animal rights activists. *Psychol. Sci.* 2:194–196.

Pluhar, E. B. 1993. Arguing away suffering: The neo-Cartesian revival. *Between the Species* 9:27–41.

Preece, R., and Chamberlain, L. 1993. *Animal Welfare and Human Values*. Waterloo, Ont.: Wilfrid Laurier Press.

Quinn, D. 1993. *Ishmael*. New York: Bantam.

Rabinowitz, A. 1986. *Jaguar: Struggle and Triumph in the Jungles of Belize*. New York: Arbor.

Rabinowitz, A. 1991. *Chasing the Dragon's Tail: The Struggle to Save Thailand's Wild Cats*. New York: Doubleday.

Rachels, J. 1989. Why animals have a right to liberty. In: T. Regan and P. Singer, eds. *Animal Rights and Human Obligations*. Second edition, pp. 122–131. Englewood Cliffs, N.J.: Prentice-Hall.

Rachels, J. 1990. *Created from Animals: The Moral Implications of Darwinism*. New York: Oxford Univ. Press.

Ramsay, M. A., and Stirling, I. 1986. Long-term effects of drugging and handling free-ranging polar bears. *J. Wildl. Mgmt.* 50:619–626.

Raynor, M. E. 1992. Necessary suffering? *N. Y. Rev. Books* 13 August:664.

Regan, T. 1983. *The Case for Animal Rights.* Berkeley: Univ. California Press.

Rollin, B. E. 1981. *Animal Rights and Human Morality.* Buffalo, N.Y.: Prometheus Books.

Rollin, B. E. 1989. *The Unheeded Cry: Animal Consciousness, Animal Pain, and Science.* New York: Oxford Univ. Press.

Rolston, H., III. 1988. *Environmental Ethics: Duties to and Values in the Natural World.* Philadelphia: Temple Univ. Press.

Rothschild, M. 1993. Thinking about animal consciousness. *J. Nat. Hist.* 27:509–512.

Rowan, A. N. 1984. *Of Mice, Models, and Men: A Critical Evaluation of Animal Research.* Albany, N.Y.: SUNY Press.

Rowan, A. N., ed. 1988. *Animals and People Sharing the World.* Hanover, N.H.: Univ. Press New England.

Ryder, R. D. 1989. *Animal Revolution: Changing Attitudes towards Speciesism.* Oxford: Basil Blackwell.

Ryder, R. D. 1992. Painism: Ethics, animal rights and environmentalism. Cardiff, Wales: Centre for Applied Ethics.

Sapontzis, S. F. 1984. Predation. *Ethics and Animals* 5:27–38.

Sapontzis, S. F. 1987. *Morals, Reason, and Animals.* Philadelphia: Temple Univ. Press.

Schaller, G. B. 1993. *The Last Panda.* Chicago: Univ. Chicago Press.

Shackley, M. 1992. Manatees and tourism in southern Florida: Opportunity or threat? *J. Env. Mgmt.* 34:257–265.

Shapiro, K. J. 1989. The death of the animal: Ontological vulnerability. *Between the Species* 5:183–194.

Sherman, P. W., and Alcock, J., eds. 1993. *Exploring Animal Behavior: Readings from American Scientist.* Sunderland, Mass.: Sinauer.

Shrader-Frechette, K. S., and McCoy, E. D. 1993. *Method in Ecology: Strategies for Conservation.* New York: Cambridge Univ. Press.

Sih, A., Crowley, P., McPeek, M., Petranka, J., and Strohmeier, K. 1985. Predation, competition, and prey communities: A review of field experiments. *Ann. Rev. Ecol. System.* 16:269–311.

Singer, P. 1975. *Animal Liberation.* New York: Avon.

Singer, P. 1990. *Animal Liberation.* Second edition. N. Y. Rev. Books.

Singer, P. 1992a. Reply. *N. Y. Rev. Books* 13 August:65.

Singer, P. 1992b. Reply. *N. Y. Rev. Books* 15 November:60–61.

Singer, P. 1993a. *Practical Ethics.* Second edition. New York: Cambridge Univ. Press.

Singer, P. 1993b. Reply. *N. Y. Rev. Books* 27 May:49–50.

Slater, P. 1993. Review of Kennedy 1992. *Anim. Welfare* 2:187–188.

Small, R. J., and Keith, L. B. 1992. An experimental study of red fox predation on arctic and snowshoe hares. *Canadian J. Zool.* 70:1614–1621.

Stamp Dawkins, M., and Gosling, L. M., eds. 1992. *Ethics in Research on Animal Behaviour: Readings from Animal Behaviour.* London: Academic Press.

Swazey, J. P., Anderson, M. S., and Louis, K. S. 1993. Ethical problems in academic research. *Amer. Sci.* 81:542–553.

Terborgh, J. 1993. Solitary enigma: A review of Schaller 1993. *N. Y. Rev. Books* 23 September:12–14.

Thorne, C., ed. 1992. *The Waltham Book of Dog and Cat Behaviour.* New York: Pergamon.

Varner, G. E., and Monroe, M. C. 1991. Ethical perspectives on captive breeding: Is it for the birds? *Endangered Species Update* 8:27–29.

Wayne, R. K., and Gittleman, J. L. 1995. The problematic red wolf. *Sci. Amer.* 273:36–39.

Webb, M. O. 1993. Why I know about as much as you: A reply to Hardwig. *J. Philos.* 90:260–270.

Wenzel, G. 1991. *Animal Rights, Human Rights, Ecology, Economy, and Ideology in the Canadian Arctic.* Toronto: Univ. Toronto Press.

White, R. J. 1990. Letter. *Hastings Center Report* November/December:43.

Wigglesworth, V. B. 1980. Do insects feel pain? *Antenna* 4:8–9.

Wilkie, T. 1993. *Perilous Knowledge: The Human Genome Project and Its Implications.* London: Faber.

Woolf, P. K. 1988. Deception in scientific research. *Jurimetrics J.* Fall.

Zimmerman, M. E., Callicott, J. B., Sessions, G., Warren, K. J., and Clark, J., eds. 1993. *Environmental Philosophy: From Animal Rights to Radical Ecology.* Englewood Cliffs, N.J.: Prentice-Hall.

CHAPTER 2

Behavioral Endocrinology and Social Organization in Dwarf Mongooses

SCOTT CREEL

Dwarf mongooses (*Helogale parvula*) are the smallest members of the Old World carnivore family Herpestidae. They are unusual among the herpestids not only for their size, but also because they live in cooperative groups (Figure 2.1). Two major factors appear to have favored sociality in dwarf mongooses and other small herpestids (Gorman 1979; Rood 1986; Waser and Waser 1985). First, they prey almost entirely on beetles and termites. Because these are evenly distributed and rapidly renewing prey, there is little cost to sharing a territory with conspecifics (Waser 1981; Waser and Waser 1985). Moreover, stealth is not critical when hunting invertebrates, so that pack mates cause little interference with prey capture (Gorman 1975, 1979). Second, dwarf mongooses are small and diurnal, and therefore highly vulnerable to predation (Rasa 1986; Rood 1986, 1990). Grouping allows for improved vigilance or defense against predators. For dwarf mongooses, improved vigilance is a major benefit, and guarding behavior is elaborate (Rood 1978; Rasa 1986).

For dwarf mongooses, guarding takes place in two contexts. The less elaborate form occurs when a pack is foraging. Mongooses often pause and move to a high point (e.g., a termite mound) to scan the area (Rasa 1986). If the guard sees a potential predator it alerts the pack with an alarm call. A guard that sees nothing alarming will normally return to foraging within a few minutes. The more costly form of cooperative vigilance is the guarding of offspring too young to forage with the pack, or babysitting (Rood 1978). When a dependent litter is confined to a den (e.g., in a termite mound), one mongoose (occasionally more) remains behind as a watchman while the pack forages. Babysitting mongooses typically remain at the den until another pack member returns to relieve them. Babysitters may remain alone with the young for several hours without foraging (Rood 1978).

In contrast to that of larger social carnivores, sociality in dwarf mongooses has little or nothing to do with cooperative hunting or defense of kills (Gittle-

Figure 2.1. Mutual grooming, a behavior that typifies the high levels of cooperation observed within dwarf mongoose packs.

man 1989; Kruuk 1972, 1975). Cooperative hunting favors sociality in some carnivores that tackle prey larger than themselves (e.g., lions, *Panthera leo* [Stander 1992a, 1992b; Stander and Albon 1993] or African wild dogs, *Lycaon pictus* [Creel and Creel 1995]), though several studies call the generality of these benefits into question (e.g., spotted hyenas, *Crocuta crocuta* [Mills 1985]; lions [Packer et al. 1990]; and cheetahs, *Acinonyx jubatus* [Caro 1994]).

Dwarf mongoose packs average 9.0 adults (range: 2–27), plus young-of-the-year from one to four sequential litters (range: 0–13). Adults of both sexes may be natal or immigrant, though dispersal is male-biased (Rood 1987; Waser et al. 1994). Subordinates of both sexes normally do not produce young of their own (Figure 2.2), but assiduously help with all aspects of raising the dominants' offspring (Rasa 1973; Rood 1978, 1980). Probably because of the energetic stress of producing up to four litters in 6 months, each averaging 22% of body mass (Creel et al. 1991), dominant females forage more than other pack members do during the breeding season (Creel and Creel 1991; Rood 1978). In most packs, subordinates perform most of the parental care other than suckling, including grooming, guarding, feeding, and carrying the young between dens (Rasa 1977, 1989; Rood 1978, 1983). When present, pregnant and pseudopregnant females nurse the dominants' young (Creel et al. 1991; Rood 1980).

Figure 2.2. A high-ranking female pins a subordinate: the dominance hierarchy leads to an unequal distribution of reproduction within a dwarf mongoose pack.

Dwarf mongooses' communal care of the young takes forms similar to that of communal care in many other social carnivores (e.g., feeding and guarding young: Macdonald and Moehlman 1982; Gittleman 1985), but the *degree* of cooperation in dwarf mongoose packs is greater than average. For example, dwarf mongooses are obligately cooperative breeders, unlike most other social carnivores: they cannot raise their young in unaided pairs (Creel 1990), probably because breeding females have adjusted their energetic investment in reproduction to take full advantage of helpers (Creel and Creel 1991).

The coefficient of relatedness among members of a dwarf mongoose pack averages 0.33 (Creel 1991; Creel and Waser 1994). Quantitative estimates of relatedness within carnivore groups are few (Gompper and Wayne, this volume), but this value is comparable to relatedness within groups of brown hyenas, *Hyaena brunnea* (0.26: Mills 1990), spotted hyenas (0.31: Mills 1985), red foxes, *Vulpes vulpes* (0.38: Macdonald 1979), and lionesses (0.25–0.50: Packer et al. 1991). This range of relatedness values is similar to that seen in communally breeding birds (Stacey and Koenig 1990 and references therein).

Although normally they are reproductively suppressed, subordinate dwarf mongooses sometimes do breed. Twelve percent of subordinate females become pregnant annually (Creel and Waser 1991, in press). The dominant female kills most offspring of subordinate females (Rasa 1973; Rood 1983), but subordinates sometimes raise offspring in a joint litter, following closely synchronized pregnancies (Creel and Waser 1991; Keane et al. 1994). Subordinate males occasionally father offspring of either dominant or subordinate mothers (Keane et al. 1994).

Social subordinates do not normally breed in most communally breeding carnivores (Macdonald and Moehlman 1982), but the degree of suppression varies widely both among and within species (Creel and Waser 1991, 1994). For example, only 6% of the African wild dog packs in Kruger National Park have more than a single breeding female (Reich 1981), compared with 38% in Masai Mara Reserve (Fuller et al. 1992). Eurasian badgers (*Meles meles*) also show wide variation among populations in the degree of female reproductive suppression (Woodroffe and Macdonald 1993).

In this chapter, I describe three features of dwarf mongooses' social organization: reproductive suppression of social subordinates, failures of reproductive suppression, and lactation by nonbreeders. (Where possible, I make comparisons to other carnivores.) I describe each facet both functionally and mechanistically, hoping to show that the two approaches shed light on one another. On the one hand, mechanistic data can help to define the traits to be considered by evolutionary analyses. For example, to address the evolution of reproductive suppression it is useful to know what aspect of reproduction has been modified in each sex. Mating rates? Access to mates? Gonadal hormone levels? Conversely, evolutionary logic often makes sense of mechanistic data. For example, knowing the endocrine mechanism underlying lactation by nonbreeders, a question remains open: why should a nulliparous female lactate at all?

The Study Population and Methods

All data I discuss come from a population of individually marked dwarf mongooses living near the Sangere River in Serengeti National Park, Tanzania. The study, which was begun in 1977 by Jon Rood, continuously monitored the demography of 12 to 20 packs on a 25 km² area of open *Acacia–Commiphora* woodland and tree savanna. Recordings of survivorship, reproduction, immigration, and emigration, made from 1977 through 1991, allowed life histories and genealogies to be assembled for 713 mongooses, most of known age. The ages of unmarked immigrants were estimated, using body mass and tooth wear. Endocrine, behavioral, and genetic data were collected in eight to ten fully marked packs from 1987 to 1991. Behavior, ecology, and methods of monitoring have been described by Rood (1978, 1980, 1983, 1986, 1987, 1990). Full methods for noninvasive endocrine, behavioral, and weight data are given by Creel et al. (1991, 1992, 1993). Genetic methods are given by Keane et al. (1994).

In brief, each mongoose was trapped, anesthetized, and marked with a unique and permanent freeze-mark using Freon-12 (Rood and Nellis 1980). Genetic analyses used muscle biopsies taken from the quadriceps. Marked packs were observed each day from 0800 to 1100 and from 1530 to 1830, when activity levels were high. Of ten fully marked packs, four were observed on foot (at and near the research station) and six were observed from a Land Rover. We used all-occurrences sampling, treating most behaviors as instantaneous. Some behaviors (e.g., duration of mating) were timed. Urine samples for radioimmunoassay were initially collected from trapped animals by keeping the trap over a stainless-steel pan. Subsequently, urine was collected without trapping, from a rubber pad (a sandal, in fact). Initially, the mongooses were given a food reward for marking the pad. The pad was washed between samples, and water blanks confirmed that cross-contamination was not a problem.

Reproductive Suppression of Social Subordinates

Serengeti dwarf mongoose packs average 9.0 adults, with the oldest individual of each sex socially dominant. Because the adult sex ratio is not skewed, subordinates outnumber dominants by 3.5 to 1 in both sexes. Despite forming only 22% of the female population, dominant females accounted for 73% of all pregnancies (219 of 302: Creel and Waser 1991). The 12% of subordinate females who become pregnant annually account for 27% of all pregnancies. Because subordinates' offspring are often killed by the dominant female, only 15% of offspring whose maternity was established by DNA fingerprinting had subordinate mothers (six of 39: Keane et al. 1994). Subordinate males are also suppressed to a large degree, and father just 24% of the offspring with known paternity (ten of 41).

Mongoose reproductive physiology is poorly known. Limited histological data (Pearson and Baldwin 1953; Hinton and Dunn 1967; Lynch 1980) suggest that ovulation is induced by mating in meerkats (suricates, *Suricata suricatta*), yellow mongooses (*Cynictis penicillata*), and small Indian mongooses (*Herpestes auropunctatus*). High mating rates (up to 50 mounts/hour: Creel et al. 1993) suggest that dwarf mongooses may also be induced ovulators.

Behavioral Mechanisms of Suppression

Females in dwarf mongoose packs enter estrus in tight synchrony (Rood 1980). Subordinate females enter estrus roughly one day later than the dominant female, which suggests that subordinates begin follicular activity in response to diestrus in the dominant female (perhaps using estrogen-dependent cues from the dominant's urine, which subordinates inspect closely). Mating periods last from 1 to 7 days, with some mating by most individuals of both sexes on most days (Rood 1980; Creel et al. 1992). Mating rates peak in the middle of the estrous period, when the dominant male stays within a meter (usually within centimeters or in contact) of the dominant female almost continuously (Rasa 1973, 1987; Rood 1980). Nonetheless, subordinate males occasionally mate with the alpha female, even during the peak of estrus (Rood 1980; Creel et al. 1992). Subordinate females may mate with males of all ranks throughout estrus (but see below).

Breeding coincides with the long rains (and abundant food: see Waser et al. 1995) between November and May (Rood 1980). Dominant females produce up to four litters in this interval (mean = 2.4: Creel and Waser 1991). Dwarf mongooses do not undergo lactational anestrus, and often conceive 1–2 weeks postpartum, while lactating heavily (Rood 1980). As a consequence, dominant females are often pregnant and lactating simultaneously, a severe energetic stress that they cannot maintain without helpers (Creel 1990; Creel and Creel 1991; Rasa 1987).

Although most adult mongooses mate, subordinates of both sexes mate less often than do the dominants (Figure 2.3). Considering all mounts (whether apparently ejaculatory or not), subordinate females mate 40% less often than dominants (2.0 ± 0.2 versus 3.3 ± 0.4 mounts/hour: $t = 3.73$, $P < 0.001$). Among males, subordinates mate 56% less often than the dominants (1.7 ± 0.1 versus 3.8 ± 0.6 mounts/hour: $t = 5.14$, $P < 0.001$). These patterns also

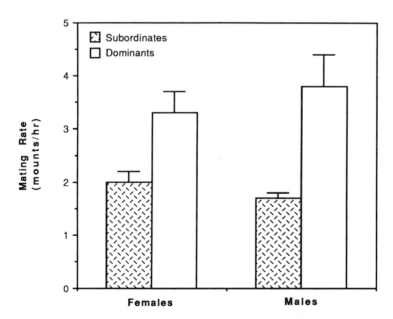

Figure 2.3. Hourly mating rates for dominant (open bars) and subordinate (shaded bars) dwarf mongooses.

hold for mounts with intromission and for ejaculatory mounts (Creel et al. 1992; see Rasa 1984, pp. 176–179, for criteria).

In addition to mating less often, subordinates are more likely to mate with other subordinates. This pattern can be seen in a significant correlation between the ranks of individuals in mating pairs (Kendall's tau = 0.39, $P <$ 0.001: Creel et al. 1992). For example, dominant males took part in 52% of 99 ejaculatory matings with females of all ranks, but garnered a higher percentage (79%) of the 33 observed ejaculatory matings with dominant females. Because only dominant females were reliably fertile, the dominant males' priority of access to them is important. Access to high-ranking males might also be important for females, because low-ranking males may have lower ejaculate volume (see below).

Subordinate males suffer a final indignity. Estrous females rarely reject mounts, but they do exhibit mate choice by mounting a male, then walking past to stand with tail averted. When females solicited matings in this way, they showed a significant preference for the dominant male, regardless of their own rank (70% of solicitations: z = 5.01, $P <$ 0.001).

Aggressive interference by dominants is an obvious mechanism by which subordinates can be precluded from mating. In gray wolves, for example, interference of this sort may be the primary cause of reproductive suppression (Packard et al. 1983). Among male dwarf mongooses, rates of aggression increase dramatically during mating periods (Figure 2.4: 0.72 ± 0.19

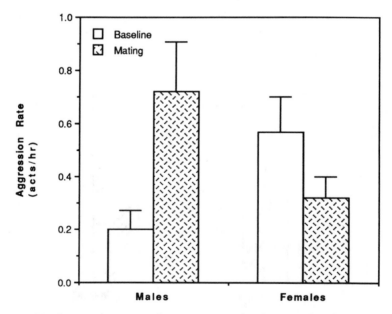

Figure 2.4. Hourly rates of aggression during nonmating (baseline) periods (open bars) and mating periods (shaded bars) for dwarf mongooses.

fights/hour, elevated from 0.20 ± 0.07 at baseline: $t = 3.39$, $P < 0.001$). This increase in aggression is almost completely due to fights over access to females, especially the alpha female (Rasa 1973; Rood 1980). During the peak of estrus, the alpha male maintains a daylong consort with the alpha female, rarely straying more than a meter, and violently attacking males who come between them (Rasa 1987; Rood 1980). Aggression also increases among high-ranking subordinate males, owing to competition for access to high-ranking subordinate females (who are more likely than low-ranking females to become pregnant: Creel and Waser 1991; Creel et al. 1992). For males, aggressive interference is an important component of reproductive success or suppression.

For females, the opposite is true. There is no increase in aggression during mating periods (Figure 2.4: 0.32 ± 0.08 fights/hour versus 0.57 ± 0.13 at baseline: $P > 0.05$). Alpha females rarely attack mating subordinates, and rates of aggression actually show a tendency to *decrease* during mating periods, perhaps owing to time taken up by mating itself.

In summary, behavioral data demonstrate that low mating rates are a component of reproductive suppression for both sexes, and that aggression directly mediates mating rates for males but not for females. Two questions arise. First, if subordinate females are not actively prevented from mating, why are their mating rates low? Second, what controls the elevation of aggression seen among males?

Endocrine Mechanisms of Suppression

Reproduction in mammals is endocrinologically controlled by the hypo-thalamic-hypophyseal-gonadal axis (Nalbandov 1976). This system has three levels linked by feedback loops. The peptide hormones of the upper two levels typically undergo rapid, pulsatile fluxes that field studies would be hard pressed to monitor. Mating and aggressive behavior, however, are controlled primarily by steroid hormones produced by the gonads, which can be assayed noninvasively using urine or fecal samples (e.g., Shideler et al. 1983). Radioim-munoassay of free or conjugated urinary steroids can be used to describe endocrine events with durations of hours (e.g., peaks of free cortisol in response to stress: Creel 1992) to days or weeks (e.g., elevated estrogen conjugates during estrus or pregnancy: Creel et al. 1992). Social carnivores' widespread use of urine and feces in scent marking (Gorman and Trowbridge 1989) can be exploited to collect samples for endocrine study of known individuals in the wild, without handling (e.g., urine from dwarf mongooses and feces from African wild dogs).

Among female dwarf mongooses, urinary estrogen conjugates (EC) remain at baseline levels outside of mating periods (Creel et al. 1992). At baseline, the EC levels of dominants (8.1 ± 1.4 ng EC/mg creatinine) are signifi-

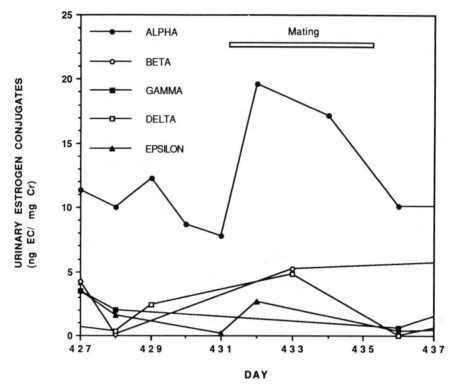

Figure 2.5. Urinary-estrogen-conjugate concentrations in five females of a single dwarf mongoose pack during a synchronized estrous period. All females mated on days 431 to 435.

cantly higher than those of subordinates (5.0 ± 0.9 ng EC/mg creatinine: Kolmogorov-Smirnov $DN = 0.56$, $P < 0.03$). There is a tendency for EC levels to increase with dominance even among subordinates (partial regression removing diurnal effects, $F = 1.74$, $P < 0.08$). During mating periods, the effect of dominance on estrogen levels becomes more exaggerated (Figure 2.5). Dominant females' EC levels peaked at 17.9 ± 3.5 ng EC/mg creatinine, far higher than the subordinates' peak values (4.8 ± 0.4 ng EC/mg creatinine: Kolmogorov-Smirnov $DN = 0.65$, $P < 0.001$). During mating periods, the EC levels of high-ranking subordinates were significantly closer to those of dominants than were the EC levels of low-ranking subordinates (partial $F = 25.29$, $P < 0.001$).

After mating, suppressed subordinates failed to establish pregnancy. Pregnant females' EC levels increased to 70-fold above baseline, but the EC levels of suppressed subordinates returned to baseline (5.0 ng EC/mg creatinine) when mating ceased. Among subordinates that did become pregnant, EC profiles were indistinguishable from those of pregnant dominant females. Never did the EC levels of a subordinate rise to a level indicating pregnancy, then fall prior to term, indicating abortion.

Taken together, the data show that reproductive suppression among female mongooses operates by blocking ovulation, fertilization, or implantation, but not by abortion of established pregnancies. Speculatively, failure of ovulation seems the most likely stage, though in rodents (McClintock 1987), mating patterns are also known to affect implantation. Because female mating behavior is estrogen-dependent (Beach 1976; Carter 1992), low baseline estrogens would be expected to lead to the low mating rates observed. Among mating-induced ovulators, estrogens and mating normally stimulate one another by positive feedback: the combination of low initial EC levels and low mating rates would impair this feedback, leading to the very low peak EC levels observed. In turn, the poor buildup of EC during estrus might preclude the pulse of luteinizing hormone that triggers ovulation. This mechanism is consistent with the data in hand, and would lead to suppression even in the absence of interference by the dominant female, as observed.

Turning to males, recall that subordinates' mating rates are low at least in part because of aggressive interference by dominants. Do differences in rates of mating and aggression depend on differences in androgens? Apparently not. Urinary androgen levels were unaffected by dominance (Figure 2.6), both during nonmating ($t = 1.03$, $P = 0.30$) and mating periods ($t = 0.33$, $P = 0.74$). Moreover, androgen levels were uncorrelated with rates of mating, aggression, or "winning" in aggressive interactions (Creel et al. 1992, 1993). The only circumstance that significantly depressed androgens was the severe and prolonged aggression accompanying immigration events. Both immigrant males and those joined by immigrants showed significant short-term declines in androgens during fierce fighting (Creel et al. 1993). Given that androgens generally promote aggression, this result is a bit paradoxical (though social instability also led to decreased androgen levels in male olive baboons, *Papio anubis:* Sapolsky 1987).

Thus, males (unlike females) are suppressed primarily by being prevented from mating with fertile females, and not by depressed gonadal steroid levels. But there is also a physiological component to suppression in males. Females mate with multiple males, and multiple paternity occurs within litters (Keane et al. 1994). Considering this result, suppression can be thought of as a kind of sperm competition in which subordinates are handicapped by low mating rates. Subordinates operate under an additional physiological handicap: low rank is associated with smaller testes, after removing effects of body mass (partial regression, $t = 2.31$, $P < 0.02$: Creel et al. 1992). Although no data on ejaculate quality exist for dwarf mongooses, smaller testes are generally associated with low mating rates and ejaculate volumes (Harcourt et al. 1981).

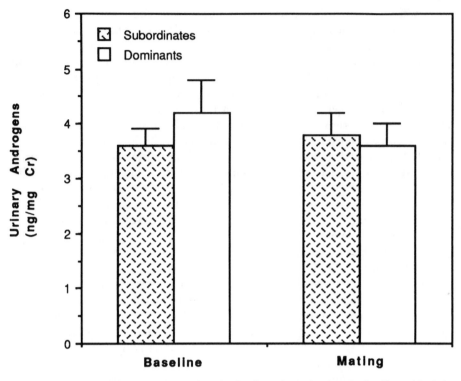

Figure 2.6. Urinary-androgen concentrations for dominant (open bars) and subordinate (shaded bars) male dwarf mongooses, during nonmating (baseline) and mating periods.

Comparison with Mechanisms of Suppression in Other Carnivores

How general are the behavioral and endocrine mechanisms of suppression seen in dwarf mongooses? Outside the Carnivora, similar mechanisms are seen in callithrichid primates (Abbott and Hearn 1978; Abbott 1987; French et al. 1984, 1989) and naked mole rats, *Heterocephalus glaber* (Abbott et al. 1989; Faulkes et al. 1990). Among carnivores, mechanistic studies of suppression have focused on captive gray wolves (Packard et al. 1985; Seal et al. 1979), African wild dogs (van Heerden and Kuhn 1985), and dingoes (Corbett 1988), and on both captive and wild spotted hyenas (Frank et al. 1985; Frank 1986, 1995, this volume; Glickman et al. 1987).

Packard et al. (1985) showed that subordinate female wolves can be suppressed at any of five stages: proestrus, ovulation, copulation, metestrus, and postparturition. Although some subordinates had suppressed endocrine function (e.g., failure of the ovulatory surge of luteinizing hormone), Packard et al. (1983, 1985) concluded that subordinate females are generally physiologically competent, and are suppressed primarily by harassment by the dominant female. In nonreproductive male wolves, seasonal development of the testes was not inhibited. Baseline testosterone levels did not correlate with reproductive activity, but response of testosterone levels to LHRH injections did (Packard et al. 1985).

The estradiol and progesterone profiles of two dominant and two subordinate African wild dog females were described by Van Heerden and Kuhn (1985). Both dominants produced litters and showed estradiol and progesterone profiles similar to those of domestic dogs and wolves (Concannon et al. 1977; Seal et al. 1979). Both subordinates had elevated estradiol during the dominants' mating periods, but their estradiol peaks were low in comparison with those of the dominants. Neither of these subordinates mated, and neither showed subsequent elevations of progesterone to indicate that they had ovulated. Both subordinates, however, came into estrus again toward the end of the dominants' gestation, then mated and ovulated (to judge from their progesterone profiles), but did not become visibly pregnant. Thus, subordinate female wild dogs are probably capable of ovulating, but did not do so at the relevant time, that is, during the pack's conceptive estrus.

All females of a captive dingo pack became pregnant, and Corbett (1988) concluded that "the major mechanism suppressing reproduction in dingoes is dominant female infanticide." Data on sex-steroids were not collected, but it seems clear that subordinates' endocrine function remains intact, at least in females. Interestingly, plasma cortisol concentrations increased during the mating season for males, but not for females (Corbett 1988). But cortisol concentrations did not correlate with rank, as would be predicted if stress played a role in suppression.

In wild spotted hyenas, plasma androgens did not correlate with rank among males within stable clans, but clan-resident males had higher androgen levels than did transient males (Frank et al. 1985). Despite the lack of correlation between androgens and rank, the mating system appears to be highly polygynous, to judge from limited mating data (Frank 1986). The dominant male mates more frequently than the subordinates, though high-ranking subordinates also mate (Frank 1986; Frank et al. 1995). Female spotted hyenas have masculinized external genitalia, are larger than males, are socially dominant to males, and have higher blood concentrations of androstenedione, a weak androgen (Glickman et al. 1987). The dominant female in one clan had androgen levels more than six times higher than those of other females (Frank et al. 1985), but this was due to pregnancy (L. Frank, pers. comm.). Controlling for pregnancy, there is no correlation between androgens and rank in females (Frank et al. 1985, pers. comm.). Most or all females of a spotted hyena clan give birth (Kruuk 1972).

Among spotted hyenas, then, subordinate males appear to reproduce little despite normal androgen concentrations (as in dwarf mongooses), whereas subordinate females breed regularly and have androgen levels indistinguishable from the dominant female's. Although subordinate females are not suppressed, they are older than dominants when they produce their first litter, and have longer interbirth intervals, creating a correlation between rank and reproductive success (Frank et al. 1995; Mills 1990). The pattern among females forms a distinct contrast to that seen in dwarf mongooses.

What general patterns in mechanisms of reproductive suppression emerge from these studies? Among females, the suppression of subordinates has a basis in ovarian function, for at least some individuals, in gray wolves, African wild dogs, and dwarf mongooses. Nonetheless, some subordinate females have normal ovarian steroid levels in all of these species. Particularly in wolves and African wild dogs, behavioral interference with subordinates' mating attempts plays a role in reproductive suppression. Infanticide is the sole documented mechanism of suppression in female dingoes, and is an important mechanism in all of the other species as well. There is enormous variation in the stage at which suppression is effected, both within and between species, but two conclusions emerge: (1) reproductive suppression of subordinate female carnivores often has an endocrine basis, as in other taxa; and (2) physiological suppression normally prevents pregnancy, rather than terminating established pregnancies.

Among male carnivores, reproductive suppression is apparently mediated by competition over access to mates and aggression. Baseline androgens do not correlate with rank or mating activity in wolves, spotted hyenas, or dwarf mongooses. Physiological suppression may occur in less overt ways, as shown by correlations between rank and testosterone response to LHRH in wolves (Packard et al. 1985), between rank and testis volume in dwarf mongooses (Creel et al. 1992), and between low androgens and transient status in spotted hyenas (Frank et al. 1985) and dwarf mongooses (Creel et al. 1993). In carnivores for which paternity has been tested, multiple paternity occurs within groups, and even within single litters—in lions (C. Packer, pers. comm.), Eurasian badgers (Evans et al. 1989), and dwarf mongooses (Keane et al. 1994)—demonstrating that subordinate males often retain their reproductive capability.

Stress physiology is affected by dominance in some species (e.g., olive baboons: Sapolsky 1985, 1987, 1992), and reproductive suppression of subordinates is sometimes attributed to the "stress of subordination" or "psychological castration" (e.g., Brown 1978; Carter 1992). In dwarf mongooses, urinary free-cortisol levels correlated with rank, but were not responsible for reproductive suppression.

Subordinate female mongooses had low baseline and peak estrogen levels (as discussed above), despite also having significantly lower urinary free-cortisol levels both at baseline ($b = 0.43 \pm 0.07$, $r = 0.40$, $t = 5.83$, $df = 176$, $P < 0.001$) and in response to stress ($b = 0.88 \pm 0.31$, $r = 0.32$, $t = 2.80$, $df = 67$, $P < 0.007$). Subordinate males, in contrast, had androgen levels indistinguishable from those of dominants (as discussed above), despite having significantly higher urinary free-cortisol levels in response to stress ($b = -1.62 \pm 0.54$, $r = $

$-0.30, t = 3.01, df = 91, P = 0.003$). Baseline cortisol levels did not correlate with rank among males. Thus, elevated cortisol was associated with normal sex-steroid levels (in males), and low cortisol levels were associated with depressed sex-steroid levels (in females). These patterns directly oppose predictions of the hypothesis that reproduction is inhibited in subordinates by chronic stress.

Elevated corticosteroids are associated with low mating rates in captive wolves (Packard et al. 1985) and perhaps in captive dingoes (Corbett 1988). But captive studies based on blood samples from animals that were immobilized repeatedly might impose stresses independent of the normal mechanisms of suppression. Packard et al. (1985) observe that in captive wolves social structure was affected by occurrences of deaths associated with immobilization. In captive African wild dogs, van Heerden and Kuhn (1985:83) state that during twice-weekly darting, the dominant female "repeatedly ran into the fence [and had] elevated peripheral serum cortisol levels." Corbett (1988) noted that dingoes in a captive pack of 12 adults received a mean of 55 wounds apiece over a single breeding season, and two were killed. Given the magnitude of disturbance and social instability seen in these captive studies, data from captive carnivores on the role of stress hormones in reproductive suppression should be treated with caution.

Evolution of Reproductive Suppression

Why do subordinates tolerate reproductive suppression? Considerable research has addressed this question, focusing mostly on birds (e.g., Brown 1987; Stacey and Koenig 1990), but also on carnivores (Jennions and Macdonald 1994; Macdonald and Moehlman 1982). Physiological and behavioral reproductive suppression probably arise where competition for limited breeding resources within a group often leads to the death of subordinates' young, either through biased provisioning (e.g., African wild dogs: Malcolm and Marten 1982) or by direct infanticide (e.g., dingoes: Corbett 1988). If subordinates' young have a sufficiently low probability of survival, selection would favor endocrine or behavioral mechanisms that preclude reproduction (but are reversible), to avoid the energetic costs of producing doomed young (Creel and Creel 1991; Kleiman 1980; Moehlman 1989).

Two classes of genetic benefit might make tolerating reproductive suppression preferable to dispersal (the main alternative). First, if subordinates increase the reproductive success of dominants to whom they are related, they will gain indirect fitness (Hamilton 1964). Second, nondispersal and toleration of suppression may yield direct genetic benefits, by increasing a subordinate's probability of ultimately attaining dominance (Woolfenden and Fitzpatrick 1984). Among dwarf mongooses, those subordinates that remain in their natal

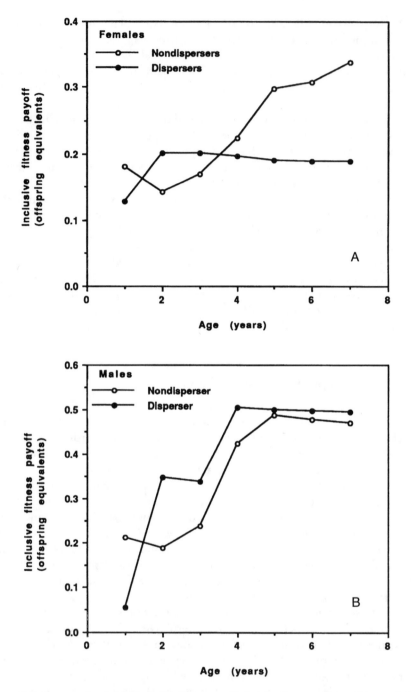

Figure 2.7. A comparison of the inclusive-fitness payoffs deriving from nondispersal (open circles) and dispersal (solid circles) in dwarf mongooses: (*A*) females, (*B*) males.

group and tolerate suppression ("nondispersers") obtain both direct and indirect payoffs (Creel and Waser 1994).

For female dwarf mongooses, calculation of the inclusive-fitness payoffs to dispersers and nondispersers of ages 1 to 7 years showed that nondispersal is favored at all ages except 2 and 3 (Figure 2.7A; Creel and Waser 1994). Genetically indirect fitness comprises 52% of the inclusive fitness of 1-year-old nondispersers. This percentage decreases steadily with age, reaching 6% among 7-year-old nondispersers. Indirect benefits decline with age primarily because relatedness between subordinates and breeders declines (from 0.35 among 1-year-olds to 0.09 among 7-year-olds) as subordinates' parents are replaced by less closely related breeders. Simultaneously, genetically *direct* benefits *increase* with age, owing to an increasing probability of being at the front of the "breeding queue" and inheriting a breeding position (rising from 2% among 1-year-olds to 17% among 7-year-olds).

The same qualitative patterns of change in direct and indirect genetic benefits occur among subordinate males, but with quantitative differences that substantially alter predictions about dispersal (Creel and Waser 1994). Among males, nondispersal and toleration of reproductive suppression is favored only among 1-year-olds (Figure 2.7B). Dispersal yields greater inclusive fitness at ages 2 to 4, and subsequently both strategies yield essentially equal inclusive fitness (Figure 2.7B). Comparison of the fitness curves of male and female dwarf mongooses (Figure 2.7A, B) predicts that males should be the more dispersive sex, and males do indeed disperse across a broader age-range than the females do, and at higher rates at all ages (Creel and Waser 1994; Rood 1987; Waser et al. 1994).

From another perspective, the fitness curves predict not only that males should be more dispersive, but also that nondispersing males should be less likely to tolerate complete reproductive suppression. Differences between the sexes in the mechanisms of suppression accord well with this prediction. Females are suppressed in a physiological manner, so that subordinates usually do not become pregnant even though the alpha female does not prevent them from mating. Subordinate males are more resistant to suppression. They have retained normal endocrine function with respect to reproduction, and they father young when not prevented by direct behavioral intervention. Genetic data suggest that these sex differences have an evolutionary impact: subordinate males father 24% of fingerprinted offspring, whereas only 15% of offspring have subordinate mothers (Keane et al. 1994).

It is commonly hypothesized that subordination may be stressful, and that endocrine stress responses underlie reproductive suppression among subordinates (Brown 1978; Mays et al. 1991; Wingfield et al. 1991; Carter 1992). Although stress undeniably can suppress reproduction by depressing gonadotropin and sex-steroid levels (Munck et al. 1984; Moore et al. 1991; Sapolsky 1992), non-stress-mediated mechanisms of reproductive suppression should be more evolutionarily stable in communal breeding systems, where suppression is a normal feature of social organization. Chronic corticosteroid elevation has a broad range of debilitating effects, from immune suppression to neural death (Munck et al. 1984; Sapolsky 1992; McEwen and Schmeck 1994). Non-stress-mediated reproductive suppression should be evolutionarily favored over stress-mediated suppression, because directly suppressed subordinates would avoid the additional costs of chronic stress responses. This hypothesis assumes that dominants stress subordinates primarily to keep them from breeding, and that subordinates who remove themselves from reproductive competition would thus gain a reduction of the "social stress" imposed by the dominants. In other words, a subordinate that can count on being stressed until its reproduction fails would benefit from shutting its reproduction off directly and cutting its losses.

This logic predicts that direct modes of suppression that do not involve corticosteroids should be the rule among cooperative breeders, and this rule holds true in the three cooperatively breeding birds studied in the wild (Mays et al. 1991; Wingfield et al. 1991). Among carnivores, suppression is not apparently mediated by corticosteroids in wild dwarf mongooses, but corticosteroids do play a role in the reproductive suppression of captive wolves (Packard et al. 1983) and perhaps captive dingoes (Corbett 1988: see above for discussion of stress in captives). I am aware of no data from the wild that contradict this hypothesis.

Failures of Reproductive Suppression

Subordinates are reproductively suppressed in the majority of communally breeding carnivores (Macdonald and Moehlman 1982), but in *all* of the species for which suppression is the norm, some subordinates do reproduce (Creel and Waser 1991): e.g., African wild dogs (Fuller et al. 1992) and gray wolves (Packard et al. 1983). Reproductive suppression is the norm among subordinate dwarf mongooses, but 27% of pregnancies were carried by subordinates (83 of 312). DNA fingerprinting shows that 6% to 30% of offspring have subordinate mothers, and 12% to 40% have subordinate fathers (95% confidence limits: Keane et al. 1994). Calculated in terms of annual fitness, subordinate reproduction yields roughly one-third of a subordinate's inclusive fitness—an appreciable proportion.

Evolution of Subordinate Reproduction

Subordinate reproduction is often regarded as simple, accidental failure of the normal mechanisms of suppression. But given its demonstrated fitness benefit in dwarf mongooses, the possibility arises that reproduction by subordinates is a form of power-sharing by dominants (Keller and Reeve 1994).

Perhaps dominants concede reproduction to subordinates when concessions are necessary to retain them within the group (and thus benefit from additional helpers: Rood 1990).

Vehrencamp (1983) developed this "power-sharing" logic to model the degree of reproductive suppression expected within a group. Her approach was to compare two quantities: (1) the indirect fitness that a subordinate accrues by nondispersal and helping and (2) the inclusive fitness that a subordinate could expect if, rather than remaining a helper, it emigrated. When the second quantity exceeds the first, then the indirect benefits of helping are not sufficient to favor nondispersal, and nondispersing subordinates should require direct reproductive success at least equal to the difference.

Vehrencamp's model can be used to predict the distribution of subordinate reproduction across ages and group sizes (Creel and Waser 1991, in press). Figure 2.8 shows the levels of direct reproductive success predicted for subordinate dwarf mongooses of given age and pack size (i.e., the minimum reproductive success required to make nondispersal the favored strategy). Qualitative predictions about subordinate reproduction are similar for both sexes. Using partial regression to control for pack size, a positive relationship is predicted between subordinates' age and direct reproduction (males: Figure 8B, $t = 4.66$, $P < 0.001$; females: Figure 8A, $t = 17.35$, $P < 0.001$). Using partial regression to control for age, no association is predicted between pack size and subordinate reproduction (males: $t = 1.05$, $P = 0.31$; females: $t = 1.08$, $P = 0.29$). Quantitatively, the slope of the partial regression of predicted reproductive success on age is steeper among subordinate females than among subordinate males. The relationship predicted between subordinate age and reproduction (for both sexes) is due mainly to declining subordinate-breeder relatedness, and to the older subordinates' improved odds of finding a breeding position in another group (Creel and Waser 1991, in press). In dwarf mongooses, age is tightly correlated with dominance ($r^2 = 69\%$: Creel et al. 1992), even following dispersal (Rood 1983).

How well do the predicted patterns of subordinate reproduction match the mongooses' actual behavior? Among subordinate females, patterns of pregnancy fit the model well. As predicted, pregnancies are more common among older subordinates (partial regression controlling pack size; $t = 6.34$, $P < 0.001$), with no systematic change across group sizes (partial regression controlling age; $t = 1.36$, $P = 0.18$). To measure the predictive power of the model, one can regress the observed frequency of subordinate pregnancy on predicted reproductive success (with each point representing an age/pack-size combination). This regression shows the model to be a good predictor, explaining 45% of the variance in the distribution of subordinate pregnancies (Creel and Waser 1991). That subordinate pregnancies fall in the pattern expected from the model suggests that subordinate reproduction does indeed constitute an evolutionary compromise by the dominant. Limited subordinate reproduction is allowed by the dominant female when necessary as an incen-

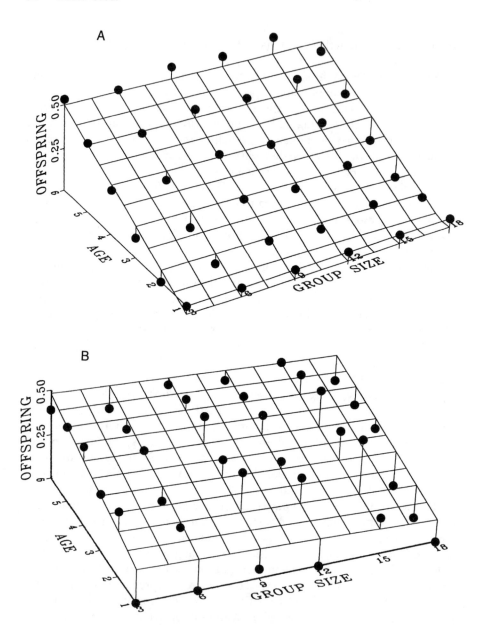

Figure 2.8. The level of direct reproductive success that subordinate dwarf mongooses should require as a "staying incentive" to make nondispersal and helping preferable to dispersal: (*A*) females, (*B*) males. (See text for details.)

tive for subordinates to remain in the pack and help (see also Keller and Reeve 1994).

For males, the results are strikingly different. The predicted relationship between reproduction and age does not emerge in the partial regression (controlling pack size) of subordinate mating rate on age ($t = 0.34$, $P = 0.70$). Furthermore, subordinate mating rate increases with group size (partial regression controlling age; $t = 2.80$, $P = 0.009$), although this was not predicted (Figure 2.8B). Unlike females, dominant males do not share reproduction preferentially with those subordinates who are predicted to require greater "staying incentives."

Why does an evolutionary model that predicts female behavior not predict male behavior as well? Mechanistic data help to resolve this question.

Mechanisms Underlying Failures of Reproductive Suppression

Do patterns of reproduction among subordinate dwarf mongooses relate to mechanisms of reproductive suppression? Among females, older subordinates (who are high-ranking: Rood 1983; Creel et al. 1992) have higher baseline and peak estrogens, and higher mating rates, than do younger subordinates (Creel et al. 1992). Because the older subordinates have reproductive function that more closely resembles the normal function of dominants, they are more likely to produce litters. That is, subordinate females' reproduction is not suppressed dichotomously. Instead, ovarian function is progressively more effective with age, in parallel with the increasing need for "staying incentives" predicted by the model (Creel and Waser 1991; Keller and Reeve 1994; Vehrencamp 1983).

Among subordinate males, patterns of mating do not match the predictions of the evolutionary model, though these patterns do make sense in light of the proximate mechanisms of suppression in males. An association between subordinates' age and mating rate is predicted, but does not emerge (Creel and Waser, in press). The lack of association reflects the mechanistic fact that even young, low-ranking male mongooses have normal androgen levels (Creel et al. 1992, 1993). The evolutionary model predicts no association between mating rate and group size, yet a positive association exists. This discrepancy reflects the fact that subordinate males retain their fertility and mate opportunistically (Creel et al. 1992; Keane et al. 1994), and in a larger pack the dominant probably has greater difficulty monitoring all the subordinates and preventing them from mating. For example, when a pack is scattered, or when matings occur out of view (both of which are more common in larger packs), dominant males monopolize matings significantly less effectively (Creel et al. 1992). In contrast, females' mating rates are not affected by scattering of the pack.

Patterns of subordinate reproduction in both sexes can be explained by mechanistic data, but only in females do the patterns match Vehrencamp's model of reproductive biases within groups. The question becomes, how do mechanistic differences between the sexes relate to the evolutionary model?

In this regard, the fundamental difference between the sexes may lie in the dominant mongoose's confidence of parentage, which would affect its ability to use infanticide to enforce suppression (Creel and Waser, in press). Vehrencamp's (1983) model *assumes* that the dominant can enforce the degree of suppression of subordinates that is optimal from its own point of view. For male dwarf mongooses, this may not hold true. Because opportunistic mating by subordinate males leads to multiple paternity even within litters (Keane et al. 1994), dominant males are poorly positioned to detect and kill the offspring of subordinate fathers. Even phenotypic self-matching could be unreliable, because the other candidates for paternity are usually relatives. If the dominant cannot ultimately use infanticide as a lever to control subordinate reproduction, then subordinates are less constrained to follow the pattern of reproduction optimal for the dominant (Creel and Waser, in press). A dominant female, having physically given birth, is more likely to know her own young (simply by keeping track of them), and thus is better positioned than the dominant male to use infanticide to enforce the degree of suppression optimal for her.

In summary, mechanistic and evolutionary analyses of subordinate reproduction mesh smoothly for females. For males they do not, but aligning the evolutionary and mechanistic approaches exposes a potential reason (lack of infanticidal leverage) for the failure of the evolutionary model to predict levels of reproduction in subordinate males.

Empirical tests of evolutionary models of reproductive skew within groups (Vehrencamp 1983; Reeve and Ratnieks 1993) have largely been restricted to social insects (Keller and Reeve 1994). Among carnivores, a parallel can be drawn to coalitions of male lions. Small coalitions often comprise nonrelatives, and their members have fairly equal reproductive success (Packer et al. 1991). Large coalitions, however, are formed exclusively by relatives, and have higher per-capita reproductive success than do small coalitions (Bygott et al. 1979). As a result, males in large lion coalitions gain greater indirect fitness benefits than do males in small coalitions. Vehrencamp's model predicts a greater likelihood of suppression in large coalitions, and this is what is seen: "one or more members of each larger coalition may best be viewed as non-breeding helpers" (Packer et al. 1991).

Nonbreeder Lactation as Alloparental Care

Pregnant subordinate mongooses normally produce a litter in synchrony with the dominant. Both females then nurse the creched young, although most of the young belong to the dominant (Rood 1980, 1983; Keane et al. 1994). Despite the presence of an additional lactating female, the dominant's production of mature offspring is reduced by 22% in joint litters, relative to that in single litters (Keane et al. 1994). Thus, dominant mongooses may allow subordinates to breed as an inducement to remain in the pack (see above), but the dominants' young do not survive better as a result of communal nursing in joint litters.

In dwarf mongooses, another form of communal nursing occurs. For every four subordinates that lactate following pregnancy, one female lactates *without* pregnancy (Rood 1980). Because lactation is usually the most energetically costly aspect of rearing young in mammals (Clutton-Brock 1991; Gittleman and Thompson 1988), nonbreeder lactation is an extreme form of alloparental care. Lactation without pregnancy has been reported in laboratory rats, domestic dogs, captive primates, one captive Asian elephant, and humans (Holman and Goy 1980; Rapoport and Haight 1987; Smith and McDonald 1974). To my knowledge, female dwarf mongooses and male Dayak fruit bats, *Dyacopterus spadiceus* (Francis et al. 1994) are the only species known to lactate spontaneously without pregnancy in the wild.

Endocrine Mechanisms of Spontaneous Lactation

Because dwarf mongooses produce relatively heavy litters ($21 \pm 2\%$ of female body mass: Creel and Creel 1991), pregnant females are easily identified, and spontaneous lactation is unlikely to be due to undetected pregnancy. As Figure 2.9 shows, pregnant subordinates produced litters of the same mass as did dominants (88.0 ± 10.3 vs. 90.9 ± 6.5 g: $t = 0.25$, $P = 0.81$). In contrast, spontaneous lactators lost 17.2 ± 4.5 g, 58% of the weight loss per offspring among pregnant females (Figure 2.9: Creel et al. 1991). Spontaneous lactators not only lost significantly less mass than pregnant females (compared to pregnant dominants, $t = 7.26$, $P < 0.001$; compared to pregnant subordinates, $t = 6.37$, $P < 0.001$), but they lost this mass gradually over a week, rather than abruptly on the day of parturition within their pack (Figure 2.10*B*). Finally, spontaneous lactators had a significant net mass gain of 36.3 ± 7.9 g in the interval from estrus to the week after parturition ($t = 2.36$, $P < 0.025$), while pregnant females showed no net change in body mass over the same interval (mean loss of 2.8 ± 7.7 g). These patterns of weight change in lactating nonbreeders (Figures 2.9, 2.10*B*) are typical of pseudopregnancy (an endocrine state similar to pregnancy, despite the lack of implanted fetuses: Nalbandov 1976).

Urinary estrogen-conjugate profiles confirm that spontaneous lactation is due to pseudopregnancy, not to undetected pregnancy. Pregnant females' EC increased to 70-fold above the baseline of 5.0 ng EC/mg creatinine (Figure 2.10*A*) (Creel et al. 1992). Spontaneous lactators' EC levels peaked at 17.7 ± 0.9 ng EC/mg creatinine, more than an order of magnitude below levels typical of pregnancy ($t = 2.80$, $P = 0.008$). Although inconsistent with pregnancy, spontaneous lactators' EC levels were significantly higher than those of non-

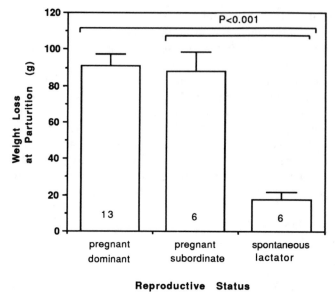

Figure 2.9. Weight lost at the time of parturition by pregnant females and spontaneous lactators, in dwarf mongooses. Numbers within bars are sample sizes. (From Creel et al. 1991; reprinted by permission.)

lactators ($t = 3.50$, $P < 0.001$), whose EC remained at baseline (Creel et al. 1991). Spontaneous lactators' EC increased, particularly in the second half of the gestation to which they were synchronized (Figure 2.10A). Like their changes in body mass, the estrogen profiles of spontaneously lactating dwarf mongooses are typical of pseudopregnancy in carnivores (Moller 1973a, b; Verhage et al. 1976; Concannon et al. 1977; Seal et al. 1979). Because dwarf mongooses' pseudopregnancies are synchronized to the true pregnancy of the alpha female, pseudopregnant females are endocrinologically conditioned to lactate when they encounter the offspring of the dominant.

Evolution of Nonbreeder Lactation

What sort of females lactate spontaneously? For most cases in which lactators' life history was known since birth, spontaneous lactation first occurred when a female was nulliparous (15 of 15 cases), young (mean = 2.8 years, range 2–5), and in the pack of her birth (17 of 18 cases). Young females still in their natal pack were normally close relatives of the pack's breeders for all cases (often their offspring), but spontaneous lactators were significantly more closely related to pack breeders ($r = 0.36 \pm 0.02$) than were subordinates that became pregnant and suckled joint litters ($r = 0.25 \pm 0.02$: $F = 15.59$, $P < 0.001$). Thus, nonbreeder lactation might be conditional upon close relatedness to the offspring to be suckled. On average, spontaneous lactators nurse young to whom they are related almost three-fourths as closely as they would

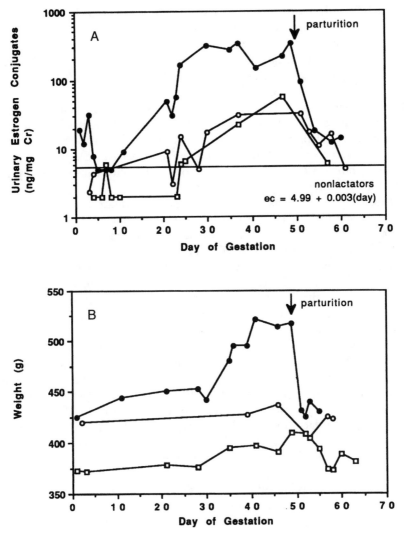

Figure 2.10. *A,* profiles of change in urinary-estrogen conjugates for three female dwarf mongooses of a single pack, during the dominant's (filled circles) gestation. The other two females (open circles and squares) lactated in synchrony with the dominant. *B,* profiles of change in body mass for the same three females over the same interval. (From Creel et al. 1991; reprinted by permission.)

be to their own young. Interestingly, the sole female who spontaneously lactated outside her natal pack nursed the young of a full sister with whom she had earlier dispersed.

Nonbreeding lactators have a pronounced effect on the survival of the young they nurse (Creel et al. 1991). At weaning, litters with spontaneous lactators were significantly larger (2.9 ± 0.4 young) than litters suckled only by the alpha female (1.6 ± 0.1 young: $t = 3.58$, $P < 0.001$). This difference did not arise from a difference in initial litter size, or as an effect of the number of helpers present. In a partial regression controlling these variables, each spontaneous lactator was associated with an increase of 1.1 weaned offspring (partial $b = 1.07 ± 0.28$, $t = 3.85$, $P < 0.005$).

Devaluing this effect (1.1 offspring) to reflect lactators' relatedness to the young they nurse (0.36) yields the inclusive-fitness benefit of spontaneous lactation: 0.79 offspring-equivalents. This fitness payoff is 33% of the mean payoff to breeding realized by an alpha female (2.4 weaned offspring). Given that the subordinates' reproductive options are highly constrained by their status, this is a substantial fitness benefit.

Although pseudopregnancy is taxonomically widespread and common among the carnivores, it is generally regarded as one of nature's jokes, an anomalous consequence of failing to become pregnant (Concannon et al. 1977; Nalbandov 1976; Verhage et al. 1976). But for dwarf mongooses, pseudopregnancy yields an evolutionary benefit, through the lactation it induces. Given its benefit, why do only 4% of nonbreeding mongooses lactate? Speculatively, the benefits might diminish as the supply of milk nears the maximum required, so that the energetic cost of lactation might exceed the benefit if several females lactated.

Finally, why has nonbreeder lactation not been reported in other social carnivores? Many species share the social preadaptations that favor nonbreeder lactation in dwarf mongooses (closely related, reproductively suppressed subordinates; e.g., canids: Moehlman 1989). Perhaps the mechanisms of reproductive suppression vary among species in ways that affect the endocrine induction of lactation. Behavior patterns may also play a role, as they do among African wild dogs, where dominant females often drive other females away from unweaned young. But pseudopregnancy *has* been documented in some social carnivores (e.g., wolves: Packard et al. 1983, 1985) and could well occur in others (e.g., other canids, lions). Perhaps nonbreeder lactation will not ultimately prove to be unique to dwarf mongooses.

Behavioral Endocrinology and Social Organization in Carnivores

With the exception of sociality itself, reproductive suppression and alloparental care of young are the two most central features of dwarf mongooses' social organization (Rood 1978, 1980, 1983). The analyses of this chapter show that these patterns of social organization influence, and are influenced by, behavioral endocrinology.

How general are the evolutionary explanations for reproductive suppression and helping behavior that we see in dwarf mongooses? A great deal is known about these issues in communally breeding birds (Brown 1987; Stacey and

Koenig 1990) and in social insects (Nonacs 1991; Keller and Reeve 1994). In these taxa, cost/benefit analyses of inclusive fitness have produced conclusions similar to those drawn for dwarf mongooses.

Cost/benefit analyses of inclusive fitness have been much less broadly applied to mammals in general (Solomon and French, in press; Jennions and Macdonald 1994) and to carnivores in particular. Levels of relatedness similar to those within dwarf mongoose groups have been reported for some species (lions: Packer et al. 1991; brown and spotted hyenas: Mills 1985, 1990; Eurasian badgers: Evans et al. 1989). Fitness benefits of helping similar to those determined for dwarf mongooses have been demonstrated for some species: black-backed jackals, *Canis mesomelas* (Moehlman 1979), red foxes, *Canis rufus* (Macdonald 1979), and African wild dogs, *Lycaon pictus* (Malcolm and Marten 1982). But dissimilar results have also been found: gray wolves, *Canis lupus* (Harrington et al. 1983), Ethiopian wolves, *Canis simensis* (Sillero-Zubiri 1994), and Eurasian badgers (da Silva et al. 1994). Still, cost-benefit analyses have been few, because the opportunity costs of tolerating suppression and helping are not known for most social carnivores. An important avenue for future research on carnivore social evolution will be to gather demographic data that allow the fitness costs and benefits of alternative reproductive strategies to be measured.

The lack of complete data sets on costs, benefits, and relatedness is partially due to the practical difficulty of studying species that are wide-ranging, elusive, and often nocturnal. But it is also due, in part, to the fact that studies on social carnivores have often been directed at other issues (e.g., predator-prey relations: Mills and Biggs 1993; or cooperative hunting: Caro 1994). Interest in "bird-style" analyses of cooperative breeding in mammals has increased recently (Jennions and Macdonald 1994; Solomon and French, in press). Current field studies of meerkats (suricates), banded mongooses (*Mungos mungo*), coatis (*Nasua nasua*), Eurasian badgers, lions, spotted hyenas, and African wild dogs are likely to reveal general patterns (if they exist) in the evolution of reproductive suppression and helping in carnivores.

How general are the physiological mechanisms described for dwarf mongooses? Many features of dwarf mongooses' social organization are common to other social carnivores (e.g., African wild dogs: Malcolm and Marten 1982; wolves: Mech 1970; golden and black-backed jackals, *Canis aureus* and *C. mesomelas*: Moehlman 1989; dholes, *Cuon alpinus*: Johnsingh 1982). Nonetheless, captive studies indicate that there is considerable variation within the carnivores in mechanisms of suppression (van Heerden and Kuhn 1985; Packard et al. 1985; Corbett 1988). Because field studies integrating evolutionary and endocrine approaches are relatively new (Carter et al. 1986; Sapolsky 1987; Wingfield and Moore 1987), considerable research will be needed to identify general patterns. By the same token, future research will be rewarded by establishing relationships between physiology and life in the real world.

Acknowledgments

This chapter is dedicated to Jon Rood, in recognition of his years of work on dwarf mongooses and in gratitude for the invitation to join his study. The research summarized here was collaborative, and I thank my colleagues for their ideas and work, both in the field and in the laboratory: Nancy Marusha Creel, Lee Elliott, Brian Keane, Dennis Minchella, Steve Monfort, Jon Rood, Peter Waser, and David Wildt. I am especially grateful to three people whose work was critical to my research: Nancy Marusha Creel, Steve Monfort, and Peter Waser. The research was supported by the National Geographic Society, the National Science Foundation, the Scholarly Studies Program of the Smithsonian Institution, Friends of the National Zoo, and a David Ross Fellowship from Purdue University. I am grateful to the Tanzania Council for Science and Technology, the Serengeti Wildlife Research Institute, and Tanzania National Parks for their permission to work in Serengeti National Park. This chapter was prepared under support from the Frankfurt Zoological Society—Help for Threatened Wildlife, Project 1112/90.

References

Abbott, D. H. 1987. Behaviourally mediated suppression of reproduction in female primates. *J. Zool. (London)* 213:455–470.

Abbott, D. H., Barrett, J., Faulkes, C. G., and George, L. M. 1989. Social contraception in naked mole-rats and marmoset monkeys. *J. Zool. (London)* 219:703–710.

Abbott, D. H., and Hearn, J. P. 1978. Physical, hormonal and behavioural aspects of sexual development in the marmoset monkey, *Callithrix jacchus. J. Reprod. Fert.* 53:155–166.

Beach, F. A. 1976. Sexual attractivity, proceptivity and receptivity in female mammals. *Horm. Behav.* 7:105–138.

Brown, J. L. 1978. Avian communal breeding systems. *Ann. Rev. Ecol. Syst.* 9:123–155.

Brown, J. L. 1987. *Helping and Communal Breeding in Birds.* Princeton, N.J.: Princeton Univ. Press.

Bygott, J. D., Bertram, B. C. R., and Hanby, J. P. 1979. Male lions in large coalitions gain reproductive advantage. *Nature* 26:839–841.

Caro, T. M. 1994. *Cheetahs of the Serengeti Plains: Group Living in an Asocial Species.* Chicago: Univ. Chicago Press.

Carter, C. S. 1992. Neuroendocrinology of sexual behavior in the female. In: J. B. Becker, S. M. Breedlove, and D. Crews, eds. *Behavioral Endocrinology,* pp. 131–142. Cambridge, Mass.: MIT Press.

Carter, C. S., Getz, L. L., and Cohen-Parsons, M. 1986. Relationships between social organization and behavioral endocrinology in a monogamous mammal. *Adv. Stud. Behav.* 16:109–145.

Clutton-Brock, T. H. 1991. *The Evolution of Parental Care.* Princeton, N.J.: Princeton Univ. Press.

Concannon, P. W., Powers, M. E., Holder, W., and Hansel, W. 1977. Pregnancy and parturition in the bitch. *Biol. Reprod.* 16:517–526.

Corbett, L. K. 1988. Social dynamics of a captive dingo pack: Population regulation by dominant female infanticide. *Ethology* 78:177–198.

Creel, S. R. 1990. How to measure inclusive fitness. *Proc. Roy. Soc. London* 241:229–231.

Creel, S. R. 1991. Reproductive suppression and communal breeding in dwarf mongooses, *Helogale parvula*. Ph.D. dissert., Purdue Univ., West Lafayette, Ind.

Creel, S. R. 1992. Cause of wild dog deaths. *Nature* 360:633.

Creel, S. R., and Creel, N. M. 1991. Energetics, reproductive suppression and obligate communal breeding in carnivores. *Behav. Ecol. Sociobiol.* 28:263–270.

Creel, S. R., and Creel, N. M. 1995. Communal hunting and pack size in African wild dogs *Lycaon pictus*. *Anim. Behav.* 50:1325–1339.

Creel, S. R., Creel, N. M., Wildt, D. E., and Monfort, S. L. 1992. Behavioral and endocrine mechanisms of reproductive suppression in Serengeti dwarf mongooses. *Anim. Behav.* 43:231–245.

Creel, S. R., Monfort, S. L., Wildt, D. E., and Waser, P. M. 1991. Spontaneous lactation is an adaptive result of pseudopregnancy. *Nature* 351:660–662.

Creel, S. R., and Waser, P. M. 1991. Failures of reproductive suppression in dwarf mongooses (*Helogale parvula*): Accident or adaptation? *Behav. Ecol.* 2:7–15.

Creel, S. R., and Waser, P. M., 1994. Inclusive fitness and reproductive strategies in dwarf mongooses. *Behav. Ecol.* 5:339–348.

Creel, S. R., and Waser, P. M. In press. Variation in reproductive suppression in dwarf mongooses: Interplay between evolution and mechanisms. In: N. Solomon and J. French, eds. *Cooperative Breeding in Mammals.* Cambridge: Cambridge Univ. Press.

Creel, S. R., Wildt, D. E., and Monfort, S. L. 1993. Androgens, aggression and reproduction in wild dwarf mongooses: A test of the challenge hypothesis. *Amer. Nat.* 141:816–825.

da Silva, J., Macdonald, D. W., and Evans, P. G. H. 1994. Net costs of group living in a solitary forager, the Eurasian badger (*Meles meles*). *Behav. Ecol.* 5:151–158.

Evans, P. G. H., Macdonald, D. W., and Cheeseman, C. L. 1989. Social structure of the Eurasian badger (*Meles meles*): Genetic evidence. *J. Zool. (London)* 218:587–595.

Faulkes, C. G., Abbott, D. H., and Jarvis, J. U. M. 1990. Social suppression of ovarian cyclicity in captive and wild colonies of naked mole rats, *Heterocephalus glaber*. *J. Reprod. Fert.* 88:559–568.

Francis, C. M., Anthony, E. L. P., Brunton, J. A., and Kunz, T. H. 1994. Lactation in male fruit bats. *Nature* 367:691–692.

Frank, L. G. 1986. Social organization of the spotted hyaena *Crocuta crocuta*. II. Dominance and reproduction. *Anim. Behav.* 34:1510–1527.

Frank, L. G., Davidson, J. M., and Smith, E. R. 1985. Androgen levels in the spotted hyaena *Crocuta crocuta:* The influence of social factors. *J. Zool. (London)* 206:525–531.

Frank, L. G., Holekamp, K. E., and Smale, L. 1995. Dominance, demography and reproductive success of female spotted hyenas. In: A. R. E. Sinclair and P. Arcese, eds. *Serengeti II: Research, Management and Conservation of an Ecosystem*, pp. 364–384. Chicago: Univ. Chicago Press.

French, J. A., Abbott, D. H., Scheffler, G., Robinson, J. A., and Goy, R. W. 1984. Cyclic excretion of urinary oestrogens in female tamarins (*Sanguinus oedipus*). *J. Reprod. Fert.* 68:177–184.

French, J. A., Inglett, B. J., and Dethlefs, T. M. 1989. The reproductive status of nonbreeding group members in golden lion tamarin social groups. *Amer. J. Primatol.* 18:73–86.

Fuller, T. K., Kat, P. W., Bulger, J. B., Maddock, A. H., Ginsberg, J. R., Burrows, R.,

McNutt, J. W., and Mills, M. G. L. 1992. Population dynamics of African wild dogs. In: D. R. McCullough and R. H. Barrett, eds. *Wildlife 2001: Populations,* pp. 1125–1139. London: Elsevier.

Gittleman, J. L. 1985. Functions of communal care in mammals. In: P. J. Greenwood, P. H. Harvey, and M. Slatkin, eds. *Evolution: Essays in Honour of John Maynard Smith,* pp. 187–205. Cambridge: Cambridge Univ. Press.

Gittleman, J. L. 1989. Carnivore group-living: Comparative trends. In: J. L. Gittleman, ed. *Carnivore Behavior, Ecology, and Evolution,* pp. 183–207. Ithaca, N.Y.: Cornell Univ. Press.

Gittleman, J. L., and Thompson, S. D. 1988. Energy allocation in mammalian reproduction. *Amer. Zool.* 28:863–875.

Glickman, S. E., Frank, L. G., Davidson, J. M., Smith, E. R., and Siiteri, P. K. 1987. Androstenedione may organize or activate sex-reversed traits in female spotted hyenas. *Proc. Natl. Acad. Sci.* 84:3444–3447.

Gorman, M. L. 1975. The diet of feral *Herpestes auropunctatus* (Carnivora: Viverridae) in the Fijian Islands. *J. Zool. (London)* 175:273–278.

Gorman, M. L. 1979. Dispersion and foraging of the small Indian mongoose, *Herpestes auropunctatus* (Carnivora: Viverridae) relative to the evolution of social viverrids. *J. Zool. (London)* 187:65–73.

Gorman, M. L., and Trowbridge, B. J. 1989. The role of odor in the social lives of carnivores. In: J. L. Gittleman, ed. *Carnivore Behavior, Ecology, and Evolution,* pp. 57–88. Ithaca, N.Y.: Cornell Univ. Press.

Hamilton, W. D. 1964. The genetical evolution of social behavior. *J. Theor. Biol.* 7:1–52.

Harcourt, A. H., Harvey, P. H., Larson, S. G., and Short, R. V. 1981. Testis weight, body weight and breeding system in primates. *Nature* 293:55–57.

Harrington, F. H., Mech, L. D., and Fritts, S. H. 1983. Pack size and wolf pup survival: Their relationship under varying ecological conditions. *Behav. Ecol. Sociobiol.* 13:19–26.

Hinton, H. E., and Dunn, A. M. S. 1967. *Mongooses: Their Natural History and Behaviour.* Edinburgh: Oliver and Boyd.

Holman, S. D., and Goy, R. W. 1980. Behavioral and mammary responses of adult female rhesus to strange infants. *Horm. Behav.* 14:348–357.

Jennions, M. D., and Macdonald, D. W. 1994. Cooperative breeding in mammals. *Trends Ecol. Evol.* 9:89–93.

Johnsingh, A. J. T. 1982. Reproductive and social behaviour of the dhole, *Cuon alpinus* (Canidae). *J. Zool. (London)* 198:443–463.

Keane, B., Waser, P. M., Creel, S. R., Creel, N. M., Elliott, L. F., and Minchella, D. J. 1994. Subordinate reproduction in dwarf mongooses. *Anim. Behav.* 47:65–75.

Keller, L., and Reeve, H. K. 1994. Partitioning of reproduction in animal societies. *Trends Ecol. Evol.* 9:98–102.

Kleiman, D. G. 1980. The socioecology of captive propagation. In: M. E. Soule and B. A. Wilcox, eds. *Conservation Biology: An Evolutionary-Ecological Perspective,* pp. 243–261. Sunderland, Mass.: Sinauer.

Kruuk, H. 1972. *The Spotted Hyena.* Chicago: Univ. Chicago Press.

Kruuk, H. 1975. Functional aspects of social hunting by carnivores. In: G. Baerends, C. Beer, and A. Manning, eds. *Function and Evolution in Behaviour: Essays in Honour of Professor Niko Tinbergen,* pp. 119–141. Oxford: Oxford Univ. Press.

Lynch, C. D. 1980. Ecology of the suricate *Suricata suricatta* and the yellow mongoose *Cynictis penicillata* with special reference to their reproduction. Bloemfontein, South Africa: Nasionale Museum.

Macdonald, D. W. 1979. 'Helpers' in fox society. *Nature* 282:69–71.

Macdonald, D. W., and Moehlman, P. D. 1982. Cooperation, altruism and restraint in the reproduction of carnivores. *Persp. Ethol.* 5:433–467.

Malcolm, J. R., and Marten, K. 1982. Natural selection and the communal rearing of pups in African wild dogs (*Lycaon pictus*). *Behav. Ecol. Sociobiol.* 10:1–13.

Mays, N. A., Vleck, C. M., and Dawson, J. 1991. Plasma luteinizing hormone, steroid hormones, behavioral role and nesting stage in the cooperatively breeding Harris's Hawk (*Parabuteo unicinctus*). *Auk* 108:619–637.

McClintock, B. 1987. A functional approach to the behavioral endocrinology of rodents. In: D. Crews, ed. *Psychobiology of Reproductive Behavior*, pp. 176–203. Englewood Cliffs, N.J.: Prentice-Hall.

McEwen, B. S., and Schmeck, H. M. 1994. *The Hostage Brain*. New York: Rockefeller Univ. Press.

Mech, L. D. 1970 *The Wolf: Ecology and Social Behavior of an Endangered Species*. New York: Natural History Press.

Mills, M. G. L. 1985. Related spotted hyaenas forage together but do not cooperate in rearing young. *Nature* 316:61–62.

Mills, M. G. L. 1990. *Kalahari Hyaenas*. London: Unwin Hyman.

Mills, M. G. L., and Biggs, H. C. 1993. Prey apportionment and related ecological relationships between large carnivores in Kruger National Park. *Symp. Zool. Soc. London* 65:253–268.

Moehlman, P. D. 1979. Jackal helpers and pup survival. *Nature* 277:382–383.

Moehlman, P. D. 1989. Intraspecific variation in canid social systems. In: J. L. Gittleman, ed. *Carnivore Behavior, Ecology, and Evolution*, pp. 143–163. Ithaca, N.Y.: Cornell Univ. Press.

Moller, O. M. 1973a. Progesterone concentrations in the peripheral plasma of the blue fox (*Alopex lagopus*) during pregnancy and the oestrus cycle. *J. Endocrinol.* 59:429–438.

Moller, O. M. 1973b. Progesterone concentrations in the peripheral plasma of the mink (*Mustela vison*) during pregnancy. *J. Endocrinol.* 56:121–133.

Moore, M. C., Thompson, C. W., and Marler, C. A. 1991. Reciprocal changes in corticosterone and testosterone levels following acute and chronic handling stress in the tree lizard, *Urosaurus ornatus*. *Gen. Comp. Endocrinol.* 8:217–226.

Munck, A., Guyre, P. M., and Holbrook, N. J. 1984. Physiological functions of glucocorticoids in stress and their relation to pharmacological actions. *Endocrinol. Rev.* 5:25–44.

Nalbandov, A. V. 1976. *Reproductive Physiology of Birds and Mammals*. San Francisco: W. H. Freeman.

Nonacs, P. 1991. Alloparental care and eusocial evolution: The limits of Queller's headstart advantage. *Oikos* 61:122–125.

Packard, J. M., Mech, L. D., and Seal, U. S. 1983. Social influences on reproduction in wolves. In: L. Carbyn, ed. *Wolves in Canada*, pp. 78–85. Canadian Wildl. Serv. Rep. No. 88.

Packard, J. M., Seal, U. S., Mech, L. D., and Plotka, E. D. 1985. Causes of reproductive failure in two family groups of wolves. *Z. Tierpsychol.* 68:24–40.

Packer, C., Gilbert, D. A., Pusey, A. E., and O'Brien, S. J. 1991. A molecular genetic analysis of kinship and cooperation in African lions. *Nature* 351:562–565.

Packer, C., Scheel, D., and Pusey, A. E. 1990. Why lions form groups: Food is not enough. *Amer. Nat.* 136:1–19.

Pearson, O. P., and Baldwin, P. H. 1953. Reproduction and age structure of a mongoose population in Hawaii. *J. Mamm.* 34:436–447.

Rapoport, L., and Haight, J. 1987. Some observations on allomaternal caretaking among captive Asian elephants (*Elaphas maximus*). *J. Mamm.* 68:438–442.

Rasa, O. A. E. 1973. Intra-familial sexual repression in the dwarf mongoose, *Helogale parvula. Naturwiss.* 60:303–304.

Rasa, O. A. E. 1977. Differences in group member response to intruding conspecifics and frightening or potentially dangerous stimuli in dwarf mongooses (*Helogale undulata rufula*). *Z. Tierpsychol.* 43:337–406.

Rasa, O. A. E. 1984. *Mongoose Watch: A Family Observed.* London: John Murray.

Rasa, O. A. E. 1986. Coordinated vigilance in dwarf mongoose family groups: The 'watchman's song' hypothesis and the costs of guarding. *Z. Tierpsychol.* 71:340–344.

Rasa, O. A. E. 1987. The dwarf mongoose: A study of behavior and social structure in relation to ecology in a small, social carnivore. *Adv. Stud. Behav.* 17:121–163.

Rasa, O. A. E. 1989. Behavioural parameters of vigilance in the dwarf mongoose: Social acquisition of a sex-biased role. *Behaviour* 110:125–145.

Reeve, H. K., and Ratnieks, F. L. W. 1993. Queen-queen conflicts in polygynous societies: Mutual tolerance and reproductive skew. In: L. Keller, ed. *Queen Number and Sociality in Insects,* pp. 45–85. Oxford: Oxford Univ. Press.

Reich, A. 1981. The behavior and ecology of the African wild dog, *Lycaon pictus*, in the Kruger National Park. Ph.D. dissert., Yale University, New Haven, Conn.

Rood, J. P. 1978. Dwarf mongoose helpers at the den. *Z. Tierpsychol.* 48:277–287.

Rood, J. P. 1980. Mating relationships and breeding suppression in the dwarf mongoose. *Anim. Behav.* 28:143–150.

Rood, J. P. 1983. The social system of the dwarf mongoose. In: J. F. Eisenberg and D. G. Kleiman, eds. *Advances in the Study of Mammalian Behavior,* pp. 454–488. Lawrence, Kansas: Amer. Soc. Mammalogists.

Rood, J. P. 1986. Ecology and social evolution in the mongooses. In: D. I. Rubenstein and R. W. Wrangham, eds. *Ecological Aspects of Social Evolution,* pp. 131–152. Princeton, N.J.: Princeton Univ. Press.

Rood, J. P. 1987. Dispersal and interpack transfer in the dwarf mongoose. In: B. D. Chepko-Sade and Z. T. Halpin, eds. *Mammalian Dispersal Patterns: The Effects of Social Structure on Population Genetics,* pp. 85–103. Chicago: Univ. Chicago Press.

Rood, J. P. 1990. Group size, survival, reproduction and routes to breeding in dwarf mongooses. *Anim. Behav.* 39:566–572.

Rood, J. P., and Nellis, D. 1980. Freeze marking mongooses. *J. Wildl. Mgmt.* 44:500–502.

Sapolsky, R. M. 1985. Stress-induced suppression of testicular function in the wild baboon: Role of glucocorticoids. *Endocrinology* 116:2273–2278.

Sapolsky, R. M. 1987. Stress, social status and reproductive physiology in free-living baboons. In: D. Crews, ed. *Psychobiology of Reproductive Behavior,* pp. 291–322. Englewood Cliffs, N.J.: Prentice-Hall.

Sapolsky, R. M. 1992. Neuroendocrinology of the Stress-Response. In: J. B. Becker, S. M. Breedlove, and D. Crews, eds. *Behavioral Endocrinology,* pp. 287–324. Cambridge, Mass.: MIT Press.

Seal, U. S., Plotka, E. D., Packard, J. M., and Mech, L. D. 1979. Endocrine correlates of reproduction in the wolf. I. Serum progesterone, estradiol and LH during the estrous cycle. *Biol. Reprod.* 21:1057–1066.

Shideler, S. E., Czekala, N. M., Kasman, L. H., Lindburg, D. G., and Lasley, B. L. 1983. Monitoring ovulation and implantation in the lion-tailed macaque (*Macaca silenus*) through urinary estrone conjugate evaluations. *Biol. Reprod.* 29:905–911.

Sillero-Zubiri, C. 1994. Behavioural Ecology of the Ethiopian wolf (*Canis simensis*). Ph.D. dissert., Oxford Univ., Oxford.

Smith, M. S., and McDonald, L. E. 1974. Serum levels of luteinizing hormone and progesterone during the estrous cycle, pseudopregnancy and pregnancy in the dog. *Endocrinology* 94:404–412.

Solomon, N., and French, J. In press. *Cooperative Breeding in Mammals.* Cambridge: Cambridge Univ. Press.

Stacey, P. B., and Koenig, W. D. 1990. *Cooperative Breeding in Birds: Long-Term Studies of Ecology and Behavior.* Cambridge: Cambridge Univ. Press.

Stander, P. E. 1992a. Foraging dynamics of lions in a semi-arid environment. *Canadian J. Zool.* 70:8–21.

Stander, P. E. 1992b. Cooperative hunting in lions: The role of the individual. *Behav. Ecol. Sociobiol.* 29:445–454.

Stander, P. E., and Albon, S. D. 1993. Hunting success of lions in a semi-arid environment. *Symp. Zool. Soc. London* 65:127–143.

van Heerden, J., and Kuhn, F. 1985. Reproduction in captive hunting dogs *Lycaon pictus. S. African J. Wildl. Res.* 15:80–85.

Vehrencamp, S. L. 1983. A model for the evolution of despotic versus egalitarian societies. *Anim. Behav.* 31:667–682.

Verhage, H. G., Beamer, N. B., and Brenner, R. M. 1976. Plasma levels of estradiol and progesterone in the cat during polyestrus, pregnancy and pseudopregnancy. *Biol. Reprod.* 14:579–585.

Waser, P. M. 1981. Sociality or territorial defense? The influence of resource renewal. *Behav. Ecol. Sociobiol.* 8:231–237.

Waser, P. M., Creel, S. R., and Lucas, J. R. 1994. Death and disappearance: Estimating mortality risks associated with philopatry and dispersal. *Behav. Ecol.* 5:135–141.

Waser, P. M., Elliott, L. F., Creel, N. M., and Creel, S. R. 1995. Habitat variation and mongoose demography. In: A. R. E. Sinclair and P. Arcese, eds. *Serengeti II: Research, Management and Conservation of an Ecosystem,* pp. 421–447. Chicago: Univ. Chicago Press.

Waser, P. M., and Waser, M. S. 1985. *Ichneumia albicauda* and the evolution of viverrid gregariousness. *Z. Tierpsychol.* 68:137–151.

Wingfield, J. C., Hegner, R. E., and Lewis, D. M. 1991. Circulating levels of luteinizing hormone and steroid hormones in relation to social status in the cooperatively breeding white-browed sparrow weaver, *Plocepasser mahali. J. Zool. (London)* 225:43–58.

Wingfield, J. C., and Moore, M. C. 1987. Hormonal, social, and environmental factors in the reproductive biology of free-living male birds. In: D. Crews, ed. *Psychobiology of Reproductive Behavior: An Evolutionary Perspective,* pp. 149–175. Englewood Cliffs, N.J.: Prentice-Hall.

Woodroffe, R., and Macdonald, D. W. 1993. Badger sociality: Models of spatial grouping. *Symp. Zool. Soc. London* 65:145–169.

Woolfenden, G. E., and Fitzpatrick, J. W. 1984. *The Florida Scrub Jay: Demography of a Cooperative-Breeding Bird.* Princeton, N.J.: Princeton Univ. Press.

CHAPTER 3

Female Masculinization in the Spotted Hyena: Endocrinology, Behavioral Ecology, and Evolution

Laurence G. Frank

Reproductive anatomy tends to be conservative among mammals, showing relatively minor variations across orders, even where the orders are characterized by marked specialization in other aspects of their biology. Probably the most anatomically extreme reproductive specialization among mammals is found in the female spotted hyena (*Crocuta crocuta*), whose external genitalia are so profoundly masculinized as to be distinguishable only with difficulty from the male's. Associated with genital masculinization is a suite of female behavioral adaptations characterized by a high level of aggressiveness, relative to that of the male, and a social system based on dominance of females over males. The unique syndrome of genital masculinization and aggressiveness in females seems to be mediated by endocrine adaptations in the pregnant female that virilize female embryos both anatomically and behaviorally. Moreover, the spotted hyena is the only living member of the Hyaenidae that is an active predator on large mammals, and the extraordinary adaptations of the female may have arisen in response to the exceptionally intense competition for food that occurs after a kill has been made.

The social system of the spotted hyena has been the subject of a 17-year study of a wild population in Kenya, and the biology of female masculinization has been intensively investigated in a laboratory colony at the University of California, Berkeley, in a study directed by Stephen E. Glickman. Through these complementary studies, we are beginning to understand this complex biological phenomenon at both proximate and ultimate levels of analysis. In this chapter, I attempt to synthesize these results. I begin with a description of the anatomical and behavioral characteristics that make the female spotted hyena so unusual among mammals, and then summarize the current state of our understanding of the underlying physiological mechanisms. I then consider some of the apparent costs of masculinization, as well as one of its more remarkable correlates, the intense neonatal aggression that appears to kill many sibling cubs in the first weeks of life. Finally, I describe the function of female aggression and dominance in the complex hyena social system, and conclude with (necessarily speculative) hypotheses regarding the origin and evolution of the system.

General and Reproductive Anatomy of the Spotted Hyena

Both sexes of the spotted hyena possess a variety of specialized features relating to diet and locomotion, which they share with the other extant members of the Hyaeninae, the striped hyena (*Hyaena hyaena*) and brown hyena (*Hyaena brunnea*). Hyaenadont dentition is characterized by massive bone-crushing premolars and powerful cranial musculature. Because the spotted hyena is an active predator on large mammals, the dentition and cranial musculature are somewhat more robust than those of the other extant hyenas (Werdelin and Solounias 1991), which supplement scavenging with predation on small vertebrates (Kruuk 1976; Mills 1990). Although still unstudied, the digestive system of hyenas seems extraordinarily long and narrow for a carnivore, apparently an adaptation for digesting bone (pers. obs.). The characteristic sloping profile of these animals (Figure 3.1) is the result of having hindlimbs

Figure 3.1. Female spotted hyena, illustrating the sloping profile due to the shorter hindlegs of these animals. (Photo by L. G. Frank.)

that are short relative to the forelimbs, apparently an adaptation that facilitates the "rocking-horse gallop" that allows the hyena to lope for hours, covering as much as 80 km in a night (Eloff 1964; Hofer and East 1993b; Tilson and Henschel 1986), with high energy efficiency (Pennycuick 1979).

In genital anatomy, however, the female spotted hyena has no parallel among mammals. For millennia, stories circulated in Europe about the hermaphroditic hyena (Aristotle 1965), and this belief is still current in Africa (Hemingway 1935). It has been variously said to be both sexes simultaneously, alternately, sequentially, or facultatively. The first scientific investigation of the spotted hyena was that of Leonard Harrison Matthews (1939), who in the mid-1930s collected over 100 individuals in Ngorongoro region of Tanganyika Territory (now Tanzania). Although his monograph (1939) showed that the hyena comprises only the usual two sexes, the hermaphrodite story was still being repeated in the scientific literature 23 years later (Deane 1962).

The male reproductive anatomy is not unusual. The female, however, is born with the clitoris so radically hypertrophied that in the adult it is nearly 90% the length of (171 vs. 193 mm) and the same diameter as (22 vs. 21 mm) the male penis (Neaves et al. 1980). The vaginal labia are fused to form a psuedoscrotum containing fat and connective tissue that give it the appearance of containing testicles. The urogenital canal traverses the length of the clitoris, and erectile tissue (the *corpus spongiosum*) allows the clitoris to become erect in the same manner as a penis. Only recently did we learn to distinguish the two sexes readily, on sight (Frank et al. 1990): the shape of the glans penis (Figure 3.2A) differs from that of the glans clitoridis (Figure 3.2B).

The female urinates, copulates, and gives birth through the clitoris. In prepubescent females (under 18 months of age), the diameter of the clitoral urogenital canal is only slightly larger than the canal of the male (Neaves et al. 1980), but at puberty it enlarges and becomes elastic, apparently under the influence of pubertal estrogens (Glickman et al. 1992a). Mating is facilitated by a pair of retractor muscles (Neaves et al. 1980), which are more robust in the female than in the male. These allow the female to retract the clitoris on itself (much like pushing up a shirtsleeve), creating an opening into which the male inserts his penis. However, because the clitoris is located in the same position as the penis, on the ventrum rather than under the tail as in other mammals, the male is forced to assume a rather awkward posture, with his rump tucked almost beneath the female, in order to achieve intromission. Once engaged, he assumes an erect posture behind her, with forepaws clasped around her lower abdomen (Figure 3.3).

Parturition through the clitoris is yet more problematical. The urogenital canal is substantially longer than that in mammals of comparable size (Matthews 1939) because of the position of the clitoris (Figure 3.4). The urogenital canal extends caudad from the uterus in the normal manner, but makes a 180-degree bend in the pelvis to enter the base of the clitoris, through which the

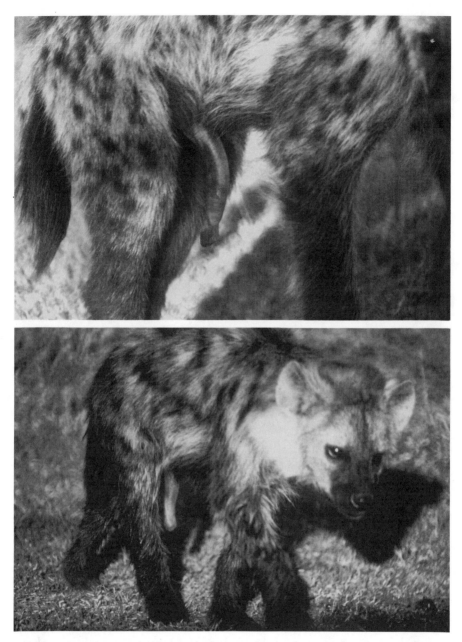

Figure 3.2. Juvenile spotted hyenas displaying phallic erection. *Above*, male. *Below*, female. (Photos by L. G. Frank.)

Figure 3.3. Mating spotted hyenas. (Photo by L. G. Frank.)

descending fetus must pass. As described below, the passage of a large fetus through such a long, narrow canal poses major difficulties, particularly in primiparous females.

The Spotted Hyena Social System

Most of the following information comes from a 17-year study of the Talek Clan, a group of about 70 individuals that inhabits a territory of about 60 km² in the Masai Mara National Reserve in southwest Kenya (Frank 1986a). Spotted hyena social organization is similar in most respects to that of baboons (*Anubis* sp.) and other terrestrial cercopithecines. Females stay in the natal group, or "clan," for life and constitute the stable core of the social group. Males disperse from their natal clan at or after puberty, eventually joining a new clan. In each sex there is a linear dominance hierarchy, and social rank is a powerful determinant of reproductive success for females (Frank et al. 1995), and probably for males as well (Frank 1986b). Juveniles attain their mothers' rank in the first 6 months of life through a complex process of social learning at the communal den (Holekamp and Smale 1993; Smale et al. 1993), and rank just below their mothers in the clan hierarchy. These ranks are relatively unchanging, and, as a result, relations between the matrilines comprising the

Figure 3.4. Reproductive tract of pregnant spotted hyena. At parturition, the fetus passes through the pelvis and then makes a 180° turn before passing through the entire length of the clitoris, a total journey of about 60 cm. The umbilical cord is only 15 cm long, and in primiparous females the fetus may become temporarily lodged in the pelvis or clitoris; because the placenta has separated previously, the resulting delay in delivery frequently results in stillbirth. (Drawing by C. M. Drea. From Frank et al. 1995; reprinted by permission.)

Talek Clan are remarkably stable over generations: the alpha female in 1995 is the great-granddaughter of the alpha female in 1979.

Adult female spotted hyenas are dominant to adult males in nearly all social interactions (Kruuk 1972; Tilson and Hamilton 1984; Frank 1986b; Frank et al. 1989; Mills 1990). Males invariably defer to females in competition over food, abandoning a kill as soon as females appear. A single subadult female can prevent five adult males from feeding on a large carcass (pers. obs.). Courting males epitomize intense approach-avoidance conflict, advancing toward the female with extreme wariness and shying away from even a mild threat. In captive peer groups, juvenile males may rank above females, but as they attain adulthood, the males gradually decline in rank until all females are dominant to all males (Frank et al. 1989).

In most social mammals, young males show higher levels of "rough-and-tumble play" than do females, which tend to play more gently (Goy and Resko 1972). In the spotted hyena, females are significantly more playful than males (Pedersen et al. 1990). Females are at the forefront in the violent interclan encounters (Henschel and Skinner 1991a) and in encounters with lions (pers. obs.).

Because males are dominant to females in most social mammals (Ralls 1976; Hrdy 1981), discussions of the phenomenon of female dominance in the spotted hyena have focused on the behavioral or ecological factors that may be responsible for the reversal of the more usual roles. It may be misleading, however, to think in terms of the advantage for females over males. As will become apparent, high-ranking females gain important advantages in female-female competition, including priority of access to food and dramatically improved reproductive success. Rank is mediated by aggression among females, and the benefits are amplified by the fact that the female rank is "inherited," ensuring that a high-ranking lineage maintains its advantages over generations. It is therefore likely that dominance over males is in fact an epiphenomenon, resulting from intense competition among females. In fact, an equally interesting question is, Why is male-male aggression so mild in a species that appears to be highly polygynous?

Another peculiarity of spotted hyena social behavior is clearly associated with female genital masculinization: phallic erections are a common social display in both sexes. This species appears to be unique among mammals in that the erect penis (or clitoris) is always an element of submissive or appeasement displays (East et al. 1993). Erections are most commonly seen in the "meeting ceremony," when animals that have been apart reunite and engage in a species-typical greeting display (Kruuk 1972; Krusko et al. 1988). The animals approach each other and stand alongside one another, head-to-tail, one or both lifting the inner hindleg and allowing inspection of its erect phallus and inguinal area. East et al. (1993), in a detailed analysis of meeting ceremonies in Serengeti spotted hyenas, show that when only one member of a greeting pair displays an erection, it is normally the subordinate. It is also the subordinate, normally, that initiates the greeting (Kruuk 1972), though a dominant animal may "force" a greeting on a reluctant subordinate. The likelihood of a subordinate female displaying an erection increases with increasing disparity between its rank and that of its partner. Greetings between adult males and adult females are rare, and only the higher-ranking adult males greet adult females. Reflecting the observation that there is generally a greater degree of association and affiliative behavior among members of high-ranking matrilines (Frank

1986b), East et al. (1993) show a similar disparity: members of high-ranking matrilines greet more often than do members of lower-ranking ones.

This unusual display is not without its costs. As Kruuk (1972) observed, each hyena puts its reproductive organs in immediate proximity to very powerful jaws. In fact, penile damage is not unusual, both in the wild (Frank 1986b) and in captivity. Greetings between captive females that have been separated for a number of weeks are very tense, and frequently end in serious fights that often begin when one member of the greeting pair bites the genital region of the other. The resulting damage may be severe enough to destroy or seriously compromise the reproductive competence of the injured party (Berger, Frank, and Glickman, unpublished data).

Endocrine Substrates of Female Masculinization

A large body of evidence demonstrates that sexual differentiation during fetal development in mammals is dependent on testosterone produced by the testis of fetal males (Jost 1972). In the absence of testosterone, a fetal mammal develops the physical and behavioral phenotype of a female, regardless of chromosomal sex. But in the presence of testosterone, and its metabolite dihydrotestosterone, the embryo develops as a male, whether the androgens are derived in the normal manner, from the testis, or experimentally administered to a chromosomal female. Exogenous testosterone can produce virtually complete male genitalia in a genetically female fetus (Short 1974), and a variety of naturally occurring endocrine anomalies can produce individuals that are genotypically one sex but phenotypically the other. The normal female spotted hyena exhibits an extreme form of female masculinization that in other mammals may be produced either experimentally, through hyperactivity of the fetal adrenal gland (which normally produces only small amounts of androgens), or through defects in maternal endocrine metabolism that produce excess androgens.

Androgens and Aggression

The behavioral effects of hormones in animals may conveniently be separated into two categories (Goy and McEwen 1980). *Organizational* effects are those that operate during embryonic development to influence the brain permanently, such that sex-specific behaviors are likely to be displayed by the adult. Androgen secreted by the fetal testis or experimentally administered to female fetuses is responsible for the display of sexual and aggressive behaviors typical of the adult male. In the absence of fetal androgens, these behavior patterns are reduced or eliminated. Organizational effects may operate independently of subsequent activation: exogenous androgen administration to female monkey fetuses causes not only virilization of the genitalia but also male-like levels of juvenile rough-and-tumble play and aggressive behavior, even in the absence of subsequent exposure to androgens (Goy and Resko 1972). *Activational* effects of hormones are those that operate reversibly in postfetal life to influence the display of behaviors or anatomical features; examples include the growth and shedding of antlers, both of which are controlled by seasonal androgen secretion, receptive behavior, which is controlled by increases in estrogen associated with estrus, and seasonal changes in rates or intensity of aggressive behavior, such as those produced by testosterone in rutting male ungulates. Alternatively, the occurrence of other behavior, such as male sexual behavior and aggressiveness, may require both prenatal organization and concurrent activation in the adult. Furthermore, social experience and milieu can interact significantly with hormonal influences. The experience of winning a fight raises a male's testosterone level, whereas losing an encounter, or experiencing social turmoil, tends to depress testosterone (Lovejoy and Wallen 1988; Bernstein et al. 1983; Sapolsky 1991). Finally, because the threshold testosterone levels necessary for the expression of aggression are low, dominance rank or rates of aggressive behavior do not correlate with circulating androgen levels. Thus, although the relationship between androgens and aggression is never simple, both organizational and activational influences of androgens are clearly important for the display of many forms of aggression (Monaghan and Glickman 1992).

In the conclusion of his monograph on spotted hyena reproductive anatomy, Matthews (1939) speculated that the female condition may be the result of testosterone produced by the ovary, which he described as showing an unusually large amount of stromal (steroidogenic) tissue and relatively few follicles compared to those of other mammals. He speculated that androgens from the maternal ovary are transferred to the embryos, thus masculinizing the genitalia in developing females. He ended his monograph with a lament that, although endocrinologically very interesting, the hyena was hardly an ideal experimental animal. Experience has proved Matthews correct on the first count: the spotted hyena is an exceptionally interesting animal in which to test hypotheses about hormonal influences on mammalian development, both anatomical and behavioral. His pessimism about its potential as an experimental animal was premature, however, for we have been able to learn a great deal about the processes that masculinize the female spotted hyena.

This review will not attempt to summarize all that has been learned about spotted hyena endocrinology, but will address primarily the central question that makes this hyena so interesting: What endocrine and developmental processes account for the extraordinary anatomy and behavior of the female spotted hyena? A more thorough account of spotted hyena endocrinology is given by Glickman et al. (1992a, b, 1993).

Field Studies of Endocrinology

The first major contribution to spotted hyena endocrinology was made by Racey and Skinner (1979), who used material obtained from animals killed in culling operations in Kruger National Park, South Africa. They found no statistically significant differences in testosterone levels between males and females, or between any of the age-sex classes they examined, including pregnant females. Taking issue with Matthews, they postulated that androgens from the fetal ovary, rather than from the maternal ovary, are responsible for masculinizing the female fetus *in utero*.

Data from the Masai Mara study failed to corroborate Racey and Skinner's key finding, however. Using blood samples from long-term study animals of known status, Frank et al. (1985a) showed that male androgen levels are significantly higher than those of females, as they are in other mammals. Social status was found to have a strong effect on male levels: transient males passing through the Talek Clan range, when captured, showed androgen levels only 25% as high as those of resident immigrant males (there was, however, high variance in the latter). Thus, the low social status of nonresident males seems to have a depressive effect on androgen levels, a finding that was subsequently corroborated by van Jaarsveld and Skinner (1991) in hyenas from the Kruger and Kalahari Gemsbok National Parks, South Africa.

Among females at Masai Mara, there was no apparent relation between social rank and circulating androgens, except that the alpha female showed levels in the male range. She was pregnant when sampled, however, and because the South African work reported no relation between pregnancy and androgen levels, we suggested a relation between her circulating androgens and her social rank. Subsequent work at Berkeley has shown that females produce very high levels of testosterone during pregnancy, and has not borne out any relation between circulating androgens and rank (Baker 1990).

In a follow-up study of 60 hyenas from Kruger National Park, Lindeque et al. (1986) again failed to find significant sex differences in circulating testosterone levels in any age-sex class of hyenas, though they tended to be higher in males. They did note, however, significantly higher testosterone levels in pregnant females than in nonpregnant parous females, but attributed this disparity to increased levels of steroid binding protein, which typically rise in pregnant female mammals.

Lindeque and Skinner (1982) reported on fetal androgens taken from twin fetuses in four pregnant females culled in Kruger National Park. They reported high levels of testosterone in the maternal circulation, the fetal circulation, and the umbilical artery, which returns blood from the fetus to the placenta, but failed to detect significant levels of androgens in either the placenta or the umbilical vein, which transports blood from the placenta to the fetus. Because androgenized blood appeared to flow from the fetus to the mother, but not the reverse, Lindeque and Skinner ruled out a maternal androgen contribution via the placenta. Following Racey and Skinner (1979), they also concluded that androgens in the fetus were synthesized in the fetal testes and ovaries, rather than being of maternal origin.

Lindeque and Skinner's was the first study to assay for androstenedione, a weak androgen that is synthesized in mammalian adrenals, testes, and ovaries (Whalen et al. 1985). In the ovaries, it serves as the precursor to estradiol, the primary estrogen, upon conversion by the enzyme aromatase. The enzyme 17 beta-hydroxysteroid dehydrogenase converts androstenedione to testosterone. All studies to date (Lindeque and Skinner 1982; Glickman et al. 1987; van Jaarsveld and Skinner 1991) have agreed that female spotted hyenas of all ages display higher levels of circulating androstenedione than do males, a pattern unusual among mammals (Feder 1985). Although some female primates (including humans) may show higher androstenedione levels than do males, during parts of the menstrual cycle (Feder 1985), the absolute values of androstenedione in spotted hyenas are higher than those reported for other female mammals.

The Berkeley Hyena Project: Endocrinology, Anatomy, and Behavior

In spite of Matthews' doubts about the feasibility of studying hyenas in captivity, a research colony was established in 1985 at the University of California, Berkeley, the purpose of which was to use the spotted hyena as a model in which to study mechanisms of sexual differentiation. Twenty infants (two cohorts of ten) were collected under permit from dens in Narok District north of the Masai Mara Reserve. The colony, now numbering about 40 animals, is housed at the Field Station for Behavioral Research on a 1/2-hectare complex of indoor-outdoor pens, separated into small social groups.

Our original plan to maintain one or two large "clans" proved infeasible when aggression escalated as the cubs entered adolescence. Owing to the tendency of spotted hyenas to redirect aggression onto lower-ranking individuals, the latter suffered frequent group attacks, and we were eventually forced to divide the original cohorts into subgroups of five (Frank et al. 1989). But although none of the original infants in our colony had more than a few weeks of social experience in the wild, all of the basic characteristics of the spotted hyena's natural social system have emerged in captivity. These include female dominance, linear dominance hierarchies, kin coalitions, and the acquisition of maternal social rank (Jenks et al. 1995).

Most of the hyena research in Berkeley has focused on the endocrine events that underlie anatomical and behavioral differences between the sexes. The first three years of the research were largely normative, coupling intensive behavioral observations of the wild-caught infants, as they matured, with monthly blood sampling and physical measurements to document the concomitant endocrine changes. Two males and four females who were bilaterally gonadectomized at four to seven months of age served as controls, individuals that would mature in the absence of endogenous steroids. Other than those due to aggression, there have been few management difficulties, and no disease problems (Berger et al. 1992).

Behavioral research took two forms: (1) nightly recording of spontaneous social behavior in the outdoor pens and (2) standardized test situations designed to elicit certain types of behavior (e.g., play or dominance interactions) under controlled conditions amenable to videotaping for subsequent detailed analysis.

As the original females began to breed, research emphasis shifted to the endocrine changes associated with pregnancy and neonatal life. The discovery that neonates fight violently at birth, and that their fighting is associated with behavioral and morphological precocity, led to an unanticipated avenue of research in both the laboratory and the field. And as more females grew to reproductive age, the emphasis shifted again, this time to the prenatal endocrine events responsible for the unusual form of female sexual differentiation seen in this species.

Postnatal Hormone Profiles

The first three years of endocrine research at Berkeley, which were based on monthly blood samples, refined the picture that had been obtained from single samples of animals shot or darted in the wild (Glickman et al. 1992b). As in other mammals, prepubertal males and females do not differ in circulating testosterone levels, but after the mean age of 26.5 months, males show significantly higher levels of the primary androgen (Figure 3.5). Chromatographic separation of plasma steroids revealed that a substantial fraction of "testosterone" in both males and females was in fact the metabolite dihydrotestosterone (DHT), which is responsible for masculinizing genital and other tissues, and that DHT constituted a higher percentage of the total androgens in females. Estrogen comprised estrone and estradiol, the fractions depending on reproductive state; uncharacterized estrogenic compounds were also present. Estrogen levels rose in females after 14 months of age, and comparison with levels in the gonadectomized females showed that estrogen is responsible for nipple growth, the increase in the diameter of the urogenital canal first described by Neaves et al. (1980), and increased clitoral elasticity, which permits mating and birth. Males and gonadectomized females showed no change in elasticity of the penile/clitoral meatus. Very low levels of testosterone, androstenedione, and estrogen in the gonadectomized individuals confirmed earlier evidence that the gonads, rather than the adrenals, are the source of androgens.

Measurements of androstenedione from plasma samples taken from the captive infants and from subadults and adults in the Mara study population confirmed that, at most ages, females have higher circulating levels of androstenedione than do males, leading Glickman et al. (1987) to suggest that

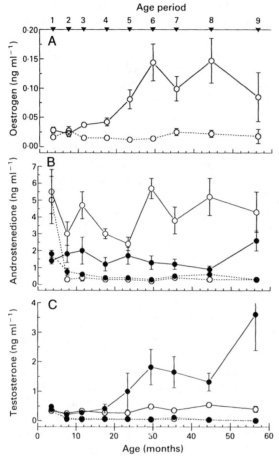

Figure 3.5. Mean concentrations (± SEM) of estrogen, androstenedione, and testosterone + dihydrotestosterone in 20 spotted hyenas between the ages of 2 months and 5 years. Closed circles are males; open circles are females. Solid lines denote intact animals; dotted lines denote subjects gonadectomized at 4–7 months of age. N = six intact males, two gonadectomized males, eight intact females, and four gonadectomized females. After surgery, levels of estrogen and testosterone fell to nondetectable levels. Androstenedione in males was not affected by castration, indicating a nontesticular source. Adrostenedione levels were significantly higher in intact females than in intact males; there was no tendency for these levels to increase with age in males. Testosterone + dihydrotestosterone was significantly higher in intact males than in intact females after the age of 36.5 months. (From Glickman et al. 1992a; reprinted by permission.)

this "weak androgen" may be implicated in the unusual levels of aggressiveness characteristic of the female. Experimental administration of androstenedione to male mice (*Mus musculus*) has shown that it may significantly increase aggressiveness (Brain 1983), and observation of human teenagers suggests that elevated levels of androstenedione may be associated with some of the emotional turmoil of adolescence (Inoff-Germain et al. 1988).

Prenatal Hormone Mechanisms

Since female hyenas are born with fully masculinized genitalia, the most interesting endocrine events occur prenatally. Thus when the captive-raised infants reached sexual maturity, we initiated studies of endocrine events during pregnancy. Every intact female went through at least two complete pregnan-

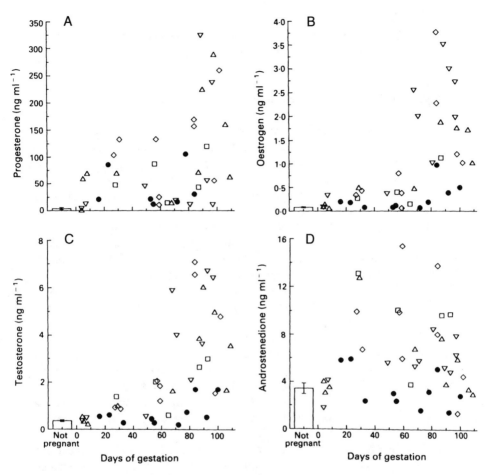

Figure 3.6. Concentrations of plasma progesterone (*A*), estrogen (*B*), testosterone + dihydrotestosterone (*C*), and androstenedione (*D*) during gestation in five spotted hyenas, representing 15 pregnancies. Each symbol represents a different animal. Bar at lower left of each graph represents mean measures and 95% confidence intervals from the same animals when nonpregnant. (From Licht et al. 1992; reprinted by permission.)

cies, during which blood samples were regularly taken from her peripheral circulation.

As in other pregnant mammals, circulating levels of the ovarian steroids progesterone and estrogen rose steadily during pregnancy, attaining levels orders of magnitude above those in the nonpregnant state (Licht et al. 1992). In contrast with what occurs in most mammals, however, levels of androgens also rose, two- to threefold in the case of androstenedione, more dramatically in the case of testosterone (Figure 3.6). Androstenedione rose above nonpregnant levels early in pregnancy and remained elevated throughout. Testosterone began to increase only about halfway through gestation, but then attained levels as high as the highest measured in adult males. About half of the "testosterone" is in fact DHT, which is normally responsible for masculinizing genital tissues, and comparable levels of testosterone were measured in plasma samples taken from neonates at birth (Frank et al. 1991).

Origins of Elevated Testosterone in Pregnancy

Although the adrenal glands are a potential source of androgens, castration experiments on adults (Frank et al. 1985b) and juveniles (Glickman et al. 1992a) showed that the gonads, rather than the adrenal glands, were (as mentioned above) the only important source of androgens, in both sexes.

To investigate the source of masculinizing androgens *in utero,* some pregnancies were interrupted. We performed caesarean sections in late pregnancy, and obtained blood samples from the component parts of the maternal-fetal circulation before and after removal of the fetal-placental unit (Licht et al. 1992). These included samples from the peripheral maternal circulation, the ovarian vein (which drains into the uterus), the uterine vein (which drains out of the uterus), the umbilical vein (which carries blood from the placenta to the fetus), the fetus itself, and the umbilical artery (which carries blood back from the fetus to the placenta).

Androstenedione in the vein draining the ovary was 35 times higher than that in the peripheral maternal circulation, but was low in the umbilical circulation. Testosterone was higher in the uterine veins than in the maternal periphery. When the placenta was detached, androstenedione rose in the maternal circulation while testosterone declined. These observations suggested that the androstenedione produced by the ovary is converted by the placenta to testosterone (and presumably dihydrotestosterone).

Testosterone levels were higher in samples taken from the umbilical vein than they were in the fetus itself or in the umbilical artery, showing a gradient from placenta to umbilicus to fetus. These studies have failed to corroborate Lindeque and Skinner's (1982) contention that the umbilical vein does not convey testosterone from the mother to the fetus. There was no difference in testosterone level between male and female fetuses. Androstenedione tended to be higher in the fetuses than in the umbilical circulation, suggesting a fetal source for fetal androstenedione. Finally, morphological evidence suggests that the spotted hyena placenta differs from those of other carnivores, and may be particularly efficient at converting substances and transporting them to the fetus (Wynn and Amoroso 1964). These observations led to a series of experiments on the steroidogenic capacity of the ovary and placenta.

The Ovary

Because in most mammals androstenedione is the normal precursor for the primary estrogen, it is produced in all mammalian ovaries, and is normally converted immediately into estradiol. When Yalcinkaya et al. (1993) incubated spotted hyena ovarian-tissue fractions with luteinizing hormone, which stimulates the production of steroids, they found that very large amounts of androstenedione were produced, particularly by the stromal tissue, as first suggested by Matthews. However, the ovary did not produce large quantities of testosterone. Thus the ovary appears to be the source of the androstenedione measured in the female circulation but cannot account for the levels of testosterone detected in both the pregnant female and the fetus.

The Placenta

Yalcinkaya et al. (1993) then tested the placenta's ability to convert androstenedione to estradiol and testosterone. In the human placenta, the enzyme aromatase normally catalyzes the conversion of androgen into estradiol, thereby protecting the female fetus from excess androgens that may be produced by the mother. In the hyena placenta, however, aromatase activity was only about 5% of that found in human placenta, indicating that maternal androgens would be passed on to the fetus largely unchanged. Moreover, they found that placentas of both hyena and human were able to convert androstenedione to testosterone efficiently, through the activity of a second enzyme, 17 beta-hydroxysteroid dehydrogenase, which was about equally active in the two species. Thus, the placenta not only fails to protect the fetus from androstenedione originating in the maternal ovary, but actively converts that relatively weak androgen to the more potent testosterone. It is not yet known how or where a significant proportion of the testosterone is converted by the enzyme 5 alpha-reductase to dihydrotestosterone, which is presumably ultimately responsible for genital masculinization.

Yalcinkaya et al. suggest an epigenetic mechanism that not only may explain hyena virilization, but also may play a role in certain human abnormalities. When immature female rats (*Rattus norvegicus*) are treated with testosterone, the number of follicles in their ovaries is greatly reduced, and stromal tissue, the source of androstenedione, increases. Thus, the high levels of testosterone delivered to the developing female hyena fetus not only may virilize her genitalia, but also may create an ovary that in turn will produce high levels of androstenedione in adulthood. This mechanism would also explain the small number of ovarian follicles noted by Matthews and subsequent investigators. A relatively minor genetic change that served to decrease aromatase production by the placenta would produce the relatively important mechanism described above. Finally, Yalcinkaya et al. note a similarity between this proposed mechanism of hyena masculinization and human polycystic ovarian syndrome, which is also characterized by high androstenedione levels and pubertal virilization. They suggest that a similar mechanism of placental testosterone synthesis may produce the characteristic ovary that secretes excessive androstenedione later in life.

A Caveat on Mechanisms of Sexual Differentiation

The endocrine mechanisms described above would appear to explain the development of male-like genitalia in female spotted hyenas. A set of preliminary observations and experiments in the Berkeley colony, however, suggests that the story may turn out to be more complex: (1) demonstrating steroid binding to androgen receptors in hyena genital tissue has proved difficult; (2) prepubertal castration, in either sex, does not prevent the development of nearly normal-sized external genitalia; and (3) administration of antiandrogens (drugs that inhibit androgen synthesis or reduce receptor activity through competitive binding) does not yield expected results. Administration of anti-androgens during pregnancy should dramatically feminize the genitalia of both male and female infants, but early results (Glickman and Licht, unpublished data) nevertheless suggest that they may not prevent the development of the phallus, in either sex. The possibility must thus be considered that a novel but as yet undiscovered mechanism may contribute to sexual differentiation in this species.

Adaptations (and Maladaptations?) of the Reproductive System

Not surprisingly, exposing a fetal female mammal to abnormally high levels of androgens would in most cases create a variety of problems tending to interfere with normal reproduction in the adult. The complex "costs" of female androgenization have been discussed at length by Glickman et al. (1993) and will be only briefly summarized here.

The anatomy of the female reproductive system is profoundly affected by exposure to androgens during critical stages of development. In experimental animals, exogenous androgens cause the clitoris to hypertrophy, the vulva to fuse, and the urogenital canal to exit the body at the base of the clitoris. Clearly, there have been adaptations in the spotted hyena to accommodate extreme genital masculinization while maintaining the patency of the urogenital canal. Concomitantly, the clitoris developed the ability to become elastic at puberty, so as to allow both copulation and parturition, and its retractor muscles enlarged, presumably also to facilitate mating.

In female mammals, ovarian cycling depends upon hypothalamic secretion of gonadotrophin-releasing hormone (GnRH), which in turn causes the pituitary to release luteinizing hormone (LH). In response to LH, the ovary secretes estrogen. In some species, fetal androgen exposure may disrupt the development of the hypothalamic-pituitary-ovarian axis such that estrus cycles are disrupted (Feder 1985). Various components of maternal behavior may also be inhibited by prenatal androgens (Glickman et al. 1993). Clearly, the developing brain of the female spotted hyena must be buffered from these maladaptive influences of the androgens that influence the development of adult aggressive behavior. Because reproductive and aggressive behaviors are influenced by different hormones at different times during fetal development, the timing of androgen exposure may be a critical factor, allowing development of enhanced aggressiveness while also maintaining normal reproductive cycling.

We can only speculate on the nature of the evolution that would yield the elaborate adaptations necessary to permit the newly masculinized female hyena to reproduce. Because all components of the system—anatomical development, endocrine development, and intact maternal behavior—needed to be present simultaneously, it is not easy to visualize a gradual adaptation to masculinization. The high frequency of birth complications (dystocia) seen in primiparous spotted hyenas, apparently deriving from the extraordinary anatomy of the reproductive tract, suggests that adaptation to masculinization may not be fully developed.

For one thing, the urogenital canal of a primiparous female is too narrow to allow easy passage of the fetus, which may become lodged in the proximal end of the clitoris at the junction with the internal reproductive tract. And at the distal end, the fetus must tear the urinary meatus to form an adequate opening. But because the urogenital canal remains permanently stretched, and the meatus tear does not close, subsequent births are relatively uneventful. The spotted hyena fetus is unusually large at birth compared to the neonates of other carnivores (Gittleman 1986), probably owing to the prolonged gestation and precocial development that are associated with the siblicidal fighting I shall describe below (Frank and Glickman 1994; Frank et al. 1995).

Furthermore, for reasons that are not understood, the umbilical cord is too short to allow the placenta to remain attached while the fetus travels the full length of the urogenital canal. Judging from the animal's size alone, one would expect the spotted hyena's birth canal, measured from the uterus to the normal position of the vulva, to be about 30 cm long. In fact, the canal makes a 180° bend in the pelvis (Figure 3.4), and then traverses the full length of the clitoris; the birth canal is in fact about 60 cm long (Matthews 1939; Neaves et al. 1980). But the umbilical cord is only 12–18 cm long. Thus, the placenta detaches from the uterine wall before the fetus is delivered; if the delivery is delayed, owing to the narrowness of the birth canal, the fetus will be stillborn. In humans, by contrast, the fetus travels only about 8 cm through the birth canal before the head is delivered, and the umbilical cord is approximately 50 cm long.

In the Berkeley colony, four out of 11 (37%) of births to primiparous females have resulted in complications that required veterinary attention, twice saving the life of the mother, and 11 out of 18 first fetuses (61%) have been stillborn. If this pattern occurs in wild hyenas as well, it may explain the high death rate seen in females three to four years old, the age of first reproduction (Glickman et al. 1993).

Labor in a primiparous hyena female is prolonged; females in Berkeley have endured labors over 48 hours long, compared to the labors of just a few hours that are typical of multiparous females. A parturient female appears to be largely oblivious to her surroundings as she paces about, rolls on her back, sits upright, and licks fluid from her clitoris. In the wild, these contortions would draw the attention of any passing lion or human, and predation on parturient hyenas may account for some of the mortality associated with females at the age of first reproduction (Glickman et al. 1993).

Using the rates of neonatal mortality and severe maternal distress observed in the laboratory, and the mortality rate among wild female hyenas at first breeding, we can estimate the fitness cost of masculinization, in terms of dead young and reduced maternal life expectancy (Frank et al. 1995). It appears that dystocia costs the average female spotted hyena between 16% and 25% of her expected lifetime production of offspring.

Siblicide in Spotted Hyenas

Although neonatal fighting is well documented in birds, it is very rare in mammals. Neonatal fighting in large litters of domestic pigs, *Sus scrofa*, may result in mortality (Fraser and Thompson 1991), and fighting has been described anecdotally in neonatal Arctic foxes, *Alopex lagopus* (Macpherson 1969), and red foxes, *Vulpes vulpes* (Henry 1975). *Habitual* neonatal siblicide, however, had not been described in any mammal until the recent discovery of intense and frequently fatal fighting in spotted hyenas (Frank et al. 1991). Because theoretical considerations suggest that siblicide is unlikely to evolve in a species with a small litter size, I present a brief introduction to siblicide theory.

There clearly are adaptive consequences of siblicidal aggression in spotted hyenas. The lack of a similar phenomenon in related mammals, however, and the close association between neonatal aggression, prenatal androgen exposure, and the entire syndrome of female masculinization lead to the possibility that neonatal aggression originally arose indirectly, as a consequence of androgenization, rather than directly, as a reproductive adaptation to ecological variability.

Spotted hyena females normally give birth to litters of twins, after a gestation period of about 110 days. The infants are highly precocial at birth, with fully erupted canines and incisors (Figure 3.7); their eyes are open, and they are capable of vigorous, directed locomotion and biting attack within minutes of birth. By comparison, other carnivores, including the other hyena species, are born in an altricial state, without teeth and capable of only very limited activity for the first weeks of life. In striped and brown hyenas, the gestation period is about 90 days (Rieger 1979; Walker 1975).

The two spotted hyena infants are typically born about one hour apart. There is no significant size difference in neonates (Smale et al. 1995). Within

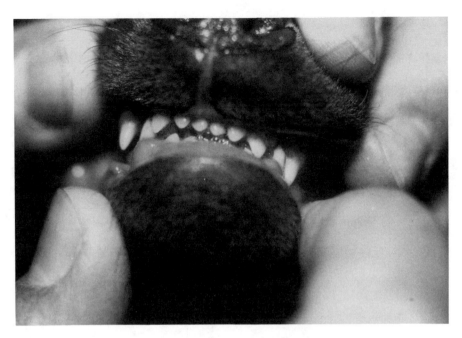

Figure 3.7. Fully erupted incisors and canines of spotted hyena on day of birth. (Photo by L. G. Frank.)

minutes of the birth of the second infant, the firstborn attacks it violently, fastening its teeth into the skin of the neck or back and shaking furiously. The second responds in kind. The attack is virtually identical to the attack behavior seen in intraspecific aggression among spotted hyena adults. The neonates fight through the first hours or days of life, until one emerges as dominant over the other. Aggression diminishes after the first day, but the dominant continues to attack and threaten the subordinate, at a lower rate, long after the latter has stopped defending itself (Figure 3.8A). Intralitter rank established by the initial fighting is stable thereafter, in both wild and captive litters (Smale et al. 1995). Play emerges as aggression declines, in the second week of life, possibly preparing the neonates for socialization with peers at the communal den (Drea et al., in press).

Hormone levels measured pre- and postnatally show that in the last month before parturition, fetal hyenas are exposed to sustained high levels of testosterone and androstenedione. Levels of both hormones start to decline at birth (Frank et al. 1991), when the neonates are separated from the maternal-placental circulation that is the source of most of these androgens (Figures 3.8B, 3.8C) (Licht et al. 1992), but both male and female infants maintain elevated levels, owing to their own endogenous production of testosterone and androstenedione, respectively.

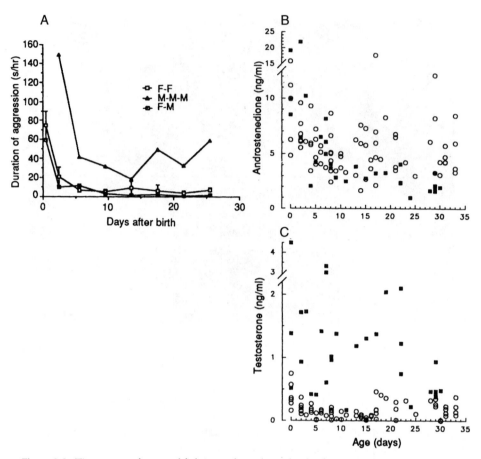

Figure 3.8. Time course of neonatal fighting and circulating levels of testosterone and androstenedione in five litters of spotted hyenas between birth and one month of age. Three litters were female-female pairs, one was mixed sex, and one was a set of male triplets. *A*, duration (secs/hour) of aggressive acts (bite-shakes, chases, and threats). The data were divided into 4-day blocks, with the first period further divided into days 0–1 and 2–3. Note that aggression is highest in the days immediately following birth, falling rapidly thereafter. The triplet litter had three potential dyads, and aggression in all three continued at a higher level than that in twin litters; the decline in aggression was similar and significant, whether measured in total duration, frequency of aggressive bouts, or mean length of bout. *B*, individual androstenedione measures in the first month of life. Closed squares are males; open circles are females. Females had significantly higher levels than males in days 20–33. *C*, individual testosterone measures in the first month of life. Levels were significantly higher in males than in females, at all ages. (From Frank et al. 1991; reprinted by permission.)

Siblicide and Sex-Ratio Adjustment

In captivity, spotted hyena neonates have been housed in pens where they have constant access to the mother, and to objects they can hide behind. Under these circumstances the neonatal aggression results in strong and stable domi-

Figure 3.9. Spotted hyena cub at 3 weeks of age, showing effects of severe neonatal aggression. This cub, severely emaciated and undersized, was the smallest of a set of triplets born in the Talek Clan. Aggression from the two dominant sibs has removed all the skin from the back of the neck and shoulders, and the resulting wound is infected. The cub was not seen again. (Photo by Paul Barber; reprinted by permission.)

nance; both members of the litter survive and grow, though the subordinate often bears extensive wounding of the back, neck, and ears as a result of attacks from the dominant.

In the wild, the consequences of sibling aggression appear to be much more severe (Figure 3.9). In eastern and southern Africa, the female normally gives birth at the mouth of a disused aardvark (*Orycteropus afer*) burrow (East et al. 1989; Frank et al. 1991; Henschel and Skinner 1991b); occasionally, other suitable cavities are used (Kruuk 1972). The neonates live in the burrow system, coming out at intervals to nurse. The mother lies in the entrance of the burrow to nurse, but the tunnel is too narrow for her to enter, and she appears to be unable to influence the behavior of her offspring within the burrow system itself.

Siblicide and Litter Sex Ratios

In about half of all spotted hyena litters, only one sibling survives the natal den stage to be brought to the communal den. Between 1980 and 1990, among those twin litters that survive intact, nearly all constituted brother-sister pairs, though occasional male-male pairs survived (Frank et al. 1991). Prior to 1991, no sets of female-female twins had been known to survive in any of the wild populations studied (Frank et al. 1995; Hofer and East 1992; Mills 1990).

If the two sexes are equally likely at conception, and since multiparous females nearly always bear twins, the expected distribution of male-male:male-female:female-female litters would be binomially distributed, and the ratio of litter types would approximate 1:2:1 (Frank et al. 1991). And in both fetal litters collected in the wild (Lindeque 1981) and litters born in the Berkeley colony (Frank 1992), the binomial distribution of litter types is in fact what is observed. Thus, litter reduction in same-sex litters appears to occur *after* birth, as a result of the starvation and trauma incurred by the subordinate cub from aggression by the dominant (Frank et al. 1991).

Long-term demographic data from the Talek Clan suggest that adult females may be able to influence the outcome of fighting in their neonates in an adaptive manner. Specifically, they may be able to offset the costs of cub mortality by manipulating the fighting in such a way as to adjust the sex ratio of their litters to existing social and demographic variables (Frank 1992).

Female Rank and Offspring Cost

Two lines of demographic evidence indicate that, at least among the highest-ranking females, daughters require greater parental investment than do sons, an effect not apparent in the lower-ranking matrilines. First, among the highest-ranking females, the interbirth interval following a singleton son was significantly shorter than the interval following twins, and the interval following a singleton daughter was the longest (Figure 3.10). In the lower ranks, there was little effect of litter composition on interlitter interval. Thus, high-ranking females tend to produce singleton sons at a high rate, and singleton daughters at a low rate.

Second, the likelihood of an infant hyena surviving to the age of one year depended in large measure on the social rank of its mother and the sex composition of the litter in which it was raised (Figure 3.11A, B). Female cubs born to high-ranking mothers survived well regardless of whether they were singletons or had twin brothers; in the lower ranks, singletons had only a slight survival advantage. Among male infants, there was a strong relationship between maternal rank and the cost or benefit of a twin sister. In the highest ranks, singleton males survived infancy significantly more often than males with twin sisters did. Having a sister improved male survival in the middle ranks, and made no difference in the lowest ranks. The interaction of maternal rank with the differential costs of male and female offspring may reflect the feeding advantage of high-ranking females and their offspring in the extraordinarily competitive feeding melees characteristic of this species. Nutritionally stressed low-ranking females and their young are likely to be subject to different costs and constraints than are the well-fed high-ranking females and young.

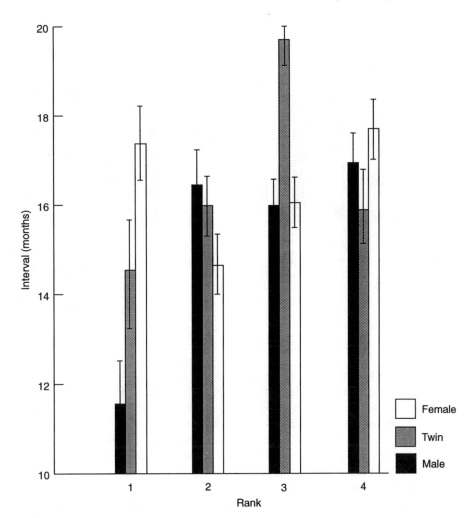

Figure 3.10. Effect of maternal rank and litter composition on interlitter intervals in 36 female spotted hyenas. Black bars are singleton male litters; open bars are singleton female litters; shaded bars are mixed-sex twin litters. Rank Level 1 comprised alpha and beta females, all of which belonged to the top-ranked matriline. Rank Levels 2–4 comprised high-middle, low-middle, and low-ranking females (N = 6, 9, 10, 11 females for Rank Levels 1–4, respectively). See Frank et al. 1995 for details of methods. For top-ranked females, singleton sons were followed by very short intervals compared to any other rank-litter type combination. (From Frank et al. 1995; reprinted by permission.)

By both measures, then, daughters are more costly than sons, at least to high-ranking mothers: they extract a longer period of investment from the mother, and they appear to decrease the probability of a twin brother surviving infancy. By preferentially investing in singleton sons, high-ranking females produce more of the cheaper sex.

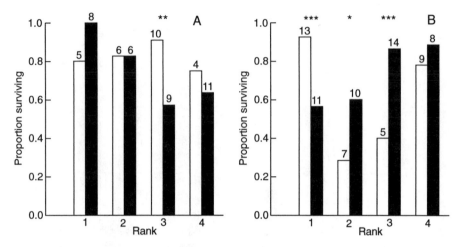

Figure 3.11. Probability of female (*A*) and male (*B*) cubs surviving to the age of 1 year, depending on maternal rank and whether they were singletons or twins. Open bars are singletons; black bars are twins. Sample size is indicated above each bar, and asterisks denote significant *z* scores comparing the proportions surviving to 1 year.

Production of Sons at High Density

Between 1980 and 1990, the adult female population of the Talek Clan was stable, at 20–23 breeding females. During this period, the mean sex ratio (proportion of males) born to females in the top half of the clan hierarchy was .67 ± .06, compared to .50 ± .06 in the lower ranks (*t* = 2.015, *df* = 1, *P* = 0.05). Since nearly all litters of multiparous females are born as twins, and since the majority of same-sex twins are apparently reduced to a singleton by fatal siblicidal fighting in the natal den, the expected distribution of singleton sons:mixed-sex twins:singleton daughters at the communal den would be approximately 1:2:1. In the lower ranks this was the observed pattern (Figure 3.12). Females in the top-ranked matriline, however, produced a significant excess of singleton sons, as compared either to twins or to singleton daughters. The relatively low proportion of twins and single females produced by top-ranked mothers further suggests that some of the singleton-male litters originated as mixed-sex twins, the sister of which died. In that period. the distribution of twin litter types was significantly different from the 1:2:1 expectation (χ^2 = 16.7, *df* = 2, *P* < 0.001).

Production of Daughters at Low Density

In early 1990, the mass poisoning of a neighboring clan by local people allowed a number of low- and mid-ranking Talek females to leave their home range and start a new clan (Holekamp et al. 1993), reducing the original study

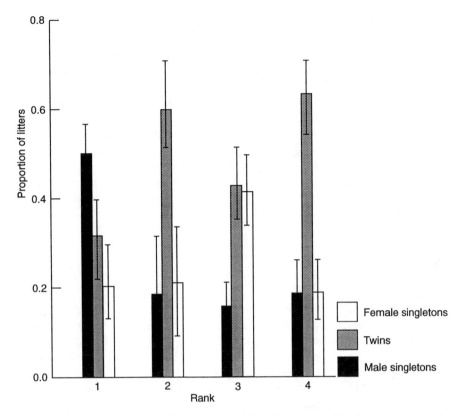

Figure 3.12. Mean proportions of litter types brought to the communal den by 21 parous females that bore at least three litters between 1979 and 1990 (N = 5, 4, 5, 7 for Rank Levels 1–4, respectively). The highest-ranked females produced significantly more singleton sons and significantly fewer singleton daughters and mixed-sex litters, compared to all lower ranks. There were no significant deviations from expected proportions of litter types in Rank Levels 2–4.

clan to 14 adult females by 1992. At the smaller size, the study clan immediately started producing an excess of daughters (Holekamp and Smale 1995). From July 1990 through November 1994, after the adult-female population of the clan had been reduced to 60% of its long-term stable level, the sex ratio of all cubs brought to the communal den declined from the pre-fission ratio of .59 (N = 138 cubs) to .41 (N = 71, χ^2 = 4.76, df = 1, $P < 0.05$). Among singleton litters of known sex, in the earlier period the sex ratio was .59 (n = 54); in the later period, the ratio was .36 (n = 11). The change in clan sex ratio was due entirely to the overproduction of daughters in the highest ranks (Holekamp and Smale 1995), among which daughters were significantly overrepresented (χ^2 = 3.88, df = 1, P = 0.05); in the lower ranks, sex ratio was not significantly different from equality (middle ranks: χ^2 = 1.20, NS; low ranks: χ^2 = 0.06, df = 1, NS).

In the 11 years prior to clan fission, only one female-female twin pair was documented. In the post-fission period, 11 pairs of twin females, constituting

25% of the 44 litters in which size-sex composition could be determined, have been brought to the communal den. Female twin litters have been distributed across all maternal ranks, but were more frequent among higher-ranking females. After clan fission the distribution of twin litter types did not differ from expectation ($\chi^2 = 3.66$, $df = 1$, NS). If most same-sex twin litters are reduced to singletons by siblicide, there should be equal numbers of singleton and twin litters brought to the communal den; this was the case in the pre-fission period ($\chi^2 = 0.5$, $df = 1$, NS). Post-fission, there was a deficit of singleton litters ($\chi^2 = 4.59$, $df = 1$, $P < 0.05$), suggesting that more litters are surviving the siblicidal period intact. In late 1994, there were 18 reproducing females in the clan, and the clan's offspring sex ratio was .50 ($n = 18$), suggesting that sex ratios may be moving toward the pre-fission pattern as the female population moves toward long-term carrying capacity.

Siblicide, Rank, and Sex-Ratio Adjustment

Thus, in a stable population at carrying capacity, high-ranking females emphasize the production of singleton sons. These males enjoy the dual advantages of rapid growth, in the absence of a sibling (Hofer and East 1992), and mother's intervention on their behalf in the highly competitive feeding melee (Frank 1986b). Moreover, their early experience of behavioral dominance may provide social skills that facilitate competition for rank—and consequent mating success—upon subsequent immigration.

But in a declining population, below carrying capacity, the emphasis appears to be on daughters. It may be to a female's advantage to produce daughters when competition for food is low (Holekamp and Smale 1995), thereby increasing the size of her matriline. Daughters of high-ranking females gain clear advantages because they retain high rank for life and are always able to feed unchallenged, even when food is scarce (Frank 1986b). If the clan then returns to its former size, a reversion to the male-biased sex ratio would be predicted.

Sex-Ratio Adjustment in Other Mammals

Male and female offspring of any species may be valued differently, depending on the form of the social system. The Trivers-Willard hypothesis (1973) predicts that in a polygynous mating system, where there is high variance in male mating success and where male quality is influenced by maternal quality, females in good condition or of high social rank should bias their offspring toward males. Poorer-quality females should produce more female offspring, since female reproductive success is less likely to be affected by physical condition. Species in which field data suggest a Trivers-Willard sex-ratio skew include red deer, *Cervus elephus* (Clutton-Brock et al. 1984, 1985); opossums, *Didelphis marsupialis* (Austad and Sunquist 1986); and spider monkeys, *Ateles paniscus* (Symington 1987).

When females are philopatric (spending their lives in the area where they were born), grown daughters may compete with their mothers for resources. The Local Resource Competition model (Clark 1978; Silk 1983, 1984) also predicts that females will then produce an excess of sons over daughters. Species in which the LRC model seems to account for skewed sex ratios include bushbabies, *Galago crassicaudatus* (Clark 1978), lemmings, *Lemmus* sp. (Fredga et al. 1977), and muskrats, *Ondatra zibethica* (Caley et al. 1988). In a different social system, however, philopatric offspring may improve the reproductive success of mothers, by helping to raise subsequent siblings or by providing coalitional support to their mother in competition with conspecifics. The Local Resource Enhancement model (Gowaty and Lennartz 1985) suggests that in these cases, females should overproduce the sex that stays and helps. In several cercopithecine monkeys where females are philopatric and the "inheritance" of female rank gives mothers greater influence over daughters' reproductive success (RS) than sons' RS, high-ranking mothers produce more daughters and attempt to limit the numbers of daughters born to other females, which results in higher production of sons by low-ranking females (see reviews by Hrdy 1987; Clutton-Brock and Iason 1986; Clutton-Brock 1991; van Schaik and Hrdy 1991).

Clearly, sex ratio may be influenced by several interacting social or biological parameters; none of the above effects need be mutually exclusive. In a comparison of red deer and rhesus monkeys (*Macaca mulatta*), Gomendio et al. (1990) showed that relative costs of sons and daughters may differ depending on the rank of the mother, that the form of the social and mating systems helps determine which sex is more expensive, and that relative costs are reflected in differing offspring sex ratios of high- and low-ranking females. Van Schaik and Hrdy (1991) have suggested that in cercopithecine primates the Trivers-Willard and LRC models can operate simultaneously, and that their effects can be distinguished depending on (1) the direction of bias in the sex ratio between high- and low-ranking mothers and (2) the rate of population growth. Specifically, they suggest that under conditions of good resource availability, the higher ranks will emphasize production of sons (Trivers-Willard), but that under more intense competition, daughters will predominate in the higher ranks (LRC). In spotted hyenas as well, the offspring sex ratio seems to show an interaction between maternal rank and competition. The interaction, however, is in the direction opposite to that proposed by van Schaik and Hrdy for cercopithecines: high-ranking females produce an excess of sons when competition is high, but emphasize production of daughters when population density—and presumably competition for food—is lower. Holekamp and Smale (1995) provide a verbal model of the process in hyenas.

These data suggest that female spotted hyenas may be able to influence their offspring sex ratio according to their own social rank and the prevailing demographic variables. Although many species of mammals are now known to bias sex ratios, the mechanisms by which they do so are generally not known. A unique form of chromosomal sex determination has been described for lemmings (Fredga et al. 1977), and various physiological processes have been suggested for other mammals (Hrdy 1987). But only two mammals are known to use postnatal mechanisms to adjust these ratios through selective mortality of neonates. Some populations of humans alter sex ratios in favor of males through selective neglect or infanticide of daughters (Hrdy 1987), and senescent female *Antechinus stuartii*, the Australian marsupial mouse, abandon investment in them (Cockburn 1994). The spotted hyena may prove to be a third case.

The Mechanism: Maternal Manipulation of Neonatal Aggression?

Given a litter size of two and the equal likelihood of their being male or female, the binomial distribution predicts male-male:male-female:female-female litters at a ratio of about 1:2:1. Primiparous spotted hyena females usually produce singletons. In the Berkeley colony, there have been 20 litters from multiparous females: two singletons, 14 pairs of twins, and four sets of triplets. In all sets of triplets, one infant would clearly have died in the first days (had we not intervened) as a result of aggression. When these triplet litters were tabulated according to the sex composition of the surviving two infants, and tallied with the 14 sets of twins, the composition of the 18 resulting litters was three sets of male twins, five sets of female twins, and ten mixed-sex twins, a mix not significantly different from chance ($\chi^2 = 0.65$, $df = 2$, NS). Ten fetal litters collected in the wild (Lindeque 1981) comprised nine sets of twins and one singleton, with a nonsignificant excess of males. Thus the majority of litters are born as twins, and there is no evidence for prenatal sex-ratio adjustment.

In the absence of evidence for such a prenatal mechanism, sex-ratio adjustment appears to occur postnatally, through manipulation of the outcomes of neonatal fighting. In the wild, females have been seen to separate neonates into different burrows within hours of birth, as well as later, during the natal den period (R. Matthews, pers. comm.; P. Barber, P. White, and L. Frank, unpublished data). The single documented occurrence of surviving triplets among wild hyenas occurred when a female split her litter between two burrows 100 m apart; when she recombined the litter at about the first month of age, the smallest disappeared, having lost most of the skin of its back owing to aggression by the larger sibs (Figure 3.9).

In a pilot study, when captive mothers and neonates were kept in a pen with access to a separate "burrow system," the mean length of agonistic bouts between neonates in the burrow was 45.6 sec, compared to 9.6 sec outside of the burrow. In one case, the burrow system had to be removed on day 4 because the subordinate cub had already sustained serious injuries in the burrow. Females whose cubs had access to the burrow were five times more likely to intervene to break up aggression between the cubs during the first six days of life than were females whose cubs had no access to a "burrow" (proportion of interactions in which mother intervened, no-burrow system: $\chi = .084 + .03$; burrow system: $\chi = .445 + .18$, $F_{1,4} = 4.35$, $P < 0.001$).

Avian Siblicide and Its Fitness Consequences

Sibling aggression broadly similar to that described for spotted hyenas is well documented in a variety of predatory birds, particularly the larger eagles and fish-eating birds such as herons, egrets, pelicans, and boobies. In siblicidal birds, incubation starts with the first egg laid; consequently, that egg hatches before the other eggs, and the first or senior chick has a significant growth advantage over subsequent chicks. The earlier-hatched chick attacks the junior chicks and quickly establishes dominance; depending on the species and the ecological situation, one or more junior chicks may die as a result of the senior's aggression. The ecological function and adaptive advantage of this behavior are reasonably well understood (reviewed by Mock 1984a, b), but the processes that seem to explain siblicide in birds do not appear to explain siblicide in spotted hyenas adequately.

Avian siblicide is obligate in some species, particularly the large eagles, and facultative in the fish-eating birds and some raptors. Two nonexclusive hypotheses seem to explain the two different forms of aggression. The Insurance Egg Hypothesis holds that *obligate* siblicide results when ecological constraints prevent the parents from successfully raising more than one chick; this is probably the case for the larger eagles that depend on relatively rare prey. In this case, the second egg is laid as insurance against the possible inviability of the first egg. *Facultative* siblicide is a more complex phenomenon, in the sense that aggression may be mediated by food availability. The Brood Reduction Hypothesis holds that the female lays more eggs than can be successfully raised in years of normal prey availability. In the blue-footed booby (*Sula nebouxii*), aggression by the senior chick becomes severe enough to kill the junior chick only when the senior's growth rate drops below a critical level (Drummond and Garcia Chavelas 1989). If there is adequate food to maintain rapid growth in the senior chick, aggression is relatively minor and most of the brood survives. This mechanism tailors brood size to available resources, maximizing reproduction in both lean and rich years. Experimental manipulations suggest that food availability may not affect the severity of aggression in siblicidal species of egrets, however (Mock and Ploger 1987).

Costs of siblicide, in terms of both parental fitness and the surviving sib's inclusive fitness, are inversely proportional to brood size, being lower in larger broods and highest in broods of two (O'Connor 1978). Fitness gains to parents and offspring depend on several variables (Parker and Mock 1987), including costs of producing a neonate, costs of fighting, gains in survival probability or reproductive potential due to brood reduction, and relatedness within the brood (whether single or multiple paternity). Potential parent-offspring conflict (POC: disagreement between the genetic interests of the parent and those of the surviving sib) over the benefits of sacrificing a sib is highest when optimal brood size is small, when brood reduction confers a substantial gain in survivorship, and when multipaternity is high. If siblicidal brood reduction optimizes brood size according to resource availability, then the reproductive interests of parents, surviving sibs, and possibly even the eliminated sib, may coincide. In this case there is no POC (O'Connor 1978). If, however, siblicide results in a brood *smaller* than could be successfully reared, and there is thus no concomitant gain in the reproductive success of the parents through improved reproduction in the survivor, then the surviving sib would appear to win in POC. In practice, however, tests of predictions (regarding parental intervention in the fighting and the withholding of feeding effort) designed to distinguish between the two outcomes have produced contradictory results (Mock 1987). Demonstration of POC requires evidence that parents and offspring are at cross-purposes (e.g., there are efforts at parental intervention) and experimental evidence of decreased fitness if one party wins. Parker and Mock (1987) conclude that, though there is likely to be only a small difference in brood size optima for parents and offspring, resolution of the difference could result in extensive siblicidal aggression, with the relative costs of offspring of different size or sex, and costs and benefits of aggression, being the determining factors. If there *is* a difference in the costs of rearing the two sexes, there may be no single optimum, and the likelihood of severe aggression would be increased.

Because avian siblicide is associated with a parentally determined asymmetry in competitive ability (asynchronous hatching), and because the parents in most species do not attempt to intervene, it would appear that siblicide may be in the interest of the parents as well as the surviving offspring. Since siblicidal behavior appears to maximize the reproductive output of the parents, by adjusting brood size to resource availability, it may be in the interest of the parents, as well as the surviving chicks, to allow the death of the "extra" chicks: siblicide may not impose an undue reproductive cost on the parents.

Spotted Hyena Siblicide and Its Fitness Consequences

Siblicide in the spotted hyena is less well understood than avian brood reduction, but in any event the hypotheses that adequately explain avian examples do not seem to fit the hyena case. The modal hyena litter size is two, the size at which the costs of siblicide are highest, in terms of both parental fitness and the surviving sib's inclusive fitness (O'Connor 1978). Potential parent-offspring conflict is also highest when litters are smallest. Thus, the costs of siblicide in hyenas appear to be high, and its adaptive value is unclear.

Siblicidal aggression in spotted hyenas does not seem to be simply an adaptation to unpredictable food supply, because under ecological conditions as disparate as the prey-rich Mara and the prey-poor southern Kalahari (Mills 1990), female hyenas normally raise mixed-sex twins and occasional litters of male twins, and they are clearly able to raise twin daughters when the clan's female population falls below carrying capacity. Significantly, the intense fighting occurs at birth, *before* the neonates have had an opportunity to assess resources, via either hunger level or growth rate. Although singletons grow faster than twins (Hofer and East 1992), the survivor appears to sacrifice inclusive fitness through killing a full sib, for twins in this species are unlikely to have different fathers (Frank 1986b; Frank et al. 1995).

It is clear that the dominant sib may have a major survival advantage under circumstances of food shortage, as is the case in the commuting hyenas of the Serengeti (Hofer and East 1993a, b, c), where singletons have a significant growth advantage over twins (Hofer and East 1989, 1992). In that population, aggression by the dominant during periods of food shortage, at ages up to three months, frequently causes starvation of the subordinate. In the Mara, where prey is more reliably available all year, siblicidal mortality normally occurs only in the first weeks of life, in the natal den (Frank et al. 1991), before levels of maternal nutrition are likely to influence cub growth. A survival advantage during periods of adversity does not explain why mortality usually occurs so early in life, nor why high-ranking females, which have priority during feeding, should so often bring singleton litters to the communal den. Further, even in the Berkeley colony, where all animals are fed ad lib, when cubs are allowed access to an artificial burrow system the fighting is sufficiently severe that mortality would result, in the absence of observer intervention. Finally, if siblicide were a facultative adaptation to food shortage, one would expect to find it in other carnivores.

Nor do the avian hypotheses explain why the outcome of the fighting should typically be fatal in same-sex but not mixed-sex litters. If one sex is energetically more expensive to raise than the other (Clutton-Brock et al. 1985; Hofer and East 1992), one would predict litter reduction in same-sex litters of the more expensive sex, not in both sexes. Twin litters of the less-expensive sex should be more common than mixed-sex litters, quite the opposite of the observed litter sex distribution. The observation that twin females survive when the adult female population of the clan is reduced strongly suggests that the outcome of siblicidal aggression is a trade-off between the potential reproductive benefits of one sex and those of the other, rather than simply a reflection of the differential costs of the two sexes. We have insufficient data on growth rates and reproductive success, particularly of males, to speculate on the trade-offs between fitness gains due to more rapid growth, gains deriving from the ability of mothers to adjust sex ratios to prevailing ecological conditions, and fitness losses due to litter reduction effected by siblicide.

Neonatal aggression in the spotted hyena may have arisen as a "side effect" of androgenization, leading to an evolutionary arms race (Frank et al. 1991). Survival of both cubs in captivity shows that the fatal outcome in the wild is greatly facilitated by the burrow-denning habit, which allows the aggression to proceed unchecked. Captive females sometimes intervene persistently when their neonates fight, but are unable to influence cub behavior in the depths of the burrow system. Ironically, the use of burrows is probably necessary to protect neonates from predation by other carnivores or from infanticide by other hyenas (East et al. 1989; Henschel and Skinner 1991b). Cub mortality due to siblicide in hyenas may be as high as 25% (Frank et al. 1991), but that figure is substantially lower than the 80% predation mortality documented in the cheetah (*Acinonyx jubatus*), which den in exposed bush lairs (Caro and Laurenson 1994).

Sex-ratio adjustment may favor males *or* females, as demographic conditions dictate. The apparent ability of at least some mothers to exploit siblicidal aggression to increase their own fitness may provide the key to understanding the evolution and maintenance of this outwardly expensive behavior. Neonatal fighting may have originally been an unselected by-product of selection for traits mediated by prenatal androgens (e.g., female aggressiveness and body size). Once early aggression was established, selection for winning may have resulted in precocial behavioral development, tooth eruption, and the extreme neonatal behavior seen today. Although winning is of course advantageous to the survivor, the apparent high cost to the mother had suggested to us that this is an instance in which parent-offspring conflict is decided in favor of the offspring. It now appears that mothers may also benefit from siblicide, by manipulating the outcome in their favor.

On average, three out of four litters offer the mother an opportunity to favor one sex, since only one-fourth of the litters will be same-sex twins of the non-preferred sex. In the absence of a mechanism for biasing the primary or secondary sex ratios, manipulation of neonatal aggression may allow a female to favor the sex that is of most value under the particular ecological or demographic conditions then extant. Even if an ability to bias sex ratio is adaptive only for high-ranking females and offsets the cost of siblicide for them, the net cost to low-ranking females may be evolutionarily unimportant because of their low reproductive success.

If it is to the mothers' advantage for one offspring to die, why does she not kill one outright instead of relying on neonatal aggression? The tendency of adult females to commit infanticide against unrelated neonates (Glickman and Frank, unpublished data; Hofer and East, unpublished manuscript) must be matched by a strong inhibition against killing their own offspring. Moreover, reliance on the more protracted process of siblicide ensures that the subordinate sib can serve as "insurance" against the early loss of the dominant, and that weak cubs are eliminated early.

If siblicidal aggression were in the interest only of the surviving cubs and served to *decrease* the fitness of females, it would be analogous to infanticide by males, which has been described for many primates and for lions (reviewed by Hausfater and Hrdy 1984). One would, then, expect female spotted hyenas to have evolved strategies for thwarting a fatal outcome in their litters, as has been the case in species where male infanticide is a common reproductive strategy (Hrdy 1979; Packer and Pusey 1983a, b; Struhsaker and Leland 1986; Smuts 1986).

Experiments with the administration of antiandrogens to pregnant females at Berkeley will test whether neonatal fighting is indeed promoted by prenatal androgen exposure. If blocking prenatal androgen successfully inhibits neonatal fighting, that finding will indicate that the fighting is androgen-dependent. Obviously, however, such a demonstration would not resolve the question of the evolutionary origin of the aggression: as suggested below, androgenization could be interpreted as an adaptation resulting *from* neonatal fighting, rather than the original *cause* of it.

Fitness Consequences of Social Rank

Dominance hierarchies are nearly universal among mammals that live in stable social groups. In the absence of hierarchies, there would be incessant squabbling over food, mates, resting sites, and other limited resources. Because higher-ranking animals typically enjoy preferential access to resources, biologists commonly assume that they also have greater fitness or reproductive success than lower-ranking individuals do. Field studies, however, have produced varied results, in some cases showing the predicted effect but in others finding little or no obvious reproductive benefit attending high social rank (Silk 1986; Altmann et al. 1988; Packer et al. 1995; Amos et al. 1995).

The results of the Masai Mara spotted hyena study are unambiguous: high social rank in this species confers a variety of advantages to individuals of both sexes. These advantages culminate in dramatically higher reproductive success among top-ranked females than for any others. High-ranking males also appear to gain reproductive benefits, but observational data can be misleading and must be corroborated with genetic data on paternity (Packer et al. 1991; Amos et al. 1995).

Social Rank and Access to Food

The most obvious advantage of dominance among females becomes clear when hyenas compete for a kill. In a high-density population, the resulting feeding melee is probably one of the most competitive situations known in any mammalian species; a 200-kg wildebeest may be reduced to blood-stained grass in a matter of minutes (Kruuk 1972).

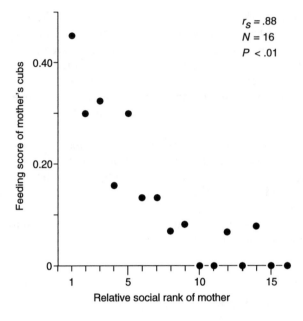

Figure 3.13. Correlation between absolute rank of females and the ability of their dependent cubs to feed on highly contested kills. Feeding score was the proportion of 5-min intervals in which a cub was seen to feed. (From Frank 1986b; reprinted by permission.)

In these melees, males are immediately excluded by females, and lower-ranking females are forced from the kill as it is reduced by the feeding throng; their dependent offspring have little opportunity to feed on the meat at all (Figure 3.13) (Frank 1986b). The cubs of high-ranking females, by contrast, are able to feed next to their mothers from an early age, and the mothers are able to sequester the diminishing carcass, thereby eating more themselves and sharing with their older kin as well as their cubs. Improved access to highly contested food appears to have important consequences for rank-related differences in reproductive parameters and recruitment (Frank et al. 1995).

Although males often make a kill, they are forced from it as soon as females arrive; even a subadult female may prevent up to five males from feeding on a large carcass (pers. obs.). Males apparently feed more frequently than females on leftover scraps such as abandoned bones, which results in a higher rate of tooth wear as compared to females (unpublished data). Transient males appear in the Talek Clan range in large numbers when the migratory wildebeest are present (Frank 1986b), but we do not see an increase in unfamiliar females at these times.

Prey Availability and Commuting

In the Serengeti, movements of migratory prey force entire hyena populations to "commute" long distances from their home range when the herds are absent (Hofer and East 1993 a, b, c). These authors speculate that there may be a relationship between an individual's rank and the likelihood that it will be forced to leave the home range in times of prey shortage. This effect is apparent in the Talek Clan even though prey fluctuations there are not as dramatic as those in the Serengeti. During periods of relative prey scarcity, low-ranking individuals of both sexes may disappear from the clan range for weeks (unpublished data), whereas high-ranking animals are nearly always to be found in the core of the territory (Frank 1986b). Males are more likely than females to abandon the clan range, presumably because they are more often excluded from kills. Thus, females in general and high-ranking ones in particular have greater access to food even in times of prey shortage, and are thus buffered against the necessity to undertake long, perilous excursions in search of food.

Social Rank and the Provisioning of Young

Provisioning of dependent young with food (carrying it away from the kill site to the cubs) is common among several social carnivores, particularly the canids, all pack members of which participate in feeding the cubs (Macdonald and Moehlman 1982). Among brown hyenas, too, provisioning by related clan members of both sexes is an important source of food for cubs (Mills 1990; Owens and Owens 1984), and Kruuk (1976) describes provisioning by a female striped hyena. As pointed out by Mills (1990), provisioning appears to be the ancestral condition in the Hyaenidae, and its relative rarity among spotted hyenas is anomalous.

In the course of extensive observations of the Talek Clan communal den, Holekamp and Smale (1990) discovered that, although rare, provisioning occurs at a higher rate than previously thought, but that it is highly constrained by rank relations within the clan. Low-ranking females attempting to bring food to their cubs at the den usually lost it to higher-ranking females, who then fed it to their own cubs. Thus, provisioning is a useful tactic only for higher-ranking mothers and cubs, but because of their access to meat at kill sites, these are the cubs least likely to derive substantial nutritional gains from provisioning.

Social Rank and Clan Fission

Given that low-ranking females have decreased access to food and display poor reproductive success (see below), dispersal or clan fission would seem, for them, to be obvious alternatives. In fact, dispersal by female spotted hyenas seems to be very rare, and only two instances of clan fissioning (Mills 1990; Holekamp et al. 1993) have been observed. Appropriate habitat is likely to be saturated with hyenas, and even if mortality has reduced a clan below carrying capacity, intolerance of female transients would discourage dispersal.

The Talek Clan split when the population of juveniles reached numbers higher than had been previously recorded, but only after adjoining habitat suddenly became available when a neighboring clan to the north had been destroyed by poisoning. The split was also coincident with a period of prey scarcity. Presumably, then, competition for food was unusually intense at the time. The ecological circumstances surrounding the fission of Mills's Kousaunt Clan are not known, but he had witnessed severe aggression against one of the clan's three matrilines, suggesting that the clan may have grown beyond its food supply.

In both cases, the split occurred along matrilineal lines: closely related adult females left the parent clan with their dependent young. In the Kousaunt Clan, the lowest-ranking matriline left, whereas in the Talek Clan several mid- and low-ranking matrilines departed. In neither case was it possible to monitor the new clans after fission, but it seems likely that both food availability and reproductive performance would improve under the less-competitive circumstances likely to obtain at the new sites.

For years, we have seen that low-ranking females may leave the clan range for extended periods during prey shortage, but we have been unable to locate them in adjacent areas. Recent data from three radio-collared low-ranking females show that some are moving distances of at least 35 km from the clan range; one such individual was known to be alive at least two years after leaving. We do not yet know whether such animals live as solitary non-breeders, attempt to breed as solitaries, or are able to join new clans. Although they have occasionally been seen in the Talek Clan range, foreign females have not attempted to join the Clan, and never seem to stay for more than a day.

Spotted hyenas are intolerant of strangers (Kruuk 1972; Hofer and East 1993b), and encounters between residents and transients may be fatal to the transients. Moreover, among both captive (Frank et al. 1989) and wild hyenas (Holekamp et al. 1993), females may be subjected to severe aggression by other females upon returning from a prolonged absence. Thus, a nomadic lifestyle, whether temporary or permanent, is likely to be both dangerous and stressful. Of the three radio-collared females that were tracked, two are known to have died far from the clan range (one was an elderly individual aged 15 years, and the other, at 9.5 years, was middle-aged).

Social Rank and Female Reproductive Success

Social rank has a powerful effect on female reproductive success (Frank et al. 1995). Females in the highest-ranking matriline started breeding at an earlier age and had shorter interlitter intervals than females of any lower rank did. Owing to these factors, and improved survival through the subadult period, these highest-ranking females produced offspring of both sexes that reached reproductive age at a rate 2.5 times that of all lower-ranking females (Figure 3.14). Differential access to food probably accounts for most of the observed differential in reproductive parameters and recruitment. Infanticide by other

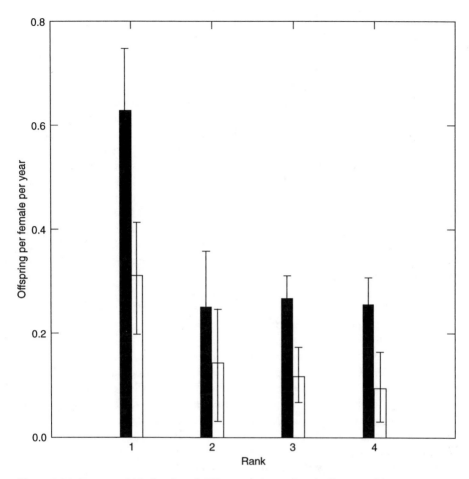

Figure 3.14. Rate at which females of different ranks produced offspring of both sexes that reached the age of sexual maturity (two years for males, three for females). Black bars are sons; open bars are daughters. $N = 36$ females for which there were at least three years of reproductive data between 1980 and 1990. (From Frank et al. 1995; reprinted by permission.)

females, however, may account for the disappearance of some litters of low-ranking females.

Because of rank "inheritance" among spotted hyenas (Frank 1986b; Holekamp and Smale 1993; Smale et al. 1993; Jenks et al. 1995), the relative ranks of the Talek Clan's matrilines have been stable over generations (Frank et al. 1995). The current alpha female is the great-granddaughter of Female 04, who was alpha in the early 1980's. Interestingly, in spite of the clear advantages of high rank, we have not seen animals move up the social hierarchy in the Mara study. Note, however, that East et al. (1993) report a case in a Serengeti clan in which an alpha matriline declined in numbers and was "overthrown" by the larger beta matriline. When the original alpha increased in size again, however, it regained its top rank. Thus, the long-term stability of matrilineal dominance relationships may be due to the high reproductive success of top-ranked females, which ensures that the highest-ranked matrilines normally have the most members.

Because of the higher rate of female recruitment in the top ranks and the inheritance of maternal rank, mid- and low-ranking females drop ever lower in rank, and show the reduced reproductive success characteristic of lower-ranking females. Over a 10-year period, the top-ranked matriline increased its numbers by 50%, while the lower-ranking matrilines either remained stable or lost members. Moreover, some lower-ranking matrilines have died out altogether over the course of the study. At this rate, the consolidation time, or time until the entire clan will have been descended from the top matriline (Altmann and Altmann 1991), will be 15 generations, or 45 years. These are conservative figures, however, because most of the members of the alpha matriline were almost certainly descendants or sisters of alpha female 04 (Frank 1986b), who was not included in this analysis because all her verified offspring since the study began were males.

Social Rank and Male Reproductive Success

Like females, male spotted hyenas acquire their mother's rank during the communal den stage (Holekamp and Smale 1993). But upon dispersal from the natal-clan home range, males become subordinate to all members, male and female, of any clan whose territory they enter. Male clan residents are intolerant of potential immigrants, and frequently chase them, although severe fights or injuries have not been witnessed in the Talek study. The processes by which dispersing males gradually integrate themselves into a new clan are currently under study by Holekamp and Smale in the Mara, but it is clear that they enter at the bottom of the male hierarchy, and that their subsequent rise in rank is at least partly a matter of residence time (Frank 1986b). For example, Male 65, an immigrant adult who ranked low in the Talek Clan male hierarchy in 1980, had risen to near the top by 1987, as higher-ranking males disappeared. Some males, however, seem able to rise more rapidly than the attrition of dominants would indicate, which suggests that factors other than tenure may occasionally be important.

As adults, males born to high-ranking mothers may have an advantage over those born to low-ranking ones. High-ranking females produce a substantial excess of singleton sons (Frank et al. 1995; Frank 1992; Holekamp and Smale 1995). Because these sons do not share maternal milk with a sibling, they enjoy the dual advantages of rapid pre-weaning growth and frequent access to meat at an early age, owing to their mothers' rank (Frank 1986b). Furthermore, their entire social experience is of high rank; current research by Holekamp and Smale tests the hypothesis that these males are more likely to enter a new clan and rise rapidly in rank there than are males of low-ranking mothers, whose social experience has been largely of submission.

Observations of hyenas mating are uncommon, and DNA fingerprinting studies of paternity in the Talek Clan are still under way. All five matings seen in the first four years of the study were by the alpha male, however, and he also had the highest rates of consortship and courtship with females. Overall, male reproductive behavior was strongly correlated with rank, as measured by male-male aggression (Frank 1986b). In subsequent years, several mid-ranking males were also seen mating (unpublished data), but generally the pattern has continued to indicate strong polygyny, the highest-ranking males gaining most of the observed matings.

Thus, if the speculation relating males' maternal rank with subsequent adult rank proves correct, the maternal rank of males would correlate strongly with adult mating success, thereby according to female rank another powerful reproductive advantage. Unfortunately, the logistical obstacles confronting such a study are formidable: it would be necessary not only to monitor natal males after they disperse, but also to become sufficiently familiar with the male social hierarchy in all their adopted clans, so as to document accurately the study individuals' social (and, therefore, reproductive) success. The alternative, complete paternity analyses of each target clan, is also logistically formidable.

Intralitter Dominance and Reverse Sexual Size Dimorphism

Several hypotheses have been advanced to explain the weight differential between male and female hyenas. Though linear measures do not differ significantly, adult females are about 10% heavier than males (Hamilton et al. 1986). Whately (1980) suggested that because of their dominance at kills, females are more likely than males to have full stomachs, which would in part account for their greater weight. Circulating androgens increase muscle mass (Glickman et al. 1992b, 1993), but this effect would tend to make females comparable to males in size, not necessarily larger. Noting that energetic efficiency increases with size, several authors have suggested that the larger size of females might improve their reproductive success—the "big mother hypothesis" (Ralls 1976; Bertram 1979; Mills 1990). Hofer and East (1993c) show that in Serengeti spotted hyenas, which are routinely forced to "commute" to follow migratory prey, cub growth is significantly curtailed during periods of maternal absence. Long-distance movements of this sort are not unusual for hyenas (Eloff 1964; Mills 1990; Tilson and Henschel 1986), and the big mother advantage might thus be greater in spotted hyenas than it is in carnivores that range over smaller distances and do not show reverse sexual size dimorphism.

The consequences of early intralitter aggression may also help to explain the size differences seen between male and female hyenas. Cubs that have experienced a severe early nutritional deficit, owing to subordinance in intralitter aggression, remain stunted for life (Glickman and Frank, unpublished data). In wild litters, the subordinate is nearly always smaller than the dominant, at all ages, but we have no indication from captive litters that there is a significant size difference between twins *at birth,* whatever their sexes. Thus, the size difference appears to arise as a result of the dominants' ability to monopolize nursing. Since brothers are usually subordinate to sisters in mixed-sex litters, the tendency for males to be smaller than females, throughout life, may be no more than an enduring effect of early aggression. If so, males maturing as singletons would not suffer from this effect. Testing this hypothesis requires data on adult weight as a function of prior litter composition; we do not yet have the requisite data for adult males.

Development of Female Dominance

In both captive and wild spotted hyenas, all adult males are subordinate to all adult females. Juvenile males, however, frequently outrank both juvenile and adult females. In the wild, cubs of both sexes acquire their mothers' rank, and male juveniles can thereby dominate lower-ranking adult females (Smale et al. 1993). In captive peer-reared groups, young males frequently outranked young females prior to puberty (Frank et al. 1989). By adulthood, however, female dominance is absolute. We believe the change to be a result of several interacting factors.

Prenatal development events, mediated at least in part by androgens, produce females that are inherently more aggressive than the males. Experiential and social factors, however, often intervene early in life to influence male-female relations. Neonatal fighting sometimes results in a brother dominating his sister, and the rank of the mother determines relations between juveniles of both sexes and all other clan members: thus, juvenile males often enjoy very high rank because of their mothers' influence. Even as juveniles, however, females are more successful than males in dominating lower-ranking adult females (Smale et al. 1993). This suggests either that juvenile females are more persistently aggressive than males, or that adult females are more likely to challenge the dominance of a higher-ranking juvenile if it is male. Since juvenile males will ultimately disperse, they do not pose the lifelong threat that a higher-ranking young female represents to lower-ranking adult females; challenging a male may have fewer long-term repercussions than challenging a female does.

Once a male disperses from his natal range, however, he finds himself in unfamiliar territory, confronted by coalitions of females and of resident males, neither of which tolerate newcomers. The dispersing male is thus subordinate to all clan members, and the stress of universal subordinance is reflected in depressed androgen production (Frank et al. 1985a; van Jaarsveld and Skinner 1991). If this male stays in his adopted clan, he gradually rises in the male hierarchy as a consequence of seniority, but he will never attain the seniority of even the lowest-ranking females, all of whom were born in the clan. Coalitions among females, in conjunction with their greater aggressiveness, serve to keep immigrant males in their place.

Female Baiting by Males

The only circumstance under which *adult* male spotted hyenas dominate adult females is "baiting" (Kruuk 1972). When several males are following a female that is approaching sexual receptivity they may attack her suddenly and vigorously. Rather than defend herself, the female crouches down, protecting her rump and emitting submissive displays and squeals, while occasionally biting ineffectually at her tormentors. The males rarely do serious damage to the female, but occasionally she suffers minor wounds. Sometimes a female relative will respond to the victim's squeals, and her arrival puts an end to the attack; otherwise, the attack may continue for several minutes, before the males lose interest and resume quietly following her.

The function of this behavior is not known, but its clear relationship to impending estrus suggests that a hormonal change in the female may alter her level of aggressiveness, or in some other way make her susceptible to male aggression. There may be a parallel to the natural phenomenon of baiting in the experimental animals of the Berkeley colony: the only robust change in behavior attributable to juvenile ovariectomy was that these females were subjected to a much higher rate of aggression from males than were intact females (Baker 1990). This difference strongly suggests that an ovarian product maintains female aggressiveness toward males, and that hormonal changes associated with estrus temporarily duplicate the effects of ovariectomy. Whether reduction in female aggression is a necessary prelude to sexual receptivity is not known, but current research in both the laboratory and the field is directed at the problem. Results of these studies may suggest hormone-replacement therapy that would restore the normal aggressiveness of ovariectomized females, thus demonstrating a direct link between a hormone and aggressiveness in a female mammal.

The Evolution of Female Masculinization and Dominance

Earlier speculations about the origin of female masculinization in spotted hyenas focused separately on the behavioral and anatomical aspects of the syndrome. By considering the established relationship between aggressiveness, androgens, and sexual differentiation, on the one hand, and field studies of hyena behavioral ecology, on the other, our (admittedly incomplete) understanding of this phenomenon attempts to synthesize ecological and physiological explanations.

The Mimicry Hypothesis

Apparently assuming that males are dominant over females, Wickler (1965) suggested that the female genitalia of the spotted hyena developed to mimic a preexisting aggressive display of the male. There is no evidence, however, for such a display in any of the extant hyenas, and, in fact, erection in either sex of the spotted hyena is an appeasement, not a dominance display (East et al. 1993; Krusko et al. 1988).

The Meeting Ceremony and Infanticide Hypotheses

Kruuk (1972) suggested that female dominance arose as a defense against male infanticide, and that the evolution of the female genitalia could be understood in the context of the meeting ceremony. Spotted hyenas are both solitary and social, coming together after long absences; he argued that the greeting ceremony allows two individuals to stay in close contact while tensions subside. By concentrating attention on a complex and conspicuous body part well-removed from the weaponry (teeth), social bonds would be reestablished peaceably. Thus, in this view the genitalia are seen strictly as a social-display organ.

Although male infanticide may have been an important selective pressure earlier in hyena evolution, there is no evidence of it today. All observations of infanticide, both in captivity (unpublished data) and in the wild (Hofer and East, unpublished manuscript) suggest that adult females are the killers. The other hyena species have similarly complex greeting ceremonies, though they involve close attention to the anal scent gland rather than the genitalia (Rieger 1978; Mills 1990). Moreover, it is difficult to visualize the evolution of a display organ that entails such major reproductive complications.

The Siblicide Hypothesis: The Chicken or the Egg?

The phenomenon of neonatal siblicide in spotted hyenas suggests an alternative hypothesis for the origin of female androgenization (East et al. 1993). It seems likely that the severe fighting at birth is facilitated by androgens that are produced in the latter half of gestation. In canids, intralitter aggression arises later in development, the age at which it appears being related to the degree of sociality of the species (Bekoff 1977). Dominance-determining aggression appears earlier in red foxes (*Vulpes vulpes*) than it does in coyotes (*Canis latrans*), and it does not occur at all in gray wolves (*Canis lupus*). Ancestral *Crocuta*, with normal endocrine profiles and normal genital dimorphism, may have exhibited a similar pattern of juvenile aggression, perhaps related to feeding priority. If the survival consequences of winning or losing the aggression were higher than in modern canids (e.g., if the loser had significantly increased likelihood of dying), selection may have favored earlier fighting, along with precocial development in size, tooth eruption, and fighting abilities conducive to winning. Androgenization of the fetuses to increase aggression may have been similarly selected, leading inevitably to the remarkable masculinizing effect on genital development seen today. In this view, male-like female genitalia and aggressiveness are a consequence of neonatal fighting, rather than the reverse.

It is even possible that enhanced female adult aggressiveness resulting from selection for neonate aggression facilitated the evolution of social predation, by obviating the deleterious effects of feeding competition on cub survival. But in order to fully explain the modern situation, which finds females more aggressive than males, it is still necessary to invoke the feeding and reproductive benefits accruing to aggressive females.

East et al. (1993) recognize the possibility that androgenization may have originated with neonatal aggressiveness, but they link selection for early aggression to the long period of maternal investment during gestation and lactation. In fact, high maternal investment would seem to select strongly *against*, rather than for, cub mortality.

The Rape Hypothesis

East et al. (1993) also argue unconvincingly against the hypothesis that links masculinization to social behavior via androgens. Addressing the evolution of the female genitalia, they appear to follow Wickler (1965) in emphasizing the undoubted signal function of the penile erection. It is not clear whether they see selection for a display organ as the original pressure for development of the female structure, but they do not cite any alternative hypothesis for the origin of the penile clitoris. Nor is it clear whether they feel that the siblicide-androgen hypothesis may account for the female genitalia, but by not citing references they deny a vast literature and state that prenatal androgen exposure cannot produce a penile structure in females.

They do, however, erect a novel hypothesis to explain the fusion of the labia to form the psuedoscrotum. They suggest that selection may have favored it as a mechanism to defeat male efforts at forced copulation, by giving the female control over mating. This hypothesis does not seem compelling, given that labial fusion is a necessary result of prenatal androgen exposure (Jost 1972). Furthermore, there is no evidence that males attempt to coerce females in this species, nor is there in any other species of hyena. Male spotted hyenas are extremely wary of females, even in the heat of passion, and females tolerate close approach by a single male only when they are ready to mate (Frank 1986b). Given their impressive weaponry and aggressiveness, and the meek nature of males, it seems highly unlikely that female hyenas need resort to a scrotal chastity belt.

East et al. (1993) reject the hypothesis that links prenatal androgens, genital masculinization, and adult aggressiveness. They argue that, because in some primate species coalitions of females are able to dominate single males, no special advantage is required to explain female dominance in spotted hyenas. That situation, however, is not comparable to the spotted hyena case, in which males are subordinate, not only to single females, but even to single infants. Their argument ignores the association between female masculinization, female aggressiveness, extraordinary feeding competition, and cub mortality. They also ignore the uncontested body of experimental evidence that unequivocally demonstrates the causal developmental links between prenatal androgens, genital masculinization, and adult aggressiveness.

Ecological Hypotheses

Recognizing the association between aggression-promoting effects of prenatal androgens, genital masculinization, and female dominance, Gould (1981) and Gould and Vrba (1982) suggested that the basis of female masculinization should be sought in "some aspect" of the species' social behavior. In the absence of field data on the ecological role of female aggressiveness, however, the selective pressures underlying this suite of adaptations were not obvious.

The key to our current understanding of the evolution of female aggressiveness and physical masculinization lay in the observation that high-ranking females and their offspring are easily able to dominate access to a limited food resource (Tilson and Hamilton 1984; Frank 1986b). In other social carnivores, cub mortality is high in times of food shortage: a large proportion of lion cubs starves (Bertram 1975), because adult males take priority at a kill, followed by adult females. Only when food is abundant do lion cubs prosper. Similarly, in wolves (van Ballenberghe and Mech 1975; Harrington et al. 1983; Delgiudice et al. 1991), the young of the year, but not adults, show physiological signs of starvation and poor growth during times of prey shortage. Thus, a significant cost of communal predation is that the young tend to suffer during the frequent periods of prey shortage.

We hypothesize that the initial selective pressure resulting in highly aggressive spotted hyena females was the development of communal predation (Figure 3.15). Both the striped and brown hyenas, which are primarily scavengers on large mammal carcasses, tend to forage and feed alone, since carcasses are rare and small prey or fruits tend to be widely dispersed (Mills 1990; Kruuk

1976). When a scavenging ancestral form of *Crocuta* developed the ability to hunt large prey communally, the new behavior produced new selective pressures. It seems likely that communal predation increased feeding competition in two ways: (1) the predators now gathered in groups to hunt large mammal prey, assuring that several hyenas would be present at a kill; and (2), because live prey is a far more abundant and rewarding resource than are scavenged carcasses, the food supply for a recently evolved communal carnivore was far greater than that available to the scavenging ancestor. Thus, the population of the newly predacious *Crocuta* would have increased rapidly, further increasing the numbers of animals that would be competing over kills. These two processes led to the situation we see today in the prey-rich savannas of Africa, where two or three spotted hyenas feeding on a freshly killed wildebeest may soon attract up to 30.

If increased aggression improved the food intake of an individual or that of its offspring, as seems likely, such intense competition would provide very strong selective pressure for aggressiveness. A relatively straightforward route to increased female aggressiveness is to expose the fetus to androgens during development. Androgens that masculinize aggressive behavior, however, will also masculinize genital development (Jost 1972; Short 1974), and in other mammals masculinized female genitalia generally preclude reproduction. Moreover, early androgen exposure may also damage normal reproductive cycling and maternal behavior (Glickman et al. 1993).

As discussed earlier, there must have been concomitant adaptations permitting successful reproduction *in spite of* the anatomical impediments, at least to the degree that the advantages of aggressiveness would come to outweigh the problems associated with masculinized genitalia. Further, reproductive and maternal behavior had to have been "buffered" from the effects of the androgens, possibly through adjustments in the timing of androgen exposure. Glickman et al. (1993) treat these problems at length and address the obvious question of why other social carnivores have not adopted the same solution to the problem of juvenile mortality caused by competitive feeding.

Thus, we need not see the original elaboration of the female genitalia as constituting the development of a display organ. It seems more likely that the erect phallus became associated with submission *after* males became subordinate; in a species without obvious dimorphism in size or ornamentation, a male can demonstrate his submissive status by displaying his maleness. Once the erection became an important display organ, there may have been elaboration of the enlarged clitoris to further resemble the penis (Hamilton et al. 1986). Unfortunately, the fossil record is mute regarding both behavior and genitalia.

According to this competition-aggression-androgen scenario, it is not necessary to hypothesize any direct pressure for females to gain dominance specifically over males. Given sufficient reproductive benefit to holding high rank, particularly when amplified by social inheritance, competition among females would provide ample selective pressure to produce highly aggressive individu-

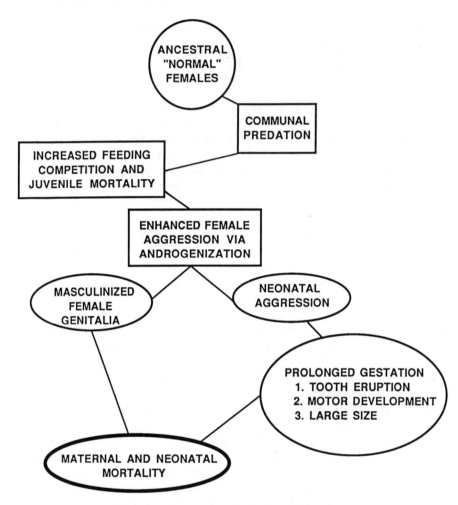

Figure 3.15. Schematic representation of the postulated evolution of female masculinization.

als. If females benefit more than males from high levels of aggressiveness, no direct male-female competition need be invoked (although dominance over males would present an additional benefit).

Given the apparently high degree of polygyny in the spotted hyena mating system, males seem remarkably nonaggressive (Frank 1986b; East and Hofer 1991). Moreover, polygyny would normally be expected to produce sexual size dimorphism favoring males. But among hyenas, discretion may be the better part of valor: the jaw strength of a hyena makes serious fighting exceptionally dangerous. If a male is likely to achieve breeding status through seniority, avoiding the risk of death or serious injury in the meantime may be the more prudent course. Thus, males may benefit by being relatively nonaggressive, whereas females clearly benefit from aggression.

Given the low probability that we will be able to generate compelling evidence bearing on the evolutionary origin of this extraordinary syndrome, speculation about the evolution of masculinization and dominance in the female spotted hyena is likely to occupy biologists for some time to come. We may never know whether the evolutionary success of the spotted hyena has occurred because of, or in spite of, its unlikely set of adaptations.

Acknowledgments

I am grateful to the Office of the President, the Department of Wildlife Conservation and Management of the Republic of Kenya, and the Narok County Council for permission to carry out field research. The Masai Mara hyena study has been supported by grants BNS 7803614, BNS 8706939, BNS 9121461, IBN 9296051, and IBN 9306912 from the National Science Foundation, and by grants from the National Geographic Society, the H. F. Guggenheim Foundation, the Center for Field Research, the Wildlife Federation, the American Association of University Women, and the Charles A. Lindbergh Fund. The Berkeley study has been supported by grant MH39917 from the National Institute of Mental Health. K. E. Holekamp and L. Smale collected demographic data from 1988 to 1992. T. Caro, S. Yoerg, and S. E. Glickman made valuable comments on an earlier draft. C. M. Drea kindly provided Figure 3.4. Most of the ideas presented in this chapter are the result of ten years of conversations with my collaborator, S. E. Glickman.

References

Altmann, M., and Altmann, J. 1991. Models of status-correlated bias in offspring sex ratio. *Amer. Nat.* 137:542–555.

Altmann, J., Hausfater, G., and Altmann, S. A. 1988. Determinants of reproductive success in savannah baboons. In: T. H. Clutton-Brock, ed. *Reproductive Success*, pp. 403–418. Chicago: Univ. Chicago Press.

Amos, B., Twiss, S., Pomeroy, P., and Anderson, S. 1995. Evidence for mate fidelity in the gray seal. *Science* 286:1897–1899.

Aristotle, 1965. *Historia Animalium*. Transl. by A. L. Peck. Cambridge, Mass.: Harvard Univ. Press.

Austad, S., and Sunquist, M. E. 1986. Sex-ratio manipulation in the common opossum. *Nature* 324:58–60.

Baker, M. 1990. Prepubertal Ovariectomy Affects Female-Male but not Female-Female Dyadic Aggression and Submission in a Colony of Peri-pubertal Spotted Hyenas (*Crocuta crocuta*). Master's thesis, Univ. California, Berkeley.

Bekoff, M. 1977. Mammalian dispersal and the ontogeny of individual behavioral phenotypes. *Amer. Nat.* 111:715–732.

Berger, D. M. P., Frank, L. G., and Glickman, S. E. 1992. Unraveling ancient mysteries: Biology, behavior, and captive management of the spotted hyena, *Crocuta crocuta*. *Proc. Joint Conf. Amer. Assoc. Zoo Vet. and Amer. Assoc. Wildlife Vet., Oakland, Calif.,* Nov. 15–19, 1992, pp. 139–148.

Bernstein, I. S., Gordon, T. P., and Rose, R. M. 1983. The interaction of hormones, behavior, and social context in non-human primates. In: B. B. Svare, ed. *Hormones and Aggressive Behavior*, pp. 535–562. New York: Plenum Press.

Bertram, B. C. R. 1975. Social factors influencing reproduction in wild lions. *J. Zool. (London)* 77:463–482.

Bertram, B. C. R. 1979. Serengeti predators and their social systems. In: A. R. E. Sinclair and M. Norton-Griffiths, eds. *Serengeti: Dynamics of an Ecosystem,* pp. 221–248. Chicago: Univ. Chicago Press.

Brain, P. F. 1983. Pituitary-gonadal influences on social aggression. In: B. B. Svare, ed. *Hormones and Aggressive Behavior*, pp. 3–23. New York: Plenum Press.

Caley, M. H. J., Boutin, S., and Moses, R. A. 1988. Male-biased reproduction and sex-ratio adjustment in muskrats. *Oecologia* 74:501–506.

Caro, T. M., and Laurenson, M. K. 1994. Ecological and genetic factors in conservation: A cautionary tale. *Science* 263:485–486.

Clark, A. B. 1978. Sex ratio and local resource competition in a prosimian primate. *Science* 201:163–165.

Clutton-Brock, T. H. 1991. *The Evolution of Parental Care*. Princeton, N.J.: Princeton Univ. Press.

Clutton-Brock, T. H., Albon, S. D., and Guinness, F. E. 1984. Maternal dominance, breeding success, and birth sex ratios in red deer. *Nature* 308:358–360.

Clutton-Brock, T. H., Albon, S. D., and Guinness, F. E. 1985. Parental investment and sex differences in juvenile mortality in birds and mammals. *Nature* 313:131–133.

Clutton-Brock, T. H., and Iason, G. R. 1986. Sex ratio variation in mammals. *Quart. Rev. Biol.* 61:339–374.

Cockburn, A. 1994. Adaptive sex allocation by brood reduction in antechinuses. *Behav. Ecol. Sociobiol.* 35:53–62.

Deane, N. N. 1962. The spotted hyaena *Crocuta crocuta crocuta. Lammergeyer* 2:26–44.

Delgiudice, G. D., Mech, L. D., and Seal, U. S. 1991. Gray wolf density and its association with weights and hematology of pups from 1970 to 1988. *J. Wildl. Diseases* 27:630–636.

Drea, C. M., Hawk, J., and Glickman, S. E. In press. Aggression decreases as play emerges in infant spotted hyaena (*Crocuta crocuta*): Preparation for joining the clan. *Anim. Behav.*

Drummond, H., and Garcia Chavelas, C. G. 1989. Food shortage influences sibling aggression in the blue-footed booby. *Anim. Behav.* 37:806–819.

East, M. L., and Hofer, H. 1991. Loud calling in a female-dominated mammalian society, II: Behavioural contexts and functions of whooping of spotted hyaenas. *Anim. Behav.* 42:651–669.

East, M. L., Hofer, H., and Turk, A. 1989. Functions of birth dens in spotted hyaenas (*Crocuta crocuta*). *J. Zool. (London)* 219:690–697.

East, M. L., Hofer, H., and Wickler, W. 1993. The erect 'penis' is a flag of submission in a female-dominated society: Greetings in Serengeti spotted hyenas. *Behav. Ecol. Sociobiol.* 33:355–370.

Eloff, F. C. 1964. On the predatory habits of lions and hyaenas. *Koedoe* 7:105–12.

Feder, H. H. 1985. Peripheral plasma levels of gonadal steroids in adult male and adult, non-pregnant female mammals. In: N. Adler, D. Pfaff, and R. W. Goy, eds. *Hand-*

book of Neurobiology. Vol. 7: Reproduction, pp. 299–370. New York: Plenum Press.

Frank, L. G. 1986a. Social organisation of the spotted hyaena (*Crocuta crocuta*), I: Demography. *Anim. Behav.* 34:1500–1509.

Frank, L. G. 1986b. Social organisation of the spotted hyaena (*Crocuta crocuta*), II: Dominance and reproduction. *Anim. Behav.* 34:1510–1527.

Frank, L. G. 1992. Female spotted hyenas may use neonatal siblicide to bias sex ratios. Talk presented at Fourth International Behavioral Ecology Congress, Princeton Univ., Princeton, N.J., Aug. 17–22. (Abstract.)

Frank, L. G., Davidson, J. M., and Smith, E. R. 1985a. Androgen levels in the spotted hyaena: The influence of social factors. *J. Zool. (London)* (A)206:525–531.

Frank, L. G., and Glickman, S. E. 1994. Giving birth through a penile clitoris: Parturition and dystocia in the spotted hyaena. *J. Zool. (London)* 234:659–665.

Frank, L. G., Glickman, S. E., and Licht, P. 1991. Fatal sibling aggression, precocial development, and androgens in neonatal spotted hyenas. *Science* 252:702–704.

Frank, L. G., Glickman, S. E., and Powch, I. 1990. Sexual dimorphism in the spotted hyaena (*Crocuta crocuta*). *J. Zool. (London)* 221:308–313.

Frank, L. G., Glickman, S. E., and Zabel, C. J. 1989. Ontogeny of female dominance in the spotted hyaena: Perspectives from nature and captivity. In: P. A. Jewell and G. M. O. Maloiy, eds. *The Biology of Large African Mammals in the Environment. Symp. Zool. Soc. London* 61:127–46.

Frank, L. G., Holekamp, H. E., and Smale, L. 1995. Dominance, demographics and reproductive success in female spotted hyenas: A long term study. In: A. R. E. Sinclair and P. Arcese, eds. *Serengeti II: Research, Management, and Conservation of an Ecosystem*, pp. 364–384. Chicago: Univ. Chicago Press.

Frank, L. G., Smith, E. R., and Davidson, J. M. 1985b. Testicular origin of circulating androgen in the spotted hyaena. *J. Zool. (London)* (A)207:613–615.

Frank, L. G., Weldele, M. L., and Glickman, S. E. 1995. Fitness costs of genital masculinization in female spotted hyenas. *Nature* 377:584–585.

Fraser, D., and Thompson, B. K. 1991. Armed sibling rivalry among suckling piglets. *Behav. Ecol. Sociobiol.* 29:9–15.

Fredga, K., Gropp, A., Winkling, I., and Frank, F. 1977. A hypothesis explaining the exceptional sex ratio of the wood lemming (*Myopus shesticolor*). *Heriditas* 85:101–104.

Gittleman, J. L. 1986. Carnivore life history patterns: Allometric, phylogenetic, and ecological associations. *Amer. Nat.* 127:744–71.

Glickman, S. E., Frank, L. G., Davidson, J. M., Smith, E. R., and Siiteri, P. K. 1987. Androstenedione may organize or activate sex reversed traits in female spotted hyenas. *Proc. Natl. Acad. Sci.* 84:3444–3447.

Glickman, S. E., Frank, L. G., Holekamp, K. E., Smale, L., and Licht, P. 1993. Costs and benefits of "androgenization" in the female spotted hyena: The natural selection of physiological mechanisms. In: P. P. G. Bateson, P. H. Klopfer, and N. S. Thompson, eds. *Perspectives in Ethology*. Vol. 10: *Behavior and Evolution*, pp. 87–117. New York: Plenum Press.

Glickman, S. E., Frank, L. G., Licht, P., Yalcinkaya, T. M., Siiteri, P. K., and Davidson, J. M. 1992b. Sexual differentiation of the female spotted hyena: One of nature's experiments. *Ann. New York Acad. Sci.* 662:135–159.

Glickman, S. E., Frank, L. G., Pavgi, S., and Licht, P. 1992a. Hormonal correlates of "masculinization" in the female spotted hyaena (*Crocuta crocuta*), I: Infancy to sexual maturity. *J. Reprod. Fert.* 95:451–462.

Gomendio, M., Clutton-Brock, T. H., Albon, S. D., Guinness, F. E., and Simpson, M. J.

1990. Mammalian sex ratios and variation in cost of rearing sons and daughters. *Nature* 343:261–263.

Gould, S. J. 1981. Hyena myths and realities. *Nat. Hist.* 90:16–24.

Gould, S. J., and Vrba, E. S. 1982. Exaptation: A missing term in the science of form. *Paleobiology* 8:4–15.

Gowaty, P. A., and Lennartz, M. R. 1985. Sex ratios of nestling and fledgling red-cockaded woodpeckers (*Picoides borealis*) favor males. *Amer. Nat.* 126:347–356.

Goy, R. W., and McEwen, B. S. 1980. *Sexual Differentiation of the Brain*. Cambridge, Mass.: MIT Press.

Goy, R. W., and Resko, J. A. 1972. Gonadal hormones and behavior of normal and psuedohermaphroditic non-human females. *Recent Progress Hormone Res.* 28:707–733.

Hamilton, W. J., III, Tilson, R. L., and Frank, L. G. 1986. Sexual monomorphism in Spotted Hyenas, *Crocuta crocuta*. *Ethology* 71:63–73.

Harrington, F. H., Mech, L. D., and Fritts, S. H. 1983. Pack size and wolf cub survival: Their relationship under varying ecological conditions. *Behav. Ecol. Sociobiol.* 13:19–26.

Hausfater, G., and Hrdy, S. B., eds. 1984. *Infanticide: Comparative and Evolutionary Perspectives*. Hawthorne, N.Y.: Aldine.

Hemingway, E. 1935. *The Green Hills of Africa*. New York: Charles Scribner's Sons.

Henry, J. 1975. *The Red Fox*. Washington, D.C.: Smithsonian Institution Press.

Henschel, J. R., and Skinner, J. D. 1991a. Territorial behaviour by a clan of spotted hyaenas *Crocuta crocuta*. *Ethology* 88:223–235.

Henschel, J. R., and Skinner, J. D. 1991b. Parturition and early maternal care of spotted hyaenas *Crocuta crocuta*: A case report. *J. Zool. (London)* 222:702–704.

Hofer, H., and East, M. L. 1989. Maternal rank and maternal care in spotted hyaenas in a system with fluctuating prey populations. Paper presented at Fifth International Theriological Congress, Rome, 19–26 Aug. (Abstract.)

Hofer, H., and East, M. L. 1992. Siblicide in spotted hyaenas is a consequence of allocation of maternal effort and a cause of skewed litter sex ratios. Talk presented at Fourth International Behavioral Ecology Congress, Princeton Univ., Princeton, N.J., Aug. 17–22. (Abstract.)

Hofer, H., and East, M. L. 1993a. The commuting system of Serengeti spotted hyaenas: How a predator copes with migratory prey, I: Social organisation. *Anim. Behav.* 46:547–558.

Hofer, H., and East, M. L. 1993b. The commuting system of Serengeti spotted hyaenas: How a predator copes with migratory prey, II: Intrusion pressure and commuters' space use. *Anim. Behav.* 46:559–574.

Hofer, H., and East, M. L. 1993c. The commuting system of Serengeti spotted hyaenas: How a predator copes with migratory prey, III: Attendance and maternal care. *Anim. Behav.* 46:575–589.

Holekamp, K. E., Ogutu, J. O., Frank, L. G., Dublin, H. T., and Smale, L. 1993. Fission of a spotted hyena clan: Consequences of prolonged female absenteeism and causes of female emigration. *Ethology* 93:285–299.

Holekamp, H. E., and Smale, L. 1990. Provisioning and food-sharing by lactating spotted hyenas (*Crocuta crocuta*). *Ethology* 86:191–202.

Holekamp, H. E., and Smale, L. 1993. Ontogeny of dominance in free-living spotted hyaenas: Juvenile rank relations with other immature individuals. *Anim. Behav.* 46:451–466.

Holekamp, H. E., and Smale, L. 1995. Rapid change in offspring sex ratio after clan fission in the spotted hyena. *Amer. Nat.* 145:261–278.

Hrdy, S. B. 1979. Infanticide among mammals: A review, classification, and examination of the implications for the reproductive strategies of females. *Ethol. Sociobiol.* 1:13–40.

Hrdy, S. B. 1981. *The Woman That Never Evolved.* Cambridge, Mass.: Harvard Univ. Press.

Hrdy, S. B. 1987. Sex-biased parental investment among primates and other mammals: A critical evaluation of the Trivers-Willard hypothesis. In: R. Gelles and J. Lancaster, eds. *Child Abuse and Neglect: Biosocial Dimensions,* pp. 97–147. New York: Aldine.

Inoff-Germain, G., Arnold, G. S., Susman, E. J., Nottelmann, E. D., Cutler, G. B., Jr., and Chrousos, G. P. 1988. Relations between hormone levels and observational measures of aggressive behavior of young adolescents in family interactions. *Developm. Psychol.* 24:129–139.

Jenks, S. M., Weldele, M. L., Frank, L. G., and Glickman, S. E. 1995. Acquisition of matrilineal rank in captive spotted hyenas: Emergence of a natural social system in peer-reared animals and their offspring. *Anim. Behav.* 50:893–904.

Jost, A. 1972. A new look at the mechanisms controlling sex differentiation in mammals. *Johns Hopkins Med. J.* 130:38–53.

Krusko, N. A., Weldele, M. L., and Glickman, S. E. 1988. Meeting ceremonies in a colony of juvenile spotted hyenas. Paper presented at annual meeting of Animal Behavior Society, Missoula, Mont. (Abstract.)

Kruuk, H. 1972. *The Spotted Hyena: A Study of Predation and Social Behavior.* Chicago: Univ. Chicago Press.

Kruuk, H. 1976. Feeding and social behaviour of the striped hyaena (*Hyaena vulgaris* Desmarest). *East African Wildl. J.* 14:91–111.

Licht, P., Frank, L. G., Pavgi, S., Yalcinkaya, T. M., Siiteri, P. K., and Glickman, S. E. 1992. Hormonal correlates of "masculinization" in the female spotted hyaena (*Crocuta crocuta*), II: Maternal and fetal steroids. *J. Reprod. Fert.* 95:463–474.

Lindeque, M. 1981. Reproduction in the spotted hyaena (*Crocuta crocuta* Erxleben). Master's thesis, Univ. Pretoria.

Lindeque, M., and Skinner, J. D. 1982. Fetal androgens and sexual mimicry in the spotted hyaena (*Crocuta crocuta*). *J. Reprod. Fert.* 67:405–416.

Lindeque, M., Skinner, J. D., and Millar, R. P. 1986. Adrenal and gonadal contribution to circulating androgens in spotted hyaenas (*Crocuta crocuta*) as revealed by LHRH, hCG and ACTH stimulation. *J. Reprod. Fert.* 78:211–217.

Lovejoy, J., and Wallen, K. 1988. Sexually dimorphic behavior in group-housed rhesus monkeys (*Macacca mulatta*) at 1 year of age. *Psychobiology* 16:348–356.

Macdonald, D. W., and Moehlman, P. D. 1982. Cooperation, altruism and restraint in the reproduction of carnivores. In: P. P. G. Bateson and P. H. Klopfer, eds. *Perspectives in Ethology,* Vol. 5, pp. 433–466. New York: Plenum Press.

Macpherson, A. H. 1969. The dynamics of Canadian arctic fox populations. *Canadian Wildl. Serv. Rep. Ser.* 8:3–52.

Matthews, L. H. 1939. Reproduction of the spotted hyaena (*Crocuta crocuta* Erxleben). *Phil. Trans. (Ser. B)* 230:1–78.

Mills, M. G. L. 1990. *Kalahari Hyaenas: Comparative Behavioural Ecology of Two Species.* London: Unwin Hyman.

Mock, D. W. 1984a. Infanticide, siblicide, and avian nest mortality. In: G. Hausfater and S. B. Hrdy, eds. *Infanticide: Comparative and Evolutionary Perspectives,* pp. 2–30. New York: Aldine.

Mock, D. W. 1984b. Siblicidal aggression and resource monopolization in birds. *Science* 225:731–733.

Mock, D. W. 1987. Siblicide, parent-offspring conflict, and unequal parental investment by egrets and herons. *Behav. Ecol. Sociobiol.* 20:247–256.

Mock, D. W., and Ploger, B. J. 1987. Parental manipulation of optimal hatch asynchrony in cattle egrets: An experimental study. *Anim. Behav.* 35:150–160.

Monaghan, E. O., and Glickman, S. E. 1992. Hormones and aggressive behavior. In: J. B. Becker, S. M. Breedlove, and D. Crews, eds. *Behavioral Endocrinology,* pp. 261–285. Cambridge, Mass.: MIT Press.

Neaves, W. B., Griffin, J. E., and Wilson, J. D. 1980. Sexual dimorphism of the phallus of the spotted hyena (*Crocuta crocuta*). *J. Reprod. Fert.* 59:506–512.

O'Connor, R. J. 1978. Brood reduction in birds: Selection for fratricide, infanticide and suicide? *Anim. Behav.* 26:79–86.

Owens, D., and Owens, M. J. 1984. Helping behaviour in brown hyaenas. *Nature.* 308:843–845.

Packer, P., Collins, D. A., Sindimwo, A., and Goodall, J. 1995. Reproductive constraints on aggressive competition in female baboons. *Nature* 373:60–63.

Packer, C., Gilbert, D. A., Pusey, A. E., and O'Brien, S. J. 1991. A molecular genetic analysis of kinship and cooperation in African lions. *Nature* 351:562–565.

Packer, C., and Pusey, A. E. 1983a. Adaptations of female lions to infanticide by incoming males. *Amer. Nat.* 121:716–728.

Packer, C., and Pusey, A. E. 1983b. Male take-overs and female reproductive parameters: A simulation of oestrous synchrony in lions (*Panthera leo*). *Anim. Behav.* 31: 334–340.

Packer, G. A., and Mock, D. W. 1987. Parent-offspring conflict over clutch size. *Evol. Ecol.* 1:161–174.

Pedersen, J. M., Glickman, S. E., Frank, L. G., and Beach, F. A. 1990. Sex differences in play of immature spotted hyenas. *Horm. Behav.* 24:403–420.

Pennycuick, P. J. 1979. Energy costs of locomotion and the concept of "foraging radius." In: A. R. E. Sinclair and M. Norton-Griffiths, eds. *Serengeti: Dynamics of an Ecosystem,* pp. 164–184. Chicago: Univ. Chicago Press.

Racey, P. A., and Skinner, J. D. 1979. Endocrine aspects of sexual mimicry in the spotted hyaena (*Crocuta crocuta*). *J. Zool. (London)* 187:315–328.

Ralls, K. 1976. Mammals in which females are larger than males. *Quart. Rev. Biol.* 51:245–276.

Rieger, I. 1978. Social behavior of striped hyenas at the Zurich zoo. *Carnivore* 1:49–60.

Rieger, I. 1979. Breeding the striped hyaena (*Hyaena hyaena*) in captivity. *Internatl. Zoo Yearb.* 19:193–198.

Sapolsky, R. M. 1991. Testicular function, social status and personality among wild baboons. *Psychoneuroendocrinology* 16:281–293.

Short, R. V. 1974. Sexual differentiation of the brain of the sheep. *Les Colloques de l'Institut National de la Sante et de la Recherche Medicale* 32:121–142.

Silk, J. B. 1983. Local resource competition and facultative adjustment of sex ratios in relation to competitive abilities. *Amer. Nat.* 121:56–66.

Silk, J. B. 1984. Local resource competition and the evolution of male-biased sex ratios. *J. Theor. Biol.* 108:203–213.

Silk, J. B. 1986. Social behavior in evolutionary perspective. In: B. B. Smuts, D. L. Cheney, R. M. Seyfarth, R. W. Wrangham, and T. T. Struhsaker, eds., *Primate Societies,* pp. 318–329. Chicago: Univ. Chicago Press.

Smale, L., Frank, L. G., and Holekamp, K. E. 1993. Ontogeny of dominance in free-living spotted hyaenas: Juvenile rank relations with adult females and immigrant males. *Anim. Behav.* 46:467–477.

Smale, L., Holekamp, K. E., Weldele, M., Frank, L. G., and Glickman, S. E. 1995. Competition and cooperation between littermates in the spotted hyaena (*Crocuta crocuta*). *Anim. Behav.* 50:671–682.

Smuts, B. B. 1986. Sexual competition and mate choice. In: B. B. Smuts, D. L. Cheney, R. M. Seyfarth, R. W. Wrangham, and T. T. Struhsaker, eds. *Primate Societies,* pp. 385–399. Chicago: Univ. Chicago Press.

Struhsaker, T. T., and Leland, L. 1986. Colobines: Infanticide by adult males. In: B. B. Smuts, D. L. Cheney, R. M. Seyfarth, R. W. Wrangham, and T. T. Struhsaker, eds. *Primate Societies,* pp. 83–97. Chicago: Univ. Chicago Press.

Symington, M. M. 1987. Sex ratio and maternal rank in wild spider monkeys: When daughters disperse. *Behav. Ecol. Sociobiol.* 20:421–425.

Tilson, R. L., and Hamilton, W. J., III. 1984. Social dominance and feeding patterns of spotted hyaenas. *Anim. Behav.* 32:715–724.

Tilson, R. L., and Henschel, J. R. 1986. Spatial arrangement of spotted hyaena groups in a desert environment, Namibia. *African J. Ecol.* 24:173–80.

Trivers, R. L., and Willard, D. 1973. Natural selection of parental ability to vary the sex ratio of offspring. *Science* 179:90–92.

van Ballenberghe, V., and Mech, L. D. 1975. Weights, growth, and survival of timber wolf pups in Minnesota. *J. Mamm.* 56:44–63.

van Jaarsveld, A. S., and Skinner, J. D. 1991. Plasma androgens in spotted hyaenas (*Crocuta crocuta*): Influence of social and reproductive development. *J. Reprod. Fert.* 93:195–201.

van Schaik, C. P., and Hrdy, S. B. 1991. Intensity of local resource competition shapes the relationship between maternal rank and sex ratios at birth in cercopithecine primates. *Amer. Nat.* 138:1555–1562.

Walker, E. P. 1975. *Mammals of the World.* Baltimore: Johns Hopkins Univ. Press.

Werdelin, L., and Solounias, N. 1991. The Hyaenidae: Taxonomy, systematics and evolution. *Fossils and Strata* 30:1–104.

Whalen, R. E., Yahr, P., and Luttge, G. G. 1985. The role of metabolism in the hormonal control of sexual behavior. In: N. Adler, D. Pfaff, and R. W. Goy, eds. *Handbook of Behavioral Neurobiology,* pp. 127–158. New York: Plenum Press.

Whately, A. 1980. Comparative body measurements of male and female spotted hyenas from Natal. *Lammergeyer* 28:40–43.

Wickler, W. 1965. Die ausseren genitalien als soziale Signalen bei einegen Primaten. *Naturwiss.* 52:268–270.

Wynn, R. M., and Amoroso, E. C. 1964. Placentation in the spotted hyena (*Crocuta crocuta* Erxleben), with particular reference to the circulation. *Amer. J. Anatomy* 115:327–362.

Yalcinkaya, T. M., Siiteri, P. K., Vigne, J., Licht, P., Pavgi, S., Frank, L. G., and Glickman, S. E. 1993. A mechanism for virilization of female spotted hyenas *in utero.* *Science* 260:1929–31.

CHAPTER 4

The Adaptive Significance of Color Patterns in Carnivores: Phylogenetic Tests of Classic Hypotheses

ALESSIA ORTOLANI AND T. M. CARO

Carnivores are an exciting taxonomic group against which to examine evolutionary theories of coloration, because they exhibit a wide variety of basic pelage colors; show a great diversity of markings on their coats, including spots, stripes, bands, and patches; have generated some of the classic hypotheses for coloration in animals; and have attracted disproportionate attention in popular literature. For example, most people in the western world know that the giant panda (*Ailuropoda melanoleuca*) is black and white, and that the leopard (*Panthera pardus*) is spotted. Moreover, the common names of a great many carnivore species are based on their color or color pattern: the brown hyena (*Hyaena brunnea*), the red fox (*Vulpes vulpes*), the striped skunk (*Mephitis mephitis*), the black-footed cat (*Felis nigripes*), the side-striped jackal (*Canis adustus*), and the ring-tailed cat (*Bassariscus astutus*).

At the turn of the century, naturalists began to speculate about the survival value of the pelage, feather, and skin colors that they saw in the growing numbers of animal specimens brought back from collecting expeditions (Wallace 1889; Poulton 1890; Beddard 1892; Thayer 1909; Roosevelt 1911; Mottram 1915, 1916). Current adaptive hypotheses for animal coloration still stem from this period, although they were subsequently given more sophistication by Cott (1940) in his benchmark book *Adaptive Coloration in Animals*. The great intuitive appeal of many of these classic hypotheses had an unfortunate consequence, however, for it caused interest in the adaptive significance of animal coloration to lapse, until just recently. Work on animal coloration since 1940 has concentrated primarily on the development of pelage patterning in mammals (e.g., Stoddart 1970; Hadley 1972; Bard 1977; Murray 1981), the mechanisms by which animals match their backgrounds (e.g., Endler 1978), and the theories underlying the evolution of coloration patterns (Mallet and Singer 1987; Endler 1988; Guilford 1988; Booth 1990). In contrast, empirical and comparative tests of evolutionary hypotheses in animal coloration are still scarce (but see Baker and Parker 1979; Lyon and Montgomerie 1986; Hoglund 1989; Dubost 1991).

To begin to redress this imbalance, we attempt in this chapter to test—both systematically and comprehensively—the classic hypotheses about the adaptive significance of coat-color patterns in carnivores (Order Carnivora). These hypotheses, which predict correlations between particular habitats or behaviors and particular coat-color patterns, usually rest on species-specific anecdotes. To test the generality of these hypotheses, we performed a comparative study across a broad spectrum of species, using quantitative tests that take phylogenetic relationships into account (Ridley 1983; Felsenstein 1985; Maddison 1990; Harvey and Pagel 1991; Gittleman and Luh 1992). Because closely related species can exhibit similar color markings and behavioral ecology as a result of shared ancestry (Ridley 1983), simply testing the correlation between coloration variables and behavioral-ecological variables would not distinguish adaptations from homologies.

Our chapter is in four sections. First, we briefly discuss extant theories of coloration in animals, highlighting hypotheses that have been advanced for carnivore coloration. Second, we outline new methods of classifying and scoring color markings, for purposes of analysis. Third, we present quantitative tests of the hypotheses (e.g., pale coloration in deserts) at the level of species within families. Finally, we discuss the advantages and shortcomings of applying comparative phylogenetic techniques to our data set, and examine the results of these comparative analyses of the adaptive significance of color markings in carnivores.

Theories of Animal Coloration

Concealing Coloration

Coloration can yield concealment through three different means: general color and pattern resemblance to background, disruptive coloration, and countershading. *General color resemblance* (Poulton 1890; Cott 1940) or *background matching* (Endler 1978) invokes the similarity between an animal's color and that of the natural background in which it lives. In *disruptive coloration* (Cott 1940), sharply contrasting colors and irregular markings break up the form of the animal, making regions of its body appear mutually discontinuous. Disruptive colors may be restricted to particular areas of the body, such as the face or the tail; the coat of the giant panda (*Ailuropoda melanoleuca*) has been described as disruptive (Pycraft 1925). *Countershading* is a third method by which animals are thought to attain concealment (Thayer 1896; Thayer 1909; Kiltie 1988). Most mammals, including carnivores, have lighter ventral surfaces. The lightening of the ventral surface and darkening of the dorsal surface may counteract the effects of shade and light, making the animal appear less visible. However, there is little empirical evidence to support this hypothesis (Kiltie 1989), and we did not test it.

Concealing coloration can also be divided into protective resemblance and aggressive resemblance (Poulton 1890), respectively reflecting the need to be concealed from predators and the need to approach prey undetected. Intuitively, aggressive background resemblance might be expected to be more prevalent in carnivores, but many species in this order are themselves subject to predation (Polis et al. 1989; Caro 1994).

Proposals for the adaptive significance of the coat patterns of some carnivore species have become all but universally accepted. For example, the polar bear (*Thalarctos maritimus*) is said to be white for concealment in arctic snow (Cott 1940). Similarly, the stoat (*Mustela erminea*) and Arctic fox (*Alopex lagopus*) are thought to change from brown in summer to white in winter to match the seasonally changing background (Poulton 1890; Beddard 1892). The fennec (*Vulpes zerda*) is presumed to have a pale coat in order to match the light, sand-colored desert ground on which it lives (Benson 1933). The jaguar (*Panthera onca*), the ocelot (*Felis pardalis*), and other felids are thought to be spotted so as to remain concealed amid forest foliage; the theory is that their coats reproduce the pattern of dappled light filtering through the leaves (Beddard 1892; Cott 1940; Kitchener 1991). Finally, the tiger (*Panthera tigris*) is supposed to have a striped coat in order to match tall grassland vegetation (Beddard 1892). Although these solutions may indeed be adaptive for the species mentioned, the actual *generality* of these hypotheses has never been tested.

Conspicuous Coloration and Interspecific Communication

Conspicuous color patterns may be used in *warning coloration* to advertise the presence of an animal across a range of different backgrounds (Hadley 1972; Guilford 1988). Warning, or aposematic, color patterns are believed to be found in animal species that have noxious defenses (Guilford 1990) and behave aggressively toward an enemy (Andersson 1976). Such patterns may advertise that an animal is toxic if eaten, or that it possesses chemical secretions or mechanical defenses that are life-threatening, or at least highly discouraging, to an enemy. In mammals, these defenses take the form of spines, quills, or anal-sac secretions. Many aposematic invertebrates and poikilothermic vertebrates advertise themselves by combinations of red, yellow, and black coloring, but among mammals, by contrast, black-and-white markings are thought to be associated with warning coloration (Wallace 1889; Pocock 1908). Black-and-white coloration is extremely conspicuous in a variety of environments, and it is particularly effective on nocturnal animals (Cott 1940; Searle 1968).

The most cited example of warning coloration among mammals is the skunk (*Mephitis, Conepatus,* and *Spilogale* species: Belt 1888; Poulton 1890; Cott, 1940; Young 1957), although polecats (*Mustela putorius*), ratels (*Mellivora capensis*), and grisons (*Galictis* spp.) have also been mentioned (Pocock 1908; Pycraft 1925). The association between conspicuousness and noxious defenses has been explored in insects (Guilford 1990), but it has not been investigated systematically in carnivores.

Conspicuous colors may also be used to attract the attention of other species and draw them closer, in order to predate them. For example, some viperid snakes (e.g., *Bothrops jararaca* and *B. jararacussu*) lure frogs by undulating their white tail tips (Sazima 1991). White-tailed mongooses, *Ichneumia albicauda,* have been observed flicking their tails to lure birds into hunting range (Estes 1991).

Conspicuous parts of the body may be used to redirect predatory attack away from the body toward distal regions. Using models of small mammals, Powell (1982) found that raptors preferentially attack marked tail tips. He suggested that the black-tipped tails of long-tailed weasels (*Mustela frenata*) and short-tailed (common, or least) weasels (*Mustela nivalis*) may deflect predatory attack from the body.

Finally, conspicuous coloration may be used in some forms of mimicry (Vane-Wright 1980; Endler 1981). In Batesian mimicry a harmless species (e.g., the viceroy butterfly) resembles in color a noxious species (in this case the monarch), thereby gaining protection because predators confuse it with the model (Rettenmeyer 1970). In Müllerian mimicry, two or more noxious species resemble each other and thereby reinforce predator avoidance for all (Turner 1971). In carnivores, two examples of Batesian mimicry have been advanced: the white dorsal and dark ventral surface of relatively harmless cheetah cubs may resemble the coat pattern of the very aggressive ratel (Eaton 1976); and the stripes on the relatively defenseless aardwolf (*Proteles cristatus*) are thought to mimic those of the more dangerous striped hyena, *Hyaena hyaena* (Gingerich 1975). Their intuitive appeal notwithstanding, these hypotheses have subsequently been refuted (Goodhart 1975; Caro 1987). Pocock (1908) has suggested that the teledu (*Mydaus javanensis*) and the ferret-badgers (*Melogale* spp.), with their striking black-and-white markings, noxious anal-sac secretions, and sympatric distribution in Southeast Asia, are Müllerian mimics.

Certain color patches on the bodies of mammals are thought to exaggerate or mimic weaponry (Guthrie and Petocz 1970). For example, white labial spots may exaggerate the length of canines, and various eye marks may increase the apparent length of horns or antlers of ungulates.

Conspicuous Coloration and Intraspecific Communication

Conspicuous coloration is also used in intraspecific communication (Wickler 1968). Patches of color on prominent areas such as the end of the tail may allow conspecifics, especially young, to locate each other more easily (Leyhausen 1979), and might therefore be prevalent in social species. Alternatively, contrasting tail-tips or patches of color behind the ears may allow individuals to follow each other in high grass, as is suggested for cheetahs (*Acinonyx jubatus:* Ewer 1973) and tigers (*Panthera tigris:* Schaller 1967; Heran 1976).

Color patches may also act as social releasers. The white throats of some carnivores, such as the red fox, are thought to orient bites from conspecifics to an area that is relatively protected by thick skin and hair (Fox 1971). Contrasting color patches may signal motivational state in the context of aggression: the chest patches of some bears, such as the Malayan sun bear (*Helarctos malayanus*), may be intimidating to rivals when the animals stand bipedally, for example (Searle 1968; Ewer 1973).

Conspicuous coloration may be used to attract mates, and thus may be sexually selected. Few mammals, however, are sexually dichromatic (notable exceptions are pinnipeds, a few primates, and some artiodactyls), and this study could find no evidence of pronounced sexual dichromatism in terrestrial carnivores.

Physical Correlates of Coloration

Pelage coloration may also play an important role in thermoregulation and ultraviolet-light protection, and so enable organisms to survive better in harsh climatic conditions (Hamilton and Heppner 1967; Hamilton 1973). Different shades of color may result in differential reflectivity of solar radiation, and may thus affect thermal balance (Burtt 1979; Schmidt-Nielsen 1979). For example, white fur may increase solar heat gain by scattering solar radiation toward the skin (Walsberg 1983), and would thus be expected in cold climates. Nevertheless, the relationship between surface coloration and solar-heat gain is complex (Walsberg 1990) and incompletely understood. For example, Gloger's rule states that species living in warm, humid regions tend to be dark (Gloger 1833; Huxley 1942); but it is not known whether this is so because warm, dark surfaces vaporize water more readily than do cool surfaces, and so keep the animal drier, or because dark animals are better concealed in dark, moist habitats, or both. Dark hair or skin also affords some protection from the harmful effects of ultraviolet radiation in the tropics (Porter 1967). The inverse relationship between the extent of external skin pigmentation and that of the pigmentation of the internal organs suggests that skin pigmentation is an adaptation against damaging ultraviolet radiation (Cole 1943).

Certain color patterns may serve purposes unrelated to those discussed above. For example, striping in zebras (e.g., *Equus burchelli*) may be an adaptation to avoid biting flies (Waage 1981). It has also been proposed that the dark markings around the eyes of many diurnal vertebrates reduce glare from the sun (Ficken et al. 1971; Rougeot 1981). The converse, that light circles around the eyes concentrate light in dark environments, has been suggested by the same authors. This hypothesis is not logical, however, since light eye rings would reflect light into the eye, light that is not image-forming, and would thus blur any image coming directly into the eye (J. Endler, pers. comm.).

The Classic Hypotheses We Tested

Given this background, we tested the validity of 15 hypotheses—hypotheses that had been formulated on specific carnivore species—across all carnivores for which comparative data are available. These are (1) white fur color and living in the arctic; (2) pale fur color and living in deserts; (3) spotted coats and arboreal lifestyle; (4) spotted coats and living in forests; (5) vertically striped coats and living in grassland; (6) black-and-white coats and having noxious anal secretions; (7) white throats and being social; (8) contrasting tail tips and being social; (9) contrasting ear marks and being social; (10–13) black tail tips or ears and living in grassland, and white tail tips or ears and living in forest; (14) dark eye contours and diurnal habit; and, finally, (15) dark fur and living in tropical forest.

Methods of Classification and Statistical Analysis

Classification of Color Patterns

In this study we classify the contrasting coat markings of adult individuals of all terrestrial carnivore species for which data were available and examine the distribution of these markings across general behavioral and ecological categories. Although new techniques for measuring the color spectrum of animal coat patterns are available now (see Endler 1990), they become impractical in a broad comparative study. Instead, Ortolani systematically classified coat-color markings for 176 carnivore species representing all extant carnivore genera (see Corbet and Hill 1991) for which the literature offers a pictorial representation and at least modest information on behavioral ecology and coat coloration. The principal sources employed were Haltenorth and Diller (1977), Kingdon (1977), Lekagul and McNeely (1977), Grzimek (1990), Macdonald (1984), Estes (1991), Nowak (1991), and Seidensticker and Lumpkin (1991).

Carnivores' coat-color patterns were classified in two main ways (see Appendix 4.1, this chapter). First, a category describing general "fur color" was constructed, using species descriptions found in the references cited above. Since descriptions of color hues can be subjective, and since fur color may vary according to geographical location, all color hues that the different sources reported for a given species were recorded and were then grouped into broader categories. Four general fur-color categories were established along a brightness gradient constructed according to the Munsell color system (Cleland 1921) (Table 4.1). The categories—dark, medium, pale, and white—included all species with uniform color coats whose variation in color shades was confined to a single brightness category, such that, for example, a species with both black and dark-brown individuals would be classified as "Dark." Three

Table 4.1. Character states and definitions of fur color

Fur color	Definition
Dark	Uniformly black, dark brown, dark gray-black, or brown-black
Medium	Uniformly brown, red, or gray
Pale	Uniformly light beige, yellow, or light gray
White	Uniformly white
Variable dark	Both medium and dark individuals found in the same species
Variable light	Both medium and pale individuals found in the same species
Variable	All uniform shades (dark, medium, pale, and white) found in the same species
Light and dark	Contrasting light and dark markings (other than black and white) over the whole body
Black and white	Contrasting solid black and solid white regions over whole body
Molt	Uniform white coat in winter changing to medium-dark in summer

Note: Melanistic forms were not included under dark or variable dark.

variable categories (Table 4.1) were constructed to accommodate uniform species that show uniform color in individuals but variation in fur color among individuals across two or more brightness categories; for example, a species with black, brown, beige, and white individuals would be classified as "Variable." Finally, species with nonuniform fur color, such as the striped skunk (*Mephitis mephitis*), were classified under ad hoc categories (i.e., "Black and white").

Second, a separate classification was constructed to score contrasting body markings found on carnivores (Table 4.2). For the purpose, we used a combination of color photographs, black-and-white photographs, and detailed drawings from the sources listed just above. Museum skins and occasionally live animals were also used, when available. Since a distinctive color marking may occur in any given region of the body (e.g., a white tail tip or a black tail tip), markings over different regions of the body may occur in different combinations when species are compared. In order to classify color markings into discrete, mutually exclusive character states, 19 areas of the body surface were initially selected, according to anatomical criteria (Figure 4.1), and each was scored for the presence or absence of specific color markings, on each species. Only ten of these areas (see Table 4.2) are discussed in this chapter, however, since these were found to be the only color markings pertinent to the hypotheses tested. A marking was defined as an area of a color contrasting with that of the rest of the body or with that of the most proximal body area. Thus a white tail tip on a white animal would not be scored as such, but a white tail tip on a black animal, or on one with a black tail, *would* constitute a marking.

Inter-Observer Reliability

Body marks were originally categorized by Ortolani, using several sources, but it was difficult to gain access to the same museum specimens and live

Table 4.2. Characters, character states, and definitions of contrasting color markings

Character and character state	Definition
1. BODY[a]	
Spots	Dark spots or rosettes cover half or more of body area[b]
Spots and uniform	Intraspecific variation: some individuals are spotted and others are uniform
Bands	Dark, broad vertical bands cover dorsum and sides of body
Blotches	Large, irregular blotches of contrasting colors cover half or more of body area
Vertical stripes	Dark vertical stripes cover half or more of body area
Horizontal stripes	Dark horizontal stripes cover half or more of body area
White cape	White stripe covers entire dorsal area, whereas flanks are dark
Dark cape	Dark cape covers entire dorsal area, whereas flanks are lighter
3. UNDERSIDES	
Dark	Ventral area is black or darker than body
Light	Ventral area is white or lighter in color than body
9. FACE	
Contrasting	Contrasting white and dark marks cover half or more of facial region
Darker	Head or face is uniformly darker than body
Lighter	Head or face is uniformly lighter than body
13. EYE CONTOUR	
Dark around	Dark fur encircles entire eye
Light above and dark below	Light fur encircles upper orbital area and dark fur encircles lower orbital region
Light above	Light fur encircles only upper orbital region
Light below	Light fur encircles only lower orbital region
Light ring	Light fur encircles entire eye
14. BELOW EYE	
White spot	Contrasting white spot below eye
Dark patch	Contrasting black patch below eye, near muzzle
17. FACIAL STRIPES[c]	
Frontal stripe	White vertical stripe between eyes extends toward nose
Forehead band	White horizontal stripe, or white hood, crosses forehead
12. BACKS OF EARS[d]	
Darker back	Back of ear darker than head
Black tip	Tip of pinna black
Lighter back	Back of ear lighter than head
White spot	White spot on center of darker pinna
White rim	Rim of pinna white
White tip	Tip of pinna white
8. THROAT AND NECK	
Darker	Throat area distinctively darker than body
Lighter	Throat area distinctively lighter than body (but not white)
White throat	Chin and throat area white
Black and white	Very distinct black and white marks on neck mark sides of neck
Chest mark	Contrasting color patch (not white) on throat or sternal region
6. TAIL	
Dark	Tail black or distinctively darker than body
Light	Tail white or distinctively lighter than body
Ringed	Black or dark rings, alternating with white or light rings, encircle tail for entire length
Dorsal stripe	Dark stripe along dorsal surface of tail

(*continued*)

Table 4.2. (*Continued*)

Character and character state	Definition
7. TAIL TIP[e]	
Black	Tail tip distinctly black
White	Tail tip distinctly white
Black or light	Tip black on some individuals, light or white on others

[a]Numbers here refer to location of body markings shown in Figure 4.1.

[b]Includes individuals with clouded spots or with spots arranged in longitudinal rows that may merge into stripes.

[c]Frontal stripe and forehead band were combined to test the association with anal-sac secretion.

[d]All categories were combined to test against sociality; darker back and black tip were combined to test the association with grassland habitat; white spot, white ear margins, and white ear tip were combined to test the association with forest habitat.

[e]All categories were combined to test against sociality.

animals a second time in order to measure inter-observer reliability. Instead, we trained two naive observers and then, for a random sample of 30 carnivores, asked them to score ten contrasting body marks for each species from a single color photograph of each. Fur color was not recorded, since it had been gleaned from written descriptions (see above). Inter-observer values were obtained by using an index of concordance, since scores were categorical (Caro et al. 1979). If one observer scored a marking as unknown, that was considered a disagreement; if both did, the marking was omitted from the analysis. The results of these efforts are shown in Table 4.3.

Most categories yielded reasonably high values; the exceptions were backs of ears, throat-neck, undersides, and eye contour. Low inter-observer reliability on the first three of the exceptions was in large part due to a high number of "unknown" scores by the naive observer, scored as such because these areas of the body were not visible in some of the photographs. This may therefore be more a reflection of using a single photograph to measure inter-observer reliability than an indication of low reliability in scoring these body parts. Eye contour, by contrast, seems to have been more difficult to score reliably from a photograph even when this body area was visible, and even though the data presented in this chapter are based in part on museum skins (which are much easier to score for this body mark than are illustrations), the analyses concerning eye contour should be treated with some caution.

Behavioral and Ecological Variables

A set of behavioral and ecological characters was scored so that we might test their association with different color patterns. For every carnivore species for which coloration data were available, social behavior, activity pattern, and type of locomotion were recorded, using the same sources used to classify color

Figure 4.1. Location of body markings scored on each species; not all are discussed in this chapter. 1, body; 2, dorsal stripe; 3, undersides; 4, shoulder mark; 5, limbs; 6, tail; 7, tail tip; 8, throat and neck; 9, face; 10, muzzle; 11, front of ears; 12, back of ears; 13, eye contour; 14, below eyes; 15, vertical eye stripe; 16, eye mask; 17, facial stripes; 18, nostril spots; 19, chin.

markings (the illustrations provided for each carnivore species, in these sources, were assumed to be independent from the descriptions of their behavioral ecology) (Appendix 4.2). Each of these three categories was then divided into discrete, mutually exclusive states (Table 4.4).

We also scored each species as being present or absent in six nonmutually exclusive habitat types broadly representing different vegetation types (Table 4.5). Although these habitats may each contain a variety of light-level environ-

Table 4.3. Inter-observer reliability on scoring color markings

Character	No. of agreements	No. of agreements and disagreements	Ratio (A/A + D)
Body	28	30	0.93
Undersides	12	24	0.50
Face	23	30	0.77
Eye contour	17	29	0.59
Below eyes	23	30	0.77
White facial stripes	30	30	1.00
Backs of ears	6	15	0.40
Throat and neck	16	27	0.59
Tail	21	26	0.81
Tail tip	15	19	0.79

Table 4.4. Behavioral characters and character states

Character	Character state
Social behavior[a]	Solitary
	Pairs
	Group
	Variable group[b]
Activity pattern[c]	Nocturnal
	Nocturnal and crepuscular
	Diurnal
	Anytime[d]
Locomotion[e]	Terrestrial
	Terrestrial but climbs trees
	Terrestrial and arboreal
	Arboreal
	Aquatic

[a]Group and variable group were combined and called "social" to test the association with white throats, contrasting tail tips, and contrasting ears.

[b]Variable group was assigned to species that show variability in grouping patterns, such that some individuals may be solitary, whereas others may be in pairs, in groups, or in any combination of these.

[c]Nocturnal and nocturnal and crepuscular were combined and called "mainly nocturnal," diurnal and anytime were combined and called "mainly diurnal," to test the association with eye markings.

[d]Denoted being active during both night and day.

[e]Terrestrial and terrestrial but climbs trees were combined and called "mainly terrestrial," arboreal and terrestrial and arboreal were combined and called "mainly arboreal," to test the association with spotted coats.

Table 4.5. Definitions of habitats
by vegetation type

Habitat[a]	Vegetation type
Temperate forest[b]	Forest
	Woodland
	Taiga
Tropical forest[b]	Tropical rainforest
	Tropical jungle
Grassland	Grassland
	Savanna
	Steppe
Arctic	Arctic
	Tundra
Riparian and aquatic	Riparian
	Aquatic
Desert	Desert
	Semidesert

[a]Species can be found in a number of habitats, and hence were scored as present or absent in each.

[b]Temperate forest and tropical forest were combined and called "forest" to test the association with spotted coats, white tail tips, and white ear markings.

ments, the aggregate difference in the extent of vegetation cover between them is almost certainly sufficient to set up different selection pressures to which coat colors may have responded over evolutionary time.

In addition, possession of a noxious anal-sac secretion was recorded. This category was difficult to score because all carnivores except ursids and four other species have anal sacs (Macdonald 1985). Moreover, the extent to which a species' anal-sac secretion is characterized as noxious or not relies heavily on subjective criteria, such as whether a human observer finds its smell highly offensive. We therefore used a conservative approach, scoring only those species that had been documented as having grossly enlarged anal sacs, and that had been observed spraying the secretion of these glands during encounters with predators (Pocock 1908; Brinck et al. 1983; Macdonald 1985). Species that met the first criterion, but for which spraying was never documented, were scored as unknown.

Phylogenetic and Statistical Analyses

We tested each classic hypothesis of adaptive coloration separately for each carnivore family. We proceeded in this manner for two main reasons. First, since carnivores exhibit different diets, hunting techniques, and locomotory adaptations, as well as social systems that often associate along family lines

(Gittleman 1989, 1993), we wanted to see if adaptive responses to similar selective pressures differed along family lines. And, second, we wanted to separate families for which good phylogenies were available, such as the felids, from those for which we had only classical taxonomies, such as the viverrids. We used the most current molecular phylogenies available for the Canidae (Geffen et al. 1996) and Ursidae (Goldman et al. 1989), and the most recent morphological phylogenies for the Mustelidae (Bryant et al. 1993) and the Hyaenidae (Werdelin and Solounias 1991). For the Felidae we used a molecular phylogeny based on the work of Collier and O'Brien (1985), Pecon Slattery et al. (1994), and Janczewski et al. (1995). Species listed in the two appendices but not included in the phylogenies (e.g., pale fox, *Vulpes pallida*) and species in the original phylogenies but not in the appendices because of lack of data were excluded from the analysis. Because no phylogenies were available for the Procyonidae, Viverridae, or Herpestidae, we followed, for them, the taxonomic classification by Wozencraft (1989). For present purposes, the red panda (*Ailurus fulgens*) was considered a procyonid (Wayne et al. 1989), and the giant panda an ursid (O'Brien et al. 1985). None of the phylogenies used were based on pelage characteristics.

We used the MacClade computer program for analysis of phylogeny and character evolution (Maddison and Maddison 1992) to map coloration and behavioral-ecological characters onto the family phylogenies. Given the character states of the terminal taxa, MacClade then reconstructs the ancestral state for the character mapped onto the specified phylogeny. All coloration and behavioral-ecological characters were assumed to be unordered.

We employed Maddison's (1990) concentrated-changes test to test the association between specific color patterns and behavioral ecology. Since Maddison's test demands the use of binary characters in the independent variable, each behavioral-ecological variable was recorded as being either present or absent by assigning it a 1 or 0 value, respectively. A 0-to-1 change on a branch of the tree represents a gain in that character, and a 1-to-0 change is considered a loss. In contrast, the reconstruction of the ancestral states of any given color pattern (e.g., dark) was performed with all the character states for that particular color variable (e.g., fur color) traced on the phylogeny. The numbers of gains and losses in the color pattern of interest (i.e., the dependent variable) and any ambiguities in the reconstruction of its ancestral states (see below) were each resolved by having all other character states traced on the phylogeny.

Dependent characters with multiple states do not have to be binary, since only the gains and losses of the character state of interest are needed for the concentrated-changes test. But in some cases the reconstruction of ancestral states in a dependent character produces ambiguities (i.e., equivocal branches) on the tree, leaving the researcher to choose among many, sometimes thousands, of equally parsimonious resolutions of the ambiguities. In the work underlying this chapter, we arbitrarily decided to test the two most extreme resolutions available: those that produced the most gains (or losses) in the character state of interest and those that produced the fewest.

The concentrated-changes test calculates the probability that gains in a particular color pattern (e.g., spotted coat) are more concentrated than expected by chance on those branches of the tree with a particular behavioral or ecological character state (e.g., forest) (Maddison and Maddison 1992). Because the concentrated-changes test assumes equal branch lengths, it supposes that change is equally likely to occur on each branch of the tree. In situations where MacClade was unable to generate exact counts of gains and losses, we ran 10,000 simulations instead, using the "actual changes" option (Maddison and Maddison 1992). The null hypothesis was rejected at the 0.1 level, given the low statistical power of the test (Maddison 1990; Sillen-Tullberg and Moller 1993).

Resolution of Problems in the Phylogenetic Analyses

Equivocal Reconstructions

Historical reconstructions of character evolution on a phylogeny may lead to ambiguity in the reconstruction of ancestral states (Maddison and Maddison 1992). In this analysis, some of the reconstructions of the ancestral character states of either coloration or behavioral-ecological characters were equivocal. An "equivocal" assignment on a branch of a phylogenetic tree means that equally parsimonious reconstructions of the character in question exist for that ancestor. When faced with ambiguities in the reconstruction of coloration characters, we examined all of the existing, equally parsimonious reconstructions using the "equivocal cycling" option in MacClade. We then chose the two most extreme, equally parsimonious reconstructions: the most-supportive (MS) reconstruction for the hypothesis tested was the one with the maximum number of gains (or losses) in the character state of interest; the least-supportive (LS) reconstruction was the one with the minimum number of gains (or losses) in the character state of interest. The concentrated-changes test was then run twice, once for each reconstruction, and the probability values obtained for the two cases were reported.

When there were ambiguities in the reconstructions of behavioral-ecological ancestral character-states, we resolved them using the ACCTRAN and DELTRAN options in MacClade. ACCTRAN accelerates changes toward the root of the tree, and thus maximizes early gains; DELTRAN delays changes away from the root, thus maximizing parallel changes on the tree (Maddison and Maddison 1992). Since all behavioral-ecological characters were coded as binary characters, using the ACCTRAN and DELTRAN options to resolve uncertainties in character reconstruction was almost always equivalent to examining the two most extreme, equally parsimonious resolutions. Probabilities obtained from the concentrated-changes test are reported for both ACCTRAN and DELTRAN reconstructions. When there were ambiguities in character reconstruction in both the coloration and the behavioral-ecological variables, four probability values are reported in the tables.

Polytomies

The felid, ursid, hyaenid, and mustelid phylogenies contained unresolved polytomies. Since the concentrated-changes test can be applied only to trees consisting of dichotomous branches (Maddison and Maddison 1992), we had to resolve all polytomies manually. If the character of interest was present in all species within the polytomy (Figure 4.2 [1]), then we resolved it randomly, since the outcome of the test would be unaffected (Figure 4.2 [2, 3]) (W. Maddison, pers. comm.). If, however, the character was present only in some species within the same polytomy but not in others (Figure 4.2 [4]), then we examined the two most extreme possible resolutions (Sillen-Tullberg and Moller 1993; W. Maddison, pers. comm.). Because the *most-parsimonious* hypothesis for the evolution of a given character is that it evolved only once, we resolved the polytomy by minimizing the number of gains for that character state (Figure 4.2 [5]). Because the *least-parsimonious* hypothesis for the evolution of a character is that it evolved independently in each species, we resolved the polytomy by making the number of gains in a given character state equal to the number of species in which that state was present (Figure 4.2 [6]). In these cases, the concentrated-changes test was run twice, and two different probability values, one for the most-parsimonious (MP) and one for the least-parsimonious (LP) resolution, are reported. Significant results under the latter resolution must be treated very cautiously, however.

Incomplete or Nonexistent Phylogenies

Since complete phylogenies were unavailable for the procyonids, viverrids, and herpestids, we considered them to be monophyletic (after Wozencraft 1989) and treated each as a large polytomy. Within the three families, genera were assumed to be monophyletic, and species within a genus were grouped together and treated as second-level polytomies. In order to run the concentrated-changes test, we then resolved the relationships within the family by using the most- and least-parsimonious method (see above). Thus each coloration character state produced two different hypotheses of ancestral relationship within a given family, and a behavioral or ecological character was then mapped onto the two trees. The most-parsimonious resolution, however, always yielded one gain, in any of the characters examined within a family, irrespective of the number of species possessing it, and never achieved signifi-

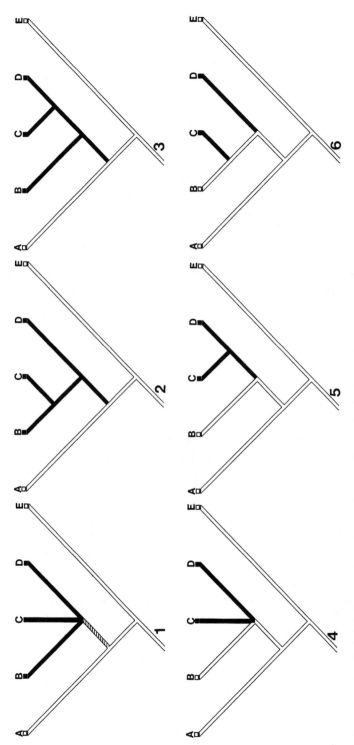

Figure 4.2. Hypothetical phylogeny of taxa A to E, showing an unresolved polytomy between taxa B, C, and D. The evolution of a hypothetical character can be traced on the tree (black branches denote presence; white branches, absence; hatched branches, equivocal assignments). *Above:* 1. Unresolved polytomy where the character traced is present in all taxa within the polytomy. 2. and 3. The only possible solutions to the polytomy; note that the character evolved only once irrespective of the resolution. *Below:* 4. Unresolved polytomy where the character traced is present in only some of the taxa within the polytomy. 5. Most-parsimonious resolution of (4): the character traced evolved only once. 6. Least-parsimonious resolution of (4): the character traced evolved twice.

cance. Therefore, we report only the probability values for the least-parsimonious resolution in these three families. We suggest great caution in the interpretation of statistically significant results for the procyonid, viverrid, and herpestid data until phylogenies for these families become available.

Significance Achieved by Chance

Because we tested several hypotheses against all families of carnivores, the number of statistical tests was large. This raised the possibility that some statistical tests achieved significance by chance. For each hypothesis, therefore, we also performed a Fisher combined-probability test (presented as χ^2 in the results section, following) across those families tested (Sokal and Rolff 1981). When more than one probability value from the concentrated-changes test was available for a carnivore family, we ran two Fisher combined-probability tests: one with the lowest and one with the highest probability value reported in the tables. This procedure reduced the number of statistical tests and corroborated results across the order.

Results of the Comparative Analyses

White Coloration in the Arctic

We tested the hypothesis that carnivores with white coats are found in treeless, snow-covered regions, as typified by arctic tundra habitats. Carnivores can assume a white coloration either by seasonal molting or by remaining white year-round (Table 4.1), and we used both categories.

The arctic fox, which turns white (or, in some variants, blue) in winter, is one of three canid species that live in the arctic tundra regions (Figure 4.3), a marginally significant association (Table 4.6). The polar bear, white year-round, lives in the arctic, but this association did not achieve significance across ursids. The stoat, least weasel, and long-tailed weasel all molt to white, and the first two live in arctic tundra regions—a significant association in the Mustelidae however the polytomy is resolved. Taking the three families together, there was a significant association between being white and living in the arctic (Fisher combined-probability test: $df = 6$, $\chi^2 = 14.742$, $P < 0.05$, or $\chi^2 = 22.53$, $P < 0.001$).

Pale Coloration in Deserts

Next, we tested the hypothesis that carnivores with light fur color tend to be found in desert regions. To be conservative, species categorized as variable

Table 4.6. Associations between white fur color and living in arctic habitat

Family	No. of species[a]	No. of gains and losses[b]	P-value[c]	Test[d]
Canidae	1 / 1	1G 0L/1G	0.060	EC
Ursidae	1 / 1	1G 0L/1G	0.214	EC
Procyonidae	na[e]			
Mustelidae	3 / 2	MP: 1G 0L/1G	0.049	EC
		LP: 2G 0L/2G	0.001	EC
Viverridae	na[e]			
Herpestidae	na[e]			
Hyaenidae	na[e]			
Felidae	na[e]			

[a]Species with white fur as against species with white fur living in arctic habitat.
[b]Gains (G) and losses (L) in white fur, as against gains (G) in arctic; MP refers to most parsimonious, LP to least parsimonious.
[c]Probability value obtained from concentrated-changes test.
[d]Exact count.
[e]Denotes not applicable, since no species in this family has white fur.

light (Table 4.1) and species that are nonuniform in fur color but have pale background coloration, such as the cheetah, were excluded from this analysis.

In only two families of carnivores do species living in desert or semidesert habitats have pale coats. Three canids—the kit fox (*Vulpes macrotis*), pale fox (*V. pallida*), and fennec (*V. zerda*)—are pale and all three live in desert or semidesert (Figure 4.3). Since the phylogenetic status of the pale fox was not specified in the canid phylogeny we used, we added this species to the "*Vulpes* clade" on the canid phylogeny and employed a most- and least-parsimonious resolution to test this hypothesis. If all three canids evolved pale coats independently, then the association with desert habitats is significant (LP test, Table 4.7). On the other hand, if the pale fox and the kit fox are really sister species, then pale coats evolved only twice in the canids, a nonsignificant association (MP test, Table 4.7). The sand cat (*Felis margarita*) has a pale coat and inhabits desert, but this association did not achieve significance across felids. Although some mustelids and herpestids are pale, none of these is found in deserts, and no ursid, procyonid, viverrid, or hyaenid was classified as being pale. Across carnivores, then, pale fur color was marginally associated with desert or semidesert environments ($df = 4$, $\chi^2 = 6.898$, $P > 0.1$, or $\chi^2 = 10.564$, $P < 0.05$).

Spotted Coats in Forest

Spotted coats are thought to provide concealment in forests because they supposedly reproduce the pattern of dappled light shining through a forest canopy (Cott 1940). To test this hypothesis, we looked for an association between spots and living in forest (temperate and tropical forest combined; Table 4.5) and, more specifically, between spots and being "mainly arboreal"

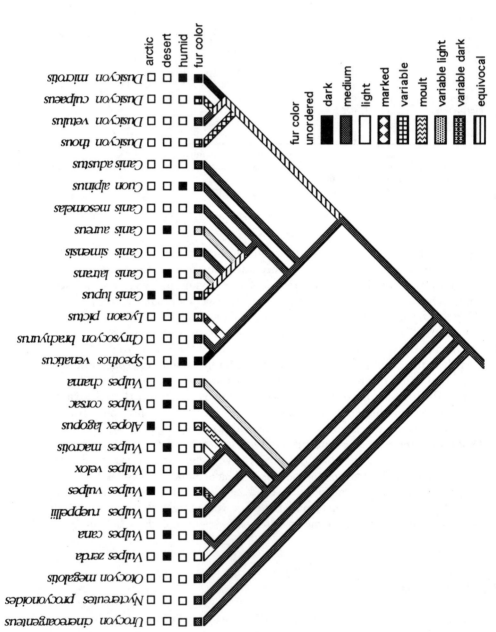

Figure 4.3. Phylogeny of the Canidae (after Geffen et al. 1996) showing the reconstructed evolution of fur color (see key); "equivocal" branches denote ambiguities in character reconstruction. The rows of boxes labeled "arctic," "desert," and "humid" denote the presence (black box) or absence (white box)

Table 4.7. Associations between pale fur color and living in desert habitat

Family	No. of species[a]	No. of gains and losses[b]	P-value[c]	Test[d]
Canidae	3 / 3	MP: 2G 0L/2G	0.095°/0.116†	EC
		LP: 3G 0L/3G	0.021°/0.033†	EC
Ursidae	na[e]			
Procyonidae	na[e]			
Mustelidae	3 / 0	nt[f]		
Viverridae	na[e]			
Herpestidae	3 / 0	nt[f]		
Hyaenidae	na[e]			
Felidae	1 / 1	1G 0L/1G	0.274°/0.242†	EC

[a]Species with pale fur, as against species with pale fur living in desert habitat.

[b]Gains (G) and losses (L) in pale fur, as against gains (G) in desert; MP refers to most parsimonious, LP to least parsimonious.

[c]Probability value obtained from concentrated-changes test: °equivocal branches in desert resolved with ACCTRAN; †equivocal branches in desert resolved with DELTRAN.

[d]Exact count.

[e]Denotes not applicable since no species in this family has pale fur.

[f]Denotes not testable, since no species with pale fur in this family lives in desert.

(arboreal, and terrestrial and arboreal, combined; see Table 4.4). Three families of carnivores include species with spotted coats. Sixteen viverrid species are spotted, and 11 of these are "mainly arboreal" (henceforth arboreal), a significant association (Table 4.8). Fourteen of the 16 live in forest, but this association was not significant. One hyaenid, *Crocuta crocuta,* is spotted, but it is strictly terrestrial and avoids tropical forest.

Nineteen felids are spotted, 16 of the 19 live in forest, and ten of the 16 are arboreal. The statistical significance of these associations could not be tested directly, because spotted coat was reconstructed as being the ancestral color pattern in felids (Figure 4.4). Thus, it was not possible to determine what kind of habitat or type of locomotion was associated with the original gain of a spotted coat in this family. But since spots are the ancestral color pattern of all extant felid species, we tested the converse hypothesis: that the "loss" of spots is associated, first, with not being in forest, and, second, with not being arboreal. Of the ten felids that have lost a spotted coat, four are not found in forest, a significant association, and three are not arboreal, again a significant association (Table 4.8). Four felid species—the North American lynx (*Lynx canadensis*), the African and Asian golden cats (*Felis aurata* and *F. temmincki*), and the pampas cat (*F. colocolo*)—show variation in their coat pattern: some individuals have spotted coats and others are uniformly colored or only sparsely spotted (Figure 4.4; indicated by a black dot). These species were not included in the analysis of losses of a spotted coat.

Across carnivores in general, there was a significant association between arboreality and having a spotted coat ($df = 4$, $\chi^2 = 20.304$, $P < 0.001$, or $\chi^2 = 14.184$, $P < 0.01$), but the association with forest living was unclear ($df = 4$, $\chi^2 = 19.180$, $P < 0.001$, or $\chi^2 = 6.984$, $P > 0.1$).

Table 4.8. Associations between spotted coats and arboreality (above) and living in forest (below)

Association and family	No. of species[a]	No. of gains and losses[b]	P-value[c]	Test[d]
SPOTTED COATS AND ARBOREALITY				
Canidae	na[e]			
Ursidae	na[e]			
Procyonidae	na[e]			
Mustelidae	na[e]			
Viverridae	16 / 11	LP: 8G 0L/7G	0.013	S
Herpestidae	na[e]			
Hyaenidae	1 / 0	nt[f]		
Felidae	19 / 10	MS: 2G 5L/3L	0.003	S
		LS: 0G 10L/3L	0.064	S
SPOTTED COATS AND LIVING IN FOREST				
Canidae	na[e]			
Ursidae	na[e]			
Procyonidae	na[e]			
Mustelidae	na[e]			
Viverridae	16 / 14	LP: 8G 0L/8G	0.342	S
Herpestidae	na[e]			
Hyaenidae	1 / 0	nt[f]		
Felidae	19 / 16	MS: 2G 5L/4L	0.003°/0.0002[†]	S
		LS: 0G 10L/4L	0.089°/0.061[†]	S

[a]Species with spotted coats, as against species with spotted coats that are arboreal (above) or living in forest (below).

[b]Gains (G) and losses (L) in spotted coat, as against gains (G) in arboreal or in forest, or losses (L), in not arboreal or in not living in forest; LP refers to least parsimonious, MS to most-supportive resolution, LS to least-supportive resolution, of equivocal branches in spotted coat.

[c]Probability value obtained from concentrated-changes test: °equivocal branches in forest resolved with ACCTRAN; [†]equivocal branches in forest resolved with DELTRAN.

[d]Simulations (10,000).

[e]Denotes not applicable, since no species in this family has a spotted coat.

[f]Denotes not testable, since no species with a spotted coat in this family is arboreal or lives in forest.

Vertically Striped Coats in Grassland

Cott (1940) wrote that "the vertical tawny-orange and black stripes of the tiger (*P. tigris*) assimilate with the tall parallel grass stems and reeds of the swamps and grassy plains where it lives." We therefore tested the hypothesis that having a vertically striped coat is associated with living in grassland habitat.

Only five carnivores have vertical stripes. Among hyaenids, the striped hyena and aardwolf are striped and both live in grassland, although this association was not significant (Table 4.9). The tiger and the European wild cat (*F. silvestris*) are vertically striped and both live in grassland, but so do another 16 unstriped felids, and the association is thus nonsignificant. Although the Andean cat (*F. jacobita*) is striped, almost nothing is known of its ecology, and

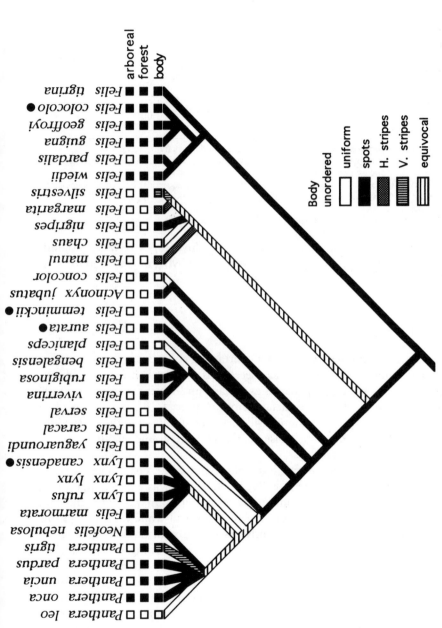

Figure 4.4. Phylogeny of the Felidae (after Collier and O'Brien 1985; Slattery et al. 1994; Janczewski et al. 1995), showing the reconstructed evolution of body color (see key); "equivocal" branches denote ambiguities in character reconstruction. Species marked with black dot show intraspecific variation in body-color patterns; both spotted and uniform individuals exist. The row of boxes immediately below the species' names denotes whether the species are mainly arboreal (black box) or not (white box); data on *Felis rubiginosa* are missing for this character. Note: *Felis iriomotensis* was not included in this phylogeny.

Table 4.9. Associations between vertically striped coats and grassland

Family	No. of species[a]	No. of gains and losses[b]	P-value[c]	Test[d]
Canidae	na[e]			
Ursidae	na[e]			
Procyonidae	na[e]			
Mustelidae	na[e]			
Viverridae	na[e]			
Herpestidae	na[e]			
Hyaenidae	2 / 2	MP: 1G 2L/1G	1.000	EC
		LP: 2G 0L/2G	1.000	EC
Felidae	2 / 2	2G 0L/2G	0.263°/0.208†	EC

[a]Species with striped coats, as against species with striped coats living in grassland.
[b]Gains (G) and losses (L) in vertical stripes, as against gains (G) in grassland; MP refers to most parsimonious, LP to least parsimonious.
[c]Probability value obtained from concentrated-changes test: °equivocal branches in grassland resolved with ACCTRAN; †equivocal branches in grassland resolved with DELTRAN.
[d]Exact count.
[e]Denotes not applicable, since no species in this family has a striped coat.

it was not included in our data set. Across all carnivores, there was no significant association between having vertical stripes and living in grassland ($df = 4$, $\chi^2 = 3.140$, $P > 0.5$, or $\chi^2 = 2.671$, $P > 0.5$).

Warning Coloration

We tested the hypothesis that highly contrasting color patterns are found in species that employ noxious anal-sac secretions. We looked for associations between six highly contrasting markings—black-and-white fur color, contrasting faces (Figure 4.5*A*), white facial stripes (Figure 4.5*B*), black-and-white neck marks (Figure 4.5*C*), dark undersides, and light tail—and the presence of a noxious anal-sac secretion. These markings were found in all carnivore families except the hyaenids.

In the mustelids, seven species have black-and-white coats and all seven produce noxious anal-sac secretions (Figure 4.6); this was a highly significant association under the most-parsimonious resolution, but was not significant under the least-parsimonious resolution of black-and-white fur color (Table 4.10). Light tails were also significantly associated with noxious anal secretions in this family. Nevertheless, dark undersides, contrasting faces, and white facial stripes were not significantly associated with anal secretions.

Viverrids exhibit four contrasting markings: dark undersides, contrasting faces, white facial stripes, and black-and-white neck marks (Table 4.10). Although 12 species have contrasting faces, and nine of these use anal-sac secretions in antipredator defense, the least-parsimonious concentrated-changes test was not significant. Similarly, white facial stripes and black-and-white necks failed to reach significance.

Figure 4.5 *Above left:* ferret badger (*Melogale moschata*), showing its contrasting face. *Above right:* grison (*Galictis vittata*), showing its forehead band. *Below:* oriental civet (*Viverra tangalunga*), showing its black-and-white neck markings. (Drawings by Otis Bilodeau.)

Herpestids show only three highly contrasting markings. Three species—the broad-striped mongoose (*Galidictis fasciata*), the white-tailed mongoose (*Ichneumia albicauda*), and the black-legged mongoose (*Bdeogale nigripes*)—have light tails, and the last two are known to spray predators with noxious secretions; this resulted in a marginally significant association (Table 4.10). But if we include Selous's mongoose (*Paracynictis selousi*), which has a long white tail tip, the association becomes highly significant (given four gains in white tail and no losses, the probability that three gains are associated with anal-sac secretions is 0.01, using an exact count). The black-legged and white-tailed

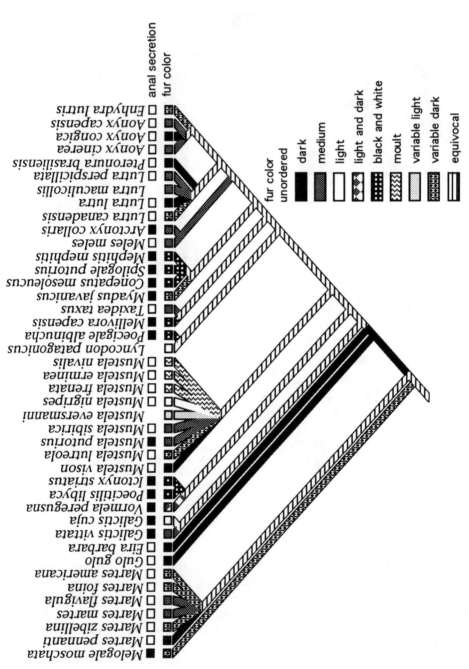

Figure 4.6. Phylogeny of the Mustelidae (after Bryant et al. 1993), showing the reconstructed evolution of fur color (see key); "equivocal" branches denote ambiguities in character reconstruction. The row of boxes labeled "anal secretion" denotes whether the species possess a noxious anal-sac secretion (black box) or not (white box). Data for *Mustela eversmanni*, *Lyncodon patagonicus*, and *Lutra maculicollis* are missing for this character.

Table 4.10. Associations between contrasting markings and use of noxious secretions in antipredator defense in mustelids, viverrids, and herpestids

Family and association	No. of species[a]	No. of gains and losses[b]	P-value[c]	Test[d]
MUSTELIDAE				
Black-and-white fur	7 / 7	MP: 4G 0L/4G	0.006°/0.014†	S
		LP: 1G 6L/1G	0.239°/0.218†	S
Light tail	10 / 9	5G 0L/4G	0.063°/0.029†	S
Dark undersides	13 / 12	MP: 1G 5L/1G	0.237°/0.274†	EC
		LP: 2G 5L/1G	0.487°/0.532†	S
Contrasting face	19 / 14	MP: 2G 2L/2G	0.110°/0.179†	EC/S
		LP: 5G 2L/2G	0.589°/0.752†	S
White facial stripes	18 / 14	MP: 2G 2L/2G	0.110°/0.179†	EC
		LP: 4G 2L/2G	0.451°/0.617†	S
Black-and-white neck	na[e]			
VIVERRIDAE				
Black-and-white fur	na[e]			
Light tail	na[e]			
Dark undersides	1 / 0	nt[f]		
Contrasting face	12 / 9	LP: 7G 0L/4G	0.201	S
White facial stripes	4 / 2	LP: 4G 0L/2G	0.481	EC
Black-and-white neck	5 / 1	LP: 2G 0L/1G	0.592	EC
HERPESTIDAE				
Black-and-white fur	na[e]			
Light tail	3 / 2	LP: 3G 0L/2G	0.073	EC
Dark undersides	2 / 2	LP: 2G 0L/2G	0.021	EC
Contrasting face	na[e]			
White facial stripes	na[e]			
Black-and-white neck	3 / 1	LP: 3G 0L/1G	0.475	EC

[a]Species with contrasting markings as against species with contrasting markings that have noxious anal secretions.

[b]Gains (G) and losses (L) in contrasting marking, as against gains (G) in anal secretions; MP refers to most parsimonious, LP to least parsimonious.

[c]Probability value obtained from concentrated-changes test: °equivocal branches in anal secretions resolved with ACCTRAN; †equivocal branches in anal secretions resolved with DELTRAN.

[d]EC = exact count; S = simulations (10,000).

[e]Denotes not applicable, since no species in this family has that contrasting pattern.

[f]Denotes not testable, since no species with that contrasting pattern in this family has noxious anal secretions.

mongooses have dark undersides, and both use anal-sac secretions in antipredator defense, producing a significant relationship. Black-and-white neck markings in this family were not associated with having anal-sac secretions, however.

The only other species of carnivore with a black-and-white coat is the giant panda, but it is not believed to have anal sacs (Macdonald 1985). Eight carnivores, such as the spectacled bear (*Tremarctos ornatus*) and raccoon (*Procyon lotor*) have contrasting faces, and one, the red panda, has a dark underside, but none of these species uses anal-sac secretions for antipredator defense. Interestingly, two procyonids, the ring-tailed cat (*Bassariscus astustus*) and the common olingo (*Bassaricyon gabbii*), have been reported to discharge noxious anal-sac secretions during encounters with enemies (Poglayen-Neuwall 1990); neither, however, possesses any of the contrasting marks analyzed, although the ring-tailed cat has a black-and-white ringed tail. Across carnivores, light tails ($df = 4$, $\chi^2 = 12.315$, $P < 0.02$, or $\chi^2 = 10.763$, $P < 0.05$) and dark undersides ($df = 4$, $\chi^2 = 10.599$, $P < 0.02$, or $\chi^2 = 8.982$, $P < 0.1$) were significantly associated with noxious anal-sac secretions; but contrasting faces ($df = 4$, $\chi^2 = 7.623$, $P < 0.1$, or $\chi^2 = 3.778$, $P > 0.3$), white facial stripes ($df = 4$, $\chi^2 = 5.878$, $P > 0.2$, or $\chi^2 = 2.430$, $P > 0.5$), and black-and-white neck marks ($df = 4$, $\chi^2 = 2.537$, $P > 0.5$) were not.

White Throats and Intraspecific Communication

Fox (1971) noticed that the white throats of canids may serve to direct bites toward this area during ritualized aggression between conspecifics. Since intraspecific conflicts are likely to be more frequent in group-living species, we tested the association between having a white throat and being social (group and variable group combined; see Table 4.4).

Of the 12 mustelids with white throats, six live in groups, which was not significant (Table 4.11). In viverrids, the African linsang (*Poiana richardsonii*) and the Congo water civet (*Osbornictis piscivora*) have white throats, and although only the former is social, the association was marginally significant. Among canids, 14 species have white throats, of which six are social. The

Table 4.11. Associations between white throats and being social

Family	No. of species[a]	No. of gains and losses[b]	P-value[c]	Test[d]
Canidae	14 / 6	MS: 3G 6L/2G	0.317	S
		LS: 0G 9L/5L	0.597	S
Ursidae	na[e]			
Procyonidae	na[e]			
Mustelidae	12 / 6	MS: 7G 0L/4G*	0.130	S
		LS: 4G 5L/2G*	0.333	S
Viverridae	2 / 1	LP: 2G 0L/1G	0.054	EC
Herpestidae	1 / 1	LP: 1G 0L/1G	0.275	EC
Hyaenidae	na[e]			
Felidae	12 / 1	MP: 7G 0L/1G	0.262	S
		LP: 8G 1L/1G	0.325	S

[a]Species with white throats, as against species with white throats that are social.

[b]Gains (G) and losses (L) in white throat, as against gains (G) in social, or losses (L), in not being social; MP refers to most parsimonious, LP to least parsimonious, MS to most-supportive resolution, LS to least-supportive resolution, of equivocal branches in white throat; *only most-parsimonious resolution tested.

[c]Probability value obtained from concentrated-changes test.

[d]EC = exact count; S = simulations (10,000).

[e]Denotes not applicable, since no species in this family has a white throat.

most-supportive (MS) resolution of equivocal branches resulted in three independent gains of white throats in the canids, which was not significantly associated with group living. Under the least-supportive (LS) resolution, white throats were reconstructed to be ancestral in the canids. Thus, we tested the association between losing a white throat and living in a group, but this association, too, was not significant (Table 4.11).

One herpestid, the yellow mongoose (*Cynictis penicillata*), has a white throat and is social, but so are another nine species. Twelve felids have white throats, but only the cheetah is found in groups. No ursids, procyonids, or hyaenids have white throats. Across all of the carnivore families, there was no association between having a white throat and being social ($df = 10$, $\chi^2 = 117.476$, $P < 0.1$, or $\chi^2 = 13.899$, $P > 0.1$).

Tail Tips and Communication

Contrasting patches of color on the tip of the tail might be used to promote social cohesiveness in group-living species, since they may allow conspecifics to follow each other through dense vegetation. Therefore, we looked for an association between having a contrasting tail tip (black, black or light, and white combined, Table 4.2) and being social (group and variable group combined).

Although ten herpestids have contrasting tail tips, only six of these are social, which was not significant (Table 4.12). Twenty-four species of canids have contrasting tail tips, but only 11 of these are social; nine procyonids have

Table 4.12. Associations between contrasting tail tips and sociality

Family	No. of species[a]	No. of gains and losses[b]	P-value[c]	Test[d]
Canidae	24 / 11	0G 3L/2L	0.485	EC
Ursidae	na[e]			
Procyonidae	9 / 2	0G 1L/1L	0.833	EC
Mustelidae	5 / 2	MS: 4G 0L/2G*	0.392	S
		LS: 3G 1L/1G*	0.666	S
Viverridae	20 / 0	nt[f]		
Herpestidae	10 / 6	LP: 10G 0L/5G	0.551	EC
Hyaenidae	2 / 1	2G 0L/1G	1.000	EC
Felidae	20 / 2	MS: 2G 2L/2L	0.977	S
		LS: 0G 4L/4L	0.852	S

[a]Species with contrasting tail tips, as against species with contrasting tail tips that are social.
[b]Gains (G) and losses (L) in tail tips, as against gains (G) in social, or losses (L), in not being social; LP refers to least-parsimonious, MS to most-supportive, LS to least-supportive resolution of equivocal branches in contrasting tail tipes; *only most-parsimonious resolution tested.
[c]Probability value obtained from concentrated-changes test.
[d]EC = exact count; S = simulations (10,000).
[e]Denotes not applicable, since no species in this family has a contrasting tail tip.
[f]Denotes not testable, since no species with contrasting tail tips in this family is social.

tail tips and two of them are social; five mustelids have tail tips and two are social; two hyaenids have tail tips and one is social; 22 felids have contrasting tail tips but only two are social. There was thus no significant association between contrasting tail tips and group living in any of these families (Table 4.12) or across carnivores in general ($df = 12$, $\chi^2 = 4.924$, $P > 0.95$, or $\chi^2 = 4.139$, $P > 0.95$). Twenty viverrids have contrasting tail tips but none is social, and no ursid has a contrasting tail tip.

Contrasting tail tips may also function as visual signals for long-distance communication, either among conspecifics or between different species. Since "black" has greater contrast in light environments and "white" has greater contrast in dark environments (Mottram 1915), we hypothesized that these patches would be found in species living in habitats in which these colors might show maximum contrast. Thus we tested the association between having a black tail tip and living in grassland habitat, and that between having a white tail tip and living in forest habitat (temperate and tropical combined). Species whose tail tips could be either black or white were not included in these tests.

In two families, black tail tips were associated with living in grassland. Five mustelids have black tail tips and all five live in grassland; and of five herpestids with black tail tips, four live in grassland. In both of these families there was a significant association only in the least-parsimonious tests (Table 4.13).

Seventeen canids have black tail tips and 12 of these live in grassland; and 17 felids have black tail tips, 11 of which are grassland species. But black tail tips were reconstructed as being ancestral in both of these families. Therefore, we tested the association between the loss of a black tail tip and not living in grassland, but the results were not significant in either canids or felids (Table 4.13).

There was no significant association between black tail tips and grassland in viverrids or in hyaenids, and across carnivores, possession of a black tail tip and living in grassland were not significantly associated ($df = 12$, $\chi^2 = 14.612$, $P > 0.2$, or $\chi^2 = 19.473$, $P > 0.1$).

There was no relationship between white tail tips and living in forest in any family (Table 4.13). In canids, mustelids, viverrids, herpestids, and felids, as well as across carnivores as a whole ($df = 10$, $\chi^2 = 2.966$, $P > 0.9$, or $\chi^2 = 4.796$, $P > 0.8$), associations were all nonsignificant.

Ears and Communication

We also tested the hypothesis that markings on the backs of ears (all character states, except uniform, combined, Table 4.2) are found in social species (group and variable group combined), since, like tail tips, they may allow conspecifics to follow each other through tall vegetation (Schaller 1967).

Twenty-two mustelids have contrasting markings on the backs of their ears and nine of them are social. This association was significant only in the most-supportive resolution (Table 4.14).

Table 4.13. Associations between black tail tips and living in grassland, and between white tail tips and living in forest

Association and family	No. of species[a]	No. of gains and losses[b]	P-value[c]	Test[d]
BLACK TAIL TIPS AND GRASSLAND HABITAT				
Canidae	17 / 12	MS: 1G 5L/2L	0.218	S
		LS: 0G 6L/2L	0.118	S
Ursidae	na[e]			
Procyonidae	na[e]			
Mustelidae	5 / 5	MP: 2G 0L/2G	0.251°/0.277†	EC
		LP: 5G 0L/5G	0.020°/0.027†	S
Viverridae	11 / 6	LP: 7G 0L/2G	0.919	S
Herpestidae	5 / 4	LP: 4G 0L/4G	0.050	EC
Hyaenidae	2 / 2	MP: 1G 0L/1G	1.000	EC
		LP: 2G 0L/2G	1.000	EC
Felidae	17 / 11	MS: 0G 6L/4L	0.242°/0.302†	S
		LS: 3G 2L/2L	0.461°/0.545†	S
WHITE TAIL TIPS AND FOREST HABITAT				
Canidae	5 / 3	MS: 3G 0L/2G	0.440°/0.498†	EC
		LS: 2G 1L/2G	0.178°/0.270†	EC
Ursidae	na[e]			
Procyonidae	na[e]			
Mustelidae	2 / 1	MS: 2G 0L/1G	0.860	EC
		LS: 1G 1L/1G	0.920	EC
Viverridae	1 / 1	LP: 1G 0L/1G	0.923	EC
Herpestidae	2 / 1	LP: 2G 0L/1G	0.709	EC
Hyaenidae	na[e]			
Felidae	2 / 1	MS: 1G 2L/1G	0.880°/1.000†	EC
		LS: 2G 0L/1G	0.962°/0.986†	EC

[a]Species with black (above) or white (below) tail tips as against species with black tail tips that live in grassland (above) or species with white tail tips that live in forest (below).

[b]Gains (G) and losses (L) in tail tips, as against gains (G) in grassland (above) or in forest (below), or losses (L), not in grassland (above) or not in forest (below); MP refers to most parsimonious, LP to least parsimonious, MS to most-supportive resolution, LS to least-supportive resolution, of equivocal branches in tail tips.

[c]Probability value obtained from concentrated-changes test: °equivocal branches in grassland or forest resolved with ACCTRAN; †equivocal branches in grassland or forest resolved with DELTRAN.

[d]EC = exact count; S = simulations (10,000).

[e]Denotes not applicable, since no species in this family has a contrasting tail tip.

There was no significant association between contrasting ear markings and sociality in either canids, herpestids, or hyaenids (Table 4.14), and few procyonids, viverrids, or felids are social. Across all families, this association was not significant ($df = 12$, $\chi^2 = 11.668$, $P > 0.3$, or $\chi^2 = 8.377$, $P > 0.7$).

We then tested the hypothesis that black marks on the backs of ears (black ear tips and darker back combined) are associated with grassland and that white marks (white spot, white tip, and white rims combined; see Figures 4.6 and 4.7) are associated with forest habitats (temperate and tropical combined), in which these color marks would be most contrasting.

Black ear marks and living in grassland were significantly associated in felids

Table 4.14. Associations between contrasting ears and sociality

Family	No. of species[a]	No. of gains and losses[b]	P-value[c]	Test[d]
Canidae	10 / 4	MS: 5G 2L/3G	0.574	S
		LS: 2G 5L/3L	0.519	S
Ursidae	1 / 0	nt[e]		
Procyonidae	6 / 1	LP: 4G 0L/1G	0.733	EC
Mustelidae	22 / /6	MS: 4G 3L/3G*	0.034	S
		LS: 0G 7L/6L*	0.195	S
Viverridae	15 / 0	nt[e]		
Herpestidae	1 / 1	1G 0L/1G	0.275	EC
Hyaenidae	1 / 1	1G 0L/1G	0.833	EC
Felidae	28 / 2	0G 3L/3L	0.893	EC

[a]Species with contrasting ears, as against species with contrasting ears that are social.

[b]Gains (G) and losses (L) in contrasting ears, as against gains (G) in social, or losses (L), in not being social; LP refers to least parsimonious, MS to most supportive resolution, LS to least-supportive resolution of equivocal branches in contrasting ears; *only most-parsimonious resolution tested.

[c]Probability value obtained from concentrated-changes test.

[d]EC = exact count; S = simulations (10,000).

[e]Not testable, since no species with contrasting ears in this family is social.

Figure 4.7. Bobcat (*Lynx rufus*), showing its white ear spot. (Drawing by Otis Bilodeau.)

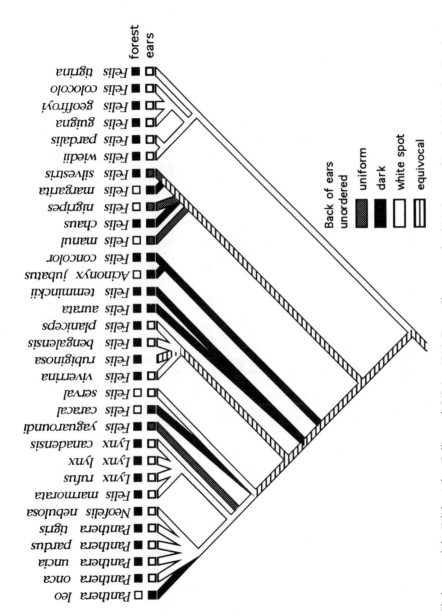

Figure 4.8. Phylogeny of the Felidae (after Collier and O'Brien 1985; Pecon Slattery et al. 1994; Janczewski et al. 1995), showing the reconstructed evolution of markings on the backs of ears (see key); "equivocal" branches denote ambiguities in character reconstruction. Data for *Felis rubiginosa* are missing, and the ancestral character state has been reconstructed as being equivocal. The row of boxes labeled "forest" denotes presence (black box) or absence (white box) of those species in forest habitat. Note: *Felis iriomotensis* was not included in this phylogeny.

Table 4.15. Associations between black ear markings and living in grassland, and between white ear markings and living in forest

Association and family	No. of species[a]	No. of gains and losses[b]	P-value[c]	Test[d]
BLACK EAR MARKS				
Canidae	9 / 7	MS: 6G 1L/6G	0.500	S
		LS: 1G 5L/1G	0.940	S
Ursidae	1 / 0	nt[f]		
Procyonidae	na[e]			
Mustelidae	7 / 5	MS: 4G 0L/3G	0.308°/0.335†	S
		LS: 2G 1L/2G	0.306°/0.342†	S
Viverridae	7 / 1	LP: 6G 0L/1G	0.425	EC
Herpestidae	1 / 1	LP: 1G 0L/1G	0.987	EC
Hyaenidae	1 / 1	1G 0L/1G	1.000	EC
Felidae	8 / 5	MS: 3G 3L/3G	0.033°/0.002†	EC
		LS: 5G 1L/4G	0.169°/0.112†	S
WHITE EAR MARKS				
Canidae	na[e]			
Ursidae	na[e]			
Procyonidae	6 / 6	LP: 4G 0L/4G	1.000	EC
Mustelidae	9 / 6	MP: 6G 0L/5G	0.137	S
		LP: 7G 0L/6G	0.073	S
Viverridae	5 / 5	LP: 3G 0L/3G	0.736	EC
Herpestidae	na[e]			
Hyaenidae	na[e]			
Felidae	20 / 18	MS: 1G 6L/5L	0.0003°/<0.0001†	S
		LS: 2G 3L/3L	0.006° / 0.0004†	EC

[a]Species with black (above) or white (below) ear marks as against species with black ear marks that live in grassland (above) or species with white ear marks that live in forest (below).

[b]Gains (G) and losses (L) in black or white ear markings, as against gains (G) in grassland (above) or forest (below), or losses (L), not in grassland (above) or not in forest (below); MP refers to most parsimonious, LP to least parsimonious, MS to most-supportive resolution, LS to least-supportive resolution, of equivocal branches in ear markings.

[c]Probability value obtained from concentrated-changes test: °equivocal branches in grassland or forest resolved with ACCTRAN; †equivocal branches in grassland or forest resolved with DELTRAN.

[d]EC = exact count; S = simulations (10,000).

[e]Denotes not applicable, since no species in this family has that colored ear mark.

[f]Denotes not testable, since no species with that colored ear mark in this family is social.

in only the most-supportive resolution (Figure 4.8, Table 4.15). Despite the widespread appearance of black ear markings in carnivores, there was no significant association with grassland habitat in any of the other families (Table 4.15), or across the carnivores ($df = 12$, $\chi^2 = 17.740$, $P > 0.1$, or $\chi^2 = 7.786$, $P > 0.7$). Parallel relationships between having white ear marks and living in forest were stronger, however: 18 of the 20 felids with white ears live in forest (Figure 4.8). The Iriomote cat (*Felis iriomotensis*) was omitted from analysis because its phylogenetic relationship with other felids is unknown. Since white ear spots were reconstructed as being ancestral in the felids (Figure 4.8), we tested the association between losing this color pattern and not living in forest, and found it to be highly significant in all of the resolutions employed (Table 4.15).

There was a marginally significant association between white ear spots and forest in the mustelids, considering the least-parsimonious test. Six procyonids and five viverrids have white ear marks, and all live in forest, although neither result was significant. No canids, ursids, herpestids, or hyaenids have white markings on their ears. Across carnivores, there was a strong association between white markings on the backs of ears and living in forest ($df = 8$, $\chi^2 = 23.009$, $P < 0.01$, or $\chi^2 = 16.078$, $P < 0.05$).

Dark Eye Contours and Diurnality

Although most carnivores have black eyelids that can be very prominent (e.g., on the polar bear), in only a few of them is the fur surrounding the eyes dark. We tested the hypothesis that having dark fur around the eyes (Figure 4.9) is associated with being "mainly diurnal" (diurnal and anytime combined, see Table 4.4), since it has been proposed that dark coloration in vertebrates reduces glare from the sun (Ficken et al. 1971).

In canids, four species have dark eye contours and two of these are diurnal, which was not significant (Table 4.16). Nor were dark eye contours significantly associated with being diurnal in the ursids, mustelids, or herpestids. Although two procyonids, the raccoon (*Procyon lotor*) and the crab-eating raccoon (*P. cancrivorus*), and 12 viverrids have dark fur around the eyes, none is diurnal. Across carnivores as a whole, there was no significant association between these variables ($df = 10$, $\chi^2 = 7.707$, $P > 0.5$, or $\chi^2 = 4.730$, $P > 0.8$).

Interestingly, some canids and some mustelids have a black patch below the eyes near or on the muzzle (Table 4.2). In canids this marking was significantly associated with crepuscular habits (given two gains and two losses, the probability that the two gains are associated with being crepuscular is 0.002, using

Figure 4.9. Suricate (*Suricata suricatta*), showing its dark eye contour. (Drawing by Otis Bilodeau.)

Table 4.16. Associations between dark eye contours and being diurnal

Family	No. of species[a]	No. of gains and losses[b]	P-value[c]	Test[d]
Canidae	4 / 2	MS: 3G 1L/2G	0.224	EC
		LS: 2G 2L/1G	0.600	EC
Ursidae	5 / 3	MP: 1G 3L/1G	0.265	EC
		LP: 5G 0L/3G	0.442	EC
Procyonidae	2 / 0	nt[f]		
Mustelidae	20 / 3	MS: 2G 3L/1G*	0.477	EC
		LS: 1G 4L/1G*	0.473	EC
Viverridae	12 / 0	nt[f]		
Herpestidae	6 / 3	LP: 6G 0L/3G	0.899	EC
Hyaenidae	3 / 1	0G 1L/1L	0.833	EC
Felidae	na[e]			

[a]Species with dark eye contours, as against species with dark eye contours that are diurnal.

[b]Gains (G) and losses (L) in dark eye contours, as against gains (G) in diurnal, or losses (L), in not being diurnal; MP refers to most parsimonious, LP to least parsimonious, MS to most-supportive resolution, LS to least-supportive resolution, of equivocal branches in eye contour; *only most-parsimonious resolution tested.

[c]Probability value obtained from concentrated-changes test.

[d]EC = exact count; S = simulations (10,000).

[e]Denotes not applicable, since no species in this family has dark eye contours.

[f]Denotes not testable, since no species with dark eye contours in this family is diurnal.

an exact count). In mustelids, it was associated with being diurnal (given three gains, the probability that three gains are diurnal is 0.06, using an exact count). Thus, in two families, there was some support for dark fur below, rather than around, the eyes and having diurnal or crepuscular habits.

Dark Fur and Humid Habitat

Gloger's rule states that species living in warm, humid conditions have darker coats (Gloger 1833). We therefore tested the association between dark fur and living in tropical forest, a humid environment. Melanistic forms, generally rare, were not included in the analysis.

Two canids, the small-eared zorro (*Dusicyon microtis*) and bush dog (*Speothos venaticus*), are dark, and both live in tropical forests, an association that proved significant (Table 4.17; Figure 4.3). Two other canids, the red fox and the gray wolf, have dark variants but do not live in tropical forest. Four ursids are dark and show little or no variation in fur color, and three of these—the spectacled bear, the Malayan sun bear, and the sloth bear (*Melursus ursinus*)— live in tropical forests, a marginally significant association. Although most North American black bears (*Ursus americanus*) and brown bears (*U. arctos*) are dark, these species show extensive geographic variation in fur color (Rounds 1987), and were therefore excluded from analysis. The cusimanse (*Crossarchus obscurus*) and the Liberian mongoose (*Liberiictis kuhni*) are dark

Table 4.17. Associations between dark fur and living in tropical forest

Family	No. of species[a]	No. of gains and losses[b]	P-value[c]	Test[d]
Canidae	2 / 3	2G 0L/2G	0.003	EC
Ursidae	4 / 3	MP: 1G 1L/1G	0.083	EC
		LP: 4G 0L/3G	0.067	EC
Procyonidae	2 / 1	LP: 2G 0L/1G	1.000	EC
Mustelidae	8 / 0	nt[e]		
Viverridae	2 / 1	LP: 2G 0L/1G	0.494	EC
Herpestidae	2 / 2	LP: 2G 0L/2G	0.060	EC
Hyaenidae	1 / 0	nt[e]		
Felidae	2 / 2	2G 0L/2G	0.345°/0.522[†]	EC

[a]Species with dark fur, as against species with dark fur that live in tropical forest.

[b]Gains (G) and losses (L) in dark fur, as against gains (G) in tropical forest; MP refers to most parsimonious, LP to least parsimonious.

[c]Probability value obtained from concentrated-changes test; °equivocal branches in tropical forest resolved with ACCTRAN; [†]equivocal branches in tropical forest resolved with DELTRAN.

[d]Exact count.

[e]Denotes not testable, since no species with dark fur in this family lives in tropical forest.

and both live in tropical forest, a marginally significant association. Although none of the other families showed a significant association between these variables, there was a significant association between dark fur and tropical forests across carnivores as a whole ($df = 12$, $\chi^2 = 24.933$, $P < 0.02$, or $\chi^2 = 26.188$, $P < 0.02$).

Advantages and Disadvantages of Comparative Analyses

Comparative analyses identify behavioral, ecological, and morphological correlates of interspecific differences that may reveal adaptations. The comparative method, and phylogenetic analyses in particular, may, however, miss adaptive patterns, for the following reasons.

Phylogenetic Analyses

First, phylogenetic analyses are employed to remove associations between traits resulting from shared ancestry. Nevertheless, the coincidence of a trait and an ecological variable in closely related species may actually be strong evidence for the adaptive significance of that trait, if it has been conserved in a particular ecological context over evolutionary time (Harvey and Pagel 1991).

Second, phylogenies may be incomplete or incorrect, which would lower the resolution of the method. To minimize the effect of incomplete phylogenies, analyses can be carried out on those parts of the tree for which phylogenies are completely known. One of the reasons we used the family level for our analyses was to separate families with good phylogenies from those for which only classical taxonomies were available.

Third, phylogenetic analyses are conservative, in the sense that they show only that a particular trait, such as vertical striping, has evolved under particular ecological circumstances, such as living in grassland, if the majority of grassland species are striped. In this example, if several nonstriped species live in grassland habitats as well, a significant association is unlikely to be revealed. These nonstriped species could nonetheless have evolved other body markings to live in that habitat, which would seem to diminish the importance of vertical striping in grassland.

Level of Resolution

Examining a wide diversity of species necessitates formulating broad ecological and behavioral categories that may mask subtle ways in which traits relate to particular environments. Although significant effects deriving from the use of broad categories are therefore likely to represent real biological phenomena, nonsignificant results may be difficult to interpret.

Second, certain analyses of correlated changes with categorical variables, such as the concentrated-changes test (Maddison 1990), demand that the independent variable, here being a behavioral-ecological variable, be coded as a binary character. This requirement of the procedure may alter the overall reconstructed evolutionary history of the independent character.

Statistical Issues

Statistical analyses that examine associations between two variables across species, such as chi-square and Fisher exact-probability tests, are very sensitive to the nonindependence of cell entries (Siegel 1956). Since closely related species may share common traits as a result of shared ancestry, and therefore may not be independent, statistics that take phylogenies into account are better tools (than those that do not) for examining associations between traits (Gittleman and Luh 1992). When phylogenies include only a few taxa, however, low sample size may make it difficult to find significant associations, a problem we encountered in the ursids and hyaenids.

Finally, the dependent character may be reconstructed as being ancestral, thus appearing at the root of the node of the clade selected. Since there are no gains in the dependent character in such cases, its association with a second character cannot be tested. In these cases, we tested the converse hypothesis: that a *loss* in the coloration character of interest was *not* associated with the same behavioral-ecological variable under which we expected the coloration character to have evolved.

Despite all these difficulties, a number of clear associations emerged from the analyses.

Implications of the Comparative Analyses

White Coloration in the Arctic

All white carnivores live in snow-covered habitats. Results of the concentrated-changes tests were significant in two of the three families tested, and across carnivores in general. Although one of the three mustelids that molt to white in the winter, the long-tailed weasel, does not live in arctic tundra regions, it changes its coat to white only in the northern part of its range (in southern Canada), which is covered by snow in winter (King 1989).

Certain carnivore subspecies, such as the tundra wolf (*Canis lupus albus*), are also white, and also live in the arctic (Macdonald 1984), and others, such as the black bear and the brown or grizzly bear, have cream or nearly white variants in arctic tundra regions (Burt and Grossenheider 1976). When these two ursid color variants are included in the analysis with the polar bear, the result is highly significant. Finally, the snow leopard (*Panthera uncia*) is white with spots, and although not inhabiting arctic regions, it lives in valleys of the Himalayas that are often snow-covered.

Despite the strong association, it is not entirely clear whether white pelage serves as background matching to facilitate prey capture or avoid predation, or alternatively (or even additionally) serves in a thermoregulatory capacity. In support of the first possibility, Hamilton (1973) pointed out that active predators and species suffering heavy predation tend to be white, whereas arctic species that do not depend so much on camouflage for survival, such as the wolverine (*Gulo gulo*) are not white.

Pale Coloration in Deserts

Pale coloration in carnivores is associated with desert living only in the canids, and only under the least-parsimonious hypothesis. Small sample size in felids may have been what resulted in nonsignificance, owing to lowered statistical power. If we had included felids with variable light coats—the caracal (*Felis caracal*), jungle cat (*Felis chaus*), and Pallas's cat (*Felis manul*)—there would have been a highly significant association with desert living in felids, despite the fact that the jungle cat is not found in desert habitats. Moreover, intraspecific variation may result in desert individuals being lighter than those living in other habitats, as has been observed in the Egyptian mongoose (*Herpestes ichneumon*), for example (but see Taylor et al. 1990). Both facts suggest that more sensitive measures of intraspecific variation in fur color and habitat may be necessary to test this association properly. Three species of mustelids and two species of herpestids are pale (Appendix 4.1) but do not inhabit deserts. All three pale mustelids—black-footed ferret (*Mustela nigripes*), little grison (*Galictis cuja*), and Patagonian weasel (*Lyncodon patagonicus*)—either are rare or have habits that are poorly known, and thus it was not possible to assess whether they are naturally found in habitats with light-colored substrates, other than deserts. In the herpestids, the suricate (*Suricata suricatta*) and the narrow-striped mongoose (*Mungotictis decemlineata*) live in dry, sandy savannas, which we classified as grassland.

It is not clear whether pale pelage in deserts is an adaptation for concealment or for thermoregulation, or, rather, is a nonadaptive consequence of bleaching from the sun (Hamilton 1973). Interestingly, all the pale carnivores that live in desert are nocturnal (Appendix 4.2). Since pale fur color is unlikely to be important for thermoregulation at night, and since sun-bleaching is unlikely to be an issue for a nocturnal animal, pale coloration is more likely to be an adaptation for concealment. Indeed, experimental studies of nocturnal predation on mice (*Peromyscus* sp.) by owls showed that there is selection against conspicuously colored prey. Light-brown mice on dark soil and dark-brown mice on light soil were captured significantly more often than were their counterparts, which suggests that protective concealment is important at night (Dice 1947; Kaufman 1974).

Spotted Coats in Forest

Among carnivores, spotted coats are found in the viverrids and felids and in one hyaenid. Eighty-three percent of all spotted carnivores live in forest habitats, and 58% of these are arboreal. These trends alone seem to lend support to Cott's (1940) original hypothesis that spots are an adaptation for remaining concealed in dense vegetation. This hypothesis, however, proved difficult to test using the comparative method, for three reasons: (1) most felid and viverrid species are forest-dwelling, arboreal creatures to begin with; (2) no viverrid phylogeny was available; (3) spots are ancestral in felids. These difficulties notwithstanding, there was a significant association in felids between the loss of a spotted coat and not living in forest. Out of ten nonspotted felids, four do not live in forest at all, and another four, including the puma (*F. concolor*), are found in both forest and grassland habitats. Thus, only two nonspotted felids, the jaguaroundi (*F. yaguaroundi*) and the flat-headed cat (*F. planiceps*), are described as being exclusively forest dwellers. Eisenberg (1989), however, writes that the jaguaroundi "frequently hunts on the boundaries of galley forests and thus may be found some distance from forest cover in savanna habitats." The flat-headed cat is rare and little is known about its habits.

At the same time, six of the 36 spotted carnivores are not found in forest, each perhaps for species-specific reasons unrelated to coloration effects. For example, the spotted hyena is believed to avoid woodland habitat because of the prevalence there of tsetse flies (*Glossina* sp.) (Kruuk 1972). Alternatively, since dappled shade can exist in almost any habitat (Mottram 1915; Endler 1993), spotted coats may provide concealment elsewhere. For example, cheetahs are difficult to see in grassland and scrub environments (Caro 1994). Indeed, many carnivores are crepuscular or nocturnal, hunting at low light intensities when image resolution is poor because of the predominance of rod vision (Ewer 1973; Endler 1978). At this time of day, it is difficult to discern spots (or stripes), since they take on a dull wash that blends in with the background (Mottram 1915).

Vertically Striped Coats in Grassland

There was no significant association between striped coats and grassland habitats in either hyaenids or felids, even though four of the five striped carnivores (all five are in these two families) are found in grassland. This suggests that striping is not specifically associated with remaining concealed in grass or swampy habitats; indeed, the tiger and European wild cat live also in forest habitats. Perhaps vertical stripes are a form of disruptive coloration, rather than background matching, as was suggested for zebras (Mottram 1915; Cott 1940). The blending of the stripes near the margins of the animal's body helps to break up its outline, particularly at dusk, when this effect is greatest (Cott 1940). In this way, vertical stripes would provide concealment in a variety of habitats, and their benefits would therefore not be confined to grassland. Indeed, many grassland-living carnivores have grizzled, or agouti, hair, which might provide a better form of background matching than do stripes.

Warning Coloration

Our analyses confirm that contrasting markings in mustelids and herpestids are associated with the use of noxious anal secretions, and therefore that these markings may advertise the fact that these species can defend themselves against predators. In mustelids, there was a one-to-one correspondence between possessing a black-and-white coat and employing this form of antipredator defense. Interestingly, although no herpestid has a black-and-white coat, the possession of a dark underside and a light tail in this family was significantly associated with having a noxious anal secretion, suggesting that different conspicuous markings evolved independently for the same function in different carnivore families.

The data for viverrids are more equivocal. Although most species with contrasting faces or white facial stripes have been reported to use noxious anal-sac secretions against predators, we found no significant association between these character states. For example, the African civet (*Viverra civetta*) has three contrasting markings (a dark underside, contrasting face, and black-and-white neck), and although it is known to have a potent anal scent, it has not been reported to use it against predators. Since viverrids are secretive and few studies have been conducted on this group, poor data rather than an absence of association may be contributing to the lack of significant effects in this family. If not, a different adaptive explanation for black-and-white markings in viverrids needs to be sought.

The fact that in mustelids, viverrids, and herpestids a contrasting face, white facial stripes, and black-and-white neck marks were not associated with noxious anal-sac secretions suggests that there may be other explanations for these markings. If they function as warning colors they might advertise other types of defenses. For example, the contrastingly marked Eurasian and American badgers (*Meles meles* and *Taxidea taxus*) are highly aggressive toward predators, although they do not possess particularly noxious anal secretions. Alternatively, these marks could be used for intraspecific communication (J. Endler, pers. comm.) or as disruptive coloration to mask conspicuous body parts, such as the eye (Cott 1940).

White Throats and Intraspecific Communication

There was no evidence that having a white throat is associated with group living in carnivores. The combined-probability test across carnivore families was not significant. In viverrids, marginal significance was driven by just one species, the African linsang; and in felids, most of the many white-throated species were solitary. Using social species to test the hypothesis that white throats are used to redirect aggressive bites to this area during intraspecific conflict may be inappropriate, since all carnivore species exhibit some level of aggressive interaction. Clearly, this hypothesis needs to be retested when behavioral data on ritualized fighting become available for a large number of carnivore species.

Tail Tips and Communication

Contrasting tail tips were not associated with sociality in any of the carnivore families. If contrasting tail tips serve principally as devices by which offspring follow their mothers (as suggested for felids: Leyhausen 1979), then they would be found in many species, which would help explain our results. Irrespective of the intended receiver, black tail tips in mustelids and herpestids were associated with grassland, suggesting that they may act as a signal in these habitats. Although 70% of canids and 65% of felids with black tail tips live in grassland habitats, the association could not be tested directly because in both families black tail tips were reconstructed as being ancestral. The converse hypothesis, that the loss of a black tail tip is associated with not living in grassland, was tested instead. Only two out of eight canids and four out of six felids exhibiting a loss of a black tail tip are not found in grassland. These associations were not significant. Failure to find parallel relationships between white tail tips and forest habitats could be due to their limited occurrence in carnivores: only 12 carnivore species have white tail tips, whereas 57 have black ones.

Testing the association between contrasting tail tips and different habitats of different light levels does not distinguish between whether black tail tips are used in intraspecific communication, whether they are used to deflect predatory attack from the body, as demonstrated for weasels (Powell 1982), or whether they act as lures to hold prey's attention, a hypothesis equally applicable to most carnivore families. Alternatively, tail tips might be used to convey to predators that they have been detected and that further advance is therefore unlikely to be successful (Hasson 1991). This function might be more prevalent in smaller species subject to a greater range of predators.

Ears and Communication

There was overwhelming evidence in felids, and some evidence in mustelids, that contrasting white ear marks are associated with living in forest. There was, however, little support for the hypothesis that black ear marks are associated with grassland habitats. Experimental studies on the conspicuousness of patterns in nature conducted by Mottram (1915) showed that both the type of background and the amount of lighting affected the visibility of both black and white objects. He showed that a white object on a black background is always much more visible than is a black object on a white or light ground, and that this difference is greater at higher illuminations (Mottram 1915). The absence of an association of either black tail tips or black ear marks with grassland habitats may result from grassland's being a poor measure of lighting conditions.

The function of contrasting ear markings is unclear. Like tail tips, ear markings could allow conspecifics to follow each other through dense vegetation (Schaller 1967; Leyhausen 1979); they could reinforce ear displays during aggressive encounters (see Fox 1971); or they could mimic fake eyes on the back of the head.

Dark Eye Contours and Diurnality

There was no evidence that having dark eye contours is associated with being diurnal. In canids, however, dark patches below the eyes were significantly associated with crepuscular species, and the same markings were marginally associated with diurnality in mustelids. Dark patches below the eyes may serve to reduce glare from the sun; American footballers claim that they mark the area below their eyes to reduce glare (pers. comm.). Dark patches could also increase facial contrast and thus aid in intraspecific signaling to conspecifics that rely on vision. This hypothesis needs to be confirmed by behavioral studies of species with dark eye masks.

Dark Fur and Humid Habitat

Dark canids, ursids, and herpestids are found in tropical forest, a humid habitat. The only two nonmelanistic dark felids, the flat-headed cat and the Iriomote cat, are also found in this habitat, but the association was not significant, since many other felids live in tropical forest as well. In the other families there was no clear pattern of association of dark species and tropical forest, perhaps because humid habitats are not confined solely to tropical forests. Nevertheless, the association was significant across all carnivores. Originally, Gloger's rule was formulated as an observation that dark coats were more prevalent in humid habitats, and the mechanism by which the association might arise was never rigorously explained. Our results suggest that carnivores do obey Gloger's rule, but for reasons that remain opaque.

Conclusions

Some of the classic hypotheses for coloration in mammals were originally generated by observing the apparent fit between the coat color of a few species and their ecological and social environments. Using a comparative approach that incorporates phylogenetic relationships, it has been possible to examine the generality of these associations across most carnivore species. But when an association between a color marking and a habitat or behavior was found to be significant, the specific *function* of that marking could only be inferred. Our results show that some of these classic hypotheses have general applicability and others do not. We confirmed that white coats are associated with arctic conditions, that pale coats may be associated with living in deserts, that spotted carnivores are arboreal and (more specifically) that spotted felids are forest dwellers, that species with black-and-white markings produce noxious anal secretions, that white ear marks are found in forest dwellers, that dark patches below the eyes are associated with crepuscular and diurnal habits, and that species with dark fur tend to live in tropical forest.

In contrast, we could not find general support for the idea that striped species live in grassland; that species with white throats, contrasting tail tips, or contrasting ears are social; that black tail tips or ear marks are related to inhabiting grassland; or that species with dark eye contours are diurnal.

These conflicting sets of findings make it clear that hypotheses about animal coloration derived from the design features of a single species' color markings need to be treated with great caution, since they may have only limited applicability (or none) in a comparative context.

Our analyses suggest three avenues for further research. First, we need experimental tests that might confirm the function of those markings that were significantly associated with particular habitats. Such tests might also distinguish between competing hypotheses. For example, the importance of contrasting tail tips in deflecting predatory attack has already been demonstrated in some species (Powell 1982), and their importance in luring prey or enabling offspring to follow their mothers in tall vegetation could be explored in a similar fashion. Experimental techniques have the advantage of testing the functional import of traits, whereas comparative associations are not adaptive explanations in themselves.

Second, the general applicability of our results can be examined by matching the same color markings to the same associated behavioral and ecological variables in other mammalian orders. For example, pale coloration could be explored in desert rodents; dark coats could be explored in forest ungulates.

Third, since the classificatory system of body marks and fur color developed here can be applied with only minor modifications to mammals other than carnivores, it provides a springboard from which to explore the behavioral and ecological correlates of other mammalian color patterns for which no adaptive explanation has heretofore been sought. A method is therefore in place to begin to understand the diversity of coloration patterns among such brilliantly colored groups as the primates, or the subtleties of coat patterns in the rodents, or the significance of body markings in the ungulates.

Appendixes

Appendix 4.1 (first four pages). Species and data used for the coloration analysis. Species names from Corbet and Hill (1991). Meanings and abbreviations: uniform, no distinct color marks present in this area; ?, missing value; LA and DB, light above and dark below; (white), in last column and only in parentheses, in some individuals only the tail tip is white, whereas in others the tail has mixed black and white hairs.

Appendix 4.2 (second four pages). Species and data used for the behavioral/ecological analyses. Species names from Corbet and Hill (1991). Abbreviations: noct. and crep., nocturnal and crepuscular; T, terrestrial; A, arboreal; T and A, terrestrial and arboreal; ?, missing value.

Appendix 4.1, p. 1

Species	Fur color	Body	Undersides	Face	Eye contour	Below eyes	Facial stripes	Back of ears	Throat and neck	Tail	Tail tip
CANIDAE											
Vulpes corsac	medium	uniform	light	uniform	light ring	dark patch	none	darker	uniform	uniform	black
Vulpes cana	medium	dark cape	light	uniform	light ring	dark patch	none	darker	white	uniform	black
Vulpes vulpes	variable dark	uniform	light	uniform	uniform	uniform	none	darker	white	uniform	black or light
Vulpes velox	medium	uniform	light	uniform	light ring	uniform	none	uniform	white	uniform	black
Vulpes macrotis	pale	uniform	light	uniform	light ring	dark patch	none	uniform	white	uniform	white
Vulpes rueppellii	medium	dark cape	light	lighter	dark around	dark patch	none	darker	lighter	uniform	white
Vulpes chama	variable light	uniform	light	lighter	uniform	dark patch	none	uniform	uniform	uniform	black
Vulpes pallida	pale	uniform	light	lighter	light ring	uniform	none	uniform	white	uniform	black
Vulpes zerda	pale	uniform	light	lighter	uniform	dark patch	none	uniform	white	uniform	black
Alopex lagopus	moult	uniform	uniform	uniform	light ring	uniform	none	uniform	uniform	uniform	uniform
Urocyon cinereoargenteus	medium	dark cape	light	uniform	uniform	dark patch	none	black tip	white	dorsal stripe	black
Dusicyon vetulus	medium	uniform	light	uniform	light ring	uniform	none	uniform	?	dorsal stripe	black
Dusicyon culpaeus	variable	dark cape	light	lighter	light ring	uniform	none	uniform	lighter	dorsal stripe	black
Dusicyon thous	variable	uniform	light	uniform	light ring	uniform	none	darker	lighter	dorsal stripe	black
Dusicyon microtis	dark	uniform	light	uniform	uniform	uniform	none	uniform	uniform	uniform	uniform
Otocyon megalotis	medium	uniform	light	contrasting	dark around	uniform	none	black tip	lighter	dorsal stripe	black
Nyctereutes procyonoides	medium	uniform	?	contrasting	dark around	uniform	none	darker	lighter	dorsal stripe	black
Speothos venaticus	dark	uniform	uniform	lighter	uniform	uniform	none	uniform	lighter	dark	white
Chrysocyon brachyurus	medium	uniform	uniform	uniform	L.A and D.B	uniform	none	uniform	lighter	uniform	white
Cuon alpinus	medium	uniform	light	darker	light ring	uniform	none	uniform	white	dark	uniform
Lycaon pictus	light and dark	blotches	uniform	darker	dark around	uniform	none	darker	darker	uniform	white
Canis adustus	medium	dark cape	light	lighter	light ring	uniform	none	uniform	white	white	white
Canis mesomelas	medium	dark cape	light	lighter	uniform	uniform	none	lighter	lighter	dark	black or light
Canis aureus	variable light	uniform	light	uniform	light ring	uniform	none	uniform	white	uniform	black
Canis latrans	variable light	uniform	light	uniform	light ring	uniform	none	uniform	white	uniform	black
Canis rufus	medium	uniform	light	uniform	light ring	uniform	none	uniform	lighter	uniform	black
Canis simensis	medium	uniform	light	uniform	light ring	uniform	none	uniform	white	dark	uniform
Canis lupus	variable	uniform	light	uniform	light ring	uniform	none	uniform	white	uniform	black
URSIDAE											
Ailuropoda melanoleuca	black&white	uniform	uniform	contrasting	dark around	uniform	none	darker	darker	uniform	uniform
Tremarctos ornatus	dark	uniform	uniform	contrasting	dark around	uniform	none	uniform	chest mark	uniform	uniform
Selenarctos thibetanus	variable	uniform	uniform	uniform	dark around	uniform	none	uniform	chest mark	uniform	uniform
Ursus americanus	variable	uniform	uniform	uniform	dark around	uniform	none	uniform	uniform	uniform	uniform
Ursus arctos	white	uniform	uniform	uniform	uniform	uniform	none	uniform	uniform	uniform	uniform
Helarctos malayanus	dark	uniform	uniform	uniform	light ring	uniform	none	uniform	chest mark	uniform	uniform
Melursus ursinus	dark	uniform	uniform	uniform	light ring	uniform	none	uniform	chest mark	uniform	uniform
PROCYONIDAE											
Bassariscus astutus	medium	uniform	light	uniform	light ring	white spot	none	white tip	lighter	ringed	black or light
Bassariscus sumichrasti	dark	uniform	light	uniform	light ring	white spot	none	white tip	lighter	ringed	black or light
Procyon cancrivorous	medium	uniform	light	contrasting	dark around	uniform	none	white tip	darker	ringed	black or light
Procyon lotor	medium	uniform	light	uniform	dark around	uniform	none	white tip	darker	ringed	black or light
Nasua nasua	medium	uniform	uniform	uniform	light ring	white spot	none	white tip	lighter	ringed	black or light
Nasua narica	medium	uniform	?	lighter	?	white spot	?	white tip	?	ringed	black or light
Nasuella olivacea	medium	uniform	light	uniform	uniform	uniform	none	uniform	uniform	ringed	black or light
Potos flavus	medium	uniform	light	lighter	?	uniform	none	uniform	uniform	uniform	uniform
Bassaricyon gabbii	medium	uniform	light	lighter	L.A and D.B	uniform	none	uniform	uniform	ringed	black or light
Ailurus fulgens	dark	uniform	dark	lighter	L.A and D.B	uniform	none	white rims	darker	ringed	black or light

Species	Fur color	Body	Undersides	Face	Eye contour	Below eyes	Facial stripes	Back of ears	Throat and neck	Tail	Tail tip
MUSTELIDAE											
Mustela erminea	moult	uniform	light	uniform	uniform	uniform	none	uniform	white	uniform	black
Mustela eversmanni	variable light	uniform	dark	contrasting	dark around	uniform	forehead band	lighter	darker	dark	uniform
Mustela frenata	moult	uniform	light	uniform	uniform	uniform	none	uniform	white	uniform	black
Mustela lutreola	variable dark	uniform	uniform	uniform	uniform	uniform	none	uniform	uniform	uniform	uniform
Mustela nivalis	moult	uniform	light	uniform	dark around	uniform	none	uniform	white	uniform	black
Mustela nigripes	pale	uniform	light	contrasting	dark around	uniform	none	uniform	lighter	uniform	black
Mustela nudipes	variable light	uniform	?	lighter	uniform	uniform	none	uniform	?	light	uniform
Mustela putorius	medium	uniform	uniform	contrasting	dark around	uniform	forehead band	white rim	chest mark	dark	uniform
Mustela sibirica	medium	uniform	light	lighter	dark around	?	?	?	lighter	uniform	uniform
Mustela strigidorsa	dark	uniform	light	uniform	?	?	?	?	uniform	uniform	uniform
Mustela vison	dark	uniform	uniform	contrasting	dark around	uniform	none	uniform	uniform	uniform	black
Vormela peregusna	light and dark	light spots	dark	lighter	light ring	dark patch	forehead band	darker	lighter	light	uniform
Martes americana	variable dark	uniform	light	darker	uniform	uniform	none	darker	darker	dark	uniform
Martes flavigula	medium	uniform	uniform	uniform	uniform	uniform	none	white rim	lighter	uniform	uniform
Martes foina	variable dark	uniform	uniform	uniform	uniform	dark patch	none	uniform	white	uniform	uniform
Martes martes	medium	uniform	?	uniform	light above	uniform	none	uniform	chest mark	uniform	uniform
Martes pennanti	dark	uniform	light	uniform	light ring	uniform	none	uniform	chest mark	dark	uniform
Martes zibellina	variable dark	uniform	light	uniform	light ring	uniform	none	white rim	?	uniform	uniform
Eira barbara	dark	uniform	uniform	lighter	uniform	uniform	none	uniform	chest mark	uniform	uniform
Galictis vittata	medium	uniform	dark	contrasting	dark around	uniform	forehead band	lighter	darker	uniform	uniform
Galictis cuja	pale	uniform	dark	contrasting	dark around	uniform	forehead band	lighter	darker	uniform	uniform
Lyncodon patagonicus	pale	uniform	dark	contrasting	dark around	uniform	forehead band	lighter	darker	uniform	uniform
Ictonyx striatus	black&white	white cape	uniform	contrasting	dark around	uniform	frontal stripe	lighter	uniform	light	uniform
Poecilictis libyca	black&white	white cape	dark	contrasting	dark around	uniform	forehead band	white rim	darker	light	uniform
Poecilogale albinucha	black&white	white cape	dark	contrasting	dark around	uniform	forehead band	white tip	uniform	light	uniform
Gulo gulo	dark	uniform	uniform	lighter	uniform	uniform	none	uniform	uniform	uniform	uniform
Mellivora capensis	black&white	white cape	dark	contrasting	dark around	uniform	forehead band	uniform	darker	dark	uniform
Meles meles	medium	uniform	dark	contrasting	dark around	uniform	frontal stripe	white tip	white	uniform	uniform
Arctonyx collaris	medium	uniform	dark	contrasting	dark around	uniform	frontal stripe	lighter	white	light	uniform
Mydaus javanensis	variable dark	white cape	uniform	contrasting	dark around	uniform	forehead band	darker	uniform	light	uniform
Taxidea taxus	medium	uniform	light	contrasting	dark around	uniform	frontal stripe	uniform	lighter	uniform	uniform
Melogale moschata	variable dark	uniform	light	contrasting	dark around	uniform	frontal stripe	darker	white	uniform	uniform
Mephitis mephitis	black&white	white cape	dark	contrasting	dark around	uniform	frontal stripe	darker	white	light	(white)
Spilogale putorius	black&white	white cape	dark	contrasting	dark around	uniform	frontal stripe	black tip	lighter	light	(white)
Conepatus mesoleucus	black&white	white cape	dark	contrasting	dark around	uniform	forehead band	black tip	darker	light	uniform
Lutra canadensis	variable dark	uniform	light	uniform	D.A and L.B	dark patch	none	uniform	lighter	uniform	uniform
Lutra lutra	dark	uniform	light	uniform	D.A and L.B	dark patch	none	uniform	lighter	uniform	uniform
Lutra maculicollis	medium	uniform	light	uniform	?	?	?	?	chest mark	uniform	uniform
Lutra perspicillata	medium	uniform	light	uniform	?	?	?	?	white	uniform	uniform
Pteronura brasiliensis	dark	uniform	uniform	uniform	uniform	uniform	none	uniform	chest mark	uniform	uniform
Aonyx capensis	medium	uniform	light	uniform	D.A and L.B	dark patch	none	white rim	white	uniform	uniform
Aonyx cinerea	medium	uniform	light	uniform	D.A and L.B	dark patch	none	white rim	white	uniform	uniform
Aonyx congica	dark	uniform	uniform	uniform	D.A and L.B	dark patch	none	white rim	white	uniform	uniform
Enhydra lutris	variable dark	uniform	uniform	lighter	uniform	uniform	none	uniform	lighter	uniform	uniform

Appendix 4.1, p. 3

Species	Fur color	Body	Undersides	Face	Eye contour	Below eyes	Facial stripes	Back of ears	Throat and neck	Tail	Tail tip
VIVERRIDAE											
Poiana richardsonii	light and dark	spots	light	?	light above	uniform	none	?	white	ringed	black or light
Genetta genetta	light and dark	spots	light	contrasting	light ring	white spot	none	lighter	uniform	ringed	black or light
Genetta angolensis	light and dark	spots	light	contrasting	light ring	white spot	none	lighter	uniform	ringed	black or light
Genetta thierryi	light and dark	spots	light	contrasting	light ring	white spot	none	uniform	uniform	ringed	black
Genetta servalina	light and dark	spots	uniform	contrasting	light ring	white spot	none	uniform	uniform	ringed	black or light
Genetta abyssinica	light and dark	spots	light	contrasting	light ring	white spot	none	uniform	uniform	ringed	black
Viverricula indica	light and dark	spots	light	lighter	light above	uniform	none	?	BW neck mark	ringed	white
Osbornictis piscivora	medium	uniform	dark	uniform	uniform	white spot	none	darker	white	dark	uniform
Viverra civetta	light and dark	spots	light	contrasting	dark around	uniform	none	white spot	BW neck mark	ringed	black
Viverra megaspila	light and dark	spots	light	uniform	dark around	uniform	none	white tip	BW neck mark	ringed	black
Viverra tangalunga	light and dark	spots	light	uniform	dark around	uniform	none	white rim	BW neck mark	ringed	black
Viverra zibetha	light and dark	spots	light	lighter	dark around	uniform	none	uniform	BW neck mark	ringed	black
Prionodon linsang	light and dark	bands	light	uniform	dark around	dark patch	none	lighter	lighter	ringed	black
Prionodon pardicolor	light and dark	spots	light	uniform	D.A and L.B	dark patch	none	uniform	lighter	ringed	black or light
Nandinia binotata	dark	spots	light	lighter	uniform	uniform	none	?	uniform	ringed	black or light
Arctogalidea trivirgata	medium	hor. stripes	light	contrasting	dark around	uniform	frontal stripe	white tip	lighter	ringed	black
Paradoxurus hermaphroditus	medium	spots	?	contrasting	dark around	white spot	none	darker	?	dorsal stripe	black
Paradoxurus zeylonensis	medium	spots	?	contrasting	light ring	white spot	none	darker	?	dark	uniform
Paguma larvata	medium	uniform	light	contrasting	dark around	uniform	frontal stripe	darker	uniform	dark	black or light
Macrogalidea musschenbroekii	medium	uniform	light	uniform	?	?	none	darker	lighter	ringed	black or light
Arctictis binturong	dark	uniform	uniform	lighter	light above	uniform	none	white rim	uniform	uniform	uniform
Fossa fossa	light and dark	spots	light	uniform	dark around	uniform	frontal stripe	?	uniform	ringed	black
Hemigalus derbyanus	light and dark	bands	light	contrasting	dark around	white spot	frontal stripe	darker	lighter	dark	uniform
Chrotogale owstoni	light and dark	bands	light	contrasting	dark around	white spot	frontal stripe	darker	chest mark	ringed	black
Cynogale bennettii	dark	uniform	light	uniform	dark around	uniform	none	uniform	uniform	uniform	uniform
Eupleres goudotii	medium	uniform	light	uniform	light ring	uniform	none	uniform	lighter	uniform	uniform
Cryptoprocta ferox	medium	uniform	light	uniform	light ring	uniform	none	uniform	lighter	uniform	uniform
HERPESTIDAE											
Galidea elegans	medium	uniform	light	lighter	light below	uniform	none	uniform	lighter	ringed	black or light
Galidictis fasciata	pale	hor. stripes	light	darker	light ring	uniform	none	uniform	lighter	light	uniform
Mungotictis decemlineata	pale	hor. stripes	light	uniform	light above	uniform	none	uniform	lighter	uniform	uniform
Salanoia concolor	medium	uniform	uniform	uniform	uniform	uniform	none	uniform	uniform	uniform	uniform
Suricata suricatta	pale	bands	uniform	lighter	dark around	uniform	none	darker	lighter	uniform	black
Herpestes auropunctatus	medium	uniform	light	uniform	uniform	uniform	none	uniform	uniform	uniform	uniform
Herpestes ichneumon	variable	uniform	light	lighter	dark around	uniform	none	uniform	uniform	uniform	black
Herpestes urva	variable dark	uniform	light	lighter	uniform	uniform	none	?	BW neck mark	uniform	uniform
Herpestes parvula	variable dark	uniform	light	lighter	uniform	uniform	none	uniform	lighter	uniform	uniform
Helogale parvula	medium	uniform	light	darker	light above	uniform	none	?	lighter	uniform	uniform
Dologale dybowskii	medium	uniform	light	lighter	uniform	uniform	none	uniform	lighter	uniform	black
Galerella sanguinea	variable dark	uniform	light	lighter	uniform	uniform	none	uniform	uniform	uniform	uniform
Atilax paludinosus	variable dark	uniform	uniform	uniform	?	uniform	none	uniform	uniform	uniform	uniform
Mungos mungo	medium	bands	uniform	uniform	dark around	uniform	none	uniform	lighter	uniform	black
Mungos gambianus	medium	uniform	light	lighter	dark around	uniform	none	?	BW neck mark	uniform	black
Crossarchus obscurus	dark	uniform	light	lighter	?	?	none	?	uniform	uniform	black
Liberiictis kuhni	dark	uniform	uniform	uniform	dark around	uniform	?	?	BW neck mark	uniform	uniform
Ichneumia albicauda	medium	uniform	dark	lighter	dark around	uniform	none	uniform	lighter	light	black or light
Bdeogale crassicauda	variable dark	uniform	light	lighter	?	?	none	?	lighter	dark	uniform
Bdeogale nigripes	medium	uniform	dark	lighter	dark around	uniform	?	?	lighter	light	uniform
Rhynchogale melleri	medium	uniform	light	lighter	?	uniform	?	uniform	lighter	uniform	black or light
Cynictis penicillata	medium	uniform	light	uniform	light below	uniform	none	uniform	white	dark	white
Paracynictis selousi	medium	uniform	light	uniform	dark around	uniform	none	uniform	lighter	uniform	white

Appendix 4.1, p. 4

Species	Fur color	Body	Undersides	Face	Eye contour	Below eyes	Facial stripes	Back of ears	Throat and neck	Tail	Tail tip
HYAENIDAE											
Proteles cristatus	light and dark	vert. stripes	light	lighter	dark around	uniform	none	uniform	lighter	uniform	black
Crocuta crocuta	light and dark	spots	light	darker	dark around	uniform	none	darker	lighter	uniform	black
Hyaena brunnea	dark	uniform	light	lighter	dark around	uniform	none	uniform	lighter	uniform	uniform
Hyaena hyaena	light and dark	vert. stripes	light	uniform	LA and D.B	uniform	none	uniform	darker	uniform	uniform
FELIDAE											
Felis aurata	medium	uniform&spots	light	uniform	light above	uniform	none	darker	white	uniform	uniform
Felis bengalensis	light and dark	spots	light	uniform	light ring	uniform	none	white spot	white	ringed	black or light
Felis caracal	variable light	uniform	light	uniform	light ring	uniform	none	darker	lighter	uniform	uniform
Felis chaus	variable light	uniform	light	uniform	light below	uniform	none	black tip	lighter	ringed	black
Felis colocolo	medium	uniform&spots	?	uniform	light ring	uniform	none	white spot	uniform	ringed	black or light
Felis concolor	medium	uniform	light	uniform	light above	uniform	none	darker	white	uniform	black
Felis geoffroyi	light and dark	spots	light	uniform	light ring	uniform	none	white spot	uniform	ringed	black or light
Felis guigna	light and dark	spots	light	uniform	light below	uniform	none	white spot	?	ringed	black
Felis iriomotensis	dark	spots	light	uniform	light below	uniform	none	white spot	uniform	ringed	black or light
Felis manul	variable light	hor. stripes	light	uniform	light ring	uniform	none	uniform	white	ringed	black
Felis margarita	pale	hor. stripes	light	uniform	light ring	uniform	none	black tip	lighter	ringed	black
Felis marmorata	light and dark	spots	light	uniform	light below	uniform	none	white spot	uniform	ringed	black or light
Felis nigripes	light and dark	spots	light	uniform	light ring	uniform	none	uniform	white	ringed	black or light
Felis pardalis	light and dark	spots	light	uniform	light ring	uniform	none	white spot	white	ringed	black or light
Felis planiceps	dark	uniform	light	lighter	light below	uniform	none	white spot	white	uniform	uniform
Felis rubiginosa	medium	spots	light	uniform	light below	uniform	none	?	white	uniform	black
Felis serval	light and dark	spots	light	uniform	light ring	uniform	none	white spot	uniform	ringed	black
Felis silvestris	variable	vert. stripes	light	uniform	light ring	uniform	none	uniform	white	ringed	black
Felis temminckii	medium	uniform&spots	light	uniform	light ring	uniform	none	darker	?	uniform	white
Felis tigrina	light and dark	spots	light	uniform	light ring	uniform	none	white spot	uniform	ringed	black or light
Felis viverrina	light and dark	spots	light	uniform	light below	uniform	none	white spot	uniform	ringed	black
Felis wiedii	light and dark	spots	light	uniform	light ring	uniform	none	white spot	uniform	ringed	black or light
Felis yaguaroundi	variable dark	uniform	light	uniform	light ring	uniform	none	uniform	lighter	uniform	uniform
Lynx canadensis	medium	uniform&spots	light	uniform	light below	uniform	none	white spot	uniform	uniform	black
Lynx lynx	light and dark	spots	light	uniform	light ring	uniform	none	white spot	uniform	ringed	black
Lynx rufus	light and dark	spots	light	uniform	light ring	uniform	none	white spot	uniform	ringed	black
Panthera leo	medium	uniform	light	uniform	light below	uniform	none	black tip	lighter	uniform	black
Panthera onca	light and dark	spots	light	uniform	light below	uniform	none	white spot	uniform	ringed	black
Panthera pardus	light and dark	spots	light	uniform	light ring	uniform	none	white spot	uniform	ringed	black or light
Panthera tigris	light and dark	vert. stripes	light	uniform	light ring	uniform	none	white spot	uniform	ringed	black
Panthera uncia	light and dark	spots	light	uniform	light ring	uniform	none	white spot	white	ringed	black
Neofelis nebulosa	light and dark	spots	light	uniform	?	uniform	none	white spot	white	ringed	black or light
Acinonyx jubatus	light and dark	spots	light	uniform	light below	uniform	none	darker	white	ringed	white

Appendix 4.2, p. 1

Species	Social behavior	Activity	Locomotion	Temperate forest	Tropical forest	Grassland	Arctic	Riparian & aquatic	Desert	Noxious anal secretion
CANIDAE										
Vulpes corsac	variable group	nocturnal	T but climbs	no	no	yes	no	no	yes	no
Vulpes cana	?	?	terrestrial (T)	no	no	yes	no	no	yes	no
Vulpes vulpes	variable group	noct. and crep.	terrestrial (T)	yes	no	yes	yes	no	no	no
Vulpes velox	pairs	nocturnal	terrestrial (T)	no	no	yes	no	no	yes	no
Vulpes macrotis	pairs	nocturnal	terrestrial (T)	no	no	yes	no	no	yes	?
Vulpes rueppellii	pairs	nocturnal	terrestrial (T)	no	no	yes	no	no	yes	no
Vulpes chama	pairs	noct. and crep.	terrestrial (T)	no	no	yes	no	no	no	no
Vulpes pallida	pairs	noct. and crep.	terrestrial (T)	no	no	yes	no	no	yes	no
Vulpes zerda	group	noct. and crep.	terrestrial (T)	no	no	no	no	no	yes	no
Alopex lagopus	variable group	anytime	terrestrial (T)	no	no	no	yes	no	no	no
Urocyon cinereoargenteus	pairs	noct. and crep.	T but climbs	yes	no	yes	no	no	no	no
Dusicyon vetulus	?	noct. and crep.	terrestrial (T)	no	no	yes	no	no	no	no
Dusicyon culpaeus	solitary	noct. and crep.	terrestrial (T)	no	no	yes	no	no	no	no
Dusicyon thous	pairs	anytime	terrestrial (T)	yes	no	yes	no	yes	no	no
Dusicyon microtis	solitary	nocturnal	terrestrial (T)	no	yes	no	no	yes	no	no
Otocyon megalotis	pairs	anytime	terrestrial (T)	no	no	yes	no	no	no	no
Nyctereutes procyonoides	pairs	noct. and crep.	terrestrial (T)	yes	no	yes	no	yes	no	no
Speothos venaticus	group	diurnal	terrestrial (T)	yes	yes	yes	no	yes	no	no
Chrysocyon brachyurus	solitary	noct. and crep.	terrestrial (T)	no	no	yes	no	no	no	no
Cuon alpinus	group	anytime	terrestrial (T)	yes	yes	yes	no	no	no	no
Lycaon pictus	group	diurnal	terrestrial (T)	yes	no	yes	no	no	no	no
Canis adustus	pairs	noct. and crep.	terrestrial (T)	yes	no	yes	no	no	no	no
Canis mesomelas	pairs	anytime	terrestrial (T)	yes	no	yes	no	no	no	no
Canis aureus	variable group	anytime	terrestrial (T)	no	no	yes	no	no	yes	no
Canis latrans	variable group	noct. and crep.	terrestrial (T)	yes	no	yes	no	no	yes	no
Canis rufus	variable group	nocturnal	terrestrial (T)	yes	no	yes	no	no	no	no
Canis simensis	solitary	anytime	terrestrial (T)	no	no	yes	no	no	no	no
Canis lupus	group	noct. and crep.	terrestrial (T)	yes	no	yes	yes	no	yes	no
URSIDAE										
Ailuropoda melanoleuca	solitary	anytime	T and A	yes	no	no	no	no	no	no
Tremarctos ornatus	solitary	noct. and crep.	T and A	no	yes	yes	no	no	no	no
Selenarctos thibetanus	solitary	nocturnal	T and A	yes	no	no	yes	no	no	no
Ursus americanus	solitary	noct. and crep.	T but climbs	yes	no	no	yes	no	no	no
Ursus arctos	solitary	anytime	T but climbs	yes	no	no	yes	yes	no	no
Ursus maritimus	solitary	diurnal	aquatic	no	no	no	yes	yes	no	no
Helarctos malayanus	solitary	nocturnal	T and A	yes	yes	no	no	no	no	no
Melursus ursinus	solitary	anytime	T and A	yes	yes	no	no	no	no	no
PROCYONIDAE										
Bassariscus astutus	solitary	nocturnal	T and A	yes	no	yes	no	yes	yes	yes
Bassariscus sumichrasti	solitary	nocturnal	T and A	yes	yes	no	no	yes	no	no
Procyon cancrivorus	solitary	noct. and crep.	T and A	no	yes	no	no	yes	no	no
Procyon lotor	solitary	noct. and crep.	T and A	yes	yes	no	no	yes	no	no
Nasua nasua	variable group	diurnal	T and A	no	yes	no	no	yes	no	no
Nasua narica	variable group	anytime	T and A	no	yes	no	no	yes	no	no
Nasuella olivacea	?	?	T and A	no	yes	yes	no	no	no	no
Potos flavus	solitary	nocturnal	arboreal (A)	no	yes	no	no	no	no	yes
Bassaricyon gabbii	solitary	nocturnal	arboreal (A)	no	yes	no	no	no	no	no
Ailurus fulgens	solitary	noct. and crep.	T and A	yes	no	no	no	no	no	yes

Appendix 4.2, p. 2

Species	Social behavior	Activity	Locomotion	Temperate forest	Tropical forest	Grassland	Arctic	Riparian & aquatic	Desert	Noxious anal secretion
MUSTELIDAE										
Mustela erminea	solitary	anytime	T but climbs	yes	no	yes	yes	no	no	no
Mustela eversmanni	solitary	nocturnal	terrestrial (T)	no	no	yes	no	no	yes	?
Mustela frenata	solitary	anytime	T but climbs	yes	no	yes	no	no	no	no
Mustela lutreola	solitary	noct. and crep.	terrestrial (T)	no	no	no	no	yes	no	no
Mustela nivalis	solitary	diurnal	terrestrial (T)	yes	no	yes	yes	no	yes	no
Mustela nigripes	solitary	nocturnal	terrestrial (T)	?	?	?	?	?	?	?
Mustela nudipes	?	?	?	?	?	?	?	?	?	?
Mustela putorius	solitary	noct. and crep.	terrestrial (T)	yes	no	yes	no	no	no	yes
Mustela sibirica	solitary	noct. and crep.	T but climbs	yes	no	yes	no	no	no	no
Mustela strigidorsa	?	?	?	?	?	?	?	?	?	?
Mustela vison	solitary	anytime	terrestrial (T)	no	no	yes	no	yes	no	no
Vormela peregusna	solitary	nocturnal	T but climbs	no	no	yes	no	no	yes	yes
Martes americana	solitary	noct. and crep.	T and A	yes	no	no	no	no	no	no
Martes flavigula	variable group	diurnal	T and A	yes	no	no	no	no	no	no
Martes foina	solitary	noct. and crep.	T but climbs	yes	no	no	no	no	no	no
Martes martes	solitary	anytime	T and A	yes	no	no	no	no	no	no
Martes pennanti	solitary	anytime	T but climbs	yes	no	no	no	no	no	no
Martes zibellina	solitary	anytime	T but climbs	yes	no	no	no	no	no	no
Eira barbara	solitary	anytime	T and A	yes	no	yes	no	no	no	yes
Galictis vittata	variable group	anytime	T but climbs	yes	no	yes	no	no	no	yes
Galictis cuja	variable group	anytime	T but climbs	yes	no	yes	no	no	no	?
Lyncodon patagonicus	?	noct. and crep.	terrestrial (T)	no	no	no	no	no	no	?
Ictonyx striatus	solitary	nocturnal	T but climbs	no	no	yes	no	no	no	yes
Poecilis libyca	solitary	nocturnal	terrestrial (T)	no	no	yes	no	no	yes	yes
Poecilogale albinucha	solitary	nocturnal	T but climbs	yes	no	yes	no	no	no	yes
Gulo gulo	solitary	anytime	T but climbs	yes	no	no	yes	no	no	no
Melivora capensis	pairs	noct. and crep.	T but climbs	yes	no	yes	no	no	yes	yes
Meles meles	variable group	noct. and crep.	terrestrial (T)	yes	no	yes	no	no	no	no
Arctonyx collaris	?	nocturnal	terrestrial (T)	no	yes	no	no	no	yes	yes
Mydaus javanensis	?	nocturnal	terrestrial (T)	no	yes	no	no	no	no	yes
Taxidea taxus	solitary	anytime	terrestrial (T)	no	no	yes	no	no	yes	no
Melogale moschata	?	noct. and crep.	T but climbs	yes	no	yes	no	no	no	yes
Mephitis mephitis	variable group	nocturnal	terrestrial (T)	yes	no	yes	no	no	yes	yes
Spilogale putorius	variable group	nocturnal	T but climbs	yes	no	yes	no	no	no	yes
Conepatus mesoleucus	solitary	nocturnal	terrestrial (T)	no	no	yes	no	no	no	yes
Lutra canadensis	solitary	anytime	aquatic	no	no	no	no	yes	no	no
Lutra lutra	solitary	noct. and crep.	aquatic	no	no	no	no	yes	no	no
Lutra maculicollis	variable group	anytime	aquatic	no	no	no	no	yes	no	?
Lutra perspicillata	pairs	?	aquatic	no	no	no	no	yes	no	no
Pteronura brasiliensis	group	diurnal	aquatic	no	no	no	no	yes	no	no
Aonyx capensis	variable group	anytime	aquatic	no	yes	no	no	yes	no	no
Aonyx cinerea	group	diurnal	aquatic	no	no	no	no	yes	no	no
Aonyx congica	variable group	noct. and crep.	aquatic	no	no	no	no	yes	no	no
Enhydra lutris	variable group	diurnal	aquatic	no	no	no	no	yes	no	no

Appendix 4.2, p. 3

Species	Social behavior	Activity	Locomotion	Temperate forest	Tropical forest	Grassland	Arctic	Riparian & aquatic	Desert	Noxious anal secretion
VIVERRIDAE										
Poiana richardsonii	variable group	nocturnal	T and A	no	yes	no	no	no	no	no
Genetta genetta	solitary	noct. and crep.	T and A	yes	no	yes	no	no	yes	yes
Genetta angolensis	?	?	?	no	yes	yes	no	no	no	?
Genetta thierryi	?	?	?	no	no	yes	no	no	no	?
Genetta servalina	?	?	T but climbs	yes	no	yes	no	no	no	?
Genetta abyssinica	?	?	?	no	no	yes	no	no	no	?
Viverricula indica	solitary	nocturnal	T and A	no	yes	yes	no	no	no	yes
Osbornictis piscivora	solitary	?	aquatic	no	yes	no	no	yes	no	?
Viverra civetta	solitary	nocturnal	terrestrial (T)	no	yes	yes	no	yes	no	?
Viverra megaspila	?	?	?	no	yes	yes	no	no	no	?
Viverra tangalunga	solitary	nocturnal	T but climbs	no	yes	yes	no	no	no	?
Viverra zibetha	solitary	nocturnal	T but climbs	no	yes	no	no	no	no	?
Prionodon linsang	solitary	nocturnal	T and A	no	yes	no	no	no	no	no
Prionodon pardicolor	solitary	nocturnal	T and A	no	yes	yes	no	no	no	?
Nandinia binotata	solitary	noct. and crep.	arboreal (A)	yes	yes	no	no	no	no	no
Arctogalidea trivirgata	solitary	nocturnal	arboreal (A)	yes	yes	yes	no	no	no	no
Paradoxurus hermaphroditus	solitary	nocturnal	arboreal (A)	no	yes	yes	no	no	no	yes
Paradoxurus zeylonensis	solitary	nocturnal	arboreal (A)	no	yes	no	no	no	no	yes
Paguma larvata	solitary	nocturnal	arboreal (A)	no	yes	no	no	no	no	yes
Macrogalidea musschenbroeki	solitary	?	T and A	no	yes	yes	no	no	no	no
Arctictis binturong	solitary	nocturnal	T and A	no	yes	no	no	no	no	no
Fossa fossa	pairs	nocturnal	T and A	yes	yes	yes	no	no	no	no
Hemigalus derbyanus	solitary	nocturnal	T but climbs	no	yes	no	no	no	no	no
Chrotogale owstoni	?	?	terrestrial (T)	no	yes	no	no	yes	no	yes
Cynogale bennettii	?	?	aquatic	no	no	no	no	yes	no	yes
Eupleres goudotii	solitary&pairs	noct. and crep.	terrestrial (T)	no	yes	no	no	yes	no	no
Cryptoprocta ferox	solitary	noct. and crep.	T and A	yes	no	yes	no	no	no	yes
HERPESTIDAE										
Galidia elegans	variable group	anytime	T and A	yes	yes	no	no	no	no	no
Galidictis fasciata	variable group	noct. and crep.	T and A	no	yes	no	no	no	no	?
Mungotictis decemlineata	variable group	diurnal	T and A	no	no	yes	no	no	no	no
Salanoia concolor	variable group	diurnal	T but climbs	yes	no	yes	no	no	no	no
Suricata suricatta	group	diurnal	terrestrial (T)	no	no	yes	no	no	no	no
Herpestes auropunctatus	solitary	anytime	T but climbs	no	no	yes	no	no	no	no
Herpestes ichneumon	variable group	diurnal	T but climbs	no	yes	no	no	yes	yes	yes
Herpestes urva	?	nocturnal	aquatic	no	no	no	no	yes	no	no
Helogale parvula	group	diurnal	T but climbs	no	?	yes	no	yes	?	?
Dologale dybowskii	solitary	diurnal	T and A	no	yes	yes	no	no	no	no
Galerella sanguinea	solitary	noct. and crep.	aquatic	no	no	no	no	no	no	no
Atilax paludinosus	group	diurnal	T but climbs	no	yes	no	no	yes	no	yes
Mungos mungo	group	diurnal	T but climbs	no	no	yes	no	no	no	no
Mungos gambianus	group	anytime	T but climbs	no	yes	yes	no	no	no	no
Crossarchus obscurus	group	diurnal	terrestrial (T)	no	yes	no	no	no	no	no
Liberiictis kuhni	solitary	noct. and crep.	T but climbs	no	yes	yes	no	yes	yes	yes
Ichneumia albicauda	solitary	nocturnal	T but climbs	yes	no	yes	no	yes	no	no
Bdeogale crassicauda	solitary&pairs	nocturnal	T but climbs	no	yes	no	no	yes	?	yes
Bdeogale nigripes	solitary&pairs	nocturnal	terrestrial (T)	no	yes	no	no	?	?	no
Rhynchogale melleri	solitary	nocturnal	terrestrial (T)	yes	no	yes	no	no	no	yes
Cynictis penicillata	group	anytime	terrestrial (T)	no	no	yes	no	no	no	no
Paracynictis selousi	solitary	noct. and crep.	terrestrial (T)	yes	no	yes	no	no	no	yes

Appendix 4.2, p. 4

Species	Social behavior	Activity	Locomotion	Temperate forest	Tropical forest	Grassland	Arctic	Riparian & aquatic	Desert	Noxious anal secretion
HYAENIDAE										
Proteles cristatus	solitary	nocturnal	terrestrial	no	no	yes	no	no	no	?
Crocuta crocuta	group	anytime	terrestrial	no	no	yes	no	no	yes	no
Hyaena brunnea	group	noct. and crep.	terrestrial	no	no	yes	no	no	yes	no
Hyaena hyaena	variable group	noct. and crep.	terrestrial	no	no	yes	no	no	yes	no
FELIDAE										
Felis aurata	solitary	anytime	T but climbs	no	yes	yes	no	no	no	no
Felis bengalensis	pairs	nocturnal	T and A	yes	yes	no	no	no	no	no
Felis caracal	solitary	nocturnal	T but climbs	no	no	yes	no	yes	yes	no
Felis chaus	solitary	anytime	T but climbs	no	yes	yes	no	no	no	no
Felis colocolo	?	nocturnal	T and A	yes	yes	yes	no	no	no	no
Felis concolor	solitary	anytime	T but climbs	yes	yes	yes	no	no	no	no
Felis geoffroyi	solitary	nocturnal	T and A	yes	no	no	no	no	no	no
Felis guigna	?	nocturnal	T and A	yes	yes	no	no	no	no	no
Felis inomotensis	solitary	nocturnal	T and A	no	no	yes	no	no	yes	no
Felis manul	solitary	nocturnal	T but climbs	no	no	no	no	no	yes	no
Felis margarita	?	noct. and crep.	terrestrial (T)	no	no	no	no	no	yes	no
Felis marmorata	solitary	nocturnal	T and A	no	yes	yes	no	no	no	no
Felis nigripes	solitary	nocturnal	T but climbs	yes	yes	yes	no	no	yes	no
Felis planiceps	?	nocturnal	terrestrial (T)	no	yes	no	no	yes	no	no
Felis rubiginosa	?	nocturnal	?	no	yes	yes	no	no	no	no
Felis serval	solitary	nocturnal	T but climbs	no	no	yes	no	yes	no	no
Felis silvestris	solitary	noct. and crep.	T but climbs	yes	yes	yes	no	no	yes	no
Felis temminckii	pairs	anytime	T but climbs	yes	yes	no	no	no	no	no
Felis tigrina	solitary	?	T and A	yes	yes	no	no	yes	no	no
Felis viverrina	solitary	?	terrestrial (T)	no	yes	no	no	no	no	no
Felis wiedii	solitary	?	T and A	no	yes	no	no	no	no	no
Felis yaguaroundi	solitary	anytime	T but climbs	yes	yes	no	no	yes	no	no
Lynx canadensis	solitary	nocturnal	T but climbs	yes	no	no	yes	no	no	no
Lynx lynx	solitary	noct. and crep.	T but climbs	yes	no	yes	no	no	no	no
Lynx rufus	solitary	anytime	terrestrial (T)	yes	no	yes	no	no	yes	no
Panthera leo	group	nocturnal	T and A	no	no	yes	no	no	yes	no
Panthera onca	solitary	nocturnal	T but climbs	yes	yes	yes	no	yes	yes	no
Panthera pardus	solitary	nocturnal	T but climbs	yes	yes	yes	no	no	yes	no
Panthera tigris	solitary	nocturnal	T but climbs	yes	yes	yes	no	yes	no	no
Panthera uncia	solitary	diurnal	T but climbs	yes	no	no	no	no	no	no
Neofelis nebulosa	pairs	nocturnal	T and A	yes	yes	no	no	no	no	no
Acinonyx jubatus	variable group	diurnal	T but climbs	no	no	yes	no	no	yes	no

Acknowledgments

The Museum of Zoology and Jim Patton at the University of California at Berkeley, the Vertebrate Museum and Ron Cole at the University of California, Davis, and the Vertebrate Museum and Karen Cebra at the California Academy of Sciences graciously allowed Ortolani access to specimens. Laurence Frank showed us his spotted hyenas at the University of California at Berkeley, and Jane Hansjorten showed us the newly arrived black-footed cats at the Sacramento Zoo. We thank Brad Shaffer for advice on phylogenetic analyses; Matt Gompper, Steve O'Brien, and Bob Wayne for forwarding manuscripts; Kirsten Christopherson and Chris Gregory for help with interobserver reliability and the latter for helping to run MacClade; Scott Creel and Bob Montgomerie for discussions; Otis Bilodeau for the line drawings he prepared on short notice; John Gittleman, Daniel Promislow, and especially John Endler for commenting on the manuscript; and, finally, John Gittleman and each other for great patience.

References

Andersson, M. 1976. *Lemmus lemmus:* A possible case of aposematic coloration and behavior. *J. Mammal.* 57:461–469.

Baker, R. R., and Parker, G. A. 1979. The evolution of bird coloration. *Proc. Roy. Soc. London* (ser. B) 287:63–130.

Bard, J. B. L. 1977. A unity underlying the different zebra striping patterns. *J. Zool.* 183:527–539.

Beddard, F. E. 1892. *Animal Coloration.* London: Hazell, Watson, and Viney.

Belt, T. 1888. *The Naturalist in Nicaragua.* London: John Murray.

Benson, S. B. 1933. Concealing coloration among some desert rodents of the southwestern United States. *United Calif. Publ. Zool.* 40:1–70.

Booth, C. L. 1990. Evolutionary significance of ontogenetic colour change in animals. *Biol. J. Linn. Soc.* 40:125–163.

Brinck, C., Erlinge, S., and Sandell, M. 1983. Anal sac secretion in mustelids. *J. Chem. Ecol.* 9:727–745.

Bryant, H. N., Russell, A. P., and Fitch, W. D. 1993. Phylogenetic relationships within the extant Mustelidae (Carnivora): Appraisal of the cladistic status of the Simpsonian subfamilies. *Zool. J. Linn. Soc.* 108:301–334.

Burt, W. H., and Grossenheider, R. P. 1976. *A Field Guide to the Mammals (North America north of Mexico),* 3rd ed. Boston: Houghton Mifflin.

Burtt, E. H., Jr., ed. 1979. *The Behavioral Significance of Color.* New York: Garland STPM Press.

Caro, T. M. 1987. Cheetah mothers' vigilance: Looking out for prey or for predators? *Behav. Ecol. Sociobiol.* 20:351–361.

Caro, T. M. 1994. *Cheetahs of the Serengeti Plains: Group Living in an Asocial Species.* Chicago: Univ. Chicago Press.

Caro, T. M., Roper, R., Young, M., and Dank, G. R. 1979. Inter-observer reliability. *Behaviour* 69:303–315.

Cleland, T. M. 1921. *A Grammar of Color.* Mittineague, Mass.: Strathmore Paper Company.

Cole, L. R. 1943. Experiments on toleration of high temperature in lizards with reference to adaptive coloration. *Ecology* 24:94–108.

Collier, G. E., and O'Brien, S. J. 1985. A molecular phylogeny of the Felidae: Immunological distance. *Evolution* 39:437–487.

Corbet, G. B., and Hill, J. E. 1991. *A World List of Mammalian Species*, 3rd ed. London: British Museum (Natural History).

Cott, H. B. 1940. *Adaptive Coloration in Mammals*. London: Methuen.

Dice, L. R. 1947. Effectiveness of selection by owls of deer-mice (*Peromyscus maniculatus*) which contrast in color with their background. *Contrib. Lab. Vertebr. Biol. Univ. Michigan* 34:1–20.

Dubost, G. 1991. Body coloration and colour alternation among tropical forest mammals. *Z. Zool. Syst. Evolutionsforsch.* 29:123–138.

Eaton, R. L. 1976. A possible case of mimicry in larger mammals. *Evolution* 30:853–856.

Eisenberg, J. F. 1989. *Mammals of the Neotropics: The Northern Neotropics*, vol. 1. Chicago: Univ. Chicago Press.

Endler, J. A. 1978. A predator's view of animal colour patterns. *Evol. Biol.* 11:319–364.

Endler, J. A. 1981. An overview of the relationships between mimicry and crypsis. *Biol. J. Linn. Soc.* 16:25–31.

Endler, J. A. 1988. Frequency-dependent predation, crypsis and aposematic coloration. *Proc. Trans. Roy. Soc. London.* (ser. B) 319:505–523.

Endler, J. A. 1990. On the measurement and classification of colour in studies of animal colour patterns. *Biol. J. Linn. Soc.* 41:315–352.

Endler, J. A. 1993. The color of light in forest and its implications. *Ecol. Monogr.* 63:1–27.

Estes, R. D. 1991. *The Behavior Guide to African Mammals*. Berkeley: Univ. California Press.

Ewer, R. F. 1973. *The Carnivores*. Ithaca, N.Y.: Cornell Univ. Press.

Felsenstein, J. 1985. Phylogenies and the comparative method. *Amer. Nat.* 125:1–15.

Ficken, R. W., Matthiae, P. E., and Horwich, R. 1971. Eye marks in vertebrates: Aids to vision. *Science* 173:936–938.

Fox, M. W. 1971. *Behaviour of Wolves, Dogs and Related Canids*. New York: Harper & Row.

Geffen, E., Gompper, M. E., Gittleman, J. L., Luh, H.-K., Macdonald, D. W., and Wayne, R. K. 1996. Size, life history traits, and social organization in the Canidae: A reevaluation. *Amer. Nat.* 147:140–160.

Gingerich, P. D. 1975. Is the aardwolf a mimic of the hyaena? *Nature* 253:191–192.

Gittleman, J. L. 1989. Carnivore group living: Comparative trends. In: J. L. Gittleman, *Carnivore Behavior, Ecology, and Evolution*, pp. 183–207. Ithaca, N.Y.: Cornell Univ. Press.

Gittleman, J. L. 1993. Carnivore life histories: A re-analysis in the light of new models. *Symp. Zool. Soc. London* 65:65–86.

Gittleman, J. L., and Luh, H-K. 1992. On comparing comparative methods. *Ann. Rev. Ecol. Syst.* 23:383–404.

Gloger, C. W. L. 1833. *Das Abändern der Vögel durch Einfluss des Klimas*. Breslau: A. Schulz.

Goldman, D., Giri, P. R., and O'Brien, S. J. 1989. Molecular genetic-distance estimates among the Ursidae as indicated by one- and two-dimensional protein electrophoresis. *Evolution* 43:282–295.

Goodhart, C. B. 1975. Does the aardwolf mimic a hyena? *Zool. J. Linn. Soc.* 57:349–356.

Grzimek's Encyclopedia of Mammals. 1990. 2nd ed. S. P. Parker, ed. New York: McGraw-Hill.

Guilford, T. 1988. The evolution of conspicuous coloration. *Amer. Nat.* 131:S7–S21.

Guilford, T. 1990. The evolution of aposematism. In: D. L. Evans and J. O. Schmidt, eds. *Insect Defenses: Adaptive Mechanisms and Strategies of Prey and Predators,* pp. 23–61. New York: State Univ. New York Press.

Guthrie, R. D., and Petocz, R. G. 1970. Weapon automimicry among mammals. *Amer. Nat.* 104:585–588.

Hadley, M. E. 1972. Functional significance of vertebrate integumental pigmentation. *Amer. Zool.* 12:63–76.

Haltenorth, T., and Diller, H. 1977. *A Field Guide to the Mammals of Africa Including Madagascar.* London: Collins.

Hamilton, W. J., III. 1973. *Life's Color Code.* New York: McGraw-Hill.

Hamilton, W. J., III., and Heppner, F. H. 1967. Radiant solar energy and the function of black homeotherm pigmentation: An hypothesis. *Science* 155:196–197.

Harvey, P. H., and Pagel, M. D. 1991. *The Comparative Method in Evolutionary Biology.* Oxford: Oxford Univ. Press.

Hasson, O. 1991. Pursuit-deterrent signals: Communication between prey and predator. *Trends Ecol. Evol.* 6:325–329.

Heran, I. 1976. *Animal Coloration: The Nature and Purpose of Colours in Vertebrates.* London: Hamlyn.

Hoglund, J. 1989. Size and plumage dimorphism in lek-breeding birds: A comparative analysis. *Amer. Nat.* 134:72–87.

Huxley, J. 1942. *Evolution: The Modern Synthesis.* New York: Harper & Bros.

Janczewski, D. N., Modi, W. S., Stephens, J. C., and O'Brien, S. J. 1995. Molecular evolution of mitochondrial 12s RNA and cytochrome b sequences in the pantherine lineage of felidae. *Mol. Biol. Evol.* 12(4):690–707.

Kaufman, D. W. 1974. Adaptive coloration in *Peromyscus polionotus:* Experimental selection by owls. *J. Mammal.* 55:271–283.

Kiltie, R. A. 1988. Countershading: Universally deceptive or deceptively universal? *Trends Ecol. Evol.* 3:21–23.

Kiltie, R. A. 1989. Testing Thayer's countershading hypothesis: An image processing approach. *Anim. Behav.* 38:542–552.

King, C. 1989. *The Natural History of Weasels and Stoats.* London: Christopher Helm.

Kingdon, J. 1977. *East African Mammals: An Atlas of Evolution in Africa,* vol. 3(A) (Carnivores). London: Academic Press.

Kitchener, A. 1991. *The Natural History of the Wild Cats.* London: Christopher Helm.

Kruuk, H. 1972. *The Spotted Hyena: A Study of Predation and Social Behavior.* Chicago: Univ. Chicago Press.

Lekagul, B., and McNeely, J. A. 1977. *Mammals of Thailand.* Bangkok: Assoc. Conserv. Wildl.

Leyhausen, P. 1979. *Behavior.* New York: Garland STPM Press.

Lyon, B. E., and Montgomerie, R. D. 1986. Delayed plumage maturation in passerine birds: Reliable signaling by subordinate males? *Evolution* 40:605–615.

Macdonald, D. W. 1984. *Encyclopedia of Mammals,* vol. 1. London: Allen and Unwin.

Macdonald, D. W. 1985. The carnivores: Order Carnivora. In: R. E. Brown and D. W. Macdonald, eds. *Social Odours in Mammals,* vol. 2, pp. 619–722. Oxford: Clarendon Press.

Maddison, W. P. 1990. A method for testing the correlated evolution of two binary characters: Are gains or losses concentrated on certain branches of a phylogenetic tree? *Evolution* 44:539–557.

Maddison, W. P., and Maddison, D. R. 1992. *MacClade: Analysis of Phylogeny and Character Evolution, Version 3.04.* Sunderland, Mass.: Sinauer.

Mallet, J., and Singer, M. 1987. Individual selection, kin selection, and the shifting balance in the evolution of warning colours: The evidence from butterflies. *Biol. J. Linn. Soc.* 32:337–350.

Mottram, J. C. 1915. Some observations on pattern-blending with reference to obliterative shading and concealment of outline. *Proc. Zool. Soc. London* 1915(49):679–692.

Mottram, J. C. 1916. An experimental determination of the factors which cause patterns to appear conspicuous in nature. *Proc. Zool. Soc. London* 1916(13):383–419.

Murray, J. D. 1981. A pre-pattern formation mechanism for animal coat markings. *J. Theoret. Biol.* 88:161–199.

Nowak, R. M. 1991. *Walker's Mammals of the World,* 5th ed., vol. II. Baltimore: John Hopkins Univ. Press.

O'Brien, S. J., Nash, W. G., Wildt, D. E., Bush, M., and Benveniste, R. E. 1985. A molecular solution to the riddle of the giant panda's phylogeny. *Nature* 317:140–144.

Pecon Slattery, J., Johnson, W. E., Goldman, D., and O'Brien, S. J. 1994. Phylogenetic reconstruction of South American felids defined by protein electrophoresis. *J. Mol. Evol.* 39(3):296–305.

Pocock, R. I. 1908. Warning colouration in the musteline Carnivora. *Proc. Roy. Soc. London* 61:944–959.

Poglayen-Neuwall, I. 1990. Procyonids. In: *Grzimek's Encyclopedia of Mammals,* 2nd ed., S. P. Parker, ed., pp. 450–451. New York: McGraw-Hill.

Polis, G. A., Meyers, C., and Holt, R. D. 1989. The ecology and evolution of intraguild predation: Competitors that eat each other. *Ann. Rev. Ecol. Syst.* 20:297–330.

Porter, W. P. 1967. Solar radiations through the living body walls of vertebrates with emphasis on desert reptiles. *Ecol. Monogr.* 37:273–296.

Poulton, E. B. 1890. *The Colours of Animals.* London: Kegan Paul, Trench, Trübner.

Powell, R. A. 1982. Evolution of black-tipped tails in weasels: Predator confusion. *Amer. Nat.* 119:126–131.

Pycraft, W. P. 1925. *Camouflage in Nature.* London: Hutchinson.

Rettenmeyer, C. W. 1970. Insect mimicry. *Ann. Rev. Entomol.* 15:43–74.

Ridley, M. 1983. *The Explanation of Organic Diversity.* Oxford: Oxford Univ. Press.

Roosevelt, T. 1911. Revealing and concealing coloration in birds and mammals. *Bull. Amer. Mus. Nat. Hist.* 30:119–231.

Rougeot, J. 1981. Déterminisme de la répartition de la pigmentation dans pelage et la peau des mammifères. *Ann. Genet. Sel. Anim.* 13(1):9–16.

Rounds, R. C. 1987. Distribution and analysis of colourmorphs of the black bear (*Ursus americanus*). *J. Biogeogr.* 14:521–528.

Sazima, I. 1991. Caudal luring in two neotropical pitvipers, *Bothrops jararaca* and *B. jararacussu. Copeia* 1991(1):245–248.

Schaller, G. B. 1967. *The Deer and the Tiger.* Chicago: Univ. Chicago Press.

Schmidt-Nielsen, K. 1979. *Animal Physiology.* London: Cambridge Univ. Press.

Searle, A. G. 1968. *Comparative Genetics of Coat Colour in Mammals.* New York: Academic Press.

Seidensticker, J., and Lumpkin, S. 1991. *Great Cats: Majestic Creatures of the Wild.* Emmaus, Pa.: Rodale Press.

Siegel, S. 1956. *Nonparametric Statistics for the Behavioral Sciences.* New York: McGraw-Hill.

Sillen-Tullberg, B., and Moller, A. P. 1993. The relationship between concealed ovulation and mating systems in anthropoid primates: A phylogenetic analysis. *Amer. Nat.* 141:1–25.

Sokal, R. R., and Rolff, F. J. 1981. *Biometry: The Principles and Practice of Statistics in Biological Research,* 2nd ed. San Francisco: W. H. Freeman.

Stoddart, D. M. 1970. Tail tip and other albinisms in voles of the genus *Arvicola* Claepede 1799. *Symp. Zool. Soc. London.* 26:271–282.

Taylor, P. J., Meester, J., and Rautenbach, I. L. 1990. A quantitative analysis of geographic colour variation in the yellow mongoose *Cynictis penicillata* (Cuvier, 1829) (Mammalia: Viverridae) in southern Africa. *Ann. Transvaal Museum* 35(11): 177–198.

Thayer, A. G. 1909. *Concealing Coloration in the Animal Kingdom.* New York: Macmillan.

Thayer, A. H. 1896. The law which underlies protective coloration. *Auk* 13:124–129.

Turner, J. R. G. 1971. Studies of Mullerian mimicry and its evolution in Bu moths and Helioconiid butterflies. In: E. B. Creed, ed. *Ecological Genetics and Evolution: Essays in Honor of E. B. Ford,* pp. 224–260. Oxford: Oxford Scientific Publications.

Vane-Wright, R. I. 1980. On the definition of mimicry. *Biol. J. Linn. Soc.* 13:1–6.

Waage, J. K. 1981. How the zebra got its stripes: Biting flies as selective agents in the evolution of zebra coloration. *J. Entomol. Soc. South Africa* 44:351–358.

Wallace, A. R. 1889. *Darwinism.* London: Macmillan.

Walsberg, G. E. 1983. Coat color and solar heat gain in animals. *Bioscience* 33:88–91.

Walsberg, G. E. 1990. Convergence of solar heat gain in two squirrel species with contrasting coat colors. *Physiol. Zool.* 63:1025–1042.

Wayne, R. K., Benveniste, R. E., Janczewski, D. N., and O'Brien, S. J. 1989. Molecular and biochemical evolution of the Carnivora. In: J. L. Gittleman, ed. *Carnivore Behavior, Ecology, and Evolution,* pp. 495–535. Ithaca, N.Y.: Cornell Univ. Press.

Werdelin, L., and Solounias, N. 1991. *The Hyaenidae: Taxonomy, Systematics and Evolution.* Fossils and Strata 30. Oslo: Universitetsforlaget.

Wickler, W. 1968. *Mimicry in Plants and Animals* (transl. from German by R. D. Martin). London: Weidenfeld and Nicholson.

Wozencraft, W. C. 1989. The phylogeny of the Recent Carnivora. In: J. L. Gittleman, ed. *Carnivore Behavior, Ecology, and Evolution,* pp. 495–535. Ithaca, N.Y.: Cornell Univ. Press.

Young, J. Z. 1957. *The Life of Mammals.* Oxford: Clarendon Press.

CHAPTER 5

Sympatry in Canids: A Review and Assessment

WARREN E. JOHNSON, TODD K. FULLER,
AND WILLIAM L. FRANKLIN

The organization of animal communities, which is influenced by many factors, has been of primary interest to ecologists for several decades. Since the seminal work of Rosenzweig (1966), however, relatively few papers have attempted to synthesize and review our knowledge of carnivore community organization. The Canidae, or dog family, is an instructive group well suited to the analysis of community structure. With members on all continents except Antarctica (Ginsberg and Macdonald 1990), it is one of the most widely distributed families of the order Carnivora. Foxes, jackals, wolves, and dogs exhibit a wide variety of behavioral and ecological adaptations. Their social systems range from loose pairs to large packs, and they are found in almost every habitat, from deserts to tropical rain forests, high country, and arctic pack ice. Canids also demonstrate considerable intraspecific variation; their social structure, habitat use, food habits, body size, and reproductive parameters can vary substantially under different ecological conditions (Macdonald 1983; Moehlman 1989; Geffen et al. 1996).

The flexibility of canid species, along with many other adaptations to environmental constraints (Bekoff et al. 1981; Bekoff et al. 1984), make them potentially strong competitors in many situations. Competition may be evidenced in the spatial distribution, resource partitioning, morphological variation, and interspecific relations between competing species. Several extensive reviews and comparisons of the ecology of canid species are available (e.g., Fox 1971; Gittleman 1986; Moehlman 1986, 1989; Ginsberg and Macdonald 1990; Sheldon 1992), but little emphasis has been placed on summarizing the interactions and relationships of sympatric species.

In this chapter we review and summarize examples of these relationships and interactions and discuss the ecological factors promoting sympatry. Specifically, we compare ecological parameters considered important in assessing potential resource competition, including body mass, social system, and degrees of prey, spatial, habitat, and temporal overlap. We also explore how competition may alter behavioral and ecological patterns, discuss the role of sympatry in the conservation and management of canid species, and identify areas of future research.

We focus our review on studies of clearly sympatric populations, conducted at the same time and place, to avoid results that may have been confounded by variations in food or habitat availability. When appropriate, we consider data from single-species studies in making causal inferences. Whenever possible, for each set of sympatric canids, we categorized the extent of diet, spatial, habitat, and temporal overlap between the sympatric species as low or high, on the basis of published studies. Body masses (from Ginsberg and Macdonald 1990) reflect ranges from throughout each species' distribution, and not necessarily from where they are sympatric with other canids.

Using the geographic regions classified by Ginsberg and Macdonald (1990), we identified, on the basis of distribution and broad habitat utilization, currently or historically sympatric canid species. A problem with this type of analysis is determining how to deal with the variety of methods and sampling procedures used in different studies. Conclusions in many cases were difficult to interpret, because the authors failed to determine whether observed differences in resource utilization resulted from spatial segregation and differential availability of resources, or reflected true mechanisms for partitioning resources.

Patterns of Geographic Distribution

There are 34–37 canid species worldwide, depending on the classification system used (see Wozencraft 1989). Africa and South America, each supporting ten species, are the two continents with the highest canid diversity. The large geographic ranges of many canids concurrently overlap with those of several other canid species (Table 5.1). For example, red foxes (all scientific names of canids are listed in Table 5.1) are sympatric with 14 other canid species (from three geographic regions); golden jackals are sympatric with 13 species (from two regions); and gray wolves are sympatric with 11 species (from three regions). At any one location, however, the diversity of canid species is usually limited, ranging from one to a high of five in East Africa.

Canid distributions have changed dramatically during the last century and continue to be in a state of flux. Comparing the historic and current distribution patterns of 34 canid species during the last 100 years (Ginsberg and Macdonald 1990), we determined that seven have increased their ranges, eight have decreased their ranges, and nine have retained relatively stable patterns of distribution (Table 5.1); changes in the distribution of the remaining ten species were difficult to establish, owing to lack of information. Although these changes in distribution have sometimes resulted in a local decrease in canid diversity, especially in areas of high concentrations of humans, in many cases one species simply replaced another (e.g., coyotes replacing gray and red

Table 5.1. Occurences of sympatry between pairs of canid species in five regions of the world and recent (<100 years) increases (+) or decreases (−) in their geographic distributions (= means relatively stable ranges; ? means lack of information). P = sympatry predicted on the basis of geographic distributions; S = sympatry confirmed in field studies; H = historical but not current sympatry.

South America	C. tho.	C. bra.	D. cul.	D. gri.	D. gym.	D. mic.	D. sec.	D. vet.	S. ven.	U. cin.	Distribution changes
Cerdocyon thous Crab-eating zorro	−	P			P	P		P	P	P	=
Chrysocyon brachyurus Maned wolf	P	−			P			P			−
Dusicyon culpaeus Culpeo zorro			−	S	P		P				=
D. griseus Gray zorro			S	−	P						+
D. gymnocercus Azara's zorro	P	P	P	P	−						=
D. microtis Small-eared zorro	P					P			P		?
D. sechurae Sechuran zorro			P				−				?
D. vetulus Hoary zorro	P	P						−			?
Speothos venaticus Bush dog	P					P			−	P	?
Urocyon cinereoargenteus Gray fox	P								P	−	=

(continued)

Table 5.1. (*Continued*)

Sub-Saharan Africa	C. adu.	C. aur.	C. mes.	C. sim.	L. pic.	O. meg.	V. cha.	Distribution changes
Canis adustus Side-striped jackal	–	S	S		P	P		=
C. aureus Golden jackal	S	–	S		P	P		+
C. mesomelas Black-backed jackal	S	S	–		P	S	S	=
C. simensis Simien jackal				–				–
Lycaon pictus African wild dog	P	P	P		–	P		–
Otocyon megalotis Bat-eared fox	P	P	S		P	–	S	+
Vulpes chama Cape fox			S			S	–	+

Holarctic	A. lag.	C. lat.	C. lup.	C. ruf.	N. pro.	U. cin.	U. lit.	V. vel.	V. vul.	Distribution changes
Alopex lagopus Arctic fox	–	S	S						S	=
Canis latrans Coyote	S	–	S			S		S	S	+
C. lupus Gray wolf	S	S	–	H				H	S	–
C. rufus Red wolf				–		H				–
Nyctereutes procyonoides Raccoon dog			S		–				S	+
Urocyon cinereoargenteus Gray fox		S		H		–		S	S	=
U. littoralis Island gray fox							–			–
Vulpes velox Swift or kit fox		S	H			S		–		–
V. vulpes Red fox	S	S	S		S	S			–	+

North Africa and the Middle East	C. aur.	C. lup.	F. zer.	V. can.	V. pal.	V. rup.	V. vul.	Distribution changes
Canis aureus Golden jackal	–	P	P	P	P	P	P	+
C. lupus Gray wolf	P	–		P	P	P	P	–
Fennecus zerda Fennec fox	P		–				P	?
Vulpes cana Blanford's fox	P	P		–		P	P	?
V. pallida Pale fox	P	P			–			?
V. rueppelli Ruppell's fox	P	P		P		–	S	?
V. vulpes Red fox	P	P	P	P		S	–	+

South and Southeast Asia and Australia	C. aur.	C. fam.	C. lup.	C. alp.	V. cor.	V. ben.	V. fer.	V. vul.	Distribution changes
Canis aureus Golden jackal	–		P	P	P	P	P	P	+
C. familiaris dingo Dingo		–						P	=
C. lupus Gray wolf	P		–	P	P	P	P	P	–
Cuon alpinus Dhole	P		P	–		P	P	P	–
Vulpes corsac Corsac fox	P		P		–		P	P	?
V. bengalensis Bengal fox	P		P	P		–		P	–
V. ferrilata Tibetan fox	P		P	P	P		–	P	?
V. vulpes Red fox	P	P	P	P	P	P	P	–	+

wolves: Litvaitis 1992; Peterson, in press). Recent changes in distribution resulting in an *increase* in canid diversity have been documented only on islands, such as Tierra del Fuego (Chile and Argentina: Jaksic and Yañez 1983), where biogeographical factors had kept the number of canid species comparatively low across the last century.

A Review of Scientific Research, by Region

Relatively few studies have specifically examined relationships between sympatric canids; those that did were concentrated in North America and sub-Saharan Africa (Table 5.2). We found 55 studies that investigated some

Table 5.2. Studies of sympatric canid species by region of the world

Region	Species	References[a]
Sub-Saharan Africa	C. mesomelas/O. megalotis/V. chama	1, 2
	C. adustus/C. aureus/C. mesomelas	3
	C. aureus/C. mesomelas	4, 5
	C. aureus/C. mesomelas/L. pictus	6
	O. megalotis/C. aureus	7
South America	D. culpaeus/D. griseus	8–11
Holarctic	C. lupus/C. latrans/V. vulpes	12
	C. lupus/V. vulpes	13
	C. lupus/C. rufus	14
	C. lupus/C. latrans	15–22
	C. latrans/V. vulpes	23–39
	C. latrans/U. cinereoargenteus	40
	C. latrans/V. vulpes/U. cinereoargenteus	41, 42
	C. latrans/V. velox	43–47
	V. vulpes/U. cinereoargenteus	48–56
	V. vulpes/A. lagopus	57–59
North Africa and the Middle East	V. ruppelli/V. vulpes	60
South and Southeast Asia and Australia	C. familiaris dingo/V. vulpes	61, 62

[a]1: Bothma et al. 1984; 2: Nel 1984; 3: Fuller et al. 1989; 4: Lamprecht 1978; 5: Moehlman 1983; 6: Schaller 1972; 7: Davis 1980; 8: Fuentes and Jaksic 1979; 9: Jaksic et al. 1983; 10: Johnson 1992; 11: Jimenez 1993; 12: Dekker 1989; 13: Macdonald et al. 1980; 14: USFWS 1989; 15: Berg and Chesness 1978; 16: Fuller and Keith 1981; 17: Carbyn 1982; 18: Cowardin et al. 1983; 19: Schmitz and Lavigne 1987; 20: Paquet 1991; 21: Paquet 1991; 22: Paquet 1992; 23: Robinson 1961; 24: Linhart and Robinson 1972; 25: Johnson and Sargeant 1977; 26: Green and Flinders 1981; 27: Sargeant 1982; 28: Dekker 1983; 29: Voigt and Earle 1983; 30: Englehardt 1986; 31: Schmidt 1986; 32: Chambers 1987; 33: Klett 1987; 34: Major and Sherburne 1987; 35: Sargeant et al. 1987; 36: Harrison et al. 1989; 37: Sargeant and Allen 1989; 38: Theberge and Wedeles 1989; 39: Thurber et al. 1992; 40: Wooding 1984; 41: King 1981; 42: Ingle 1990; 43: O'Farrell 1984; 44: O'Farrell and Gilbertson 1986; 45: Carbyn 1989; 46: Covell 1992; 47: Cypher and Scrivner 1992; 48: Errington 1935; 49: Scott 1955; 50: Wood et al. 1958; 51: Follman 1973; 52: Ashby 1974; 53: Yearsley and Samuel 1980; 54: Hockman and Chapman 1983; 55: Schloeder 1988; 56: Sunquist 1989; 57: Schmidt 1986; 58: Smits et al. 1989; 59: Bailey 1992; 60: Lindsay and Macdonald 1986; 61: Brown 1990; 62: Marsack and Campbell 1990.

ecological aspect of two sympatric canids, seven studies of three sympatric canids, and none that concurrently examined relationships of four or more sympatric canids. Because there are large gaps in our knowledge of the Canidae, even about many of the more common, more widely distributed species, we present here a brief review of just seven of the more extensively studied assemblages of coexisting canids (Figure 5.1), summarizing what is known and what is unknown.

East Africa

Black-backed jackals, golden jackals, side-striped jackals, African wild dogs, and bat-eared foxes are sympatric in the Serengeti ecosystem and Rift Valley of Tanzania and Kenya, but have never been studied concurrently in a comprehensive manner. This is the largest group of sympatric canid species in the world, probably reflecting the high diversity and abundance of food in the region (Van Valkenburgh 1985; Wayne et al. 1989). The spatial relationship among the five species is unclear, but African wild dogs and bat-eared foxes do simultaneously overlap spatially with each other and with at least one of the three jackal species. Bat-eared foxes, the smallest of the five species, differ ecologically from the other canids (Table 5.3) in preying principally on a wide variety of insects, especially termites (*Hodotermes:* Lamprecht 1978; Malcolm 1986; Nel 1990; Nel and Mackie 1990). African wild dogs, the largest of the five species, differ from the other canids in living in packs of related adults and in preying on large herbivores such as Thompson's gazelle (*Gazella thomsoni*), wildebeest (*Connochaetes taurinus*), and impala (*Aepyceros melampus*) (Schaller 1972; Frame 1986; Fuller and Kat 1990).

Of the mechanisms of resource partitioning operant among the five species, those among the three jackals are the least understood. Although these three species diverged from one another more than 2 million years ago, they remain behaviorally and morphologically very similar (Wayne et al. 1989). Jackals of each species weigh 7–15 kg, are omnivorous, and have a social system centered around a mated pair (Lamprecht 1978; Moehlman 1983, 1986; Fuller et al. 1989). Although there are no obvious behavioral or physiological traits that determine the distribution of particular jackal species, when sympatric they seem to segregate spatially—the side-striped jackals inhabiting areas of more dense undergrowth, the black-backed jackals inhabiting woodland areas, and the golden jackals inhabiting open plains—as well as temporally (Moehlman 1983; Fuller et al. 1989). Still, all of the three jackal species can be trapped at the same site (Fuller et al. 1989), and golden and black-backed jackals are often seen at the same carcasses (Schaller 1972:350; Lamprecht 1978). The observed sympatry among these three species may be facilitated by their opportunistic foraging habits, flexible social systems (including cooperative hunting and pup-rearing), and, perhaps most important, a diverse and abundant

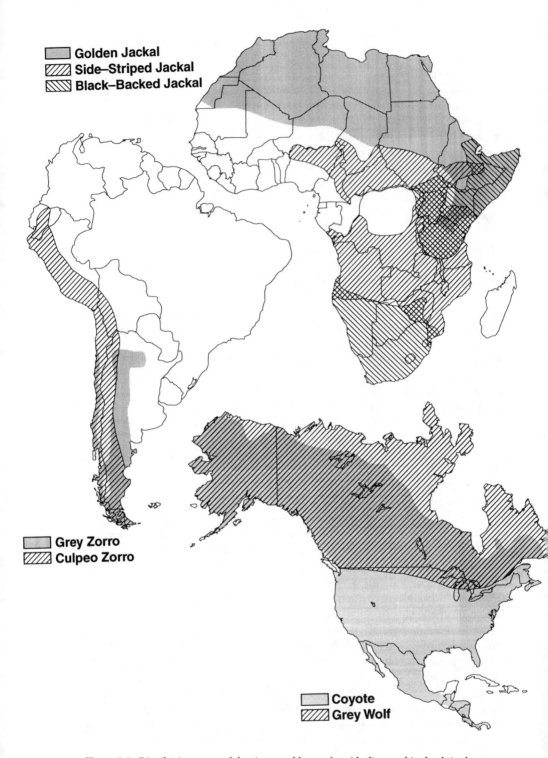

Figure 5.1. Distribution maps of the six assemblages of canids discussed in depth in the text.

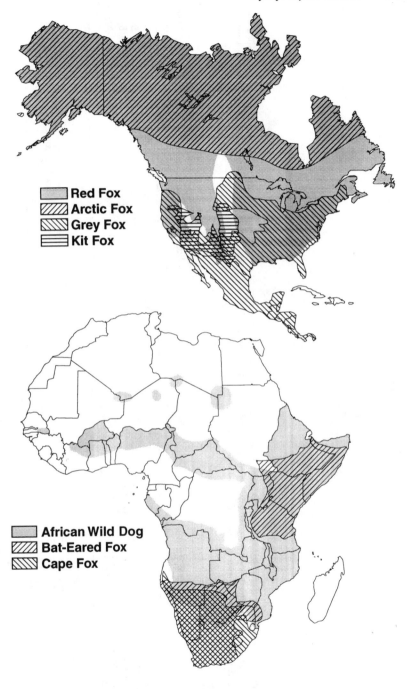

Figure 5.1 (Cont.)

Table 5.3. Body mass, diet class, and diet breadth of canid species. Resources that are partitioned (no overlap in use) are listed below the dashed lines, and those that are not partioned (overlap) are listed above the dashed lines. Single dashes denote areas that have not been addressed, and question marks indicate suppositions from one-species studies.

EAST AFRICA

Species	Body mass (kg)	Diet	Diet breadth	Bat-eared fox	Black-backed jackal	Side-striped jackal	Golden jackal	African wild dog
Bat-eared fox	4	Insectivore	High	- - - - - -	Temporal	Temporal, habitat	—	Spatial
Black-backed jackal	6–12	Omnivore	High	Food	- - - - - -	Temporal	Temporal	Spatial
Side-striped jackal	6–14	Omnivore	High	Food	- - - - - -	- - - - - -	Temporal	Spatial
Golden jackal	7–15	Omnivore	High	Food	Habitat, spatial	Habitat, spatial	- - - - -	Spatial
African wild dog	17–36	Carnivore	Narrow	Food	Habitat, spatial	Food	Food	- - - - -

SOUTHERN AFRICA

Species	Body mass (kg)	Diet	Diet breadth	Cape fox	Bat-eared fox	Black-backed jackal	African wild dog
Cape fox	3	Omnivore	High	- - - - -	Spatial?	Spatial?	Spatial
Bat-eared fox	4	Insectivore	High	Food	- - - - - -	Spatial	Spatial
Black-backed jackal	6–12	Omnivore	High	Food?	Food?	- - - -	Spatial
African wild dog	17–36	Carnivore	Narrow	Food	Food	Food	- - - - - -

Southern South America

Species	Body mass (kg)	Diet	Diet breadth	Gray zorro	Culpeo zorro
Gray zorro	3–4	Omnivore	High	- - - - - -	Temporal, food
Culpeo zorro	8–12	Omnivore	High	Spatial, habitat	- - - - - -

Northern North America

Species	Body mass (kg)	Diet	Diet breadth	Red fox	Coyote	Gray wolf
Red fox	3–11	Omnivore	High	- - - - - -	Food, temporal	Spatial, temporal
Coyote	9–16	Omnivore	High	Spatial, habitat	- - - - - -	Spatial, temporal
Gray wolf	16–60	Carnivore	Low	Food	Food	- - - - - -

Southeastern North America

Species	Body mass (kg)	Diet	Diet breadth	Gray fox	Red fox	Coyote
Gray fox	3–7	Omnivore	High	- - - - - -	Spatial	Spatial
Red fox	3–11	Omnivore	High	Habitat?, food?	- - - - - -	Food, temporal
Coyote	9–16	Omnivore	High	Temporal, micro-habitat?	Spatial	- - - - - -

(continued)

Table 5.3. (*Continued*)

SOUTHWESTERN NORTH AMERICA

Species	Body mass (kg)	Diet	Diet breadth	Kit fox	Red fox	Coyote
Kit fox	2–3	Omnivore	High	- - - - -	—	Spatial
Red fox	3–11	Omnivore	High	—	- - - - -	Food, temporal
Coyote	9–16	Omnivore	High	—	Spatial	- - - - -

NORTHERN NORTH AMERICA

Species	Body mass (kg)	Diet	Diet breadth	Arctic fox	Red fox
Arctic fox	3–11	Omnivore	High	- - - - -	Food
Red fox	3–11	Omnivore	High	Spatial?	- - - - -

food supply (Stuart 1976; Lamprecht 1978; Macdonald 1979; Moehlman 1983; Rowe-Rowe 1983).

Southern Africa

Black-backed jackals, African wild dogs, Cape foxes, and bat-eared foxes are sympatric in Namibia, Botswana, and the Republic of South Africa. The only concurrent ecological information on these species is from incidental observations on black-backed jackals, bat-eared foxes, and Cape foxes by Bothma et al. (1984) in the Namib Desert, Namibia, and by Nel (1984) in the southwestern Kalahari. Bothma et al. (1984) concluded that these three species showed food-niche and temporal separation, but not spatial separation (Table 5.3). The results of Bothma et al. (1984) are difficult to interpret because the food-habits data presented are from different years, and the conclusions on temporal and spatial separation stem only from visual sightings and not from radiotelemetry data. Nel (1984) came to similar conclusions concerning resource partitioning after 9 years of short (1- to 2-week), biannual observation periods. Although African wild dog densities in this area have been reduced drastically since historic times, the animals are now probably found in sympatry with one or more of the other three species (and side-striped jackals) where they still occur.

Southern South America

The culpeo zorro and the gray zorro are sympatric in southern South America, especially in the foothills between the Andes and the lowlands to the east and west. Size differences between the two increase with latitude, the culpeo zorro weighing 7–12 kg and the gray zorro weighing 3–4 kg in the southern portion of their ranges. Fuentes and Jaksic (1979) attributed this phenomenon to increasing sympatry between the two species, resulting in different body sizes to more efficiently partition food resources. But they did not find a correlation between size differences between the two species and variations in mean size of potential prey species at different latitudes; prey size remained constant. Other means of resource segregation, such as spatial and/or temporal differences in habitat use, were not addressed.

The most detailed radiotelemetry study of the two species where they are sympatric was conducted in Torres del Paine National Park in the Patagonia of southern Chile (Johnson 1992; Johnson and Franklin 1994a; Johnson and Franklin 1994b; Johnson and Franklin 1994c). In this study, differences were found in food habits and in habitat utilization. Although they fed on the same prey items, the culpeo zorro preyed significantly more on European hare (*Lepus capensis*) and less on carrion than did the gray zorro, which, for its part, fed more on insects and berries. Gray zorro were found in more open, upland grassland and shrub transition habitats, whereas culpeo zorro used more dense, matorral shrubland or *Nothofagus* thicket. But since gray and culpeo zorros were distributed in nonoverlapping, interspersed home ranges, Johnson and Franklin (1994c) concluded that the observed differences in prey habits and habitat utilization could be attributed to differences in resource availability in the home ranges (Table 5.3). Within its home range each species showed selection for the same habitat types and fed on vertebrate prey items in proportion to prey availability. Johnson and Franklin (1994a) hypothesized, on the basis of differential energy needs related to body mass, that the ranges of both species were determined by the more dominant culpeo zorro. Culpeo zorro would occur where densities of medium-sized prey species were sufficient, whereas the gray zorro, being able to exploit smaller food items, would occupy other areas.

These findings were supported by Jimenez (1993) in a study on the two species in north-central Chile. He found that culpeo zorro excluded gray zorro from the areas of higher densities of small rodents, a disparity that was reflected in their diets. Although both species consumed the same prey items, culpeo zorro consumed comparatively more mammals and gray zorro more reptiles, insects, and fruit.

Northern North America

Gray wolves, coyotes, and red foxes are currently sympatric throughout parts of the northern United States and much of southern Canada. They have been the most intensively studied species of sympatric carnivores (26 of the 61 studies) and the subject of several reviews (Carbyn 1987; Litvaitis 1992; Peterson, in press). Factors influencing the distribution of the three species, however, are incompletely quantified, and the degree of competition among them, or precisely what limiting resources might be factors in such competition, is not fully known. As is true of other sympatric canids, the distribution of these three species has changed dramatically during the last 200 years. The present distribution of wolves, coyotes, and red foxes is influenced strongly by altered habitats, human intervention, and prey availability. Red foxes have expanded their range south and west from their original distribution in the northern parts of the continent. Coyotes, previously confined to mainly arid areas in western North America, are currently found in every state, province, and country north of Panama. Gray wolves, once distributed across most of North America, have been severely reduced (though they are now expanding) south of the 48th parallel (Schmidt 1986).

These changes in distribution have been attributed to habitat alterations by humans and persecution by humans, which have generally favored red foxes and coyotes and not gray wolves, and to strong interspecific competition among the three canids (Table 5.3). There is in fact substantial evidence of an inverse relationship between the population densities of gray wolves and coyotes (Berg and Chesness 1978; Carbyn 1982), and between the densities of coyotes and red foxes (Robinson 1961; Linhart and Robinson 1972; Sargeant 1982; Cowardin et al. 1983; Sargeant et al. 1984). One of the proximate causes of this numerical response may be the exclusion or displacement of coyotes by gray wolves (Berg and Chesness 1978; Fuller and Keith 1981) and of red foxes by coyotes (Voigt and Earle 1983; Major and Sherburne 1987; Sargeant et al. 1987; Harrison et al. 1989; Ingle 1990) through active aggression by the larger canid and/or avoidance by the smaller one (Fuller and Keith 1981; Carbyn 1982; Dekker 1983, 1989; Sargeant and Allen 1989). Wolves have often been reported to kill coyotes (Berg and Chesness, 1978; Carbyn 1982; Paquet 1992), and there are anecdotal accounts of coyotes evicting red foxes from dens (Hilton 1978) or killing them, especially when the foxes are caught in hunters' traps (Voigt and Earle 1983; Sargeant and Allen 1989). Nonetheless, the complete exclusion of red foxes by coyotes has been demonstrated only in small areas (Dekker 1983; Sergeant et al. 1987; Harrison et al. 1989).

Alternatively, the apparent spatial segregations observed could result from differential preferences in habitat or prey selection (Todd et al. 1985), especially between coyotes and red foxes, whose home ranges do not seem to overlap as readily as those of gray wolves and coyotes do (Sargeant and Allen 1989; Paquet 1991a; Paquet 1991b). In support of this contention, some studies have found that coyotes and red foxes select different habitats within the same area (Major and Sherburne 1987; Theberge and Wedeles 1989). Although coyotes and red foxes generally feed on the same food items, they do so in different proportions, with red foxes feeding to a greater extent on small mammals, insects, and fruit, and coyotes feeding to a greater extent on medium-sized mammals (Green and Flinders 1981; Chambers 1987; Major and Sherburne 1987; Theberge and Wedeles 1989; Dibello et al. 1990). It is unclear, however, whether differences in feeding habits are related to prey availability or to differences in selection. For example, Theberge and Wedeles (1989) found that coyotes and red foxes used the same major prey species in the same order of importance during a peak in the snowshoe hare (*Lepus americanus*) cycle, but that during a decline in hare populations the red foxes, more than the coyotes, shifted their diet to alternative prey such as small rodents.

Using a slightly different argument, Paquet (1992) proposed that the association of gray wolves and coyotes was influenced largely by prey availability. He hypothesized that gray wolf distribution was determined by the availability of large ungulate prey species, and that coyotes were excluded from or avoided these areas unless they were able to scavenge on carrion from the kills of gray wolves. Paquet (1992) and Meleshko (1986) suggested that food-resource partitioning, the larger gray wolves preying on the larger prey species, also facilitated the spatial overlap of these two canids. Similarly, Schmitz and Lavigne (1987) concluded that differences in the sizes of gray wolves and coyotes could be accounted for by differences in their food habits, and not by evolutionary pressures between these two competing carnivores.

Only a few observations have been made on the relationship between gray wolves and red foxes (Andriashek et al. 1985; Dekker 1989), but Mech (1970) concluded that red foxes probably benefited by association with gray wolves because of the food provided them by carrion and abandoned kills. There is also some indirect evidence that red fox numbers are generally greater in areas where wolves are found, perhaps because of a corresponding decrease in coyotes (Peterson, in press), even though gray wolves sometimes kill red foxes.

Southeastern United States

Coyotes, red foxes, and gray foxes are sympatric in portions of southern and eastern United States. Although the three are economically and ecologically important furbearers, there is no general agreement on their relationships to one another, and relatively few empirical studies have been done on sympatric populations of all three species. Over the last 300 years, as the coyote has expanded its range eastward and as red foxes were introduced from Europe and became established in eastern North America in the seventeenth century (Mansueti 1955), these three canids have become more sympatric.

There have been conflicting opinions, each based largely on circumstantial evidence, on whether or not there is an inverse relationship in population density between coyotes and gray foxes (Davis 1966; Small 1971; Stoudt 1971; Sargeant 1982; Wooding 1984), or one between red and gray foxes (Trapp and Hallberg 1975). There is some evidence that gray foxes dominate red foxes in interspecific interactions (Fox 1971; Schloeder 1988).

Several studies have found overlap in red and gray fox home ranges (Follman 1973; Schloeder 1988; Sunquist 1989; Ingle 1990), with varying degrees of differential habitat use; gray foxes have generally been found to use the more wooded areas and red foxes to use more open habitats (Table 5.3).

As discussed previously, home-range overlap between red foxes and coyotes seems to be rare. In contrast, overlap of gray fox home ranges with those of coyotes seems to be more common, the gray foxes using slightly different habitats (Wooding 1984) than do the coyotes, or gray foxes avoiding coyote activity loci, both temporally and spatially (Ingle 1990). Unfortunately, no studies have included an adequate examination of the food habits of these sympatric canids.

Concurrent food-habits studies of coyotes, red foxes, and gray foxes are rare. King (1981) found that these three carnivores fed on the same food types in winter in northeastern Arkansas, but in different proportions. In an apparent contradiction, King (1981) concluded that although diet overlap was high among the three species, their diets were only superficially similar and would not result in significant competition. Hockman and Chapman (1983) analyzed the stomachs of red and gray foxes collected during two years in Maryland and also found high diet overlap. They concluded that the probability of competition is reduced because red foxes feed primarily on small mammals, whereas gray foxes are more omnivorous, feeding more on plants such as persimmon fruit (*Diospyro virginiana*) and corn (*Zea mays*). Using a different approach, Jaslow (1987) concluded that morphological and physiological differences between red and gray foxes may facilitate sympatry by allowing each species to exploit different foods more effectively.

Southwestern United States

Kit foxes (or swift foxes) and coyotes are sympatric in southwestern United States and, recently, non-native red foxes have extended their range over large parts of this region. Kit fox densities have been reported either to increase after coyote population reductions (Robinson 1961; Covell 1992) or to show no detectable change (Cypher and Scrivner 1992). Coyotes are known to kill kit foxes (Egoscue 1956; Scott-Brown et al. 1987) and, more rarely, red foxes kill kit foxes (Ralls and White 1995). As was discussed earlier, coyotes and red foxes do not overlap spatially; in contrast, coyotes and kit foxes do have overlapping home ranges (White et al. 1994).

Northwestern North America

Red foxes and arctic foxes are sympatric in Alaska and northwestern Canada. Smits et al. (1989), in the only concurrent study on sympatric populations in the region, determined that the two species have similar summer food habits, leading to potential competition. Similarly, Hersteinsson and Macdonald (1982), on the basis of similarities in morphology, social organization, and resource utilization, determined that the two species would be direct competitors where sympatric.

In an artificial enclosure, arctic foxes were subordinate to red foxes (Rudzinski et al. 1982) and appeared to avoid close encounters with red foxes in the wild (Schamel and Tracy 1986). In spite of their apparent behavioral dominance, red foxes have not been observed excluding arctic foxes from any area where the two were naturally sympatric. Sterile red foxes, however, have been employed to eliminate introduced arctic foxes from islands in Alaska (Bailey 1992). The potential ability of red foxes to outcompete and exclude arctic foxes from an area may be attenuated in some places by the greater capacity of arctic foxes to withstand colder temperatures and to acquire winter food (Chesemore 1975; Smits et al. 1989).

Generalizations from Studies of Sympatric Canids

Factors Influencing the Numbers of Sympatric Canids

It is obvious that various canid species can successfully coexist in sympatry in many situations. The number of canids that share a given area seems to be determined by a combination of biogeographical history, the impact of human intervention, and environmental diversity and productivity. For example, large numbers of sympatric canids occur in portions of eastern and southern Africa because of the heterogeneous environments characterizing the regions, and because large, diverse populations of ungulates are available. Canids, which are coursing predators, are especially well-adapted to these open terrains. South America also supports a high number of canid species, but does not have as many sympatric canids in any given area. This pattern may reflect the relative scarcity of large ungulate species in South America (see Keast 1972). In contrast, Australia, because of its unique biogeographical isolation, has only two canid species, dingoes and red foxes, both of which are introduced and thriving in spite of human persecution. For these, Australia offers suitable habitat, apparently minimal competition, and abundant prey.

Determinations of canid diversity and the numbers of sympatric canid species are also dependent on the scale employed to assess the degree of overlap. Many canid species may be able to maintain exclusive home ranges, thus excluding other canid species on a small scale or at a local level. Generally, however, one species does not completely exclude another from a region.

Humans can play important roles in canid distribution, diversity, and sympatry as a "keystone species," by altering habitats and disrupting community structure (Peterson, in press). Humans have actively removed certain canids, especially the larger ones (e.g., gray wolves) from certain broad areas. In contrast, some canids may be more tolerant of humans than others are, and may use areas of human settlement as refuges from their competitors. Although nonphysical to lethal aggression has been well-documented between several canid species, most notably by gray wolves toward coyotes, coyotes toward red foxes, and dingoes toward red foxes, the predation of one canid on another does not seem to be common (but coyotes regularly kill kit foxes; Ralls and White 1995).

Although there are no examples in canids that the transmission of disease has determined the distribution of a species, rabies and distemper, for example, are important sources of mortality in some populations, and canids are often disease vectors. In Alaska, wolves acquire rabies only in areas of overlap with arctic foxes (Brand et al., in press). Red foxes are also a leading carrier of rabies, and its prevalence can be linked to the density and distribution of red fox populations (Macdonald 1980; Carey 1982).

Although epizootic outbreaks can severely diminish canid densities, populations generally recover after densities become too low to sustain the outbreak

(Macdonald 1980). Such outbreaks may have lasting effects on small, relic populations of endangered species, however, permanently altering community structure. Isolated populations of African wild dogs and Simien jackals (also known as Ethiopian wolves), for example, could be especially vulnerable to epizootics. African wild dog populations have been exposed to or reduced by several diseases that are transmitted by canids, including rabies, canine distemper, parvo, and anthrax (Schaller 1972; Fanshawe et al. 1991; Alexander et al. 1993). Simien jackals also are vulnerable to these same diseases, especially via domestic dogs (Gottelli and Sillero-Zubiri 1992).

Competitive Interactions in Canids

The occurrence of three canid species in general sympatry is the predominant pattern throughout most of North America, Eurasia, and Africa. Moreover, these species usually fill one of three broadly defined ecological roles based on size-related diets: a large-sized (> 20 kg), mainly carnivorous species (e.g., gray wolves, African wild dogs, or dholes); a medium-sized (10–20 kg), omnivorous species (e.g., golden jackals or coyotes); and a small-sized, more omnivorous species with more flexible food habits (e.g., Cape foxes and red foxes). Situations including four sympatric canid species occur in parts of Africa by the addition of the bat-eared fox, which avoids potential competition with the other canids by specializing on insects. The presence of five sympatric species in East Africa is an anomaly, perhaps explained by the unusually large numbers of herbivore prey in the region.

Some canid communities lack a large-sized canid species, owing either to historic biogeographic causes, to the absence of sufficient prey, or to extirpation caused by humans or by habitat loss. In the absence of a large-sized canid species, two general patterns are observed. First, as in northern North America (with arctic foxes and red foxes), Australia (with dingoes and red foxes), and much of South America (e.g., culpeo zorros and gray zorros), only two canid species coexist, and the two interact in a way similar to that between the smaller two species in the three-species scenario. Second, a third species (e.g., coyotes, red foxes, or golden jackals) invades and partially fills the role of the missing large-sized canid and/or disrupts the populations and interrelationships of the smaller two species.

Interactions between pairs of canids produce several outcomes, including complete exclusion, partial exclusion, scattered interspecies territories, or complete overlap. Although these outcomes and their mechanisms for partitioning resources are complex, some patterns emerge that can potentially be explained in terms of energetics. We will make three assumptions: (1) that the larger canid species has higher energy requirements deriving from higher basal metabolic rates (McNab 1980); (2) that the larger canid species has the potential to impose a substantial energetic cost on the smaller species, through either aggression or predation; and (3) that there are two interspersed habitats in the same region, one with greater food availability.

Where there is little diet overlap, as usually occurs between the largest canid and the smallest canid in an assemblage, then the two (e.g., gray wolves and red foxes) usually overlap spatially, and may or may not use different habitats. When food habits are more similar between two canid species of different sizes (e.g., coyotes and red foxes), then interference competition (Park 1962) is common, and the larger species tends to be distributed wherever there are sufficient food resources to meet its energy requirements and there are no other limiting factors (e.g., gray wolves, frigid winters, or human extirpation). The smaller species in this scenario tends to occupy adjacent areas offering lower amounts of food resources, and thus often appears to be selecting different types of habitat or food. Where the two species are more closely matched in size (e.g., various jackal species, or red foxes and gray foxes), habitat partitioning is more common, and although some degree of food-resource partitioning may occur, exploitative competition (Park 1962) for food may be more likely.

In summary, these patterns suggest that, at least in those cases where food habits are similar, competition for food resources may help determine spatial distributions of sympatric canids. In these cases, the mechanisms of suppression or displacement are still unproven, but are likely to involve a combination of exclusion by the dominant species and avoidance by the subordinate, depending on the ecological circumstances in the area of potential overlap (Persson 1985). Distribution of the dominant species (and thus indirectly the subordinate species) is probably determined by energetic restraints (food, resources, temperature, etc.) and, in some cases, by human extirpation.

Indirect Measures of Community Structure

Differential morphology of sympatric canids, both between and within species, has been used to provide evidence for resource partitioning and potential competition. Wayne et al. (1989) demonstrated that, with the exception of the three East African jackal species (all *Canis*), sympatric Canidae differ from one another in body size by a factor of two or more, and also differ on one or more of the parameters measuring cranial and dental shape. Dayan et al. (1989a) demonstrated that the slope of the regression of red fox lower carnassial length against mean ambient temperature differed from that of Ruppell's sand fox between regions of sympatry and regions of allopatry. Such morphological differences, or character displacements, often have been used to infer differences in diet, foraging strategy, or life history patterns, the degree of such differences potentially reflecting levels of competition and/or resource stability (e.g., Cody 1974; Ricklefs et al. 1981; Van Valkenburgh 1985, 1989, 1991; Dayan et al. 1989a, b). These studies tend to support the importance of food-resource partitioning in forming canid community structure.

But there may be problems with this approach (see Dayan and Simberloff, this volume). First, in many cases the choice of measured characters can be somewhat arbitrary, and characteristics that do not fall into neat patterns may be ignored or unreported. Thus, more research may be needed on the functional roles of the measured attributes in order to understand their ecological and evolutionary importance. Second, alternative explanations for the observed patterns often need to be more thoroughly addressed, since the occurrence of patterns of size differences does not alone prove that these are the causes of different foraging strategies. Third, the choice of species to compare, the combining or separating of males and females in the analysis, and the choice of statistical methods employed can influence one's conclusions concerning whether size differences reflect evolutionary pressures in canid communities. Nevertheless, careful studies that do include appropriate measurements of adequate samples of sympatric canid species, along with documented ecological interactions and evolutionary relationships, can elucidate the subtle but convincing relationships posited between character displacement and competition (e.g., the various African jackals: Van Valkenburgh and Wayne 1994); such studies are rare, and should be emulated.

Management and Conservation Implications

Interspecific relationships between sympatric canids must be considered when formulating management strategies. The success of efforts to conserve a threatened or endangered canid species, or to mitigate the effect of a "pest" species, may depend on the interactive role of other sympatric canid species, too.

There are several ways in which the interactions between sympatric canid species can have a strong influence on canid conservation and management efforts. Changes in canid community structure usually follow a species' range expansion, extirpation, or reintroduction. The best-documented case is that of coyotes, which may have been able to expand their range in North America in part because of the elimination of gray wolves. The coyote has also rapidly expanded its range south into Central America, following changes in habitat associated with increased human development of the land and agricultural clearing of primary forest (Vaughan 1983). This southerly expansion could in time affect the taxonomically distinct South American canid communities (except for gray foxes), especially if coyotes prove to be successful in naturally or artificially crossing the barrier of the humid tropics of Panama into South America.

Red foxes, which may also have benefited from the removal of gray wolves in parts of their range (especially Europe), and which are expanding their distribution worldwide, could have a similar effect on local Canidae. Hersteinsson et al. (1989) believe that red foxes, which have become more common as gray wolves have been eradicated, may represent the greatest threat to arctic foxes in Scandinavia. Similarly, Hersteinsson and Macdonald (not 1982; cited in Ginsberg and Macdonald 1990) believe that range expansion of red foxes may be the most critical threat to the long-term survival of many of the lesser-known small canid species of North Africa and the Middle East. Red fox expansion in California can have, potentially, a similar detrimental effect on kit fox populations (Jurek 1992; Lewis et al. 1993).

Interspecific relationships among canids should also be considered before canid reintroduction programs are implemented. Reintroductions will probably become increasingly important to canid conservation, and because most reintroduction efforts will attempt to restore past community structure, interspecific relationships will be of concern when long-established canids reduce, either directly or indirectly, the probability of successful reintroduction. For example, the reintroduction of swift foxes (kit foxes) into parts of Canada may be hampered by red foxes. In other instances, such as the possible reintroduction of gray wolves into parts of western United States, the concern will be what effect the reintroduction will have on resident coyote and red fox populations, and on the prey community (R. Crabtree, pers. comm.). Wolf reintroduction may result in some decrease in numbers of the large prey species (Cook 1993) and, owing to concomitant reductions in coyote or red fox populations, some increase in small and medium-sized prey species.

Reintroduction efforts may also be jeopardized by the potential of hybridization. For example, in parts of southeastern United States newly introduced red wolves (*Canis rufus*) or their offspring may eventually hybridize with resident coyotes, as has apparently happened in the past (Schmitz and Kolenosky 1985; Lehman et al. 1991; Wayne and Jenks 1991; Jenks and Wayne 1992). Potential hybridization between canid species complicates conservation efforts (O'Brien and Mayr 1991; Gittleman and Pimm 1991; Geist 1992), especially when populations are at low densities. In addition to that which has occurred between red wolves and coyotes, hybridization has been noted between gray wolves and coyotes (Lehman et al. 1991), gray wolves and domestic dogs (Boitani 1982), and Simien jackals and domestic dogs (Gottelli and Sillero-Zubiri 1992).

Needed Research

In spite of the evidence supporting the prevailing idea that sympatric canids compete for available resources, the notion that canid communities or assemblages are structured solely by interspecific relations remains poorly documented. In addition to the descriptive studies that are now fairly common, we clearly need more detailed and more carefully planned studies on sympatric canids, and there are several approaches that might be particularly useful.

At the very least, studies of sympatric canids need to focus on how species are selecting and utilizing resources within individual home ranges, as opposed to selection within a study area at large. Perceptions of habitat and food preferences might change dramatically with a more detailed and more accurate accounting of resource availability for, and use by, each individual. This level of precision is especially important when there is no spatial overlap of home ranges between species, or when there is considerable variability among individuals. Ideally, this information should be combined with data on resource utilization by the species when or where populations are *not* sympatric. This could be accomplished through studies at different sites or times that have different canid assemblages, or under conditions of experimental removal or introduction of one or more species. Monitoring the effects of canid reintroduction programs, such as may occur with wolves in western United States, provides a splendid opportunity to examine canid community structure and competition experimentally.

Other experimental studies that would be of great value include purposefully increasing or decreasing the availability of particular food to see if (and how) community structure changes. A sufficient number of large prey items may be important for maintaining the populations of larger canids such as wolves or the culpeo zorro, which, in turn, may modify the distributions of smaller canids such as coyotes or South American gray zorros, respectively. Changes in ungulate hunting regulations or ungulate protection can produce large changes in food availability, and canid populations could be monitored as these adjustments occur. Similarly, studies of sympatric canids undertaken during large cyclic fluctuations of smaller mammals would likely prove productive.

Canids are wide-ranging and often difficult to observe and catch, but there are several family characteristics that make these experimental studies particularly worthwhile and particularly challenging. Because most canids are generalists and adapt to a wide variety of environmental conditions, their responses to natural changes or human perturbations may initially not be obvious. And because canids are highly mobile and have flexible socioecologies, the individuals within a species are likely to respond differently to experimental manipulation. Finally, recent and continuing changes in canid distribution imply that, in some cases, the results of competitive interactions may not yet be apparent and may still be evolving. These concerns only emphasize the planning, funding, and initiative required to quantify and understand the interactions of sympatric canids adequately, as well as the degree to which each influences the other's density, distribution, and behavior.

Acknowledgments

We thank L. Best, W. Clark, E. Klass, and an anonymous reviewer for their valuable comments on drafts of the manuscript. Preparation of this report was supported by the National Geographic Society, Patagonia Research Expeditions (Iowa State University), the Organization of American States, the International Telephone and Telegraph Corporation Fellowship Program, and the National Wildlife Federation. This is Journal Paper No. J-14873 of the Iowa Agriculture and Home Economics Experiment Station, Ames, Iowa (Project No. 2519).

References

Alexander, K. A., Conrad, P. A., Gardner, I. A., Parrish, C., Appel, M., Levy, M. G., Lerche, N., and Kat, P. W. 1993. Serologic survey of selected canine pathogens in African wild dogs (*Lycaon pictus*) and sympatric domestic dogs in Masai Mara, Kenya. *J. Zoo Animal Medicine* 24:140–144.

Andriashek, D., Kiliaan, H. P. L., and Taylor, M. K. 1985. Observations on foxes, *Alopex lagopus* and *Vulpes vulpes,* and wolves, *Canis lupus,* on the off-shore sea ice of northern Labrador. *Canadian Field-Nat.* 99:86–89.

Ashby, J. R. 1974. Food habits and distribution of the red fox (*Vulpes fulva* Desmarest) and gray fox (*Urocyon cinereoargenteus* Schreber) on the Catoosa WMA, Tennessee. Master's thesis, Tennessee Technological University.

Bailey, E. P. 1992. Red foxes, *Vulpes vulpes,* as biological control agents for introduced Arctic foxes, *Alopex lagopus,* on Alaskan islands. *Canadian Field-Nat.* 106:200–205.

Bekoff, M., Daniels, T. J., and Gittleman, J. L. 1984. Life history patterns and the comparative social ecology of carnivores. *Ann. Rev. Ecol. Syst.* 15:191–232.

Bekoff, M., Diamond, J., and Mitton, J. B. 1981. Life history patterns and sociality in canids: Body size, reproduction, and behavior. *Oecologia* 50:386–390.

Berg, W. E., and Chesness, R. A. 1978. Ecology of coyotes in northern Minnesota. In: M. Bekoff, ed. *Coyotes: Biology, Behavior, and Management,* pp. 229–246. New York: Academic Press.

Boitani, L. 1982. Wolf management in intensively used areas of Italy. In: F. H. Harrington and P. C. Paquet, eds. *Wolves of the World: Perspectives of Behavior, Ecology, and Conservation,* pp. 158–172. Park Ridge, N.J.: Noyes Publications.

Bothma, J., Du, P., Nel, J. A. J., and Macdonald, A. 1984. Food niche separation between four sympatric Namib Desert carnivores. *J. Zool. (London)* 202:327–340.

Brand, C. J., Pybers, M. J., Ballard, W. B., and Peterson, R. O. In press. Infectious and parasitic diseases of the gray wolf and their potential effect on wolf populations in North America. In: L. N. Carbyn, S. H. Fritts, and D. Seit, eds. *Wolves in a Changing World.* Edmonton, Alta.: Canadian Circumpolar Institute, University of Alberta.

Brown, G. W. 1990. Diets of wild canids and foxes in East Gippsland, 1983–1987, using predator scat analysis. *Australian Mammal.* 13:209–213.

Carbyn, L. N. 1982. Coyote population fluctuations and spatial distributions in relation to wolf territories in Riding Mountain National Park, Manitoba. *Canadian Field-Nat.* 96:176–183.

Carbyn, L. N. 1987. Gray wolf and red wolf. In: M. Novak, G. A. Baker, M. E. Osbard, and B. Malloch, eds. *Wild Furbearer Management and Conservation in North America,* pp. 358–376. Toronto: Ontario Trappers Association, Ministry of Natural Resources.

Carbyn, L. N. 1989. Status of the swift fox in Saskatchewan. *Blue Jay* 47:41–52.

Carey, A. B. 1982. The ecology of red foxes, gray foxes, and rabies in the eastern United States. *Wildl. Soc. Bull.* 10:18–26.

Chambers, R. E. 1987. Diets of Adirondack coyotes and red foxes. *Trans. NE Sect. Wildl. Soc.* 44:90 (abstract).

Chesemore, D. L. 1975. Ecology of the arctic fox (*Alopex lagopus*) in North America. In: M. W. Fox, ed. *The Wild Canids,* pp. 143–163. New York: Van Nostrand Reinhold.

Cody, M. L. 1974. *Competition and the Structure of Bird Communities.* Princeton, N.J.: Princeton Univ. Press.

Cook, R. S., ed. 1993. Ecological issues on reintroducing wolves into Yellowstone National Park. Sci. Monog. NPS/NRYELL/NRSM-93/22. Denver: U.S. Natl. Park Service.

Covell, D. F. 1992. Ecology of the swift fox (*Vulpes velox*) in southeastern Colorado. Master's thesis, University of Wisconsin, Madison.

Cowardin, L. M., Sargeant, A. B., and Duebbert, H. F. 1983. Problems and potentials for prairie ducks. *Naturalist* 34:4–11.

Cypher, B. L., and Scrivner, J. H. 1992. Coyote control to protect endangered San Joaquin kit foxes at the Naval Petroleum Reserves, California. In: J. E. Borrecco and R. E. Marsh, eds. *Proc. 15th Vertebrate Pest Conference,* pp. 42–47. Davis: Univ. California.

Davis, G. K. 1980. Interaction between bat-eared fox and silver-backed jackal. *East Africa Nat. Hist. Soc. Bull.* Sep/Oct.:79.

Davis, W. B. 1966. The mammals of Texas. *Texas Parks and Wildl. Dept. Bull. No. 41.* Austin, Tex.

Dayan, T., Simberloff, D., Tchernov, E., and Yom-Tov, Y. 1989a. Inter- and intraspecific character displacement in mustelids. *Ecology* 70:1526–1539.

Dayan, T., Tchernov, E., Yom-Tov, Y., and Simberloff, D. 1989b. Ecological character displacement in Saharo-Arabian *Vulpes:* Outfoxing Bergmann's rule. *Oikos* 55:263–272.

Dekker, D. 1983. Denning and foraging habits of red foxes, *Vulpes vulpes,* and their interactions with coyotes, *Canis latrans,* in Central Alberta, 1972–1981. *Canadian Field-Nat.* 97:303–306.

Dekker, D. 1989. Population fluctuations and spatial relationships among wolves, *Canis lupus,* coyotes, *Canis latrans,* and red foxes, *Vulpes vulpes,* in Jasper National Park, Alberta. *Canadian Field-Nat.* 103:261–264.

Dibello, F. J., Arthur, S. M., and Krohn, W. B. 1990. Food habits of sympatric coyotes, *Canis latrans,* red foxes, *Vulpes vulpes,* and bobcats, *Lynx rufus,* in Maine. *Canadian Field-Nat.* 104:403–408.

Egoscue, H. J. 1956. Preliminary studies of the kit fox in Utah. *J. Mammal.* 37:351–357.

Engelhardt, D. B. 1986. Analysis of red fox and coyote home range use in relation to artificial scent marks. Master's thesis, University of Maine, Orono.

Errington, P. L. 1935. Food habits of mid-west foxes. *J. Mammal.* 16:192–200.

Fanshawe, J. H., Frame, L. H., and Ginsberg, J. R. 1991. The wild dog: Africa's vanishing carnivore. *Oryx* 25:137–146.

Follman, E. H. 1973. Comparative ecology and behavior of red and gray foxes. Ph.D. dissert., Southern Illinois University, Carbondale.

Fox, M. W. 1971. *Behavior of Wolves, Dogs and Related Canids.* New York: Harper and Row.

Frame, G. W. 1986. Carnivore competition and resource use in the Serengeti ecosystem of Tanzania. Ph.D. dissert., Utah State University, Logan.

Fuentes, E. R., and Jaksic, F. M. 1979. Latitudinal size variation of Chilean foxes: Tests of alternative hypotheses. *Ecology* 60:43–47.

Fuller, T. K., Biknevicius, A. R., Kat, P. W., Van Valkenburgh, B., and Wayne, R. K.

1989. The ecology of three sympatric jackal species in the Rift Valley of Kenya. *African J. Ecol.* 27:313–323.

Fuller, T. K., and Kat, P. W. 1990. Movements, activity, and prey relationships of African wild dogs (*Lycaon pictus*) near Aitong, southwestern Kenya. *African J. Ecol.* 28:330–350.

Fuller, T. K., and Keith, L. B. 1981. Non-overlapping ranges of coyotes and wolves in northeastern Alberta. *J. Mammal.* 62:403–405.

Geffen, E., Gompper, M. E., Gittleman, J. L., Lu, H.-K., Macdonald, D. W., and Wayne, R. K. 1996. Size, life history traits, and social organization in the Canidae: A reevaluation. *Amer. Nat.* 147:140–160.

Geist, V. 1992. Endangered species and the law. *Nature* 357:274–276.

Ginsberg, J. R., and Macdonald, D. W. 1990. *Foxes, Wolves, Jackals, and Dogs: An Action Plan for the Conservation of Canids.* Gland: International Union for Conservation of Nature and Natural Resources.

Gittleman, J. L. 1986. Carnivore life history patterns: Allometric, phylogenetic, and ecological associations. *Amer. Nat.* 127:744–771.

Gittleman, J. L., and Pimm, S. L. 1991. Crying wolf in North America. *Nature* 351:524–525.

Gottelli, D., and Sillero-Zubiri, C. 1992. The Ethiopian wolf: An endangered endemic canid. *Oryx* 26:205–214.

Green, J. S., and Flinders, J. T. 1981. Diets of sympatric red foxes and coyotes in southeastern Idaho. *Great Basin Nat.* 41:251–254.

Harrison, D. J., Bissonette, J. A., and Sherburne, J. A. 1989. Spatial relationships between coyotes and red foxes in eastern Maine. *J. Wildl. Mgmt.* 53:181–185.

Hersteinsson, P., Angerbjorn, A., Frafjord, K., and Kaikusalo, A. 1989. The arctic fox in Fennoscandia and Iceland: Management problems. *Biol. Conserv.* 49:67–81.

Hersteinsson, P., and Macdonald, D. W. 1982. Some comparisons between red and arctic foxes, *Vulpes vulpes* and *Alopex lagopus*, as revealed by radio tracking. *Symp. Zool. Soc. London* 49:259–289.

Hilton, H. 1978. Systematics and ecology of the eastern coyote. In: M. Bekoff, ed. *Coyotes: Biology, Behavior, and Management,* pp. 567–585. New York: Academic Press.

Hockman, J. G., and Chapman, J. A. 1983. Comparative feeding habits of red foxes (*Vulpes vulpes*) and gray foxes (*Urocyon cinereoargenteus*) in Maryland. *Amer. Midl. Nat.* 110:276–285.

Ingle, M. A. 1990. Ecology of red foxes and gray foxes and spatial relationships with coyotes in an agricultural region of Vermont. Master's thesis, University of Vermont, Burlington.

Jaksic, F. M., and Yañez, J. L. 1983. Rabbit and fox introductions in Tierra del Fuego: History and assessment of the attempts at biological control of the rabbit infestation. *Biol. Conserv.* 26:367–374.

Jaksic, F. M., Yañez, J. L., and Rau, J. R. 1983. Trophic relations of the southernmost populations of *Dusicyon* in Chile. *J. Mammal.* 64:697–700.

Jaslow, C. R. 1987. Morphology and digestive efficiency of red foxes (*Vulpes vulpes*) and grey foxes (*Urocyon cinereoargenteus*) in relation to diet. *Canadian J. Zool.* 65:72–79.

Jenks, S. M., and Wayne, R. K. 1992. Problems and policy for species threatened by hybridization: The red wolf as a case study. In: D. McCullough and R. Barret, eds. *Wildlife 2001: Populations,* pp. 237–251. London: Elsevier Press.

Jimenez, J. E. 1993. Comparative ecology of *Dusicyon* foxes at the Chinchilla National Reserve in Northeastern Chile. Master's thesis, University of Florida, Gainesville.

Johnson, D. H., and Sargeant, A. B. 1977. Impact of red fox predation on the sex ratio of prairie mallards. *U.S. Fish Wildl. Ser. Rep.* 6:1–56.

Johnson, W. E. 1992. Comparative ecology of two sympatric South American foxes, *Dusicyon griseus* and *D. culpaeus*. Ph.D. dissert., Iowa State University, Ames.

Johnson, W. E., and Franklin, W. L. 1994a. The role of body size in the diets of sympatric grey and culpeo foxes. *J. Mammal.* 75:163–174.

Johnson, W. E., and Franklin, W. L. 1994b. Conservation implications of South American grey fox (*Dusicyon griseus*) socioecology in the Patagonia of Southern Chile. *Vida Silvestre Neotropical* 3:16–23.

Johnson, W. E., and Franklin, W. L. 1994c. Partitioning of spatial and temporal resources by sympatric *Dusicyon griseus* and *D. culpaeus* in the Patagonia of southern Chile. *Canadian J. Zool.* 72:1788–1793.

Jurek, R. M. 1992. Non-native red foxes in California. State of California, The Resources Agency, Department of Fish and Game, Non-game Bird and Mammal Section Report 92–04.

Keast, A. 1972. Comparisons of contemporary mammal faunas of southern continents. In: A. Keast, F. C. Erk, and B. Glass, eds. *Evolution, Mammals, and Southern Continents,* pp. 433–501. Albany: State Univ. New York Press.

King, A. W. 1981. An analysis of the winter trophic niches of the wild canids of northeast Arkansas. Master's thesis, Arkansas State University, Fayetteville.

Klett, S. 1987. Home ranges, movement patterns, habitat use, and interspecific interaction of red foxes and coyotes in northwest Louisiana. Master's thesis, Southeast Louisiana University.

Lamprecht, J. 1978. On diet, foraging behaviour and interspecific food competition of jackals in the Serengeti National Park, East Africa. *Z. Saugetierk.* 43:210–233.

Lehman, N., Eisenhawer, A., Hansen, K., Mech, L. D., Peterson, R. O., Gogan, P. J. P., and Wayne, R. K. 1991. Introgression of coyote mitochondrial DNA into sympatric North American gray wolf populations. *Evolution* 45:104–119.

Lewis, J. C., Sallee, K. L., and Goligtly, R. T. 1993. Introduced red fox in California. State of California, The Resources Agency, Department of Fish and Game, Non-game Bird and Mammal Section Report 93-10.

Linhart, S. B., and Robinson, W. B. 1972. Some relative carnivore densities in areas under sustained coyote control. *J. Mammal.* 53:880–884.

Litvaitis, J. A. 1992. Niche relations between coyotes and sympatric Carnivora. In: A. H. Boer, ed. *Ecology and Management of the Eastern Coyote,* pp. 73–85. Wildlife Research Unit, Univ. New Brunswick, Canada.

Macdonald, D. W. 1979. The flexible social system of the golden jackal, *Canis aureus. Behav. Ecol. Sociobiol.* 5:17–38.

Macdonald, D. W. 1980. *Rabies and Wildlife: A Biologist's Perspective.* Oxford: Oxford Univ. Press.

Macdonald, D. W. 1983. The ecology of carnivore social behaviour. *Nature* 301:379–384.

Macdonald, D. W., Boitani, L., and Barrasso, P. 1980. Foxes, wolves and conservation in the Abruzzo Mountains. In: E. Zimen, ed. *The Red Fox,* pp. 223–235. The Hague: Junk.

Major, J. T., and Sherburne, J. A. 1987. Interspecific relationships of coyotes, bobcats, and red foxes in western Maine. *J. Wildl. Mgmt.* 51:606–616.

Malcolm, J. R. 1986. Socio-ecology of bat-eared foxes (*Otocyon megalotis*). *J. Zool. (London)* 208:458–517.

Mansueti, R. 1955. Case of the displaced fox. *Maryland Nat.* 25:3–8.

Marsack, P., and Campbell, G. 1990. Feeding behaviour and diet of dingoes in the Nullarbor Region, Western Australia. *Australian Wildl. Res.* 17:349–357.

McNab, B. K. 1980. Food habits, energetics, and the population biology of mammals. *Amer. Nat.* 116:106–124.

Mech, L. D. 1970. *The Wolf: The Ecology and Behavior of an Endangered Species.* New York: Natural History Press.

Meleshko, D. W. 1986. Feeding habits of sympatric canids in an area of moderate ungulate density. Master's thesis, University of Alberta, Edmonton.

Moehlman, P. D. 1983. Socioecology of silver-backed and golden jackals, *Canis mesomelas* and *C. aureus.* In: J. F. Eisenberg and D. G. Kleiman, eds. *Recent Advances in the Study of Mammalian Behavior,* pp. 423–453. Amer. Soc. Mammalogists Spec. Pub. 7.

Moehlman, P. D. 1986. Ecology of cooperation in canids. In: D. I. Rubenstein and R. W. Wrangham, eds. *Ecological Aspects of Social Evolution: Birds and Mammals,* pp. 64–86. Princeton, N.J.: Princeton Univ. Press.

Moehlman, P. D. 1989. Intraspecific variation in canid social systems. In: J. L. Gittleman, ed. *Carnivore Behavior, Ecology, and Evolution,* pp. 143–163. Ithaca, N.Y.: Cornell Univ. Press.

Nel, J. A. J. 1984. Behavioural ecology of canids in the south-western Kalahari. *Koedoe* (supplement) 1984:229–235.

Nel, J. A. J. 1990. Foraging and feeding by bat-eared foxes *Otocyon megalotis* in the southwestern Kalahari. *Koedoe* 33:9–16.

Nel, J. A. J., and Mackie, A. J. 1990. Food and foraging behaviour of bat-eared foxes in the south-eastern Orange Free State. *South African J. Wildl. Res.* 20:162–166.

O'Brien, S. J., and Mayr, E. 1991. Bureaucratic mischief: Recognizing endangered species and subspecies. *Science* 251:1187–1188.

O'Farrell, T. P. 1984. Conservation of the endangered San Joaquin kit fox, *Vulpes macrotis mutica,* on the Naval Petroleum Reserves, California. *Acta Zool. Fennica* 172:207–208.

O'Farrell, T. P., and Gilbertson, L. 1986. Ecology of the desert kit fox, *Vulpes macrotis arsipus,* in the Mojave Desert of southern California. *Southern Calif. Acad. Sci.* 85:1–15.

Paquet, P. C. 1991a. Winter spatial relationships of wolves and coyotes in Riding Mountain National Park, Manitoba. *J. Mammal.* 72:397–401.

Paquet, P. C. 1991b. Scent-marking behavior of sympatric wolves (*Canis lupus*) and coyotes (*C. latrans*) in Riding Mountain National Park. *Canadian J. Zool.* 69:1721–1727.

Paquet, P. C. 1992. Prey use strategies of sympatric wolves and coyotes in Riding Mountain National Park, Manitoba. *J. Mammal.* 73:337–343.

Park, T. 1962. Beetles, competition, and populations. *Science* 138:1369–1375.

Persson, L. 1985. Asymmetrical competition: Are larger animals competitively superior? *Amer. Nat.* 126:261–266.

Peterson, R. O. In press. Wolves as interspecific competitors. In: L. N. Carbyn, S. H. Fritts, and D. Seit, eds. *Wolves in a Changing World.* Edmonton, Alta.: Canadian Circumpolar Institute, Univ. Alberta.

Ralls, K., and White, P. J. 1995. Predation on endangered San Joaquin kit foxes by larger canids. *J. Mammal.* 76:723–729.

Ricklefs, R. E., Cochran, D. C., and Pianka, B. R. 1981. *Ecology and Natural History of Desert Lizards.* Princeton, N.J.: Princeton Univ. Press.

Robinson, W. D. 1961. Population changes of carnivores in some coyote-control areas. *J. Mammal.* 42:510–515.

Rosenzweig, M. L. 1966. Community structure in sympatric carnivora. *J. Mammal.* 47:602–612.

Rowe-Rowe, D. T. 1983. Black-backed jackal diet in relation to food availability in the Natal Drakensberg. *South African J. Wildl. Res.* BB13:17–23.

Rudzinski, D. R., Graves, H. B., Sargeant, A. B., and Storm, G. L. 1982. Behavioral interactions of penned red and arctic foxes. *J. Wildl. Mgmt.* 46:877–884.

Sargeant, A. B. 1982. A case history of a dynamic resource: The red fox. In: G. C. Sanderson, ed. *Midwest Furbearer Management,* pp. 121–137. Wichita, Kans.: Proc. 1981 Symp., Midwest Fish Wildl. Conf.

Sargeant, A. B., and Allen, S. H. 1989. Observed interactions between coyotes and red foxes. *J. Mammal.* 70:631–633.

Sargeant, A. B., Allen, S. H., and Eberhardt, R. T. 1984. Red fox predation on breeding ducks in midcontinent North America. *Wildl. Monogr.* 89:1–41.

Sargeant, A. B., Allen, S. H., and Hastings, J. O. 1987. Spatial relations between sympatric coyotes and red foxes in North Dakota. *J. Wildl. Mgmt.* 51:285–293.

Schaller, G. 1972. *The Serengeti Lion: A Study of Predator-Prey Relations.* Chicago: Univ. Chicago Press.

Schamel, D., and Tracy, D. M. 1986. Encounters between arctic foxes, *Alopex lagopus,* and red foxes, *Vulpes vulpes. Canadian Field-Nat.* 100:562–563.

Schloeder, C. A. 1988. Home ranges, habitat selection and activity patterns of gray and red foxes in West Virginia. Master's thesis, West Virginia University, Morgantown.

Schmidt, D. H. 1985. Controlling arctic fox populations with introduced red foxes. *Wildl. Soc. Bull.* 13:592–594.

Schmidt, R. H. 1986. Community-level effects of coyote population reduction. In: J. Carnes, Jr., ed. *Community Toxicity Testing,* pp. 49–65. Amer. Soc. Testing Materials Special Technical Publication.

Schmitz, O. J., and Kolenosky, G. B. 1985. Wolves and coyotes in Ontario: Morphological relationships and origins. *Canadian J. Zool.* 63:1130–1137.

Schmitz, O. J., and Lavigne, D. M. 1987. Factors affecting body size in sympatric Ontario *Canis. J. Mammal.* 68:92–99.

Scott, T. G. 1955. Dietary patterns of red and gray foxes. *Ecology* 36:366–367.

Scott-Brown, J. M., Herrero, S., and Reynolds, J. A. 1987. Swift Fox. In: M. Novak, J. A. Baker, M. E. Obbard, and B. Malloch, eds. *Wild Furbearer Management and Conservation in North America,* pp. 432–441. Toronto: Ontario Ministry of Natural Resources.

Sheldon, J. W. 1992. *Wild Dogs: The Natural History of the Nondomestic Canidae.* San Diego: Academic Press.

Small, R. L. 1971. Interspecific competition among three species of Carnivora on the Spider Range, Yavapai Co., Arizona. Master's thesis, University of Arizona.

Smits, C. M. M., Slough, B. G., and Yasui, C. A. 1989. Summer food habits of sympatric arctic foxes, *Alopex lagopus,* and red foxes, *Vulpes vulpes,* in the northern Yukon Territory. *Canadian Field-Nat.* 103:363–367.

Stoudt, J. H. 1971. Ecological factors affecting waterfowl production in the Saskatchewan parklands. *U.S. Fish and Wildl. Ser. Res. Pub.* 99:1–58.

Stuart, C. T. 1976. Diet of the black-backed jackal *Canis mesomelas* in the central Namib Desert, South West Africa. *Zool. Africa* 11:193–205.

Sunquist, M. E. 1989. Comparison of spatial and temporal activity of red foxes and gray foxes in north-central Florida. *Florida Field Nat.* 17:11–18.

Theberge, J. B., and Wedeles, C. H. R. 1989. Prey selection and habitat partitioning in sympatric coyote and red fox populations, southwest Yukon. *Canadian J. Zool.* 67:1285–1290.

Thurber, J. M., Peterson, R. O., Woolington, J. D., and Vucetich, J. A. Coyote coexistence with wolves on the Kenai Peninsula, Alaska. *Canadian J. Zool.* 70:2494–2498.

Todd, A. W., Keith, L. B., and Fischer, C. A. 1985. Population ecology of coyotes during a fluctuation of snowshoe hares. *J. Wildl. Mgmt.* 45:629–640.

Trapp, G. R., and Hallberg, D. L. 1975. Ecology of the gray fox (*Urocyon cinere-oargenteus*): A review: In: M. W. Fox, ed. *The Wild Canids*, pp. 164–178. New York: Van Nostrand Reinhold.

U.S. Fish and Wildlife Service. 1989. *Red Wolf Recovery Plan*. Atlanta: U.S. Fish and Wildlife Service.

Van Valkenburgh, B. 1985. Locomotor diversity within past and present guilds of large predatory mammals. *Paleobiology* 11:406–428.

Van Valkenburgh, B. 1989. Carnivore dental adaptations and diet: A study of trophic diversity within guilds. In: J. L. Gittleman, ed. *Carnivore Behavior, Ecology, and Evolution*, pp. 410–436. Ithaca, N.Y.: Cornell Univ. Press.

Van Valkenburgh, B. 1991. Iterative evolution of hypercarnivory in canids (Mammalia: Carnivora): Evolutionary interactions among sympatric predators. *Paleobiology* 17:340–362.

Van Valkenburgh, B., and Wayne, R. K. 1994. Shape divergence associated with size convergence in sympatric East African jackals. *Ecology* 75:1567–1581.

Vaughan, C. 1983. Coyote range expansion in Costa Rica and Panama. *Brenesia* 21:27–32.

Voigt, D. R., and Earle, B. D. 1983. Avoidance of coyotes by red fox families. *J. Wildl. Mgmt.* 47:852–857.

Wayne, R. K., and Jenks, S. M. 1991. Mitochondrial DNA analysis implying extensive hybridization of the endangered red wolf *Canis rufus*. *Nature* 351:565–568.

Wayne, R. K., Van Valkenburgh, B., Kat, P. W., Fuller, T. K., Johnson, W. E., and O'Brien, S. J. 1989. Genetic and morphological divergence among sympatric canids. *J. Heredity* 80:447–454.

White, P. J., Ralls, K., and Garrott, R. A. 1994. Coyote–kit fox spatial interactions as revealed by telemetry. *Canadian J. Zool.* 72:1831–1836.

Wood, J. E., Davis, E. E., and Komarek, E. V. 1958. The distribution of fox populations in relation to vegetation in southern Georgia. *Ecology* 39:160–162.

Wooding, J. B. 1984. Coyote food habits and the spatial relationship of coyotes and foxes in Mississippi and Alabama. Master's thesis, Mississippi State University, State College.

Wozencraft, W. C. 1989. Classification of the recent Carnivora. In: J. L. Gittleman, ed. *Carnivore Behavior, Ecology, and Evolution*, pp. 569–593. Ithaca, N.Y.: Cornell Univ. Press.

Yearsley, E. F., and Samuel, D. E. 1980. Use of reclaimed surface mines by foxes in West Virginia. *J. Wildl. Mgmt.* 44:729–734.

PART II

ECOLOGY

INTRODUCTION

All facets of ecology are developing rapidly today, partly because methods of study (e.g., radiotelemetry analysis) continue to improve and partly because many ecological data bases are becoming more complete. That these trends have come to play such a major role in the study of carnivores is undoubtedly linked to the fact that, for carnivores to continue to survive and receive the benefits of informed decisions concerning their conservation and management, we need a solid understanding of their ecological context.

The importance of field methodology in studying carnivores has always been paramount; our ability to peer into the secretive and often dangerous lifestyles of carnivores hinges on developing more effective, more accurate, but less intrusive methods. Mills reviews new techniques for capture, census, and determination of dietary make-up, especially for large carnivores; he emphasizes that different methods should be selected for different species-specific and ecological questions. For example, conclusions about dietary patterns based on scat analysis will be biased toward the kinds of food that can be identified in fecal matter, the pass rate of different foods, and levels of biomass. Studies of the brown hyena (*Hyaena brunnea*) reveal that up to 22% of the hair in scat analysis is unidentifiable. Mills offers a number of guidelines for determining the types of questions and field techniques that should be considered prior to undertaking field study of carnivores.

The theme of population structure is assessed in the following two chapters, one by Dayan and Simberloff, the other by Waser. Both chapters raise issues about mechanisms (dispersal) and functional results (size differences) of community dynamics and ecology. Dayan and Simberloff analyze a unique feature of carnivores: body size spans a greater range in the Carnivora than in any other mammalian order (Gittleman 1985). Specifically, the authors critically review various selective factors that affect variation in size among carnivore communities, partly in the context of the hot debate surrounding the role of competition (Lewin 1983; Pimm and Gittleman 1990). Using a data base on 16 carnivore species living in Israel, Dayan and Simberloff show that among three guilds (mustelids/viverrids, felids, and canids), there is remarkably strong statistical support for character displacement in canine size. Such a firm cross-species result calls for detailed study of the causal mechanisms (e.g., type and size of prey; sexual selection) driving these observed patterns.

Particular patterns of dispersal are inherent in the lifestyles of every carnivore species, yet the subject is very poorly known, and Waser's is the first comprehensive review. As he points out, one of the reasons that dispersal has failed to receive comparative treatment is that there are many ways to measure dispersal and an almost endless number of independent variables that can affect dispersal. Working from an original comparative data base, collected directly from many field workers, Waser systematically reviews carnivore dispersal patterns in relation to the types of dispersal observed, the mortality consequences of dispersal, the impact of dispersal on reproduction, and the indirect effects of fitness on dispersal. He concludes with a list of trenchant unanswered questions calculated to focus attention on testing critical hypotheses about carnivore dispersal.

The remaining chapters in Part II address conservation. As a united front, they show forcefully that there are no simple recipes for conserving endangered carnivore species, nor are there easy solutions for most management issues. Thus, a broad amalgam of practical methods, biological knowledge, and theoretical ideas, sensitive to species-specific problems, is what must be sought in developing individual conservation programs. At a general level, these two chapters bring together different levels of analysis on captive populations, on natural populations, and in the biopolitical arena of human/carnivore interactions. Reading and Clark reevaluate different aspects of reintroduction, ranging from ethical considerations to different techniques to issues of cost. Selecting two case studies, the river otter (*Lutra canadensis*) and the North American lynx (*Lynx canadensis*), they demonstrate that continuous pre-release and post-release studies are crucial, and that, to a large extent, the value of this effort is in the eye of the beholder: what will most influence the success of reintroduction will be the soundness of the organizational framework established. Thus, as George Schaller writes, the success of our efforts depends to a great degree on how well we serve as mediators between science and culture. This point is nowhere more compelling than in efforts to conserve the giant panda (*Ailuropoda melanoleuca*) in China. Johnson, Yao, You, Yang, and Shen show that conservation research on the giant panda necessarily requires at least the following goals: obtaining sufficient funds; establishing a captive-breeding program; pursuing field behavioral-ecology research; conducting a census survey of panda populations, bamboo, suitable habitat, and the region's socioeconomic carrying capacity; constructing a cohesive, results-oriented conservation-management plan; and securing appropriate personnel training. Each of these goals involves many other considerations, such as political and policy considerations, economic change, and resource depletion, to name a few. In many ways, the giant panda is a worst-case scenario of what it takes to preserve an endangered carnivore species. Still, these are precisely the factors that, at some level, are involved in *every* case.

Collectively, the chapters on ecology and associated issues in the conservation biology of carnivores elucidate how and why success in conservation relies on a multidisciplinary approach. As E. O. Wilson (1992:36) forewarns, "Carnivores . . . are predestined by their perch at the apex of the food web to be big in size and sparse in numbers. They live on such a small portion of life's available energy as always to skirt the edge of extinction, and they are the first to suffer when the ecosystem around them starts to erode." We, as students and stewards of carnivores, therefore have a most difficult task. Whether we succeed or fail only time will tell.

John L. Gittleman

References

Gittleman, J. L. 1985. Carnivore body size: Ecological and taxonomic correlates. *Oecologia* 67:540–554.

Lewin, R. 1983. Santa Rosalia was a goat. *Science.* 221:636–639.

Pimm, S. L., and Gittleman, J. L. 1990. Carnivores and ecologists on the road to Damascus. *Trends Ecol. Evol.* 5:70–73.

Wilson, E. O. 1992. *The Diversity of Life.* Cambridge, Mass.: Belknap Press of Harvard Univ. Press.

CHAPTER 6

Methodological Advances in Capture, Census, and Food-Habits Studies of Large African Carnivores

M. G. L. MILLS

The study of large carnivores poses particular methodological problems for the researcher. Because they live at low densities and many are nocturnal, they are often difficult animals to contact and observe. The reliable and relatively inexpensive radiotelemetry systems that are available today have revolutionized studies of free-ranging birds and mammals and it is now possible to locate and monitor the movements of elusive animals on a regular and predictable basis. Carnivore biologists have been quick to capitalize on this development (Mech 1974). In addition to, or in combination with, radiotelemetry, a range of other study methods has been developed, for specific objectives and particular species. Here, I review and evaluate some of the methods that have been used to capture, census, and study the food habits of brown hyena (*Hyaena brunnea*), spotted hyena (*Crocuta crocuta*), African wild dog (*Lycaon pictus*), lion (*Panthera leo*), leopard (*Panthera pardus*), and cheetah (*Acinonyx jubatus*) in Africa.

Capture Methods

A prerequisite for most carnivore studies—and often for management purposes as well—is an efficient and safe method of capture of the animals, for the purposes of fitting radio transmitters, marking individuals, collecting blood samples, and obtaining baseline allometric data. The development of safe immobilizing drugs, a significant development in the field of game capture, is not within the scope of this chapter. Recent reviews of the relevant immobilizing drugs can be found in Swan (1993) and McKenzie and Burroughs (1993).

With some species of relatively open habitats, and with exceptionally well-habituated individuals of other species in more closed habitats, the free-ranging animal can be immobilized simply by approaching it slowly in a vehicle and shooting it with an immobilizing dart when it is within a suitable distance, usually 20 m or less (Kruuk 1972; Bertram 1978; pers. obs.). When this direct approach is impossible, as it usually is with shy and nocturnal species, or ones that live in closed or otherwise inaccessible habitats, other methods of capture must be used.

Trapping

Of the species I shall deal with, the two that are particularly difficult to catch are the nocturnal, solitary, and generally shy brown hyena and leopard. The most successful method for catching these species, and often for others, is by means of a baited drop-door trap.

In the southern Kalahari, I caught brown hyenas in traps measuring 1800 by 850 by 775 mm (long, wide, and high), made from scrap-steel bars 30 mm thick placed 90 mm apart (Figures 6.1 and 6.2). A sheet of conveyer belting

Figure 6.1. A set and baited drop-door trap for a brown hyena, Kalahari Gemsbok National Park, South Africa. Note the conveyer-belting sheet on the door, which prevents tooth damage by hyenas trying to escape, and the rope and pulley for the safe release of unwanted animals.

Figure 6.2. A brown hyena entering a baited drop-door trap.

was fastened to the door to prevent the hyenas from damaging their teeth while trying to escape. The release mechanism consisted of a pin holding the gate open and a length of thin rope, attached to the pin, that extended to the back of the cage, passed over a wheel, and was fastened to a large piece of meat, usually springbok (*Antidorcas marsupialis*), left hanging in the back of the cage (Figures 6.1 and 6.2). Whenever possible I dragged the carcass from which the bait was cut 1–2 km each side of the trap, at the beginning of a trapping session, to establish a scent trail. This tactic increased the trapping success rate (Mills 1981). A rope was attached to the trapdoor and passed through a pulley tied to a branch in the tree under which the trap was positioned. This arrangement enabled the safe release of unwanted animals such as leopards and lions, as well as previously captured brown hyenas. In 666 trapnights 29 different brown hyenas were caught 169 times, as were six lions and four leopards (Mills 1990). Black-backed jackals (*Canis mesomelas*) were the only carnivores to interfere regularly with the traps, and once they had sprung the trap and eaten the bait they were able to escape between the bars.

A drawback to this trapping method is that it is nonselective, in the sense that some individuals are more likely to enter the trap than others. For example, four hyenas belonging to the same social group were caught 17, 11, 8, and 0 times, respectively, in 160 trap-nights (Mills 1990).

Hamilton (1981) trapped leopards in Tsavo National Park, Kenya, as well as stock-raiding leopards in central Kenya, in metal box traps measuring 2000 by 600 by 900 mm. The leopards were initially attracted to the capture site by a bait hung in a suitable tree above the trap. After a leopard had fed on a bait for one or two nights, the bait was moved from the tree and into the trap. Taking this step prevented the trapping of nontarget species. In Tsavo, 12 leopards were caught or recaught a total of 22 times over a 26-month period, although the number of trap-nights was not given.

In Kruger National Park, South Africa, Bailey (1993) trapped leopards in metal-mesh cage traps, some measuring 2400 by 800 by 900 mm, others 2000 by 600 by 800 mm (Figure 6.3). These traps were sprung by a foot-release mechanism that was set off by the weight of the leopard attempting to reach the bait tied at the back of the cage. The smaller traps were found to be as efficient as the larger ones and easier to maneuver. The traps were set in trees 1–3 m above ground on a solid wooden platform in order to prevent the capture of lions and spotted hyenas, and the traps were camouflaged with branches of evergreen trees. Selection of trap site was important, and before a trap was set in an area, time was taken looking for leopard signs. The main baits used were buffalo (*Syncerus caffer*) and elephant (*Loxodonta africana*) meat, and the average age of the bait at time of capture was 3.5 days. In 2,412

Figure 6.3. A leopard caught in a baited drop-door trap in Kruger National Park, South Africa. The trap has been placed in a tree to prevent the capture of nontarget species.

trap days, 30 different leopards were caught a total of 112 times. Significantly more leopards were caught during dry seasons than during wet seasons, probably because leopard hunting success was higher during wet seasons when vegetation cover was more dense. Male leopards were caught more frequently than females, and never was a female with dependent young captured. Old males and emaciated leopards were caught more frequently than were younger, more healthy animals.

In some situations considerable patience is required to trap leopards. Norton (1986) attempted to trap leopards in the mountainous areas of the Cape Province, South Africa. The traps, made of metal frame covered with 25-mm diamond mesh, measured 1800 by 600 by 600 mm. The baits used were dead chickens, dead or live helmeted guineafowl (*Numida meleagris*), dead rock dassies (*Procavia capensis*), or parts of other carcasses, such as baboons (*Papio ursinus*), springbok, and sheep (Norton and Lawson 1985; Norton and Henley (1987). In one forest and its adjacent mountains, where it was known that at least three leopards were present, none was trapped in over 1000 trap-nights. In another area more than 500 trap-nights each were needed to catch three leopards.

Spotted hyenas were caught in similar cage traps in Aberdare National Park, Kenya. Eighteen hyenas were caught in 112 trap-nights (Sillero-Zubiri and Gottelli 1993). In Kruger National Park, however, I have trapped spotted hyenas in a slightly different manner (unpublished). Here, a large trap measuring 10 by 10 by 2 m, constructed from expanded metal and employing two large drop doors, was used (Figure 6.4). Hyenas were attracted to the trapping site by dragging the bait several kilometers each side of the trap before placing it in the trap and playing tape recordings of selected hyena calls over loudspeakers. The doors were closed by an electrical switch operated from a vehicle parked 30 m away. In this way several (one to three) hyenas could be caught simultaneously. The cage was then approached on foot and each hyena was immobilized by a dart fired from a Capchur pistol. This method is suitable for catching large numbers of spotted hyenas in bush country, 23 having been caught in 21 trap-nights. The major drawback of the method is that the hyenas panic when approached to be darted, and about half of them damaged themselves about the face when running into the side of the cage. For this reason expanded metal is not a good material to use in any carnivore trap, and a suitably strong wire mesh is preferable.

Although cheetahs can sometimes be darted from a vehicle during the day, in the agricultural areas of Namibia (D. Morsbach, pers. comm.) and in Suikerbosrand Nature Reserve, South Africa (Pettifer 1981), they have been trapped. The cheetah traps were not baited, however. Rather, a thorn fence was erected around a latrine-tree, and an unbaited two-door walk-in trap was sprung by a foot-release mechanism placed in a gap in the fence. Placing fresh cheetah feces in the trap as a lure is also reported to be successful in catching cheetahs (de Wet 1993), although this method is more likely to catch males than females. In

Figure 6.4. A spotted hyena about to enter a multiple-cage trap in Kruger National Park, South Africa. The cable connects the electrical gate-closing mechanism to a switch located in a vehicle 30 m away.

Kruger National Park I had no success with these methods in over 100 trap-nights. But when a small rag liberally doused with urine from a female in estrus was hung in the middle of the trap, a male was caught on the second day (Figure 6.5).

On some occasions a live lure, such as a goat, might be useful for trapping particularly wary predators, such as problem animals on farmland (de Wet 1993). The ethics of this technique must be balanced against the extent of the damage inflicted by the problem animal. Alternatively, rubber-padded steel foothold traps set at regular crossing places in a fence, or at the site of a kill, have been used to catch stock-raiding lions (van der Meulen 1977), which were then translocated. This method may also be useful with stock-raiding leopards.

Baiting

Smuts et al. (1977a, b) describe a mass-capture technique for lions at baits in Kruger National Park, a method that was also successfully deployed in the southern Kalahari (Smuts et al. 1977b; Mills et al. 1978).

Lions were lured to baits at night by playing amplified tape recordings of the

Figure 6.5. A two-door trap set for cheetahs at a latrine-tree in Kruger National Park, South Africa. Access to the tree is only through the trap.

growling-snarling sounds of feeding lions (not lion roars, which often caused lions to move away), and the vocalizations associated with feeding spotted hyenas. In addition, scent trails were laid by dragging the carcass several kilometers each side of the capture site. The use of recordings in conjunction with a carcass and scent trails reduced the time lions took to arrive by a factor of 2.6, when compared with using only a carcass and scent trails. The effective luring range of lions was about 4 km. The lions were darted with immobilizing drugs at night, either from the back of a parked vehicle or from inside a portable blind or caravan. Eighty-two percent of the 775 lions darted became recumbent within 10 m of the carcass. Animals of both sexes and all age classes, except cubs under 2 months of age, were caught by the method. The mortality rate due to capture-related problems was just 0.52%, which, considering the large number of animals caught, can be deemed satisfactory.

Whately (1979) describes a method whereby difficult-to-catch spotted hyenas were caught using meat impregnated with phencyclidine hydrochloride at a dosage rage of 1.7–2.2 mg/kg of meat. He attracted hyenas to a carcass by playing amplified recordings of hyena calls. Here, selective administration of the drug was achieved by suspending a small piece of meat containing the drug from a tree above the carcass. An observer could then lower the drugged mouthful in front of a particular hyena. Because it was then able to feed on part of the bait, the hyena was encouraged to stay in the vicinity. The first signs of ataxia were noticed about 4 h after the hyena had taken the drug, and after 6–7 h the animal was recumbent. This method was also reported, without giving details, to have been used on lions, with varying results. In particularly difficult cases, leopards and brown hyenas may also be caught by this method.

Sometimes a fresh bait can be used to lure species like cheetahs and wild dogs that have already been located to within darting distance (pers. obs.).

Boma Capture

A common method for catching herbivores (Densham 1974) is to drive them with a helicopter into a plastic capture boma or corral. English et al. (1993) adopted this method to catch a pack of 17 African wild dogs that had to be translocated. The terrain was a broken granitic landscape that made approaching the dogs by vehicle impossible. The dogs could be driven by the helicopter for distances of up to 10 km, but not all of them were caught in the first attempt, and it became progressively more difficult to drive them into the boma. Some were eventually caught from the helicopter by means of a net-gun, and others were corraled in a half-circle of nets tensioned with a top-and-bottom cable.

These are expensive techniques, but they may be justified when an endangered species such as the African wild dog is involved and the terrain is impassable by vehicle. It is imperative that as many dogs as possible are caught at the first attempt, since it becomes progressively more difficult to drive them into the boma with each succeeding attempt (English et al. 1993).

Conclusions concerning Capture Methods

Table 6.1 shows some methods that have been used to capture the large carnivores under discussion, and offers some recommendations concerning which methods are likely to be most successful. This information should not be taken as implying that methods not recommended here for a particular species should not be used. Situations vary, and the researcher should be flexible and inventive in order to achieve maximal success. The particular method employed will depend on a number of factors, such as the behavior of the carnivores, the objectives of the researcher or manager, and the habitat involved.

Although all large-carnivore species, with the exception of African wild dogs, may be caught in a cage trap, it usually takes many days of trapping before success is achieved. If a free-ranging animal can be approached close enough that it may be darted from a vehicle, this is a good option, although some species, particularly spotted hyenas, tend to run quite far when darted,

Table 6.1. Recommendations on some methods of capturing large African carnivores
(+ = recommended; ± = recommended only under certain conditions; − = not recommended)

Method of capture	Brown hyena	Spotted hyena	Wild dog	Lion	Leopard	Cheetah
Darting from a vehicle	−	±	+	+	−	+
Darting on bait	−	−	±	+	−	±
Cage trap	+	+	−	−	+	+
Leg-hold trap	−	−	−	±	±	−
Drugged meat	±	±	−	±	±	−
Boma trapping	−	−	±	−	−	−

and care must be taken not to lose contact with the animal once it has been darted. In thicker bush areas this is an important consideration. Darts incorporating radio transmitters are available, but are expensive and are liable to be damaged by hyenas. Lions are best caught at baits. The use of leg-hold traps, drugged carcasses, and capture bomas are recommended only in exceptional cases. It is these situations that require innovative ideas and experimentation. McKenzie (1993) provides practical guidelines for a wide range of capture methods for all species in a variety of conditions.

Census Methods

Large carnivores are notoriously difficult to census, since conventional air and ground transect methods, used on large herbivores, are usually not appropriate, although the spotted hyena population on the Serengeti Plain has been sampled by ground transects (Anon. 1977). Often, population densities have been calculated by extrapolating from observations of known or radio-collared individuals made during studies not primarily concerned with monitoring population trends (Makacha and Schaller 1969; Schaller 1972; Rodgers 1974; Elliott and McTaggart Cowan 1977; Whately and Brooks 1978; Hanby and Bygott 1979; Tilson and Henschel 1986; Mills 1990; Bailey 1993). Some studies, however, have utilized special methods for censusing carnivores (Table 6.2).

Table 6.2. Methods that have been used successfully to census large African carnivores

Method	Brown hyena	Spotted hyena	Wild dog	Lion	Leopard	Cheetah
Lincoln index	No	Yes	No	Yes	No	No
Intensive survey	No	No	No	Yes	No	No
Audiotape	No	Yes	No	No	No	No
Photographic record	No	No	Yes	No	No	Yes

Lincoln Index

The Lincoln index is a widely used and most useful method for estimating animal abundance (Seber 1973). It is a mark-recapture method that relies on a number of underlying assumptions, the most important of which are that marked and unmarked animals have the same probability of being caught in the second sample, and that the population is closed, with no recruitment or mortality during the sampling. Several workers have made use of a modified Lincoln index for censusing large carnivores.

Kruuk (1972), for example, calculated the number of spotted hyenas in the Ngorongoro Crater, Tanzania, by marking a sample of animals with ear notches, then establishing the proportion of marked hyenas in each clan range during nine visits to the crater over 3 years, and comparing this proportion with the number of marked hyenas assumed to be present at that time, that is, after discounting marked hyenas that had either died or emigrated. Interestingly, by applying a less elaborate method, where merely the proportion of marked to unmarked hyenas seen during an observation period, regardless of the sites of marking and resighting, were noted, a total population similar to the one derived by the more detailed method was derived (Kruuk 1972). In the Serengeti, where the hyenas move over a much larger area, and where they do not mix randomly, the area was arbitrarily split into a number of smaller regions and a modified Lincoln index was calculated for each. The population was then assessed as the sum of the populations in the smaller regions (Kruuk 1972). Similarly, amplified tape recordings were used to attract spotted hyenas to bait sites in Aberdare National Park. Using sightings of known, ear-notched hyenas, population size was estimated over a 4-month study period (Sillero-Zubiri and Gottelli 1993).

The Lincoln index was also employed to census lions in Kalahari Gemsbok National Park (Mills et al. 1978). A sample of 54 lions was marked by branding unobtrusive symbols on shoulder or rump, and follow-up observations were made for 6 months thereafter by field staff engaged on routine patrols. The figure obtained by this method was of the same order of magnitude as the predicted population in this area, and the method was considered sound.

Provided that the assumptions pertaining to the Lincoln index can be met, this is a useful method for censusing large carnivores, and with innovative thinking can even be used in areas of thick bush (Sillero-Zubiri and Gottelli 1992). In order that the assumption of a closed population be borne out, the time allowed to elapse for the follow-up observations of marked and unmarked animals should be kept to the minimum.

Intensive Surveys

Smuts (1976) describes a lion survey of the 5560-km² central district of Kruger National Park. Employing the mass-capture technique outlined above, two teams gradually worked their way through the area, attempting to locate and mark a number of individuals from each pride. Although the results obtained gave an accurate picture of the population and its structure, it was a time-consuming and expensive operation, requiring several months of intensive field work.

The Use of Sound

Spotted hyenas have been surveyed over the entire 20,000 km² of Kruger National Park by the use of sound. An amplified, 6-minute-long tape of sounds known to attract spotted hyenas—the bleating of a blue wildebeest (*Connochaetes taurinus*) calf, spotted hyenas mobbing lions, an interclan fight, and hyenas squabbling over a kill—was played twice at each of 173 calling stations, and the number of hyenas attracted within 30 min of the commencement of the tape was noted. The calling stations were situated more than 10 km apart. Experiments determined that hyenas were attracted to the sound from a maximum distance of 3.5 km, took an average of 21 min to appear, and responded in groups; that is, if one responded, all of them did. Within a 3.5-km radius of the calling station, the response was independent of distance and was estimated to be 0.55, with the 95% confidence intervals being 0.25–0.60. This information can be combined with the census counts from a given habitat to form a probability model that can then be used to estimate the expected number of hyenas per unit area. The model also adjusts for nonresponse, and offers the possibility of comparisons between years and between habitats (Mills and Juritz, in prep.).

Photographic Methods

Species that are individually identifiable and diurnal may be censused through photographic methods. Areas that receive many tourists and have a good network of roads are particularly well suited to this type of survey. Maddock and Mills (1993) were able to census the African wild dog population in Kruger National Park by asking tourists and staff members to photograph dogs whenever they were encountered. The distinctive coat patterns of these animals make them particularly suitable to this method. Over 5000 photographs were received during a 1-year period, from which 357 dogs in 26 packs were identified. Other demographic data were also collected, such as pack size, sex ratios, litter size, and pup survival. Recently, the cheetah population in Kruger National Park has also been censused in this manner (Bowland 1993).

Leopards have been photographed for individual identification by means of automatic cameras with built-in flashes attached to trigger plates and hidden two-way switches, but these attempts have met with little success (Smith 1977; Stuart and Stuart 1991). In the latter case, interference by hikers and by baboons and other animals crossing the trigger plate were the reasons given for failure.

Conclusions concerning Census Methods

Provided that animals can be individually identified, large-carnivore populations can be censused through a Lincoln-index approach. It is a robust method that can be modified to suit most circumstances. The use of sound is applicable to the censusing of spotted hyenas, because they respond so quickly to the taped sounds. Lions also respond, but at a far slower pace, making this method less suitable for this species. Diurnal species with variable coat patterns, such as wild dogs and cheetahs, are suitable for photographic surveys. In game parks, the enlistment of tourists in these surveys may significantly increase the data base, as well as provide a good public-relations operation.

Food-Habits Studies

A knowledge of the food habits of a carnivore is central to understanding many aspects of its behavior and ecology (Mills 1992). It is therefore not surprising that much attention has been given to this aspect of carnivore behavior, and many methods for studying food habits have been utilized.

Stomach Analysis

Smuts (1979) assessed the diet of lions and spotted hyenas by analyzing remains in the stomach contents of animals killed during a predator-control experiment in Kruger National Park. But besides the fact that the stomachs of large carnivores are seldom available in adequate numbers, the thicker skin of large animals tends to remain undigested for longer than that of smaller prey, thus biasing the results toward larger prey. Furthermore, although the presence of maggots in a stomach is indicative of scavenging, it is not always possible to say whether the food in question was killed by the carnivore or scavenged.

Fecal Analysis

The identification of food remains found in feces is the most common method for analyzing carnivore food habits. Amongst large African carnivores this has been done in a number of studies, particularly with hyenas (Kruuk 1972, 1976; Bearder 1977; Mills and Mills 1978; Tilson et al. 1980; Henschel and

Skinner 1990; Sillero-Zubiri and Gottelli 1992; Skinner et al. 1992), but also with leopards (Grobler and Wilson 1972; Smith 1977; Norton et al. 1986; Bailey 1993) and cheetahs (Mills 1992). Food remains such as those from insects, reptiles, birds, and wild fruits, as well as teeth, nails, and some hairs of mammals, can be identified macroscopically. Other hairs require microscopic identification techniques and comparison with known specimens (Mills and Mills 1978; Henschel and Skinner 1990).

Fecal analysis has been found to be useful for constructing a basic description of a carnivore's diet, particularly where other types of observation are impossible, or where time does not permit detailed observations. The method may also supplement other types of observation, for example in the identification of small food items eaten by brown hyenas, but not identified, during direct observations (Mills and Mills 1978). A third application is for inter- and intraspecific comparisons of diets (Kruuk 1976; Mills and Mills 1978; Henschel and Skinner 1990; Skinner et al. 1992).

For hyenas, which leave prominent white scats at latrines and dens, an adequate fecal sample can usually be collected; 25 scats per month from any number of individuals were found to be an adequate basis for determining the basic food habits of spotted hyenas (Henschel and Skinner 1990). With other species, sampling is more difficult. Schaller (1972) concluded that it was difficult and unproductive to attempt to collect adequate samples of lion feces, mainly because they do not defecate at regular sites, and scavengers often consume their feces. It may also be difficult to collect adequate samples from low-density species such as wild dogs, or secretive ones such as leopards. Cheetah males defecate on prominent objects, whereas females defecate randomly, and males tend to take larger prey than do females (Mills 1992). Therefore, male-biased fecal samples, which do not truly reflect the diet of the species, are likely to be collected. A cheetah fecal sample I collected (Mills 1992) also illustrates the risk of inadequate sampling: a high occurrence of zebra (*Equus burchelli*) in this sample (almost certainly small foals) probably reflected the fact that nearly half the scats were collected in August, September, and October, when small zebra foals are most abundant.

Even with an adequate fecal sample there are problems and limitations with fecal analysis that must be borne in mind when interpreting results. First, the identification of hair is often difficult, particularly in a species with a diverse diet, such as the brown hyena, where 22% of scats examined contained some unidentified hair (Mills and Mills 1978). Second, the amount of hair ingested per kilogram of food eaten varies with the species and the part eaten, making quantitative assessment of the data unreliable (Kruuk 1972; Bearder 1977; Mills and Mills 1978; Henschel and Skinner 1990). This difficulty is particularly serious in species with a wide diet. Experiments on the passage rates of different prey species, which have been conducted with gray wolves (*Canis lupus:* Floyd et al. 1978), servals (*Felis serval*), and black-backed jackals (Bowland and Bowland 1991) could lead to more reliable quantitative analyses for species like hyenas (Bearder 1977). Third, the contribution in terms of biomass of the various food items cannot be measured, and data can be analyzed only on a presence-or-absence basis. Accordingly, food items such as rodents, insects, and reptiles may be overrepresented. Fourth, for most prey species it is not possible to differentiate between the hair of adults and that of juveniles. Finally, it is impossible to determine whether food was scavenged or killed by the predator, which, particularly in the case of hyenas, is an important consideration.

Clearly, there is little that can be done to overcome most of these biases. Whenever possible, therefore, fecal-analysis data should be supplemented with data from other sources. Some of these are taken up next.

Tracking Spoor

Soft sand and sparse vegetation, such as occur in the dunes of the Kalahari, and the expertise of Bushman trackers, make it possible to follow the tracks of carnivores for long distances, and to reconstruct much of their behavior. The method is similar to following tracks in snow (Mech 1980).

The spoor of lions (Eloff 1973, 1984), cheetahs (Labuschagne 1979), leopards (Bothma and le Riche 1984, 1990), and spotted hyenas (Mills 1990) have revealed much useful data on these species' feeding habits in the southern Kalahari. Tracking brown hyena spoor, however, was less productive, because they move erratically and solitarily, and frequently consume food items completely, leaving no evidence of their presence in the diet (Mills and Mills 1978). Mills (1992) showed that the frequency of feeding observations was significantly lower when tracking brown hyena spoor than when visually following them, but that this was not the case when following spotted hyena spoor. The most serious drawback to tracking spoor in the southern Kalahari, affecting data on all species, is that it cannot be done along riverbeds, where the ground is hard. Although they cover only about 5% of the area, the riverbeds are an important habitat for all the large carnivores, since many prey species concentrate there (Mills and Retief 1984; Mills 1990).

In Rhodes Matopos National Park, Zimbabwe (Smith 1977), and in Londolozi Game Reserve, South Africa (Hes 1991), the skills of local trackers have been used to reconstruct the behavior of leopards by searching along tracks and sandy riverbeds. Although it was not possible to obtain as complete a picture in these areas as can be assembled in the southern Kalahari, valuable data were collected.

Opportunistic Observations

Observations of kills encountered opportunistically are easy to make, and have been utilized in a number of studies on the food habits of large African carnivores (Wright 1960; Kruuk and Turner 1967; Pienaar 1969; Schaller

1972; Tilson et al. 1980; Mills 1984). Such observations are clearly biased toward large animals, however, and must be treated with caution when used to describe the diet of a species. Where large samples can be collected over long periods, and if additional data such as the sex and age of the prey are included, opportunistic observations can be useful for studying such aspects of predation as the sex- and age-class selection of adult prey (Mills 1984). Long-term data sets can reveal switches in diet under changing ecological conditions (Mills et al. 1995).

Radio-Location Observations

Mills (1992) defined a radio-location observation as one where, at the moment when it was located, a radio-collared carnivore was feeding on a kill, a method that has been used in studying lions (Whyte 1985; Mills 1992), leopards (Bertram 1982), cheetahs, and wild dogs (Mills 1992). Depending on the feeding habits of the carnivores concerned, radio-location data, like various other data, may give results biased toward larger prey. Data from cheetahs appear to be less biased toward large animals than are data from other large African carnivores, since they consume their prey more slowly than do the others, thus increasing the chances of finding them on a carcass (Mills 1992).

Direct Observations

Following carnivores in a vehicle and then observing them in the hunt is an important method of studying the food habits of large African carnivores (Estes and Goddard 1967; Kruuk 1972, 1976; Schaller 1972; Elliot et al. 1977; Mills 1978, 1990, 1992; van Orsdol 1982, 1984, 1986; FitzGibbon and Fanshawe 1989; Fuller and Kat 1990, 1993; Stander 1991; Mills and Shenk 1992; Fanshawe and FitzGibbon 1993). Provided that sufficient care is taken not to disturb the prey by approaching too close during the day, by dazzling them at night with a spotlight, or by allowing the predators to use the vehicle as cover when approaching their prey (Mills 1992), this method gives the most accurate results and the most complete information on the food habits of a carnivore—which is particularly important for studies aimed at assessing predator/prey relationships. It is, however, a labor-intensive method, requiring time and patience, and can be carried out only in reasonably open habitats and on unbroken substrates. It also puts vehicles at risk, since trees, large bushes, and concealed stumps and rocks not only damage the bodywork but may also damage tires, wires, and other working parts beneath the vehicle.

A disciplined approach to data collection is important if direct observations are to yield good results. So that observation periods may not inflate kill frequency, their duration and frequency should be predetermined and carried out irrespective of the observer's expectations of when kills are going to be made. Mills (1992) showed that, for both lions and cheetahs, observations made where the time spent during each session was not fixed beforehand, but depended largely on whether the observer expected that the hunters would hunt or not, tended to exaggerate kill rate. Moon phase, too, can affect results, since lions are less likely to hunt during full-moon periods (Schaller 1972; van Orsdol 1984; pers. obs.). In Kruger National Park, South Africa, an observation period of three nights, chosen irrespective of moon phase, was found to be both practical and unbiased for lion-feeding studies.

Conclusions concerning Food-Habits Studies

Each method for analyzing the food habits of large carnivores has its advantages and disadvantages (Table 6.3). The method chosen will depend largely on the objectives of the study, the circumstances, and the habitat. Direct observations give the best results and should normally be a researcher's first choice.

Table 6.3. Advantages and disadvantages of some methods for analyzing the food habits of large African carnivores

Method	Advantages	Disadvantages
Stomach analysis	1. Quick and easy method for diet profile	1. Adequate sample unlikely to become available 2. Not always possible to determine if food was killed or was scavenged
Fecal analysis	1. Relatively quick and inexpensive for a basic description of the diet 2. Can supplement other types of observation 3. Useful for comparison purposes	1. May be difficult to collect an adequate sample 2. Identification of hair sometimes difficult 3. Data can be analyzed only on a presence-or-absence basis 4. Not possible to determine if food was scavenged or was killed
Tracking spoor	1. Not necessary actually to see the animals	1. Can be done only on soft substrates and in open vegetation 2. Not good for brown hyena
Opportunistic encounter and radio location	1. Relatively easy to collect data 2. Can measure sex and age selection of certain adult prey 3. Useful for long-term studies.	1. Usually heavily biased toward large prey
Direct observation	1. Least-biased method 2. Can measure prey selection, kill frequency, and consumption rates	1. Labor-intensive 2. Possible only in certain habitats 3. May disturb animals 4. Hard on vehicles

Fecal analysis can often provide the necessary background for broad-based behavioral-ecology studies, but the limitations of this method must be borne in mind. Tracking spoor, opportunistic observations, and radio-location observations have limited application, but can be useful in pursuing certain objectives. Stomach analysis may be used where carnivores are being culled.

Frequently, combining the data of several methods has been used to analyze carnivore food habits. This approach makes comparing the results from different studies difficult, and may lead to erroneous conclusions. For example, the mixing of opportunistic observations with direct observations may increase the sample size, but it is bound to increase the contribution of large mammals and decrease the contribution of smaller mammals in the apparent diet of the carnivore concerned. Results obtained from separate methods should be treated separately, unless it can be shown that the two sets of data are compatible. If not, accurate data may become contaminated by more biased data, thus invalidating conclusions. The best yardstick for the accuracy of the data is direct observations in a planned regime. This method has been compared with results from other methods in Mills (1992).

Acknowledgments

I thank the National Parks Board, South Africa, for facilities provided.

References

Anon. 1977. Census of predators and other animals on the Serengeti plains, May 1977. Report no 52. Arusha: Tanzania National Parks.

Bailey, T. N. 1993. *The African Leopard*. New York: Columbia Univ. Press.

Bearder, S. K., 1977. Feeding habits of spotted hyaenas in a woodland habitat. *East African Wildl. J.* 15:263–280.

Bertram, B. C. R. 1978. *Pride of Lions*. London: Dent.

Bertram, B. C. R. 1982. Leopard ecology as studied by radio tracking. *Symp. Zool. Soc. London* 49:341–352.

Bothma, J. du P., and le Riche, E. A. N. 1984. Aspects of the ecology and behaviour of the leopard *Panthera pardus* in the Kalahari Desert. *Koedoe* 27(supplement):259–279.

Bothma, J. du P., and le Riche, E. A. N. 1990. The influence of increasing hunger on the hunting behaviour of southern Kalahari leopards. *J. Arid Environ.* 18:79–84.

Bowland, A. E. 1993. The 1990/1991 cheetah photographic survey. Unpublished manuscript. Skukuza: Kruger National Park.

Bowland, J. M., and Bowland, A. E., 1991. Differential passage rates of prey components through the gut of serval *Felis serval* and black-backed jackal *Canis mesomelas*. *Koedoe* 34:37–39.

Densham, W. D. 1974. A method of capture and translocation of wild herbivores using opaque plastic material and a helicopter. *Lammergeyer* 21:1–25.

De Wet, T. 1993. Physical capture of carnivores. In: A. A. McKenzie, ed. *The Capture and Care Manual*, pp. 255–262. Pretoria: Wildlife Decision Support Services and South African Veterinary Foundation.

Elliott, J. P., and McTaggart Cowan, I. 1977. Territoriality, density, and prey of the lion in Ngorongoro Crater, Tanzania. *Canadian J. Zool.* 56:1726–1734.

Elliot, J. P., McTaggart Cowan, I., and Holling, C. S. 1977. Prey capture by the African lion. *Canadian J. Zool.* 55:1811–1828.

Eloff, F. C. 1973. Lion predation in the Kalahari Gemsbok National Park. *J. South African Wildl. Mgmt. Assoc.* 3:59–64.

Eloff, F. C. 1984. Food ecology of the Kalahari lion *Panthera leo vernayi*. *Koedoe* 27(supplement):249–258.

English, R. A., Stalmans, M., Mills, M. G. L., and van Wyk, A. 1993. Helicopter-assisted boma capture of African wild dogs *Lycaon pictus*. *Koedoe* 36:103–106.

Estes, R. D., and Goddard, J. 1967. Prey selection and hunting behaviour of the African wild dog. *J. Wild. Mgmt.* 31:52–70.

Fanshawe, J. H., and FitzGibbon, C. D. 1993. Factors influencing the hunting success of an African wild dog pack. *Anim. Behav.* 45:479–490.

FitzGibbon, C. D., and Fanshawe, J. H., 1989. The condition and age of Thomson's gazelles killed by cheetahs and wild dogs. *J. Zool.* 218:99–107.

Floyd, T. J., Mech, L. D., and Jordan, P. A., 1978. Relating wolf scat content to prey consumed. *J. Wildl. Mgmt.* 42:528–532.

Fuller, T. K., and Kat, P. W. 1990. Movements, activity, and prey relationships of African wild dogs (*Lycaon pictus*) near Aitong, southwestern Kenya. *African J. Ecol.* 28:330–350.

Fuller, T. K., and Kat, P. W. 1993. Hunting success of African wild dogs in southwestern Kenya. *J. Mamm.* 74:464–467.

Grobler, J. H., and Wilson, V. J. 1972. Food of the leopard *Panthera pardus* (Linn.) in the Rhodes Matopos National Park, Rhodesia, as determined by faecal analysis. *Arnoldia* 5:1–9.

Hamilton, P. H. 1981. *The Leopard* Panthera pardus *and the Cheetah* Acinonyx jubatus *in Kenya*. Report for the U.S. Fish and Wildlife Service, the African Wildlife Leadership Foundation, and the Government of Kenya.

Hanby, J. P., and Bygott, J. D. 1979. Population changes in lions and other predators. In: A. R. E. Sinclair and M. Norton-Griffiths, eds. *Serengeti: Dynamics of an Ecosystem*, pp. 249–262. Chicago: Uni. Chicago Press.

Henschel, J. R., and Skinner, J. D. 1990. The diet of the spotted hyaena (*Crocuta corcuta*) in Kruger National Park. *African J. Ecol.* 28:69–82.

Hes, L. 1991. *The Leopards of Londolozi*. Cape Town: Struik Winchester.

Kruuk, H. 1972. *The Spotted Hyena*. Chicago: Univ. Chicago Press.

Kruuk, H. 1976. Feeding and social behaviour of the striped hyaena (*Hyaena vulgaris* Desmarest). *East African Wildl. J.* 14:91–111.

Kruuk, H., and Turner, M. 1967. Comparative notes on predation by lion, leopard, cheetah and wild dog in the Serengeti area, East Africa. *Mammalia* 31:1–27.

Labuschagne, W. 1979. 'n Bio-ekologiese en gedragstudie van die jagluiperd *Acinonyx jubatus jubatus* (Schreber, 1776). Master's thesis, University of Pretoria, South Africa.

Maddock, A. H., and Mills, M. G. L. 1993. Population characteristics of African wild dogs *Lycaon pictus* in the Eastern Transvaal Lowveld, South Africa, as revealed through photographic records. *Biol. Conserv.* 67:57–62.

Makacha, S., and Schaller, G. B. 1969. Observations on lions in the Lake Manyara National Park, Tanzania. *East African Wildl. J.* 7:99–103.

McKenzie, A. A., ed. 1993. *The Capture and Care Manual*. Pretoria: Wildlife Decision Support Services and South African Veterinary Foundation.

McKenzie, A. A., and Burroughs, R. E. J. 1993. Chemical capture of carnivores. In: A.

A. McKenzie, ed. *The Capture and Care Manual*, pp. 224–244. Pretoria: Wildlife Decision Support Services and South African Veterinary Foundation.

Mech, L. D. 1974. Current techniques in the study of elusive wilderness carnivores. *Proc. XI Int. Cong. Game Biol., Stockholm*, 315–322.

Mech, L. D., 1980. *The Wolf*. New York: Natural History Press.

Mills, M. G. L. 1978. Foraging behaviour of the brown hyaena (*Hyaena brunnea* Thunberg, 1820) in the southern Kalahari. *Z. Tierpsychol.* 48:113–141.

Mills, M. G. L. 1981. The socio-ecology and social behaviour of the brown hyaena Hyaena brunnea, Thunberg, 1820 in the southern Kalahari. D.Sc. dissert., University of Pretoria.

Mills, M. G. L. 1984. Prey selection and feeding habits of the large carnivores in the southern Kalahari. *Koedoe* 27(supplement):281–294.

Mills, M. G. L. 1990. *Kalahari Hyaenas*. London: Unwin Hyman.

Mills, M. G. L. 1992. A comparison of methods used to study food habits of large African carnivores. In: D. R. McCullough and R. H. Barret, eds. *Wildlife 2001: Populations*, pp. 1112–1124. London & New York: Elsevier Applied Science.

Mills, M. G. L., Biggs, H. C., and Whyte, I. J. 1995. The relationship between lion predation, population trends in African herbivores and rainfall. *Wildl. Res.* 22:75–87.

Mills, M. G. L., and J. M. Juritz. In prep. The use of sound for censusing spotted hyenas.

Mills, M. G. L., and Mills, M. E. J. 1978. The diet of the brown hyaena *Hyaena brunnea* in the southern Kalahari. *Koedoe* 21:125–149.

Mills, M. G. L., and Retief, P. F. 1984. The response of ungulates to rainfall along the riverbeds of the southern Kalahari. *Koedoe* 27(supplement):129–142.

Mills, M. G. L., and Shenk, T. M. 1992. Predator-prey relationships: The impact of lion predation on wildebeest and zebra populations. *J. Anim. Ecol.* 61:693–702.

Mills, M. G. L., Wolff, P., le Riche, E. A. N., and Meyer, I. J. 1978. Some population characteristics of the lion (*Panthera leo*) in the Kalahari Gemsbok National Park. *Koedoe* 21:163–171.

Norton, P. M. 1986. Ecology and conservation of the leopard in the mountains of the Cape Province. Unpublished report, Cape Department of Nature Conservation.

Norton, P. M., and Henley, S. R. 1987. Home range and movements of male leopards in the Cedarberg Wilderness Area, Cape Province. *South African J. Wildl. Res.* 17:41–48.

Norton, P. M., and Lawson, A. B. 1985. Radio tracking of leopards and caracal in the Stellenbosch area, Cape Province. *South African Wildl. Res.* 15:17–24.

Norton, P. M., Lawson, A. B., Henley, S. R., and Avery, G. 1986. Prey of leopards in four mountainous areas of the south-western Cape Province. *South African J. Wildl. Res.* 16:47–52.

Pettifer, H. L. 1981. Aspects of the ecology of cheetahs (*Acinonyx jubatus*) on the Suikerbosrand Nature Reserve. In: J. A. Chapman and D. Pursley, eds. *Worldwide Furbearer Conf. Proc.*, Vol. 2., pp. 1121–1142. Falls Church, Virginia: R. R. Donneley.

Pienaar, U. de V. 1969. Predator-prey relationships amongst the larger mammals of the Kruger National Park. *Koedoe* 12:108–176.

Rodgers, W. A. 1974. The lion (*Panthera leo*, Linn.) population of the eastern Selous Game Reserve. *East African Wildl. J.* 12:313–317.

Schaller, G. B. 1972. *The Serengeti Lion*. Chicago: Univ. Chicago Press.

Seber, G. A. F. 1973. *The Estimation of Animal Abundance*. London: Charles Griffen.

Sillero-Zubiri, C., and Gottelli, D. 1992. Feeding ecology of spotted hyaena (Mammalia: *Crocuta crocuta*) in a mountain forest habitat. *J. African Zool.* 106:169–176.

Sillero-Zubiri, C., and Gottelli, D. 1993. Population ecology of spotted hyaena in an equatorial mountain forest. *African J. Ecol.* 30:292–300.

Skinner, J. D., Funston, P. J., van Aarde, R. J., van Dyk, G., and Haupt, M. A. 1992. Diet of spotted hyaenas in some mesic and arid southern African game reserves adjoining farmland. *South African J. Wildl. Res.* 22:119–121.

Smith, R. M. 1977. Movement patterns and feeding behaviour of leopard in the Rhodes Matopos National Park, Rhodesia. *Arnoldia* 8:1–16.

Smuts, G. L. 1976. Population characteristics and recent history of lions in two parts of the Kruger National Park. *Koedoe* 19:153–164.

Smuts, G. L. 1979. Diet of lions and spotted hyaenas assessed from stomach contents. *South African J. Wildl. Res.* 9:19–25.

Smuts, G. L., Whyte, I. J., and Dearlove, T. W. 1977a. A mass capture technique for lions. *East African Wildl. J.* 15:81–87.

Smuts, G. L., Whyte, I. J., and Dearlove, T. W. 1977b. Advances in the mass capture of lions (*Panthera leo*). *Proc. XIII Int. Cong. Game Biol., Atlanta,* 420–431. Washington, D.C.: The Wildlife Society and Wildlife Management Institute.

Stander, P. E. 1991. Foraging dynamics of lions in a semi-arid environment. *Canadian J. Zool.* 70:8–21.

Stuart, C. T., and Stuart, T. 1991. A leopard in the wilderness. *African Wildl.* 45:251–254.

Swan, G. E. 1993. Drugs used for the immobilization, capture, and translocation of wild animals. In: A. A. McKenzie, ed. *The Capture and Care Manual,* pp. 2–64. Pretoria: Wildlife Decision Support Services and South African Veterinary Foundation.

Tilson, R., and Henschel, J. 1986. Spatial arrangement of spotted hyaena groups in a desert environment, Namibia. *African J. Ecol.* 24:173–180.

Tilson, R., von Blotnitz, F., and Henschel, J. 1980. Prey selection of spotted hyaena (*Crocuta crocuta*) in the Namib Desert. *Madoqua* 12:41–49.

Van der Meulen, J. H. 1977. Notes on the capture and translocation of stock raiding lions in north eastern and north western Rhodesia. *South African J. Wildl. Res.* 7:15–17.

Van Orsdol, K. G. 1982. Ranges and food habits of lions in Rwenzori National Park, Uganda. *Symp. Zool. Soc. London* 49:325–340.

Van Orsdol, K. G. 1984. Foraging behaviour and hunting success of lions in Queen Elizabeth National Park, Uganda. *African J. Ecol.* 22:79–99.

Van Orsdol, K. G. 1986. Feeding behaviour and food intake of lions in Rwenzori National Park, Uganda. In: S. D. Miller and D. D. Everett, eds. *Cats of the World: Biology, Conservation and Management,* pp. 377–388. Washington, D.C.: National Wildlife Federation.

Whately, A. 1979. Selective capture of spotted hyaenas using orally administered sernylan. *Lammergeyer* 27:25–27.

Whateley, A., and Brooks, P. M. 1978. Numbers and movements of spotted hyaenas in Hluhluwe Game Reserve. *Lammergeyer* 26:44–52.

Whyte, I. J. 1985. The present ecological status of the blue wildebeest (*Connochaetes taurinus taurinus* Burchell, 1823) in the central district of the Kruger National Park. Master's thesis, University of Natal, South Africa.

Wright, B. S. 1960. Predation on big game in East Africa. *J. Wildl. Mgmt.* 44:1–15.

CHAPTER 7

Patterns of Size Separation
in Carnivore Communities

TAMAR DAYAN AND DANIEL SIMBERLOFF

Carnivores often range quite widely, and many species occur in a wide variety of biotic and abiotic environments. Different populations of the same species are often found under different climatic conditions, and may also coexist, in different communities, with different suites of potential prey and competitors. Thus, different carnivore populations may be subject to different selective pressures. These differences in selective regimes are probably reflected in differences in a variety of biological traits, the most conspicuous of which is body size. Many ecological and evolutionary studies have therefore dealt with morphological variation among carnivores.

Autecology: Ambient Temperature as a Selective Force

Many studies have treated the effect of climate on carnivore body size. Much research has demonstrated that body size in some carnivores is correlated with mean ambient temperature or latitude, the latter taken to be a surrogate for temperature (e.g., Davis 1981; Klein 1986). Temporal size change related to paleoclimatic change has also been demonstrated (e.g., Kurtén 1965; Davis 1977, 1981; Klein 1986). These phenomena accord with Bergmann's rule (Bergmann 1847), redefined by Mayr (1956, 1963), which states that the body size of homeotherms from cooler regions tends to exceed that of conspecific populations from warmer regions. Bergmann's explanation of this rule was that, because body volume (which produces heat) increases more rapidly than does corresponding body surface area (which dissipates heat), larger mammals produce more heat and lose less to the environment than do smaller ones. Large body size must then be adaptive for existence in cold climates.

Bergmann's rule is only one of several ecogeographic rules, empirical generalizations that define relationships between morphological variation and the physical environment, but it is the best studied of those rules and the only one that has been intensively explored for carnivores. Thus the importance of the role of ambient temperature as an evolutionary force that affects the body size of mammals in general, and of carnivores in particular (see, for example, Klein 1986), is now widely accepted. A cursory review of the published literature, however, reveals that two-thirds of the carnivore species studied display some deviation from Bergmann's rule, at least in parts of their ranges (Dayan et al. 1991; see also Gittleman 1985). These deviations may derive, in part, from differences in the composition of the communities in which the different carnivore populations are embedded: coexistence with different potential competitors may have affected the sizes of these carnivores (Dayan et al. 1991).

Synecology: Competition as a Selective Force

Research on the possible evolutionary effect of interspecific competition on carnivore morphology has a venerable history. For many years numerous studies of various animal taxa have interpreted morphological size relationships among sympatric species of similar ecological requirements as indirect evidence for competition. Coexistence of competitors has been viewed as leading to coevolutionary change. Brown and Wilson (1956) defined character displacement as a phenomenon that may arise when two similar animal species overlap geographically: the differences between them may be accentuated in the zone of sympatry and weakened or lost entirely in parts of their ranges outside this zone. Character displacement might be generated by selection against hybridization or by selection to reduce competition for limiting resources. It is the latter possibility that concerns us. Hutchinson (1959) theorized that competition for food might impose a limit on the similarity between trophic apparati of potentially competing species. He saw different sizes as enabling different species to use different sizes of food. Starting with Holmes and Pitelka (1968), many researchers on groups of three or more species reported overdispersion of the mean sizes—a tendency toward equal size ratios between species adjacent in a size ranking (references in Simberloff and Boecklen 1981). Strong et al. (1979) termed this phenomenon "community-wide character displacement."

The positive relationship between body size and prey size among carnivores (Gittleman 1985) makes the hypothesis of food-size partitioning among them a reasonable one. Generally, larger predators take a larger range of prey sizes, primarily because of an increase in the upper size limit (Schoener 1969; Wilson 1975; Gittleman 1985). This evidence, coupled with a rich literature documenting aggressive interference among carnivores competing for prey (discussion and references in Van Valkenburgh 1984; Van Valkenburgh and Hertel 1993), enhances the suitability of carnivores for the study of character displacement. Therefore, among studies of character displacement in mammals, carnivores have been the most prominent, even though they are difficult to study in the field.

Fuentes and Jaksic (1979) argued that character displacement occurs between two Chilean foxes (the colpeo fox, *Dusicyon culpaeus,* and the Argentine gray fox, *D. griseus*) that diverge in body length with increasing latitude, a phenomenon they attributed to increased habitat overlap. Jiménez (1993), however, failed to confirm this result, finding both fox species increasing in size with increasing latitude, in accord with Bergmann's rule. Rosenzweig (1968), McNab (1971), and Ralls and Harvey (1985) all studied the possibility of character displacement in North American weasels and arrived at conflicting conclusions. Rosenzweig (1968) attributed little of the geographical size variation of the stoat, or ermine (*Mustela erminea*), to competition with other weasels, or to size of potential prey, and saw key roles for mean annual temperature and latitude. He studied males and females separately and a mixed population of both. McNab (1971) interpreted the size variation within the same species as resulting from the combined effect of the size of the available prey and character displacement. In both studies the character measured was head plus body length. McNab (1971) arrived at similar conclusions for South American felids. In fact, he suggested that Bergmannian size clines, where found, were only a by-product of selection resulting from interspecific competitive interactions.

Ralls and Harvey (1985) used the condylobasal length of the skull as a measure of body size for the North American weasels and found no evidence of character displacement between any two species. They found that, for *M. erminea,* size increased with latitude and that there were no obvious correlates of size in the longtailed weasel (*M. frenata*) or the common or least weasel (*M. nivalis*). For the ermine, Ralls and Harvey (1985) also examined a number of other measures, including several they considered directly related to feeding, and found such high correlations between these and condylobasal skull length that they restricted their analyses to the latter. Because of the relatively high degree of sexual size dimorphism in weasels, McNab (1971) and Ralls and Harvey (1985) analyzed the two sexes separately and compared males to males and females to females when seeking character displacement.

Until recently, no comparably thorough analyses had been carried out on Old World weasels. Hutchinson (1959) and Williamson (1972), however, had pointed out that *M. erminea* in Ireland, where it occurs alone, is smaller than *M. erminea* in Britain, where *M. nivalis* is also present. Dayan and Tchernov (1988), further, suggested that the large size of *M. nivalis* in Egypt and on some Mediterranean islands results from character release (*sensu* Grant 1972) in the absence of larger congeners.

Kiltie (1984) studied size ratios among sympatric Neotropical cats using several measures—body weight, body length, relative maximum bite force, and relative maximum gape (estimated partly from jaw length). For each trait, ratios were computed between pairs of species adjacent in a size ranking. Thus, for *n* species, there are *n* − 1 ratios. When all six species were analyzed, he found that none of the four sets of ratios was statistically distinguishable from those that would have been expected if the species sizes had been independently and randomly chosen between the smallest and largest observed sizes. Ratios for maximum gape among just the four largest species were significantly more similar to one another, and the minimum ratio was significantly greater than expected by chance, patterns he thought might have been generated by interspecific competition. He noted that the two smallest species, the margay (*Felis wiedii*) and the jaguarundi (*F. yagouraoundi*), were very similar in maximum gape, but that this similarity may have been permitted by different habitat use—where they are sympatric, the former species is more arboreal and the latter more terrestrial. Despite the sexual size dimorphism in these species, Kiltie (1984) studied a mixed sample of males and females, using averages or midpoints of ranges.

Later Kiltie (1988) extended his analysis of size relationships in Neotropical cats to field assemblages in tropical Africa and Asia. He suggested that jaw length correlates well with modal prey size. Using means of this character for sexually mixed samples, he found that all assemblages contained two species that are indistinguishable in jaw length. When those were treated as a single species he found assemblage-wide ratio regularity. The ratios were more similar to one another than would have been expected if the sizes were independent and random. Again, he viewed this pattern as conceivably generated by competition, and he noted that the two species that were very similar in size in each assemblage always differed in some other way: one was terrestrial and the other arboreal, or one used mesic and the other xeric habitats, or one had a dappled coat and the other a plain one.

The Controversy Surrounding the Role of Competition in Selecting for Size

Recently, much of the ecological literature on size relationships has been reconsidered and contested. Controversy surrounds the community-ecology literature that deals with morphological coevolution among competitors (e.g., Lewin 1983; Pimm and Gittleman 1990). This controversy casts doubt upon both the morphological/evolutionary role of competition and its role in structuring ecological communities.

Statistical Tests

Grant (1972, 1975) criticized some published cases of ecological character displacement and suggested that they could be otherwise interpreted. When studies claiming size-ratio equality and unusually large minimum size ratios were subjected to rigorous statistical analysis, it appeared that, for a large fraction of claimed instances of size displacement, observed sizes did not falsify a hypothesis of independence between species (Roth 1981; Simberloff and Boecklen 1981). Schoener (1984) has questioned the power of these tests to detect deviation of a set of sizes from a random arrangement, and several authors (references in Simberloff 1989) questioned whether the main test used for ratio equality, the Barton-David test (Barton and David 1956; Simberloff and Boecklen 1981), is the most powerful available. There has been no full test of relative power, but the most extensive examination to date (Arita 1993) suggests that Barton-David tests are generally at least as powerful as others suggested and often more powerful. Arita (1993) finds, however, that no test is very powerful when there are very few sympatric species. Tonkyn and Cole (1986) pointed out that the Barton-David test estimates the largest and smallest possible sizes by those observed, and that this procedure introduces some error into the calculation of tail probabilities for observed deviation. This problem seems unavoidable and obtains for alternative tests as well (Simberloff 1989), but it should not lead to great errors unless species numbers are very small.

Guild Composition

Studies dealing with ecological character displacement and community-wide character displacement were criticized not only for lack of suitable statistical tests but also for choice of potential competitors (Simberloff and Boecklen 1981). Different choice of possible competitors as guild members may, of course, lead to conflicting results. Morphological studies dealt with members of a guild, a concept that has been interpreted in several ways (reviewed by Simberloff and Dayan 1991). Root (1967) defined ecological guilds as groups of species using the same class of resources in a similar way. The major discrepancy among various researchers' studies and interpretations has been the significance of the "similar way" in which resources were supposed to have been garnered. Although Root (1967) emphasized its significance, others have questioned it: "To state the question with a bit of hyperbole, does it matter that a particular insect species is captured by a silken spider web as opposed to a bird's beak? The ecosystem and community consequences are similar—one less insect of that species—and manner is irrelevant from the specific perspective" (Hawkins and MacMahon 1989:443).

Jaksic et al. (1981) and Jaksic and Delibes (1987) viewed the carnivores in their studies as part of a general predatory guild that also included owls, raptors, and snakes, clearly underplaying the role of "similar manner" in the way prey were sought and taken. Van Valkenburgh (1985) used a narrower definition of predatory guilds, including only extant and extinct mammalian predators. In her guild of large land predators, she included the nonaquatic, nonvolant mammal species within a community that take prey and potentially compete for food. Van Valkenburgh (1985) limited guild membership to species above jackal size (7 kg), because she expected increased competition among large predators and also because these are better represented in the fossil record. None of these studies directly measured interactions among species within the designated guilds.

Other investigators (e.g., McNab 1971; Fuentes and Jaksic 1979) have simply studied congeneric species, assuming that closely related species are the most likely to compete with one another. One advantage of the guild concept stressed by Root (1967) is that it focuses attention on all sympatric species involved in a potentially competitive interaction, regardless of their taxonomic relationships. Guild designations, however, still tend to include closely related species, collectively termed "taxon-guilds" by Schoener (1986).

Another concern was that, although studies that were perceived to demonstrate character displacement and community-wide character displacement were readily published, it is more than likely that many others that *failed* to detect any morphological pattern were *not* reported, and accordingly there is a publication bias toward supportive evidence (Simberloff and Boecklen 1981). Even within the taxa of a given region there would be a tendency to emphasize the morphological patterns in certain guilds while neglecting those in others.

Choice of Characters

Another problem with studies of morphological relationships has been that different researchers choose different characters to measure (Simberloff and Boecklen 1981). Choice of different measures may, of course, yield different results. Hutchinson (1959) specified that minimum size ratios should occur between trophic apparati, but for mammals, surprisingly, he suggested skull length as a measure. Later studies used this measure or used measures of body size such as head plus body length. In recent years there has been a growing realization that trophic apparati more directly related to feeding should be studied. Thus Ralls and Harvey (1985) measured several "feeding" characters—length of the lower jaw, upper carnassial length, and others—for North American weasels, as did Kiltie (1984) for Neotropical cats. In his study of sexual size dimorphism in the genus *Mustela*, Holmes (1987), too, chose several characters related to feeding ecology. The characters chosen for these studies were generally related to the feeding behavior, but not primarily the killing behavior, of these carnivores.

In sum, there has been a frequent failure to justify the species viewed as guild associates, especially with respect to how prey are taken. Traits taken to represent size have also often been rather casually selected. Both of these problems cast doubt on the robustness of many conclusions. The use of statistical tests has introduced an element of objectivity and rigor into the study of size separation, but more research is needed on the power of key tests.

Morphological Relationships among Israeli Carnivores

Israel supports a diverse carnivorous fauna of 16 extant terrestrial, nonaquatic species. This diversity stems from both geographical and historical factors (Yom-Tov and Tchernov 1988). Israel is situated at a biogeographic crossroads, a meeting point for fauna and flora from different biogeographic origins and of different biogeographic distribution (see, for example, Tchernov 1988). For many species Israel is at either the northernmost or the southernmost end of the range. Thus for several carnivore species, Israel and, perhaps, contiguous areas form a narrow zone of overlap, where carnivores that have not coevolved over most of their distributional ranges must coexist. If morphological coevolution among carnivores does occur, one should ask how these carnivores react to each other's presence in sympatry and whether ecological character displacement is seen in this zone. Our studies of Israeli carnivores (Dayan et al. 1989a, b, 1990, 1992a) differed from previous carnivore studies in several significant respects.

Composition of Israeli Guilds

We divided Israeli carnivores into guilds based on their limb morphology, which reflects and affects their locomotor behavior (cf. Van Valkenburgh 1985). The locomotor abilities of a carnivore reveal much about its way of life, preferred habitat, foraging behavior, and escape strategy (Van Valkenburgh 1984; see also Taylor 1989). We reasoned that carnivores that are similar in their locomotor behavior seek and take prey in a similar manner and are thus likely to encounter and to take similar spectra of prey species (Dayan et al. 1992b). We thus attached much significance to the way in which resources were taken, in the original spirit of Root's (1967) guild definition. Using these criteria we divided all of Israel's terrestrial nonaquatic carnivores into three guilds (Dayan et al. 1992b), excluding the striped hyena (*Hyaena hyaena*), a scavenger that feeds largely on carcasses and specializes in crushing bones (Ewer 1973). Hyenas have highly developed forelimbs and disproportionately small hindquarters, a specialized limb morphology that may be associated with the mechanics of bone crushing (Ewer 1973). Its ecology, morphology, and diet set the hyena apart from other Israeli carnivores.

Thus we studied morphological relationships among the other Israeli carnivores within their respective guilds. We found that morphological criteria provide a clearcut guild definition. The three guilds are:

1. Mustelids/viverrids—elongated, small to medium-sized, short-limbed pentadactyl plantigrades to semiplantigrades (Novikov 1962; Harrison 1968;

Osborn and Helmy 1980). There is no reduction of the ulna, and the radius retains the primitive ability to rotate around it. This guild comprises five species: four mustelids (the European badger, *Meles meles;* the ratel, or honey badger, *Mellivora capensis;* the beech marten, or stone marten, *Martes foina;* and the marbled polecat, *Vormela peregusna*) and one viverrid (the Egyptian mongoose, *Herpestes ichneumon*) (Dayan et al. 1989a).

2. Felids—carnivores having the feet digitigrade, with five toes on the forefeet but only four on the hindfeet (see, for example, Harrison 1968). The radius retains the ability to rotate around the ulna, thus enabling cats to grip with their forefeet, which is essential for climbing but also enables them to manipulate prey. No material was available for the now extirpated lion (*Panthera leo*) and cheetah (*Acinonyx jubatus*). Thus we studied only the small felids—the European wild cat (*Felis silvestris*), jungle cat (*F. chaus*), and caracal (*F. caracal*) (Dayan et al. 1990). The only large felid currently found in Israel, the leopard (*Panthera pardus*), is intermediate in size between the cheetah and the lion.

3. Canids—carnivores of open areas highly adapted for cursorial locomotion, the feet digitigrade, with five toes on the forefeet but only four on the hindfeet (Clutton-Brock et al. 1976). Canids are unable to rotate the foreleg, and accordingly cannot use them to manipulate prey (Taylor 1989). The gray wolf (*Canis lupus*), golden jackal (*C. aureus*), red fox (*Vulpes vulpes*), Ruppell's sand fox (*V. rueppelli*), and Blanford's fox (*V. cana*) are members of this guild in Israel (Dayan et al. 1992a). For the Middle Paleolithic, specimens of the larger three species, but not of the smaller two foxes, are available.

Choice of Characters for Israeli Guilds

Like earlier researchers, we emphasized that trophic apparati and not general measures of body size should be studied when one seeks ecological character displacement. We studied the trophic apparati directly related to the killing behavior of carnivores: their specialized dentitions. In the Carnivora, the emphasis is on the canine and carnassial teeth (e.g., Gromova 1968; Savage and Long 1986). In the mustelid/viverrid and small-felid carnivore guilds, we focused particularly on the maximal diameter of their upper canines. These carnivores have often been reported to kill their prey by inserting an upper canine between two particular vertebrae in the region where the thoracic and cervical vertebrae meet, then dislocating them, thus lacerating the spinal chord (see, for example, Gossow 1970; Heidt 1972; Leyhausen 1979). We therefore hypothesized that the maximal diameter of the upper canine, used for pushing the vertebrae apart, would be the feature most likely to show size displacement if resource partitioning by different prey size does indeed occur. In these small predators, canines may well reflect the typical size of prey; they would be adapted to the particular vertebral size and structure most frequently encountered (Leyhausen 1965, 1979).

Canids, by contrast, do not kill their prey by a stylized death bite. They are generally more omnivorous, and they do not possess morphological specializations enabling them to bring down large prey as quickly and efficiently as do cats and mustelids (Kleiman and Eisenberg 1973). Their canine teeth are relatively large but not particularly sharp. They are not specifically adapted to deliver a highly oriented, lethal bite (Ewer 1973); neither are they as strong as the canines of felids (Van Valkenburgh and Ruff 1987). Canids kill their prey through a series of slashing bites (Ewer 1973; Leyhausen 1979), and small prey are killed by violent shaking (Seitz 1950). Because the rostrum of canids is long, the shearing power of the jaws is quite reduced, except near the angle of the jaw itself (Kleiman and Eisenberg 1973). This morphology dictates much less bite strength in the canines as compared with the shearing power of the carnassials. In carnivores with shorter rostra, the difference between bite strength at the carnassials and that at the canines is not as pronounced.

Sexual Size Dimorphism

Two of our Israeli carnivore guilds contained highly dimorphic species. One possible interpretation of the evolution of sexual size dimorphism is that it reduces resource overlap and competition between males and females (reviewed by Hedrick and Temeles 1989; Shine 1989). An extensive literature deals specifically with the marked sexual size dimorphism of mustelids (for review, see King 1989).

Several studies of sexual size dimorphism in weasels concluded that it is adaptive for reducing intersexual food competition. Brown and Lasiewski (1972) argued that elongate mustelids are more dimorphic than less elongate species of equal body mass because elongation increases energy needs and consequently increases intersexual competition for food. Simms (1979) used an estimate of body diameter as an index of size and demonstrated a good match between estimated body diameter (and therefore minimum tunnel width) of weasels and minimum tunnel width of likely available prey. He argued that sexual dimorphism in the three North American weasels is at least partly a consequence of intraspecific resource partitioning. Gliwicz (1988), however, suggested that the sexual size dimorphism stems from the fact that pregnancy substantially changes the female body diameter. Since the maximum diameter of weasels should not exceed that of the burrows of their basic prey, females (when not pregnant) are thinner than males and consequently shorter and much smaller. Powell (1979), Moors (1980), and Powell and Leonard (1983) suggested that smaller size was favored in females because it enabled the female to channel more energy into reproduction because of her own reduced needs. A similar model, based on the different energetic needs of males and females, predicts different optimal body sizes for the two sexes during the breeding season, thereby providing an energetic explanation for sexual size dimorphism (Sandell 1989).

Erlinge (1979) and Moors (1980) suggested that the larger size of the polygynous male weasels results from intrasexual selection favoring large males over small ones in competition for mates. The smaller size of females was seen as adaptive because it permitted them lower maintenance costs and food requirements, especially important during the reproductive period (Erlinge 1979; Moors 1980). Likewise, Moors (1984), in a study of various mustelid species, concluded that sexual dimorphism in body size is not primarily an adaptation to competition. Holmes (1987) examined sexual dimorphism in a number of tooth characteristics in North American weasels and arrived at a similar conclusion. He suggested that sexual dimorphism is not mainly a means of partitioning resources and that the degree of sexual dimorphism in a population of any species is unaffected by the presence or absence of other weasel species.

The ecological interpretation of sexual size dimorphism as a means for reducing resource overlap between the sexes is similar to the idea of size displacement as a means for reducing interspecific resource overlap. Previous studies of sexually dimorphic carnivores, however, did not try to study both hypotheses, but compared males to males and females to females, or else used means of the mean sizes of the two sexes or midpoints of their ranges. Two of our carnivore guilds—the mustelid/viverrid guild and the small-felid guild—comprise species that are dimorphic, albeit to varying degrees, for body size and skull length (see also Moors 1984) and even more dimorphic for canine diameter. For these guilds we viewed each sex as a separate morphospecies and analyzed all morphospecies at once. We reasoned that, if competition for food is a prime selective pressure, then each sex of each species is competing with its conspecific opposite sex plus both sexes of all other species. Thus character displacement should act between sexes of one species just as it does between each sex of that species and the males and females of other species (Dayan et al. 1989a). Intraspecific competition has not been explored as assiduously as interspecific competition has, but where it has been sought it usually seems to be at least as strong as the latter (see, for example, Connell 1983; Abrams 1987a, b).

Canids are not very dimorphic and are less dimorphic than are most mustelids, viverrids, and felids (Ewer 1973). This difference may be related to the fact that monogamy and paternal care are typical of canids but extremely rare in the other families (Kleiman 1977). Ralls (1977) found that monogamous mammals typically have little sexual size dimorphism, whereas extremely polygynous species tend to be very dimorphic. Moors (1980) reported that highly dimorphic species tend to be strongly carnivorous, whereas those with low dimorphism tend to be omnivorous or to eat relatively small prey. Kleiman and Eisenberg (1973) also suggested that the tendency toward omnivory in canids helps to permit pair-bonding, which Ralls (1977) correlates with low sexual size dimorphism. Whatever the evolutionary reason for the low sexual size dimorphism in canids, we used mixed populations of males and females.

Morphological Patterns

In the mustelid/viverrid and small-felid guilds of Israel (Figures 7.1 and 7.2, respectively), we found a tendency toward size-ratio equality both inter- and intraspecifically in one trait, maximum canine diameter (Dayan et al. 1989a, 1990). This pattern suggests a coevolutionary morphological response of the different species (and sexes) to one another. The species of each guild exhibit some geographic size variation, and their zone of overlap is quite limited geographically, yet where all coexist, in Israel, there is a regular pattern— community-wide character displacement in canine diameter. Skull length, a general measure of size, exhibited no regular pattern, nor did upper carnassial length (Dayan et al. 1989a, 1990; Dayan and Simberloff, unpublished data).

In the Israeli fossil record we find the common weasel (*Mustela nivalis*) (Dayan and Tchernov 1988), which by limb morphology should be grouped with the mustelids/viverrids. The size of the fossil specimens appears quite similar to that of the Recent Egyptian population, and in canine diameter the Recent Egyptian population fits well with the general trend (Dayan et al. 1989a).

For the three members of the Israeli Middle Paleolithic canid guild for which material was available, we found size-ratio equality for lower first molar (car-nassial) length, the only trait available for measurement. For the five-member

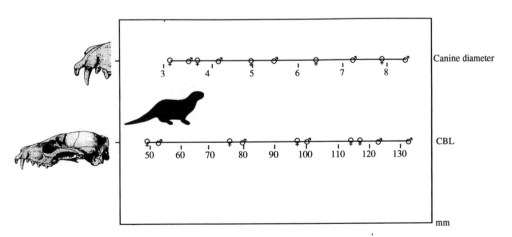

Figure 7.1. Canine diameters and condylobasal skull lengths of species in the mustelid/viverrid guild of Israel.

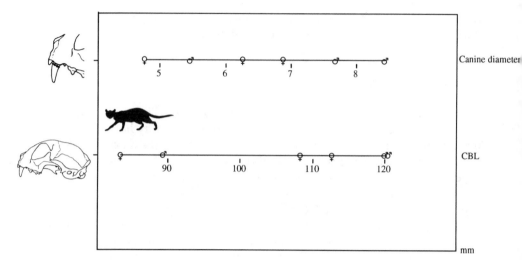

Figure 7.2. Canine diameters and condylobasal skull lengths of species in the small-felid guild of Israel.

Recent guild (Figure 7.3), we found size-ratio equality not only in lower carnassial length but in upper canine diameter and skull length (Dayan et al. 1992a). Thus, the different outcomes for the canids, on the one hand, and the mustelids/viverrids and the felids, on the other, are only partially consistent with our hypotheses. The latter two guilds use their canines in a stylized way, and the canines alone show the predicted regular pattern. The canids do *not* use their canines in this way, and all three traits show a regular pattern. Geographic variation in Israel and North Africa is complex and also not easily interpreted; we discuss these patterns below.

When all three guilds in Israel, or even any pair of them, are viewed together (Figure 7.4), there is no tendency toward size-ratio equality (Simberloff and Dayan 1991). For example, canine diameters of golden jackals, male caracals, and male honey badgers (each in a different guild) are almost identical, as are those of red foxes, male wild cats, and male Egyptian mongooses (also of different guilds). Our hypothesis is that, because of their different locomotor and killing behaviors, the three guilds we posit have characteristically different ways of getting their food and thus qualify as distinct. Unfortunately, there are insufficient dietary data on these species in Israel to determine empirically whether these different behaviors do result in resource partitioning between the designated guilds (Simberloff and Dayan 1991).

Dayan et al. (1989a, b, 1990, 1992a) suggested that ecological character displacement is best manifested by the trophic apparati directly related to feeding ecology, and that other traits, such as body size, are probably more affected by other selective pressures, such as autecological ones. It is quite

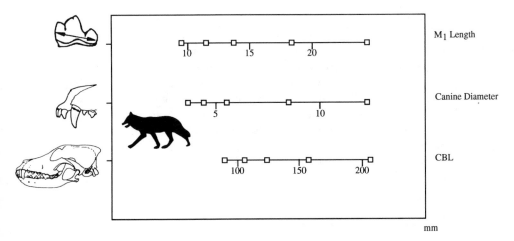

Figure 7.3. Lower carnassial lengths, canine diameters, and condylobasal skull lengths in the canid guild of Israel.

possible that the pattern depicted by body size or skull length is only a passive correlated response to selection acting on the trophic apparati and that choice of unsuitable morphological characters has generated part of the controversy over ecological character displacement and community-wide character displacement. The patterns we found indicate ecological character displacement and community-wide character displacement among Israeli carnivores, but

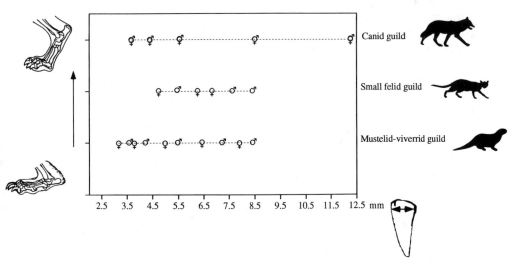

Figure 7.4. Canine diameters of the three carnivore guilds of Israel, as a function of increasing adaptation to cursorial locomotion.

there is currently little direct evidence to verify the ecological interpretation of these phenomena. Because there are very few data in Israel on the food habits of carnivores, we do not know whether niche partitioning through taking different prey-size groups does occur. There has been little research on the functional role of tooth size and skull size in the ability of an individual to bring down prey of different sizes or types (but see Leyhausen 1979). We have only circumstantial evidence that prey are limiting to these species. Clearly, the ecological and evolutionary basis for this phenomenon remains to be studied. The problem of scanty ecological literature on food habits led us to seek other groups of species, elsewhere in the world, for which food habits *have* been reported.

Other Carnivore Guilds and Studies

Several other carnivore guilds in various parts of the world—embracing mustelids, felids, canids, and even hyaenids—have been studied. We, too, examined some New World and Old World mustelids whose food habits have been researched. We studied the three North American weasels that have been the subject of several ecomorphological studies (e.g., NcNab 1971; Ralls and Harvey 1985), and again we found size ratios tending toward equality in upper canine diameters at seven of eight localities studied, but a regular pattern for skull lengths at only four of the eight localities (Dayan et al. 1989a). The ecological literature supports a hypothesis of prey-size selection among the different morphospecies. These weasels, however, are only part of the larger guild of North American mustelids.

We also examined the entire mustelid guild of the British Isles (Dayan and Simberloff 1994). The British mustelids, a subset of mainland European mustelids (from which their populations derive), are the Eurasian badger (*Meles meles*), European pine marten (*Martes martes*), polecat, or European ferret (*Mustela putorius*), stoat (*Mustela erminea*), and least weasel (*Mustela nivalis*). Irish mustelids, whose populations derive from the British populations, form a still smaller subset: the badger, marten, and stoat. Thus the mustelid populations of these two islands form an interesting "natural experiment" (Diamond 1986) for the study of character displacement. In addition to the native mustelid species, in the past few decades farmed American mink (*Mustela vison*) have established feral populations in the British Isles.

Community-wide character displacement was found for canine diameter among mustelids of Great Britain. For skull length it was seen only when the largely vermivorous badger was excluded. When we added feral mink, the regular pattern disappeared, but when we substituted the mink for the polecat, which is now restricted to Wales and adjacent English counties, community-wide character displacement was manifest. For Irish mustelids, size ratios were not equal, but the pattern for canines was more regular than that for skull lengths. Adding the local feral mink did not result in a regular pattern, but the addition of the mink and exclusion of the badger yielded equal ratios for skull length, though not for canines (Dayan and Simberloff 1994).

A considerable literature is available on the food habits of the European mustelids. The morphological patterns plus the published empirical data on these species support a hypothesis of prey-size partitioning. With the possible exception of the badger, the larger species take somewhat larger prey items, and males take larger prey than do the females, at least during certain times of the year. The significant differences in size between some of the British and Irish populations of the same morphospecies suggest the possibility of ecological release among Irish mustelids, whose populations originally derive from British ones (Dayan and Simberloff 1994). In particular, canine sexual size dimorphism is greater for Irish pine martens, stoats, and mink, as would be expected if there were fewer competitors (Selander 1966). On Ireland, the female pine marten and female stoat are substantially smaller than their counterparts in Great Britain, approaching the sizes of males of missing species (the polecat for the pine marten and the weasel for the stoat). For skull length there is no consistent pattern (Dayan and Simberloff 1994).

We also examined variation in skull length and canine diameter for the native British mustelids and found that five of the six morphospecies occurring on both Ireland and Great Britain had greater coefficients of variation on Ireland (Dayan and Simberloff 1994); most of the differences were significant. This result accords with the "niche-variation" hypothesis (Van Valen 1965; Grant 1967) that morphological variation should increase in the absence of competing species, since competitive release allows different individuals within a population to specialize on different resources. Of course, increased sexual dimorphism could accomplish the same result and is also much greater (for canine diameter) for the pine marten and stoat on Ireland than it is on Great Britain (for the badger, it was lower on Ireland).

Reig (1992) studied morphological variation in the pine marten (*Martes martes*) and the stone marten (*M. foina*) in Europe. Variation in 23 craniodental measurements of pine martens across 15 populations indicated a north-south gradient of increasing skull size. The skulls of stone martens across 13 populations increased in size from west to east. Patterns of variation in sympatric and allopatric samples, and variation in relative size between species within their area of sympatry, suggested no character displacement between species (Reig 1992). Reig studied several characters related to feeding (e.g., length of first lower molar, length of lower canine) but did not measure the upper canines that are used to kill prey. It would be interesting to see whether character displacement between the two species occurs for this trait.

We studied the small felids of the Sind desert of Pakistan, looking at the same characters that we chose for Israeli small felids (Dayan et al. 1990). This guild comprises the same species as does the Israeli guild (wild cat, jungle cat, and caracal) plus another medium-sized cat—the fishing cat (*Felis viverrina*).

For this guild we found a regular pattern of size ratios tending toward equality in canine diameter. The sequence of the morphospecies, however, was different from that of Israel, and this difference reflected an almost twofold difference in sexual size dimorphism between the wild cat of Israel and the wild cat of the Sind desert. For skull lengths and upper carnassial lengths, there was no tendency toward size-ratio equality. This result strongly supports the hypothesis of an ecological role for sexual size dimorphism, at least in the feature used to kill prey (Dayan et al. 1990). The size decreases in the wild cat, jungle cat, and caracal in the Sind, in comparison to their conspecifics in Israel, were also consistent with the hypothesis that the presence of another, slightly larger medium-sized cat in the Sind desert felid guild resulted in ecological character displacement.

North American canids from the Upper Pleistocene were also briefly analyzed. Three large canid species were found in the La Brea tar pits: dire wolves (*Canis dirus*), gray wolves (*C. lupus*), and coyotes (*C. latrans*) (see, for example, Nowak 1979). Size ratios between mean carnassial lengths of these species, determined from data published by Nowak (1979), tend toward equality (Dayan et al. 1993), but new datings of the tar pits are now available (Marcus and Berger 1984), and this research should be carried out thoroughly, without time-averaging of the specimens (Dayan et al. 1993). Such a study is currently under way.

On the other hand, Van Valkenburgh and Wayne (1994), who examined three sympatric jackal species, arrived at conclusions very different from those we have reached in our research. The geographic ranges of the three species—the side-striped jackal (*Canis adustus*), the golden jackal (*C. aureus*), and the black-backed jackal (*C. mesomelas*)—overlap only in East Africa. Van Valkenburgh and Wayne (1994) measured 62 cranial, dental, and mandibular traits, including lengths of upper and lower carnassials and upper canine length. They tested four predicted results of competition: (1) divergence in size should be pronounced in locations where all three species are present, (2) the addition of a third species in East Africa should result in character displacement either in the invading species or in the resident species most similar to it, in characters related to the diet, (3) intraspecific morphological variability should be higher in areas with fewer coexisting species, and (4) species should be less sexually dimorphic when sympatric with more different species. Thus, sexual dimorphism should be least in East Africa. Van Valkenburgh and Wayne (1994) found no evidence for size divergence among these jackals; in fact, the three species appear to converge in size. But they did find character divergence and probable niche compression of black-backed jackals, as evidenced by adaptations for increased carnivory, reduced sexual dimorphism, and reduced variability. The response was asymmetric and affected this species alone (Van Valkenburgh and Wayne 1994).

Werdelin (1993) studied late Miocene and earliest Pliocene hyaenids from several localities in Eurasia and Africa. These hyenas were for the most part small to medium-sized carnivores with no strong adaptations to the type of bone-cracking scavenging seen in modern-day hyenas. Rather, they have been characterized as "doglike" by Werdelin and Solounias (1991), showing remarkable resemblances to canids in their skull and dental morphology and probably in postcranial morphology as well (Werdelin 1993). Sexual size dimorphism for these species was low. Because of their resemblance to extant canids, Werdelin (1993) explored the possibility that they would show the same pattern of community-wide character displacement in carnassial length that Dayan et al. (1989b, 1992a) found for canids. He found size ratios tending toward equality among the carnassials of sympatric late Miocene hyenas. This pattern was not exhibited by the lower third premolar. This finding indicates the process of community-wide character displacement among these species, operating on their specialized dentitions, although the question of how the pattern arose is even more difficult to resolve for extinct taxa than for extant ones (Werdelin 1993).

Alternative Hypotheses and Unresolved Issues

The results of studies of ecological character displacement in carnivores leave us with several issues that still require careful study. It is also important that we search for alternative hypotheses that might explain the morphological patterns described. For canine diameter, one alternative hypothesis has been suggested, but it is unlikely that it holds.

An Alternative Behavioral Hypothesis

Canines may be used for interspecific communication in threat behavior; in carnivores, at least part of such behavior consists of showing the teeth by opening the mouth and, in some species, by a stylized raising of the lip to bare the canines further (Ewer 1973). Dayan et al. (1989a) tentatively raised the alternative hypothesis that selection entailed aggressive behavior rather than partitioning of prey but found the existing data insufficient to test this idea.

Bekoff (1989), reviewing the literature on the behavioral development of terrestrial carnivores, found few studies of mustelids. Poole (1966) described a "defensive threat," in which a polecat stands with arched back and raised head, facing its aggressor with bared teeth and hissing or screaming. This behavior is carried out generally after, not before, combat, by the individual wanting to *avoid* further aggression, not by the aggressor, who is given wide berth (Poole 1967). Opening the mouth and baring the teeth is a behavior also described, but not discussed, for weasels (Gossow 1970). The discussion by Neal (1986) of badger social behavior mentioned no display of teeth, whereas the extensive description by Kruuk (1989) of the aggressive and defensive behavior of badgers included a depauperate set of visual cues and no tooth display. Generally, mustelids have well-developed scent glands that are used for marking territories (Gorman and Trowbridge 1989). In fact, Lockie (1966) suggested that the idea of visual display derives largely from avian studies, whereas small carnivores lay scent or use other territorial markers, avoiding face-to-face encounters. This very limited literature may indicate that baring teeth is relatively unimportant in mustelids, except perhaps in defensive behavior, and that it is unlikely that evolution would have calibrated canine size differences among mustelids in response to this selective pressure.

Canid and felid behaviors have received more study. Threat behavior in cats has been carefully described and includes various visual cues, among them a variety of facial expressions (Leyhausen 1979), which are graphically represented by Bradshaw (1992); these include opening the mouth. Kleiman and Eisenberg (1973) suggest that the facial expressions of canids appear more exaggerated than do those of felids. During display, some canids, including the gray wolf (Kleiman 1967) and the golden jackal (Seitz 1959), curl up their lips to bare the canine teeth. The longer muzzle of canids may contribute to the appearance of greater expression in canids than in felids, especially during threat, when the canid muzzle becomes wrinkled and the teeth are bared (Kleiman and Eisenberg 1973).

A basic problem with a hypothesis relating canine size ratios to behavior is that no behavioral models treat interspecific size ratios. Furthermore, the fact that we are dealing with different species and differing social systems would greatly complicate a possible behavioral model (Dayan et al. 1989a). But even were such a model to be erected, we would still have to explain why canids, in which baring the canines is so prominent, exhibit a regular pattern not only in the canine teeth but also in carnassials and skull lengths, whereas mustelids, in which this visual cue appears to be unimportant, nevertheless exhibit canine diameter regularity but do not have the sizes of the carnassials or skulls equally spaced. Finally, it should be noted that, for felids, this pattern occurs for canine diameter and not for canine length (Dayan et al. 1990), which might be of more obvious visual significance. For other groups, canine lengths were not measured. In sum, it appears to us that the most parsimonious explanation for the canine pattern we found so consistently in all of our carnivore studies is that it is generated by selective pressure to partition prey according to prey size.

Remaining Problems

If this explanation is correct, then the fact that all three traits show interspecific size-ratio equality in canids is an important problem for future study. Why should there be size-ratio equality for canine diameter in the canids, which do not use their canines as the felids and mustelids/viverrids do? Related puzzling observations come from North Africa, where the largest (wolf) and smallest (Blanford's fox) of the five Israeli canid species are missing (Dayan et al. 1992a). The golden jackal subspecies in Egypt, *Canis aureus lupaster,* is very large in canine diameter, lower carnassial length, and skull length. For all three traits it is roughly midway in size between the Israeli populations of wolves and golden jackals. In other words, it appears to have undergone ecological release in the absence of the wolf. The wolf, however, is missing from all of North Africa, yet the golden jackal shows clines of decreasing canine diameter and skull length from east to west, whereas the carnassials remain unchanged (and very large relative to those of the Israeli population). In Algeria, for example, jackal size is identical to that in Israel, as measured for both canines and skulls, whereas the carnassials equal those of the Egyptian population. What selective pressures are at work here?

In contrast, carnassial length does *not* decrease from Israel to Egypt for Ruppell's sand fox, despite the absence of the smaller Blanford's fox, and it appears to stay constant all across North Africa (Dayan et al. 1992a). Of course, one might not have expected diminution, given the resource bases in the different sites, but these are unknown. At the same time, the canine diameter of Ruppell's sand fox *does* seem to decrease in Egypt, as does skull length. Both increase again in Algeria, though not to the sizes in Israel (Dayan et al. 1992a). An ANOVA shows no significance differences in either trait, but sample sizes are very small. It would be interesting to determine whether, with adequate samples, these results hold.

Another unstudied problem is the difference in sexual dimorphism between canids and the other groups. As noted above, the reduced sexual dimorphism in canids may reflect the monogamy and paternal care that typify canids but are rare in the other families. Moreover, the more omnivorous habits of canids, relative to those of felids, mustelids, and viverrids, may also contribute to reduced sexual dimorphism. A study of a wide range of carnivores and a phylogenetic analysis of the wide range of carnivores would help to determine the causes of this pattern (see Gittleman and Van Valkenburgh, in prep.).

Finally, we have not established with field data that the three guilds we posit are actually partitioning resources in Israel, though the hypothesis is reasonable on the basis of their locomotor behavior. The patterns of size-ratio equality obtain within each guild in Israel, but not when the three guilds are combined into one overarching carnivore guild (Simberloff and Dayan 1991). But Van Valkenburgh and Wayne (1994:1577), following Wayne et al. (1989), imply that competition is likely between members of each of the different guilds:

In our previous study on jackals we hypothesized that a constraint on body size divergence is the unique richness in East Africa of other carnivore taxa, both larger and smaller than a typical jackal (6–10 kg) (Wayne et al. 1989). There are at least 20 species smaller and seven species larger than the jackals in the Serengeti ecosystem (Kingdon

1977) and a similar diversity exists in South Africa (Smithers 1983). Thus any significant increase or decrease in size might bring the jackals into greater competition with other species.

Van Valkenburgh (1985) similarly defined carnivore guilds containing several families. Clearly, more research is needed on likely bases for guild assignment, and this research will have to be field-oriented.

In sum, we hope that our recent papers have contributed to a clarification of morphological coevolution among carnivores and that carnivores will continue to have a major role in questions of ecomorphology. If the questions that still need to be addressed attract further study, the joint efforts of field ecologists, mammalian morphologists, and paleontologists in these areas should pave the way for a fuller understanding of evolution in carnivore communities.

Acknowledgments

We thank B. Van Valkenburgh, R. K. Wayne, and L. Werdelin, for access to unpublished manuscripts, and I. Zohar, for drawing the figures.

References

Abrams, P. A. 1987a. An analysis of competitive interactions between three hermit crab species. *Oecologia (Berlin)* 72:233–247.

Abrams, P. A. 1987b. Resource partitioning and competition for shells between intertidal hermit crabs on the outer coast of Washington. *Oecologia (Berlin)* 72:248–258.

Arita, H. T. 1993. Tests for morphological competitive displacement: A reassessment of parameters. *Ecology* 74:627–630.

Barton, D. E., and David, F. N. 1956. Some notes on ordered intervals. *J. Roy. Stat. Soc. B* 18:79–94.

Bekoff, M. 1989. Behavioral development of terrestrial carnivores. In: J. L. Gittleman, ed. *Carnivore Behavior, Ecology, and Evolution*, pp. 89–124. Ithaca, N.Y.: Cornell Univ. Press.

Bergmann, C. 1847. Über die Verhaltnisse der Warmekonomie der Thiere zu ihrer Grosse. *Göttingen Studien* 3:595–708.

Bradshaw, J. W. S. 1992. *The Behavior of the Domestic Cat*. Wallingford, U.K.: CAB International.

Brown, J. H., and Lasiewski, R. C. 1972. Metabolism of weasels: The costs of being long and thin. *Ecology* 53:939–943.

Brown, W. L., and Wilson, E. O. 1956. Character displacement. *Syst. Zool.* 5:49–64.

Clutton-Brock, J., Corbet, G. B., and Hills, M. 1976. A review of the family Canidae with a classification by numerical methods. *Bull. British Mus. (Nat. Hist.) Zool.* 29(3):1–199.

Connell, J. H. 1983. On the prevalence and relative importance of interspecific competition: Evidence from field experiments. *Amer. Nat.* 122:661–696.

Davis, S. 1977. Size variation in the fox, *Vulpes vulpes,* in the Palaearctic region today, and in Israel during the Quaternary. *J. Zool. (London)* 182:343–351.

Davis, S. J. M. 1981. The effects of temperature change and domestication on the body size of Late Pleistocene to Holocene mammals of Israel. *Paleobiology* 7:101–114.

Dayan, T., and Simberloff, D. 1994. Character displacement, sexual dimorphism, and morphological variation among British and Irish mustelids. *Ecology* 75:1063–1073.

Dayan, T., Simberloff, D., and Tchernov, E. 1993. Morphological change in Quaternary mammals: A role for species interactions. In: R. A. Martin and A. Barnosky, eds. *Morphological Change in Quaternary Mammals of North America: Integrating Case Studies and Evolutionary Theory*, pp. 71–83. Cambridge: Cambridge Univ. Press.

Dayan, T., Simberloff, D., Tchernov, E., and Yom-Tov, Y. 1989a. Inter- and intraspecific character displacement in mustelids. *Ecology* 70:1526–1539.

Dayan, T., Simberloff, D., Tchernov, E., and Yom-Tov, Y. 1990. Feline canines: Community-wide character displacement among the small cats of Israel. *Amer. Nat.* 136:39–60.

Dayan, T., Simberloff, D., Tchernov, E., and Yom-Tov, Y. 1991. Calibrating the paleothermometer: Climate, communities, and the evolution of size. *Paleobiology* 17:189–199.

Dayan, T., Simberloff, D., Tchernov, E., and Yom-Tov, Y. 1992a. Canine carnassials: Character displacement in the wolves, jackals, and foxes of Israel. *Biol. J. Linn. Soc.* 45:315–331.

Dayan, T., and Tchernov, E. 1988. On the first occurrence of the common weasel (*Mustela nivalis*) in the fossil record of Israel. *Mammalia* 52:165–168.

Dayan, T., Tchernov, E., Simberloff, D., and Yom-Tov, Y. 1992b. Tooth size: Function and coevolution in carnivore guilds. In: P. Smith and E. Tchernov, eds. *Structure, Function and Evolution of the Teeth*, pp. 215–222. London: Freund.

Dayan, T., Tchernov, E., Yom-Tov, Y., and Simberloff, D. 1989b. Ecological character displacement in Saharo-Arabian *Vulpes:* Outfoxing Bergmann's Rule. *Oikos* 55:263–272.

Diamond, J. M. 1986. Overview: Laboratory experiments, field experiments, and natural experiments. In: J. Diamond and T. J. Case, eds. *Community Ecology*, pp. 3–22. New York: Harper and Row.

Erlinge, S. 1979. Adaptive significance of sexual dimorphism in weasels. *Oikos* 33:233–245.

Ewer, R. G. 1973. *The Carnivores*. Ithaca, N.Y.: Cornell Univ. Press.

Fuentes, E. R., and Jaksic, F. M. 1979. Latitudinal size variation of sympatric foxes: Tests of alternative hypotheses. *Ecology* 60:43–47.

Gittleman, J. L. 1985. Carnivore body size: Ecological and taxonomic correlates. *Oecologia* 67:540–554.

Gittleman, J. L., and Van Valkenburgh, B. In prep.

Gliwicz, J. 1988. Sexual dimorphism in small mustelids: Body diameter limitation. *Oikos* 53:411–414.

Gorman, M. L., and Trowbridge, B. J. 1989. The role of odor in the social lives of carnivores. In: J. L. Gittleman, ed. *Carnivore Behavior, Ecology, and Evolution*, pp. 57–88. Ithaca, N.Y.: Cornell Univ. Press.

Gossow, H. 1970. Vergleichende Verhaltenstudien an Marderartigen I. Über Lautausserungen und zum Beuteverhalten. *Z. Tierpsychol.* 27:405–480.

Grant, P. R. 1967. Bill length variability in birds of the Tres Marias Islands, Mexico. *Canadian J. Zool.* 45:805–815.

Grant, P. R. 1972. Convergent and divergent character displacement. *Biol. J. Linn. Soc.* 4:39–68.

Grant, P. R. 1975. The classical case of character displacement. *Evol. Biol.* 8:237–337.

Gromova, G. I. 1968. *Fundamentals of Palaeontology*, Vol. XIII. Jerusalem: Israel Program for Scientific Translations.

Harrison, D. L. 1968. *The Mammals of Arabia*. London: Ernest Benn.

Hawkins, C. P., and MacMahon, J. A. 1989. Guilds: The multiple meanings of a concept. *Ann. Rev. Entomol.* 34:423–451.

Hedrick, A. V., and Temeles, E. J. 1989. The evolution of sexual dimorphism in animals: Hypotheses and tests. *Trends Ecol. Evol.* 4:136–138.

Heidt, G. A. 1972. Anatomical and behavioral aspects of killing and feeding by the least weasel, *Mustela nivalis* L. *Proc. Arkansas Acad. Sci.* 26:53–54.

Holmes, R. T., and Pitelka, F. A. 1968. Food overlap among coexisting sandpipers on northern Alaska tundra. *Syst. Zool.* 17:305–318.

Holmes, T. 1987. Sexual dimorphism in North American weasels with a phylogeny of Mustelidae. Ph.D. dissert., University of Kansas, Lawrence.

Hutchinson, G. E. 1959. Homage to Santa Rosalia, or why are there so many kinds of animals? *Amer. Nat.* 93:145–159.

Jaksic, F. M., and Delibes, M. 1987. A comparative analysis of food-niche relationships and trophic guild structure in two assemblages of vertebrate predators differing in species richness: Causes, correlations, and consequences. *Oecologia* 71:461–472.

Jaksic, F. M., Greene, H. W., and Yanez, J. L. 1981. The guild structure of a community of predatory vertebrates in central Chile. *Oecologia* 49:21–28.

Jiménez, J. E. 1993. Comparative ecology of *Dusicyon* foxes at the Churchill National Reserve in north central Chile. Master's thesis, University of Florida, Gainesville.

Kiltie, R. A. 1984. Size ratios among sympatric Neotropical cats. *Oecologia* 61:411–416.

Kiltie, R. A. 1988. Interspecific size regularities in tropical felid assemblages. *Oecologia* 76:97–105.

King, C. M. 1989. The advantages and disadvantages of small size to weasels, *Mustela* species. In: J. L. Gittleman, ed. *Carnivore Behavior, Ecology, and Evolution,* pp. 302–334. Ithaca, N.Y.: Cornell Univ. Press.

Kingdon, J. 1977. *East African Mammals,* Vol. 3, Part A (Carnivores). London: Academic Press.

Kleiman, D. G. 1967. Some aspects of social behavior in the Canidae. *Amer. Zool.* 7:365–372.

Kleiman, D. G. 1977. Monogamy in mammals. *Quart. Rev. Biol.* 52:39–69.

Kleiman, D. G., and Eisenberg, J. F. 1973. Comparison of canid and felid social systems from an evolutionary perspective. *Anim. Behav.* 21:637–659.

Klein, R. G. 1986. Carnivore size and Quaternary climatic change in southern Africa. *Quaternary Res.* 26:153–170.

Kruuk, H. 1989. *The Social Badger.* Oxford: Oxford Univ. Press.

Kurtén, B. 1965. The Carnivora of the Palestine caves. *Acta Zool. Fennica* 107:1–74.

Lewin, R. 1983. Santa Rosalia was a goat. *Science* 221:636–639.

Leyhausen, P. 1965. Über die funktion der relativen stimmungshierarchie (Dargestellt am Beispiel der phylogenetischen und ontogenetischen Entwicklung des Beutfangs von Raubtieren). *Z. Tierpsychol.* 22:412–494.

Leyhausen, P. 1979. *Cat Behavior.* New York: Garland STPM Press.

Lockie, J. D. 1966. Territory in small carnivores. *Symp. Zool. Soc. London* 18:143–165.

Marcus, L., and Berger, R. 1984. The significance of radiocarbon dates for Rancho La Brea. In: P. S. Martin and R. G. Klein, eds. *Quaternary Extinctions,* pp. 159–183. Tucson: Univ. Arizona Press.

Mayr, E. 1956. Geographical gradients and climatic adaptation. *Evolution* 10:105–108.

Mayr, E. 1963. *Animal Species and Evolution.* Cambridge: Harvard Univ. Press.

McNab, B. K. 1971. On the ecological significance of Bergmann's rule. *Ecology* 52:845–854.

Moors, P. J. 1980. Sexual dimorphism in the body size of mustelids (Mammalia: Carnivora): The role of food habits and breeding systems. *Oikos* 34:147–158.

Moors, P. J. 1984. Coexistence and interspecific competition in the carnivore genus *Mustela. Acta Zool. Fennica* 172:37–40.

Neal, E. 1986. *The Natural History of Badgers.* London: Croom Helm.

Novikov, G. A. 1962. *Carnivorous Mammals of the U.S.S.R.* Jerusalem: Israel Program for Scientific Translations.

Nowak, R. 1979. *North American Quaternary Canis.* Monographs of the Museum of Natural History, Univ. Kansas, No. 6.

Osborn, D. J., and Helmy, I. 1980. The contemporary land mammals of Egypt, including Sinai. *Fieldiana Zool.* (new ser.), No. 5.

Pimm, S. L., and Gittleman, J. L. 1990. Carnivores and ecologists on the road to Damascus. *Trends Ecol. Evol.* 5:70–73.

Poole, T. B. 1966. Aggressive play in polecats. *Symp. Zool. Soc. London* 18:23–44.

Poole, T. B. 1967. Aspects of aggressive behaviour in polecats. *Z. Tierpsychol.* 24:351–369.

Powell, R. A. 1979. Mustelid spacing patterns: Variations on a theme by *Mustela. Z. Tierpsychol.* 50:153–165.

Powell, R. A., and Leonard, R. D. 1983. Sexual dimorphism and energy expenditure for reproduction in female fisher *Martes pennanti. Oikos* 40:166–174.

Ralls, K. 1977. Sexual dimorphism in mammals: Avian models and unanswered questions. *Amer. Nat.* 111:917–938.

Ralls, K., and Harvey, P. H. 1985. Geographic variation in size and sexual dimorphism of North American weasels. *Biol. J. Linn. Soc.* 25:119–167.

Reig, S. 1992. Geographic variation in pine marten (*Martes martes*) and beech marten (*M. foina*) in Europe. *J. Mamm.* 73:744–769.

Root, R. B. 1967. The niche exploitation pattern of the blue-gray gnatcatcher. *Ecol. Monogr.* 37:317–350.

Rosenzweig, M. L. 1968. The strategy of body size in mammalian carnivores. *Amer. Midl. Nat.* 80:299–315.

Roth, V. L. 1981. Constancy of size ratios of sympatric species. *Amer. Nat.* 118:394–404.

Sandell, M. 1989. Ecological energetics, optimal body size and sexual size dimorphism: A model applied to the stoat *Mustela erminea* L. *Funct. Ecol.* 3:315–324.

Savage, R. J. H., and Long, M. R. 1986. *Mammal Evolution.* London: British Museum (Natural History).

Schoener, T. W. 1969. Models of optimal size for solitary predators. *Amer. Nat.* 103:277–313.

Schoener, T. W. 1984. Size differences among sympatric, bird-eating hawks: A worldwide survey. In: D. R. Strong, Jr., D. Simberloff, L. G. Abele, and A. B. Thistle, eds. *Ecological Communities: Conceptual Issues and the Evidence,* pp. 254–281. Princeton, N.J.: Princeton Univ. Press.

Schoener, T. W. 1986. Overview: Kinds of ecological communities—ecology becomes pluralistic. In: J. Diamond and T. J. Case, eds. *Community Ecology,* pp. 467–479. New York: Harper and Row.

Seitz, A. 1950. Untersuchungen über angerborene Verhaltweisen bei Caniden I und II. *Z. Tierpsychol.* 7:1–46.

Seitz, A. 1959. Beobachtungen an handaufgezogenen Goldschakalen (*Canis aureus algirensis* Wagner 1843). *Z. Tierpsychol.* 16:747–771.

Selander, R. K. 1966. Sexual dimorphism and differential niche utilization in birds. *Condor* 68:113–151.

Shine, R. 1989. Ecological causes for the evolution of sexual dimorphism: A review of the evidence. *Quart. Rev. Biol.* 64:419–461.

Simberloff, D. 1989. Pattern analysis and the detection of interactions in natural communities. In: G. Montalenti, A. Renzoni, and A. Anelli, eds. *Proc. III Congresso Societa Italiana di Ecologia, Siena, October 21–24, 1987*, pp. 19–29. Parma: Edizioni Zara.

Simberloff, D., and Boecklen, W. J. 1981. Santa Rosalia reconsidered: Size ratios and competition. *Evolution* 35:1206–1228.

Simberloff, D., and Dayan, T. 1991. Guilds and the structure of ecological communities. *Ann. Rev. Ecol. Syst.* 22:115–143.

Simms, D. A. 1979. North American weasels: Resource utilization and distribution. *Canadian J. Zool.* 57:504–520.

Smithers, R. H. N. 1983. *The Mammals of the Southern African Subregion.* Pretoria, South Africa: Univ. Pretoria Press.

Strong, D. R., Szyska, L. A., and Simberloff, D. 1979. Tests of community-wide character displacement against null hypotheses. *Evolution* 33:897–913.

Taylor, M. E. 1989. Locomotor adaptations by carnivores. In J. L. Gittleman, ed. *Carnivore Behavior, Ecology, and Evolution*, pp. 382–409. Ithaca, N.Y.: Cornell Univ. Press.

Tchernov, E. 1988. The paleobiogeographical history of the southern Levant. In: Y. Yom-Tov and E. Tchernov, eds. *The Zoogeography of Israel*, pp. 159–250. Dordrecht, Netherlands: W. Junk.

Tonkyn, D. W., and Cole, B. J. 1986. The statistical analysis of size ratios. *Amer. Nat.* 128:66–81.

Van Valen, L. 1965. Morphological variation and the width of the ecological niche. *Amer. Nat.* 99:377–390.

Van Valkenburgh, B. 1984. A morphological analysis of ecological separation within past and present carnivore guilds. Ph.D. dissert., The Johns Hopkins University.

Van Valkenburgh, B. 1985. Locomotor diversity within past and present guilds of large predatory mammals. *Paleobiology* 11:406–428.

Van Valkenburgh, B., and Hertel, F. 1993. Tough times at La Brea: Tooth breakage in large carnivores of the Late Pleistocene. *Science* 261:456–459.

Van Valkenburgh, B., and Ruff, C. B. 1987. Canine tooth strength and killing behaviour in large carnivores. *J. Zool. (London)* 212:379–397.

Van Valkenburgh, B., and Wayne, R. K. 1994. Shape divergence associated with size convergence in sympatric East African jackals. *Ecology* 75:1567–1581.

Wayne, R. K., Van Valkenburgh, B., Kat, P. W., Fuller, T. K., Johnson, W. E., and O'Brien, S. J. 1989. Genetic and morphological divergence among sympatric canids. *J. Hered.* 80:447–454.

Werdelin, L. 1993. Community-wide character displacement in the carnassials of Miocene hyenas. Unpublished manuscript.

Werdelin, L., and Solounias, N. 1991. The Hyaenidae: Taxonomy, systematics and evolution. *Fossils Strata* 30:1–104.

Williamson, M. 1972. *The Analysis of Biological Populations.* London: Edward Arnold.

Wilson, D. S. 1975. The adequacy of body size as a niche difference. *Amer. Nat.* 109:769–784.

Yom-Tov, Y., and Tchernov, E., eds. 1988. *The Zoogeography of Israel.* Dordrecht, Netherlands: W. Junk.

CHAPTER 8

Patterns and Consequences of Dispersal in Gregarious Carnivores

PETER M. WASER

Most young carnivores, like most young mammals of any sort, leave home at sexual maturity and, with luck, find a stable home range elsewhere. Emigration from the natal home range is more likely, and the distance traveled before settling tends to be greater, if the young carnivore is male. A few individuals disperse great distances, but most settle within a few home-range diameters of their birthsites. Cases presenting this general pattern have been documented in all carnivore families. Nevertheless, the pattern is a caricature, concealing enormous variation in the details. Dispersal may be short- or long-distance, common or rare, early in life or late, sex-biased or not, an overnight departure or the culmination of months of tentative exploration.

A recurring phenomenon within all carnivore families is natal philopatry: some individuals remain on their natal home range far past the onset of reproductive maturity. Natal philopatry appears to be at the base of sociality in all carnivore groups. It has been argued that high fitness costs associated with dispersal set the stage for the evolution of carnivore gregariousness, even when the advantages of social hunting are nonexistent or equivocal (Messier and Barrette 1982; Waser and Jones 1983; Packer 1986). As this chapter will describe, gregariousness also has the potential to complicate dispersal's costs and benefits in interesting ways.

This chapter will focus on dispersal in six well-studied species, in each case a relatively gregarious member of its family: the gray wolf (*Canis lupus*), the lion (*Panthera leo*), the spotted hyena (*Crocuta crocuta*), the coati (*Nasua narica*), the European badger (*Meles meles*), and the dwarf mongoose (*Helogale parvula*). "Dispersal," a heterogeneous phenomenon, will be given a very general definition: animals that leave either their natal home ranges or their natal social groups will be viewed as emigrating or attempting dispersal; and if emigrants settle on a stable home range or immigrate into a new social group, dispersal will be viewed as successful. I will describe the patterns of dispersal exhibited by each species, but focus on the consequences: how does emigration influence the survival, and the number and quality of breeding opportunities, of dispersing individuals?

A full accounting of the fitness consequences of dispersal versus philopatry is an extremely complex undertaking, because dispersal is likely to influence survival, access to mates, survival of young, and survival of nondescendant kin. All of these effects may be immediate, or they may occur years later. Dispersal's fitness consequences are likely to differ between sexes, among age classes, as a function of the relatedness of dispersers to other group members, and as a function of their group size and social rank. Since the methodology for such accounting is still being developed (Creel and Waser 1994; Lucas et al., 1996), the best that this review can do is to point to the kinds of fitness effects that have been identified.

Nevertheless, we can be optimistic about the prospects for understanding the ultimate causes of carnivore dispersal. The data we need would describe the life-history trajectories of large numbers of individual carnivores throughout their lives: under what conditions they dispersed (or, just as important, failed to disperse), and how well they survived and reproduced in the following years. These are exactly the kinds of data that most studies of gregarious carnivores are beginning to produce.

The Studies

Perhaps the most extensive and longest long-term studies of any carnivore population are those of the Serengeti lions, in Tanzania. This study, pursued continuously since the late 1960s, involves hundreds of individually recognizable animals, including the complete population of Ngorongoro Crater. Because lions are conspicuous, largely oblivious to human observers, and individually recognizable from natural marks, longitudinal data are available for many animals (Schaller 1972; Bertram 1975; Pusey and Packer 1987; Packer et al. 1988; Packer and Pusey 1993). These advantages are shared by spotted hyenas, the subject of multi-year studies in Kenya, 1979 to present (Frank 1986a, b, this volume; Holekamp and Smale 1993; Holekamp et al. 1993; Smale et al. 1993; Frank et al. 1995), Tanzania, 1987 to present (Hofer and East 1995), Botswana, 1978–84 (Mills 1990), and South Africa, 1982–84 (Henschel and Skinner 1987).

Wolves have been less cooperative subjects; because virtually all wolf populations are subject to human predation and/or live in habitats affording poor visibility, studies have relied heavily on radio-collared animals, and no populations have been completely marked. Nevertheless, the number of animals radio-tagged has been extensive, and multi-year data are available from three sites in Minnesota, 1966 to present (Mech 1987; Gese and Mech 1991), 1972–76 (Fritts and Mech 1981), 1980–86 (Fuller 1989), two sites in Quebec, 1980–84 (Messier 1985a, b; Potvin 1987), and two sites in Alaska, 1975–82

(Van Ballenberghe 1983a; Ballard et al. 1987) and 1976–81 (Peterson et al. 1984).

Dwarf mongooses, though difficult to recognize from natural marks, are diurnal and live in open country; a large Serengeti population has been freeze-marked, habituated, and under observation since 1974 (Rood 1987, 1990; Creel and Waser 1991, 1994). Coatis, though diurnal, live in dense forest where observation conditions are poor, and dispersal data for them are thus less complete; animals on Barro Colorado Island, Panama, have been marked and followed in 1976–78 (Russell 1982) and since 1989 (Gompper 1994). Finally, dispersal data for the European badger, whose nocturnal habits make it still more difficult to monitor, come primarily from four sites in Britain: Bristol, 1978–85 (Cheeseman et al. 1988), Gloucestershire, 1976–85 (Cheeseman et al. 1988), Oxford, 1972–75 (Kruuk 1989; 1987–92, Woodroffe et al. 1995), and Speyside, 1976–81 (Kruuk 1989). In each of these studies substantial numbers of animals were radio-collared or individually marked, but it was usually not possible to census entire populations at regular intervals.

Some Comments on the Data

Emigration is a rare event, and hard to track; problems of logistics, as well as the interests of investigators, mean that dispersal in different species has been viewed through different filters. In trying to compare aspects of dispersal across species, or even across populations within species, it becomes clear that the available comparisons are not entirely quantitative. For example, nearly complete cohorts of dwarf mongooses can be marked and their fates monitored through time, but when an animal disappears it may be difficult to determine with certainty whether it emigrated and, if it did, what its fate was (Waser et al. 1994a). In contrast, radio-tracking studies of wolves tell us unambiguously when an animal has left its pack, but we may not know its exact age, or whether it is leaving its *natal* pack, or whether the radio-collared individuals are a random sample of the population.

Paradoxically, the data that can be most easily compared across studies are often the least quantitative. Virtually all studies report the proportions of males and females that emigrate while under observation; these numbers can therefore be compared (Table 8.1). The numbers cannot, however, be translated directly into per-capita probabilities of emigration, and are not always directly comparable across studies. Studies differ in the length of time that individuals are under observation, as well as the composition of the sex and age classes that have been followed. Moreover, animals that do *not* emigrate while under observation include those that die in the group and those that disappear under unknown circumstances, as well as those that survive the observation period. For example, Table 8.1 reports that only 43% of male hyenas disperse while under observation; but it should be remembered that many of the re-

Table 8.1. Percentages of individuals that emigrate while under observation

Carnivore species	Males	Females	G
Gray wolves	29% (101/353)	26% (87/340)	0.58
Dwarf mongooses	36% (126/348)	23% (71/309)	39.85***
European badgers	32% (66/206)	13% (37/280)	76.16***
Lions	87% (115/132)	38% (56/148)	24.88***
Spotted hyenas	43% (52/122)	6% (6/94)	13.76***
Coatis	46% (31/67)	23% (18/78)	5.54**

Notes: Emigration probabilities were significantly higher for males in all species except wolves. Animals were considered to emigrate if they left their original group or home range while under observation, or if they were initially marked or recognized as transients, or if they disappeared under conditions thought by the authors to be suggestive of emigration rather than death (for instance, simultaneously with same-sexed group members, in lions and mongooses). Thus, the numbers include both social and locational dispersal. Where multiple studies exist, numbers of animals observed and numbers of animals dispersing are summed across studies. As described in the text, methods and intensities of observation differ among species, and many animals counted as "nondispersing" may have died. Therefore, the percentages should not be read as per-capita probabilities of emigration, and may not be comparable across species, although sex ratios should be. (The G statistic for the log likelihood test indicates whether the proportions of males and females that disperse differ: ** = $P < 0.01$, *** = $P < 0.001$.)

Sources: Gray wolves: Fritts and Mech 1981; Peterson et al. 1984; Ballard et al. 1987; Potvin 1987; Fuller 1989. Dwarf mongooses: Waser et al. 1994a. European badgers: Cheeseman et al. 1988; Kruuk 1989; Woodroffe et al., in press. Lions: Pusey and Packer 1987, pers. comm. Spotted hyenas: Frank 1986a; Henschel and Skinner 1987; Mills 1990; K. Holekamp, pers. comm. Coatis: M. Gompper, pers. comm.

maining 57% died while under observation, and others may have dispersed after observations terminated. The lack of standard approaches to presenting dispersal data places strong limits on cross-species comparisons. In the future, data reported in the form of "time lines" (e.g., Figure 8.1) will be particularly useful, because time lines make it possible to determine how many animals of each age and sex are under observation, as well as their fates, and to allow comparison of the rates and consequences of both dispersal and philopatry.

The Patterns

Dispersal starts with a simple yes/no decision—to emigrate or to stay in the natal group—but individual carnivores' decisions to leave may not be independent; some are forced out, and some emigrate in coalitions. Once out of the group, most carnivores also leave the natal home range, but some do not. Following a distinction that has been useful in the description of primate dispersal (Isbell et al. 1990), carnivores sometimes disperse "socially" but not "locationally" (more on that distinction below). Dispersal, if successful, usually terminates with immigration into a new group, but can end with settle-

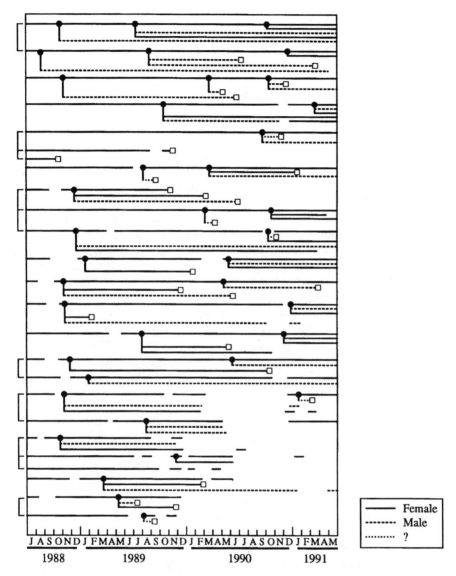

Figure 8.1. Representative time line (for female spotted hyenas, modified from Holekamp et al. 1993). Vertical lines descending from solid circles denote births; squares denote known deaths; gaps in lines and unterminated lines denote disappearances. Particular classes of dispersers can be identified, and many of the fitness consequences of dispersal and philopatry classes can be estimated, through the use of time lines, which are particularly useful if they indicate the sex of each animal, the periods it was under observation, the dates of its immigration and emigration, and its eventual fate.

ment on an unoccupied site. Finally, if an emigrating animal is not immediately successful at finding such an opportunity, it may either pursue a transient existence or return, perhaps repeatedly, to its natal group.

Emigration Alone or in Coalitions

Most mammals, and many members of the six species to be taken up here, emigrate independently, as individuals. Joint dispersal, however, is not uncommon in male dwarf mongooses, and it is the rule among male lions. In both species, males that emigrate together are often close relatives, remain together during dispersal, and compete jointly, as a coalition, for opportunities to join other groups. In both species, there are also cases in which males that emigrate together later split up, or are joined by unrelated males while in transit. Some female dwarf mongooses, wolves of both sexes, and females in one population of European badgers (Figure 8.2) also emigrate in pairs or trios, and members of these coalitions are believed to be related (Ballard et al. 1987; Rood 1987; Woodroffe et al. 1995).

For female lions, spotted hyenas, and coatis, another version of emigration in coalitions is the predominant form of dispersal. In these three species, female philopatry is the rule, but when females leave their natal group they do so predominantly through the departure of a matriline (Russell 1982; Mills 1990; Holekamp et al. 1993; Gompper 1994) or of a cohort (Hanby and Bygott 1987; Pusey and Packer 1987). Group fission can be viewed as emigration by a coalition of females, sometimes accompanied by immatures of both sexes.

Social versus Locational Dispersal

Group fission in lions and coatis is distinctive in that the dispersing coalitions may not choose to leave their original home range. For example, in 23 out of 24 cases of group fission in lions, the new pride's range was entirely or partly within the females' natal area (Pusey and Packer 1987). Recent genetic evidence suggests that coati males also exhibit social, but not locational, dispersal. Solitary males captured in the vicinity of a group are closely related to each other and to the members of that group, which suggests that emigrating males remain on their natal ranges, making brief forays into other home ranges during the breeding season (Gompper and Wayne, this volume).

Emigration without Immigration

Dispersal by female coalitions of lions, spotted hyenas, and coatis, and also by solitary male coatis, is distinctive for another reason: it does not terminate in immigration. Male coatis remain solitary after emigration, entering groups only to breed. Female lion, spotted hyena, and coati coalitions usually form the

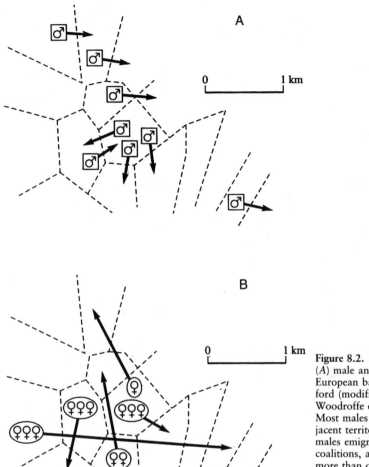

Figure 8.2. Dispersal of (*A*) male and (*B*) female European badgers at Oxford (modified from Woodroffe et al. 1995). Most males disperse to adjacent territories. Most females emigrate in coalitions, and many move more than one territory away before settling.

nucleus of new groups that form by accretion, growing through reproduction or as they are joined by males.

Emigration without immigration is also exhibited by dwarf mongooses: some females leave their natal home ranges to settle, without a male, in previously unoccupied areas (Rood 1987). New groups form later, when the female is joined by an emigrating male. The same behavior may be exhibited by some female wolves: emigrating females settle more than twice as long before pairing as do emigrating males (Gese and Mech 1991).

Emigration to Known Destinations

A small proportion of dwarf mongooses, especially juveniles, disperses when adjacent groups are in contact with each other, and emigration and immigration are thus virtually simultaneous events (Rood 1987; S. Creel, pers. comm.). In these cases, it seems likely that individuals assess the potential for successful immigration without leaving their natal groups. Perhaps similarly, a handful of male lion coalitions move directly into adjacent prides (Pusey and Packer 1987). Some wolves and spotted hyenas emigrate into adjacent groups under circumstances that suggest they can detect opportunities for immigration without leaving their natal pack (Mech 1987; K. Holekamp and H. Hofer, pers. comm.).

Emigration and Transience

In other cases, animals that emigrate appear to strike out into the unknown; after leaving, they pursue an itinerant existence, searching for groups to enter or unoccupied areas in which to settle. Some emigrants disperse in a straight line, but others follow highly circuitous paths; there can be tremendous variation even within a single population (Mech 1987). Little is known about the ways animals assess their chances of joining new groups, or how many groups they try to enter, but where data are available (mongooses, male badgers, wolves), most successful dispersers end up in immediately adjacent groups (Figure 8.2). In lions, males in large coalitions usually enter adjacent groups, whereas those in small coalitions disperse farther, reflecting the greater ease with which males in large coalitions displace resident males (Pusey and Packer 1987). In wolves, yearlings and pups may have to traverse more territories than adults do before they are able to settle (Gese and Mech 1991). Interestingly, emigrating female dwarf mongooses and European badgers (Figure 8.2) tend to settle farther from their natal ranges than do males, which suggests that those females that do emigrate are more selective, or have a more difficult time immigrating, than do the males (Waser et al. 1994a; Woodroffe et al. 1995).

When emigrants are not immediately successful at immigration, they may wander for long periods. The mean time between emigrating and settling is less than a month for female wolves, but more than 4 months for males; the range is 1 week to 12 months (Gese and Mech 1991). "Lone wolves" of both sexes have been radio-tracked for a year or more (Messier 1985a; Fuller 1989). For male spotted hyenas the process of finding and immigrating into another group may also take many months, and a few animals are truly nomadic (K. Holekamp, pers. comm.). Emigrating male lions remain nomadic for a mean of 17 months (range 0–44 months; Pusey and Packer 1987). Emigrating dwarf mongooses can remain solitary for more than a year (J. Rood, unpubl. data), though the median duration of dispersal is on the order of a month.

Repeated Forays

Leaving the natal group and home range is often an irrevocable, one-shot decision, but there is increasing evidence that final departure can be preceded by exploratory sorties. For example, initial descriptions of male spotted hyena dispersal suggested that males near reproductive maturity simply left—but some males were resighted, perhaps months later, revisiting the natal group (Frank 1986a). More recent radiotelemetry data suggest that it is common for male hyenas to make repeated forays, lasting from just a few hours up to 8 months, into other territories; such males have been sighted up to 55 km from their natal group (Mills 1990; K. Holekamp, pers. comm.). Females may make forays as well (Holekamp et al. 1993). In Serengeti populations, both sexes regularly "commute" long distances to follow migratory prey, during which they may gain information on reproductive opportunities (Hofer and East 1993a, b). East and Hofer (1991) have shown how dispersing males use long-distance calls ("whoops") to assess the length of other groups' queues for reproductive opportunities.

In wolves, exploratory forays out of the natal group and home range are the rule, for both males and females. "Extraterritorial" forays generally last for a few weeks, and extend over a few tens of kilometers; in Minnesota, the average young wolf made about one foray before dispersing, but in Quebec, yearlings made on average three forays per year, and females made more forays than did males (Figure 8.3) (Messier 1985a; Gese and Mech 1991). Forays by males and, to a slightly lesser extent, by females are also common in most European badger populations (Cheeseman et al. 1988; Woodroffe et al. 1995). A small proportion of dwarf mongooses of both sexes make forays to nearby groups (Rood 1987).

In these species, patterns of movement during "forays" and "dispersal" may be difficult to distinguish. In several wolf populations, the rates, durations, and distances of forays are identical between animals that eventually disperse and those that do not (Messier 1985a; Gese and Mech 1991). A "dispersing" animal may be simply one whose foray ends in success. Some animals, however, almost certainly mate while on forays, even without settling in the group they mate in. For example, female European badgers have been trapped making forays while in estrus (Woodroffe et al. 1995), and genetic evidence suggests more outbreeding than is expected from dispersal rates (Evans et al. 1989; da Silva et al. 1994). Some male wolves have been reported to associate with females during forays and mate with them, as have a very few male dwarf mongooses (Mech 1987; S. Creel, pers. comm.). Female lions occasionally leave their groups and mate with nomadic males (Pusey and Packer 1987).

Male-Female Differences

For each species, a crude first approximation of the probability of emigration is the proportion of recognizable animals that leave their groups while under observation (Table 8.1). Such data confirm that males are generally more likely to emigrate than females. In most populations of lions, spotted hyenas, and European badgers, and in Barro Colorado coatis, males are more

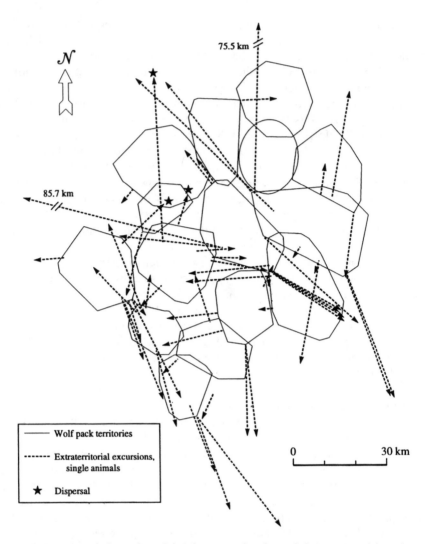

Figure 8.3. Forays made by male and female gray wolves beyond their territorial boundaries (modified from Messier 1985a). Both the animals that returned to their original groups and those that settled or immigrated (three of them, indicated with stars) had similar movement patterns.

than twice as likely to emigrate as females. In dwarf mongooses, males are about half again as likely to leave their natal groups. Only in wolves are the emigration probabilities of the two sexes similar.

Where more than one population of a species has been studied, the proportions of males that emigrate while under observation do not differ significantly from one population to the next. The probability of female emigration, however, is much more labile (Table 8.2). In lions, wolves, European badgers, and

Table 8.2. Interpopulation variation in dispersal tendencies

Species and population	Males	Females
Lions	G = 0.75	G = 10.36***
Serengeti	89% (67/75)	27% (22/85)
Ngorongoro	84% (48/75)	52% (34/65)
Gray wolves	G = 5.06	G = 11.45*
Minnesota[a]	24% (39/162)	23% (36/154)
Alaska[b]	34% (28/83)	15% (10/67)
Minnesota[c]	33% (12/36)	37% (16/43)
Alaska[d]	38% (13/34)	30% (9/30)
Minnesota[e]	25% (4/16)	26% (6/23)
Quebec[f]	23% (5/22)	43% (10/23)
European badgers	G = 6.38	G = 24.98***
Gloucestershire	30% (32/75)	9% (15/163)
Bristol	44% (24/54)	30% (16/43)
Speyside	26% (6/23)	0% (0/37)
Oxford	18% (4/22)	22% (6/27)
Spotted hyenas	G = 5.16	G = 10.11*
Kenya[g]	43% (23/53)	0% (0/40)
Kenya[h]	37% (16/43)	9% (3/35)
Botswana	33% (5/15)	22% (3/13)
South Africa	73% (8/11)	0% (0/6)

Notes: For males, contingency tests show no significant differences among populations in tendency to emigrate while under observation, in any species. For females, there are significant differences among populations in tendencies to emigrate, for all four species. (The G statistic for the log likelihood test indicates whether the proportions of animals that disperse differ among populations: * = $P < 0.05$, *** = $P < 0.001$.)

Data: [a]Gese and Mech 1991; [b]Ballard et al. 1987; [c]Fuller 1989; [d]Peterson et al. 1984; [e]Fritts and Mech 1981; [f]Potvin 1987; [g]Frank 1986a; [h]K. Holekamp, pers. comm. Other sources as in Table 8.1.

spotted hyenas, contingency analyses indicate that there are significant differences among populations in female emigration tendencies. The reasons for this apparently higher lability of female tendencies are not yet clear. Such differences might reflect differences in demography. Lionesses emigrate more readily in the Ngorongoro population, where juvenile survival is higher (Pusey and Packer 1987). Female wolves emigrate least often in an Alaskan population that is heavily hunted (Ballard et al. 1987); opportunities to obtain breeding positions in the natal group may be unusually high under these conditions. Badger females emigrate at high rates in a low-density Bristol population where vacant areas between groups are common (Cheeseman et al. 1988), and not at all in higher-density populations; in the highest-density population, females emigrated, but only in coalitions (Woodroffe et al. 1995).

Further patterns are evident where studies provide residence histories of known-aged animals from which age-specific emigration probabilities can be estimated. For example, such data suggest that emigration rates are remarkably similar, age by age, for male and female wolves; rates are highest at age two, around the period of reproductive maturity (Fritts and Mech 1981; Peterson et al. 1984; Potvin 1987; Fuller 1989). In contrast, male spotted hyenas begin to emigrate at age two, around the time of reproductive maturity, but females rarely leave at any age (Frank 1986a; Mills 1990; Hofer et al. 1993). In dwarf mongooses, both males and females are most likely to emigrate at ages one to two, when they are mature but often reproductively suppressed; and males emigrate over a broader range of ages than do females (Waser et al. 1994a). Young European badgers can begin to make forays prior to maturation, but dispersal usually occurs afterward. In contrast to mongooses, badger females but not males can emigrate at any age (Woodroffe et al. 1995).

A third way to examine dispersal patterns is by their consequences: what proportions of animals remain in their natal groups at each age? Through a combination of mortality and emigration, no male spotted hyenas remain in their natal groups by age five. Although most females die in their first 5 years, the 20% that remain are all still in natal groups (Figure 8.4) (Frank et al. 1995). Only $^2/_{57}$ (3%) of male coatis are still alive in their natal groups by age three, when their testes have descended; in contrast, nearly all females still alive at this age remain philopatric (Gompper 1994). Only $^1/_{132}$ (1%) of male lions remain in their natal group by age four, while $^{76}/_{148}$ (51%) of females are still there (Pusey and Packer 1987). Similarly, in cohorts of one-year-old dwarf mongooses, only $^7/_{210}$ males (3%) remain alive and in their natal group by age

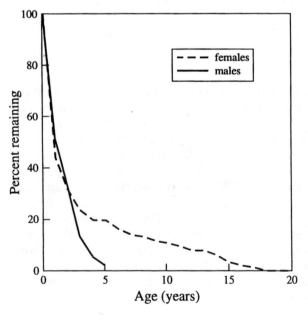

Figure 8.4. Percentages of male and female spotted hyenas remaining in their natal group, as a function of age (adapted from Frank et al. 1995). All males have either died or emigrated by age five. Female disappearances are presumed to represent mortality only. $N = 194$ individuals (sex unstated).

four, compared with $^{50}/_{211}$ females (24%) (J. Rood, S. Creel, and P. Waser, unpubl. data).

Secondary Dispersal

After immigrating into a new group, some dwarf mongooses disperse again later in life. Secondary dispersal is much less common than natal dispersal: only a third of mongoose natal dispersers emigrate again. Most animals that leave again are males, and many of these are males evicted by immigrating male coalitions. Some secondary dispersers emigrate from groups whose opposite-sexed breeder has died. These animals are all males; females in groups whose breeding male dies remain on the group home range (Rood 1987).

Similar patterns occur in lions: males often emigrate secondarily when evicted by other male coalitions. Some secondary dispersal is voluntary, however, and males move into groups with more females, or away from groups in which they have mature daughters (Hanby and Bygott 1987; Pusey and Packer 1987). Wolves, too, can emigrate secondarily when the opposite-sexed breeder dies (Fritts and Mech 1981; Fuller 1989). Secondary dispersal is known in spotted hyena males (Hofer et al. 1993; K. Holekamp and L. Frank, pers. comm.), but too few animals have been followed longitudinally to allow one to generalize about its frequency and contexts. No data are as yet available regarding secondary dispersal in badgers or coatis.

Mortality Consequences

When male coalitions take over lion prides, they often drive out young animals of both sexes. For these young lions, the alternative to emigration would be death (Hanby and Bygott 1987). Some wolves disperse during the breeding season, when rates of aggression within groups are known to be elevated; perhaps here, too, emigration is the better part of valor (see also Zimen 1976). And $^6/_{43}$ male and $^1/_{31}$ female dwarf mongooses emigrated when they were evicted by coalitions of immigrants (Rood 1987). Clearly, there are circumstances in which emigration, at least in the short term, may decrease mortality risk.

In the longer term, however, emigration almost certainly increases mortality. The evidence for increased mortality associated with emigration is often indirect: mortality rates are elevated in the sex or age classes that emigrate most often. In South Africa, 38% of female spotted hyenas, but only 9% of males, are more than 6 years old (Henschel and Skinner 1987). Likewise, adult hyena sex ratios in Kenya and Tanzania are strongly skewed toward females, probably reflecting male mortality during dispersal (Frank et al. 1995; Hofer, pers. comm.) For lions and coatis, as well, a strong male bias in emigration rate is associated with higher male death rates (Packer et al. 1988; Gompper 1994).

Mortality during Dispersal

Estimates of the mortality risk associated with emigration can be obtained in two ways. In gray wolves, enough dispersers have been radio-collared to allow direct observation of mortality. In Quebec, wolves making extraterritorial forays die at rates five times higher than residents (1.5 vs. 0.3 deaths per wolf-year; Messier 1985b). In north-central Minnesota, nonresident wolves are reported to have annual survival rates of 0.52, compared with 0.67 for residents (Fuller 1989). In Kenai, Alaska, survival rates are 0.38 for nonresidents and 0.73 for residents (Peterson et al. 1984). In northeast Minnesota, only $^{44}/_{61}$ (72%) of wolves that emigrate succeed in settling (Gese and Mech 1991).

Where animals have not been radio-tagged, the upper and lower bounds on dispersal risk can be calculated from standard census data as long as the study population is neither a source nor a sink (Waser et al. 1994a). Such calculations indicate that between 36 and 59% of emigrating male dwarf mongooses, and between 35 and 80% of emigrating females, die without immigrating into another group. Similarly, the demography of European badgers suggests that 19–61% of males and 19–47% of females die during dispersal (R. Woodroffe, unpublished data).

Emigrants may incur mortality either because they are alone or because they are in unfamiliar territory; they may also incur mortality when they attempt to immigrate into a new group. For large, well-armed carnivores like lions, spotted hyenas, and wolves, it seems likely that most dispersal-caused mortality is actually associated with intrasexual, or at least intraspecific, competition. For example, nomadic male lion coalitions usually engage in repeated contests with resident males before deposing them; these contests can result in serious injury or death (Packer and Pusey 1982). Similarly, most studies of wolves report that the second largest cause of mortality (after human aggression) is aggression from other wolves, usually associated with movements off the normal territory (e.g., Fritts and Mech 1981; Ballard et al. 1987). European badgers acquire scars and broken teeth around the time of dispersal, and immigrant males are found to have higher rates of scarring than do philopatric males, where data are controlled for age (Woodroffe et al. 1995; Woodroffe and Macdonald 1995b; Kruuk 1989). Immigrating male spotted hyenas receive considerable aggression and are intensely submissive, dominated even by pups in the groups they are trying to join; the process of assimilation into a new group may take months (Henschel and Skinner 1987; Mills 1990; Smale et al. 1993). Roughly three times as many transient as immigrating males are sighted (Holekamp et al. 1993), suggesting either that males visit several groups before successfully immigrating, or that many transient males die.

There is also evidence that emigrants incur greater risk because they are in unfamiliar territory. The largest source of wolf mortality is *Homo sapiens,* and emigrating animals are less careful to avoid areas settled by humans than are residents (Fritts and Mech 1981; Peterson et al. 1984; Fuller 1989). In Tanzania, spotted hyenas foraging off their normal home ranges are more vulnerable to human predation (Hofer et al. 1993).

Emigrating small carnivores, such as dwarf mongooses, would seem to be highly vulnerable to predation. Nevertheless there is some suggestion that even here a major portion of dispersal-caused mortality is associated with immigration. Rood (1987) found that males attempting to immigrate could be met with intense aggression by resident males; immigrating males might face such aggression for more than a month before being accepted into a group. Immigrant males often have visible injuries, poor coat condition, and low testosterone levels, conditions that are known not to be present at emigration (Creel et al. 1993; S. Creel, pers. comm.). Older male mongooses are more likely to survive dispersal, as would be expected if their rank relative to resident males influences survival when they try to immigrate (Waser et al. 1994a).

Mortality after Dispersal

After immigration or settlement on a new home range, dispersers may still pay a mortality cost. Lionesses that successfully disperse subsequently have lower probability of surviving from ages four to eight than those that do not: 86% versus 96% in Serengeti, 44% versus 100% in Ngorongoro (Pusey and Packer 1987). Dispersing dwarf mongooses tend to immigrate into smaller groups, where mortality rates are higher (Rood 1990). Dispersing male spotted hyenas immigrate at low rank, which may decrease their access to food and thus threaten their survival (Smale et al. 1993).

Reproductive Consequences

Like mortality consequences, reproductive consequences of dispersal occur not only while dispersing animals are between groups, but also after they have successfully immigrated.

Opportunity Costs

If females mate preferentially with resident males, one might assume that males do not mate during dispersal and thus that they pay a reproductive opportunity cost (Alberts and Altmann 1994). Observations of mating in some species tend to support this assumption: female dwarf mongooses have almost never been seen to mate with dispersing males (Rood 1990; S. Creel, pers.

comm.). Lionesses and female spotted hyenas mate with nomadic males infrequently (Hanby and Bygott 1987; Henschel and Skinner 1987; M. East and K. Holekamp, pers. comm.), and available genetic evidence indicates that among lions all offspring are fathered by pride males ($N = 78$; Gilbert et al. 1991).

By contrast, all male coatis mate without immigrating into another group (Gompper and Wayne, this volume). Genetic evidence of outbreeding beyond that expected from observed badger dispersal rates (Evans et al. 1989; da Silva et al. 1994) might reflect mating by transient animals. Even among dwarf mongooses and spotted hyenas, females have become pregnant in groups having no resident male; whether they mated with dispersing males, or made forays to meet residents in other groups, is unknown (Rood 1987; Mills 1990). Anecdotal accounts of mating during forays, mentioned earlier, suggest that the possibility of mating during dispersal warrants further investigation.

Escape from Groups without Mates

In contrast to the potential loss of mating opportunities during dispersal, reproductive consequences after immigration are likely to be positive. As has already been noted, some cases of emigration in dwarf mongooses and wolves are precipitated by the death of a mate. Leaving a group that contains no reproductive opportunities at all is an infrequent but clearcut example of a general pattern: carnivore dispersal may, quite generally, be out of a group presenting few reproductive opportunities and into one presenting many.

For example, male lions sometimes disperse voluntarily (that is, without having been evicted) to groups containing more females. Similarly, lionesses and female badgers disproportionately leave large groups in which per-capita reproductive success is low (Hanby and Bygott 1987; Pusey and Packer 1987; Woodroffe et al. 1995). Dwarf mongooses of both sexes emigrate disproportionately from both very small and very large groups; in small groups the survival and reproductive rates of all animals are low, whereas in large groups the length of the queue for the dominant breeding position is long (Rood 1987; J. Rood, S. Creel, and P. Waser, unpublished data).

It seems likely that dispersing carnivores often gain reproductively, because they leave groups in which they are (1) unlikely to mate or produce offspring because they are subordinate; (2) unlikely to mate because of inbreeding avoidance; or (3) are unlikely to produce offspring because of inbreeding depression. But the degree to which dispersers garner these different reproductive advantages differs considerably from one species to another.

Escape from Reproductive Suppression

Dwarf mongooses and wolves exemplify species in which socially subordinate animals of both sexes are unlikely to reproduce. Among dwarf mongooses, dominance is highly correlated with age; animals remain subordinate for up to 10 years when older individuals of the same sex are present, even though reproduction is possible among yearlings in groups with no older members (Rood 1980; Creel and Waser 1994). Dominant males father 76%, and dominant females are the mothers of 85%, of the young (Keane et al. 1994). Among wolves, social suppression of subordinate females is the primary reason for delayed breeding. Females can breed as yearlings in captivity but rarely breed before age three in the wild; one exception to this generalization was a disperser to a pack lacking more dominant females (Peterson et al. 1984).

In dwarf mongooses and wolves, there are multiple reasons for failure of reproduction among subordinates. Among males of both species, the primary means of reproductive suppression is through behavioral interference. Females, suppressed like males by behavioral means, are suppressed endocrinologically, as well (Packard et al. 1985; Creel et al. 1992), and this endocrinological suppression may well reflect the low probability of survival for subordinates' offspring (Creel and Waser 1996). In the field, subordinate female wolves rarely have offspring (Van Ballenberghe 1983b; Peterson et al. 1984; Ballard et al. 1987); when they do, the offspring may be lost (Peterson et al. 1984). Similar trends have been noted in captive wolf packs (Packard and Mech 1980). In dwarf mongooses, subordinate females that become pregnant asynchronously with the dominant females never produce surviving offspring (Rood 1980; Keane et al. 1994). The potential causes of low reproductive success by subordinate females include abandonment by other group members and infanticide (Creel, this volume; Peterson et al. 1984; Rood 1990).

Whatever the mechanisms of suppression, there is substantial evidence in both species that dispersing animals gain more reproductive opportunities. Among Alaskan wolves, dispersers left packs averaging 7.6 animals and ended up in packs of mean size 2.8 (Ballard et al. 1987). Dispersers thus increased the probability of being the dominant members of their sex. In other wolf populations, the observation that most dispersers attempt to found new packs rather than immigrating into existing ones (Fritts and Mech 1981; Mech 1987; Fuller 1989; Gese and Mech 1991) suggests that dispersing animals attain breeding positions they would otherwise have to wait for. Dominant males have been observed to disperse only after they fell in rank (Peterson et al. 1984).

Among dwarf mongooses, males and especially females disperse to groups that contain fewer, older same-sexed animals (Rood 1987; Creel and Waser 1994). Even when dispersers do not immigrate into dominant positions, by "shortening the queue" they increase their future prospects for reproduction (Figure 8.5).

Lions and coatis contrast strongly with dwarf mongooses and wolves in offering no evidence of reproductive suppression and no suggestion that either males or females escape the costs of social subordination by dispersing. In fact, emigrating lionesses reproduce later in life than do their philopatric counterparts (Pusey and Packer 1987).

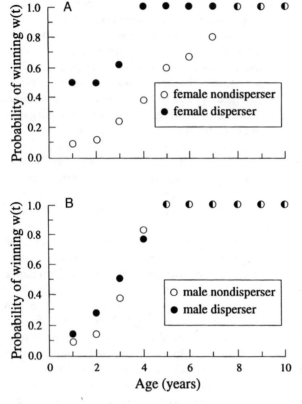

Figure 8.5. Probability of winning a contest for the dominant, breeding position in a dwarf mongoose group, as a function of age for (*A*) female and (*B*) male dispersers and nondispersers. Because dispersers (especially females) immigrate into groups with few older, same-sexed competitors, they are more likely than philopatric animals to acquire a breeding position, and to acquire it sooner. (From Creel and Waser 1994; reprinted by permission.)

The pattern for spotted hyenas and European badgers is intermediate. Male badgers that immigrate have higher testosterone titers than do males that have not dispersed, but this disparity could be either a cause or a consequence of dispersal (Woodroffe et al. 1995). For females, the degree and perhaps the causes of suppression differ interestingly among populations (Woodroffe and Macdonald 1995a). In some populations, only a single dominant female breeds, whereas in others, multiple females may do so; where the latter is the case, the factors that lead some females to be more successful at bringing pregnancies to term appear to differ between populations and between years (da Silva et al. 1994; Woodroffe and Macdonald 1995a). In any case, the data available indicate that European badgers of both sexes tend to disperse from large groups to small ones (males left groups containing on average 2.9 males and moved into groups containing 1.2; females emigrated from groups containing 6.0 females and entered groups with only one: Woodroffe et al. 1995). Since per-capita reproductive success is higher in smaller groups (da Silva et al. 1994), females gain reproductively from dispersal.

In studies of spotted hyenas, there is as yet no clear evidence of reproductive suppression, but there are clear dominance hierarchies among males and among females (Frank 1986b, this volume; Hofer and East 1993a; Holekamp and Smale 1993). Females in high-ranking matrilines breed earlier and recruit more offspring per pregnancy, presumably because of differential access to prey (Frank et al. 1995), or perhaps because they can kill offspring of low-ranking females (H. Hofer and M. East, pers. comm.). Female dispersal in the form of group fission is precipitated by low- to mid-ranking animals that increase their rank by dispersing (Holekamp et al. 1993). Among male spotted hyenas, reproduction is closely related to dominance, but only *after* dispersal. Males in their natal groups never mate, regardless of rank, whereas those that have dispersed either join groups with no resident males (Mills 1990) or rise in rank as they remain longer in their new packs (K. Holekamp, pers. comm.). Thus, dispersing males gain a reproductive benefit, but not because they are avoiding suppression in their natal group.

Escape from Inbreeding

The reason that dispersing hyena males gain reproductively presumably reflects the costs of inbreeding. Natal males rarely attempt mating, and when they do, females ignore them (K. Holekamp, pers. comm.; Henschel and Skinner 1987). In addition, captive siblings have never been observed to mate with one another (L. Frank, pers. comm.). The most straightforward interpretation is that the adverse effects of close inbreeding have selected males that avoid mating with related females, or females that avoid mating with related males. In the first case, philopatric males would lose because of inbreeding depression; in the second, they would lose because of female mate preferences, the preferences themselves generated by inbreeding depression. Unfortunately, no data documenting the magnitude of inbreeding depression in hyenas exist.

Our knowledge of inbreeding depression and avoidance in the other species is similarly incomplete. No data are available on the costs of parent-offspring inbreeding or sibling inbreeding in coatis or badgers, nor do we know anything about the behavior of females approached by related males.

There is considerable speculation in the literature that wolves are inbred (Woolpy and Eckstrand 1979). If so, inbreeding depression should be slight. Close inbreeding has been shown to decrease several measures of reproductive success in captive wolves (Laikre and Rynam 1991), but remarkably few field data are available. In Alaska, the offspring of a grandfather-granddaughter mating survived to at least 6 months (Peterson et al. 1984). Low reproductive success in the isolated Isle Royale population in Michigan could reflect inbreeding problems, but might also reflect food shortage or disease (Wayne et al. 1991). Data on avoidance among wolves are similarly spotty. Genetic data from Canadian populations show no excess of homozygotes (Kennedy et al. 1990). In captivity, where dispersal is not an option, close inbreeding may occur after the death of a dominant breeder (Packard et al. 1985). In the field, father-daughter and sibling mating apparently do not occur, but the extent to which this reflects female rejection of related males' advances is unknown.

Lionesses, like female spotted hyenas, tend not to cooperate with occasional mating attempts by close male kin (Hanby and Bygott 1987). Among lions, two types of evidence from population comparisons are consistent with the presence of strong inbreeding depression. Males in the Ngorongoro population, which is small and relatively inbred, have smaller testes, lower testosterone levels, and higher levels of sperm deformities than do males in the larger Serengeti population (Packer and Pusey 1993). Reproductive success in Ngorongoro has gradually declined, coincident with a calculated gradual increase of the average inbreeding coefficient (Figure 8.6) (Packer et al. 1991). But at the same time, there appear to be no data on the outcome of matings among close kin.

Dwarf mongooses show a pattern markedly different from that in lions and spotted hyenas. Fingerprinting results confirm that both parent-offspring and

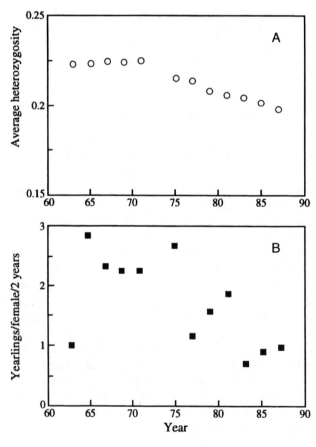

Figure 8.6. Estimated average heterozygosity of breeders (*A*) and observed yearling production per female (*B*) in Ngorongoro lions. Genetic variation in this small, inbred population has gradually declined since it was founded in the 1960s; presumably as a result, reproductive success has also declined. (Modified from Packer et al. 1991.)

sibling matings result in offspring, and pedigree analyses suggest that such pairings produce about 14% of all litters. At weaning, litters of dominants that are first-degree relatives are slightly, but not significantly, smaller than litters of less closely related dominants (1.60 ± 0.30, $n = 17$, vs. 1.70 ± 0.20, $n = 43$; $t = 0.18$, $P = 0.86$; Keane et al., in press.).

Overall, evidence that philopatric animals would forfeit reproductive success through inbreeding depression, or through loss of matings because of behavioral inbreeding avoidance, ranges from negative (in dwarf mongooses) to strongly suggestive (in lions and spotted hyenas). In most species there are at least a few cases in which animals have been observed to emigrate from groups containing opposite-sexed adult close relatives, adults that might otherwise be potential mates. Male lions sometimes leave prides voluntarily for groups containing fewer daughters, and daughters may leave prides at reproductive maturity if their fathers are still present (Hanby and Bygott 1987; Pusey and Packer 1987; Packer and Pusey 1993). Male wolves have been observed to emigrate rather than mate with their daughters (Mech 1987). Male European badgers have left groups in which the only remaining females were their sisters (R. Woodroffe, pers. comm.).

Although dispersal may decrease the likelihood of mating with parents, offspring, or siblings, it is important to note that it does not guarantee that mates will be unrelated. Repeated dispersal among neighboring groups has the potential to set up situations in which dispersers enter groups that contain relatives (Cheney and Seyfarth 1983). The extent to which this scenario occurs differs among carnivore species. In lions, genetic data confirm that patterns of dispersal nearly always result in mates that are distantly related; in only a few cases do dispersing males join females that have earlier fissioned from the same natal group (Packer and Pusey 1993). One known case of mating among relatives in wolves occurred through repeated dispersal by relatives into the same group (Peterson et al. 1984), but how often such situations arise is unknown. In contrast, genetic evidence demonstrates that male dwarf mongooses move preferentially into groups whose members are *more* related to them, on average, than members of other groups are; even parent-offspring and full-sibling matings sometimes result (Keane et al., in press).

Indirect Fitness Consequences

Because most gregarious carnivores live surrounded by close relatives, it has been natural to suggest that philopatric carnivores could gain indirect fitness. Dwarf mongooses have become a classic example: Rood (1978) showed that the presence of subordinates markedly increases the number of surviving yearlings produced by a group's dominant breeding pair, and subordinates are often closely related to the yearlings they raise. Subsequent studies confirmed that the indirect fitness acquired by philopatric subordinates is substantial, constituting more than half of a one-year-old nondisperser's inclusive fitness (Creel and Waser 1994). Although the recipient of a subordinate mongoose's help is sometimes unrelated, and although there is enough interchange among neighboring groups that even dispersers gain some indirect fitness, the loss of indirect fitness attendant upon dispersal represents a clear cost (Creel and Waser 1994; Lucas et al. 1996).

Concerning wolves, the suggestion has been made that subordinate individuals could gain indirect fitness by "helping at the nest," but available data suggest that subordinates increase the survival of the dominants' young only under conditions of food abundance. When food is scarce, offspring in larger groups survive *less* well (Harrington et al. 1983). In other carnivores, dispersal may influence indirect fitness in other ways. Among lions, larger coalitions of males are more successful at evicting pride males and have longer reproductive tenures; but when pride males are related, all males do not reproduce equally. Male lions that emigrate with their brothers, even if they do not mate when their coalition joins a pride, presumably gain indirect fitness when their brothers mate (Packer et al. 1991). Nonbreeding female European badgers have been reported to guard and feed other females' young, but their relatedness to these young and their effect on the young's survival are not yet known (Woodroffe 1993).

Conclusions

Dispersal is an enormously variable phenomenon. Even among the six species selected for discussion here, there are great differences in the forms, rates, distances, and consequences of emigration. Population comparisons give us every reason to believe that the phenomenon is similarly variable within species. Nevertheless, what we know suggests a number of hypotheses and speculations.

1. *Exploratory forays are a common prelude to dispersal.* Particularly in the larger species, it appears to be an oversimplification to regard emigration as a single decision in which a young animal breaks its ties with its natal group and strikes out, never to return. Although this picture certainly characterizes some individuals, it seems possible that the forays that are reported are but the tip of the iceberg.

2. *Females are more likely than males to engage in social, but not locational, dispersal.* Departure by matrilines or large female cohorts, usually described as group fission, is a form of dispersing from a social network without necessarily leaving a familiar area. Coati males also emigrate socially, but not locationally; still, for coatis, group fission without leaving the natal area is primarily a tactic seen among females. This finding is consistent with the possibility that habitat quality (and the effect of familiarity on foraging success) tends to contribute more to female than to male fitness.

3. *Mating by transient animals is a possibility.* Genetic data and behavioral anecdotes suggest the possibility that emigration need not be a reproductive dead end, even when not followed by immigration. Male coatis, in fact, seem to have made a career of such a possibility. Data from studies of other carnivores also support the idea that females sometimes mate preferentially with nonresident males (brown hyenas: Mills 1982; Ethiopian wolves: Gottelli et al. 1994).

4. *Intraspecific aggression may be a primary cause of dispersal-associated mortality.* More detailed monitoring of animals during dispersal will be needed if we are to determine the conditions under which they mate; such monitoring will be even more important in determining the conditions under which they die. Existing data on mortality during dispersal, such as they are, tell us only two things: first, dispersal can markedly increase mortality; second, aggression by residents during the immigration process can be a major source of that mortality, more significant in some carnivores than the predation or starvation suffered by solitary animals dispersing through unfamiliar territory.

5. *Gregariousness complicates the "stopping rules" for dispersers.* For nongregarious carnivores, the conditions that lead selection to favor settlement or attempted immigration rather than continued search presumably include the quality of the habitat an emigrant is traversing, or perhaps the number and quality of the potential mates that that habitat contains. For gregarious carnivores, an emigrant's stopping decision is also influenced, potentially, by the dominance position the emigrant would acquire if it attempted to immigrate, by how strongly the group residents would resist its immigration, and (for species with strong dominance hierarchies) by the length of the queue the emigrant must wait in for mating opportunities. In addition, the willingness of the emigrant's natal-group members to tolerate its return will influence the degree to which it can engage in repeated forays rather than accepting suboptimal immigration opportunities.

6. *Gregariousness also complicates dispersal's costs and benefits by creating conflicts of interest.* Where survival and reproductive success depend on group size, as they do in lions, European badgers, and dwarf mongooses, selection has favored emigration from groups of suboptimal size. Where dominance interactions skew the fitnesses of group members, as they do in mongooses, such selection should act particularly strongly on subordinates. And where dominants benefit from the presence of the subordinates in the group, the resulting conflict of interests between dominants and subordinates may lead the dominants to cede enough fitness to tip the selective balance *against* subordinate dispersal (Vehrencamp 1983). "Power-sharing" of this sort has been demonstrated in dwarf mongooses (Creel and Waser 1991; Keane et al. 1994).

7. *Yet another complication introduced by gregariousness is the opportunity for escalated mate competition.* Gregariousness can increase the variance in the payoffs to philopatry and dispersal, especially for males. In gregarious species, particularly those in which mating success is skewed strongly by dominance

status, males may gain access to matings with multiple females if they disperse, but only a minority of these males will succeed. In such circumstances, selection may more strongly favor emigration, but it may also more strongly favor resistance to immigration by same-sexed group residents. Even among species that are not gregarious, this situation appears to have led to escalation in the form of male coalition formation, that is, males dispersing together and contesting jointly for breeding positions elsewhere (cheetahs: Caro and Collins 1987; slender mongooses: Waser et al. 1994b). Among gregarious species like dwarf mongooses and lions, as well as banded mongooses (*Mungos mungo*) and African wild dogs (*Lycaon pictus*) (Rood 1986; Frame et al. 1979), the process has gone a step further: the dispersing coalitions are in turn resisted by coalitions of resident males.

8. *The influence of reproductive suppression is highly variable from one species to the next.* Subordinates stand to gain more from emigration if they are more completely suppressed in the natal group, as is the case among wolves and dwarf mongooses but not among lions. It is tempting to posit a feedback loop to explain these differences: where there are benefits to group life but dispersal is risky, selection may force subordinates to accept greater reproductive suppression at home than where dispersal is safe. Where suppression is strong, selection should favor rank-dependent dispersal into shorter queues elsewhere. Mortality rates during dispersal are consistent with the idea that dispersal is riskier for mongooses and wolves than it is for lions.

9. *The influence of inbreeding depression also varies greatly from one species to the next, and may be tied to the role of reproductive suppression.* The magnitude of inbreeding depression in carnivores is less than clear: several species' lack of genetic variation has been highly publicized (e.g., cheetahs: O'Brien et al. 1987), but the few data relating offspring survival to inbreeding coefficient suggest little inbreeding depression (Ralls et al. 1988; Lacy et al. 1993; Caro and Laurenson 1994). For the six species discussed here, the degree of inbreeding depression seems highly variable. In lions, the evidence of inbreeding depression is reasonably strong, and behavioral evidence implicates clear avoidance of kin matings. In dwarf mongooses, inbreeding depression appears to be slight, kin matings are not uncommon, and repeated dispersal into particular nearby groups creates the possibility of inbreeding even after dispersal. Perhaps high dispersal risk, the same condition that might set the stage for strong reproductive suppression, also leads to more frequent inbreeding, and thus, eventually, to lower costs of inbreeding.

10. *For gregarious carnivores, emigration is a tactic carrying mortality risks but offering reproductive advantages.* The great differences among populations and species in the forms of dispersal they pursue present a challenge, but also a promise. These differences provide the raw material that will eventually let us measure the fitness consequences of the behavioral decisions involved in carnivore dispersal: to leave your natal group or stay, to leave with others or alone, to continue moving or stop, to attempt immigration or not. The details

of those fitness consequences will undoubtedly be species- or even population-specific, but across all the variants the data show that two generalizations will remain. Across all species, the primary advantages of philopatry have to do with mortality: to remain in a familiar area, with familiar groupmates, is safer than leaving. At the same time, across all species the primary advantage of dispersal is reproductive: the emigrant that survives will move to a situation with more mates, with unrelated mates, with less competition for mates, and with better resources for raising young.

Acknowledgments

I thank John Gittleman for encouraging this review and Scott Creel, Marion East, Laurence Frank, Todd Fuller, Eric Gese, John Gittleman, Matt Gompper, Heribert Hofer, Kay Holekamp, Brian Keane, David Macdonald, David Mech, François Messier, Anne Pusey, Rolf Peterson, and Rosie Woodroffe for helpful comments and patient responses to queries.

References

Alberts, S. C., and Altmann, J. 1995. Balancing costs and opportunities: Dispersal in male baboons. *Amer. Nat.* 145:279–306.

Ballard, W. B., Whitman, J. S., and Gardner, C. L. 1987. Ecology of an exploited wolf population in south-central Alaska. *Wildl. Monogr.* 98:1–53.

Bertram, B. C. R. 1975. Social factors influencing reproduction in wild lions. *J. Zool. (London)* 177:463–482.

Caro, T. M., and Collins, D. A. 1987. Male cheetah social organization and territoriality. *Ethology* 74:52–64.

Caro, T. M., and Laurenson, K. 1994. Ecological and genetic factors in conservation: A cautionary tale. *Science* 263:485–486.

Cheeseman, C. L., Cresswell, W. J., Harris, S., and Mallinson, P. J. 1988. Comparison of dispersal and other movements in two badger (*Meles meles*) populations. *Mamm. Review* 18:51–59.

Cheney, D. L., and Seyfarth, R. M. 1983. Nonrandom dispersal in free-ranging vervet monkeys: Social and genetic consequences. *Amer. Nat.* 122:392–412.

Creel, S. R., Creel, N. M., Wildt, D. E., and Monfort, S. L. 1992. Behavioral and endocrine mechanisms of reproductive suppression in Serengeti dwarf mongooses. *Anim. Behav.* 43:231–245.

Creel, S. R., and Waser, P. M. 1991. Failures of reproductive suppression in dwarf mongooses: Accident or adaptation? *Behav. Ecol.* 2:7–16.

Creel, S. R., and Waser, P. M. 1994. Inclusive fitness and reproductive strategies in dwarf mongooses. *Behav. Ecol.* 5:339–348.

Creel, S. R., and Waser, P. M. 1996. Variation in reproductive suppression in dwarf mongooses: Interplay between mechanisms and evolution. In: J. French and N. Solomon, eds. *Cooperative Breeding in Mammals.* New York: Cambridge Univ. Press.

Creel, S. R., Wildt, D. E., and Monfort, S. L. 1993. Aggression, reproduction and androgens in wild dwarf mongooses: A test of the challenge hypothesis. *Amer. Nat.* 141:816–823.

da Silva, J., Macdonald, D. W., and Evans, P. G. H. 1994. Net costs of group living in a solitary forager, the Eurasian badger (*Meles meles*). *Behav. Ecol.* 5:151–158.

East, M. L., and Hofer, H. 1991. Loud calling in a female-dominated society II: Behavioral contexts and functions of whooping of spotted hyenas *Crocuta crocuta. Anim. Behav.* 42:651–670.

Evans, P. G. H., Macdonald, D. W., and Cheeseman, C. L. 1989. Social structure of the Eurasian badger (*Meles meles*): Genetic evidence. *J. Zool. (London)* 218:587–595.

Frame, L. H., Malcolm, J. R., Frame, G. W., and Lawick, H. van. 1979. Social organization of African wild dogs *Lycaon pictus* on the Serengeti plain, Tanzania. *Z. Tierpsychol.* 50:225–249.

Frank, L. G. 1986a. Social organization of the spotted hyaena (*Crocuta crocuta*) I: demography. *Anim. Behav.* 34:1500–1509.

Frank, L. G. 1986b. Social organization of the spotted hyaena *Crocuta crocuta* II: Dominance and reproduction. *Anim. Behav.* 34:1510–1527.

Frank, L. G., Holekamp, K. E., and Smale, L. 1995. Dominance, demography and reproductive success of female spotted hyaenas. In: A. R. E. Sinclair and P. Arcese, eds. *Serengeti, II: Dynamics, Management, and Conservation of an Ecosystem,* pp. 364–384. Chicago: Univ. Chicago Press.

Fritts, S. H., and Mech, L. D. 1981. Dynamics, movements, and feeding ecology of a newly protected wolf population in Northwestern Minnesota. *Wildl. Monogr.* 80:1–79.

Fuller, T. K. 1989. Population dynamics of wolves in north-central Minnesota (USA). *Wild. Monogr.* 105:1–41.

Gese, E. M., and Mech, L. D. 1991. Dispersal of wolves (*Canis lupus*) in northeastern Minnesota, 1969–1989. *Canadian J. Zool.* 69:2946–2955.

Gilbert, D. A., Packer, C., Pusey, A. E., Stephens, J. C., and O'Brien, S. J. 1991. Analytical DNA fingerprinting in lions: Parentage, genetic diversity, and kinship. *J. Hered.* 82:378–386.

Gompper, M. E. 1994. The importance of ecology, behavior, and genetics in the maintenance of coati (*Nasua narica*) social structure. Ph.D. dissert., University of Tennessee, Knoxville.

Gottelli, D., Sillero-Zubiri, C., Applebaum, G. D., Roy, M. S., Girman, D. J., Garcia-Moreno, J., Ostrander, E. A., and Wayne, R. K. 1994. Molecular genetics of the most endangered canid: The Ethiopian wolf, *Canis simensis. Mol. Ecol.* 3:301–312.

Hanby, J. P., and Bygott, J. D. 1987. Emigration of subadult lions. *Anim. Behav.* 35:161–169.

Harrington, F. H., Mech, L. D., Fritts, S. H. 1983. Pack size and wolf pup survival: Their relationship under varying ecological conditions. *Behav. Ecol. Sociobiol.* 13:19–26.

Henschel, J. R., and Skinner, J. D. 1987. Social relationships and dispersal patterns in a clan of spotted hyaenas *Crocuta crocuta* in the Kruger National Park. *South African J. Zool.* 22:18–24.

Hofer, H., and East, M. L. 1993a. The commuting system of Serengeti spotted hyaenas: How a predator copes with migratory prey, I: Social organization. *Anim. Behav.* 46:547–558.

Hofer, H., and East, M. L. 1993b. The commuting system of Serengeti spotted hyaenas: How a predator copes with migratory prey, II: Intrusion pressure and commuters' space use. *Anim. Behav.* 46:559–574.

Hofer, H., and East, M. L. 1995. Population dynamics, population size, and the commuting system of Serengeti spotted hyaenas. In: A. R. E. Sinclair and P. Arcese, eds. *Serengeti, II: Dynamics, Management, and Conservation of an Ecosystem,* pp. 332–363. Chicago: Univ. Chicago Press.

Hofer, H., East, M. L., and Campbell, K. L. I. 1993. Snares, commuting hyaenas and migratory herbivores: Humans as predators in the Serengeti. *Symp. Zool. Soc. London* 65:347–366.

Holekamp, K. E., Ogutu, J. O., Dublin, H. T., Frank, L. G., and Smale, L. 1993. Fission of a spotted hyena clan: Consequences of prolonged female absenteeism and causes of female emigration. *Ethology* 93:285–299.

Holekamp, K. E., and Smale, L. 1993. Ontogeny of dominance in free-living spotted hyaenas: Juvenile rank relations with other immature individuals. *Anim. Behav.* 46:451–466.

Isbell, L. A., Cheney, D. L., and Seyfarth, R. M. 1990. Costs and benefits of home range shifts among vervet monkeys (*Cercopithecus aethiops*) in Amboseli National Park, Kenya. *Behav. Ecol. Sociobiol.* 27:351–358.

Keane, B., Creel, S. R., and Waser, P. M. In press. No evidence for inbreeding avoidance or inbreeding depression in a social carnivore. *Behav. Ecol.*

Keane, B., Waser, P. M., Creel, S. R., Creel, N. M., Elliott, L. F., and Minchella, D. J. 1994. Subordinate reproduction in dwarf mongooses. *Anim. Behav.* 47:65–75.

Kennedy, P. K., Kennedy, M. L., Clarkson, P. L., and Liepins, I. S. 1990. Genetic variability in wild populations of the gray wolf, *Canis lupus*. *Canadian J. Zool.* 69:1183–1188.

Kruuk, H. 1989. *The Social Badger*. Oxford: Oxford Univ. Press.

Lacy, R. C., Petric, A., and Warneke, M. 1993. Inbreeding and outbreeding in captive populations of wild animal species. In: N. W. Thornhill, ed. *The Natural History of Inbreeding and Outbreeding*, pp. 352–374. Chicago: Univ. Chicago Press.

Laikre, L., and Ryman, N. 1991. Inbreeding depression in a captive wolf population. *Conserv. Biol.* 5:33–40.

Lucas, J. R., Creel, S. R., and Waser, P. M. 1996. Dynamic optimization and cooperative breeding: An evaluation of future fitness effects. In: N. Solomon and J. French, eds. *Cooperative Breeding in Mammals*. New York: Cambridge Univ. Press.

Mech, L. D. 1987. Age, season, distance, direction, and social aspects of wolf dispersal from a Minnesota pack. In: B. D. Chepko-Sade and Z. T. Halpin, eds. *Mammalian Dispersal Patterns: The Effects of Social Structure on Population Genetics*, pp. 55–74. Chicago: Univ. Chicago Press.

Messier, F. 1985a. Solitary living and extraterritorial movements of wolves in relation to social status and prey abundance. *Canadian J. Zool.* 63:239–245.

Messier, F. 1985b. Social organization, spatial distribution, and population density of wolves in relation to moose density. *Canadian J. Zool.* 63:1068–1077.

Messier, F., and Barrette, C. 1982. The social system of the coyote *Canis latrans* in a forested habitat. *Canadian J. Zool.* 60:1743–1753.

Mills, M. G. L. 1982. The mating system of the brown hyaena, *Hyaena brunnea*, in the southern Kalahari. *Behav. Ecol. Sociobiol.* 10:131–136.

Mills, M. G. L. 1990. *Kalahari Hyaenas*. London: Unwin Hyman.

O'Brien, S. J., Wildt, D. E., Bush, M., Caro, T. M., and Fitzgibbon, C. 1987. East African cheetahs: Evidence for two population bottlenecks? *Proc. Natl. Acad. Sci.* 84:508–511.

Packard, J. M., and Mech, L. D. 1980. Population regulation in wolves. In: M. Cohen, R. S. Malpas, and H. G. Klein, eds. *Biosocial Mechanisms of Population Regulation*, pp. 135–150. New Haven, Conn.: Yale Univ. Press.

Packard, J. M., Seal, U. S., Mech, L. D., and Plotka, E. D. 1985. Causes of reproductive failure in two family groups of wolves *Canis lupus*. *Z. Tierpsychol.* 68:24–40.

Packer, C. 1986. The ecology of sociality in felids. In: D. I. Rubenstein and R. J. Wrangham, eds. *Ecological Aspects of Social Evolution: Birds and Mammals*, pp. 429–451. Princeton, N.J.: Princeton Univ. Press.

Packer, C., Gilbert, D. A., Pusey, A. E., and O'Brien, S. J. 1991. A molecular genetic analysis of kinship and cooperation in African lions. *Nature* 351:562–565.

Packer, C., Herbst, L., Pusey, A. E., Bygott, J. D., Hanby, J. P., Cairns, S. J., and Borgerhoff-Mulder, M. 1988. Reproductive success of lions. In: T. H. Clutton-Brock, ed. *Reproductive Success,* pp. 363–383. Chicago: Univ. Chicago Press.

Packer, C., and Pusey, A. E. 1982. Cooperation and competition within coalitions of male lions: Kin selection or game theory? *Nature* 296:740–742.

Packer, C., and Pusey, A. E. 1993. Dispersal, kinship and inbreeding in African lions. In: N. W. Thornhill, ed. *The Natural History of Inbreeding and Outbreeding,* pp. 375–391. Chicago: Univ. Chicago Press.

Packer, C., Pusey, A. E., Rowley, H., Gilbert, D. A., Martenson, J., and O'Brien, S. J. 1991. Case study of a population bottleneck: Lions of the Ngorongoro crater. *Conserv. Biol.* 5:219–230.

Peterson, R. O., Woolington, J. D., and Bailey, T. N. 1984. Wolves of the Kenai Peninsula, Alaska. *Wildl. Monogr.* 88:1–52.

Potvin, F. 1987. Wolf movements and population dynamics in Papineau-Labelle Reserve, Quebec. *Canadian J. Zool.* 66:1266–1273.

Pusey, A. E., and Packer, C. 1987. The evolution of sex-biased dispersal in lions. *Behaviour* 101:275–310.

Ralls, K., Ballou, J. D., and Templeton, A. R. 1988. Estimates of lethal equivalents and the cost of inbreeding in mammals. *Conserv. Biol.* 2:185–193.

Rood, J. P. 1978. Dwarf mongoose helpers at the den. *Z. Tierpsychol.* 48:277–287.

Rood, J. P. 1980. Mating relationships and breeding suppression in the dwarf mongoose. *Anim. Behav.* 28:143–150.

Rood, J. P. 1986. Ecology and social evolution of the mongooses. In: D. I. Rubenstein and R. J. Wrangham, eds. *Ecological Aspects of Social Evolution: Birds and Mammals,* pp. 131–152. Princeton, N.J.: Princeton Univ. Press.

Rood, J. P. 1987. Dispersal and intergroup transfer in the dwarf mongoose. In: B. D. Chepko-Sade and Z. T. Halpin, eds. *Mammalian Dispersal Patterns: The Effects of Social Structure on Population Genetics,* pp. 85–103. Chicago: Univ. Chicago Press.

Rood, J. P. 1990. Group size, survival, reproduction, and routes to breeding in dwarf mongooses. *Anim. Behav.* 39:566–572.

Russell, J. K. 1982. Timing of reproduction by coatis (*Nasua narica*) in relation to fluctuations of food resources. In: E. G. Leigh, Jr., A. S. Rand, and D. M. Windsor, eds. *The Ecology of a Tropical Forest,* pp. 413–432. Washington, D.C.: Smithsonian Institution Press.

Schaller, G. 1972. *The Serengeti Lion.* Chicago: Univ. Chicago Press.

Smale, L., Frank, L., and Holekamp, K. E. 1993. Ontogeny of dominance in free-living spotted hyaenas: Juvenile rank relations with adult females and immigrant males. *Anim. Behav.* 46:467–477.

Van Ballenberghe, V. 1983a. Extraterritorial movements and dispersal of wolves in southcentral Alaska. *J. Mamm.* 64:168–171.

Van Ballenberghe, V. 1983b. Two litters raised in one year by a wolf pack. *J. Mamm.* 64:171–172.

Vehrencamp, S. L. 1983. Optimal degree of skew in cooperative societies. *Amer. Zool.* 23:327–335.

Waser, P. M., Creel, S. R., and Lucas, J. R. 1994a. Death and disappearance: Estimating mortality risks associated with philopatry and dispersal. *Behav. Ecol.* 5:135–141.

Waser, P. M., and Jones, W. T. 1983. Natal philopatry in solitary mammals. *Quart. Rev. Biol.* 50:355–390.

Waser, P. M., Keane, B., Creel, S. R., Elliott, L. F., and Minchella, D. J. 1994b. Possible male coalitions in a solitary mongoose. *Anim. Behav.* 47:289–294.

Wayne, R. K., Gilbert, D. A., Lehman, N., Hansen, K., Eisenhawer, A., Griman, D., Peterson, R. O., Mech, L. D., Gogan, P. J. P., Seal, U. S., and Krumenaker, R. J. 1991. Conservation genetics of the endangered Isle Royale gray wolf. *Conserv. Biol.* 5:41–51.

Woodroffe, R. 1993. Alloparental behaviour in the European badger. *Anim. Behav.* 46:413–416.

Woodroffe, R., and Macdonald, D. W. 1995a. Female/female competition in European badgers (*Meles meles*): Effects on breeding success. *J. Anim. Ecol.* 64:12–20.

Woodroffe, R., and Macdonald, D. W. 1995b. Costs of breeding status in the European badger, *Meles meles. J. Zool. (London)* 235:237–246.

Woodroffe, R., Macdonald, D. W., and da Silva, J. 1995. Dispersal and philopatry in the European badger, *Meles meles. J. Zool. (London)* 237:227–240.

Woolpy, J. H., and Eckstrand, I. 1979. Wolf pack genetics, a computer simulation with theory. In: E. Klinghammer, ed. *The Behavior and Ecology of Wolves*, pp. 206–224. New York: Garland STPM Press.

Zimen, E. 1976. On the regulation of pack size in wolves. *Z. Tierpsychol.* 40:300–341.

CHAPTER 9

Carnivore Reintroductions:
An Interdisciplinary Examination

RICHARD P. READING AND TIM W. CLARK

In the past few centuries, carnivore species have declined dramatically in numbers and range throughout the world (Dunlap 1988; Hummel and Pettigrew 1991). As the pace of decline quickens and the number of carnivore species threatened with extinction grows, the need for employing reintroduction as a conservation tool increases (Jones 1990). We define reintroduction as returning species to, and reestablishing them in, areas they once inhabited, and, more generally, translocations that involve establishing species in areas well-suited to them regardless of whether they once inhabited those areas. On that basis, we review case studies of carnivore reintroductions that have recently been undertaken or planned (Table 9.1). It should be noted that reintroductions are costly, however, and most efforts fail (Griffith et al. 1989; Ounsted 1991; Short et al. 1992; Yalden 1993; Schaller, this volume). As Scott and Carpenter (1987:545) state,

Because of the high costs associated with release programs and the endangered status of many of the animals, we cannot afford to introduce individuals to new environments without a high probability of their surviving and contributing genetically to a wild population. The techniques used to establish [new populations] or augment endangered populations are experimental and unproved methods.

It is clearly desirable to improve approaches to reintroduction.

This chapter describes a holistic, interdisciplinary paradigm for carnivore reintroductions. This model was primarily developed, and refined, on the basis of a proposed reintroduction of the endangered black-footed ferret (*Mustela nigripes*) in Montana (see Reading et al. 1991; Reading 1993) and includes biological/technical, valuational, and organizational variables. Evidence from past carnivore-reintroduction efforts suggests that many such efforts suffer from a narrow concentration on the biological, ecological, and technical aspects of recovery, to the neglect of several important nonbiological, nontechni-

Table 9.1. Selected carnivore reintroduction and translocation programs

Family and species	Location	Years	Outcome	Selected references
PHOCIDS				
Hawaiian monk seal (*Monachus schauinslandi*)	Hawaii, USA	1992–present	Uncertain	Anonymous 1992b
MUSTELIDS				
Fisher (*Martes pennanti*)	Nova Scotia, Canada	1955	Success	Berg 1982
	Montana, USA	1959–60	Success	Weckwerth and Wright 1968
	Oregon, USA	1959–63	Uncertain	Berg 1982
	Minnesota & Wisconsin, USA	1956–63	Success	Berg 1982
	Ontario, Canada	1956	Unknown	Berg 1982
	Maine, USA	1972	Failure	Berg 1982
	New York, USA	1976–77	Uncertain	Brown and Parsons 1983
American pine marten (*Martes americana*)	Alaska, USA	1934–52	Success	Berg 1982
	Saskatchewan, Canada	1954	Uncertain	Berg 1982
	Michigan, USA	1957	Failure	Berg 1982
	Ontario, Canada	1969–70	Failure	Berg 1982
	Wisconsin, USA	1975–76	Unknown	Berg 1982
European pine marten (*Martes martes*)	England	1981–82	Success?	Anonymous 1993
Eurasian otter (*Lutra lutra*)	England	1983–85	Success	Jefferies et al. 1986
River otter (*Lutra canadensis*)	Missouri, USA	1982–89+	Success	Erickson and McCullough 1987; Erickson and Hamilton 1988
	Pennsylvania, USA	1983–84	Success	Serfass and Rymon 1985
Sea otter (*Enhydra lutris*)	Alaska, USA	1965–69	Success	Jameson et al. 1982
	British Columbia, Canada	1969–72	Success	Bigg and MacAskie 1978
	Washington, USA	1969–70	Success	Jameson et al. 1986
	Oregon, USA	1970–71	Failure	Jameson et al. 1982
	California, USA	1987–90	Uncertain	Booth 1988; Anonymous 1992a
Black-footed ferret (*Mustela nigripes*)	Wyoming, USA	1991–present	Uncertain	Oakleaf et al. 1991; Miller et al. 1994, in press; Reading and Miller 1994

Table 9.1. (*Continued*)

Families and species	Location	Years	Outcome	Selected references
Siberian polecat (*Mustela eversmanni*)	Colorado & Wyoming, USA	1989–90	Trial	Biggins et al. 1990, 1991
FELIDS				
Canadian lynx (*Lynx canadensis*)	New York, USA	1989–91	Uncertain	Brocke et al. 1990, 1991
Eurasian lynx (*Lynx lynx*)	Germany	1970	Failure	Breitenmoser and Breitenmoser-Wursten 1990
	Switzerland	1971	Success	Breitenmoser and Breitenmoser-Wursten 1990; Breitenmoser and Haller 1993
	Slovenia	1973	Success	Gossow and Honsig-Erlenburg 1986; Breitenmoser and Breitenmoser-Wursten 1990
	Austria	1976–78	Success	Gossow and Honsig-Erlenburg 1986; Breitenmoser and Breitenmoser-Wursten 1990
	Czech Republic	1982	Success	Breitenmoser and Breitenmoser-Wursten 1990
	France	1983–88	Uncertain	Breitenmoser and Breitenmoser-Wursten 1990
Bobcat (*Lynx rufus*)	Georgia, USA	1988	Uncertain	Warren et al. 1990
Mountain lion (*Felis concolor*)	Florida, USA	1989	Failure	Belden and Hagedorn 1993
Serval (*Felis serval*)	Transvaal, South Africa	1982–83	Success	van Arde and Skinner 1986
Cheetah (*Acinonyx jubatus*)	Hluhluwe/Umfolozi, South Africa	1966–69	Success	Anderson 1992; Rowe-Rowe 1992
	Mkuzi, South Africa	1968–70	Failure	Anderson 1992; Rowe-Rowe 1992
	Kruger, South Africa	1973	Failure	Anderson 1992
	Itala, South Africa	1979	Removed	Anderson 1992; Rowe-Rowe 1992
	Bophuthatswana, South Africa	1981–84	Success	Anderson 1986

Species	Location	Years	Outcome	References
Leopard (*Panthera pardus*)	Mkuzi, South Africa	1975–76	Success	Anderson 1992
	Itala, South Africa	1980–91	Success	Anderson 1992
	Songimvelo, South Africa	1990	Success	Anderson 1992
Lion (*Panthera leo*)	Uttar Pradesh, India	1958	Failure	Sale 1986
	Hluhluwe/Umfolozi, South Africa	1963	Failure	Anderson 1981
	Hluhluwe/Umfolozi, South Africa	1965	Success	Anderson 1981
	Eastern Shores, South Africa	1978–81	Rare	Anderson 1992; Rowe-Rowe 1992
	Natal, South Africa	1991	Success	Rowe-Rowe 1992
URSIDS				
Black bear (*Ursus americanus*)	Montana, USA	1967–77	Variable	MacArthur 1981
	Pennsylvania & Idaho, USA	1973–83	Variable	Alt and Beecham 1984
Brown bear (*Ursus arctos*)	Austria	1992	Uncertain	Anonymous 1993
	Montana, USA	1990–present	Uncertain	Anonymous 1993
CANIDS				
Gray wolf (*Canis lupus*)	Alaska, USA	1972	Success	Henshaw et al. 1979
	Michigan, USA	1974	Failure	Weise et al. 1975, 1979
	Minnesota, USA	1975–78	Failure	Fritts et al. 1984
	Wyoming, USA	Proposed	N/A	Bath 1991; Singer 1991; Thompson 1993
Red wolf (*Canis rufus*)	North Carolina, USA	1987–90	Success	Phillips and Parker 1988; Gittleman and Pimm 1991; Moore and Smith 1991; Parker and Phillips 1991
Wild dogs (*Lycaon pictus*)	Hluhluwe/Umfolozi, South Africa	1980–81	Success	Anderson 1992; Rowe-Rowe 1992
	Venetia, South Africa	1983	Success	Anderson 1992
Swift fox (*Vulpes velox*)	South Dakota, USA	1980	Failure	Sharps and Whitcher 1982; Fauna West Wildlife Consultants 1991
	Saskatchewan & Alberta, Canada	1983–present	Uncertain	Carbyn 1989, 1992; FaunaWest Wildlife Consultants 1991
Red fox (*Vulpes vulpes*)	Iowa, USA	1966–70	Success	Andrews et al. 1973
HYAENIDS				
Brown hyena (*Hyaena brunnea*)	Hluhluwe/Umfolozi, South Africa	1977	Failure	Anderson 1992; Rowe-Rowe 1992
	Itala, South Africa	1980	Failure	Anderson 1992; Rowe-Rowe 1992
	Eastern Shores, South Africa	1983	Success	Anderson 1992; Rowe-Rowe 1992
	Transvaal, South Africa	1981–82	Success	Skinner and van Aarde 1987

cal elements (Reading 1993). Biologists and managers are increasingly realizing that wildlife management is strongly influenced and often dominated by sociopolitical forces (Kellert 1985a; Decker and Goff 1987; Moore 1992; Tear and Forester 1992; Thompson 1993). We hope here to encourage managers and conservationists to look well beyond the narrow confines of the biological sciences when developing and implementing reintroductions (see also K. Johnson et al., this volume).

Because reintroduction success is the ultimate goal, it is the focus of our model (Figure 9.1). Carnivore-reintroduction success is dependent on key-actor behaviors, which in turn are influenced by several variables. Although variables influencing key-actor behavior form a continuum of factors, we have arbitrarily delineated three groups of them as salient: (1) biological/technical

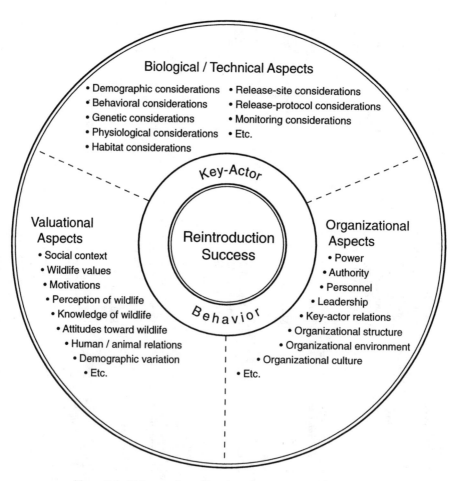

Figure 9.1. Universe of considerations for reintroduction paradigm.

aspects, (2) valuational aspects, and (3) organizational aspects. Several of these variables, previously described by Clark (1989) and by Kellert and Clark (1991), are discussed in more detail below. In examining the biological/technical aspects, we shall look briefly at two case studies: river otters (*Lutra canadensis*) and North American lynx (*Lynx canadensis*). Our examination of nonbiological (valuational and organizational) aspects again considers two case studies: red wolves (*Canis rufus*) and the black-footed ferrets. We then close with some general conclusions.

Biological/Technical Aspects of Reintroductions

Of all the factors influencing carnivore-reintroduction success, the biological and the technical are the most obvious, and are those most often stressed (Bixby 1991). Evidence suggests that biological and technical issues are less often associated with reintroduction failure than are valuational and organizational variables (Reading 1993; Yalden 1993); there is no doubt, however, that all of these issues must be successfully addressed, or the result will be almost certain failure (Griffith et al. 1989; Kleiman 1989; Reading 1993). Here we examine a range of biological/technical and reintroduction techniques and protocols and review two cases, river otter and lynx reintroductions, to illustrate various points.

Biological/Ecological Considerations

Traditionally, reintroductions have focused on the biological and ecological aspects of the species, and plans must indeed consider carefully the prospects for the species' survival in the release area, given the characteristics of both the organism and the ecosystem with which it is (or will be) associated (Griffith et al. 1989; Stanley-Price 1989; Yalden 1993). Important considerations include genetics, demographics, behavior, physiology, and habitat requirements (Cade 1988; Stanley-Price 1989; Kleiman 1989; Yalden 1993). Knowledge of these considerations informs the development of effective release protocols and, more important, provides baseline data against which the success of the reintroduction can be measured (Beck and Miller 1990).

Because small populations may be subject to significant losses of genetic variability, an understanding of genetic considerations is crucial (but see also Lacy and Clark 1989; Yalden 1993). Limited genetic variability can reduce a small population's ability to adapt to environmental changes and subject it to inbreeding and associated deleterious effects (e.g., reduced vigor or smaller litters: Soulé 1980; Lacy 1987). From a genetic standpoint, researchers have suggested that effective populations of 50 or 500 animals are needed to maintain genetic variance in the short or long term, respectively (Soulé 1980; Frankel and Soulé 1981). Effective populations are always smaller than actual populations, because demographic factors, such as population fluctuations and sex ratios, and behavioral factors, such as juvenile dispersal, prevent random mating and lead to unequal representation of individual genomes in future generations (Lande 1988). Although the validity and utility of the "50/500 rule" has been debated, many geneticists recommend using these numbers as a general guideline for most species and most populations (Lande and Barrowclough 1987; Shaffer 1987).

It is also important that projects attempt to ascertain genetic variation among species, subspecies, or populations and the pedigree of the reintroduced individuals (Templeton 1990). If the species exhibits a high degree of local adaptation and animals are reintroduced far from their original source population, reintroduction success may be limited. Furthermore, if a remnant population exists, matings with an introduced population that lacks the local adaptations could yield outbreeding depression (i.e., the disruption of coadapted gene complexes: Templeton 1990). Finally, the presence of a large population of some closely related species capable of interbreeding could result in genetic introgression or even swamping of the reintroduced population (Kleiman 1989; Peek et al. 1991; Yalden 1993).

Demographic parameters, such as fecundity, mortality, population growth rate, age structure, sex ratio, and life expectancy of carnivore populations are crucial to understanding population dynamics (Berg 1982; Kleiman et al. 1986; Erickson and Hamilton 1988; Stanley-Price 1989). Indeed, current concern over "minimum viable population" sizes (MVPs), including population-viability analyses (PVAs), are based primarily on demographics (Goodman 1987; Lacy 1987; Lacy and Clark 1993). Knowledge and understanding of these demographic variables can facilitate reintroduction by permitting, with the aid of PVAs and other demographic models, more rigorous prediction of desirable (1) captive population sizes and age structures, (2) target population sizes, (3) sizes and/or numbers of reintroduction sites, and (4) numbers, age structures, and sex ratios of the animals to be released.

The success of a carnivore reintroduction requires that the released animals find prey, escape from other predators, and reproduce fertile offspring. Therefore, behavioral considerations are also vital to a successful reintroduction. Foraging includes prey identification (e.g., search images), hunting skills (e.g., search patterns, cooperative hunting), food-processing ability (e.g., killing method, handling time), and prey caching (Fentress 1979; Henshaw et al. 1979; Kleiman et al. 1986; Stanley-Price 1989; Miller et al. 1990a; Yalden 1993). Because reintroduced animals must also avoid being predated, they must recognize and avoid other predators and potentially dangerous species—including humans (Holloway and Jungius 1973; Kleiman 1989; Miller et al. 1990b; Brocke et al. 1991). Carnivore offspring often require a high degree of care, including protection, shelter, and provisioning from parents, relatives, or helpers (Kleiman et al. 1986; Peek et al. 1991; Yalden 1993). Those needs, in turn, often necessitate the formation of complex social organizations (Henshaw et al. 1979; Beck and Miller 1990; Peek et al. 1991; Yalden 1993). Intraspecific behaviors involving communication, territoriality, dominance/ submission, avoidance, and courtship/mating are also important (Kleiman 1989; Stanley-Price 1989; Peek et al. 1991). Finally, the ability to locomote and navigate properly are crucial (Holloway and Jungius 1973; Henshaw et al. 1979; Kleiman 1989). Addressing behavioral concerns is particularly important for reintroductions of captive animals that may have lost many important behavioral skills (Fentress 1979; Mech 1979; Derrickson and Snyder 1992). Several species imprint on food, parents, natal sites, etc., during a critical phase of their development, and failure to provide the appropriate cues during this critical phase could doom reintroduction success (Stanley-Price 1989; Derrickson and Snyder 1992; Miller et al. 1994).

The reproductive physiology, nutrition, pathology, and general health and fitness of the animals are important physiological considerations. Knowledge of estrous cycles, gestation, timing of parturition, and other aspects of the reproductive physiology of the species are obviously important for captive-breeding programs, but may also be important for the timing of reintroductions (Kleiman et al. 1986; Kleiman 1989; Derrickson and Snyder 1992). Understanding the nutritional needs and pathological susceptibilities of the organism is important for assessing the suitability of release sites (see below; see also Holloway and Jungius 1973; Mech 1979; Thorne and Williams 1988; Kleiman 1989; Yalden 1993). Carnivore-reintroduction programs should also ensure that no diseases are introduced into the wild with the reintroduced animals, and, if possible, released animals should be inoculated against known wild diseases (Woodford and Kock 1991; Derrickson and Snyder 1992). Finally, the general health and physical fitness of released animals is undoubtedly important to survival following reintroduction (Holloway and Jungius 1973). Levels of fat and nutrient reserves are thus important for reproductive success and survival through adverse climatic periods.

The evaluation of potential reintroduction sites, pursued on the basis of habitat requirements, spatial characteristics, and management considerations, should be undertaken prior to program implementation. Several habitat considerations are obvious; others are less so. For example, a release site must have adequate prey, refugia, cover, denning sites, etc. (Mech 1979; Sale 1986; Kleiman 1989; Brocke et al. 1991; Belden and Hagedorn 1993; Yalden 1993). The status of prey species and other predators in the area, and the impact of the reintroduction on these species, should also be assessed, especially where rare, sensitive, threatened, or endangered species occur (Carbyn 1989; Griffith et al. 1989; Singer 1991; Yalden 1993). The presence and history of diseases and parasites in the area may be crucial to site analyses and selection (Belden et al. 1986; Stanley-Price 1989). Variables that are less intuitively apparent and thus more difficult to assess are ecosystem resilience and stability, which are especially important for systems characterized by frequent natural disturbances (e.g., fires, storms, droughts, volcanoes: Kleiman 1989; Stanley-Price 1989).

Spatial considerations, exceedingly important for the small populations typical of carnivore reintroductions, are rarely given adequate consideration (Mech 1979; Gilpin 1987). A potential release site's important spatial characteristics include its degree of insularization, its size and shape, and its location in relation to the historic range of the species (Wilcox 1980; Panwar and Rodgers 1986; Simberloff 1988; Kleiman 1989; Yalden 1993). The degree of isolation from other potential habitat and release sites in part dictates the importance of the size and shape of the release site. Often, carnivore reintroductions will require several discrete release sites linked by natural corridors or paths for artificial migration. In such cases, a successful reintroduction will result in a set of small, local populations, or a metapopulation (see Gilpin 1987; Hanski and Gilpin 1991).

Metapopulation theory and the closely related theories of insular biology and island biogeography are often not clear regarding optimal reserve designs, though for any individual site, larger is almost always better (MacArthur and Wilson 1967; Simberloff 1988; Hanski and Gilpin 1991). Usually the biology, ecology, and genetic history of a species will suggest the appropriate design strategy. Species with low genetic variance, such as cheetahs (*Acinonyx jubatus*: O'Brien et al. 1983), may benefit from population subdivision if patches have different selection pressures (Gilpin 1987). With large carnivores, habitat patches will often be quite large (e.g., thousands to tens of thousands of hectares: Mech 1979; Peek et al. 1991; Yalden 1993). The relative importance of edge effects and refugia will also influence shape considerations in reintroduction sites (Forman and Gordon 1986). For example, species subject to mortality along habitat edges may require sites with a smaller edge-to-area ratio (Yahner 1988), whereas species subject to frequent disease epizootics may survive only in relatively isolated habitat "peninsulas" (Cade 1988).

Potential release sites may also require significant habitat restoration prior to reintroduction (Kleiman 1989; Stanley-Price 1989). Certainly, the original cause of the species' local extirpation must be eliminated (Yalden 1986, 1993; Kleiman 1989; Stanley-Price 1989). As Wiley et al. (1992:175) state, "Reintroductions in former range in the absence of any actions to modify limiting factors are difficult to justify." In addition, the site should receive adequate formal (i.e., legal) and informal (i.e., respect for the law) protection (Holloway and Jungius 1973; Kleiman 1989; Peek et al. 1991; Yalden 1993). Finally, the kinds, degree, and trends of human disturbances (e.g., resource extraction, outdoor recreation, pollution) must be assessed and, if necessary, mitigated (Mech 1979; Sharps and Whitcher 1982; Saunder 1984; Gossow and Honsig-Erlenburg 1986; Breitenmoser and Breitenmoser-Wursten 1990).

Although it is obviously desirable to obtain and use as much information about the biology and ecology of the species as possible, the rarity of most endangered species dictates that pertinent information in statistically significant samples is often lacking and difficult to obtain. Nevertheless, time is at a premium for endangered-species recovery, and conservationists must proceed in the face of uncertainty, using the best available data (Frankel and Soulé 1981). Often, data from studies of closely related species are available and should be used (Snyder and Snyder 1989; Miller et al. 1990a, b; Belden and Hagedorn 1993).

Reintroduction Techniques and Protocols

In addition to ecological considerations, reintroduction programs must address techniques and protocols. These include site, animal, and management considerations, such as release techniques, animal selection, and species and data management. Kleiman (1989) and Stanley-Price (1989) identify a host of important technical aspects of reintroductions, including release-site preparation, proper management of a self-sustaining source population, preparation and training of the animals to be released, and demographic and genetic considerations in animal selection. The reintroduction must also be well-managed and monitored, with detailed record-keeping. Attention to reintroduction techniques, protocols, and management is essential because reintroductions are risky, complex, and costly in terms of both money and such other resources as time and personnel (Mech 1979; Peek et al. 1991; Yalden 1993), and it is important that we attempt to document *why* a reintroduction fails or succeeds.

Release sites may require preparation prior to reintroduction, notwithstanding that the species formerly inhabited the site. Preparation may include short-term control of other predators in the region, the gradual release of animals using shelters and/or food supplementation, the cooperation of local officials, and education programs for local residents (Belden et al. 1986; Phillips and Parker 1988; Kleiman 1989; Peek et al. 1991; Carbyn 1992). As discussed earlier, larger habitat-restoration programs should be completed well before reintroduction begins.

A well-managed source population is crucial to reintroduction success (Kleiman 1989; Yalden 1993). This population can be either wild or captive, but wild source populations are preferable because of the costs and problems associated with captive populations, like the erosion of feeding behaviors and the need for pre- and even post-release training (Mech 1979; Griffith et al. 1989; Miller et al. 1990b; Derrickson and Snyder 1992). Removing animals for reintroduction should neither jeopardize nor stress the source population. This is particularly true when translocating animals from wild carnivore populations. Programs should strive to select demographically and genetically representative, but unimportant, or "surplus," animals from the source population. Because of the implications of founder effects (Templeton 1990), it is important to release a group of individuals carrying as much genetic variability as possible (Griffith et al. 1989).

Animal preparation may entail little or much effort (Griffith et al. 1989; Kleiman 1989; Stanley-Price 1989; Miller et al. 1990a, b; FaunaWest Wildlife Consultants 1991). The former (preparations requiring little effort) are referred to as "hard" releases, the latter as "soft" releases. The amount of acclimatization, behavioral training, inoculations, provisioning, and socialization required should be determined by assessing the associated costs, feasibility, and contribution to survivorship (Holloway and Jungius 1973; Jefferies et al. 1986; Griffith et al. 1989; FaunaWest Wildlife Consultants 1991). Post-release training may be necessary or desirable, and may be simpler than behavioral training beforehand (Kleiman 1989; Beck and Miller 1990). The sex ratios, age structure, and social groups of released animals should mimic those of natural populations.

The number of animals per release, number of releases per year, number of years of releases, and the timing of releases are also important (Erickson and Hamilton 1988; Griffith et al. 1989; Kleiman 1989; Stanley-Price 1989). Some of these variables may be determined by additional factors, such as the numbers of surplus animals available for release (Cade 1988). Other variables, such as season, weather, and time of day of the release, should be based on knowledge of the species' biology and ecology (Erickson and Hamilton 1988; Kleiman 1989; Stanley-Price 1989).

Important management concerns include monitoring and research techniques and priorities, species management, record-keeping, and data management (Mech 1979; Kleiman et al. 1986; Kleiman 1989; Beck and Miller 1990; Yalden 1993). Participants should engage in contingency planning, and should clarify decision-making authority. Several difficult management decisions must be addressed and resolved prior to program implementation (Stanley-Price 1989). For example, it is important that all parties concerned agree on the amount and timing of intervention likely to influence the survivorship of released animals, which may include recapturing injured or starving animals and preventing predation. In addition, participants must agree on program goals and the amount and intrusiveness of monitoring and research. This involves choosing between more intrusive, more costly, and more informative experimental releases or less intrusive, less costly, and less informative operational releases (Erickson and Hamilton 1988; Frey and Walter 1989).

Radiotelemetry is a particularly useful and widely used monitoring technique that can provide quality data on dispersal, mortality, home ranges, activity cycles, habitat use, and more (Mech 1979; Carbyn 1989; Peek et al. 1991; Brocke et al. 1991). The knowledge of an animal's location thus obtained also facilitates visual observation or capture to assess its physical condition. Complete and accurate records should be maintained for every aspect of the reintroduction, including nutrition, behavior, and reproduction. Such records are important even for failed projects, in order to learn what went wrong and avoid the same problems in future reintroductions. These results, discouraging or not, should be published, so that the widest group of people can profit from the lessons of reintroductions (Kleiman et al. 1986; Scott and Carpenter 1987; Stanley-Price 1989; Griffith et al. 1989; Beck and Miller 1990). Studies using closely related but more common species as "surrogates" can provide valuable data about pre-release training benefits, release techniques, and the efficacy of monitoring techniques (e.g., Belden et al. 1986; Biggins et al. 1990, 1991; Miller et al. 1990a, b; Belden and Hagedorn 1993).

Two case studies—river otter reintroduction in Missouri from 1981 to 1989 and lynx reintroduction in New York from 1982 to 1992—illustrate various aspects of the biological/technical dimension of reintroduction. We have limited our analyses to specific time frames, owing to the ongoing nature of the reintroduction efforts, and because those time frames are in any case sufficient to illustrate our major points. The river otter case appears to be a success, whereas the lynx case remains problematic.

Case Study 1: River Otter

River otters were once widespread across North America, from the Arctic to northern Mexico, but during the 1800s and early 1900s they were decimated in large portions of their range by fur trappers (Berg 1982; Erickson and McCullough 1985). In Missouri, otters were eliminated from all but two locations within the state (Erickson and McCullough 1985; Erickson and Hamilton 1988). Habitat loss and degradation since the turn of the century is thought to have prevented natural recovery. The plight of the otter was noted in 1937, and legal protection was quickly passed in the state legislature that year. The desirability of reintroducing otters to Missouri has been periodically mentioned since that time, but until recently little was done, owing to the lack of a large, healthy source population (Erickson and Hamilton 1988).

The source-population problem was solved in 1981 with an agreement between the Missouri Department of Conservation and the Kentucky Department of Fish and Wildlife Resources. Missouri traded eastern wild turkeys (*Meleagris gallopavo sylvestris*) for river otters that Kentucky had acquired from commercial trappers in Louisiana (Erickson and McCullough 1985; Schwartz 1989). Not only were the Louisiana otters available, but they are believed to be from the same subspecies (*L. c. lataxima*) as those otters originally found in Missouri (Erickson and Hamilton 1988). Trades of this sort with Kentucky and with other agencies delivered 511 animals to Missouri from 1982 to 1989, with exchanges for 400–500 more otters planned (Schwartz 1989). Priority is given to trades for otters of the *lataxima* subspecies. Although reintroductions of river otters have been undertaken in several other states and at least one Canadian province, none of these has come close to the numbers or the scientific rigor involved in the Missouri program (Erickson and McCullough 1987; Schwartz 1989).

Research in the Missouri restoration effort began with a statewide survey of otters and potential otter habitat. Two small populations of otters and numerous potential release sites were located (Erickson and McCullough 1987). Next, a comprehensive management plan was developed, one that called for the reintroduction of about 800 otters at 40 sites from 1982 to 1991 (Erickson and Hamilton 1988). Because of the great popularity of river otters, public interest in the program has been high and easily encouraged. Indeed, Missourians approved the first conservation-associated sales tax in the nation to help pay for such projects (Schwartz 1989).

The Missouri otter restoration program also included intensive research and the monitoring of reintroduced animals, in an effort to assess the feasibility and effectiveness of release techniques (Erickson and McCullough 1985). Important variables that were measured from radiotelemetry data in the first two releases included post-release survival, movements, intraspecific interactions, and reproduction. The high survivorship (81% for the first year) and relatively modest movements by most animals were positive signs (Erickson and McCullough 1985, 1987; Erickson and Hamilton 1988). Most mortality was associated with translocation stress or human conflicts (e.g., animals trapped, or tangled in commercial fish net). Mortality due to trapping led to prohibitions and restrictions on state lands during subsequent releases. Reproduction did not occur during the first post-release breeding season, possibly because spring releases interrupted breeding, but reintroduced individuals bred successfully in subsequent years (Erickson and McCullough 1985, 1987; Erickson and Hamilton 1988).

Following the success of the earlier reintroductions, Missouri gained the potential for obtaining animals for reintroduction from within the state. The release managers elected not to do so, so as to avoid stressing the new populations and to save money by capitalizing on the state's large turkey population (Erickson and Hamilton 1988). Releases initially included a 5-day acclimation and observation period following arrival and processing (i.e., tagging, weighing, measuring), but the program has since moved to true hard releases in which animals are released 1 day after processing (Erickson and McCullough 1987; Erickson and Hamilton 1988). This move has saved money and permitted the release of large numbers of animals into several sites over a relatively short period of time.

Long-term monitoring of the reintroduced otter population has been difficult because of their "low density, secretive nature, and poorly accessible habitat" (Erickson and Hamilton 1988:410). Nevertheless, winter track surveys, recoveries of carcasses, and observational records have been used in an effort to assess trends. Numerous sightings have been made of otters and their winter tracks, and reproduction has been confirmed at several sites since the program was initiated (Erickson and Hamilton 1988). These positive results have encouraged the state to continue its program. Aside from the more obvious benefits to wildlife viewers (and, possibly, fur trappers, once the species has recovered), Missouri views otters as an important indicator species. As D. Hamilton (quoted in Schwartz 1989:148) states, "There may be no better indicator of the quality of aquatic habitat than river otters. . . . Toxic chemicals, heavy metals, pesticides, agricultural wastes—they all end up in the river otter. So if we can manage a river otter population, we're doing a pretty good job at managing the aquatic habitat and providing a good environment for everything else in the system." The status of river otters and that of the state's waterways are at stake with this program.

Erickson and Hamilton (1988) offer several recommendations for improving the chances for successful reintroduction, on the basis of the results of the Missouri program (see also Berg 1982). Release-strategy recommendations include: (1) releasing sufficient numbers of otters (≥ 20) at each release site, so as to ensure persistence and eventual population growth, (2) using female-biased sex ratios to promote reproductive encounters, (3) spacing releases in such a way as to permit populations to merge (see below), (4) releasing all animals at the same location, so as to encourage intraspecific interactions and permit behavior- and habitat-dictated spacing, (5) prohibiting trapping on public lands within the release area, and (6) prioritizing potential release sites using the criteria discussed below.

Several habitat considerations should be addressed in evaluating potential reintroduction sites and selecting actual release points. Erickson and Hamilton (1988) offer specific recommendations for evaluating, comparing, and selecting reintroduction sites. These evaluative variables include spacing new sites at least 160 km away from native or previously reestablished populations, and emphasizing localities with low human disturbance, high water quality, diverse habitat features, responsible land tenure, and abundant and diverse prey. The future of the river otter in Missouri is bright; with more reintroductions planned and with increasing numbers of sightings in the areas of past reintroductions, it appears that otters will become established throughout much of their former range in the state.

Case Study 2: North American Lynx

Lynx are medium-sized (~ 10 kg) North American felids that prey principally on snowshoe hares (*Lepus americanus*) (Brocke 1990). The species was extirpated from the northeastern United States, including upstate New York, in the late 1800s (Brocke and Gustafson 1992). Brocke and Gustafson suggest that logging severely altered the habitat, permitting the bobcat (*Lynx rufus*) to expand its range and displace the lynx in competition for resources. Trapping and incidental human-caused mortality may also have contributed to lynx declines.

Protection of the Adirondack region of the Northeast began in 1855, with the establishment of Forest Preserve Lands and the "Wild Forever" provision of the New York State Constitution (Brocke et al. 1990). New York continued to acquire additional property and afforded protection to these lands, culminating with the 1971 legislation that created the Adirondack Park Agency (Wargo 1984; Brocke et al. 1990). Throughout this period reforestation was pursued, and the Park's bobcat population was declining (Brocke and Gustafson 1992; Brocke et al. 1991). During the 1970s, interest in restoring extirpated species to Adirondack Park grew steadily (Brocke et al. 1990). A study of the feasibility of restoring lynx to the Park, completed in 1982, determined that there was an adequate prey population of snowshoe hares. The study also found that bobcats, an important lynx competitor, were concentrated below about 800 m in elevation. Modeling suggested the Park could support about 70+ lynx (Brocke and Gustafson 1992; Brocke et al. 1991), and a core area of largely roadless habitat was already designated as "wilderness" (Brocke et al. 1991). This total habitat, however, is not continuous, but is instead a patchwork of public lands interspersed among private lands, with varying land uses and land-use restrictions (Wargo 1984; Brocke et al. 1990).

Preparation for the introduction spanned almost 10 years, giving participants time to conduct additional feasibility studies and develop public support (Brocke et al. 1990). The reintroduction plan that was developed included acquisition of lynx from supportive trappers in the Yukon Territory of Canada, with the cooperation of the Yukon Department of Renewable Resources. Lynx were held for 2–4 weeks and radiotelemetry collars were affixed on the animals before they were released into the Adirondacks (Brocke 1990; Brocke et al. 1990; Brocke and Gustafson 1992). Release sites were selected on the basis of topography, conifer cover, prey density, and human-disturbance potential (Brocke and Gustafson 1992). Reintroduction was "semi-soft," with animals held on site for 2–5 days, during which disturbance was minimized and adequate food for the acclimation period was provided (Brocke 1990; Brocke and Gustafson 1992). Feces and urine from the animals were spread in a circle 0.8 km in diameter around the release site in an attempt to delay dispersal (Brocke 1990).

Reintroductions began in January 1989, with the release of 18 animals (12 males: 6 females) over a 5-month period. An additional 32 lynx (9: 23) were released during the winter of 1989–90, and 33 (14: 19) were reintroduced during the winter of 1990–91, for a total of 83 animal (35: 48) releases (Brocke 1990; Brocke and Gustafson 1992; Brocke et al. 1991). No animals were released in 1991–92.

Reintroduced lynx were monitored weekly using radiotelemetry. Although movement varied greatly, most animals traveled relatively long distances (Brocke et al. 1991). Some lynx dispersed hundreds of kilometers, including one animal that was shot in New Brunswick, Canada, some 720 km from its release point (Brocke 1990; Brocke et al. 1991). Telemetered lynx used large areas, averaging 421 km² among 29 females, 1760 km² among 21 males (Brocke and Gustafson 1992). Daily straight-line distance averaged 2.6 km for males and 2.0 km for females (Brocke et al. 1991). A female released with young dispersed the least, suggesting the potential value of using such animals for release (Brocke 1990). The lynx use areas were greater than expected, exceeding the home ranges of wild lynx and bobcats recorded in other studies (27–251 km²: Brocke et al. 1991). Evidence suggests that some breeding has occurred, but this remains unverified (Brocke et al. 1991).

Mortality, too, was higher than expected, and of the 32 known mortalities, 12 were attributable to automobile collisions and five to out-of-state shootings (Brocke 1990; Brocke and Gustafson 1992). The high road mortality suggests that colonizing lynx are vulnerable to road densities and potential future development. Vulnerability to other human impacts may also increase with greater road densities and development, as has been found for other species (Knight et al. 1988; Mech et al. 1988; Mattson 1990). But it is hoped that collisions will decrease as lynx begin to travel less and settle into more permanent home ranges (Brocke 1990).

Whether or not the lynx reintroduction program will ultimately succeed in reestablishing a population in Adirondack Park is still far from clear. Nevertheless, valuable insights have been collected from the releases made thus far. Large dispersal distances, stemming perhaps from the lack of large, continuous core areas with low road densities, appear to be particularly detrimental to lynx released in the Adirondacks. Restricting the movements of reintroduced animals may be especially crucial, and it has become obvious, in any case, that large, continuous blocks of undeveloped land, preferably protected under some form of public ownership, are crucial to large-carnivore population conservation, and particularly to reintroduction attempts (Brocke et al. 1991).

These two case studies focused largely on the biological/technical dimensions of carnivore reintroduction. Both cases demonstrate the importance of both pre-release studies of the habitat at potential reintroduction sites and post-release studies of survivorship, dispersal, behavior, and reproduction. Among other things, the river otter case also illustrates the value of releasing large numbers of animals, the importance of the timing of releases, and the potential benefit of using reintroduced carnivore populations as indicator species. The lynx case demonstrates the significance of human impact and fragmented habitat to a carnivore reintroduction, the importance of limiting the dispersal of released animals, and the value of testing different release strategies and techniques. Both cases also referred directly and indirectly to the nonbiological dimensions of reintroduction efforts, alluding to their potential importance. For example, both cases discussed the importance of local support and worked actively to achieve it. The otter case also involved an innovative, cooperative arrangement between different state wildlife agencies, and the lynx case illustrated the importance of land-ownership patterns. Far more nonbiological variables than these, however, are associated with any reintroduction effort.

Nonbiological/Nontechnical Aspects of Reintroductions

As reintroductions become more important to carnivore conservation and management, the need for more systematic, holistic reintroduction efforts grows. The implications of "systematic" and "holistic" are explicit and detailed considerations of both biological and nonbiological dimensions. Only from this perspective are reintroductions most likely to be successful. We recommend, therefore, that reintroduction programs address the full array of factors influencing the conservation outcome (Reading et al. 1991; Tear and Forester 1992; Reading 1993), including the valuational and organizational aspects of carnivore reintroductions. Here we examine a range of valuational and organizational considerations and review two cases—red wolf and black-footed ferret reintroductions—to illustrate various points.

Valuational Considerations

The values people ascribe to carnivores, and the factors affecting how those values are formed, factor heavily into the success or failure of carnivore-reintroduction programs. As Glass and Moore (1990:548) state, "While biophysical interrelationships set limits of [reintroduction] capability, managerial objectives for wildlife species also are determined by attitudes, perceptions, and the economic value placed upon the species." Systematic examinations of valuational considerations have usually been lacking or insufficient in carnivore-management efforts (Kellert 1985a; Bath 1991; Reading and Kellert 1993; Thompson 1993; see also K. Johnson et al., this volume).

Local support is especially crucial to reintroduction efforts (Mech 1979; Bath 1991; Reading and Kellert 1993; Yalden 1993). Moore (1992:6) suggests that "Carnivore re-introductions need preliminary habitat and prey assessment, but more importantly they need assessment and majority consensus of the local human population." It is crucial that local values and attitudes be examined prior to reintroducing carnivores into a region, so as to understand local concerns and opinions and permit the development of pertinent and effective public-relations campaigns designed to educate people and develop support for carnivore recovery (Reading and Kellert 1993). If the local public is unsupportive or, worse, antagonistic, the consequences may be disastrous (e.g., Hook and Robinson 1982; Belden et al. 1986; Breitenmoser and Breitenmoser-Wursten 1900).

Inappropriate valuation lies at the heart of many problems facing carnivore conservation. Several values attributable to wildlife, such as ecological, aesthetic, or existential, fall outside the precincts of traditional economic markets (Rolston 1981; Kellert 1987). During policy formulation these values are usually ignored or overridden by the more easily measured costs of carnivore conservation, such as livestock predation (Clark et al. 1993; Yalden 1993).

Incorporating the full array of values into the policy arena is imperative if one hopes to convince both decision-makers and the public of the necessity of preserving carnivores.

Values imply a preferred mode of conduct or end-state of existence (Rokeach 1972). Values have affective, cognitive, and directional components, and serve as the criteria for judgments, preferences, choices, and behaviors (Williams 1979). Values guide actions, attitudes, comparisons, evaluations, and justifications (Williams 1979). Attitudes are affinities for, or aversions to, objects or situations based on beliefs; and beliefs are perceived relationships between things or between a thing and its characteristics (Rokeach 1972). Beliefs and attitudes have their foundations in feeling, behaving, and thinking, and in interactions with others (Kellert 1980).

Values and their hierarchical orderings, or value systems, arise from the influence of several factors (Reading 1993), including (1) a person's understanding of experiences, interactions, and information (knowledge); (2) motivations, perceptions, and attitudes; (3) the context of the perception or valuation; (4) the cultural and social setting; and (5) the influence of social institutions (Rokeach 1972, 1979; Williams 1979; Brown 1984). Values, just as attitudes and behaviors, are not fixed, but change with changing contexts, knowledge, influences, and experiences (Sinden and Worrell 1979; Williams 1979).

The differential influence of education, experiences, culture, social institutions, and such on individuals results in differential values (see also Johnson et al., this volume). These differences are reflected among demographic groups. Kellert (1994) discusses three demographic patterns in values and attitudes toward wildlife. First, people with greater dependence on land and natural resources, as reflected in rural versus urban residency, land ownership, and occupation (in particular, one dependent on natural resources), tend to express strong utilitarian and dominionistic values, and weak moralistic and humanistic values and attitudes, toward wildlife. As a result, they are likely to favor the exploitation and subjugation of wildlife for human benefit. Second, there are patterns in demographic variables associated with socioeconomic status: people with more formal education and higher incomes tend to display higher naturalistic and ecologistic values and attitudes toward wildlife, suggesting a strong interest in outdoor recreation and support for wildlife conservation and protection. Finally, ascriptive demographic factors, especially age and gender, display characteristic patterns: younger people and females are more likely to demonstrate highly moralistic and humanistic values and attitudes toward wildlife, and would therefore likely demonstrate great affection for individual animals and oppose direct, consumptive uses.

Changing existing attitudes, behaviors, and especially values is difficult, but not impossible. Change is most easily accomplished when individuals become aware of internal contradictions among their own values, or between their values, their attitudes, and their behaviors (Rokeach 1979; Williams 1979).

"Dissonance reduction theory" provides the most widely invoked explanation for the fact that these changes are possible. It suggests that individuals will seek to reduce the discomfort they experience from holding inconsistent values, attitudes, and behaviors by changing their more dissonant, peripheral values, attitudes, and behaviors to better reflect their core values (Rokeach 1979; Williams 1979; Chaiken and Stangor 1987; Tessler and Shaffer 1990). Generally, social, cultural, and institutional forces inhibit significant value change (Rokeach 1979), and change in values is accordingly slow, especially in small, long-established communities or social groups (Williams 1979).

Understanding valuation can be important to carnivore-reintroduction success. A number of factors influence attitudes and values toward carnivores, including many real and perceived characteristics of the species (e.g., ferociousness, size, sentient capacity, phylogeny, degree of relatedness to humans, the animal's symbolic nature, the relationship between the animal and humans, and how the animal is used by humans: Kellert 1994). It is far easier to garner support for more charismatic species like bears and otters than for smaller, lesser-known species (Westman 1990), and many species, such as giant pandas (*Ailuropoda melanoleuca*) and gray wolves (*Canis lupus*), project powerful cultural or symbolic values (see K. Johnson et al., this volume).

Important human/carnivore relationships that influence attitudes include the population and conservation status of a particular species (e.g., rare, threatened, or endangered), human/wildlife conflicts (e.g., threats to human safety), wildlife-utilization opportunities (e.g., hunting and trapping), and human/animal property relations (e.g., land-use restrictions: Bath 1991; Reading and Kellert 1993; Kellert 1994: see also Gittleman and Pimm 1991). Finally, values and attitudes toward wildlife species are strongly influenced by the perceived worth of the animals. The perceived worth an individual assigns to a carnivore is influenced by all of the factors mentioned above and more, including knowledge, moral and ethical issues (e.g., animal rights), traditional market values (e.g., pelts, tourist attractions), and extra-market values (e.g., rodent control, ecosystem indicator roles). For these reasons, carnivores are valued in many different ways and in complex ways (see Clutton-Brock, this volume).

Other aspects of valuation associated with carnivore reintroductions include values and attitudes toward conservation programs and pertinent laws and regulations. For example, status as threatened or endangered under the U.S. Endangered Species Act (ESA) elicits fear and hostility among certain sectors of society (e.g., agricultural interests) and compassion and support among others (e.g., members of conservation organizations: Bath 1989, 1991; Bath and Buchanan 1989; Thompson 1993). Negative attitudes are often based on real or perceived fears of the restrictive components of the ESA, which many people view as a threat to their livelihoods and lifestyles, on negative attitudes toward wildlife in general, and on the real or perceived effects of past recovery programs (Kellert 1991; Reading and Kellert 1993; Thompson 1993). Positive attitudes are often rooted in recognition of, and concern for, the loss of biodiversity, and in positive values and attitudes toward wildlife (Kellert 1985b, 1986; Gossow and Honsig-Erlenburg 1986; Yalden 1993).

Values placed on carnivores, and on wildlife in general, are often divided into several categories (see Kellert 1980; Steinhoff 1980; Reading 1993):

1. *Aesthetic values:* Carnivores inspire and stimulate creativity, contributing to artistic expression and nature appreciation.
2. *Cultural values:* Carnivores provide a source of cultural, historic, religious, and symbolic values as symbols of group identity, as expressions of unique social experiences, or as objects of special attachment.
3. *Domination values:* Carnivores provide a challenge to human mastery, control, and domination because of their strength, speed, cunning, and well-honed senses. They are valued as metaphors for the "wilderness" that is to be tamed, subjugated, and ordered by humanity.
4. *Ecological values:* Carnivores provide ecosystems with ecological services, diversity, and stability, often through complex, dependent interrelationships.
5. *Humanitarian values:* Humanitarian values of carnivores are closely related to moral/ethical values. People often develop strong emotional attachments to carnivores and a strong interest in their humane treatment.
6. *Moral/ethical values:* Carnivores provoke an appreciation of existence values, reverence of life, character-building, and intrinsic values.
7. *Negative values:* Negative values toward carnivores are based on fear and hostility, often as a result of limited exposure or bad experiences, or owing to fear of property damage and physical harm to humans or to domestic animals and a more general unease with the unknown or unfamiliar.
8. *Recreation values:* Carnivores evoke reflection, a sense of awe and wonderment, escapism, challenge, and relaxation for many people, through consumptive (e.g., hunting and trapping), nonconsumptive (e.g., viewing and photography), and vicarious (e.g., stories and television) recreation.
9. *Scientific values:* Carnivores are a source of intellectual stimulation and understanding through research.
10. *Utilitarian values:* Carnivores provide utilitarian products, such as hides.

Several of these values, the utilitarian and the dominionistic, for example, are often closely linked. But whatever the variable and complex ways in which carnivores are valued, economics factor heavily into almost any discussion.

Many of the values enumerated above have in fact been quantified with varying degrees of success by economists, but wildlife production and consumption take place largely outside traditional, organized markets (Thompson 1993). This is so partly because we choose to make wildlife an extra-market good (Davis 1985), but also because many of the values attributable to wildlife (e.g., intrinsic values, cultural values, ecological values) are difficult, if not impossible, to quantify. Sociological and psychological research have demonstrated the importance of these less-tangible values, but because of their normative nature, they are rarely incorporated into policy and management. Therefore, in spite of their importance, most of the benefits of carnivore reintroductions are often ignored. In contrast, costs associated with reintroduction are more easily ascertained and receive more emphasis.

Program performance and effectiveness can be improved by considering and addressing the valuational aspects of carnivore-reintroduction programs. Understanding values and valuation permits recognition of strongly held values, of the major factors and forces creating and maintaining values and value systems, and of the attitudes, beliefs, and behaviors arising from them (Reading and Kellert 1993; Thompson 1993). This, in turn, facilitates the development of more effective education and public-relations programs and conservation strategies, and enables managers and conservationists to recognize and stress the full array of values and potential values that wildlife offers.

Organizational Considerations

Because carnivore conservation is carried out by various government, non-profit, and private organizations, it is also essential to understand the organizational dimension of recovery efforts. A major variable in the success or failure of a carnivore reintroduction effort is the kind of organizational system used, and this dimension is the least appreciated variable in carnivore conservation. As Clark and Harvey (1988:35) state,

Professionals working in [endangered-species recovery] programs often view recovery primarily as a biological problem. . . . Yet the organizational arrangements, decision-making processes, and other policy variables affecting recovery programs can be as critical to success as technical and biological tools. A better understanding of the policy and organizational dimensions of endangered species work could greatly enhance the effectiveness of many recovery programs.

Of all the variables hypothesized as influencing key-actor behavior and, ultimately, successful reintroduction (Figure 9.1), organizational variables appear to be the least explicitly perceived and managed. Still, understanding organizations permits description, diagnosis, and prescription of situations and problems encountered within them (Gordon 1983).

We suggest that carnivore-reintroduction failures often occur for a variety of organizational reasons, including poor matching of organizational structures to the reintroduction task, delegating implementation to an individual or organization lacking the necessary expertise, issuing conflicting directives, depending too heavily on other organizations, exercising weak leadership or overly wide (i.e., excessive) discretion, and many other factors (Yaffee 1982; Clark et al. 1993, 1994). Understanding the environments, structures, and cultures of organizations, patterns and sources of power and authority, and factors influencing key-actor relations helps explain behavior and implementation failures in carnivore-reintroduction programs.

Organizations are influenced by the environments in which they operate. The external environment of an organization includes the larger social context, the operating environment, and the power setting of the organization (Warwick 1975). These factors combine to influence goals and value systems. Perrow (1986) suggests that most organizations move continually toward centralized control, and that a hostile environment, greater complexity, and more rapid change generally lead to more rigid organizations with more rules. But to be most effective in an environment of high complexity, diversity, uncertainty, and rapid change, an organization should be less rigidly structured and less centralized (Gordon 1983; Perrow 1986). Such external environments are typical of many carnivore-reintroduction programs.

The internal environment of an organization comprises three main elements: structure, culture, and power/authority. Structural elements include the way an organization is set up, the way it operates, and valuational/perceptual elements, which describe the culture of an organization and include things like ideologies, goals, tasks, work atmosphere, management philosophies, and traditions (Warwick 1975).

Organizational structure is perhaps the most important factor influencing organizational performance (Clark et al. 1989). Structure refers to the pattern of interacting relationships, units, and practices in an organization, including hierarchy and underlying rules. Structures dictate task assignments, resource allocation, information channels, control of communication, and more, all of which influence program effectiveness (Warwick 1975). Both formal and informal structures operate within each organization (Weinstein 1984; Perrow 1986). Often the two are dramatically different; frequently, informal hierarchies represent the more common, or routine, path of operation and decision-making.

To enhance performance, organizations should be well matched to their tasks (Clark and Westrum 1989; Clark et al. 1989). Meyer (1968) suggests there are two basic organizational strategies. Organizations characterized by direct supervision and a wide span of control promote flexibility and rapid response. Such structures are desirable in the uncertain, complex, and time-urgent environments of carnivore-reintroduction programs, because they are conducive to individual and organizational learning, maintain rapid and open communication, and remain task-oriented, innovative, and self-critical (Westrum 1986; Clark and Westrum 1989; Clark et al. 1989; Clark and Cragun 1994). Other organizations, characterized by a narrow span of control, long hierarchies, and top-down control, promote predictable behavior and stable performance. Such organizations are useful for routine tasks, for task specialization, and to establish and clarify lines of communication (Perrow 1986). But both internal and external pressures generally drive organizations toward greater bureaucratization (i.e., long hierarchies: March and Simon 1958; Warwick 1975). Given the bureaucratic pressures operating on many agencies involved in carnivore reintroductions, and the inter-organizational nature of such programs, developing an effective structure often requires the establishment of a parallel organization, such as a recovery team (Clark et al. 1989).

Gordon (1983) promotes the formation of special, multidisciplinary groups or teams, especially for highly complex, uncertain, and urgent tasks like reintroductions. Clark and Westrum (1989) suggest that effective teams are those that are creative, have strong internal support, communication, information-processing, and learning capabilities, and are properly staffed, led, and buffered. Conflict, if properly managed, can be functional and productive. The team should comprise well-trained, experienced professionals with strong problem-solving skills and the ability to work as part of an interdisciplinary team with little supervision. To maximize effectiveness and flexibility, a team should be an independent unit, not simply a separate hierarchical level of the organizations, departments, or branches its members represent. There are potential dangers with smaller working groups and teams (e.g., two to ten members), which include "satisficing" (in which adequate but suboptimal solutions are adopted), slower decision-making, the exclusion of external expertise, and "group think" (in which confrontation and dissension are avoided for the sake of consensus) (March and Simon 1958; Janis 1972; Gordon 1983). Small groups may also make riskier decisions, but in carnivore-reintroduction programs it is usually better to be "risk-embracing" (i.e., to acknowledge and confront risks) than "risk-aversive" (i.e., to avoid risky decisions: March and Simon 1958; Clark and Westrum 1989).

Strong, open, objective, and competent leadership is another critical variable for any organizational or inter-organizational effort (Clark and Westrum 1989; Clark and Cragun 1991). Strong leaders are able to buffer organizations from their external environments and to help subordinates resolve conflicts and overcome challenges. Open, objective leaders are willing to consider divergent views and novel ideas. Westrum (1988:12) suggests that a leader's role is to be an "executive, structurer, and exemplar." Competent leaders possess human-relations, administrative, and technical skills and are responsible for building and maintaining healthy, productive, task-oriented teams (Clark and Westrum 1989; Clark et al. 1989).

Organizations are usually characterized by a culture, or set of shared ideologies, philosophies, norms, and goals, that influences the behavior of members and is passed on over time (Tichy 1983). Some organizational cultures (e.g., the U.S. Fish and Wildlife Service, conservation organizations) are supportive of carnivore-reintroduction and -conservation efforts while others (e.g., Animal Damage Control, stockgrowers associations) are antagonistic. Organizational culture is influenced and maintained by several factors, including history, tradition, organizational leaders, external constraint, and powerful internal and external groups (Warwick 1975; Wilson 1980; Westrum 1988). Once established, organizational cultures tend to propagate themselves through selective recruiting and retention of individuals with similar personalities and ideologies, and reorientation or elimination of individuals who differ. Similarly, individuals choose to work for organizations with certain cultures. Over time, then, well-established organizational cultures become resistant to change (Westrum 1988).

Three types of organizations can be distinguished on the basis of their management philosophies (Westrum 1986, 1988). For instance, *pathological* organizations characteristically deny or suppress new ideas, information, or evidence of problems, shirk responsibility, and punish or conceal failures. *Calculative/rational* organizations use knowledge gained from past experiences to develop future strategies, which leads to mechanical, rule-driven bureaucracies. Calculative rationality works well in relatively simple, predictable environments, but fails to deal effectively with rapid change and high complexity. Environments characterized by change and complexity require more "organic," less hierarchical organizations that stress creativity, information flow, self-examination, wide participation, and task-oriented goals. This kind of an organization, called a *generative* organization, focuses on generating ideas, evaluation, and future performance.

The goals driving organizational behavior result from several factors, including knowledge or assumptions about alternatives and their consequences, the goals and values of groups, and the skills of individuals, especially leaders (March and Simon 1958; Yaffee 1982; Westrum 1988). As an organization attempts to survive or grow, the original task goals are often weakened or perverted (Perrow 1986). "Goal displacement" occurs when survival or control goals (e.g., maintaining lead-agency status in a carnivore-reintroduction program) come to predominate over the original task goals (e.g., reestablishing a carnivore population) (Sills 1957). Differences among individuals and groups within an organization can also lead to multiple, conflicting goals. As a result, organizations, especially government agencies, often have vague, nonoperational goals and poor methods of evaluating performance (Warwick 1975; Weimer and Vining 1989). Success then becomes hard to define and measure, let alone achieve.

Finally, the struggle for power and authority is differentially important to the culture of an organization. Carnivore-reintroduction programs commonly consist of several actors, including local people, local, state, and national governmental and nongovernmental organizations, and university affiliates. In any situation where multiple actors are working toward a common goal, issues of authority and power potentially dominate interactions. Power and authority relationships among key actors evolve as programs are carried out, although in many instances the traditional organizational relations and preexisting laws, regulations, and mandates strongly influence the development of relationships among actors.

Power is an attribute of both individuals and organizations, and stems from authority and control of resources (Wilson 1980). Authority arises from legitimacy based on either: (1) formal laws, rules, and regulations, (2) tradition, custom, or loyalty, or (3) perceptions of exceptional qualities of leaders by their followers or subjects (Weber 1968; Wargo 1984). Power rests on things like money, personnel, equipment, knowledge or expertise, land tenure, political clout, and, for carnivore-reintroduction programs, control of the animals. Conflict in carnivore reintroductions often develops over control of resources, including the animals themselves, over the real or perceived transfer of private-property rights to the state without adequate compensation (Wargo 1984; Reading and Kellert 1993; Thompson 1993), and over state versus federal power struggles (e.g., Clark 1989; Reading 1993).

Understanding how organizations operate, the factors influencing their performance, and techniques for improvement can lead to improved organizational arrangements, and thus program effectiveness. But in carnivore-reintroduction efforts, organizational dimensions are rarely addressed or even understood. Many common organizational problems affect carnivore-reintroduction success, and the application of organizational theory, especially government-bureaucracy theory, to carnivore-reintroduction programs could therefore vastly improve effectiveness and efficiency.

In short, knowing the biological and technical facts of a carnivore-reintroduction program is not sufficient if we hope to maximize efficiency and success rates. Wildlife biologists must also understand the important social-science dimensions of the reintroduction challenge. Acquiring that understanding requires an interdisciplinary approach in which both valuational and organizational considerations figure prominently. By better understanding how these variables influence reintroduction success, wildlife practitioners can streamline their programs, reduce conflicts, and ultimately improve the overall performance of carnivore-reintroduction programs. Two case studies—red wolf reintroduction efforts, from 1973 to 1990, and black-footed ferret recovery and reintroduction efforts, from 1981 to 1993—illustrate various aspects of the nonbiological (valuational and organizational) dimensions of carnivore reintroduction. We have been involved in black-footed ferret reintroduction and recovery efforts since the 1970s. As before, we will limit our analyses here to specific time frames, owing to the ongoing nature of the reintroduction efforts and because those time frames were in any case sufficient to illustrate our major points.

Case Study 3: Red Wolf

Red wolves are the first carnivore species to be reintroduced after going extinct in the wild (Phillips 1990). At around 18-36 kg, red wolves are intermediate in size between gray wolves (*C. lupus*) and coyotes (*C. latrans*) (Cohn 1987). Controversy surrounds the taxonomic status of the red wolf and its relationship to coyotes and gray wolves, some studies suggesting that red wolves are nothing more than wolf-coyote hybrids (Nowak 1979; Wozencraft 1989; Peek et al. 1991; Kidder 1992). This debate has important implications, and some conservationists question the wisdom of expending the vast human and financial resources necessary to reintroduce an animal whose taxonomic distinctiveness is at best questionable (Gittleman and Pimm 1991).

Putting aside the taxonomic question for the moment, conventional wisdom suggests the following recent history of decline for red wolves. These wolves once ranged across the southeastern United States from central Texas east to the Atlantic, and from the Ohio River south to the Gulf of Mexico (Moore and Smith 1991). Direct eradication efforts and habitat alteration following European settlement led to declines in the range and numbers of red wolves (Banks 1988; Moore and Smith 1991). More recently, "pure" red wolves faced genetic swamping as numbers declined, the species' social structure was disrupted, and interbreeding with coyotes led to significant genetic introgression (Phillips and Parker 1988; Peek et al. 1991). In 1973 the U.S. Fish and Wildlife Service (FWS) began capturing animals for inclusion in a captive-breeding program; the vast majority of these animals, however, were eliminated as hybrids, and only 14 of the 40 "pure" animals produced viable offspring (Cohn 1987; Reese 1989; Kidder 1992).

Still, the captive population grew, and by the late 1970s FWS began reintroduction-feasibility studies (Moore and Smith 1991). This work included short-term, experimental releases on Bull Island, South Carolina, in 1976 and 1978, and ecological investigations of possible reintroduction sites (Taylor 1986; Cohn 1987; Banks 1988). These releases demonstrated the feasibility of reintroduction, but a larger reintroduction site was required (Moore and Smith 1991). After some searching, the FWS selected a site on the border of Kentucky and Tennessee called the Land Between the Lakes. The site consisted of a National Recreation Area of 68,000+ ha created by two Tennessee Valley Authority (TVA) projects (Cohn 1987; Phillips and Parker 1988). Although the site looked promising biologically and a technical reintroduction plan was developed, the actual release of red wolves was not forthcoming. Opposition to the proposed reintroduction surfaced almost immediately. Local farmers feared livestock depredations, hunters voiced concerns about hunting restrictions, and conservation organizations worried about both large numbers of hunters in the area and the presence of coyotes (Taylor 1986; Cohn 1987; Phillips and Parker 1988; Kidder 1992). According to Moore and Smith (1991:267), "The heart of this opposition stemmed from an inadequate job by the U.S. Fish and Wildlife Service, TVA, and cooperating state wildlife organizations in working with local residents and concerned environmentalists to allay fears and hearsay." Both state wildlife agencies withdrew their support, and the reintroduction was called off (Phillips and Parker 1988). These results are not surprising, given the widespread recent and historical antagonism of farmers, ranchers, and hunters toward predators, especially wolves (Hook and Robinson 1982; Kellert 1985b, 1986, 1991; Bath 1989; Bath and Buchanan 1989).

Cooperators in the red wolf program readily admit that their failure to address adequately the valuational considerations led to the failure of the reintroduction attempt (Taylor 1986; Cohn 1987). As program leader W. Parker stated, "The plan was feasible, but we were not able to carry it off because it became a political issue" (Cohn 1987:315). But the story did not end there; program participants learned from that first attempt. "We got a real education out there," said Parker (Taylor 1986:315). Thus, when a second opportunity for reintroduction developed with the donation of almost 48,000 ha of undisturbed land in Dare County, North Carolina, participants brought with them the lessons learned from the Land Between the Lakes experience (Kidder 1992). "Fully aware that the earlier reintroduction plan failed because the public had not been adequately prepared, Parker went out of his way this time to inform the local gentry," according to Banks (1988:105).

The donated land was designated Alligator River National Wildlife Refuge (ARNWR) in 1984, and officials began an intensive effort to contact, educate, and listen to the concerns of, local residents (Banks 1988; Phillips and Parker 1988; Phillips 1992). In ARNWR, wrote Moore and Smith (1991:268), "FW personnel . . . helped sway local public opinion in favour of the wolf through numerous informal contacts with friends and neighbors . . . [and] . . . red wolf project personnel worked carefully with the local populace and national environmental groups to develop the strategy to reintroduce wolves." To facilitate local acceptance, the reintroduced population was designated "experimental, nonessential" under Section 10(j) of the Endangered Species Act (see Parker and Phillips 1991). This provision, included in 1982 amendments to the Act, increases management flexibility of reintroduced populations in an effort to gain acceptance and encourage cooperation by local publics (Bean 1983; Parker and Phillips 1991). The flexibility afforded by the experimental, nonessential designation helped garner local support (Moore and Smith 1991). Finally, media involvement was high (Phillips 1990).

The result of these efforts paid off (Phillips 1992; but see Wilcove 1987). Although some concerns were noted, most publics and organizations expressed support for the reintroduction, their numbers including local hunters and trappers, the North Carolina Wildlife Resources Commission, and environmentalists (Banks 1988; Phillips and Parker 1988; Moore and Smith 1991; Parker and Phillips 1991; Kidder 1992). Ever since the releases began, cooperation has been high, and has included the Department of Defense (which shares the peninsula with the refuge), local civic groups, local landowners, and other citizens (Phillips 1990; Parker and Phillips 1991).

From 1987 to 1990, 32 red wolves were released on 12 occasions. Several of these animals survived to late 1991, reproduction in the wild has been noted, and all mortalities through 1990 were natural or accidental (Phillips 1990; Parker and Phillips 1991). These results bode well for the ability of the program to reestablish the population and maintain local support and cooperation. Public enthusiasm for the reintroduction program has remained high (Moore and Smith 1991). Indeed, the success of the Alligator River National

Wildlife Refuge reintroduction program has spawned similar red wolf introductions in several other locations, including Bull Island, South Carolina, and Horn Island, Mississippi (Moore and Smith 1991). Program participants have used the publicity surrounding red wolf reintroduction as an opportunity to educate the public about conservation (Kidder 1992:15): "The red wolf and other 'glamour' species can educate people about the need to preserve biological diversity, says Phillips. 'People get excited about wolves and that gives us a chance to point out other important things that people don't get excited about.' " After reviewing the red wolf and other carnivore translocations and reintroductions, Moore and Smith (1991:264) "concluded that public acceptance of such re-introductions . . . is vital for success."

Case Study 4: Black-footed Ferret

Black-footed ferrets are endangered, mink-sized mustelids (~1 kg) that depend on prairie dogs (*Cynomys* spp.) and their burrows for food and shelter, respectively (Forrest et al. 1985). (For a review of ferret biology and ecology, see Clark 1989, 1994; Seal et al. 1989; Miller et al., in press.) Ferret numbers presumably declined as prairie dogs fell victim to massive, government-sponsored eradication programs in the first half of this century (Miller et al. 1990c). These programs, which continue on a smaller scale today, have been successful in eliminating prairie dogs from most of their historic range (Reading and Miller 1994). As their prey base was being systematically destroyed, ferret populations became fragmented and the animals began dying from various deterministic and stochastic factors (Harris et al. 1989). Important among these factors was canine distemper, to which black-footed ferrets are highly susceptible (Thorne and Williams 1988).

In the late 1970s, many people believed black-footed ferrets were extinct, but a combination of diligent searching and good luck resulted in the discovery of a population of ferrets in Meeteetse, Wyoming, in 1981 (Carr 1986). The Meeteetse ferret population was monitored and studied until 1985, when plague in their white-tailed prairie dog (*C. leucurus*) prey base and canine distemper in the ferrets themselves began driving the species to extinction (Thorne and Williams 1988; Clark 1989). Eighteen individuals captured from 1985 to 1987 became the nucleus of a captive colony (Miller et al. 1988). No ferrets survived in the wild, but after a slow start, captive propagation has produced hundreds of ferrets, and the population now numbers well over 300 individuals in seven facilities (Thorne et al. 1993).

Reintroduction of ferrets bred in captivity began in 1991. The Black-footed Ferret Recovery Plan calls for the establishment of a minimum of ten separate, self-sustaining populations comprising at least 1,500 individuals prior to breeding (U.S. Fish and Wildlife Service 1988). The first ferret reintroduction site was Shirley Basin, Wyoming; and Meeteetse, Wyoming, was designated as the alternative. Two releases of ferrets have taken place in Shirley Basin: 49 animals (32 males: 17 females) in 1991 and 90 (56: 34) in 1992 (Thorne et al. 1993). A third release of a planned 50 animals is currently (autumn 1993) under way. At least four animals from the 1991 releases and eight from the 1992 releases successfully overwintered, and, of these, at least two in 1991 and four in 1992 were females that reproduced.

Since black-footed ferrets were rediscovered in 1981, ferret recovery efforts have suffered from a host of organizational problems at the state and federal levels (Carr 1986; May 1986; Weinberg 1986; DeBlieu 1991). Clark and colleagues have described the suboptimal organizational arrangements that were at the heart of these problems (Clark and Harvey 1988; Clark and Westrum 1989; Clark 1989; Clark and Cragun 1991). Despite recognition of these problems by several program participants (Clark 1989; Thorne and Oakleaf 1991; Miller 1992; Reading and Miller 1994; Miller et al., in press) and by outside interests as well (Carr 1986; May 1986; Weinberg 1986; Greater Yellowstone Coalition 1990; Alvarez 1993), organizational considerations have apparently received little explicit attention in current or past black-footed ferret recovery efforts.

Soon after the discovery of the Meeteetse ferret population, the U.S. Fish and Wildlife Service (FWS) transferred legal authority over the program to the Wyoming Game and Fish Department under the authority of the Endangered Species Act (Clark 1989). Over 100 individuals from more than 20 organizations participated in the Meeteetse ferret recovery efforts (Clark and Harvey 1988). Often, relations between actors, especially the conservation community and Wyoming, were poor and conflict-laden (Weinberg 1986; Clark 1989). Although programmatic problems were widely acknowledged, the underlying reasons for these problems were differentially recognized by the participants (e.g., Clark and Harvey 1988; Anonymous 1991; Thorne and Oakleaf 1991). The reasons for the organizational arrangements, conflicts, and associated organizational weaknesses were complicated, diverse, and not easily addressed under the conditions that existed at the time or those that remain today.

Attempting to shed light on these problems and remedy them, Clark and colleagues (Clark and Westrum 1989; Clark and Harvey 1988; Clark 1989; Clark and Cragun 1991) used theory and tools from the organizational and policy sciences. They described many of the programmatic weaknesses encountered in the ferret recovery program in Wyoming, as well as their underlying causes, explanations for the misinterpretation of these weaknesses by several program participants, and recommendations for improvements (see works cited above), but their recommendations have not been adopted by either state or federal agencies.

Two contrasting styles of organization and management had been adopted within the larger black-footed ferret recovery program in Wyoming, according to Clark (1989). He described the nongovernmental conservation community's "organic" style of organization and management (Gordon 1983) and contrasted this with the government agencies' more bureaucratic approach. Recognizing the uncertainty and urgency typical of endangered-species programs, the nongovernmental organizations adopted a "risk-embracing" style that was designed to be flexible, adaptable, and task-oriented. Professionalism, critical review, rapid and open communication, limited hierarchy, and task orientation were stressed. Clark (1989) contrasted the government agencies' approach (Wyoming Game and Fish Department and the FWS), which exhibited traditional bureaucratic styles of organization and management (Warwick 1975; Yaffee 1982). These "risk-aversive" agencies, characterized by bureaucratic and scientific conservatism, stressed authority, standard operating procedures, rigid hierarchies, and fixed rules, roles, and regulation. These two modes of organizing and operating inevitably clashed.

In the end, the more bureaucratic style of organization and management of the agencies came to dominate the ferret program. Wyoming Game and Fish Department, the lead agency in Wyoming, has apparently been unable and/or unwilling to correct recurrent organizational weaknesses, and the state has seemed to focus on maintaining and strengthening its control over the program. The result has been high levels of conflict and distrust among actors, communication breakdowns and delays, and goal displacement toward control issues, all of which led to the near extinction of the species (see Carr 1986; May 1986; Weinberg 1986; Clark and Westrum 1989; Clark and Harvey 1988; Clark et al. 1989; DeBlieu 1991; Alvarez 1993). The many problems of the program have been well documented. Nevertheless, the public-relations branches of both the state Game and Fish Department and the federal government have consistently avoided mention of programmatic problems. Instead, most people outside the program know only of the increased ferret numbers and reintroductions and conclude that the program is a successful model. But the way the program is organized and the manner in which it operates do not present a model effort. Recommendations from a number of analyses of the organizational arrangements have been little used, and recovery efforts continue to be plagued by weak organizational performance and conflict (Greater Yellowstone Coalition 1990; Miller 1992; Miller et al. 1994, in press; Reading and Miller 1994).

The number of organizational actors involved in the ferret recovery efforts in Wyoming has continued to decline. By 1993, even the FWS was only marginally involved in the reintroduction efforts, which had come to be dominated by Wyoming Game and Fish. Reading and Miller (1994) discuss current problems with the reintroduction program, including many of the same rigid bureaucratic structures described by Clark and Harvey (1988) and Clark (1989), inadequately constituted and poorly operating working groups, lack of rigorous research and monitoring, displacement of task-recovery goals by program-control goals, and narrow measures of program success that fail to consider important considerations such as program efficiency, gains in reliable knowledge, local public relations, and inter-organizational relations.

The irony is that the reintroduction program may still succeed in restoring ferrets to the wild, but the opportunity costs to other ferret reintroductions and other programs could still be significant, because the ferret program has: (1) failed to initiate reintroduction at a second site, despite the presence of excess captive animals and prepared release sites; (2) gained little reliable knowledge for improving reintroduction techniques; (3) failed to monitor the status of released animals; (4) failed to determine the genetic relatedness of the captive population's founders; (5) cost millions of dollars more than it could have; and (6) strained relations among many individuals and organizations.

The red wolf and black-footed ferret case studies illustrate the importance of valuational and organizational considerations to carnivore-reintroduction programs. The red wolf case demonstrates the importance of working to assess and address local values, attitudes, and concerns early in a carnivore-reintroduction effort, knowing when to avoid reintroduction into a potentially polemic situation, using management flexibility to garner local support and quell fears, and learning from past experiences. The black-footed ferret case illustrates the importance of organizational considerations to effective and efficient performance. On the basis of organizational analyses of our experiences in ferret recovery efforts, we suggest that strong bureaucratic controls, poor communication, unproductive conflict, displacement of recovery goals by control goals, and other organizational problems have acted to limit expertise and restrict involvement in ferret reintroduction activities, strain relationships among key actors, inhibit learning, waste resources, and slow the species' recovery. In short, the program remains problematic.

Conclusions

Carnivore reintroductions can be complex, for a host of biological and nonbiological reasons. For success to be the rule, rather than the exception, the full range of variables affecting carnivore reintroductions must be explicitly addressed. The model in Figure 9.1 and our associated descriptions and cases provide a framework that can serve as a guide for individuals and programs undertaking carnivore reintroductions. The dimensions of this model can be divided into biological/technical and nonbiological variables.

Although better understood than other variables, the biological and technical aspects of carnivore reintroductions nonetheless offer a formidable challenge. Carnivore reintroductions—especially reintroductions of endangered carnivores for which important biological data are often lacking—are risky and uncertain. To increase the probability of success, reintroduction programs should address a wide range of biological and ecological considerations, including demographics, behavior, genetics, physiological needs, and habitat requirements. Programs should also address technical considerations, such as release techniques, site preparation and management, animal selection and preparation, monitoring, and data management. New advances in the rapidly expanding discipline of conservation biology have much to offer in these areas.

But addressing biological and technical concerns is only part of the challenge; carnivore-reintroduction programs must also address valuational and organizational considerations. The importance of these considerations to carnivore-reintroduction programs is increasingly being recognized. Public support, especially from local people, is often crucial to reintroduction efforts. Still, valuational concerns are generally poorly understood by most people working on carnivore-reintroduction programs, and are rarely adequately addressed. Reintroductions should assess public perceptions, attitudes, values, and knowledge as they relate to the species and the reintroduction program. It is equally important to understand the full array of factors influencing stasis and dynamism in attitudes and values. Such understanding facilitates the development of more effective public-relations and education programs.

Although no less important than biological/technical or valuational aspects, the organizational aspects of carnivore reintroductions are perhaps the least understood and least commonly addressed. Because the reach of organizations and organizational control is pervasive, understanding organizations and how they operate can increase the efficiency and effectiveness of almost any endeavor, including reintroductions. Understanding important organizational dimensions, including environments, structures, personnel, leadership, ideologies, philosophies, goals, and power and authority, enables participants both to improve, and to work more effectively within, existing organizational arrangements.

The three classes of variables are all interrelated, they all influence each other, and the boundaries between them thus become fuzzy. For example, the values, attitudes, knowledge, and technical abilities of personnel working on a carnivore reintroduction are important from a biological/technical, a valuational, and an organizational perspective. Similarly, designing effective release protocols and data management depend upon both biological/technical and organizational considerations. All of the variables discussed above also affect key-actor behavior, which in turn affects reintroduction success. It is therefore crucial that carnivore-reintroduction plans strive to address the full array of variables influencing success. Such plans could help ensure the orientation of all the actors toward carnivore-reintroduction success and rapid, efficient movement toward that goal.

Acknowledgments

Ann Peyton Curlee, John L. Gittleman, and two anonymous reviewers offered excellent comments and suggestions on the manuscript. U. and C. Breitenmoser-Wursten (Swiss Lynx Project), R. Panaman (International Carnivore Protection Society), M. Rahbar (Re-introduction Specialist Group of the World Conservation Union), D. T. Rowe-Rowe (Natal Parks Board), V. Sabarwal (Yale University), G. Schaller (Wildlife Conservation International), M. Stanley-Price (Re-introduction Specialist Group of the World Conservation Union), R. Wallace (Marine Mammal Commission), and R. Wirth (Mustelid, Viverrid, and Procyonid Specialist Group of the World Conservation Union) provided assistance in locating case material.

References

Alt, G. L., and Beecham, J. J. 1984. Reintroduction of orphaned black bear cubs into the wild. *Wildl. Soc. Bull.* 12:169–174.

Alvarez, K. 1993. *Twilight of the Panther: Biology, Bureaucracy and Failure in an Endangered Species Program.* Sarasota, Fla.: Myakka River Publ.

Anderson, J. L. 1981. The re-establishment and management of a lion *Panthera leo* population in Zululand, South Africa. *Biol. Conserv.* 19:107–117.

Anderson, J. L. 1986. Restoring a wilderness: The reintroduction of wildlife to an African national park. *Int. Zoo Yearb.* 24/25:192–199.

Anderson, J. 1992. South African carnivores. *Re-introduction News* 4:8–9.

Andrews, R. D., Storm, G. L., Phillips, R. L., and Bishop, R. A. 1973. Survival and movements of transplanted and adopted red fox pups. *J. Wildl. Mgmt.* 37:69–72.

Anonymous. 1991. The land ethic. *Wyoming Wildl.* 55(9):2.

Anonymous. 1992a. Southern sea otter translocation to San Nicolas Island, August 1991–July 1992. Unpublished; annual report. Washington, D.C.: Marine Mammal Commission.

Anonymous. 1992b. Juvenile monk seals collected at French Frigate Shoals for rehabilitation and release at Midway Island. *SW Fish. Sci. Center Rept. Act.* July-Aug:7–8.

Anonymous. 1993. *Carnivore Re-introductions.* Nairobi, Kenya: IUCN/SSC Re-introduction Specialist Group.

Banks, V. 1988. The red wolf gets a second chance to live by its wits. *Smithsonian* 18(12):100–107.

Bath, A. J. 1989. The public and wolf restoration in Yellowstone National Park. *Soc. Nat. Res.* 2:297–306.

Bath, A. J. 1991. Public attitudes in Wyoming, Montana, and Idaho toward wolf restoration in Yellowstone National Park. *Trans. N. A. Wildl. and Nat. Res. Conf.* 56:91–95.

Bath, A. J., and Buchanan, T. 1989. Attitudes of interest groups in Wyoming toward wolf restoration in Yellowstone National Park. *Wildl. Soc. Bull.* 17:519–525.

Bean, M. J. 1983. *The Evolution of National Wildlife Law.* New York: Praeger.

Beck, B. B., and Miller, B. 1990. Implications for black-footed ferrets from the reintroduction of the golden lion tamarin. In: T. Thorne, ed. *Proceedings from a Workshop on the Reintroduction of Black-footed Ferrets, March 20–21, 1990.* Cheyenne: Wyoming Game and Fish Department.

Belden, R. C., and Hagedorn, B. W. 1993. Feasibility of translocating panthers into northern Florida. *J. Wildl. Mgmt.* 57:388–397.

Belden, R. C., Hines, T. C., and Logan, T. H. 1986. Florida panther re-establishment: A discussion of the issues. *Proc. 6th Natl. Wildl. Rehabilitators Assoc. Symp. Clearwater Beach, Fla.* 6:115–123.

Berg, W. E. 1982. Reintroduction of fisher, pine marten, and river otter. In: G. C. Sanderson, ed. *Midwest Furbearer Management*, pp. 159–173. Wichita: Kansas Chapter of The Wildlife Society.

Bigg, M. A., and MacAskie, I. B. 1978. Sea otters reestablished in British Columbia. *J. Mamm.* 59:871–874.

Biggins, D. E., Hanebury, L. R., and Miller, B. J. 1991. Trial release of Siberian polecats (*Mustela eversmanni*). Unpublished progress report, U.S. Fish and Wildlife Service.

Biggins, D. E., Hanebury, L. R., Miller, B. J., and Powell, R. A. 1990. Trial release of Siberian polecats (*Mustela eversmanni*). Unpublished progress report, U.S. Fish and Wildlife Service.

Bixby, K. 1991. Predator conservation. In: K. A. Kohm, ed. *Balancing on the Brink of Extinction: The Endangered Species Act and Lessons for the Future*, pp. 199–213. Washington, D.C.: Island Press.

Booth, W. 1988. Reintroducing a political animal. *Science* 241: 156–158.

Breitenmoser, U., and Breitenmoser-Wursten, C. 1990. *Status, Conservation Needs and Reintroduction of the Lynx (Lynx lynx) in Europe*. Nature and Environment Series, No. 45. Strasbourg: Council of Europe.

Breitenmoser, U., and Haller, H. 1993. Patterns of predation by reintroduced European lynx in the Swiss Alps. *J. Wildl. Mgmt.* 57:135–144.

Brocke, R. H. 1990. The return of the lynx: A progress report. *The Conservationist* 44(5):9–13.

Brocke, R. H., and Gustafson, K. A. 1992. Lynx (*Felis lynx*) in New York State. *Reintroduction News* 4:6–7.

Brocke, R. H., Gustafson, K. A., and Fox, L. B. 1991. Restoration of large predators: Potentials and problems. In: D. J. Decker, M. E. Krasny, G. R. Goff, C. R. Smith, and D. W. Gross, eds. *Challenges in the Conservation of Biological Resources: A Practitioner's Guide*, pp. 303–315. Boulder, Colo.: Westview Press.

Brocke, R. H., Gustafson, K. A., and Major, A. R. 1990. Restoration of lynx in New York: Biopolitical lessons. *Trans. N. A. Wildl. Nat. Res. Conf.* 55:590–598.

Brown, M. K., and Parsons, G. R. 1983. Movement of a male fisher in southern New York. *N.Y. Fish and Game J.* 30:114–115.

Brown, T. C. 1984. The concept of value in resource allocation. *Land Econ.* 60:231–246.

Cade, T. J. 1988. Using science and technology to reestablish species lost in nature. In: E. O. Wilson, ed. *Biodiversity*, pp. 279–288. Washington, D.C.: National Academy Press.

Carbyn, L. N. 1989. Status of the swift fox in Saskatchewan. *Blue Jay* 47:41–52.

Carbyn, L. N. 1992. Swift fox comeback in Canada. *Re-introduction News* 5:3–4.

Carr, A., III, 1986. Introduction. *Great Basin Nat. Mem.* 8:1–7.

Chaiken, S., and Stangor, C. 1987. Attitudes and attitude change. *Ann. Rev. Psychol.* 38:575–630.

Clark, T. W. 1989. Conservation biology of the black-footed ferret (*Mustela nigripes*). *Wildl. Preserv. Trust Internat. Spec. Sci. Rept.* 3:1–175.

Clark, T. W. 1994. Restoration of the endangered black-footed ferret: A 20-year overview. In M. L. Bowles and C. J. Whelan, eds. *Restoration and Recovery of Endangered Species*, pp. 272–297. Cambridge: Cambridge Univ. Press.

Clark, T. W., and Cragun, J. R. 1991. Organization and management of endangered species programs. *Endang. Species Update* 8(8):1–4.

Clark, T. W., and Cragun, J. R. 1994. Organizational and managerial guidelines for endangered species restoration programs and recovery teams. In: M. L. Bowles and C. J. Whelan, eds. *Restoration and Recovery of Endangered Species*, pp. 9–33. Cambridge: Cambridge Univ. Press.

Clark, T. W., Crete, R., and Cada, J. 1989. Designing and managing successful endangered species recovery programs. *Environ. Mgmt.* 13:159–170.

Clark, T. W., and Harvey, A. H. 1988. Implementing endangered species recovery policy: Learning as we go? *Endang. Species Update* 5:35–42.

Clark, T. W., Reading, R. P., and Clarke, A. L., eds. 1994. *Endangered Species Recovery: Finding the Lessons, Improving the Process.* Washington, D.C.: Island Press.

Clark, T. W., Reading, R. P., and Curlee, A. P. 1993. Conserving threatened carnivores: Developing interdisciplinary, problem-oriented strategies. Unpublished report to the Geraldine R. Dodge Foundation. Jackson, Wyo.: Northern Rockies Conservation Cooperative.

Clark, T. W., and Westrum, R. 1989. High-performance teams in wildlife conservation: A species reintroduction and recovery example. *Environ. Mgmt.* 13:663–670.

Cohn, J. P. 1987. Red wolf in the wilderness: Captive breeding program strives to keep a rare predator pure. *Bioscience* 37:313–316.

Davis, R. K. 1985. Research accomplishments and prospects in wildlife economics. *Trans. N. A. Wildl. Nat. Res. Conf.* 50:392–404.

DeBlieu, J. 1991. *Meant to be Wild: The Struggle to Save Endangered Species through Captive Breeding.* Golden, Colo.: Fulcrum.

Decker, D. J., and Goff, G. R., eds. 1987. *Valuing Wildlife: Economic and Social Perspectives.* Boulder, Colo.: Westview Press.

Derrickson, S. R., and Snyder, N. F. R. 1992. Potentials and limits of captive breeding in parrot conservation. In: S. R. Beissinger and N. F. R. Snyder, eds. *New World Parrots in Crisis: Solutions from Conservation Biology,* pp. 133–163. Washington, D.C.: Smithsonian Institution Press.

Dunlap, T. R. 1988. *Saving America's Wildlife: Ecology and the American Mind, 1850–1990.* Princeton, N.J.: Princeton Univ. Press.

Erickson, D. W., and Hamilton, D. A. 1988. Approaches to river otter restoration in Missouri. *Trans. N. A. Wildl. Nat. Res. Conf.* 53:404–413.

Erickson, D. W., and McCullough, C. R. 1985. Doubling our otters: Encouraging restoration results mean a more ambitious program for 1985. *Missouri Conservationist* 46(3):4–7.

Erickson, D. W., and McCullough, C. R. 1987. Fates of translocated river otters in Missouri. *Wildl. Soc. Bull.* 15:511–517.

FaunaWest Wildlife Consultants. 1991. An ecological and taxonomic review of the swift fox (*Vulpes velox*) with special reference to Montana. Report to the Montana Department of Fish, Wildlife, and Parks. Bozeman, Mont.

Fentress, J. C. 1979. Behavior mechanisms and preparing wolves for life in the wild. In: E. Klinghammer, ed. *The Behavior and Ecology of Wolves,* pp. 307–315. New York: Garland STPM Press.

Forman, R. T., and Gordon, M. 1986. *Landscape Ecology.* New York: John Wiley & Sons.

Forrest, S. C., Clark, T. W., Richardson, L., and Campbell, T. M., III. 1985. Black-footed ferret habitat: Some management and reintroduction considerations. *Wyoming BLM Mgmt. Wildl. Techn. Bull.* 2:1–49.

Frankel, O. H., and Soulé, M. E. 1981. *Conservation and Evolution.* New York: Cambridge Univ. Press.

Frey, H., and Walter, W. 1989. The reintroduction of the bearded vulture *Gypaetus barbatus* into the Alps. In: B. U. Meyburg and R. D. Chancellor, eds. *Raptors in the Modern World,* pp. 341–344. London: World Working Group on Birds of Prey and Owls.

Fritts, S. H., Paul, W. J., and Mech, L. D. 1984. Movements of translocated wolves in Minnesota. *J. Wildl. Mgmt.* 48:709–721.

Gilpin, M. E. 1987. Spatial structure and population vulnerability. In: M. E. Soulé, ed., *Viable Populations for Conservation*, pp. 125–139. New York: Cambridge Univ. Press.

Gittleman, J. L., and Pimm, S. L. 1991. Crying wolf in North America. *Nature* 351:524–525.

Glass, R. J., and More, T. A. 1990. Public attitudes, politics, and extramarket values for reintroduced wildlife: Examples from New England. *Trans. N. A. Wildl. Nat. Res. Conf.* 55:548–557.

Goodman, D. 1987. The demography of chance extinction. In: M. E. Soulé, ed. *Viable Populations for Conservation*, pp. 11–34. New York: Cambridge Univ. Press.

Gordon, J. R. 1983. *A Diagnostic Approach to Organizational Behavior*. Boston, Mass.: Allyn and Bacon.

Gossow, H., and Honsig-Erlenburg, P. 1986. Management problems with reintroduced lynx in Austria. In: S. D. Miller and D. D. Everett, eds. *Cats of the World: Biology, Conservation and Management*, pp. 77–83. Washington, D.C.: National Wildlife Federation.

Greater Yellowstone Coalition. 1990. GYC's first 1990 program progress report. Bozeman, Mont.: Greater Yellowstone Coalition.

Griffith, B., Scott, J. M., Carpenter, J. W., and Reed, C., 1989. Translocation as a species conservation tool: Status and strategy. *Science* 245:447–480.

Hanski, I., and Gilpin, M. 1991. Metapopulation dynamics: Brief history and conceptual domain. *Biol. J. Linnean Soc.* 42:3–16.

Harris, R. H., Clark, T. W., and Shaffer, M. L. 1989. Extinction probabilities for isolated black-footed ferret populations. In: U. S. Seal, E. T. Thorne, M. A. Bogan, and S. H. Anderson, eds. *Conservation Biology and the Black-footed Ferret*, pp. 69–82. New Haven, Conn.: Yale Univ. Press.

Henshaw, R. E., Lockwood, R., Shideler, R., and Stephenson, R. O. 1979. Experimental release of captive wolves. In: E. Klinghammer, ed. *The Behavior and Ecology of Wolves*, pp. 319–345. New York: Garland STPM Press.

Holloway, C. W., and Jungius, H. 1973. Reintroduction of certain mammal and bird species into the Gran Paradiso National Park. *Zool. Anzeiger* 191:1–44.

Hook, R. A., and Robinson, W. L. 1982. Attitudes of Michigan citizens toward predators. In: F. H. Harrington and P. C. Paquet, eds. *Wolves of the World: Perspectives of Behavior, Ecology, and Conservation*, pp. 382–394. Park Ridge, N.J.:

Hummel, M., and Pettigrew, S. 1991. *Wild Hunters: Predators in Peril*. Toronto: Key Porter Books.

Jameson, R. J., Kenyon, K. W., Jefferies, S., and VanBlaricom, G. R. 1986. Status of a translocated sea otter population and its habitat in Washington. *Murrelet* 67:84–87.

Jameson, R. J., Kenyon, K. W., Johnson, A. M., and Wright, H. M. 1982. History and status of translocated sea otter populations in North America. *Wildl. Soc. Bull.* 10:100–107.

Janis, I. L. 1972. *Victims of Groupthink: A Psychological Study of Foreign-Policy Decisions and Fiascoes*. Boston, Mass.: Houghton Mifflin.

Jefferies, D. J., Wayre, P., Jessop, R. M., and Mitchell-Jones, A. J. 1986. Reinforcing the native otter *Lutra lutra* population in East Anglia: An analysis of the behavior and range development of the first release group. *Mamm. Rev.* 16:65–79.

Jones, S. R. 1990. *Captive Propagation and Reintroduction: A Strategy for Preserving Endangered Species?* Endang. Species Update Spec. Issue: Ann Arbor: University of Michigan School of Natural Resources.

Kellert, S. R. 1980. Contemporary values of wildlife in American society. In: W. W. Shaw and E. H. Zube, eds. *Wildlife Values: Center for Assessment of Noncommodity Natural Resource Values*, pp. 31–60. Institute Ser. Report Number 1.

Kellert, S. R. 1985a. Social and perceptual factors in endangered species management. *J. Wildl. Mgmt.* 49:528–536.

Kellert, S. R. 1985b. Public perceptions of predators, particularly the wolf and coyote. *Biol. Conserv.* 31(2):167–189.

Kellert, S. R. 1986. The public and the timber wolf. *Trans. N. A. Wildl. Nat. Res. Conf.* 51:193–200.

Kellert, S. R. 1987. The contributions of wildlife to human quality of life. In: D. J. Decker and G. R. Goff, eds. *Valuing Wildlife: Economic and Social Perspectives*, pp. 22–229. Boulder, Colo.: Westview Press.

Kellert, S. R. 1991. Public views of wolf restoration in Michigan. *Trans. N. A. Wildl. Nat. Res. Conf.* 56:152–161.

Kellert, S. R., in press. Public attitudes toward bears and their conservation. In: J. J. Clear, C. Servheen, and L. J. Lyon, eds. *Proceedings of the Ninth International Bear Conference,*

Kellert, S. R., and Clark, T. W. 1991. The theory and application of a wildlife policy framework. In: W. R. Mangun and S. S. Nagel, eds. *Public Policy Issues in Wildlife Management*, pp. 17–36. New York: Greenwood Press.

Kidder, C. 1992. Return of the red wolf. *Nature Conservancy* Sept./Oct.:10–15.

Kleiman, D. G. 1989. Reintroduction of captive mammals for conservation: Guidelines for reintroducing endangered species into the wild. *Bioscience* 39:152–161.

Kleiman, D. G., Beck, B. B., Dietz, J. M., Dietz, L. A., Ballou, J. D., and Coimbra-Filho, A. F. 1986. Conservation program for the golden lion tamarin: Captive research and management, ecological studies, educational strategies, and reintroduction. In: K. Benirschke, ed. *Primates: The Road to Self-Sustaining Populations*, pp. 959–979. New York: Springer-Verlag.

Knight, R. R., Blanchard, B. M., and Eberhardt, L. L. 1988. Mortality patterns and population sinks for Yellowstone grizzly bears, 1973–1985. *Wildl. Soc. Bull.* 16:121–125.

Lacy, R. C. 1987. Loss of genetic diversity from managed populations: Interacting effects of drift, mutation, immigration, selection, and population subdivision. *Conserv. Biol.* 1:143–158.

Lacy, R. C., and Clark, T. W. 1989. Genetic variability in black-footed ferret populations: Past, present, and future. In: U. S. Seal, E. T. Thorne, M. A. Bogan, and S. H. Anderson, eds. *Conservation Biology and the Black-footed Ferret*, pp. 83–103. New Haven, Conn.: Yale Univ. Press.

Lacy, R. C., and Clark, T. W. 1993. Simulation modeling of American marten populations: Vulnerability to extinction. *Great Basin Nat.* 53:282–292.

Lande, R. 1988. Genetics and demography in biological conservation. *Science* 241:1455–1460.

Lande, R., and Barrowclough, G. F. 1987. Effective population size, genetic variation, and their use in population management. In: M. E. Soulé, ed. *Viable Populations for Conservation*, pp. 87–123. New York: Cambridge Univ. Press.

MacArthur, K. L. 1981. Factors contributing to effectiveness of black bear transplants. *J. Wildl. Mgmt.* 45:102–110.

MacArthur, R. H., and Wilson, E. O. 1967. *The Theory of Island Biogeography*. Princeton, N.J.: Princeton Univ. Press.

March, J. G., and Simon, H. 1958. *Organizations*. New York: John Wiley & Sons.

Mattson, D. J. 1990. Human impacts on bear habitat use. *Internat. Conf. Bear Res. and Mgmt.* 8:35–56.

May, R. M. 1986. The cautionary tale of the black-footed ferret. *Nature* 320:13–14.

Mech, L. D. 1979. Some considerations in re-establishing wolves in the wild. In: E.

Klinghammer, ed. *The Behavior and Ecology of Wolves*, pp. 445–457. New York: Garland STPM Press.

Mech, L. D., Fritts, S. H., Radde, G. L., and Paul, W. J. 1988. Wolf distribution and road density in Minnesota. *Wildl. Soc. Bull.* 16:85–87.

Meyer, M. 1968. Two authority structures of bureaucratic organizations. *Admin. Sci. Quart.* 13:211–228.

Miller, B. J. 1992. History of black-footed ferret captive propagation. Unpublished presentation manuscript. Providence, R.I.: University of Rhode Island.

Miller, B. J., Anderson, S. H., DonCarlos, M. W., and Thorne, E. T. 1988. Biology of the endangered black-footed ferret and the role of captive propagation in its conservation. *Canadian J. Zool.* 66:765–773.

Miller, B. J., Biggins, D., Hanebury, L., and Vargas, A. 1994. Reintroduction of the black-footed ferret (*Mustela nigripes*). In: P. T. S. Olney, G. M. Mace, and A. T. C. Feistner, eds. *Creative Conservation: Interactive Management of Wild and Captive Animals*, pp. 455–464. London: Chapman & Hall.

Miller, B. J., Biggins, D., Wemmer, C., Powell, R., Calvo, L., Hanebury, L., Horn, D., and Wharton, T. 1990b. Development of survival skills in captive-raised Siberian polecats (*Mustela eversmanni*) II: Predator avoidance. *J. Ethol.* 8:95–104.

Miller, B. J., Biggins, D., Wemmer, C., Powell, R., Hanebury, L., Horn, D., and Vargas, A. 1990a. Development of survival skills in captive-raised Siberian polecats (*Mustela eversmanni*) I: Locating prey. *J. Ethol.* 8:89–94.

Miller, B. J., Reading, R. P., and Forrest, S. R., in press. *In the Dying Light: Recovery of the Black-footed Ferret.* Washington, D.C.: Smithsonian Institution Press.

Miller, B. J., Wemmer, C., Biggins, D. E., and Reading, R. P. 1990c. A proposal to conserve black-footed ferrets and the prairie dog ecosystem. *Environ. Mgmt.* 14:763–769.

Moore, D. 1992. Re-establishing large predators. *Re-introduction News* 4:6.

Moore, D. E., III, and Smith, R. 1991. The red wolf as a model for carnivore reintroductions. In: J. H. W. Gipps, ed. *Beyond Captive Breeding: Re-introducing Endangered Mammals to the Wild*, pp. 263–278. Oxford: Clarendon Press.

Nowak, R. M. 1979. North American Quaternary *Canis*. *Univ. Kansas Mus. Nat. Hist. Mon.* 61:1–154.

Oakleaf, B., Luce, B., Thorne, E. T., and Biggins, D. 1991. Black-footed ferret reintroduction in Wyoming: Project description and 1991 protocol. Cheyenne, Wyo.: Wyoming Game and Fish Department, U.S. Bureau of Land Management, and U.S. Fish and Wildlife Service.

O'Brien, S. J., Wildt, D. E., Goldman, D., Merril, C. R., and Bush, M. 1983. The cheetah is depauperate in genetic variation. *Science* 221:459–462.

Ounsted, M. L. 1991. Re-introducing birds: Lessons to be learned for mammals. In: J. H. W. Gipps, ed. *Beyond Captive Breeding: Re-introducing Endangered Mammals to the Wild*, pp. 75–85. Oxford: Clarendon Press.

Panwar, H. S., and Rodgers, W. A. 1986. The re-introduction of large cats into wildlife protected areas. *Indian Forester* 112:939–944.

Parker, W. T., and Phillips, M. K. 1991. Application of the experimental population designation to recovery of endangered red wolves. *Wildl. Soc. Bull.* 19:73–79.

Peek, J. M., Brown, D. E., Mech, L. D., Shaw, J. H., and Van Ballenberghe, V. 1991. Restoration of wolves in North America. *Wildl. Soc. Tech. Rev.* 91–1.

Perrow, C. 1986. *Complex Organizations: A Critical Essay*, 3rd ed. New York: McGraw–Hill.

Phillips, M. K. 1990. Red wolf restoration is first-ever reintroduction of wild-extinct carnivore (North Carolina). *Restoration and Mgmt. Notes* 8(2):131.

Phillips, M. K. 1992. Red wolves (*Canis rufus*) in N. Carolina. *Re-introduction News* 4:7–8.

Phillips, M. K., and Parker, W. T. 1988. Red wolf recovery: A progress report. *Conserv. Biol.* 2:139–141.

Reading, R. P. 1993. Toward an Endangered Species Reintroduction Paradigm: A Case Study of the Black-footed Ferret. Ph.D. dissert., Yale University, New Haven, Conn.

Reading, R. P., Clark, T. W., and Kellert, S. R. 1991. Toward an endangered species reintroduction paradigm. *Endang. Species Update* 8(11):1–4.

Reading, R. P., and Kellert, S. R. 1993. Attitudes toward a proposed reintroduction of black-footed ferrets (*Mustela nigripes*). *Conserv. Biol.* 7:569–580.

Reading, R. P., and Miller, B. J. 1994. The black-footed ferret recovery program: Unmasking professional and organizational weaknesses. In: T. W. Clark, R. P. Reading, and A. L. Clarke, eds. *Endangered Species Recovery: Finding the Lessons, Improving the Process*, pp. 73–100. Washington, D.C.: Island Press.

Reese, M. D. 1989. Red wolf recovery effort intensifies. *Endang. Species Techn. Bull.* 14(1–2):3.

Rokeach, M. 1972. *Beliefs, Attitudes, and Values: A Theory of Organization and Change.* San Francisco, Calif.: Jossey-Bass.

Rokeach, M. 1979. Introduction. In: M. Rokeach, ed. *Understanding Human Values: Individual and Societal*, pp. 1–11. New York: The Free Press.

Rolston, H., III. 1981. Values in nature. *Environ. Ethics* 3(2):113–128.

Rowe-Rowe, D. T. 1992. *The Carnivores of Natal.* Pietermaritzburg, South Africa: Natal Parks Board.

Sale, J. B. 1986. Reintroduction in Indian wildlife management. *Indian Forester* 112:867–873.

Saunder, R. T. 1984. San Nicolas: Island of hope for the California sea otter. *The Otter Raft* Summer:4–7.

Schwartz, D. M. 1989. The comedian of America's wetlands stages a comeback. *Smithsonian* 20(3):138–149.

Scott, J. M., and Carpenter, J. W. 1987. Release of captive-reared or translocated endangered birds: What do we need to know? *Auk* 104:540–545.

Seal, U. S., Thorne, E. T., Bogan, M. A., and Anderson, S. H., eds. 1989. *Conservation Biology and the Black-footed Ferret.* New Haven, Conn.: Yale Univ. Press.

Serfass, T. L., and Rymon, L. M. 1985. Success of river otter introduced in Pine Creek drainage in northcentral Pennsylvania. *Trans. Northeastern Sect. Wildl. Soc.* 41:138–149.

Shaffer, M. L. 1987. Minimum viable populations: Coping with uncertainty. In: M. E. Soulé, ed. *Viable Populations for Conservation*, pp. 69–86. New York: Cambridge Univ. Press.

Sharps, J. C., and Whitcher, M. F. 1982. Swift fox reintroduction techniques. Unpublished report. Rapid City, S.D.: South Dakota Department of Game, Fish, and Parks.

Short, J., Bradshaw, S. D., Giles, J., Prince, R. I. T., and Wilson, G. R. 1992. Reintroduction of macropods (Marsupialia: Macropodoidea) in Australia—A review. *Biol. Conserv.* 61:189–204.

Sills, D. L. 1957. *The Volunteers.* New York: Free Press.

Simberloff, D. 1988. The contribution of population and community biology to conservation science. *Ann. Rev. Ecol. Syst.* 19:473–511.

Sinden, J., and Worrell, A. 1979. *Unpriced Values: Decisions without Market Values.* New York: John Wiley & Sons.

Singer, F. J. 1991. Some predictions concerning a wolf recovery into Yellowstone

National Park: How wolf recovery may affect park visitors, ungulates, and other predators. *Trans. N. A. Wildl. and Nat. Res. Conf.* 56:567–583.

Skinner, J. D., and van Arde, R. J. 1987. Range use by brown hyaenas *Hyaena brunnea* relocated in an agricultural area of the Transvaal. *J. Zool. (London)* 212:350–352.

Snyder, N. F. R., and Snyder, H. A. 1989. Biology and conservation of the California condor. *Current Ornith. Res.* 6:175–267.

Soulé, M. E. 1980. Thresholds for survival: Maintaining fitness and evolutionary potential. In: M. E. Soulé and B. A. Wilcox, eds. *Conservation Biology: An Evolutionary-Ecological Perspective*, pp. 151–169. Sunderland, Mass.: Sinauer Associates.

Stanley-Price, M. R. 1989. *Animal Re-introductions: The Arabian Oryx in Oman.* New York: Cambridge Univ. Press.

Steinhoff, H. W. 1980. Analysis of major conceptual systems for understanding and measuring wildlife values. In: W. W. Shaw and E. H. Zube, eds. *Wildlife Values*, pp. 11–21. Center for Assessment of Noncommodity Natural Resource Values, Inst. Ser. Rept. No. 1.

Taylor, R. E. 1986. Plan to give wolves new home may be a howling success. *Endang. Species Techn. Bull. Reprint* 3(8):1–2.

Tear, T. H., and Forester, D. 1992. Role of social theory in reintroduction planning: A case study of the Arabian oryx in Oman. *Soc. Nat. Res.* 5:359–374.

Templeton, A. R. 1990. The role of genetics in captive breeding and reintroduction for species conservation. *Endang. Species Update* 8(1):14–17.

Tessler, A., and Shaffer, D. R. 1990. Attitudes and attitude change. *Ann. Rev. Psychol.* 41:479–523.

Thompson, J. G. 1993. Addressing the human dimensions of wolf reintroduction: An example using estimates of livestock depredation and costs of compensation. *Soc. Nat. Res.* 6:165–179.

Thorne, E. T., Kwiatkowski, D. R., and Oakleaf, B. 1993. Management and background of black-footed ferrets selected for reintroduction in 1992. In: *1992 Annual Completion Report, April 15, 1992–April 15, 1993: Black-footed Ferret Reintroduction, Shirley Basin, Wyoming*, pp. 1–25. Cheyenne: Wyoming Game and Fish Department.

Thorne, E. T., and Oakleaf, B. 1991. Species rescue for captive breeding: Black-footed ferret as an example. In: J. H. W. Gipps, ed. *Beyond Captive Breeding: Re-introducing Endangered Mammals to the Wild*, pp. 241–261. Oxford: Clarendon Press.

Thorne, E. T., and Williams, E. 1988. Diseases and endangered species: The black-footed ferret as a recent example. *Conserv. Biol.* 2:66–73.

Tichy, N. M. 1983. *Managing Strategic Change: Technical, Political, and Cultural Dynamics.* New York: John Wiley & Sons.

U.S. Fish and Wildlife Service. 1988. *Black-footed Ferret Recovery Plan.* Denver, Colo.: U.S. Fish and Wildlife Service.

van Arde, R. J., and Skinner, J. D. 1986. Pattern of space use by relocated *Felis serval. South African J. Ecol.* 24:97–101.

Wargo, J. P. 1984. *Ecosystem Preservation Policy.* Unpublished Ph.D. dissert., Yale University.

Warren, R. J., Conroy, M. J., James, W. E., Baker, L. A., and Diefenbach, D. R. 1990. Reintroduction of bobcats on Cumberland Island, Georgia: A biopolitical lesson. *Trans. N. A. Wildl. Nat. Res. Conf.* 55:580–589.

Warwick, D. 1975. *A Theory of Public Bureaucracy: Politics, Personality, and Organization.* Cambridge, Mass.: Harvard Univ. Press.

Weber, M., 1968. *Economy and Society: An Outline of Interpretive Sociology.* Vol. I. New York: Bedminster Press.

Weckwerth, R. P., and Wright, P. L. 1968. Results of transplanting fishers in Montana. *J. Wildl. Mgmt.* 32:977–980.

Weimer, D. L., and Vining, A. R. 1989. *Policy Analysis: Concepts and Practice.* Englewood Cliffs, N.J.: Prentice-Hall.

Weinberg, D. 1986. Decline and fall of the black-footed ferret. *Nat. Hist.* 2/86:63–69.

Weinstein, D. 1984. Bureaucratic opposition: Whistle-blowing and other tactics. In: R. Westrum and K. Samaha, eds. *Complex Organizations: Growth, Struggle, and Change,* pp. 254–268. Englewood Cliffs, N.J.: Prentice-Hall.

Weise, T.F., Robinson, W. L., Hook, R. A., and Mech, L. D. 1975. An experimental translocation of the eastern timber wolf. *Audubon Conserv. Rept.* 5:1–28.

Weise, T. F., Robinson, W. L., Hook, R. A., and Mech, L. D. 1979. Experimental release of captive wolves. In: E. Klinghammer, ed. *The Behavior and Ecology of Wolves,* pp. 346–419. New York: Garland STPM Press.

Westman, W. E. 1990. Managing for biodiversity: Unresolved science and policy questions. *Bioscience* 40:26–33.

Westrum, R. 1986. Management strategies and information failures. Unpublished paper delivered at NATO Advanced Research Workshop on "Failure analysis of information systems," August 1986, Bad Winsheim, Germany.

Westrum, R. 1988. Organizational and interorganizational thought. Unpublished paper delivered at World Bank Conference on "Safety Control and Risk Management," October 1988, Washington, D.C.

Wilcove, D. S. 1987. Recall to the wild: Wolf reintroduction in Europe and North America. *TREE* 2(6)146–147.

Wilcox, B. A. 1980. Insular ecology and conservation. In: M. E. Soulé and B. A. Wilcox, eds. *Conservation Biology: An Evolutionary-Ecological Perspective,* pp. 95–117. Sunderland, Mass.: Sinauer Associates.

Wiley, J. W., Snyder, N. F. R., and Gnam, R. S. 1992. Reintroduction as a conservation strategy for parrots. In: S. R. Beissinger and N. F. R. Snyder, eds. *New World Parrots in Crisis: Solutions from Conservation Biology,* pp. 165–200. Washington, D.C.: Smithsonian Institution Press.

Williams, R. M., Jr. 1979. Change and stability in values and value systems: A sociological perspective. In: M. Rokeach, ed. *Understanding Human Values: Individual and Societal,* pp. 15–46. New York: The Free Press.

Wilson, J. Q. 1980. *The Politics of Regulation.* New York: Harper.

Woodford, M. H., and Kock, R. A. 1991. Veterinary considerations in re-introduction and translocation projects. In: J. H. W. Gipps, ed. *Beyond Captive Breeding: Reintroducing Endangered Mammals to the Wild,* pp. 101–110. Oxford: Clarendon Press.

Wozencraft, W. C. 1989. Classification of recent Carnivora. In: J. L. Gittleman, ed. *Carnivore Behavior, Ecology, and Evolution,* pp. 569–593. Ithaca, N.Y.: Cornell Univ. Press.

Yaffee, S. L. 1982. *Prohibitive Policy: Implementing the Federal Endangered Species Act.* Cambridge, Mass.: MIT Press.

Yahner, R. H. 1988. Changes in wildlife communities near edges. *Conserv. Biol.* 2:333–339.

Yalden, D. W. 1986. Opportunities for reintroducing British mammals. *Mamm. Rev.* 16:53–63.

Yalden, D. W. 1993. The problems of reintroducing carnivores. *Symp. Zool. Soc. Lond.* 65:289–306.

CHAPTER 10

Human/Carnivore Interactions: Conservation and Management Implications from China

KENNETH G. JOHNSON, YIN YAO, CHENGXIA YOU,
SENSHAN YANG, AND ZHONGMIN SHEN

Human/carnivore interactions are loosely classified into two primary and four secondary categories. Primary interactions include (1) human impacts, mainly from habitat destruction and killing, and (2) human programs, such as habitat protection and natural-resource management intended to balance ecological and socioeconomic needs. Secondary interactions include (1) the projection of human attitudes onto carnivores, (2) the incorporation of carnivores into human culture, (3) depredations by carnivores, and (4) human conflicts and conservation politics. Our chapter will focus on the primary interactions and make specific suggestions for addressing the conservation and management problems inherent in these interactions. The four secondary interactions will also be briefly discussed, because these factors usually interplay within the complex arena of carnivore conservation. Our approach was based on an interdisciplinary evaluation of a cooperative research and conservation project on giant pandas (*Ailuropoda melanoleuca*) and other carnivores in China. Pandas and other large carnivores generate exceptional human interest and are ideal species for the investigation of human/carnivore interactions and their conservation and management implications.

Conservation Programs and the CMR

Giant pandas and several sympatric carnivores are threatened because a rapidly expanding human population has reduced their distribution to six remote, rugged mountain ranges in the Central Mountain Region (CMR) of China. The CMR, which straddles subtropical lowlands and temperate highlands, harbors an extraordinary diversity of habitats and species, because the oriental and palearctic realms overlap these complexly zoned mountains. Species of oriental origin include golden monkeys (*Rhinopithecus roxellanae*), stump-tailed macaques (*Macaca thibetana*), takin (*Budorcas taxicolor*), tufted deer (*Elaphodus cephalophus*), serow (*Capricornis sumatraensis*), golden pheasants (*Chrysolophus pictus*), and clouded leopards (*Neofelis nebulosa*). Palearctic species include pika (*Ochotona thibetana*), musk deer (*Moschus berezovski*), snow cock (*Tetraogallus tibetanus*), steppe cat, or Pallas's cat (*Felis manul*), and snow leopards (*Panthera uncia*), to mention a few of the better-known species. Sympatric interactions have been investigated in the field among giant pandas and red pandas (*Ailurus fulgens:* Johnson et al. 1988a), Asiatic black bears (*Ursus thibetanus:* Schaller et al. 1989), and leopards (*Panthera pardus:* Johnson et al. 1993). Most other carnivore communities face similar threats to their continued existence and, thus, their conservation and management face similar implications. Many animals are protected by Chinese law, and most of those protected are included in international listings of threatened and endangered species.

Giant pandas, a national treasure in China and a worldwide symbol for the conservation of endangered species, have served effectively to draw attention to serious ecological problems. Since their introduction to the Western world by a French missionary in 1869, pandas have continually increased in cultural, socioeconomic, educational, scientific, and other values. The use of pandas as ambassadors of friendship can be traced back to the Tang Dynasty (A.D. 618–907), and they still play an important role in diplomacy today. Two giant pandas were given to the United States as the highest gesture of friendship when diplomatic relations were normalized in 1972. Since then, 17 pandas have been given to zoos in eight countries. In 1986, permanent gifts were discontinued, and attempts were made to address the insatiable world demand for pandas with exhibition loans.

As a biological enigma, giant pandas have continually tested our understanding of evolutionary mechanisms, phylogenetic relationships, taxonomic classifications, and behavioral ecology (Davis 1964; Gould 1980; O'Brien et al. 1985; Schaller et al. 1985, 1989; Johnson et al. 1988a, b; Gittleman 1994). Many of the biological characteristics of pandas, such as their population ecology and habitat fragmentation, represent extremes, and therefore are also instructive for the conservation and management of wildlife and natural resources in other areas. Pandas and large carnivores in general are sensitive ecological and political indicator species, because of their specialized behaviors, spatial needs, diverse habitat requirements, sparse numbers, high potential for human conflicts, and the intricate political situations usually surrounding their conservation and management. Large carnivores are typically the first ecosystem components to show the effects of excessive human impact, and their popularity can often rally support for conservation organizations and resource-management programs.

Since 1980, the World Wide Fund for Nature (WWF–International) and China have been conducting a giant panda research and conservation project. The project has six main goals: (1) to raise funds, (2) to establish a captive-breeding program, (3) to study behavioral ecology in the wild so as to learn basic habitat needs, (4) to survey giant pandas, forest, bamboo, and socio-economic factors throughout the remaining habitat, (5) to write conservation plans, and (6) to train nature-reserve personnel.

Sufficient funds have been raised to build a U.S. $1.5-million captive-breeding and research center in Wolong Reserve, conduct several international exchanges in captive technology, and make plans to release captive-bred pandas into the wild (Mainka and Qiu 1992). Results of field research show that pandas generally need large, interconnected areas of relatively undisturbed forest supporting at least two bamboo species, and that habitat destruction, poaching, and removal for zoos and captive breeding are their major threats to survival; neither the natural cycle of bamboo die-offs nor reproductive deficiencies are serious threats to giant pandas (Schaller et al. 1985; Johnson et al. 1988a, b; Pan et al. 1988; Schaller et al. 1989; Reid et al. 1989; Schaller 1993; Johnson, in prep.). Survey results indicate that current panda populations and habitat are sufficient that the management strategy employed should be the protection and management of wild populations and natural habitat throughout the CMR (Johnson, in prep.). A draft conservation plan has been published by WWF–International (MacKinnon et al. 1989); the document will undoubtedly serve as a successful academic and training exercise and publicity and fund-raising tool, but, as written, its implementation in China is unlikely.

Another effort, entitled "China's National Conservation Project for the Giant Panda and Its Habitat," has subsequently been developed by the Ministry of Forestry according to procedures acceptable within China, and was approved by State Council and the State Planning Commission in June 1992. Several joint training courses have been given in China, one study tour was conducted in Southeast Asia, and one Chinese student has received a graduate degree in natural-resource management in the United States.

Although progress has been made in panda conservation, the need to protect natural resources has become more urgent in the last few years as the actual problems and management implications have become clear. Our objectives here are (1) to identify and discuss major ecologically related problems and potential solutions and (2) to suggest an integrated and adaptive resource-management system to facilitate wildlife protection and ecological and socio-economic sustainability in the CMR of China. Many of the human/carnivore interactions and associated conservation and management implications discussed here may serve as an outline for evaluating problems and developing resource-management programs in other areas.

Major Problems

Major ecologically related problems were ranked, from more general to more specific and from national to local level, in order to reflect their interrelated nature and the administrative hierarchy that is necessary to address them realistically. The ranking also represented the relative importance and scope of the various problems that must be resolved.

Political and Policy Considerations

Political and policy issues are of foremost importance in China and should be given primary consideration when beginning conservation projects in any country. Social unrest, which peaked in Beijing on 4 June 1989, likely affects conservation and resource management in China by lowering priorities and funding for domestic programs and yielding reduced or erratic economic production, financial investment, foreign cooperation in science and technology, and tourist revenues, and an increase in the national debt. The long-term implications for China's political stability and policy consistency are unknown (see Spence 1990). Social unrest, political instability, and policy inconsistency relating to resource management are constant problems, particularly in developing countries, and such conditions must be dealt with realistically if effective international conservation programs are to be conducted.

Chinese accomplishments in protecting the nation's environment, wildlife, and natural resources include a system of at least 420 nature reserves embracing a variety of habitats and covering about 4.5% (432.8 million ha) of China's vast landscape (Jia 1992). By comparison, about 22% of the current panda distribution (14 reserves) is under reserve designation (Johnson, in prep.), and the new Chinese management project calls for the establishment of 14 additional reserves, demonstrating the strong commitment of the Chinese government to the protection and management of pandas and China's diverse habitat community. Among the 420 reserves already established in China, 30 have been designated as national priorities, and, of these 30, four have been included in the UN's Man and the Biosphere reserve network. Of the 14 reserves currently in panda habitat, three are national reserves and one (Wolong) has received recognition by the UN as an International Biosphere Reserve. Each reserve in China is designated as one of three types: forest ecosystem, wildlife, and natural or historical site (Zhu 1989).

In terms of policy and legislation for nature conservation in China, there are 12 major acts, regulations, and government orders in place, ranging from the decree in 1962 from State Council, for maintaining and rationally utilizing wildlife, to the Environmental Protection Law and the Law of Wild Animal Protection, both adopted in 1989. These provisions form a legal foundation that is generally stronger than that of the U.S., primarily because of constitutional guarantees in China (see Proceedings of the Sino-American Conference on Environmental Law 1987).

Although substantial progress has been made in the development of policy and legislation for nature conservation in China, many problems remain, especially regarding effective implementation and enforcement. One important problem is that the administrative systems, whether in the central government or in the provincial and local governments, that are in charge of implementing these acts and regulations are not as effective as they could be, because the acts are relatively recent and their effective implementation and enforcement are still developing, under rather severe resource constraints.

Another important problem is that many people whose behaviors have direct or indirect impacts on natural resources more or less lack confidence in government policy. The success of natural-resource management depends largely on the cooperation of the people who live in the areas, or whose activities affect natural resources. But frequent changes in government policies, such as the Great Leap Forward and the Cultural Revolution, have left the general public with negative impressions. Such a history of inconsistent policies presents a major long-term problem, and places ecology, environmental protection, and resource management into their proper perspective as important and interrelated matters of national security.

Population Growth

Rapid population growth is a historic and continuing problem in China and many other countries. China is addressing population problems through a rigorous, visionary National Family Planning Policy (NFPP) that limits families to one child and ties personal income levels to progress in achieving population goals (Figure 10.1). The NFPP of China has been hailed internationally as an appropriate example for other countries, and China has played a prominent role in several international conferences on solutions to world population problems (see Green 1988; Russell 1989).

Incomplete evaluations of China's NFPP indicate a reduced rate of increase, especially in the cities. Less success, however, has been achieved in rural areas and particularly in remote areas dominated by minorities. The initial population goals were about 1.5 billion people around 2050, with an ultimate goal of steady reductions to an optimal level of 750 million, but the largest age class of the current population consists of those born during the 1960s when the whole country was pursuing its population policy known as "more people makes more strength." This cohort of the population is, and will be through the next several years, at its reproductive peak. The national population increase in the last few years has already broken the initial and revised projections, continually raising the goal of national population growth by the year 2000 from 1.2 to 1.3 billion. Even the new goal is uncertain, with the most pessimistic predictions suggesting about 1.5 billion people by the year 2000.

People in many reserves consist primarily of Tibetan, Yi, or Qiong minorities. Because they are exempt from the one-child family plan, local populations could grow unchecked in the future. Requirements of food and shelter can be expected to rise accordingly. Unless other solutions are provided to

Figure 10.1. A billboard using the positive image of a baby giant panda to promote the one-child National Family Planning Policy.

meet their needs, their struggle for basic subsistence will make reserve regulations very difficult to implement. The population of Wolong Reserve, for example, has increased from approximately 3600 to 4000 in just 5 years. A new system of modified quotas for minorities is currently being implemented on an experimental basis. Results of the NFPP will be a dominant factor in China's future.

Ecological Sustainability

Ecological sustainability is difficult to assess, but China appears to be close to the carrying capacity of her land and resource base, as is perhaps indicated by the serious famines of as recently as the 1960s, recent warnings about possible grain shortages, the westward movement of people into marginal lands, and the use and even outright destruction of natural resources faster than their production or restoration. Currently, China has about 1.2 billion people, or approximately 25% of the world's population, but only 7% of the world's arable land and 3% of the world's forest. These figures clearly illustrate the severity of present resource constraints. For comparison, the U.S. has about 5% of the world's population, 14% of the world's arable land, and 7% of the world's forest (United Nations 1991; U.S. Dept. of Agriculture 1991).

Present ecological evaluations indicate that only about 750 million people, half the projected peak, can be supported optimally within China's projected resource capacity. Ecological sustainability has been integrated with economic considerations into the National Family Planning Policy of China.

Economic Sustainability

Economic sustainability seems feasible for China because an annual increase of about 10% in gross national product (GNP) has been maintained over the past 10 years, and a measure of prosperity exists in the agricultural sectors, which have exported food to Africa in recent years. Free-market incentives seem to be successfully priming the socialism-based system, but the economy is probably close to maximum production, and inflation, which was essentially nonexistent in the previously controlled economy, has increased to around 20% per year. GNP is a poor and incomplete measure of economic production, but as a convenience is still used for broad comparisons. Economic sustainability in forest resources is at present doubtful, owing to excessive, long-term degradation, and will likely be possible only through massive injections of capital into the restoration of forest ecosystems. Ecological and economic sustainability are difficult to project in the long term and are closely related to political considerations and population growth; and the underlying support for economic and ecological sustainability is uncertain (see Smil 1993).

For China, as a developed but still developing country with a history spanning several thousand years, economic improvement is of crucial importance, and serious resource constraints are a reality. In the past 10 years, "economic reforms" and "open-door" policies have brought many changes and improvements in China, including a high-percentage increase in GNP, greater foreign investment, and substantial increases in living standards, especially in rural areas. Still, ideological barriers confront the appropriate goals of "economic reform," and the local people lack psychological preparation for responding to the dramatic changes to come. In 1988, for example, China had an 11-percent increase in real GNP, but also experienced a 27-percent annual rate of inflation (Davidson 1989). The chronic confrontation between "market-oriented policy" and "planning-oriented policy" will continue to be intense. Planning-oriented policy, which has incorporated a series of retrenchment measures for more centralized control, has apparently prevailed recently. Although the influences of these measures on long-term economic production cannot be assessed accurately at the moment, China is certainly in a critical period of economic transition (see Davidson 1989; Russell 1989).

The current economic situation thus has many implications for nature conservation and environmental protection. In China, the idea of protecting natural resources and the environment even while accelerating the pace of economic production has been recognized as a basic national policy since 1983 (Qu

1987). Issues of nature conservation and environmental protection will likely remain in the headlines, but strong financial investments in these areas may be unrealistic, because priorities will probably be given to economic investments.

Inadequate planning and poor management of natural resources have seriously deteriorated China's ecological systems. Even though policies unfavorable to natural resources, such as the Great Leap Forward and the Cultural Revolution, have been of relatively brief duration, they have nonetheless had serious long-term impacts on natural resources and human attitudes. Thus, the principle that long-term economic sustainability rests in ecological sustainability is of fundamental importance in China as it is elsewhere.

Endangered Species and Ecological Decline

Increasing numbers of endangered species, reduced biotic diversity, and ecological decline are especially acute problems in China, owing to loss of natural habitat from centuries of intense land use, resource depletion, and environmental degradation. Pandas, perhaps the world's best-known endangered species, have served effectively to call attention to serious ecological problems in a region that is unparalleled in the temperate zone for biological diversity (Figure 10.2). Protecting pandas in the wild also ensures the continued existence of

Figure 10.2. The giant panda Zhen Zhen eating arrow bamboo in Wolong Reserve.

thousands of other valuable animals, plants, other resources, and essential ecological and hydrological processes, and enhances the prospects for ecological and socioeconomic sustainability.

Pandas and large carnivores in general are sensitive ecological-indicator species because of their specialized behaviors, extensive spatial needs, diverse habitat requirements, sparse numbers, potential conflicts with humans, and need for interrelated management actions to assure their continued existence. Large carnivores, which are typically the first ecosystem components to show the effects of excessive human impact, can rally support for conservation organizations and resource-management programs. Pandas have served to call attention to interrelated ecological problems, and with the help of an energetic management initiative they may serve as management-indicator species for a sustained quality of life for the CMR as a whole.

Objective monitoring of large carnivores as management-indicator species is particularly relevant to problems such as habitat loss and fragmentation, minimum critical size of habitat, minimum viable population size, interactions between socioeconomic and ecological processes, and integrated and adaptive management of landscapes on regional scales. The status of large carnivores is becoming more precarious, and they will be one of the most difficult orders to conserve as human impacts on the environment intensify in the future (see Schaller, this volume). At the current state of the evolution and use of concepts such as endangered species, biotic diversity, sustainability, and integrated and adaptive management, an effort to consider all ecological and socioeconomic components on a regional or even a reserve scale would be impractical and unnecessary. But objective monitoring of indicator species and an integrated management and feedback system offer a workable alternative to traditional approaches that have focused more on individual components than on overall ecological and management processes.

Pandas and other large carnivores are also ideal political-indicator species, because of intense human interest and the special administrative, legal, and financial considerations necessary to assure their continued existence. Conservation programs may originate from the moral, aesthetic, and personal or professional interests of the people involved, but too often projects to save endangered species begin in developed countries as enthusiastic, short-sighted campaigns with publicity and fund-raising as their main motivation. Funding appeals and supposed actions concentrate on the symptoms that are palatable to the general public, and little effort is made to solve or even address the underlying ecological and socioeconomic problems. Conservation efforts rooted in such a shallow and self-serving approach quickly lose momentum, effectiveness, and therefore credibility, and usually come to be dominated by special interests, public-relations and fund-raising considerations, and conflicts among different agendas. Using endangered species as publicity and fund-raising symbols has allowed the vested interests and public image of conservation organizations and corporations to become more important than the actual results of conserving wildlife and natural resources (Figure 10.3).

Figure 10.3. The giant panda Hua Hua being carried through the Yingxinggou gorge on the first leg of his journey back to the wild.

Extinction is finality. The irretrievable loss of links in the chain of life can only lead to a dangerous downward spiral. The implications are larger than giant pandas, other endangered species, international symbolism, or conservation politics, and have come to represent the ecological health and economic sustainability of our entire life-support system—and the political will of human culture. In more practical terms, endangered species are tangible benchmarks for tracking the vulnerability of the other species in their natural habitat. If giant pandas and their habitat, so deep in the mountains of central China, cannot be conserved, what species, habitat, or community is secure?

Water Resources

Water is an especially essential resource tied directly to panda habitats, which constitute some of the most critical upper and middle reaches of the Yangtze River watershed. The Yangtze is of profound importance to China, and State Council is currently discussing a strategic plan to protect, reforest, and manage the watershed. Forested watersheds provide dependable, high-quality water supplies for drinking, irrigation, transportation, hydropower, fisheries, and other uses, and responsible management of watersheds reduces threats to human life and crops from flooding, landslides, drought, and famine. But deforestation has led to serious erosion, which removes irreplaceable topsoils, reduces water quality and produces erratic flows, reduces the life of hydroprojects, including the Three-Gorges Dam, through siltation, and lowers potential revenues from tourist cruises through the popular Three-Gorges waterway. A significant portion of China's water originates in the CMR, and ecological and socioeconomic implications clearly extend far beyond the immediate vicinity of giant panda habitats (see Proceedings of a Workshop on Forest Hydrological Resources in China 1990).

Water is the lifeblood of China's agricultural and industrial economy, and flows along the Yangtze have direct implications for human health and safety, transportation, commerce, industry, hydropower, pollution, irrigation, and other hydrological problems and features, all the way from western China to Shanghai, China's most populous and most highly industrialized city. The impacts of water and air pollution on agricultural, industrial, and forest systems are poorly understood, but even slight changes—due to maximized production, inelastic demand, inflexible resource constraints, and biological sensitivities—could yield gravely detrimental effects (see Proceedings of the Symposium on Effects of Acid Precipitation on Comprehensive Agriculture and Its Countermeasures 1988).

Deforestation

In China, critical shortages of wood, energy sources, land area and especially arable land, and other resources have led to serious levels of deforestation. Rapid deforestation produces severe impacts on water resources, agriculture, endangered species, and ecological and socioeconomic sustainability. Deforestation and habitat fragmentation are characterized generally by a steady progression from roadbuilding, timber extraction, human settlement, firewood cutting, hunting, grazing, and farming to the complete conversion of large, mountainous areas to direct human utilization, even at high elevations and on steep slopes. Deforestation is a major cause of most conservation and management problems in the CMR, and is thus the logical place to begin a reversal of the synergistic trend toward ecological and socioeconomic collapse.

The subalpine forest in the Aba Zang Autonomous Prefecture of western Sichuan, where Wolong Reserve is located, has been disturbed or destroyed to such an extent that forest cover has dropped from 29.5% to 14.0% since 1956 (Yang 1985). What forest remained was cut apart into several isolated areas, without consideration for animal movements. Within forest clear-cutting sites, the ecological condition of the land changed drastically, across many dimensions. Daily temperature began to fluctuate widely. Bushes and shrubs colonized rapidly and formed dense root systems (2.8 times more dense than that within forest) that were unfavorable for the regeneration of forest (Yang 1985). The most critical aspect of the problem is hydrology: surface runoff from the new, lower vegetation became more than double that in forest, and the evaporation rate was 3.8 times greater. The CMR's water-supply capacity was greatly weakened (Yang 1985).

Lugo (1985) has noted that forests in many regions of the world have not recovered from deforestation. Hundreds of years may be required to restore ecologically fragile regions to their original forest and productivity, if that will indeed be possible. With rugged mountainous topography and interspersion of transitional zones among flora and fauna, many parts of the CMR are inherently fragile, both physically and ecologically. Once forest ecosystems in those parts are altered, their restoration will be difficult, if possible at all. The impact already has serious negative consequences not only for pandas but also for many other species, water resources, and other ecological and socioeconomic interactions.

Deforestation occurs mainly from (1) conversion of forested land to agricultural use, under pressure of population increases, (2) high-impact, short-term commercial logging by organized government units that are required to achieve annual timber quotas under the policies of central, provincial, county, and township governments (most timber units now no longer have available timberlands in which to relocate), and (3) lower-impact, long-term cutting of firewood and selective timber by local people. More than 2 million m³ of wood per year may be used for cooking and heating in the mountainous regions of Sichuan (Liu and Lu 1985). If deforestation is to be reduced, alternative energy resources and alternative means of timber, agricultural, and other production must be found.

Socioeconomics

Political and socioeconomic realities dictate that the basic needs of local people be met. Conservation is largely a human-rights issue involving the management of people and natural resources, especially in the CMR. Most local people in the CMR are poor by national standards; most are minorities (mostly Tibetan, Yi, and Qiong), and as such receive special considerations from government, as addressed by constitutional law. For example, minorities are exempt from the National Family Planning Policy and can be relocated only on a voluntary basis (Figure 10.4).

Some 54 minorities account for approximately 8% of China's total population. The basic constitutional principle regarding minorities is that all people are equal, and that all should unite in the search for prosperity. The constitution also provides that the government give minorities financial, material, and technical help for local and regional development. The Minority Autonomous Region Act was issued by the People's Congress in 1984, and because most minorities are scattered in remote and undeveloped areas, the government has adopted many special policies to see that their needs are met in practice.

Figure 10.4. A child in the traditional dress of the Yi minority, in Mabian county, experiencing what is likely her first view of foreigners.

Although some parts of the CMR fall into minority autonomous areas, others do not, but many people living in the latter areas are nonetheless minorities. This pattern presents a sensitive, complex situation that must be considered carefully.

Most land in the CMR is too marginal and too steep for agriculture, most of the accessible forests have been removed, transportation is limited, and as currently utilized and managed these areas appear to fall below ecological and socioeconomic sustainability. Limited socioeconomic data indicate that local townships usually operate with large deficits and require substantial annual subsidies from county, provincial, and/or central governments, which in turn encourage further unprofitable expansion. Many people starved in this region in the 1950s and 1960s and again after the major earthquake and flooding of 1976. Far from human expansion, then, the trend should be toward gradual movement of people *away* from this region, for reasons of human welfare as well as healthier socioeconomics and ecological stability.

Potential Solutions and Suggested Strategies

Given continuing political stability, consistent policies, economic growth, and foreign investments, the prospects for integrated and adaptive resource management in the CMR appear favorable. A comprehensive, regional approach is necessary: because conservation and resource-management problems are interrelated within a complex political, socioeconomic, and biological arena, they cannot be solved independently. The protection of biotic integrity and the continuation of healthy ecological processes cannot be assured on a reserve or even county scale, and only long-term regional approaches will be viable. Those approaches, however, should not be allowed to detract from current efforts toward halting immediate ecological declines by fully protecting key habitat areas, eliminating poaching, and limiting other human intrusions. Regional approaches are viewed by some as idealistic, but the chances of success on the regional level can be enhanced by developing a functional management system in one key area as a pilot program and a basis for further refinement and expansion. Under conditions of high human population pressure, such as occur in China and many other countries, integrated and adaptive regional management may provide the only realistic approach to the balancing of long-term ecological and socioeconomic needs.

Sustained Funding and Effort

First, the necessary long-term funding must be secured from a cooperative and carefully coordinated fund-raising network, to ensure that near-term efforts are sustained until a suitable management system is in operation. The power of the panda's public appeal is probably sufficient to generate the necessary funds, provided a compatible financial relationship can evolve among (1) nongovernmental organizations and other fund-raising, conservation, and special-interest groups, (2) joint economic ventures within sustainable ecological guidelines, and (3) multinational government appropriations. The panda's future must be reoriented from the primary domain of fund-raising and conservation organizations to the governmental and economic sectors assuming primary responsibility and control.

Funds from a unified, worldwide fund-raising network may be consolidated into a public trust and endowment fund, one that is legally established, internationally administered, and periodically audited by independent financial institutions. Funding requests should be both honest and realistic. A long-term, trial-and-error process must be accepted from the outset, but emergency campaigns for short-term projects must over time be eliminated in favor of a more comprehensive and progressive approach.

Appropriate uses of funds raised outside China to benefit pandas inside China are a sensitive and unsettled issue. Attempts to gain conservation concessions within China, through pressure applied from outside, especially by way of negative press campaigns and lawsuits, are counterproductive and questionable. Such pressure has never substantially changed China's policies and has not worked to resolve financial conflicts between conservation interests, panda exhibition, and captive-breeding loans. Outside pressure often worsens the situation by closing off avenues of dialogue and change that can arise only from within China. History shows that China is a world unto herself, one that has largely been impervious to outside pressure; Spence (1969) documented clearly the numerous failures of Western influence in the face of staunchly ingrained Chinese attitudes.

Success in cooperative efforts in China can be hastened by a healthy respect and genuine concern for the Chinese people and the political, socioeconomic, administrative, and temporal context within China. Western conservation attitudes and agendas, planning strategies, time frames, and financial and administrative conditions cannot simply be imposed upon China and expected to bear fruit. Field research has shown the possibilities of working *with* China, in the Chinese context. Giant pandas can be saved from extinction only through a long-term effort led by China herself.

Integrated and Adaptive Management Systems

We propose an organization made up of relevant committees to facilitate the development, implementation, and refinement of an integrated and adaptive management system (Ma et al. 1983; He et al. 1983; Hollings 1984; Walters 1986) for the CMR (Figure 10.5). The purpose of the organization would be to oversee operations and put money directly into the hands of local people for economic investments, habitat restoration, and forest management. The suggested committees are not intended to be new organizations superimposed upon the existing government structure, but, rather, a regrouping and redefining of authority, responsibilities, and personnel within the existing structure. Such a reorganization would foster the evolution of an optimal management system for the CMR.

The committees would be representative of all main government levels, from production teams and village leaders to working directors of related departments at provincial and national levels. An administrative committee, empowered with strong authority by State Council, would act as the coordinator among government units, and would provide overall guidance in the management process. An operational committee would be primarily responsible for developing, implementing, and evaluating fieldwork projects through consultation with other committees and related government units. Because of their especially important role, these people should be experienced in the region they serve and have a demonstrated working knowledge of resource management. A consultative committee would provide scientific guidance, help monitor progress, suggest revisions, and, where feasible, collect data and publish results in peer-reviewed, management-oriented outlets. The financial committee

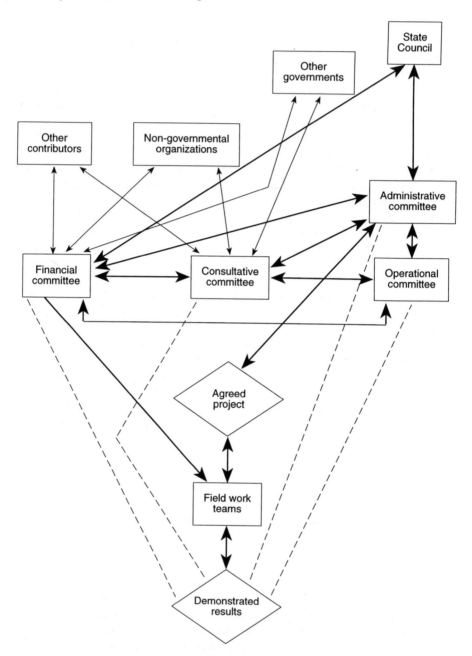

Figure 10.5. Suggested components (square boxes), goals and objectives (diamonds), linkages (light and dark lines, reflecting the relative intensity of the interactions), and feedback (dashed lines) for an integrated and adaptive resource management system in the Central Mountain Region of China (CMR).

would raise money, supervise the allocation of these funds, and file periodic and independently audited financial reports. Foreign involvement would consist of financial and scientific contributions within the consultative committees. Visits to fieldwork sites could be organized occasionally for foreign collaborators and contributors. Membership on committees would rotate gradually over several years so as to ensure both continuity and effective long-term operations.

Ideally, the organization would work as an interactive system of checks and balances, with financial and work allocations administered directly to local people for the pursuit of specific, high-priority projects that are primarily under their responsibility and commissioned for their benefit. The feasibility of actions should be tested and publicly demonstrated as pilot projects before large-scale implementation is undertaken. Progress should proceed on the basis of demonstrated results and continue under the close supervision of the committees, the responsible government leaders, and interested and involved local citizens.

Evaluations of site-specific patterns in the distributional ecology and relative abundances of giant pandas have shown that socioeconomic factors such as village densities, timber-removal quotas, road impacts, and size of habitat fragments are dominant influences within the current panda range (Johnson, in prep.). Analyzing these ecological and socioeconomic interactions in a stepwise procedure led to the prioritization of 15 areas for habitat restoration and special management (Figure 10.6). Proceeding from these actual conditions, in a realistic and practical manner, will maximize the long-term survival chances of pandas and ecological processes by maintaining the distributional area and integrity of the current biological communities in the CMR. Forest restoration and management actions should begin by reconnecting larger habitat fragments, then proceeding in descending priority to smaller and more outlying fragments as funds, village relocations, and road closures permit (Figure 10.6). Because the socioeconomic concerns of local people are dominant considerations, the outlined process includes investment opportunities and the direct involvement of local people (Johnson, in prep.).

Experimental management systems can adapt to local needs and serve as an intermediate step toward developing a special ecological zone similar to the special economic zones along the coast of China. Special ecological zones should correspond to biogeographical regions, with management focusing on ecological and economic sustainability within the balanced capacity of the local land and resource base. These special ecological zones could complement the special economic zones by ensuring the latter a steady flow of resources, and the benefits deriving from the success of the two types of zones may also provide for a better balance of wealth and resources between east and west China. Special ecological zones functioning with a high degree of autonomy may help to minimize upheavals from political movements and policy changes,

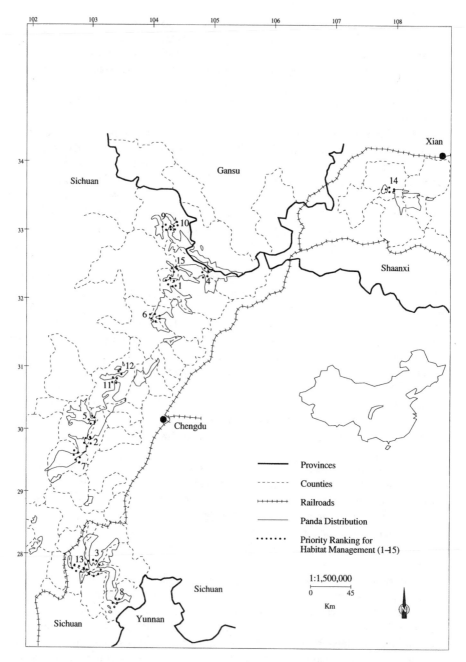

Figure 10.6. Distribution of giant pandas and priority rankings for habitat restoration and forest management in 15 areas that are intended to optimize the distributional area and integrity of biological communities in the Central Mountain Region (CMR) of China. (From Johnson, in prep.)

and perhaps to enhance the quality and stability of life for the region and the nation as a whole.

Educational and Occupational Involvement

The support of local people, paramount in these matters, may best be achieved through an educational work program focusing on ecological/economic sustainability, a harmonious land ethic, occupational training, tailored work assignments, and vested economic and ecological returns for local people. Experience with the Civilian Conservation Corps and the Young Adult Conservation Corps in the U.S. (Lorin 1941; Holland and Hill 1942; Salmond 1967) may be adaptable to a public-works program revolving around people, pandas, and conservation in the CMR of China. Because the Cultural Revolution produced a shortage of leaders educated in ecology and resource management, professional education of Chinese inside and outside China needs also to be expanded. The support and organized involvement of local people will foster the gradual development of an ecologically compatible and self-sustaining economy. In fact, managing pandas and habitats should afford the local people a unique opportunity for developing their remote mountain region in a wise, prudent, and satisfying manner.

Building an Ecologically Compatible Economy

Economic strategies with high potential for improving standards of living and reducing ecological decline in the CMR include (1) regulated forestry; (2) bolstered aquaculture and fisheries; (3) augmented hydropower for local use and export, and possibly for selected light industry; (4) due consideration of traditional medicine, especially from plants; (5) the growth of horticulture; (6) development of local arts and crafts; and (7) regulated tourism free of the promotion of habituated pandas.

Economic improvement and environmental protection in the CMR may seem contradictory goals. On the one hand, the economy in the area is poorly developed and there is currently insufficient funding for environmental protection. Local people must exploit natural resources simply to make a living. But on the other hand, when economic development leads the way, the issue of environmental protection is easily ignored. Long-term and short-term interests and local and national interests are often in conflict. People's needs for food and living space must be met, but natural resources must be used widely and maintained for future generations. In order to meet current demands, trees have been cut, crops are grown on steep hillsides, and hard living has evolved as a praiseworthy way of life. Conservation is not merely a publicity campaign in China, and wise use of limited resources must continue as a natural way of life in this isolated region.

Areas rich in natural resources usually offer splendid scenery, and if used wisely can generate significant tourist revenues. Investments in roads, facilities, and an efficient service industry should be offered in carefully selected areas. With adequate planning and facilities, some ecologically less sensitive areas can be opened to tourism so as to relieve pressures on natural resources in other areas. Habituated pandas have reduced chances of survival and reproduction, and usually end up being brought into captivity. Pandas that lose their natural fear of people are more vulnerable to poachers and can also injure people, especially where food, naive tourists, and zealous tourguides are involved. Natural conditions are at a premium in China and, with proper promotion, will draw as many tourists as can safely be accommodated.

China derives traditional medicines from many species of plants and animals. Tigers and bears are highly valued for traditional-medicine recipes, and leopard and even panda skins are a lucrative and illegal item too often found in the local and international fur trade. An organized industry could export selected plant products on a sustained-yield basis, and some wild species could be placed in cultivation, yielding profits without detriment to the environment or populations. Plant varieties could be developed and exported as ornamentals. Many undiscovered species, especially of plants, likely occur in the CMR, and may harbor unknown and valuable uses. Local arts and crafts may constitute another valuable export to other areas of China and abroad. Abundant, clean water affords an opportunity to produce high-quality native fish for local use or export, and increases the likelihood that forested slopes will be maintained for water resources and other purposes.

Reforestation Programs

Reforestation is necessary for environmental protection, for creating linkages between fragmented landscapes, and for economic production. Since the current land base in the CMR is not economically sustainable and is especially unsuited to agriculture, the best solution is to restore the natural forest ecosystem and restrict the forest industry to long-term sustained yield. Ecological and economic balance may be restored through highly regulated rotations, fixed land allocations, lower-impact selective cutting, simultaneous replanting, and limited roadbuilding and human settlement. Funds to achieve this ambitious goal may come partly from joint economic ventures with foreign organizations, as is now done routinely in other industrial segments in China. The juxtaposition of about 25% of the world's population and 3% of the world's forest in China will ensure the profitability of financial investments in this region, which is currently China's second most important timber-production area.

An aspect of forest management needing special attention is the use of original forest types to restore core areas and key watersheds (Huang and

Yang 1990). Early efforts of this sort failed, owing to grazing by livestock (Campbell 1985), and grazing accordingly must be forbidden in tree and bamboo planting areas. In buffer zones, forest cover should be at least 60%, and in timber-production areas clear-cuts must be less than 3 ha in extent (Yang 1985), if employed at all. Selective cutting is a preferred alternative for lower-impact timber harvesting.

Strengthening the Nature Reserves

The 12 designated panda reserves, two scenic reserves (Dayi and Huanglong), and 14 planned new reserves in the CMR constitute an immense investment in a conservation network that covers a significant representation of the region's habitats (see Figure 10.7). Three minor changes would enable the reserves to serve more effectively as a core system of strictly protected habitats and watersheds. First, the reserves need to be repositioned to protect key connecting habitats, and to conform more closely to the irregularly shaped panda distribution, forest fragments, and concentrated biological refuges, which are typically long, thin sections running along rugged and remote ridgecrest (Figure 10.7). The high-elevation border of the panda distribution is usually defined by distinctive heath and alpine habitats, while the low-elevation border is defined by the encroaching zone of human impact (Figure 10.8). The establishment of strong physical boundaries following topographically distinct features and encompassing whole watersheds is basic to enhancing protection, security, and enforcement, providing buffer zones of alpine and restoration forest and bamboo around core panda habitats, and encompassing the full elevational and slope gradient to maximize the protection of biotic diversity.

Second, reserve boundaries also need to be clearly marked, and made to exclude major villages and other zones of heavy human use that absolutely cannot be relocated, such as Gengda township in Wolong Reserve. Equitable exchanges of production lands currently in reserves for presently unprotected and relatively undisturbed natural habitats could be negotiated on a regional scale. Options for land exchanges are limited on a reserve or county scale, but a regional perspective should yield more opportunities for equitable exchanges.

Third, reserves need to be more effectively managed within an integrated regional landscape of strictly protected lands, buffer zones, and production lands, complete with connective habitats of replanted native trees and native bamboo. Case studies and professional exchanges may help.

A reserve strategy evaluated in the Southern Appalachian Region of the southeastern United States (Miller et al. 1987) may be helpful in the CMR. The strategy was derived from analyses of rare and endangered vascular-plant records and extant topographic heterogeneity patterns. Computer analyses of bounded areas simulating nature reserves indicated that size, elevational diver-

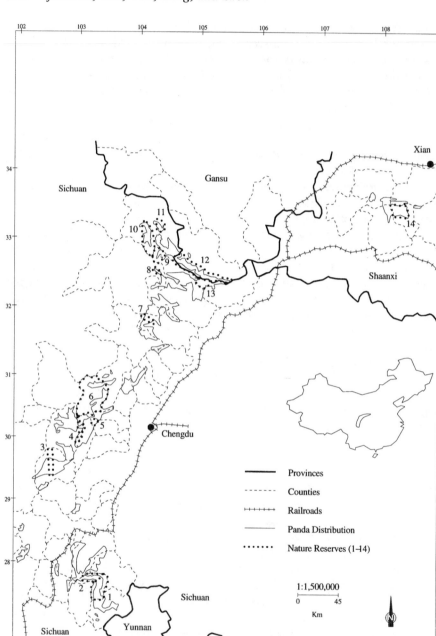

Figure 10.7. Distribution of giant pandas and the 14 existing nature reserves in China's Central Mountain Region (CMR).

Figure 10.8. A village and cornfields in the foreground, with a forest "island" in the middle-ground and the alpine habitat on the mountaintops.

sity, and slope diversity were important predictors of richness in rare vascular-plant species. Analysis of the accumulation patterns of species also indicated the likelihood of unrecorded rare and endangered vascular plants. The strategy proposed that the nature reserves selected be conformable to particular biogeographic regions, and that land costs be compared in terms of expense per species acquired, for reserves of different sizes and shapes. Economic considerations need to be integrated with social and biological concerns to achieve the most tractable regional reserve strategy (Miller et al. 1987).

Resettlement Programs

Humane relocation of people will be necessary in some areas to safeguard human welfare, restore sustainability, and protect key habitats and watersheds. An important precedent was set when 61 families were successfully relocated from Tangjiahe Reserve with an appropriation of 1.1 million yuan (U.S. $367,000) from the Sichuan People's Government (Machlis and Johnson 1987). Such moves are sensitive, expensive, and difficult, because vacant land is extremely scarce, but in certain situations this management option will be necessary.

A standard legal concept often referred to as the "grandfather clause" could be added to all Chinese legislation establishing and regulating nature reserves and other important habitats. The grandfather clause essentially functions as a lifetime lease for the head of the family, and is routinely used in reserves in Western countries to gradually and humanely relocate people living within recently designated reserves. The grandfather clause states that families living in a reserve at the time of its legal establishment may remain until the registered head of the family retires. The land-use rights cannot be passed on to the next generation or sold, but will then revert to the reserve so that it can be restored to its natural state. In the transition period between the initiation of the grandfather clause and the retirement of the first generation, the children and other family members could be given special opportunities for education, training, jobs, and other positive incentives outside the areas designated for nature protection. Once these family members have been resettled in other areas, the aging seniors in the reserves will be allowed to choose between retiring with their relocated relatives or living out their retirement in the reserve. Successful implementation of the grandfather clause can gradually resettle people as well as improve their standard of living.

Multidisciplinary Research Programs

A coordinated, multidisciplinary research program would substantially support integrated and adaptive resource management and monitoring in the CMR. Research priorities should focus on four central concerns. First, socioeconomic surveys and policy assessments that address (a) local people's opinions and attitudes about current and proposed policies and practices relating to resource management; (b) social-impact assessment; and (c) development and periodic refinement of suitable policy to reduce the pressure of local people on forest ecosystems and to reorient human uses to regionally sustainable strategies. All surveys and policy development should carefully consider the local cultures and customs.

Second, landscape planning and management, on a regional scale, that addresses such considerations as (a) locating core habitat areas and key watersheds that must remain forested under any circumstance; (b) determining which areas must be reforested immediately so as to establish landscape linkages; (c) determining which local people should be relocated first and where they should be resettled; (d) determining which lands are suited to afforestation and reforestation and establishing optimal procedures for maximizing multiple benefits, such as watersheds, timber, and firewood; and (e) assessing land suitability and the carrying capacity of development areas, particularly in terms of forestry, agricultural production, and human-population levels.

Third, the monitoring of ecosystem dynamics through (a) objective population-trend surveys of pandas as indicator species; (b) the development of projections for the long-term dynamics and interrelationships of natural and altered habitats, wildlife populations, and ecological and socioeconomic processes; (c) the use of remote-sensing and geographic-data systems to monitor and help manage landscapes; (d) the selection of atmospheric and climatic parameters to assess acid deposition, air pollution, global climate change, and other broad-scale and nonlocalized environmental threats to natural habitat areas; and (e) an evaluation of the relationships of hydrological and ecological features to management policies and actions.

Finally, research and experimentation in development zones concentrating on (a) economic- and ecological-relationship research employing, for example, the economico-ecological regionalization model, the ecological agrosystem model, forestry-rotation models, and other dynamic systems models (Ma et al. 1983; He et al. 1983); (b) learning to manage lands for multiple uses and within the limits of ecosystem tolerance to stress, so as to minimize negative impacts; (c) evaluating energy alternatives such as small-scale hydropower or coal to decrease the demand for wood; and (d) identifying specialized technologies that may hold special merit, such as accelerating forest regeneration by using pioneer species for both watershed restoration and succession to original forest types (Zhang 1985).

International Cooperation

Effective international cooperation, especially among scientific and management agencies, would be mutually conducive to integrated and adaptive resource management. The CMR and the Southern Appalachian Region of the southeastern United States pose similar ecological and management challenges, and just as exchanges in research have been appropriate and mutually beneficial, so can similar exchanges in resource management. Foreign involvement needs to focus more on the development and comparative evaluation of integrated management systems and less on traditional research, conservation, and management approaches that focus on species rather than on processes. Systems approaches to resource management hold research merit as well as direct and simultaneous benefits to people and the environment. Other creative initiatives should be developed to facilitate innovations at the interface between research, management, and financial investment.

The currently available science and technology can be used to mount valuable socioeconomic and ecological programs on an unprecedented scale if the restrictive attitudes and traditional barriers among different conservation and fund-raising interests, between basic and applied research, among the physical, social, and biological sciences, and between political ideologies and international financial alliances can be resolved through improved international cooperation. Cross-cultural exchanges and the infusion of Western science and technology into China's long history and cultural and biological diversity may yield fertile offshoots. Science and technology have largely subdued nature, and perhaps the time has come to adapt science and technology to Chinese philosophy, so that we may finally learn, as Confucius and Lao Tze expressed over 2000 years ago, "to fit into nature's way through respect, obedience, and mutual responsibility."

Secondary Interactions

The four categories of human-carnivore interactions addressed in the foregoing encompass secondary factors that may also be involved in the conservation and management of carnivores.

Projection of Human Attitudes onto Carnivores

Increasing human interest in carnivores, and in wildlife in general, has stimulated an inappropriate imposition of human attitudes and ethics on wildlife (Rolston 1992; Bekoff and Jamieson, this volume). Mass media and the urgency to "save" species infused by emotional appeals from fund-raising organizations have furthered the cause of conservation, but have often confused the conservation issues. Concern for the welfare of wildlife is usually misdirected toward the extension of human attitudes and rights to animals, without due respect to their wildness and the ecological processes that are the true basis for their well-being (Rolston 1992). Medical interventions and infatuations with captive breeding, genetics, and other dramatic technologies also focus on individual animals and biological subcomponents, further severing wildlife from its natural world. Scientific inquiry, too, tends to dissect and reduce species and ecological processes, with insufficient regard for synergistic interrelationships in natural biotic communities.

Pandas are a prime example of a popular species capable of generating such an abundant outpouring of human concern that the best of intentions can easily become clouded by the limitations of human perception. The tendency is to want to bring pandas into the human domain, to coddle them, and to reproduce them artificially. Such impulses are understandable and have been profitably exploited, but they lead to misunderstandings of pandas as simply an image shaped by human attitudes. The phenomenon of a unique assemblage of species interacting within their unique natural habitat is easily overshadowed by the shallow and shortsighted nature of human emotion and understanding. A more appropriate response would be to reverse this tendency by adopting the view, so succinctly captured by Thoreau, that "In wildness is the preservation of the world."

Captive breeding has consumed substantial resources and has indeed played a prominent role in panda conservation. Still, an excessive emphasis on the potential for artificial propagation and an unrealistic infatuation with ad-

Figure 10.9. The giant panda Hua Hua being returned to the exact site in the Wolong field-study area (Wuyipeng) from which he had been removed for captive-breeding purposes.

vanced technologies have fostered the dangerous perception that captive breeding is a sufficient solution to panda conservation. The estimated 75 pandas in captivity apparently will not be self-sustaining anytime in the foreseeable future, despite extensive efforts since the first panda was bred in captivity in 1963 and artificially inseminated in 1978; and to date no captive-bred pandas have been released into the wild. Captive-breeding programs have removed numerous pandas from the wild (Figure 10.9), and unrealized expectations have been a major distraction from the management efforts necessary to enable ecological and evolutionary processes to continue in natural habitats. Habitat protection and resource management can negate the need for the expensive and artificial interventions, of unknown consequences, that are associated with captive breeding, thus avoiding ethical, legal, and financial concerns that are easily distorted when transferred across cultural and political boundaries. Captive breeding of giant pandas shows limited potential for success and does not solve the problem of habitat loss and ecological decline. Experiences with the captive breeding of pandas provide a poignant opportunity for international conservation organizations to reevaluate and clarify the goals, priorities, and programs that are being exported to developing countries. The developing countries, for their part, need to consider carefully the relevance and practicality of the technological support being requested in joint projects.

The legal basis for conservation and endangered-species management also needs to focus more on community and ecosystem processes at regional scales. Politically and financially motivated efforts to preserve a few high-profile endangered species, often through artificial means, do little to resolve the serious problems underlying a more widespread ecological decline. As our responsibilities and challenges in wildlife protection expand, we will need a new ethical and legal basis for the guidance of programs, and, ironically, the most fitting basis may lie in natural history, ecology, and evolutionary biology. "Once we can discriminate the differences between wild nature and human culture, the *is* in nature and the *ought* in ethics are not so far apart after all" (Rolston 1992:622).

Carnivores in Human Culture

Humans have long admired and likely feared carnivores and attempted to depict their attributes in such cultural rituals as stories, art, dance, and burial (Gittleman 1989:vii–x; Clutton-Brock, this volume). The lion (*Panthera leo*) is known as "The King of Beasts," and the dog (*Canis familaris*) is "Man's Best Friend." Such positive attributes of carnivores as agility, strength, elegance, and loyalty are inspirational to humans. Mayans and peoples of other cultures believe that jaguars (*Panthera onca*) and other large carnivores are gods. Many cultures have a version of "The Big Bad Wolf" (*Canis lupus*) story. A giant panda skull was buried with one Chinese emperor. Such symbolism has mixed consequences for human attitudes and actions, and carnivores tend to become inaccurately assimilated in most interactions with humans, usually losing their unique character and coming to be thought of exclusively in human terms.

Carnivores may kill livestock or even people, causing financial losses and provoking rage and fear. Broad generalizations and misunderstandings allow bounties and other damage-control programs to become inappropriately entrenched in society. The coevolution of carnivores and humans may have helped shape human culture, and human culture has definitely seen to the domestication or extinction of several carnivore species, but human culture is currently poised to play a dominant role in altering the survival chances of many wild carnivores. Healthy human/carnivore interactions thus depend on our understanding wild animals, the human domain, and the context of responsibilities and consequences for crossing these boundaries, whether in attitude or action.

Depredations by Carnivores

Carnivore depredations can bring serious financial losses to human enterprises adjacent to natural habitat, and can provoke a negative human attitude toward carnivores in general. Depredations by giant pandas are rare, but

Asiatic black bears and other wildlife sharing protected panda habitats often eat corn, apples, and other crops from adjacent farmland. Financial-compensation programs usually permit some alleviation of verified losses, and most depredation problems can be solved with minimal management effort, focused only where needed. The severity of depredation problems is usually overestimated, particularly when the human imagination has been stimulated by generalizations about carnivorous lifestyles.

Educating local people about the actual costs and benefits of maintaining natural habitat and wild populations in their areas may help minimize perceived and real problems. Local people could more effectively solve animal-depredation problems and realize some supplementary income on a contract basis. Establishing large government programs to address scattered, sporadic problems with depredating individuals should not be necessary in dealing with most negative human/carnivore interactions. Where large carnivores must be maintained in fragmented and human-dominated landscapes, the killing of particular individuals that have lost their wildness may be necessary, in order that humans may coexist with wild carnivore populations in general.

Human Conflicts and Conservation Politics

Disputes relating to conservation money and loans of giant pandas to Western zoos, which were reduced to litigation in 1988, have raised fundamental questions concerning the appropriate role of conservation groups in the management of endangered species (Roberts 1988). A similar lawsuit was settled out of court in 1992, with the only result being payment of lawyer fees for the plaintiff (WWF). As with gorillas (*Gorilla gorilla*) and most other high-profile endangered species, outside conservation interests conflict with the local realities, because of intense political infighting for the control and management of endangered species by organizations that are staffed and driven by fund-raising interests (Foster 1992). The great amounts of money involved simply exacerbated the conservation politics surrounding giant pandas.

Though wild pandas occur exclusively within China, the well-known giant panda logo is the copyrighted symbol of WWF. Fund-raising rights for panda conservation are an unresolved issue, and legal confrontations have resulted only in settlements that do great harm to human relations and the spirit of cooperation. Lawsuits cannot resolve these conflicts, because endangered-species legislation and international trade conventions present a weak legal basis for interpretation in national or international courts of law. Attempts by special-interest groups to invoke conservation laws for their own agendas not only weaken the intended purpose and implementation of legislation and international conventions, but also seriously detract from those long-term efforts that actually do conserve wildlife and natural resources.

Current international conservation programs, such as the Convention on

International Trade of Endangered Species (CITES) have been weakened by a political process that allows special-interest conservation groups to dictate rules to the countries that are responsible for the species of concern, and there is little scientific basis for setting rules or monitoring success (Aldhous 1992). In general, the processes and weaknesses reflected in CITES characterize many other international conservation efforts. Regardless of the intentions of conservation groups, legal and financial appeals set forth in a confrontational and antagonistic manner are counterproductive, at least in China and usually in most other countries. Confrontational tactics may serve short-term publicity, propaganda, and fund-raising purposes, especially when applied within national boundaries with a common political and legal system. But attempts to use such tactics and media campaigns across international and cultural boundaries render such efforts ineffective, except in their capacity to raise more and more funds to conduct conservation projects that are unrelated to the management efforts actually necessary to conserve wildlife and natural resources. If they are to have salutary effect, conservation organizations must abandon their current tactics and motives and look to a cooperative, long-term process oriented to resource-management professionals who live in the field and maintain collaborative working relationships with local management authorities.

Summary and Conclusions

An interdisciplinary team evaluated information on giant pandas; other carnivores; ecological, socioeconomic, and political processes; conservation; and natural resource management. Pandas and carnivores generate exceptional human interest and are ideal for investigating human/carnivore interactions and conservation and management implications. Human/carnivore interactions of primary importance include: (1) human impacts arising mainly from habitat destruction, killing, and removal for zoos and captive breeding and (2) programs such as habitat protection and resource management. Our main objectives were (1) to identify and discuss major ecologically related problems and potential solutions and (2) to suggest an integrated and adaptive resource management system for the facilitation of wildlife protection and ecological and economic sustainability in the Central Mountain Region of China.

Major interrelated problems included: (1) political instability and especially inconsistency in long-term policies, (2) population growth and the results of the National Family Planning Policy, (3) ecological sustainability and the carrying capacity of China's resource base, (4) the prospects for economic sustainability under market or planning-oriented policy, (5) ecological decline and the continued threat to endangered species, (6) the deterioration of water resources that are the lifeblood of China's (7) agricultural and industrial economy, with implications from the headwaters of the Yangtze River in western China all the way to the east coast, (8) deforestation and its synergistic impacts, and (9) the basic socioeconomic concerns of local people in the CMR, including Tibetan, Yi, and Qiong minorities.

Potential solutions and a suggested strategy involve (1) sustained funding and sustained effort from a carefully coordinated worldwide network, with China's governmental and economic sectors assuming primary responsibility and control; (2) an integrated and adaptive management system to assist current government operations by putting funds directly into the hands of local people for forest and habitat management, and ultimately for the development of a special ecological-management zone; (3) gaining the support of local people through an educational work program focusing on a harmonious land ethic, occupational training, and other direct benefits to local people through (4) an ecologically compatible economy based mainly on regulated forestry, fisheries and aquaculture, hydropower, horticulture, traditional medicine, local arts and crafts, and regulated tourism free of the promotion of habituated pandas; (5) reforestation and regulated forest management for environmental protection, landscape linkages, and economic production; (6) refining and integrating the 14 existing reserves and 14 planned reserves into a regional network of core protected habitats and watersheds, complete with buffer zones and production lands; (7) humane relocation of people from key habitats through positive incentives and the grandfather clause; (8) a multidisciplinary research program focusing on socioeconomics, policy assessment, regional planning, and integrated and sustained ecosystem management; and (9) effective international cooperation directed toward resolving restrictive attitudes and traditional barriers among different conservation interests, between basic and applied research, among the physical, social, and biological sciences, and between international financial and political alliances that may enable unprecedented progress in learning, from cross-cultural exchanges, to "fit into nature's way through respect, obedience, and mutual responsibility."

Human/carnivore interactions that were considered of secondary importance included (1) the projection of human attitudes onto carnivores, (2) the incorporation of carnivores into human culture, (3) depredations by carnivores, and (4) legal and financial conflicts among conservation interests and responsible management authorities. Misdirected human intentions and conservation politics actually hindered the conservation and management of carnivores and other natural resources.

Endangered species such as giant pandas and other carnivores are ideal indicators of the degree of human impact on natural ecosystems and can call attention to the urgent needs of resource management. To protect carnivores, in fact, means to protect vegetation, other animals, water supplies, soil, and much more, and thus to sustain the quality of human life as well. In the protection of carnivores, traditional knowledge and methods are strongly challenged by questions regarding how to conserve resources effectively and how to use them to meet human requirements efficiently. Although these problems are of worldwide concern, the key to progress is site-specific adaptation of ecological principles and resource-management processes to the biological and cultural diversity and complexity inherent in our world. Integrated and adaptive resource management appears to be the most effective and perhaps the only way to protect carnivores and to achieve ecological and economic sustainability in an area of extreme importance to both China and our global environment.

Acknowledgments

Our field work was financed through an agreement between China's Ministry of Forestry, the Forestry Bureaus of Sichuan, Gansu, and Shaanxi provinces, and the World Wide Fund for Nature (WWF–International, Gland, Switzerland). Arrangements for a special-topics class were made through Frank McCormick and the University of Tennessee, Graduate Program in Ecology. Guidance and encouragement were received from many officials, including Wang Menghu (Ministry of Forestry, Beijing), Fu Daozhen, Yu Xiwen, Gong Tongyang, Hu Tieqing, Fu Chengjun, Bi Fengzhou, Jiang Jijian, Cui Yangtao, Pu Tao, Shao Kaiqing and the Sichuan Wildlife Survey Team (Forest Bureau, Chengdu), Lai Binghui and Li Chengyu (Wolong Natural Reserve), Yue Zhishun, Jiang Mingdao, and Teng Qitao (Tangjiahe Natural Reserve), Jiang Fuqian (Gansu Forest Bureau, Lanzhou), and Fan Chuandao (Shaanxi Forest Bureau, Xian). Hu Jinchu and Qin Zisheng (Nanchong Normal College) and Xie Wenzhi and Li Guohui (Shaanxi Zoological Institute) provided scientific advice. Figures 10.5–10.7 were prepared by the Geographic Information Systems Laboratory at the Environmental Management Technical Center of the National Biological Survey in Onalaska, Wisconsin. John Gittleman, George Schaller, Don Reid, Gary Machlis, Sam Wallace, and Dave Maehr reviewed the manuscript. We are deeply grateful to these people and institutions for their assistance.

References

Aldhous, P. 1992. Critics urge reform of CITES endangered list. *Nature* 355:758–759.
Campbell, J. J. N. 1985. Giant panda conservation and bamboo forest destruction. In: *Proceedings of the International Symposium on Ecology and Development of Tropical and Subtropical Mountain Areas*, 1:131–148. Chengdu, China: China Academic Publishers.
Davidson, P. 1989. China: Swimming against the rising tide. In: *Survey of Business: Faculty Perspectives on the Global Village*, 25:20–22. Knoxville: Tennessee.
Davis, D. D. 1964. The giant panda: A morphological study of evolutionary mechanisms. *Fieldiana: Zool. Mem.* 3:1–339.
Foster, J. W. 1992. Mountain gorilla conservation: A study in human values. *J. Amer. Vet. Med. Assoc.* 200:629–633.
Gittleman, J. L. 1989. Preface. In: J. L. Gittleman, ed. *Carnivore Behavior, Ecology, and Evolution*, pp. vii–x. Ithaca, N.Y.: Cornell Univ. Press.

Gittleman, J. L. 1994. Are the pandas successful specialists or evolutionary failures? *BioScience* 44:456–464.

Gould, S. J. 1980. *The Panda's Thumb: More Reflections in Natural History.* New York: W. W. Norton.

Green, L. W. 1988. Promoting the one-child policy in China. *Amer. J. Public Health Policy* 78:273–283.

He, S. Y., Ma, S. J., Huang, Z. L., and Zou, S. J. 1983. *Mathematical Methods in Economico-Ecological Regionalization.* Academia Sinica, Inst. of Automation, Beijing, Vol. 62. (In Chinese.)

Holland, K., and Hill, F. 1942. *Youth in the CCC.* Philadelphia: Ayer.

Hollings, C. S. 1984. *Adaptive Environmental Assessment and Management.* New York: John Wiley and Sons.

Huang, L. L., and Yang, Y. P. 1990. On the conservation function of natural forest in western Sichuan. In: P. F. Ffolliott and D. P. Guertin, eds. *Proceedings of a Workshop on Forest Hydrological Resources in China,* p. 121. Springfield, Va.: U.S. Department of State Publication 9829.

Jia, J. 1992. Laws and regulations help protect reserves. *China Environment News* 39:1.

Johnson, K. G. In prep. *Giant Pandas of China: Community Processes and Conservation of Ecological Integrity.* London: Chapman and Hall.

Johnson, K. G., Schaller, G. B., and Hu, J. C. 1988a. Comparative behavior of red and giant pandas in the Wolong Reserve, China. *J. Mammal.* 69:552–564.

Johnson, K. G., Schaller, G. B., and Hu, J. C. 1988b. Responses of giant pandas to a bamboo die-off. *Natl. Geogr. Res.* 4:161–177.

Johnson, K. G., Wang, W., Reid, D. G., and Hu, J. C. 1993. Food habits of Asiatic leopards (*Panthera pardus fusea*) in Wolong Reserve, Sichuan, China. *J. Mammal.* 74:646–650.

Liu, Z. G., and Lu, R. S. 1985. Study on exploitation potentialities of the subtropical mountain land of Sichuan. In: *Proceedings of the International Symposium on Ecology and Development of Tropical and Subtropical Mountain Areas,* 1:193–197. Chengdu, China: China Academic Publishers.

Lorin, L. L. 1941. *Youth Work Programs: Problems and Policies.* Philadelphia: Ayer.

Lugo, A. 1985. A primer on tropical deforestation. *Tropical Forest* 2:12–19.

Ma, S. J., He, S. Y., Zou, Z. J., and Huang, Z. L. 1983. On the economico-ecological system regionalization: Theory and method. *Acta Ecologica Sinica, Beijing,* Vol. 3. (In Chinese.)

Machlis, G. E., and Johnson, K. G. 1987. Panda outpost. *National Parks* 61:14–16.

MacKinnon, J. J., Bi, F. Z., Qiu, M. J., Fu, C. D., Wang, H. B., Yuan, S. J., Tian, A. S., and Li, J. G. 1989. *National Conservation Management Plan for the Giant Panda and Its Habitat.* Gland, Switzerland: World Wide Fund for Nature.

Mainka, S., and Qiu, X. M. 1992. Preparing for the re-introduction of giant pandas. *Proc. Amer. Assoc. Zoo and Wildl. Veterinarians* 19:65–69.

Miller, R. I., Bratton, S. P., and White, P. S. 1987. A regional strategy for reserve design and placement based on an analysis of rare and endangered species distribution patterns. *Biological Conservation* 39:255–268.

O'Brien, S., Nash, W., Wildt, D., Bush, M., and Benveniste, R. 1985. A molecular solution to the riddle of the giant panda's phylogeny. *Nature* 317:140–144.

Pan, W. S., Gao, Z. S., and Lu, Z. 1988. *The Giant Panda's Natural Refuge in the Qinling Mountains.* Beijing: Beijing Univ. Press. (In Chinese with English summary.)

Proceedings of a Workshop on Forest Hydrological Resources in China. 1990. Springfield, Va.: U.S. Department of State Publication 9829.

Proceedings of the Sino-American Conference on Environmental Law. 1987. Boulder, Colo.: Natural Resources Law Center, Univ. Colorado, School of Law.

Proceedings of the Symposium on Effects of Acid Precipitation on Comprehensive Agriculture and Its Countermeasures. 1988. Beijing: China Ocean Press. (In Chinese.)

Qu, G. P. 1987. *Environmental Problems and Policies in China.* Beijing: China Environmental Sciences Press. (In Chinese.)

Reid, D. G., Hu, J. C., Sai, D., Wang, W., and Huang, Y. 1989. Giant panda *Ailuropoda melanoleuca* behavior and carrying capacity following a bamboo die-off. *Biol. Conserv.* 49:85–104.

Roberts, L. 1988. Conservationist in Panda-monium. *Science* 241:529–531.

Rolston, H., III. 1992. Ethical responsibilities toward wildlife. *J. Amer. Vet. Med. Assoc.* 200:618–622.

Russell, M. 1989. Population, poverty and prosperity: China's environmental woes. In: *Survey of Business: Faculty Perspectives on the Global Village,* 25:23–27. Knoxville: Univ. Tennessee.

Salmond, J. A. 1967. *The Civilian Conservation Corps 1933–1942: A New Deal Case.* Durham, N.C.: Duke Historical Publications Service.

Schaller, G. B. 1993. *The Last Panda.* Chicago: Univ. Chicago Press.

Schaller, G. B., Hu, J., Pan, W., and Zhu, J. 1985. *The Giant Pandas of Wolong.* Chicago: Univ. Chicago Press.

Schaller, G. B., Teng, Q. T., Johnson, K. G., Wang, X. M., Shen, H. M., and Hu, J. C. 1989. The feeding ecology of giant pandas and Asiatic black bears in the Tangjiahe Reserve, China. In J. L. Gittleman, ed. *Carnivore Behavior, Ecology, and Evolution,* pp. 212–240. Ithaca, N.Y.: Cornell Univ. Press.

Smil, V. 1993. *China's Environmental Crisis: An Inquiry into the Limits of National Development.* New York: M. E. Sharpe/East Gate Books.

Spence, J. D. 1969. *To Change China: Western Advisors in China 1620–1960.* Boston: Little, Brown.

Spence, J. D. 1990. *The Search for Modern China.* New York: W. W. Norton.

United Nations. 1991. *Food and Agricultural Organization Production Yearbook.* Rome.

United States Department of Agriculture. 1991. *Foreign Agriculture 1990–1991.* Washington, D.C.: Foreign Agriculture Service.

Walters, C. 1986. *Adaptive Management of Renewable Resources.* New York: MacMillan.

Yang, Y. P. 1985. Great importance should be attached to the ecological balance of the subalpine forest of western Sichuan. *Proceedings of the International Symposium on Ecology and Development of Tropical and Subtropical Mountain Areas,* 1:91–98. Chengdu, China: China Academic Publishers.

Zhang, S. W. 1985. The possibility of recovering ecological equilibrium in hill areas of middle Sichuan red basin from the enlightenment of a typical revived ecosystem in a county. *Proceedings of the International Symposium on Ecology and Development of Tropical and Subtropical Mountain Areas,* 1:167–172. Chengdu, China: China Academic Publishers.

Zhu, J. 1989. Nature Conservation in China. *J. Applied Ecology* 26:825–833.

PART III

EVOLUTION

Introduction

This assemblage of chapters on evolution encompasses various topics linked by levels and modes of selection. Recent discussion about phylogenetic hierarchy in adaptive trait variation, selection at the species level versus the individual level, developmental constraints on evolutionary change, and effects of hybridization on speciation (e.g., Eggleton and Vane-Wright 1994; Harrison 1990; Harvey and Pagel 1991; McKinney and McNamara 1991; Williams 1992), all at the forefront of debate in evolutionary biology, are dealt with here in various contexts of carnivore biology.

Dating back to Darwin's (1868) *The Variation of Animals and Plants under Domestication*, artificial selection has been seen to be informative in determining the factors that influence trait evolution. Thus in reviewing the domestication process, Clutton-Brock examines why and how carnivores, especially the dog and the cat, have adapted so extraordinarily well to human cultures. Beginning with a description of the long history of human/carnivore relationships, she develops both biological and cultural arguments for the domestication process in carnivores; specifically, she argues that the process probably originated independently many times in many places, and that various independent factors ranging from food-stores to shelter in human settlements may have operated with dogs and cats. Clutton-Brock also pursues an interesting discussion of "aborted" domestics in such species as cheetahs (*Acinonyx jubatus*), the Egyptian mongoose (*Herpestes ichneumon*), and various South American canids.

One of the most profound characteristics of carnivores is their ability, morphologically endowed, to capture and consume large and potentially dangerous prey. Biknevicius and Van Valkenburgh present a critical review of the considerable literature on the functional morphology of carnivory and offer a classification scheme for analyzing how morphological characters (jaws, teeth, skull, muscular strength) are involved in feeding. Extending these functional patterns of extant species to three extinct carnivorans—the sabertooth cat (*Smilodon fatalis*), the giant American lion (*Panthera atrox*), and the dire wolf (*Canis dirus*)—the authors show that biomechanical models can be usefully applied to historical problems.

The remaining chapters evaluate evolutionary pattern and process at individual (Gompper and Wayne), population (Wayne and Koepfli), and cross-species (Hunt; Flynn; Werdelin) levels. Methodologically, the authors use various approaches, ranging from new techniques in molecular biology to modern phylogenetic analysis of morphology. Gompper and Wayne summarize allozyme and DNA fingerprint analyses within the context of the evolution of social organization in carnivores. In general, applications of new molecular-genetic techniques in behavior and ecology show that paternity can be detected with great resolution, dispersal patterns are sometimes counterintuitive, classical "solitary" carnivores are certainly not asocial, and genetic relatedness within semipermanent groups can be remarkably variable. Using similar molecular techniques at the population level, Wayne and Koepfli reveal that, after widespread differences in dispersal are accounted for, differences in population size and fragmentation of populations significantly affect population structure and genetic variability. Moreover, historical bottlenecks in some carnivore species (e.g., the black-footed ferret, *Mustela nigripes*, and the island fox, *Urocyon littoralis*) show their effects for many generations and should be carefully assessed in terms of phylogeny, particularly when the genetic variability within a population or species is being examined for conservation purposes (see Wayne and Gittleman 1995).

Hunt presents a thorough synthesis of the biogeography of fossil and extant carnivore species, including each major familial group. Using key morphological characters, he shows what specific postcranial, basicranial, auditory, and dental structures were involved in each of the major carnivore radiations; his figures illustrating many of the primary biogeographical shifts in carnivore history will undoubtedly be influential in guiding future work in this area. Flynn unites molecular and morphological approaches in a synthesis of higher-level phylogenetics in carnivores and then shows how the fossil record can be instructive for evaluating rates of evolution. He demonstrates that in certain groups (e.g., Hyaenidae), estimated evolutionary rates differ depending on whether a fossil record is used to calibrate a time clock. As morphological, genetic, and fossil material coalesce into systematics, there will be continued debate about what the "correct" answers to evolutionary questions might be. But as Flynn shows, there is little doubt that merging these materials is a productive strategy. Finally, Werdelin applies hypotheses of evolutionary process derived from extant forms to fossil assemblages to show that adaptive trends can be fully understood only through historical (phylogenetic) origins; the goal is to integrate phylogeny and functional morphology in such a way as to pinpoint why and how some characteristics have changed less than others. For example, stalk-and-ambush carnivores are relatively more prevalent both now and in the fossil record than are the scavenger/bone-cracker species. The point is convincingly made that morphological constraints on entering these and other ecological niches are critical in determining the phylogeny of carnivores.

In studies of carnivore evolution, more emphasis is now being placed on how the elements of evolutionary biology can be directed toward problems of conservation and biodiversity. Eldredge (1992) presents a useful scheme suggesting three problem areas: (1) ecological diversity, the number of organisms in an ecosystem; (2) genealogical diversity, the number of taxa within a monophyletic clade; and (3) phenotypic diversity, the amount of variation within and among populations, species, or higher taxa. Focusing on the evolutionary pattern and process of these areas will help immeasurably in preserving carnivore biodiversity.

John L. Gittleman

References

Darwin, C. 1868. *The Variation of Animals and Plants under Domestication*. London: John Murray.

Eggleton, P., and Vane-Wright, R., eds. 1994. *Phylogenetics and Ecology*. London: Academic Press.

Eldredge, N. 1992. Where the twain meet: Causal intersections between the genealogical and ecological realms. In: N. Eldredge, ed. *Systematics, Ecology, and the Biodiversity Crisis*, pp. 1–14. New York: Columbia Univ. Press.

Harrison, R. G. 1990. Hybrid zones: Windows on evolutionary process. *Oxford Surveys in Evolutionary Biology* 7:69–128.

Harvey, P. H., and Pagel, M. 1991. *The Comparative Method in Evolutionary Biology*. Oxford: Oxford Univ. Press.

McKinney, M. L., and McNamara, K. J. 1991. *Heterochrony: The Evolution of Ontogeny*. New York: Plenum Press.

Wayne, R. K., and Gittleman, J. L. 1995. The problematic red wolf. *Sci. Amer.* 273: 26–31.

Williams, G. C. 1992. *Natural Selection: Domains, Levels, and Challenges*. Oxford: Oxford Univ. Press.

CHAPTER 11

Competitors, Companions, Status Symbols, or Pests: A Review of Human Associations with Other Carnivores

JULIET CLUTTON-BROCK

It is probable that humans were beginning to displace the large carnivores as top predators well before the end of the last ice age. By the early Holocene the supremacy of human hunters was irreversible and was already causing world-wide changes in faunal compositions, ecosystems, and the behavior of the large carnivores. My aim here is to provide a sense of the increasing intensity and magnitude of these changes over the last 10,000 years, and to describe the impact that associating with humans has had on many species of carnivore, ranging from the lion to the ferret. Domestication and the transformation of the wolf and the wild cat into the most highly valued of all animal companions is the obvious manifestation of human dominance, but there is no species of carnivore that has escaped exploitation in one way or another.

Of necessity, much of the information on the history of human associations with animals must come from written sources that cannot be scientifically attested. The subject has, therefore, to be more discursive than perhaps would be expected from a biological work titled *Carnivore Behavior, Ecology, and Evolution*. The scene was set, however, in the preface to the first volume of this series, which begins with how "the great hero Hercules' most famous feat was killing the 'invulnerable' lion with his bare hands."

The relationships and close understanding that can develop between people and their dogs and cats are based on the capacity—of people and these animals alike—for interspecific communication and the animals' ability to reciprocate demonstrations of affection (a topic I shall return to). Within recent years there have been many new studies on the development of human/animal interactions, and on the processes of domestication. Current hypotheses on the origins of domestication and its effects on carnivores are summarized here. For further information, see the references at the end of this chapter, in particular Berry (1969), Clutton-Brock (1987, 1993a, b, 1994), Clutton-Brock and Jewell (1993), Hafez, ed. (1969), Hemmer (1990), and Turner and Bateson (1988).

Human Hunters and the
Symbolism of Large Carnivores

The belief that animals intentionally present themselves to be killed is prevalent among many northern hunters, and Ingold (1994) cites the Cree people of northern Canada, who believe that hunting "is a rite of regeneration" and that animals will not return to hunters who have treated them badly in the past. Bad treatment includes unnecessary killing, causing undue pain or suffering, and failing to observe the proper ceremonies and rituals of burial.

Bears

Like other carnivores, many species of bears have always been a part of human societies, and folk memories of bears remain with us in the legends, nursery tales, and cherished toys of today. Perhaps the best-known association with bears to survive into modern times is that of the Ainu people of northern Japan, where bear cubs are reared in the villages and even suckled by women. When they are between two and three years old, and beginning to be aggressive, the bears are killed during a great festival that lasts several days. Hallowell (1926:126) wrote that, before its death, the bear is told, "We have given you food and joy and health; now we kill you in order that you may in return send riches to us and to our children."

Through all the mythology of bear ceremonies and the hunting of bears runs the belief that the soul of the bear must be appeased in death by propitiations and special burials. In their account of the Lappish bear graves in northern Sweden, Zachrisson and Iregren (1974:14) describe the Nordic practice of burying the bones of bears that had been killed for their highly esteemed meat. They quoted the seventeenth-century author Niurenius, who wrote:

When one asks the reason for this [burying the bones, after eating the meat], they say that it is done so that they may, without injury and danger to life, catch and kill the dangerous bears. Formerly when all the animals could speak, the bear said and proclaimed that he would always allow himself to be killed without injuring any man, provided that after his death, they honoured him.

Beliefs like these, of indigenous hunter-gatherer societies, are in direct opposition to the wanton carnage practiced by more "highly developed" societies, say of the Roman period, when countries were invaded and wild animals were captured in their thousands to be slaughtered in the Circuses. As Toynbee (1973) commented, the deep sensitivity the Romans felt toward the natural world on an individual basis contrasted starkly with the pleasure they took at the sight of huge numbers of magnificent animals being ignominiously slaughtered in the public spectacles. It is equally difficult for us to understand the appalling sport of bear-baiting that was so popular in the Middle Ages, only to be replaced by bull-baiting after the brown bear had become a rare animal. The baiting of animals was not outlawed in Britain until 1835, and, as reported by the Libearty Campaign (World Society for the Protection of Animals 1993), the torment of "dancing" bears, as well as bear-baiting, continues in eastern Europe and Asia today.

Large Cats

In the ancient world it was the lion more than any other animal that symbolized the power and grandeur of the conquering hero. Beginning with the Assyrian Empire in the ninth century B.C. and continuing to the present day in the heraldic arms of European families, lions have been portrayed, not only as the reserved prey of kings, but also as guardians of their palaces.

So many lions were required for the royal hunts and for sacrifices in Assyria that they had to be captured and held in cages to await their fate. Ashurnasirpal II (883–859 B.C.) claimed:

In those days on the opposite bank of the Euphrates I have killed 50 wild bulls and 80 I caught. I killed 30 elephants with my bow and arrow, in royal combat I killed 215 powerful wild bulls. By stretching out my hand, and with a stout heart, I caught 15 strong lions in mountains and forests. (Quoted from Hobusch 1980:35)

The famous stone bas-reliefs (now mostly in the British Museum) from the Assyrian palaces of Nimrud and Nineveh (800–600 B.C.) show lions being released from cages to be shot at close range by the king. At the time, the lion was still a common predator of domestic livestock throughout western Asia, as is shown by portrayals on Sumerian seals and in many references in the Old Testament. In Isaiah (31:4), for example, we read, "Like as the lion and the young lion roaring on his prey, when a multitude of shepherds is called forth against him." It may be that the lion and the wolf had both increased in numbers with the ready availability of domestic livestock, even while the relentless pressure from human hunters was steadily increasing.

As described by Toynbee (1973), lions were first displayed in Rome in 186 B.C. In 55 B.C. Pompey displayed about 20 elephants, 600 lions, 410 female leopards, a lynx, a rhinoceros, and some apes. Lions were brought to the Circuses from Africa, Arabia, Syria, and Mesopotamia. The first Roman emperor, Augustus (63 B.C.–A.D. 14), in order to enhance his position as ruler of the world, employed more than 1000 gladiators, and more than 3500 animals were killed in combats in the arenas during his reign. Besides those that were killed, thousands of wild animals that had been captured in North Africa and Asia were brought to Rome by boat for display in the spectacles. These included hundreds of lions, leopards, bears, and some tigers.

The fighting of gladiators was forbidden by Roman law in A.D. 325, but the "amusements" with wild animals continued, even though lions were becoming scarce. In A.D. 414 the common people were still allowed to kill them for their own safety, but hunting and trade in lions was allowed only by special license, as shown by the following decree (Pharr 1952:436):

We allow everyone the right to kill lions, and We permit no one at any time to fear malicious prosecution therefor, for the safety of Our provincials necessarily shall take precedence over Our amusements; nor does that appear at all to hinder Our amusements, since We have given license only to kill wild beasts, not to hunt and sell them.

In the ancient world, as in modern times, the taming of lions was relatively common, and there are several contemporary accounts of lions recognizing and sparing human friends in the arena, the best-known tale being that of Androcles and the lion, which was poignantly related by the Roman writer Aelian (born ca. A.D. 170: VII.48, trans. Scholfield 1971), and mention may be made here of the earlier Biblical tale of Daniel in the lion's den (Daniel 6:16–24).

Though throughout history the lion has been the ultimate symbol of majesty and power, the tiger has been the most feared of all carnivores, and the target for concerted attack by the British in India, who were determined to rid the Empire of this "intolerable pest." But they had first to overcome the ancient traditions of the Indian people, who considered it wrong to kill a tiger unless it attacked first. For this reason, in 1888, "the reward for killing a Royal Tiger was raised to two hundred florins." It may be noted that this information is given in a students' textbook on mammals written 100 years ago by Flower and Lydekker (1891:514). These authors, both of them acclaimed biologists, clearly believed that the sooner the tiger was exterminated, the better for the world. About the leopard they wrote, "In habits the Leopard resembles the other Cat-like animals, yielding to none in the ferocity and bloodthirstiness of its disposition."

The Bonds between Humans and Carnivores

As discussed by Ingold (1993), the modern Western view of the human hunter is of someone who seeks to capture or kill animals in the wild, animals previously untouched by human society, not for food but for the challenge and excitement of the hunt. Integral to this view is the belief that humans are outside and above nature, and that the animal prey will be unaware of the hunter except in the final chase. But as Ingold (1993) and Thomas (1990) make clear, in human societies based on hunting and gathering there is no distinction between the human and the natural worlds; the two merge into "a single field of relationships." Thus the account by Thomas (1990) of how, as recently as the 1950s, lions and the Juwa hunters of the Kalahari lived in mutual trust, assessing each other's strengths and understanding each other's motives, seems highly plausible. That mutual trust must have been the way of the Kalahari for the last 5 million years, since the emergence of the earliest hominids.

The few remaining hunter-gatherer societies—like those of the Kalahari, scattered in environmental refuges today—are relics of the way of life of humans all over the world until roughly 10,000 years ago. Until the late Pleistocene, throughout the world, human hunters lived in ecological balance both with other carnivores and with their prey. Inevitably, however, the ever-increasing success of the long-distance projectile and the hidden pitfall trap or snare, combined with control of fire, led to human supremacy in the hunt. Probably nearly everywhere (except in southern Africa) by 10,000 years ago this supremacy was enhanced with the aid of the domestic dog (Clutton-Brock 1984; Clutton-Brock and Jewell 1993). And so the alienation of humanity from nature began. Even so, the division between the human and animal worlds is emphasized only in those societies that are strongly influenced by Western philosophy and religion.

It is generally believed that one of the remarkable, if not unique, characteristics of humans is their ability to command and communicate with so many other species of the higher vertebrates. I should like to argue that this is a fallacy, which, like many other aspects of the relationships between humans and animals, has deep-seated roots in the Eurocentric view of the supremacy of mankind. The belief that animals are not sentient beings, that they have a threshold of pain higher than that of humans, and that they lack powers of interspecific communication probably had its beginnings with the Roman domination of the Western world. Speech is not necessary for communication, and the powers of different species of animals to understand each other are greatly underestimated. Understanding of this sort is especially keen between carnivores: a person may be able to tell a dog to sit, but the dog can just as easily dominate a cat with a look, and a lion can fully express its feelings to a hyena.

As has been repeatedly noted over the last 30 years, human society is very like wolf society, and this should not be surprising, since both of these hunting species evolved highly social, intelligent, and versatile behavioral patterns (Scott and Fuller 1965; Hall and Sharpe 1978; Clutton-Brock 1987). Both wolf and human depend on their family groups for sustenance and companionship and for help with rearing their young. The great strength of the bond between human and dog (wolf) derives from their understanding each other's dominant and submissive postures, their many common emotional responses, such as love, hate, and jealousy, and their similar methods of communication. When in close contact, humans speak, laugh, or cry, while dogs bark, growl, whine, or yap; humans smile, while dogs "look pleased" and wag their tails. The two hunting species even have the same ability to communicate over long distances, the human with a whistle or yodel, the wolf with a howl. It is little wonder, then, that the dog has been the universal companion of humans, worldwide, for the last 10,000 years.

The close bond between people and cats has a different basis, and it is often postulated that it was the cat that initiated the relationship, for the cat is "the animal that walks by itself" (Clutton-Brock 1993b). There are rather few remains of early domestic cats from archaeological sites, but all evidence points to their origin in Egypt or western Asia about 5000 years ago. Although it is true that the most likely ancestor of the domestic cat, the African wild cat (*Felis silvestris lybica*) is a solitary hunter, it appears that its behavior is so versatile that it can readily adapt to living in the social environment of the human home (Clutton-Brock 1993b; Turner and Bateson 1988). And, like the dog, the cat has the ability to understand and communicate with its human companions. Finally, both dog and cat have a great capacity for prolonged sleep and contentment, which enables them to remain relaxed while the humans around them carry on with their own activities.

Domestication of Carnivores

The domestication of carnivores is both a biological and a cultural process, one that begins when a small number of individuals are separated from the wild species and become habituated to humans. If these animals then breed, they form a founder group, but the progeny of that group changes over successive generations both in response to natural selection under the new regime of the human community and its environment, and by artificial selection for economic, cultural, and aesthetic reasons (Bökönyi 1989; Clutton-Brock 1987, 1992, 1993a, 1993b; Ducos 1989; Hemmer 1990; Turner and Bateson 1988). The relationship between human and animal is transformed from one of mutual trust, in which the environment and its resources are shared, to total human control and domination (Ingold 1994).

In the past, biologists and ethologists have put domestication aside, as a phenomenon that is part of human culture and outside nature. I should like to argue that domestication is evolution, and that breeds of domestic animals (and cultivated plants) should be considered as local ecotypes or demes with special adaptations to particular micro-environments (Berry 1969). Thus the dog can be seen as an ecotype separate from the wolf (*Canis lupus*), or, as I prefer, it can be considered as a separate species (*Canis familiaris*), since dogs are (normally) reproductively isolated from wolves. And in answer to those who say that the dog cannot be a separate species because the interbreeding of dog with wolf will produce fully fertile offspring, so will, I hasten to point out, the interbreeding of dog with jackal, dog with coyote, and coyote with wolf (Gray 1972; Clutton-Brock and Jewell 1993), though these canids are never collectively classified as a single species.

There are, as I see it, two routes to domestication. The first is through the enfolding of tamed animals into a human community, where they form a breeding population that in time becomes isolated from, and different from, the wild species. This process can occur only when the natural behavioral patterns of the originating animal species match those of humans; that is, the animals must be gregarious or social, breed readily, have a wide home range and no particular territorial restrictions, and require only a short flight distance. The wolf fulfills all these conditions, and it is easy to see why it was the earliest animal to be domesticated. The cat, however, does not, and I believe that its domestication (as with other "solitary" carnivores like the ferret) followed the second route, through commensalism. The commensal bond may be either temporary or permanent; it is an association between individuals of two species in which one species benefits from the other, without untoward impact on the other.

The most plausible explanation for the early associations of the wild cat (*Felis silvestris*) with humans is that the cat became a tolerated commensal, attracted to settlements, buildings, and food-stores by rodents, lizards, and other small, noncommensal animals already established there as pests.

The significant difference between commensalism and domestication is that the commensal animal benefits from the human environment but lives and breeds in the wild, whereas the domesticated animal is owned, and its breeding and whereabouts are controlled. It is often claimed that fully domestic animals benefit from their association with their human owners, and are thus commensals, but I believe that only wild animals can live as commensals, because, as written by Darwin (1868:4), domestic animals, "have been modified not for their own benefit, but for that of man." It is a moot point whether in terms of genetic success the dog benefits more than the wolf.

The same species can associate with humans either as fully domesticated animals or as commensals, common examples being dogs, cats, domestic ferrets, black and brown rats, mice, and pigeons (Clutton-Brock 1993b). All these species could have been domesticated in the first instance through the taming route or the commensal route, or through a mixture of the two. Even after generations of breeding under domestication, animals can revert to living as commensals, or even as truly "wild," in which case they are said to be *feral*; that is, they form a breeding group of animals that has derived from domestic stock but now lives in a self-sustaining population.

It is probable that for most mammals that were domesticated, including the wolf and the wild cat, which had very widespread distributions, the process occurred many times over in different parts of the world, and that the initial domestic populations were bred from very small founder numbers (Clutton-Brock 1992). In later times, individuals with unusual characteristics were preferred, and the domestic populations took on a look different from that of the wild species. Consider for example the single-colored, yellow coat of the dingo or an up-standing tail or lop ears. Selective breeding would enhance characters that originally appeared as chance mutations, and the "new" animals would spread rapidly over whole continents. For example, dingoes are very likely all descended from a few individual dogs that were taken in boats to Australia by aboriginal people 12,000 to 5000 years ago (I shall return to the dingoes shortly).

Once domestication is established, new breeds are produced by further reproductive isolation, leading to genetic drift, as occurs in the founder populations of new subspecies in the wild. The founders of the new breed contain only a small fraction of the total variation of the parent species (Clutton-Brock 1992; Hemmer 1990). The new breed is thus a genetically unique population which, by shifting of gene frequencies, may, in Mayr's words, "give rise to an evolutionary novelty" (Mayr 1970:361). A certain amount of (usually controlled) inbreeding of the founders will stabilize the phenotype, and we can see in this process how in time the wolf gave rise to both the Irish wolfhound and the chihuahua.

The first stages of domestication are almost always accompanied by a reduction in body size and a shortening of the jaws. This trend is so marked that it is taken as the criterion for recognizing the first signs of domestication in the remains of most mammals from archaeological sites (Clutton-Brock 1987, 1992).

The perceptual world of the animal under human protection is very different from that of its wild progenitor, which must search for its own food and escape from predators, and there is strong selective pressure, under domestication, for animals that are placid and not fearful. This pressure has led, in the domestic carnivores, to a reduction in the complexity of the neocortex (and hence in the size of the brain) and in the lowering of responses by the sensory organs (Hemmer 1990). The auditory bullae, as well as the cranial capacities, of domestic dogs and cats are relatively much smaller than those in the wolf and the wild cat. What is more, these reductions appear to be irreversible, since feral dogs and cats (that is, animals that have returned to live in the wild, like the pariah dog) retain the size range for bullae and brain of the domestic forms (Hemmer 1990).

Concomitant with the effects of artificial selection on the domesticate are the same forces of natural selection that occur in wild species. For example, the effects of climatic selection on domestic animals appear to parallel the well-known correlations in size and body shape that can be seen in subspecies of wild animals across a geographical cline. For example, the Scottish wild cat (*Felis silvestris grampia*), with its small ears, stocky limbs, and heavy body, contrasts with the more gracile African wild cat (*Felis silvestris lybica*), and domestic cats of the same regions tend to reflect these body shapes (Clutton-Brock 1993b).

It has been shown for a variety of mammals, including foxes and cats, that there is a close link between coat color and temperament. Black and blotched tabby cats are more placid than the wild-type striped tabby, and in fact the former are found more commonly in the urban environment, whereas striped tabbies are more common on farms (Clutton-Brock 1993b; Todd 1977).

Domesticated Carnivores in Times Past

Cheetah

Since ancient Egyptian times (ca. 1450 B.C.), the cheetah (*Acinonyx jubatus*) has been a symbol of nobility and high status, and were it not for the fact that, in the past, cheetahs rarely bred in captivity, they could be common domestic animals today. The custom of employing cheetahs, the fastest of all carnivores, to pursue gazelle and other quarry appears to have begun in Egypt, to have spread from there to ancient Assyria, and to have moved on to India. It was during the time of the Moghul emperors, however, that the cheetah became an indispensable adjunct in the royal hunts, and Akbar (A.D. 1542–1605) is said to have kept 9000 cheetahs (Clutton-Brock 1987). All of these were caught in the wild, and it is therefore not surprising that the cheetah became extinct in Asia.

Mongoose

In ancient Egypt the ichneumon, or Egyptian mongoose (*Herpestes ichneumon*), may have been as important a commensal as the cat. According to Diodorus Siculus (80–20 B.C.) and other classical writers, this was because the mongoose

goes about breaking the eggs of the crocodiles, since the animal lays them on the banks of the river, and—what is most astonishing of all—without eating them or profiting in any way it continually performs a service which, in a sense, has been prescribed by nature and forced upon the animal for the benefit of men. (I:35.7, trans. Oldfather 1989)

The crocodiles were of course very dangerous, but they were not killed, because they prevented robbers from swimming across the Nile from Arabia and Libya to invade Egypt (I:88.3). Thus by looking upon both mongooses and crocodiles as sacred animals, the rulers maintained a fine balance between the two with little effort from the people.

Mongooses were also valuable for killing snakes, and from New Kingdom times (1567 B.C.) onward they were allowed to roam around the temples (Zeuner 1963:404). They were fed, like the cats, on milk and raw fish, and when they died their bodies were mummified in very large numbers (Diodorus Siculus I:82.3–83.4–8, trans. Oldfather 1989).

The introduction of the small Indian mongoose (*Herpestes auropunctatus*) to the West Indies in 1872, with the aim of reducing rodent pests in the sugar plantations, has had a less salutary outcome. In the words of Lever (1985:78): "As in the case of feral cats in New Zealand the havoc wrought by mongooses on wildlife in the Caribbean reads like a tale of disaster."

Anubis

Representations of Anubis are among the most common of all depictions from ancient Egypt. With the head of a "dog" and the body of a man, Anubis was god of the dead and chief companion of the great god Osiris, whom the Egyptians believed went on a campaign to spread cultivation throughout the world, traveling first to Ethiopia. According to a little-quoted account by Diodorus Siculus (I:18.2, trans. Oldfather 1989), Osiris was accompanied on his travels by his two sons, Anubis wearing a dog's skin and Macedon wearing the foreparts of a wolf. They were distinguished for their valor, and it was for this reason that these two animals were held in special honor by the Egyptians.

Anubis is of interest to biologists because of the anomalies in the systematics of the so-called Egyptian jackal (*Canis lupaster*) and the Simien jackal (*Canis simensis*). The question being: are these canids really jackals, or are they wolves, or hybrids, or even anciently feral domestic dogs? To me, the head of the typical Anubis looks very like the head of a Simien jackal, with its long, thin snout and large ears. It does not seem improbable that this canid was known to the Egyptians, and because it was doglike but exotic it became especially revered.

Ferguson (1981) has made a strong case for *Canis lupaster* being a small wolf rather than a large jackal, and his biological criteria may be supported by the new molecular evidence of Wayne (1993) that *Canis simensis* is closer to the gray wolf (*Canis lupus*) than to the jackals. If so, could it be possible that the Simien "jackal," rather than being a wild canid that was known to the ancient Egyptians and revered as a god, is actually the relic of an anciently feral population of a special breed of "Anubis" dogs, dating to ancient Egyptian times, and perhaps descended from *Canis lupaster*? This suggestion may sound wildly improbable, but it is supported at least in part by the single-colored yellow coat of *Canis simensis*, which is found in no other wild canid but is a characteristic of early domestic dogs, for example the dingo. Unfortunately, the question cannot be resolved with breeding experiments, as all species of jackal are interfertile with both *Canis lupus* and the domestic dog (Gray 1972).

Dingo

That the dingo is a feral domestic dog of ancient origin is undisputed, and it can be postulated that dogs were first taken to Australia by aboriginal people, by boat, between 12,000 and 5000 years ago (Barker and Macintosh 1979). These parameters of time are based on the reasoning that because no remains of pre-European dogs have been found on Tasmania, it is unlikely that dogs were in Australia before this island became geographically isolated. Tasmania's separation occurred by the breaking through of the Bass Straits around 12,000 years ago. It also seems safe to conclude that dogs had reached the continent before the pig became a widespread domestic animal in Southeast Asia about 5000 years ago, because the pig never reached Australia. The first dingoes may well have been taken to Australia as a source of meat rather than as hunting companions. Like the singing dogs in New Guinea (another ancient feral "breed") the dingo is most probably a direct descendant of dogs that were originally domesticated from tamed Indian wolves (Corbett 1985). The earliest radiocarbon date for remains of dingo is 3450 ± 95 years before present (Milham and Thompson 1976).

Early accounts by Europeans of the aborigines and the dingo describe a way of life that had probably changed little for thousands of years. In the nineteenth century A.D. the great majority of dingoes lived and hunted as wild carnivores. The aboriginal families kept some dingoes as pets, used some as hunting partners, ate them when meat was scarce, and also lavished affection on them.

Today, the dingo is treated as vermin and is in great danger of extinction, both from being slaughtered whenever possible and through interbreeding with European feral dogs (Ginsberg and Macdonald 1990). The extermination of the dingo would be a great loss, because it is part of the living heritage of aboriginal culture and part of Australian history.

South American Canids

Until the arrival of the Spanish in the eighteenth century the East and West islands of the Falklands were inhabited by only one terrestrial mammal. This was a fairly large canid that was described by early travelers as either foxlike or wolflike. It appeared to be a unique species, and it became known as the Falkland Island wolf, but unfortunately its thick fur became popular with fur traders, and the "wolf" was extinct by the close of the nineteenth century. When he visited the islands from the *Beagle* in 1836, Darwin (1845:194) recorded that the "wolf" fed on birds, particularly the upland goose. He called the canid *Canis antarcticus* and wrote: "As far as I am aware, there is no other instance in any part of the world, of so small a mass of broken land, distant from a continent, possessing so large an aboriginal quadruped peculiar to itself."

Hamilton Smith, who saw large numbers of skins of the canid in a New York fur store in 1839, said that the skins were indistinguishable from those of the coyote, *Canis latrans* (Hamilton Smith 1839). Later, on the basis of the few skulls and skins collected by Darwin, the canid was classified as *Dusicyon australis* by Pocock (1913), who saw it as closely related to the South American foxes of this genus. But in a general review of the family Canidae, it transpired that the skull characters of the Falkland Island wolf were closer to the North American coyote and the Australian dingo in the genus *Canis* than to the South American foxes of the genus *Dusicyon* (Clutton-Brock, Corbet, and Hills 1976). Recently, there has been some support for this finding in a preliminary molecular analysis by Wayne (pers. comm. 1990), who found that the mitochondrial DNA sequence of the Falkland Island wolf is closest to that of the coyote.

Because there have been no records of any member of the genus *Canis* in South America since the end of the Pleistocene, if the Falkland Island wolf was indeed a member of the genus *Canis*, rather than *Dusicyon*, the question arises how it got to the islands. The Falklands are some 480 km from the mainland of South America, and it is unlikely that a large carnivore could have survived on the islands during the periglacial conditions of the last ice age. So perhaps the "wolf" was an anciently feral domestic dog taken there by humans thousands of years ago, and it does share with the dingo the characters that are associated with domestication: white markings on the pelage, a wide muzzle with large, compacted teeth in the premolar region, and expanded frontal sinuses (Clutton-Brock 1977).

On the mainland of South America there is some evidence that the indigenous foxes belonging to the genus *Dusicyon* were also domesticated, for we have the word of Hamilton Smith (1839) that the native South Americans kept "Aguara dogs," which were tamed foxes. He wrote, "Several [members of the genus *Dusicyon*] can be sufficiently tamed to accompany their masters to hunt in the forest. . . . They are in general silent and often dumb animals; the cry of some is seldom but faintly heard in the night, and in domestication others learn a kind of barking." Hamilton Smith further observed that these Aguara "dogs" were being supplanted by imported European dogs, which, being descended from the social wolf, were much more companionable and would breed in the domestic environment.

Carnivores and the Fur Trade

One of the main functions of imperialism is to exploit the resources of the invaded territory, and for the last 3000 years, until the last decade or so, furs and ivory have counted as among the most valuable of these resources. As every "new" continent was "discovered," its wildlife was desecrated or exterminated.

Beginning in the seventeenth century the Dutch, French, and then the English (with the founding of the Hudson's Bay Company in 1666) began an onslaught on the carnivores of the North American continent. In the words of Davey (undated, ca. 1895:60):

The virgin forests and waters of America furnish countless beavers, rich sables, the pine and stone martins, the beautiful mink, lynx, badger, [raccoon], the choicest white and black fox, the cross, blue, red, and white fox, the seal, and sea otter, the opossum, the bison, the black and grisly bear, besides others too numerous to mention.

Davey quotes the *yearly* import of pelts to England alone as 50,000 wolves, 30,000 bears, 22,000 American otters, 750,000 raccoons, 40,000 cats, 50–100,000 American "sable" [American pine marten, *Martes americana*], and 265,000 foxes of various kinds (including North American and Siberian subspecies). Toward the end of his book, written to promote sales from The International Fur Store in London, Davey expounded as follows:

The introduction of late years of the use of the skins of the larger animals, such as the Polar bear, the buffalo, the bison, the tiger, the leopard, and even the lion, as decorations in the furnishing of large apartments, has led to the increased commerce in these skins, which are unquestionably beautiful, not only as mats and rugs, but also for wall decoration. The skin of the leopard makes exceedingly pretty chairbacks. (Davey undated, ca. 1895:102)

This huge demand for furs was bound to exceed the available supply, and from the beginning of this century the breeding of captive carnivores on fur farms became a new and very profitable industry in many countries (see, for example, Catchpole [1930], who was official inspector to the Silver Fox Breeders Association of Great Britain). Inevitably, there were numerous escapes from the fur farms, with the result that exotic carnivores became naturalized in new countries, causing disruption to the already precarious balance of the indigenous wild carnivores and their prey. This has been particularly serious with the American mink (*Mustela vison*), which by the 1950s had become naturalized in the British Isles, Iceland, Scandinavia, Netherlands, France, Spain, and Russia (Lever 1985).

Another carnivore that has spread across central and eastern Europe, as a result of introductions from the Far East to the USSR in the 1920s and 1930s, for the value of its fur, is the raccoon dog (*Nyctereutes procyonoides*). Lever (1985:45) warned that if any of the 140 raccoon dogs imported from Europe in 1980 by a fur farmer in Ontario, Canada, were to escape they could become a serious threat to the wildlife of North America.

A by-product of fur-farming has been information about the process of domestication and the link between temperament and coat color, which was revealed by the research of Clyde Keeler between the 1940s and 1960s on wild American mink and ranch mink and the red fox (*Vulpes vulpes*). In both mink and fox Keeler (1975) found that the paler-colored varieties bred on fur farms were heavier, had smaller adrenal glands, and were notably less aggressive and easier to tame than were the wild type.

The weight of the foxes increased, and the relative weight of their adrenal glands decreased, in the order of wild color to silver to pearl to amber. Tame behavior of the foxes also correlated with these weights, for while red and silver foxes reacted in fright when a stranger approached to within about 180 m, pearl foxes did not react until the observer came within about 130–150 m, and amber foxes could be approached to within 5 m (Hemmer 1990; Keeler 1975).

In other experiments on the silver variety of the red fox, on a fur farm in the former USSR, Belyaev and Trut (1975) described how they succeeded in breeding foxes that they claimed behaved in every way like domestic dogs. They began by selecting for their breeding stock those 6- to 8-week-old fox cubs that were most responsive to being called, to being hand-fed, and to being handled. After 15 years of selection for docile behavior, the foxes would come when they were called, the most responsive would wag their tails in greeting, and they would bark on seeing people. And what is perhaps most significant, since there are many records of tamed foxes, the breeding cycles were altered so that the females came into estrus earlier and sometimes twice a year.

Conclusions

The science of ethology, as it has developed over this century, has been based on the study of behavioral patterns in populations of mainly wild animals. The vast numbers of observations and investigations that have been carried out on behavior are of course of fundamental importance for the understanding of all living animals. As I see it, however, there has been a serious lack of research into personality and temperament, that is, in the study of individual variations in behavior, which are especially important in carnivores. Everyone who lives with a dog or a cat knows that the animal is an individual with a unique personality whose responses to other animals, people, and situations are variable and often unexpected. Yet the flexibility of behavior and individual personalities of wild carnivores, which ostensibly live in much more demanding environments than do the domestic breeds, seldom receive any comment from the biologists who study these animals. The behavior of the individual is too often assumed to typify the behavior of the species, and individual temperaments are ignored. This failing must be a legacy of the fear of being labeled anthropomorphic, which was the bugbear of the 1960s, but surely it is time to see that the learned behavior (what I call culture) of individual animals is of great value for the understanding of animal societies and interactions.

Study of the learned behavior of the highly intelligent carnivores must also have relevance for understanding the development of learning in humans. I strongly believe, with Thomas (1993) and Bekoff and Jamieson (this volume), that there is also a great need for research into the consciousness of animals, for I find it as extraordinary to deny that animals are conscious beings with thoughts and emotions as to claim that they do not feel pain.

As shown by Thomas (1993), observations on the personalities and variable behavior of individual animals can be made on domestic dogs easily and with little cost. Until very recently, however, the behavior of domestic animals has been a neglected field of study. Indeed, it has been considered unworthy of the biologist's time, because domestic animals are thought of as outside the natural world and artifacts of human culture. I believe, to the contrary, that there is no dichotomy between the natural and the human environment, and that as much can be learned about consciousness in animals from living with a family of dogs as from observing a family of wolves, though the latter will accord the observer higher status among his or her peers.

It is evident that the hunting and reproductive behavior, ecology, and distributions of the canids, bears, and large cats has been interwoven with human activity over at least the last 500,000 years. Humans must have been killing carnivores for food, for their thick furs, and as competitors for game since the first hunters learned how to survive the freezing cold of the northern hemisphere during the ice ages. The effects of human hunting meant that the large carnivores that had evolved as the top of the ecological pyramid became, themselves, prey that had to learn to outwit the master predator (Laughlin 1968). This must have affected the hunting, feeding, and breeding behavior of all species in a multitude of ways, but perhaps most obviously in transforming daytime hunters, such as the wolf (see Mech 1970:152), jackals, and coyote, into nocturnal predators. The avoidance of humans thus becomes part of the culture of all wild carnivores, large or small, social or solitary. If the carnivore does not learn this from its elders it is fearless: Darwin was able to walk up to a Chiloe fox, or gray zorro (*Dusicyon fulvipes*), on the island of Chiloe, in southern Chile, and kill it with his geological hammer (Darwin 1845:268).

To the early hunter-gatherers the large carnivores must indeed have been dreaded foes. Geist (1989) has even postulated that humans arrived so recently in North America because fear of its large predators (and in particular of the huge, extinct bear *Arctodus simus*) kept them on the other side of Bering Strait. In modern times the wolf, the large cats, and the bears are still considered by perhaps most people to be ruthless killers and symbols of the wild. And yet, is it not strange that the dog and cat (and teddy bear) are the most familiar of all animals and the beloved companions of millions of humans. The paradox of the feared carnivore and its domesticated counterpart must have been felt since the first wolves began living around human settlements, at least 12,000 years ago. It is paralleled by all human conflicts and wars, where strangers are killed simply because of their unfamiliar language, behavior, religion, and customs. So huntsmen will use the dogs that they know to kill the wolves or the foxes that they do not know.

In early times the killing of large carnivores was necessary for human survival, but gradually, with the spread of agriculture into Europe and Asia over the last 3000 years, hunting became the prerogative of the nobility, and the high status of the huntsman was enhanced by tales of the extreme ferocity of his prey. Throughout the Medieval period in Europe, and for at least 600 years, there were strict laws to control hunting and to ensure that it remained an elitist pursuit. From childhood, the sons of feudal lords had to learn the skills of horsemanship and the complicated language and protocol of the huntsman as an essential element of training for battle. The carnivores that were considered to be beasts of venery (the most highly valued quarry) were the wolf and the bear. The wild cat and the fox were beasts of the chase, and the badger and otter were classed as vermin.

As recently as three hundred years ago the slaughter of these animals, on a huge scale, persisted in central Europe. For example, in Prussia in 1700, 4300 wolves, 147 bears, and 229 lynxes were killed (Hobusch 1980:152). According to the same author (p. 101) there was a surprising (at least to modern Europeans) demand for the meat of hunted carnivores. In 1669, carcasses of the following animals were for sale as smoked or salted meat from a provision store in Dresden: 15 bears, 74 wolves, 15 lynxes, 55 badgers, 170 foxes, and 27 otters; another store, meanwhile, supplied the court at Dresden with 12 whole foxes for roasting and 52 for stewing.

The faunal compositions of entire continents have been molded not only by the killing of carnivores in the millions for their furs but also by the introductions of carnivores to new countries, the earliest of which may be the dingo, the most recent the American mink.

My study of the history of human associations with wild and domestic animals has made clear that there are no wild places on the habitable earth, and almost no species of animal or plant, that have not been influenced by human activity. In the future, human interference with what is considered to be the "natural world" is bound to increase. As all wildlife becomes restricted to reserves with ever-diminishing boundaries, populations of wild animals, and especially the large carnivores, will have to be managed. Their breeding will have to be controlled: those on the edge of extinction will have to be taken into schemes of captive breeding, and those that are too numerous will have to be culled.

In the management of wildlife reserves there will indeed come a time when it is difficult to draw a line between the wild, the captive, and the tame. But even in this situation the question of whether hunting for sport should be allowed will continue to be a central issue. We should remember the kings of ancient Mesopotamia, who shot at lions as they were released from cages, and ponder on the progress of humanity.

References

Barker, B. C. W., and Macintosh, A. 1979. The dingo: A review. *Archaeol. and Phys. Anthropol. in Oceania* 14(1):27–42.

Belyaev, D. K., and Trut, L. 1975. Some genetic and endocrine effects of selection for domestication in silver foxes. In: M. W. Fox, ed. *The Wild Canids*, pp. 416–426. New York: Van Nostrand Reinhold.

Berry, R. J. 1969. The genetical implications of domestication in animals. In: P. J. Ucko and G. W. Dimbleby, eds. *The Domestication and Exploitation of Plants and Animals*, pp. 207–218. London: Duckworth.

Bökönyi, S. 1989. Definitions of animal domestication. In: J. Clutton-Brock, ed. *The Walking Larder Patterns of Domestication, Pastoralism, and Predation*, pp. 22–27. London: Unwin Hyman.

Catchpole, G. G. 1930. *The Silver Fox.* London: Metropolitan Press Agency.

Clutton-Brock, J. 1977. Man-made dogs. *Science* 197:1340–1342.

Clutton-Brock, J. 1984. Dog. In: I. L. Mason, ed. *Evolution of Domesticated Animals.* pp. 198–211. London & New York: Longman.

Clutton-Brock, J. 1987. *A Natural History of Domesticated Mammals.* Cambridge: Cambridge Univ. Press; and London: British Museum (Natural History).

Clutton-Brock, J. 1992. The process of domestication. *Mamm. Rev.* 22(2):79–85.

Clutton-Brock, J. 1993a. Domestication of animals. In J. S. Jones, R. D. Martin, and D. R. Pilbeam, eds. *Cambridge Encyclopedia of Human Evolution,* pp. 380–388. Cambridge: Cambridge Univ. Press.

Clutton-Brock, J. 1993b. The animal that walks by itself. *1994 Yearbook of Science and the Future,* 156–177. Chicago: Encylopaedia Britannica.

Clutton-Brock, J. 1994. The unnatural world: Behavioural aspects of humans and animals in the process of domestication. In A. Manning and J. Serpell, eds. *Animals and Human Society: Changing Perspectives,* pp. 23–35. London: Routledge.

Clutton-Brock, J., Corbet, G. B., and Hills, M. 1976. A review of the family Canidae with a classification by numerical methods. *Bull. British Museum (Nat. Hist.) (Zool.)* 9(3):119–199.

Clutton-Brock, J., and Jewell, P. A. 1993. Origin and domestication of the dog. pp. 21–31. In H. E. Evans, ed. *Miller's Anatomy of the Dog,* 3rd ed. Philadelphia: W. B. Saunders/Harcourt Brace.

Corbett, L. K. 1985. Morphological comparisons of Australian and Thai dingoes: A reappraisal of dingo status, distribution, and ancestry. *Proc. Ecol. Soc. Australia* 13:277–291.

Darwin, C. 1845. *Naturalist's Voyage around the World.* London: John Murray.

Darwin, C. 1868. *The Variation of Animals and Plants under Domestication,* 2 vols. London: John Murray.

Davey, R. ca. 1895 (undated). *Furs and Fur Garments.* London: The International Fur Store and Roxburghe Press.

Ducos, P. 1989. Defining domestication: A clarification. In: J. Clutton-Brock, ed. *The Walking Larder Patterns of Domestication, Pastoralism, and Predation,* pp. 28–30. London: Unwin Hyman.

Ferguson, W. F. 1981. The systematic position of *Canis aureus lupaster* (Carnivora: Canidae) and the occurrence of *Canis lupus* in North Africa, Egypt, and Sinai. *Mammalia* 45(4):459–465.

Flower, W. H., and Lydekker, R. 1891. *An Introduction to the Study of Mammals Living and Extinct.* London: Adam & Charles Black.

Geist, V. 1989. Did large predators keep humans out of North America? In: J. Clutton-Brock, ed. *The Walking Larder Patterns of Domestication, Pastoralism, and Predation,* pp. 282–294. London: Unwin Hyman.

Ginsberg, J. R., and Macdonald, D. W. 1990. *Foxes, Wolves, Jackals, and Dogs: An Action Plan for the Conservation of Canids.* Gland, Switzerland: IUCN.

Gray, A. P. 1972. *Mammalian Hybrids.* Slough, England: Commonwealth Agricultural Bureaux.

Hafez, E. S. E., ed. 1969. *The Behavior of Domestic Animals.* Baltimore: Williams & Wilkins.

Hall, R. L., and Sharp, H. S. 1978. *Wolf and Man: Evolution in Parallel.* New York and London: Academic Press.

Hallowell, A. I. 1926. Bear ceremonialism in the northern hemisphere. *Amer. Anthropol.* (N.S.)28(1):1–175.

Hamilton Smith, C. 1839. *Dogs.* Vol. 1 In: W. Jardine, ed. *Naturalist's Library.* Edinburgh: Lizars.

Hemmer, H. 1990. *Domestication: The Decline of Environmental Appreciation.* Cambridge: Cambridge Univ. Press.

Hobusch, E. 1980. *Fair Game: A History of Hunting, Shooting, and Animal Conservation*. New York: Arco.

Ingold, T. 1994. From trust to domination: An alternative history of human-animal relations. In: A. Manning and J. Serpell, eds. *Animals and Human Society: Changing Perspectives*, pp. 1–2. London: Routledge.

Keeler, C. 1975. Genetics of behavior variations in color phases of the red fox. In: M. W. Fox, ed. *The Wild Canids*, pp. 399–415. New York: Van Nostrand Reinhold.

Laughlin, W. S. 1968. Hunting: An integrating biobehavior system and its evolutionary importance. In R. B. Lee and I. De Vore, eds. *Man the Hunter*, pp. 304–320. Chicago: Aldine.

Lever, C. 1985. *Naturalized Mammals of the World*. London: Longman.

Mayr, E. 1970. *Populations, Species, and Evolution*. Cambridge: Belknap Press of Harvard Univ. Press.

Mech, L. D. 1970. *The Wolf: The Ecology and Behavior of an Endangered Species*. New York: Amer. Mus. Nat. Hist.

Milham, P., and Thompson, P. 1976. Relative antiquity of human occupation and extinct fauna at Madura Cave, south-eastern Australia. *Mankind* 10:175–80.

Oldfather, C. H. (transl.) 1989. *Diodorus of Sicily*. Loeb Classical Library. Cambridge: Harvard Univ. Press.

Pharr, C. (transl.) 1952. *The Theodosian Code and Novels and the Sirmondian Constitutions*. Princeton, N.J.: Princeton Univ. Press.

Pocock, R. I. 1913. The affinities of the antarctic wolf (*Canis antarcticus*). *Proc. Zool. Soc. (London)*:382–393.

Scholfield, A. F. (transl.) 1971. *Aelian on Animals*. Loeb Classical Library. Cambridge: Harvard Univ. Press.

Scott, J. P., and Fuller, J. L. 1965. *Genetics and Social Behavior of the Dog*. Chicago: Univ. Chicago Press.

Thomas, E. M. 1990. Reflections: The old way. *The New Yorker* 15 October:78–110.

Thomas, E. M. 1993. *The Hidden Life of Dogs*. Boston: Houghton Mifflin.

Todd, N. B. 1977. Cats and commerce. *Sci. Amer.* 237(5)100–107.

Toynbee, J. M. C. 1973. *Animals in Roman Life and Art*. London: Thames & Hudson.

Turner, D. C., and Bateson, P., ed. 1988. *The Domestic Cat: the Biology of Its Behaviour*. Cambridge: Cambridge Univ. Press.

Wayne, R. K. 1993. Phylogenetic relationships of canids to other carnivores. In H. E. Evans, ed. *Miller's Anatomy of the Dog*, 3rd ed, pp. 15–20. Philadelphia: W. B. Saunders/Harcourt Brace.

Zachrisson, I., and Iregren, E. 1974. Lappish bear graves in northern Sweden: An archaeological and osteological study. *Early Norland 5*. Univ. Stockholm.

Zeuner, F. E. 1963. *A History of Domesticated Animals*. London: Hutchinson.

CHAPTER **12**

Design for Killing:
Craniodental Adaptations of Predators

AUDRONE R. BIKNEVICIUS AND
BLAIRE VAN VALKENBURGH

Predaceous carnivorans (members of the order Carnivora) are those whose diets are composed principally of vertebrate prey that they capture and kill. All predators rely on strong skulls, jaws, and teeth to kill and dismember their prey. Killing techniques, however, may differ; felids kill with a single, penetrating bite, whereas canids and hyaenids are more likely to kill with several, shallower bites. Not surprisingly, differences among carnivoran species in killing and feeding behavior are evident in the morphology and biomechanics of their skulls and teeth. Since the publication of Ewer's (1973) superb book, there have been numerous studies of form and function in the carnivoran skull. Most of these studies are more quantitative than those reported in the past; most use the tools of the engineer to model bones and teeth as beams of various dimensions and on that basis calculate expected stresses. Below, we review much of this more recent work, discussing it within the context of the two tasks that are critical for a predator's success, making a kill and consuming a carcass. In addition, we present an example of the application of biomechanical analysis to the inference of function in three extinct carnivorans, the sabertooth cat (*Smilodon fatalis*), the giant American lion (*Panthera atrox*), and the dire wolf (*Canis dirus*). We begin with a discussion of the feeding problems that confront a terrestrial mammalian predator and the anatomical building blocks that compose their skulls, jaws, and teeth.

Struggling Prey and Tough Food

Prey are characteristically uncooperative sources of nutrition; at a minimum, they resist capture, and some threaten their attacker with horns, hooves, or teeth. Attempts by the prey to escape from its assailant may place large, unpredictable loads on the skulls and teeth of the predator, loads that can result in serious injury or death (Hornocker 1970; Packer 1986; Van Valkenburgh 1988). Faced with these dangers, many predaceous carnivorans choose prey much smaller than themselves, which are easier to subdue. Other species, however, habitually and successfully hunt large prey, their efforts enabled by adaptations for prey control such as protractile claws and the ability to lock their jaws (felids and mustelids, respectively: King 1989; Kitchener 1991) or by the cooperation of hunting partners (some canids and one hyaenid among living carnivorans: Mech 1970; Mills 1990).

Because struggling prey present a serious risk to a predator, kills are best made quickly. For a solitary predator, each bite—preferably the first—must be made with precision, positioned so as to take a significant or definitive toll on the prey. As a result, solitary hunters tend to be rather stereotypical in their approach to prey, focusing their attacks on the craniofacial or cervical regions of the prey in order to effect death quickly by neural distress or, more slowly, by suffocation (Figure 12.1A) (Leyhausen 1979; Kruuk and Turner 1967; Rowe-Rowe 1978). By contrast, bites are individually less crucial for predators cooperating on a kill (Figure 12.1B); some individuals of the group hold or distract the prey while others deliver repeated, debilitating slashes to the abdomen (Estes and Goddard 1967; Mech 1970).

Once the prey dies, predators are faced with a new challenge, that of opening and consuming the carcass. Vertebrate carcasses consist largely of skin, muscle, and bone, three materials with different textures and different mechanical properties. Skin, for example, is a composite structure composed principally of collagen and elastin, and is characterized by great pliancy and toughness (Vogel 1988), a combination that ensures that even when the skin is pierced, crack propagation is difficult. The textures and crack-propagation problems of internal meats (e.g., striated muscles and viscera) are similar to those posed by skin. By contrast, bone is a very hard and more brittle material, owing to the presence of hydroxyapatite crystals embedded within the collagen matrix. Large bite forces are necessary to initiate a crack in bone, and the resulting stresses on the teeth involved can be substantial. Whereas muscle and skin are best comminuted with sharp, bladelike teeth, bone-cracking is better accomplished by stout, blunt, cone-shaped teeth (Lucas and Luke 1984). Given these different requirements, carnivorans tend to be relatively heterodont, displaying a battery of teeth that differ from one another in both shape and function.

The effort made by large-bodied carnivorans to capture prey must often be matched by that needed to retain possession of a carcass. As carcass size increases, so does the probability of its theft and the difficulty and value of its defense. Success is largely a matter of the relative mass of the opponents: the larger or more numerous species wins. Interspecific competition for carcass control, which can be fierce, has resulted in the evolution of compensatory behaviors and morphology (Bertram 1979). For example, as relatively slender, mostly solitary hunters, African cheetahs (*Acinonyx jubatus*) commonly yield possession of their kills to larger or more social carnivores (Schaller 1972;

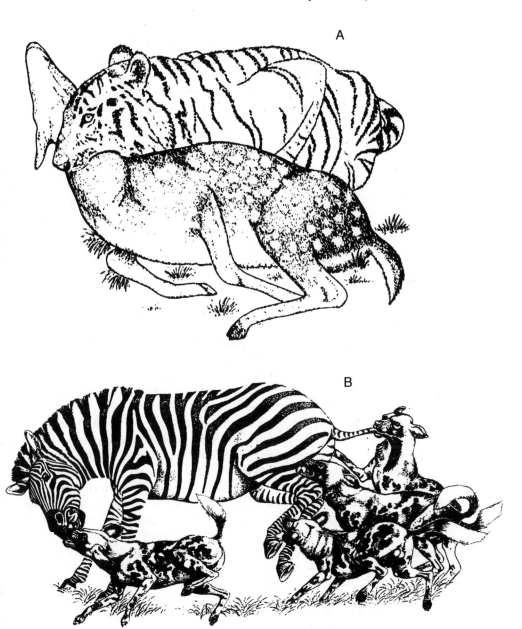

Figure 12.1. Prey capture and killing techniques in predaceous carnivorans. *A*, solitary hunting by felids: the tiger (*Panthera tigris*) is using a throat bite in order to suffocate a deer. *B*, group hunting by large canids: cooperative effort by pack members of hunting dogs (*Lycaon pictus*) in subduing a zebra. (*A*, reprinted by permission from Kitchener 1991; *B*, adapted from Whitfield 1978.)

Mills 1990; Caro 1994). Cheetahs, however, which have comparatively high hunting-success rates, minimize encounters with other predators by seeking their prey during the day. Similarly, leopards (*Panthera pardus*) that are sympatric with wholly terrestrial predators and scavengers cache carcasses in trees, thereby effectively reducing or eliminating competitive interspecific interactions (Schaller 1972; Seidensticker 1976). In addition to competition between species, that within species can be substantial in social predators. For a lion (*Panthera leo*), for example, loss of meat to conspecifics may be more critical than loss to spotted hyenas (*Crocuta crocuta*) (Packer 1986). Such losses can be reduced by aggressive actions toward conspecifics, a potentially dangerous strategy, or by rapid feeding. The one that eats the fastest is likely to get the most. For example, in their rush to consume elements of the carcass quickly, hyenas are known to swallow whole the ribs and articular surfaces of limb bones (Marean et al. 1992). Rapid feeding can benefit solitary hunters as well, if there is a significant risk of carcass theft.

The Anatomical Building Blocks

Given the problems of prey capture, carcass consumption, and carcass retention, predaceous carnivorans must be equipped with appropriate tools for efficiently subduing prey and for rapidly ingesting large quantities of meat. These tools are represented in carnivorans in the form of teeth and the bone in which they are rooted, as well as in the muscles and ligaments that both enable and restrict movements of the craniofacial skeleton.

Vertebrate teeth are composed of a thin cap of enamel over a thick dentin base. Although enamel is the hardest and stiffest of all skeletal materials (Wainwright et al. 1976), repetitive use will score its surface. When the enamel cap is breached, the softer underlying dentin erodes quickly. Teeth must be designed to perform their functions efficiently, since mature enamel degenerates with use and cannot be repaired. Moreover, unlike most fish and reptiles, mammals do not replace their teeth continuously during their lifetimes, and the teeth they take to the hunt are built to last and to fracture only rarely.

The heterodont dentition of carnivorans can be structurally and functionally subdivided into grasping incisors, penetrating canines, and food-processing cheek teeth (Figure 12.2*A*), the cheek teeth comprising premolars and molars. By convention, incisors, canines, premolars, and molars are designated as I, C, P, and M, respectively. Numbers following the designation, usually as superscripts or subscripts, indicate position in the array; for example I^2 and P_4 represent the second upper incisor and the fourth lower premolar, respectively. The carnassial teeth (the upper fourth premolars, P^4, and lower first molars, M_1), which have bladelike (sectorial) modifications in most carnivorans, are used in dividing soft but tough foods such as skin and muscle. Postcarnassial

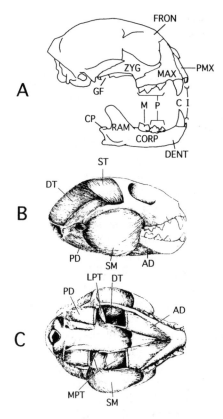

Figure 12.2. Components of the craniofacial anatomy associated with killing and feeding, illustrated in the domestic cat. *A*, skull and teeth (compare molars with those of the canid in Figure 12.8). The temporomandibular joint is composed of CP and GF and the soft tissues surrounding these bony structures. *B, C*, major masticatory muscles in lateral and ventral views. Abbreviations: AD, anterior digastric; I, incisor; C, canine; CORP, mandibular corpus; CP, condyloid process; DENT, dentary; DT, deep temporalis; FRON, frontal; GF, glenoid fossa; M, molar; MAX, maxilla; MPT, medial pterygoid; P, premolar; PD, posterior digastric; PMX, premaxilla; RAM, (ascending) ramus; SM, superficial masseter; ST, superficial temporalis; ZYG, zygomatic (jugal). (*B* and *C* adapted by permission from Gorniak and Gans 1980.)

molars are often modified for crushing bone and other hard food items in the predaceous carnivorans that retain them (e.g., canids, mustelids). Carnivorans that lack postcarnassial molars may process hard foods using appropriately buttressed precarnassial cheek teeth (e.g., the robust premolars in hyaenids).

Teeth are rooted in bone, in the lower jaw (two dentaries comprising the mandible) and the upper jaw (paired premaxillae and maxillae). The dentary is often considered as having two functional regions, the mandibular corpus (or horizontal ramus), which holds the teeth, and the ascending ramus, which serves as a jaw-muscle attachment area (Figure 12.2*A*). Bone is a living tissue capable of adjusting its mass and distribution by growth, and of remodeling itself in response to its surrounding mechanical environment (Lanyon and Rubin 1985). Features of the mechanical environment known to affect the mammalian craniofacial skeleton, and the form of individual bones in it, include food consistency (e.g., Bouvier and Hylander 1984) and differences in feeding behaviors (e.g., Radinsky 1981a; Smith 1993). The response of jaw bones to different loading regimes *in vivo* have been relatively little studied in carnivorans, and analyses on other (mainly herbivorous) mammals provide the bulk of our present understanding; if one can assume some uniformity in bone response to stress within mammals, then one can draw some reasonable expectations for load-stress relationships in carnivoran jaws (see below).

Bones are united at joints. In the masticatory systems of carnivorans, there is only one synovial (freely movable) joint, the temporomandibular joint (TMJ), composed of the mandibular condyle of the dentary (CP, Figure 12.2A), the glenoid fossa of the temporal bone (GF, Figure 12.2A), an interposing intra-articular disc, and surrounding connective tissue. The TMJ is a load-bearing joint during most killing and feeding activities (Scapino 1965; Badoux 1972; Buckland-Wright 1978; Dessem 1989; *contra* Smith and Savage 1959). The mobility of the TMJ of most predaceous carnivorans is most free about a transverse axis (the hinge-like depression and elevation of the lower jaw), and has limited movements about the other axes (Gorniak and Gans 1980; Dessem and Druzinsky 1992). The two dentaries are united anteriorly at the mandibular symphysis, a fibrocartilaginous joint whose mobility is limited by the surrounding soft tissues. In some carnivorans (e.g., some mustelids and ursids: Scapino 1981), the mandibular symphysis ossifies. All other joints between the bones of the face and those of the braincase are sutural, and thus allow little movement (Buckland-Wright 1978).

Jaw closure is accomplished by raising the mandible with the temporalis, masseter, and medial and lateral pterygoid muscles (Figure 12.2B, C), and, in felids at least, there is a simultaneous ventral flexion of the skull over the vertebral column (Gorniak and Gans 1980; Dessem 1989; Gans et al. 1990; Dessem and Druzinsky 1992). Thus the skull is pulled down as the jaws are elevated. Predaceous carnivorans possess a larger temporalis complex than do herbivorous mammals, but a smaller masseter complex (Turnbull 1970; Ewer 1973). The principal jaw-opening muscle is the digastric (AD and PD, Figure 12.2C), which acts by lowering the mandible. The digastric is assisted by nuchal (neck) muscles that raise the head (e.g., semispinalis capitis) (Scapino 1976; Gorniak and Gans 1980). Maximum tension is produced in the jaw-closing muscles at large gapes and in the jaw-opening muscles at close to dental contact (MacKenna and Türker 1978). Consequently, carnivorans are able to generate considerable force when taking large bites, a feature necessary both for prey capture and for rapid ingestion of meat and bones of large diameter. The generation of great force in the jaw adductors has the potential to damage both hard and soft tissues in the mouth; a jaw-opening reflex, however, which is driven by the digastric muscles with concurrent inhibition of the jaw-closing muscles when the jaws are near centric occlusion, serves to protect these structures (Sherrington 1917; Lund et al. 1984).

Mastication occurs unilaterally in carnivorans; that is, meat or bone is positioned and compressed between the cheek teeth on one side, the working- or biting-side jaw. The contralateral (nonbiting) jaw is referred to as the balancing side. Relative to more herbivorous mammals, carnivorans chew rapidly, the chewing aided in part by the high percentage of fast-twitch fibers in their adductor muscles (Baker and Laskin 1968; Tamari et al. 1973; Taylor et al. 1973; MacKenna and Türker 1978; Gorniak 1986). Mandibular movements during mastication are principally sagittal, with small mediolateral and rotational components for the alignment of opposing teeth on the working side (Gorniak and Gans 1980; Dessem 1989; Dessem and Druzinsky 1992). Mediolateral movements of the lower jaw are limited by the shape of the TMJ and its capsular ligaments, as well as by the interlocking of the canines (Scapino 1965; Gorniak and Gans 1980).

Making the Kill

Prey are subdued using two primary organs of prehension, claws and the anteriormost teeth (the incisors and canines). As the killing bite or bites are applied, the teeth, jaws, and skulls of predators are probably stressed at near-maximum levels. Moreover, because the prey may struggle, the direction of loading can vary unpredictably, increasing the probability of accidental fracture (Alexander 1981). To avoid such accidents, bones and teeth are constructed with high safety factors; that is, they should be able to sustain loads much greater than those they usually experience. Notably, a study of the incidence of tooth fracture in nine extant species of carnivorans ($N = 1458$ specimens) revealed that approximately 25% of all adult individuals surveyed had at least one broken tooth (Van Valkenburgh 1988). On all species, the teeth broken most often were the canines, constituting over 50% of all observed breaks. Given that the canines are essential tools in prey capture, and that teeth are incapable of healing, their apparent vulnerability to fracture was not expected. Safety factors for canines appear to be limited by functional requirements: shorter canines with greater external diameters will have higher safety factors but are likely to perform more poorly as stabbing weapons.

To date, the study of functional design for killing in the carnivoran skull has emphasized two aspects: (1) the strength and function of the anterior dental battery and (2) the strength of the mandible and skull during strong biting. Just as claw morphology reflects claw use (Van Valkenburgh 1985; MacLeod and Rose 1993), these anterior teeth and the jaws in which they are rooted bear structural and mechanical indicators of the external loadings imposed on them. Biomechanical analyses of the facial skeleton often model teeth and the upper and lower jaws as cantilevered beams (Figure 12.3*A*, *B*: e.g., Radinsky 1981a; Demes 1982; Van Valkenburgh and Ruff 1987; Thomason 1991b; Biknevicius and Ruff 1992a). In order to qualify as a beam, a structure should be longer than it is wide. A mechanical beam model allows the analysis of tooth or jaw design using standard engineering principles of beam strength and

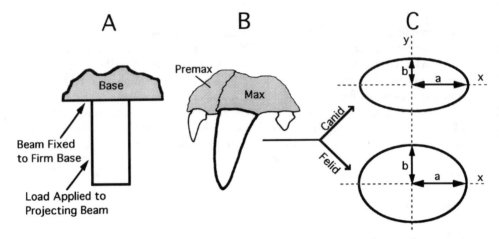

Figure 12.3. Modeling canine teeth as cantilever beams. *A*, idealized cantilever beam with a single fixed end and a free, projecting segment to which loads are applied. *B*, upper canine cantilevered to the maxilla. Canines are tapered beams that function to pierce flesh, but their design is such that dental material is concentrated where stresses are greatest, that is, at the base. *C*, cross section of the canines of a canid (the gray wolf, *Canis lupus*, top) and a felid (the leopard, *Panthera pardus*, bottom) at the dentino-enamel junction, idealized as solid ellipses having regular perimeters and composed of a homogeneous material. Computation of cross-sectional geometric properties is simplest with these assumptions; for example, the section modulus about the anteroposterior axis (Zx, or mediolateral bending strength) is computed as $Zx = Ix/b$, where Ix (the second moment of area about the same axis) equals $\pi ab^3/4$.

rigidity under loading. It is typical to assume that the tissue representing the biological beam is homogeneous and isotropic (although more complex models are possible), and most analyses accordingly concentrate on the distribution of tissue within the beam. Common cross-sectional properties analyzed are cortical cross-sectional area (CA, proportional to axial rigidity), second moments of area (I, proportional to bending rigidities perpendicular to the axes about which they are computed), and polar movements of area (J, proportional to torsional rigidity and computed as the sum of any two orthogonal second moments of area) (see Biknevicius and Ruff 1992b for equations). Since maximum bending stress is defined by Fdc/I, where F is the applied force, d is the moment arm length, and c is the distance from the neutral axis of bending to the outer perimeter in the plane of bending, and since maximum stress is inversely proportional to strength, then I/c (also known as the section modulus, Z) is directly proportional to bending strength. I, J, and Z are cross-sectional parameters that are determined not just by the amount of tissue in the cross-section but also by its distribution about neutral axes of bending and torsion. Consequently, the farther a mass of tissue is positioned from the neutral axes, the greater its impact on strength and rigidity under nonaxial loading. In the example shown in Figure 12.3C, the canines of gray wolves (*Canis lupus*) and leopards have subequal anteroposterior diameters (14.16

and 14.36 mm, respectively), but the mediolateral diameter in gray wolves is smaller than that in leopards (7.60 vs. 10.40 mm, respectively). This modest difference in the size of the minor axis results in a substantial difference in bending strength in that axis, as estimated by Zx (see the formula in the figure legend), because the mediolateral radius (b) is cubed in the computation of second moments of area. For example, the mediolateral diameter of the gray wolf canine is 73% that of the leopard, but the former's strength in bending about the anteroposterior axis is only 53% that of the latter.

In what follows, we will expound on the value of beam theory for describing patterns of structural differences in craniofacial skeletons, but it is important to realize that biological structures are not ideal engineering beams. For example, bones are composed of living tissue that is capable of responding to external stimuli by deposition and resorption, and the soft tissue surrounding bones (as well as interposing at sutures) is generally ignored. One of the key problems is that *in vivo* testing has rarely been used to substantiate the functional inferences derived from beam analyses in carnivorans.

The Anterior Dental Battery

The anterior dentition of carnivorans, particularly the canines, is loaded heavily during prey capture. Teeth contact the prey when the jaws are closed by elevation of the lower jaw and, at least in felids, depression of the upper jaw (Gorniak and Gans 1980). This action results in compressive loads on the teeth, but complex off-axis loadings to the canines and incisors during killing and feeding, as well as the curved shape of the teeth, dictate that bending and torsion are also likely to be important loading regimes. Normal shear stresses are considered to be less critical, at least in the canines (see Appendix I in Van Valkenburgh and Ruff 1987). Incisors and canines have been modeled as cantilevered beams with elliptical cross sections at their dentino-enamel junctions (Figure 12.3) (Van Valkenburgh and Ruff 1987; Biknevicius et al., in press). Simplifying assumptions employed in these analyses are that dental cross sections are solid and that they are composed of a homogeneous material.

CANINE SHAPE AND STRENGTH

The canines of living carnivorans inflict stab and slice wounds. Stabbing deeply with the canines, which is effective for securing strong holds on struggling prey, is typical of felids and mustelids (Ewer 1973; Leyhausen 1979). Canids use their canine teeth to produce slice or slash wounds that result in large, gaping lacerations, which bleed profusely and can result in rapid evisceration of the prey (Mech 1970). These different killing techniques are reflected by the shape of the upper canines in living large-bodied predators, especially in those features that confer strength in bending (Van Valkenburgh and Ruff 1987). Although dental material in the upper canines of both felids and canids is aligned preferentially along the anteroposterior axis, the upper canines of felids are relatively round in cross section when compared with the more knifelike upper canines of canids (Figure 12.3C). A beam of round cross section is equally resistant to bending stresses about all axes. Such a design is appropriate for felids, since their deep and prolonged killing bites, sustained while they maintain their hold on struggling prey, are likely to result in stresses across the teeth that are unpredictable in magnitude and direction. Any incidental contacts with bone can further exacerbate stresses to the teeth. Thus, felid canines must be resistant to loads applied from various directions. Elliptical cross sections, by contrast, suggest preferential adaptation for loads applied in one direction, that of the long axis of the ellipse. The mediolaterally compressed canines of canids are thus less able to resist bending when loaded in a mediolateral direction. This may be less critical, however, for an animal that uses its canine teeth to create shallow, slashing wounds; the shallow canine bites of canids are less likely to contact bone and may be less affected by prey movements because they are relatively brief in duration.

The canine teeth and killing behavior of spotted hyenas do not follow the same pattern. Spotted hyena kills are canidlike in the sense that they often depend on a cooperative effort among clan members, during which multiple, relatively shallow and quick canine bites are administered to the prey (Kruuk 1972; Mills 1990). Nevertheless, their canines are shaped more like those of felids. The apparently excessive buttressing of hyena canines might be related to the animals' habit of cracking bones with their premolars; given their proximity to the premolars, the canines might be subject to incidental loads of high magnitudes when bones are manipulated and crushed. Notably, the canine teeth of several extinct canids that have been proposed as bone-cracking specialists (on the basis of the robustness of their premolars; e.g., subfamily Borophaginae) are also rounder in cross section than are those of modern canids (Van Valkenburgh and Ruff 1987).

In some cases, it appears that phylogenetic history has constrained canine design. For example, the canines of domestic dogs are more similar, both morphometrically and mechanically, to those of wolves than to those of foxes (Plishka and Bardack 1992), reflecting the lupine ancestry of all domestic breeds (Wayne 1986). But membership within a lineage does not always dictate uniform canine morphologies; for example, the canines of extinct sabertooth felids are characterized by mediolateral compression, and are, therefore, somewhat more canidlike than felidlike (Van Valkenburgh and Ruff 1987).

INCISOR DESIGN AND ARCADE SHAPE

Morphological differences in incisors or in the configuration of incisor arcades have been used as phylogenetic characters (Flynn et al. 1988), but their potential functional significance remained largely unexplored until recently (Biknevicius et al., in press). As was found in the upper canines, the dental material of the upper medial incisors (I^1, I^2) in felids and canids is preferentially arranged along the anteroposterior axis at the dentino-enamel junction; that is, incisor bases are mediolaterally compressed. Consequently, incisors are constructed to better resist bending in the anteroposterior plane than in the mediolateral plane. This is not surprising, given that incisal biting is often associated with a rearward pull of the head, an action that will load the anterior teeth from back to front. Relative to body size, the incisors of canids have greater mediolateral and anteroposterior bending strengths than do those of felids (Figure 12.4)—the reverse of the situation for the upper canines, in which bending strength in both planes is greater in felids than in canids.

Incisor mediolateral strength appears to be associated with specific incisor arcade contours (Figure 12.4), such that, in felids, straight, coronally aligned incisor arcades are paired with mediolaterally weak incisors, whereas parabolic arcades in canids are coupled with incisors that are stronger in mediolateral bending. The linear arrangement of incisors in felid premaxillae results in a nearly complete mediolateral buttressing of the medial incisors by the neighboring teeth. By contrast, the medial incisors of canids are less well buttressed by adjacent teeth, because of their parabolic arrangement, and are therefore likely to encounter bending stresses from all directions.

The functional significance of particular incisor arcade configurations is unclear. The condition in most primitive carnivorans is parabolic, and the straight arcade of extant felids thus appears to be derived. In modern felids, it appears that canine biting has been greatly favored over incisal biting. This was probably less true in some sabertooth cats, such as *Smilodon fatalis*, that displays a more curved incisor arcade (see below). Parabolic arcades probably

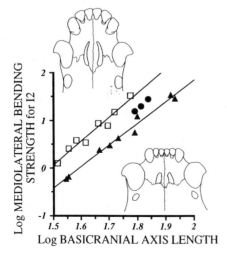

Figure 12.4. Plot of mediolateral bending strength (Zx in mm³) of the second incisor (I^2) against basicranial axis length (in mm); all variables $\log_{10}$-transformed. X-axis defined as in Figure 12.3. Symbols: squares, canids; triangles, felids; circles, hyaenids. Least-squares regression lines are shown for canids (*upper*) and felids (*lower*). The superimposed arcade at the upper left is typical of a canid; that at the lower right is typical of a felid.

confer on the incisors some functional independence from the canines; and parabolic arcades in herbivores are associated more with species that are more selective feeders (browsers) than with those with straight arcades (grazers) (Solounias et al. 1988). In canids, a parabolic arcade allows the incisors to be used apart from the canines to pull or nip at prey. It is also plausible that canids, which are omnivores, benefit from having more selective anterior tooth use when feeding, for example, on berries or invertebrates, than do the more strictly carnivorous felids.

Cranial and Mandibular Strength

The struggle between prey and predator during prey capture is likely to cause complex loading regimes of considerable magnitudes within the predator's jaw bones. Undoubtedly, combined dorsoventral and mediolateral bending as well as torsion (twisting) are common. In order to understand the impact of these loads on bony tissue, jaws have been analyzed with some success as simple mechanical beams.

THE LOWER JAW

Beam theory, supported by the results of strain-gauge studies on herbivorous and omnivorous mammals (e.g., Hylander 1981), suggests that bilateral use of the canines during killing bites should cause dorsoventral bending deformation in the mandibular corpus behind the canines (Figure 12.5A). As the canines encounter resistance and bite force increases, compressive stresses will be produced in the ventral mandibular border and tensile stresses will be produced in the dorsal border. The ideal biomechanical solution to pure dorsoventral bending loads is to increase the amount of bone tissue at the dorsal and ventral margins, producing a deeper corpus. Along the long axis of the corpus, shear stresses due to torsion are likely during canine biting because the resultant adductor-muscle force is lateral to the long axis of the corpus (Figure 12.5B). The effect of cortical-bone distribution patterns on torsional strength in the mandibular corpus is still unclear (Daegling et al. 1992), but the expectation is that deposition of bony tissue at the outer perimeter closest to the longitudinal axis, increasing both the smallest external diameter and cross-sectional circularity, should improve torsional strength (Hylander 1988).

Mandibular length, and thus snout length, varies greatly among predaceous carnivorans: mustelids and felids are characterized by short snouts, whereas canids and hyaenids are typified by elongate snouts (Radinsky 1981a, b; Van Valkenburgh and Ruff 1987). Short snouts are associated with increased force when biting with the anterior dentition because, all other things being equal,

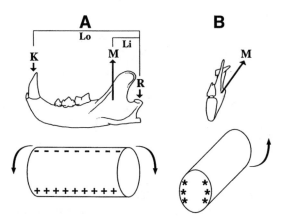

Figure 12.5. Hypothesized loading diagrams for canine killing bites, illustrated on a left mandible. *A*, the combination of ventrally directed canine killing bite (K) and temporomandibular-joint reaction force (R) with a dorsally directed masticatory muscle force (M) results in dorsoventral bending of the mandibular corpus. Lo and Li are the out-lever to the canine and in-lever to the muscle resultant, respectively, shown from the sagittal plane only (compare with Figure 12.9). The idealized cylindrical corpus shown below indicates the distribution of tensile stresses dorsally and compressive stresses ventrally. Stress magnitudes are greatest at the outer perimeter in the plane of bending, that is, at the dorsal (———) and ventral (+++) borders; these are the areas where bone should ideally be added in order to increase dorsoventral bending strength. *B*, the lateral component of the adductor muscle force tends to evert the ventral border of the mandibular corpus, and thus to impose shear stresses in the corpus. Shear stresses due to torsion are predicted to be greatest at the subperiosteal perimeter of the minor axis, that is, at the buccal and lingual borders (∗∗∗), where bone should ideally be added in order to increase torsional strength.

the out-lever (resistance moment arm) to these teeth is shorter than that found in longer snouts.

Bite strength at the anterior teeth varies with the relative proportions of the masticatory musculature. The temporalis muscles of felids, hyaenids, and mustelids have greater mass and mechanical advantage than do those of canids (Smith and Savage 1959; Radinsky 1981a, b; Van Valkenburgh and Ruff 1987). The combination of short dentaries and mechanically superior masticatory muscles results in an enhancement of biting force with the anteriormost teeth in felids and mustelids (adaptations for bite strength in hyaenids are more likely for bone-cracking with the anterior cheek teeth; see below).

The magnitude of bending moments experienced by the mandibular corpus is a function of both dentary length (the moment arm to the anterior teeth) and the size of the forces exerted at the teeth (bite and external forces). Long dentaries or high applied forces alone should result in large bending moments in the precarnassial corpus. One measure of the resistance of the lower jaw to deformation during loading is the amount and distribution of cortical bone in the corpus between the canine and carnassial teeth, as reflected by the cross-sectional geometric properties of that region (Badoux 1968; Radinsky

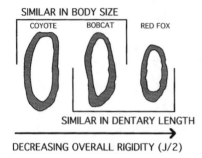

SIMILAR IN BODY SIZE

COYOTE   BOBCAT   RED FOX

SIMILAR IN DENTARY LENGTH

DECREASING OVERALL RIGIDITY (J/2) →

Figure 12.6. Tracings of the endosteal (*inner*) and sub-periosteal (*outer*) cortical bone perimeters of mandibular corpus cross sections at the P_4-M_1 interdental gap of (*left to right*) coyote (*Canis latrans*), bobcat (*Lynx rufus*), and red fox (*Vulpes vulpes*). Mean bending and torsional rigidities, represented by $J/2$ (where J is the polar moment of area computed as the sum of any two orthogonal second moments of area), are 1771.6, 1098.4, and 428.6 mm^4 for the three species, respectively. Strength measures in the precarnassial corpus of canids exceed those of felids of similar body weight; relative to dentary length, felid values are greater than canid values.

1981a, b; Biknevicius and Ruff 1992a). Prey capture in solitary predators, particularly those that customarily kill prey larger than themselves (e.g., felids, predaceous mustelids), likely results in the imposition of great bending stresses on the corpus. Accordingly, these carnivorans have more strongly buttressed corpora than do carnivorans of similar dentary length that are social killers, or those that rely on small prey (e.g., canids; Figure 12.6). Buttressing of the mandible for powerful anterior tooth use in felids is associated with great dorsoventral and mediolateral bending strength; the latter also provides indirect evidence for strength in torsion. Bending-moment magnitudes are further controlled in felids, in spite of large canine bite forces, by the abbreviations of their snouts (thereby reducing the moment arm: Biknevicius and Ruff 1992a).

A study of the cross-sectional geometric properties of the mandible in canids revealed a correspondence between typical prey size and mandibular strength; canids that routinely pull down prey much larger than themselves displayed enhanced buttressing of the mandibular corpus anterior to the carnassial (Van Valkenburgh and Koepfli 1993).

The Cranium

Because the strength of a muscle is determined in part by its size and cross-sectional area (Thomason 1991b), several cranial features have been used to estimate the size of jaw-closing muscles, in both fossil and extant species. For example, muscle leverage as well as the length and width of the temporal fossa and zygomatic arch have been used to approximate the size and strength of the temporalis. Comparisons of these measures among predaceous carnivorans reveal a greater potential for large temporalis forces in felids, mustelids, and hyaenids than in canids (Radinsky 1981a, b; Werdelin 1988).

The modeling of carnivoran skulls as hollow beams has revealed a less precise correspondence between killing behavior and bony form than was found for the lower jaw. That the skull does behave mechanically as a single structure is indicated by the presence of stress-distributing tracts of bone, many of which are continuous across the cranial sutures (Tucker 1954a–d, f, 1955;

Buckland-Wright 1971). Classic beam theory suggests that the stress patterns in the rostrum should be comparable in magnitude but inverse in orientation to that found in the dentary (e.g., compressive and tensile stresses should be produced in the dorsal and ventral borders, respectively). These expectations have not been fully supported by empirical data. For example, strength estimates for the rostrum commonly exceed those for the mandible by a factor of 1 to 8, depending on the type of load (Thomason 1991a). Estimates of bending strength of the rostrum far exceed those expected solely on the basis of bending loads generated by the masticatory system (Thomason and Russell 1986; Thomason 1991a). Furthermore, those regions of the skull that appear strongest in bending rarely correspond to areas where bending moments are expected to be greatest (Demes 1982; Thomason 1991a, b). There do appear to be constraints on skull design associated with the need to resist torsional loads (Covey and Greaves 1994): no ratios of jaw width to length fall significantly below $\pi/2$; and the large ratios of some carnivorans (e.g., felids, mustelids, and ursids) indicate that their rostra are overbuilt for resisting pure torsional loads.

Clearly, the facial skeleton is overdesigned for typical loads. This suggests that the rostrum is not constructed exclusively to minimize bone tissue and maximize strength, as expected by beam theory, but rather is a compromise between mechanical and nonmechanical factors (Hylander and Johnson 1992; Russell and Thomason 1993). An example of a nonmechanical factor here is the need to accommodate and protect neural and respiratory structures adequately.

Another aspect of the cranium that bears on killing performance in carnivorans is the temporomandibular joint (TMJ) (Figure 12.7). Unlike most herbivorous mammals, which have relatively flat and open TMJ's, carnivorans have a glenoid fossa that is bounded caudally by a well-developed postglenoid process at its posteromedial margin and a more variably developed preglenoid process at its anterolateral margin. These processes, together with the surrounding soft tissue (muscle and ligaments), resist dislocation of the TMJ as a result of loadings to the mandible during prey capture and mastication (Smith and Savage 1959; Badoux 1964; Noble 1979; Scapino 1981). Many mustelids have particularly well-developed processes that appear to function, in part, as mechanical stops (Dessem and Druzinsky 1992), enabling them to lock their jaws effectively in adduction with great muscular forces when biting active, large-bodied prey.

Carcass Consumption

Whereas killing bites are made primarily or solely with the canines and incisors, the tasks of subdividing and consuming prey are performed by all teeth. As mentioned earlier, the different shapes of the cheek teeth correspond

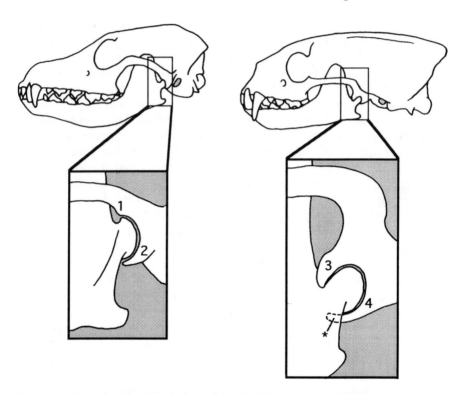

Figure 12.7. Lateral views of the skulls, and details of the temporomandibular joints, of a mustelid (the wolverine, *Gulo gulo, right*) and a nonmustelid carnivoran (the coyote, *Canis latrans, left*). Nonmustelids generally have a small preglenoid process (1) and a modest postglenoid process (2) relative to those found in many predaceous mustelids, in which the mandibular condyle is surrounded by an enlarged preglenoid process around its lateral aspect (3) and a postglenoid process (4) that extends anteroventrally around its medial aspect (*).

to their different functions. Bladelike carnassials are critical for slicing skin, and blunt-cusped molars or premolars are necessary for cracking bones. Selection appears to have favored the evolution of teeth constructed for specific functions, such as cutting skin, and, in many carnivorans, for the ability to perform such functions rapidly. The need to feed rapidly is likely to act as a potent selective force, favoring greater bite forces, larger gapes, and strong, well-designed teeth (cf. Bakker 1983; Van Valkenburgh 1991). Rapid feeding is probably most important in large-bodied predaceous carnivorans, such as jackals, wolves, hyenas, and lions, where carcass theft occurs regularly and/or group members feed together. Small carnivorans, such as weasels and mongooses, are solitary hunters of relatively small prey and probably less driven to feed rapidly. Below, we review previous work on the functional morphology of carnivoran cheek teeth and adaptations of the mandible and cranium for greater bite force during chewing.

Functional Morphology
of Postcanine Teeth

A diagnostic characteristic for the Carnivora is the presence of sectorial (bladelike) modifications at P^4 and M_1 (Figure 12.8). These modifications include reduction of the protocone together with fusion (or near fusion) of the paracone and metacone and enlargement of the metastyle in the upper carnassial; in the lower carnassial, there is a reduction or loss of the metaconid and a strong development of an oblique shearing paracristid between the paraconid and protoconid, often paired with a reduction or loss of the talonid

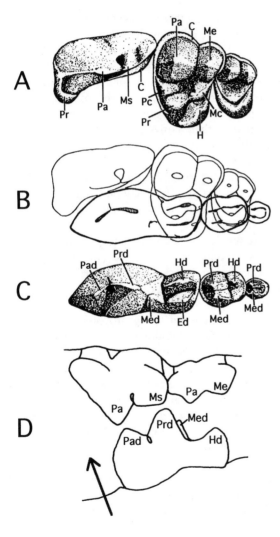

Figure 12.8. Occlusal morphology of the cheek teeth of a canid. *A*, occlusal view of upper left $P^4M^{1,2}$. *B*, occlusal relations of the teeth shown in *A* and *C* in centric occlusion. *C*, occlusal view of lower right $M_{1,2,3}$. *D*, buccal view of the left carnassial teeth (P^4 and M_1) and M^1, approaching centric occlusion (arrow indicates net movement of the lower jaw toward the upper jaw). The paracristid is the dental crest extending from the paraconid to the protoconid on M_1. Abbreviations: C, cingulum; Ed, entoconid; H, hypocone; Hd, hypoconid; Mc, metaconule; Me, metacone; Med, metaconid; Ms, metastyle; Pa, paracone; Pad, paraconid; Pc, protoconule; Pr, protocone; Prd, protoconid. (*A*–*C* adapted by permission from Scapino 1965.)

basin (Butler 1946; Flynn et al. 1988). The critical role of the carnassials in carnivoran success is suggested by the low intraspecific variability seen in carnassial tooth size relative to that observed for other postcanine teeth (Gingerich and Winkler 1979; but see Pengilly 1984). Over evolutionary time, the effective blade length of the carnassials has been increased in some highly predaceous species by the conversion of the talonid of M_1 into a bladelike cusp known as a trenchant heel (Bakker 1983; Van Valkenburgh 1991). Within carnivorans, this conversion has occurred repeatedly, appearing in members of the Felidae, Canidae, Hyaenidae, and Viverridae. The iterative appearance of this feature suggests that it is beneficial; perhaps a longer blade increases feeding rate and is advantageous for species where competition for food is significant (Bakker 1983; Van Valkenburgh 1991).

The orientation of the carnassial blades in highly predaceous carnivorans is nearly in line with the long axis of the jaw, whereas the blades are set at an angle to the jaw among carnivorans that are more omnivorous or insectivorous (Crusafont-Pairo and Truyols-Santonja 1956; Van Valen 1969). Carnassial blades are not singular structures, but, rather, each is subdivided into two shearing edges by a V-shaped notch at the free edge of the blade (Figure 12.8D). During jaw closure, the upper and lower blades shear past one another, trapping and stressing food between their converging notches (Greaves 1974; Lucas and Luke 1984). This effective method of dividing soft and tough materials is enhanced by the fact that the carnassials come into occlusion in a scissors-like fashion, from back to front (Figure 12.8D) (Greaves 1974). This allows a concentration of forces at the cusp tips, effectively piercing and dividing soft, tough materials. Sharpness of the blade edges is maintained by tooth-on-tooth abrasion (thegosis: Mellett 1981).

Carnassial blades are likely to become blunted, and thus to perform poorly as meat and skin cutters, if they are used regularly for breaking up large bones. Consequently, the durophagous carnivorans crack bone primarily with the other cheek teeth. Postcarnassial molar surfaces are routinely used by canids and mustelids for crushing bones (Ewer 1973; Riley 1985; Van Valkenburgh 1989). In both groups, crushing occurs between the protocones and talon basins of the upper molars and the talonid basins and hypoconids of the lower molars, respectively (Figure 12.8A–C). Predaceous mustelids show further specializations for crushing: their M^1 possesses an enlarged inner lobe formed by a particularly strong lingual cingulum. Because both canids and mustelids crush foods with their postcarnassial molars, their lower carnassial retains a talonid basin for occlusion with M^1. Typical hyaenids, by contrast, have greatly reduced or lacking postcarnassial molars (see Figure 12.10A, below). In these species, bone-cracking is accomplished with their robust premolars ($P^{2,3}$ and $P_{3,4}$), since they lack functional crushing surfaces caudal to the carnassials (Van Valkenburgh 1989). Hyaenids, therefore, exhibit a combination of long carnassial blades and relatively stout premolars for efficient meat-slicing and bone-cracking, respectively. Similarly, although felids are less frequent bone-chewers, some are known to crush small to medium-size bones using the premolars anterior to the carnassial (Schaller 1972; Blumenschine 1987).

Maximizing Gape and Bite Force

The ability to produce a large bite force at a large gape is important to carnivores, and yet both parameters cannot be maximized simultaneously. In theory, the mechanical advantage of jaw-adductor muscles and, consequently, the magnitude of bite force is greatest near the jaw joint, because the resistance moment arm of the object being bitten is minimal. Gape, however, clearly decreases as the jaw point is approached, meaning that larger objects cannot be positioned where bite force should be maximal (Ewer 1973; Greaves 1983). This dilemma forces a compromise: the carnassials must be positioned far enough forward in the jaw to allow complete separation between the upper and lower blades at full gape, but they must also be positioned rearward enough to produce substantial bite forces. Notably, in spite of varying diets among living carnivorans, the position of the carnassial blade relative to the ipsilateral temporomandibular joint (TMJ) is remarkably uniform: the carnassials are typically located at approximately the midpoint of the distance from the canine to the mandibular condyle along the mandibular corpus (Gaunt 1959; Radinsky 1981a; Greaves 1985).

Greaves (1983, 1985) analyzed the functional significance of the midpoint position of the carnassials, using a model of jaw mechanics to predict the location of maximum bite forces along the tooth row. The forces produced by the jaw-closing muscles are resisted by reaction forces at the biting tooth and the TMJ's, on both working and balancing sides; these three force locations delineate a "triangle of support" (Figure 12.9A). If the resultant of all the adductor musculature lies outside the triangle of support, the mandible will rotate about a line extending from the balancing-side TMJ to the biting tooth and create tension in the working-side TMJ. Tension is disadvantageous, because jaw joints are poorly designed to counter high tensile forces (Greaves 1983). The precise location of the adductor-muscle resultant is determined by muscle attachments and is unknown for most species, but is unlikely to be located more anterior than the rearmost molars (Spencer and Demes 1993). Assuming equal and maximal working- and balancing-side muscle effort, the muscle resultant lies at the midpoint of a line passing immediately behind these teeth (R on the dashed line, Figure 12.9A). The anteriormost location of high bite-force potential along a toothrow is theoretically identified as the point where a line that originates from the contralateral temporomandibular joint and passes through the midline muscle resultant intersects the working-side jaw (BF on the line originating at BS, Figure 12.9A). This is the position where gape is greatest, given maximal bite force. In most carnivorans, it is at or near the carnassial tooth.

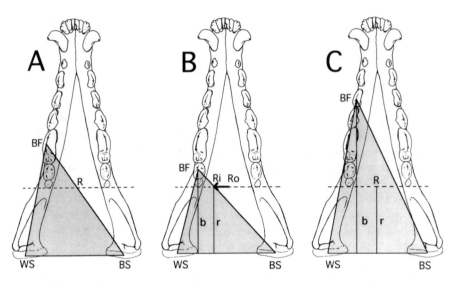

Figure 12.9. A model of biting biomechanics that analyzes forces in both frontal and sagittal planes. *A*, adductor muscles are bilaterally and maximally active, so that the muscle resultant (R) lies at the midpoint of a line that lies no farther forward than immediately behind the caudalmost molars. The anteriormost position of maximum bite force (BF) is the intersect of a line that originates at the balancing-side (BS) condyle and passes through R and the working-side mandibular corpus. The shaded area is the "triangle of support" in which R must lie in order to avoid rotation of the jaw about the line BS-BF, resulting in tension in the working-side (WS) temporomandibular joint. *B*, biting with the more posterior molars requires lateral movement of R from a position outside of the triangle of support (R_o) toward the working side and into the triangle of support (R_i). This movement is facilitated by a decrease in balancing-side muscle activity; if this does not occur, rotation of the mandible about the line BS-BF will occur. Bite-force magnitudes remain high, since $BF = R_i r/b$ (where b and r are lever-arm lengths measured from a line connecting the condyles to the biting tooth and the muscle resultant, respectively) as the requisite decrease in R (from R_o to R_i) is compensated by a decrease in the out-lever b. *C*, although biting with anterior teeth retains R within the triangle of support, and therefore allows R to be maximal, it results in decreasing BF with increasing b, since $BF = Rr/b$. (Based on Greaves 1983.)

It also follows that when biting with teeth caudal to the intersect, equal and maximal recruitment of working- and balancing-side muscles should produce tension in the working-side TMJ, because the muscle resultant then lies outside the triangle of support (Ro, Figure 12.9*B*). This tension can be avoided by reducing balancing-side muscle activity such that there is a lateral shift of the muscle resultant closer to the working side and into the triangle of support (Ri, Figure 12.9*B*) (Dessem 1989). Equally high bite forces can be generated by most caudal teeth by matching smaller muscle resultants with shorter out-lever arm lengths to the biting teeth.

One noteworthy exception to the midpoint carnassial position is the dental configuration of the spotted hyena, in which M_1 is located closer to the TMJ than it is in most other carnivorans (Radinsky 1981b; Werdelin 1989; Biknevicius and Ruff 1992a; Biknevicius 1993). Concomitant caudal shifting of the premolars effectively positions the bone-cracking teeth (P_3, P_4) within or nearer to the region of maximum bite force while still maintaining substantial gape. The position of the primary bone-cracking teeth in hyaenids is a compromise between the need for large gape, in order to accommodate large-diameter bones, and the need for strong bites, to accomplish the task.

Mandibular and Cranial Adaptations

Although feeding on a carcass may place smaller, more predictable loads on craniofacial structures than does killing live prey, substantial stresses are probably generated during rapid feeding and bone-cracking. Strain-gauge analyses in primates have shown that powerful isometric biting with the cheek teeth will twist the working-side dentary about its long axis and bend the balancing-side dentary about its mediolateral axis (dorsoventral bending: Hylander 1981). One implication of these results for carnivoran feeding is that since shear stresses due to torsion will be concentrated in the mandible between the bite point and the ascending ramus, the use of caudal teeth for bone-chewing should effectively limit stresses due to torsion in the entire corpus. Moreover, the crowns of caudal teeth tend to be in line with the vertical height of the corpus, further reducing torsion by directing loads straight through the corpus (Biknevicius 1990). Consequently, the magnitudes of shear stresses due to torsional moments are controlled to some degree by the preferential use of caudal teeth during bone-cracking. The effect of stresses resulting from dorsoventral bending during bone ingestion, however, are additive to those imposed during canine killing bites (Figure 12.10A), and are thus ideally countered by the preferential distribution of bone tissue dorsally and ventrally; this distribution effectively increases corpus height and improves dorsoventral bending strength.

The great magnitude of cross-sectional geometric properties (e.g., section moduli) computed for the mandible at or behind the site of bone-crushing reflects bone-chewing behaviors in carnivorans (Biknevicius and Ruff 1992a). Still, values for individual locations along the corpus may not distinguish between the buttressing associated with powerful anterior tooth use (e.g., for the canine killing bite) and that associated with bone-cracking. Higher stresses within the mandible due to localized bone-processing activities are better revealed by a comparison of cortical-bone cross-sectional properties at successive interdental gaps (see Table 12.2). An increase in the vertical height of the corpus and a thickening of the cortical bone along its ventral border have been observed caudal to the contrasting sites of bone-processing in hyaenids and canids (Figure 12.10B). This additional bone tissue produces a mandible that is more resistant to dorsoventral bending in the region of maximum stress (e.g., behind P_4 in hyaenids and behind M_1 in canids). Such localized thickenings are

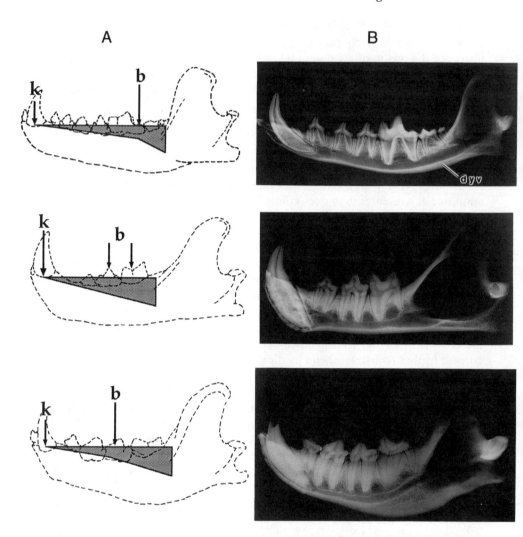

Figure 12.10. *A,* hypothesized dorsoventral bending-moment diagrams for (*top to bottom*) canid, felid, and hyaenid dentaries. Vertical arrows represent sagittal forces due to killing (k) and bone-processing (b). Arrow length indicates relative size of force. Bone-processing in felids occurs with all of the cheek teeth (Van Valkenburgh, in press), as indicated by the multiple but short arrows for bone-processing at the premolars and molars. *B,* lateral radiographs of carnivoran dentaries for (*top to bottom*) gray wolf, *Canis lupus;* puma, *Puma concolor;* and spotted hyena, *Crocuta crocuta.* All dentaries scaled to the same dentary length; dyv represents the thickness of the ventralmost portion of the basal bone of the corpus. (Adapted from Biknevicius and Ruff 1992a.)

useful indicators of the site of bone-cracking in specialized premolar or molar bone-chewers, but may not be apparent in some durophagous carnivores that use their cheek teeth more indiscriminately for bone ingestion, such as fe-lids (Figure 12.10A) and Tasmanian devils (*Sarcophilus harrisii:* Biknevicius 1990).

A final mandibular feature that has been evaluated for its correspondence to diet is the mandibular symphysis. Although fused mandibular symphyses are found in a few carnivorans, patent symphyses are typical (Scapino 1981). Patent symphyses can be quite rigid, depending on the degree of interdigitation between the bony processes on the surfaces of the symphyseal plates and the arrangement of cruciate ligaments between the plates. The function of symphyseal fusion in carnivorans, however, is still obscure, since its presence does not correlate well with body size, dental morphology, or diet (Scapino 1981). Nor is fusion essential for effective transmission of muscle force from the balancing-side dentary to the working-side dentary (*contra* Scapino 1981); the unfused symphyses of raccoons (*Procyon lotor*), domestic cats (*Felis catus*), and dogs (*Canis familiaris*) are capable of transmitting 82–100% of vertical bite force (Dessem 1985).

Although the functional consequences of chewing for the crania of carnivorans have been examined, precise correspondence between form and function in a broad range of predaceous carnivorans has not yet been demonstrated. High masticatory forces during biting with the carnassials may be transmitted through the rostrum to the craniofacial apex (between the orbits in the frontal bone) by way of a thickened arch of bone in the maxilla (Tucker 1954c, 1955; Badoux 1966; Buckland-Wright 1978; Roberts 1979). Accordingly, the skulls of some bone-cracking carnivorans, such as living hyaenids and the fossil canid *Osteoborus*, have high, vaulted foreheads (Werdelin 1989). Strains tend to be greater in the bones of the working-side rostrum than in those of the balancing side during unilateral biting (Kakudo and Amano 1968). The observation of minute movements between cranial bones in chewing cats (Buckland-Wright 1972, 1978) suggests that the aggregate effect of movements in facial sutures may be significant, and that sutures may have a shock-absorbing function.

Temporomandibular joint morphology has great bearing on the forces that can be generated during postcanine biting. As mentioned earlier, strong unilateral biting with the cheek teeth can generate tension in the balancing-side jaw joint that can result in a ventral dislocation of the TMJ if unchecked (Greaves 1983). According to Greaves' (1983, 1985) model and some empirical studies in canids and felids (Liebman and Kussick 1965; Vitti 1965; Dessem 1989), this tension is minimized by a reduction in balancing-side muscle activity. European ferrets (*Mustela putorius*), however, are an exception to this pattern, in showing nearly equal and simultaneous bilateral activity of the jaw muscles during bone-crushing (Dessem and Druzinsky 1992). The skulls of ferrets display a bony glenoid fossa that virtually wraps around the mandibular condyle. In some mustelids (e.g., the wolverine, *Gulo gulo*), the glenoid processes must be broken if one is to disarticulate the mandibles and skull (Fig. 12.7). It has been suggested that the wrap-around morphology of the ferret jaw joint allows maximal recruitment of the balancing- and working-side muscles (Bramble 1978; Dessem and Druzinsky 1992). But this begs the question of why other nonmustelid species do not have such secure jaw joints.

Three Pleistocene Megacarnivorans
of North America

Success in inferring the functional significance of observed morphologies in extinct species often depends on how familiar these morphologies are. A cranio-dental complex in an extinct species that is similar morphologically to some closely related, living form is relatively easy to interpret. More difficult are the paleocarnivorans that are morphologically unparalleled among extant carni-vorans, however closely or distantly related they may be to extant forms. These interpretive possibilities are illustrated below by three sympatric large-bodied predators from Rancho La Brea, a late Pleistocene locality in California: *Canis dirus* (dire wolf), *Panthera atrox* (American lion), and *Smilodon fatalis* (saber-tooth cat) (Figure 12.11A). The first two species have close morphological counterparts among living members of their family (gray wolf and African lion, respectively: Kurtén and Anderson 1980; Akersten 1985). The dire wolf was similar in size to or slightly larger than the largest extant gray wolves, but has been described as a more specialized bone-cracker (Merriam 1912; Stock and Harris 1992). The American lion looks very much like a huge African lion; recent mass estimates suggest it was nearly twice the size (Anyonge 1993). *Smilodon*, which was approximately 50% larger than living lions, represents a distinct form that has been the subject of prolonged discourse among paleon-tologists, the discourse fueled by its fascinating blend of familiar felid features with sabertooth specializations. Perhaps because there are no living analogs for sabertooths, the predatory behavior of *Smilodon* has been argued to have resembled that of animals as different from one another as African lions (Emerson and Radinsky 1980) and varanid lizards (Akersten 1985).

In order to gain insights into their killing and feeding behaviors, loading distribution patterns were reconstructed for the lower jaws of these late Pleis-tocene carnivorans (see legend for Figure 12.11). Relative bending strength was compared among fossil and modern species, using the magnitude of sec-tion moduli at the P_4-M_1 interdental gap relative to dentary length (Table 12.1; Figure 12.11B). This comparison minimizes the effects of different body proportions and sizes. As noted above, among living carnivorans the precar-nassial corpora of felids are more strongly buttressed for both dorsoventral and mediolateral bending than are those of canids, and hyaenid corpora are markedly buttressed for strength in dorsoventral bending (Biknevicius and Ruff 1992a). This pattern is largely upheld by the fossil taxa. Both the Ameri-can lion and *Smilodon* have mandibles that exceed those of the dire wolf in dorsoventral bending strength, and the values for all three fall within the ranges of their respective families (Zx, Figure 12.11B). In mediolateral bending strength, the extinct lion appears weaker than modern cats, but *Smilodon* and dire wolf are similar to or slightly stronger than their extant relatives in this respect (Zy, Figure 12.11B). The overall similarity between the fossil and modern species in the scaling of relative bending strength at the P_4-M_1 inter-dental gap suggests that the two extinct felids differed from the extinct canid in

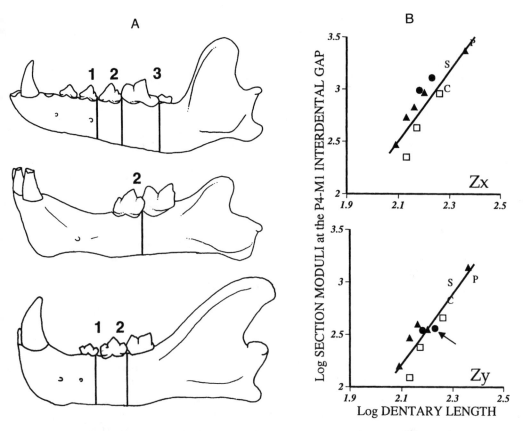

Figure 12.11. *A,* lateral view of the left dentaries of (*top to bottom*) the dire wolf (*Canis dirus*), sabertooth cat (*Smilodon fatalis*), and American lion (*Panthera atrox*). Each paleospecies is represented by eight or nine individuals housed in the George C. Page Museum (Natural History Museum of Los Angeles County). All specimens were retrieved from the asphalt traps of Rancho La Brea and are thus remarkably well preserved and minimally deformed by fossilization. Cross-sectional geometries of the corpora were reconstructed at the interdental gaps between (1) P_3-P_4, (2) P_4-M_1, and (3) M_1-M_2, where possible, using biplanar radiographs and an asymmetrical hollow model of bone from which section moduli were computed (Biknevicius and Ruff 1992a, b). Data for extinct species and large-bodied living carnivorans are listed in Table 12.1. *B,* plots of strength of the mandibular corpus in dorsoventral and mediolateral bending (section moduli Zx and Zy, respectively, in mm³) at the P_4-M_1 interdental gap against dentary length (DL, in mm) in extant large-bodied canids, felids, and hyaenids and three Pleistocene carnivores; all variables $\log_{10}$-transformed. Symbols: C, *Canis dirus;* P, *Panthera atrox;* S, *Smilodon fatalis;* all others as in Figure 12.4. Arrow indicates the jaguar (*Panthera onca*). Ordinary least-squares regression lines based on a combined data set of living canids and felids: log Zx = −4.577 + 3.367(log DL), r = 0.91; and log Zy = −4.865 + 3.372(log DL), r = 0.91. (Data from Table 12.1.)

Table 12.1. Species means of dentary length (DL) and section moduli about the transverse and sagittal axes (Zx and Zy, respectively) at the P_3-P_4 (PP), P_4-M_1 (PM), and M_1-M_2 (MM) interdental gaps; data are log_{10}-transformed. Dentary length in mm; section moduli in mm^3. Paleospecies indicated by (†).

Family and species	N	DL	ZxPP	ZyPP	ZxPM	ZyPM	ZxMM	ZyMM
CANIDAE								
Canis latrans	10	2.13	2.17	1.88	2.35	2.09	2.31	1.92
Canis lupus	10	2.26	2.85	2.57	2.96	2.66	2.94	2.46
Lycaon pictus	10	2.17	2.58	2.35	2.63	2.38	2.78	2.32
Canis dirus (†)	9	2.29	2.91	2.65	3.01	2.83	3.12	2.58
FELIDAE								
Acinonyx jubatus	10	2.09	2.42	2.21	2.47	2.20	—	—
Panthera leo	9	2.36	3.27	3.06	3.37	3.14	—	—
Panthera onca	10	2.20	2.91	2.62	2.97	2.55	—	—
Panthera pardus	9	2.16	2.80	2.54	2.83	2.60	—	—
Puma concolor	10	2.13	2.63	2.38	2.73	2.47	—	—
Panthera atrox (†)	8	2.39	3.38	2.98	3.45	3.03	—	—
Smilodon fatalis (†)	8	2.29	—	—	3.24	3.00	—	—
HYAENIDAE								
Crocuta crocuta	9	2.23	2.94	2.71	3.11	2.56	—	—
Hyaena hyaena	9	2.18	2.88	2.64	2.99	2.54	—	—

Table 12.2. Family and species means (and standard errors) of Zx/Zy ratios at the P_3-P_4, P_4-M_1, and M_1-M_2 interdental gaps. Paleospecies indicated by (†). Differences in Zx/Zy by taxon and by location were tested using ANOVA and, then, when differences were found, using Tukey's multiple comparison test (adjusted for unequal sample sizes: Zar 1984).

Family and species	Zx/Zy ratios of interdental gaps		
	P_3-P_4	P_4-M_1	M_1-M_2
CANIDAE			
Canis latrans	1.97 (0.06)	1.83 (0.08)	2.45 (0.27)[a]
Canis lupus	1.92 (0.06)	2.01 (0.05)	3.19 (0.83)[a]
Lycaon pictus	1.68 (0.03)	1.78 (0.05)	3.41 (1.52)[a]
Extant Canidae Mean:	1.85 (0.04)	1.87 (0.04)	2.97 (0.87)[a]
Canis dirus (†)	1.83 (0.07)	1.82 (0.04)	3.50 (0.72)[a]
FELIDAE			
Acinonyx jubatus	1.65 (0.05)	1.86 (0.05)	—
Panthera leo	1.66 (0.07)	1.72 (0.04)	—
Panthera onca	2.34 (0.14)	2.62 (0.09)	—
Panthera pardus	1.85 (0.07)	1.72 (0.05)	—
Puma concolor	1.78 (0.05)	1.82 (0.05)	—
Extant Felidae Mean:	1.86 (0.05)	1.96 (0.06)	—
Panthera atrox (†)	2.55 (0.18)	2.70 (0.19)	—
Smilodon fatalis (†)	—	1.80 (0.12)	—
HYAENIDAE			
Crocuta crocuta	1.72 (0.05)	3.62 (0.28)[b]	—
Hyaena hyaena	1.73 (0.04)	2.89 (0.12)[b]	—
Extant Hyaenidae Mean:	1.72 (0.03)	3.26 (0.17)[b]	—

[a]Ratio at M_1-M_2 significantly different from those at P_3-P_4 and P_4-M_1; $P < 0.05$.

[b]Ratio at P_4-M_1 significantly different from that at P_3-P_4; $P < 0.001$.

their killing behavior in much the same way as do modern big felids and canids. That is, the dire wolf probably used relatively shallow, slashing bites, whereas the two felids used more prolonged, deeper bites. The relative weakness of the mandible of the American lion in mediolateral bending is curious, but the mandible of the living jaguar (*Panthera onca*, arrow) is also relatively weak, and the mandible is thus unlikely to reflect distinctive killing behaviors in the fossil lion.

To explore their potential as bone-crackers, we estimated the distribution of loads along the mandibular corpus in the dire wolf and American lion by comparing the ratios of relative bending strength about mediolateral versus dorsoventral planes (Zx/Zy) at successive interdental gaps (Table 12.2). *Smilodon* is not included in the analysis because it lacks lower postcarnassial molars and therefore has only one postcanine interdental gap, the ratio of which does not differ significantly from that of extant felids (Table 12.2). As noted earlier, hyaenids and canids exhibit significant thickening of the mandible adjacent to the bone-cracking teeth, which are the premolars in hyaenids and the molars in canids (Biknevicius and Ruff 1992a). The dire wolf also shows a marked increase in Zx/Zy values between the P_4-M_1 and M_1-M_2 junctions that is similar to but exceeds that observed in the gray wolf (Table 12.2). Thus, like gray wolves and unlike hyenas, dire wolves probably used their postcarnassial molars to crack bones. Overall, the mandibular corpus of dire wolves is essentially wolflike, showing no special adaptations to distinguish it from that of gray wolves except that, by virtue of their larger absolute size, dire wolves were likely able to crush larger bones and gain access to valuable marrow within the bones of larger herbivores. Unlike those of the canids or hyaenids, the mandibles of the fossil and extant lions show no regional thickening of the mandibular corpus, and thus the American lion appears to have been similarly nonspecialized for bone-cracking (Table 12.2).

The analysis of cross-sectional geometry of the jaws of *Smilodon* contributes important new data concerning the killing and feeding behaviors of this sabertooth cat. The cross-sectional data indicate that the mandibular corpora of *Smilodon* were as strong as or stronger than those of extant felids of similar dentary length. This suggests that, relative to extant large-bodied felids, *Smilodon* did not have a weak bite, as has been suggested by some (Matthew 1910; Merriam and Stock 1932; Tucker 1954e). The hypothesis of a strong bite is consistent with more recent studies of sabertooth function. Emerson and Radinsky (1980) argued for large bite forces in sabertooths on the basis of temporalis muscle orientation and the placement of the carnassials near the jaw joint (but see Bryant, in press). Similarly, Akersten (1985) asserted that the robust character of *Smilodon*'s skull and jaws indicated a powerful bite, and Van Valkenburgh and Ruff (1987) presented data on canine strength in *Smilodon* that indicated that large stresses occurred within the canines. If the new data on jaw strength are combined with previous data on craniodental function in *Smilodon*, a more complete picture of the killing and feeding behaviors of this unusual animal emerges.

The issue of bite force with the postcarnassial teeth is important. Although it is unlikely that the masticatory system of *Smilodon* was insufficient for shearing meats with the carnassials, the ability of *Smilodon* to crack fairly large bones, as lions do today, is an issue of debate. A study of dental microwear features on the carnassials of *Smilodon* suggests that bone was rarely consumed (Van Valkenburgh et al. 1990), but no comparable analysis has been conducted on the premolars, the teeth most likely to have been used for bone-cracking. Furthermore, because dental microwear features are transient and seem largely to reflect diet in the previous month or two (Teaford and Oyen 1989), evidence of occasional bone-eating might be missed in an analysis of enamel microwear. The fourth lower premolars (P_4) of *Smilodon* are compressed mediolaterally relative to those of modern lions (Van Valkenburgh 1991), making them presumably more vulnerable to fracture on contact with bone. Indeed, a more recent analysis of tooth-fracture frequencies in *Smilodon* revealed numerous cheek teeth broken in life, an indication of regular bone consumption (Van Valkenburgh and Hertel 1993). To judge from the tooth-breakage and jaw-strength analyses, it seems likely that *Smilodon* did eat bones, at least occasionally.

Because of its sabers, the killing behavior of *Smilodon* has been a subject of greater interest than has its feeding behavior. Simpson (1941) associated the elongate canines in *Smilodon* with the capacity to inflict deep (puncture) wounds. Although the stabbing hypothesis for *Smilodon* has its champions (e.g., Miller 1969, 1980), most paleontologists favor saber bites that resulted in relatively shallower wounds, in order to minimize potentially disastrous contact with bone by the mediolaterally compressed upper canines (Kurtén 1952; Emerson and Radinsky 1980; Akersten 1985; Van Valkenburgh and Ruff 1987). The source of the power for these bites would have been the jaw-adductor muscles acting together with the cervical musculature. These wounds would have been longer (but less deep) than those inflicted by the canines of living felids, owing to the great anteroposterior diameter of the sabers and the sweeping, arclike trajectory of the canines during jaw closure (Figure 12.12A) (Akersten 1985). Significantly, the reconstructed canine shear-bite also requires that the lower jaw be stabilized against the body of the prey as the upper canines were driven through its flesh (Figure 12.12A). Such bites administered to contemporary large-bodied prey, such as the giant camelids, bison, and perhaps proboscideans, would have required great force, and would have stressed the mandibular corpus greatly. This level of stress may explain the great buttressing found in the corpora of *Smilodon*.

Many researchers maintain that canine bites in *Smilodon* must have been short in duration in order to minimize mediolateral bending stresses to the sabers (Akersten 1985; Van Valkenburgh and Ruff 1987). Certainly, high mediolateral bending loads applied to the canine tip could have had tragic consequences. But this oft-cited inability of *Smilodon* to maintain canine bites, owing to the potential failure of these teeth, may be exaggerated. At full

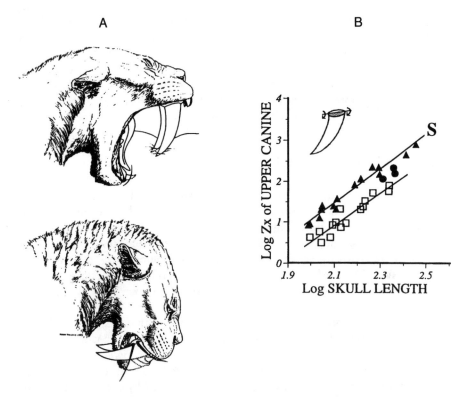

Figure 12.12. Killing bite of *Smilodon*. *A*, reconstructed canine shear bite. At full closure (*bottom*), most of the canine length is external to the body of the prey. *B*, plot of mediolateral bending strength of the canine base (Zx, in mm³) on skull length (SL, in mm); all variables log₁₀-transformed. Symbols as in Figure 12.11. Ordinary least-squares regression lines for felid canines: log Zx = −7.41 + 4.23(log SL), *r* = 0.98; equation for canid canines: log Zx = −7.48 + 4.00(log SL), *r* = 0.95. (Data from Van Valkenburgh and Ruff 1987.) (*A* adapted by permission from Akersten 1985.)

closure during a canine shear-bite, while the prey was being held firmly by the incisor battery, most of the canine (except its base) would have been located external to the prey (Figure 12.12*A*) (Akersten 1985). Significantly, the base of the sabers in *Smilodon* is similar in mediolateral bending strength to that found in the upper canines of living felids when loads are locally administered (Figure 12.12*B*). Furthermore, the teeth comprising the incisor battery (upper and lower incisors plus lower canines) are particularly robust in *Smilodon*, and these teeth interdigitate when the jaws are closed (Merriam and Stock 1932; Miller 1969; Akersten 1985). The shape of the upper incisor arcade is more parabolic than that found in living felids (Biknevicius et al., in press), providing spatial, and possibly functional, separation of the incisors from the upper canines. It appears that the incisors of living felids function more in feeding than in killing, whereas those of *Smilodon* were probably equally important in both activities.

It is not unreasonable, then, to envision that *Smilodon* could have maintained its hold on the prey, not with penetrating canine bites, as do living felids (since this is likely to result in contact with bone), but with the combined effort of saber bases and incisor battery. This might have been an effective way for *Smilodon* to control a weakened prey that had already received several quick but debilitating canine shear-bites. Prey death may have occurred by a throat hold not unlike that used by felids today.

Current and Future Research

Several topics are likely to be the foci of current and future research on carnivoran craniofacial function, and some of them have been inspired by the availability of new imaging technologies. For example, the crack-resisting construction of enamel can now be studied at a microstructural level with the use of electron microscopy (Koeningswald and Pfretzschner 1991; Rensberger 1995). Bone density and distribution within skulls and mandibles can be more accurately quantified using computerized imaging systems, such as CT scans (Biknevicius 1993; Thomason 1991b).

One important aspect of carnivoran killing behavior that has been neglected is that of neck mechanics. Net jaw adduction in carnivorans often involves flexion of the head on the spine as well as elevation of the lower jaws (Gorniak and Gans 1980), and neck function is thus likely to be a critical component of feeding biomechanics. Relative neck strength, estimated by the maximum moment arm of the head-turning muscles (Radinsky 1918a), is greater in solitary, large-prey predators (mustelids and felids) than in group or small-prey hunters (canids and hyaenids). The fact that hyaenid neck musculature has not been judged powerful by this technique, in spite of substantial contrary evidence from gross morphological studies (Spoor and Badoux 1986), indicates that more research is needed on the biomechanics of the cervical structures (Selbie et al. 1993).

A second topic in need of study concerns the ontogeny of feeding adaptations in carnivorans. The transition from suckling to feeding and then killing results in substantial changes in the functional requirements of skull and teeth (Emerson and Bramble 1993). Studies of these transitions will likely yield significant insights into the process of bone growth and remodeling. In addition, the observable design of bone should be recognized as an interplay between biomechanical and ancestral (genetic) factors. Separation of these factors may be possible through microstructural studies and ontogenetic analyses of craniofacial structures (e.g., Biknevicius 1993).

Finally, remarkably little effort has been given to amassing detailed qualitative and quantitative descriptions of carnivoran feeding behaviors in the wild (Van Valkenburgh, in press). Do hyenas usually or always crack bones with their premolars? How does the use of incisors with parabolic arcades differ from that of incisors with straight arcades? Is the function of the carnassials restricted to slicing skin and meat? Are there differences among species in typical feeding rates that correspond to differences in craniodental morphology? The answers to these and similar questions might provide us with the whys and wherefores of the morphologies we observe, and thus might suggest the adaptive pressures that are the driving forces in craniodental evolution.

Acknowledgments

We thank M. Spencer for his assistance with Greaves' model; L. Betti-Nash, T. Fox, and D. Pratt for providing assistance with the illustrations; W. Anyonge, J. L. Gittleman, S. M. Reilly, and J. J. Thomason for offering invaluable comments and suggestions on the chapter. Portions of this work were supported by the Department of Anatomical Sciences at the State University of New York at Stony Brook (ARB), the Lucille P. Markey Charitable Trust (ARB), the National Science Foundation (RPG 92-11214, to ARB), Research Enhancement Fund of Ohio University (ARB), and the Academic Senate of the University of California (BVV).

References

Akersten, W. A. 1985. Canine function in *Smilodon* (Mammalia; Felidae; Machairo-dontinae). *Contrib. Sci., Nat. Hist. Mus. Los Angeles County*, 356:1–22.

Alexander, R. M. 1981. Factors of safety in the structure of animals. *Sci. Prog.* 67:109–130.

Anyonge, W. 1993. Body mass in large extant and extinct carnivores. *J. Zool. (London)* 231:339–350.

Badoux, D. M. 1964. Lines of action of masticatory forces in domesticated dogs. *Acta Morphol. Neerl.-Scand.* 5:347–360.

Badoux, D. M. 1966. Framed structures in the mammalian skull. *Acta Morphol. Neerl.-Scand.* 6:239–250.

Badoux, D. M. 1968. Statics of the mandible. *Acta Morphol. Neerl.-Scand.* 6:251–256.

Badoux, D. M. 1972. Some notes on the stress in the caudal part of the lower jaw in domestic dog. *Proc. K.Ned. Akad. Wet. Amsterdam, Series C* 75:34–43.

Baker, G. I., and Laskin, D. M. 1968. Histochemical characterization of the muscles of mastication. *J. Dent. Res.* 48:97–104.

Bakker, R. T. 1983. The deer flees, the wolf pursues: Incongruencies in predator-prey coevolution. In: D. J. Futuyma and M. Slatkin, eds. *Coevolution*, pp. 350–382. Sunderland, Mass.: Sinauer Associates.

Bertram, B. C. R. 1979. Serengeti predators and their social systems. In: A. R. E. Sinclair and M. Norton-Grifiths, eds. *Serengeti: Dynamics of an Ecosystem*, pp. 221–248. Chicago: Univ. Chicago Press.

Bikneviçius, A. R. 1990. Biomechanical design of the mandibular corpus in carnivores. Ph.D. dissert., The Johns Hopkins University, Baltimore.

Bikneviçius, A. R. 1993. Ontogeny of feeding adaptations in carnivorans. *Amer. Zool.* 33:102A.

Biknevicius, A. R., and Ruff, C. B. 1992a. The structure of the mandibular corpus and its relationship to feeding behaviours in extant carnivorans. *J. Zool. (London)* 228:479–507.

Biknevicius, A. R., and Ruff, C. B. 1992b. Use of biplanar radiographs for estimating cross-sectional geometric properties of mandibles. *Anat. Rec.* 232:157–163.

Biknevicius, A. R., Van Valkenburgh, B., and Walker, J. In press. Incisor size and shape: Implications for feeding behavior in sabertoothed "cats." *J. Vert. Paleontol.*

Blumenschine, R. J. 1987. Characteristics of an early hominid scavenging niche. *Curr. Anthrop.* 28:383–407.

Bouvier, M., and Hylander, W. L. 1984. The effect of dietary consistency on gross and histologic morphology in the craniofacial region of young rats. *Amer. J. Anat.* 170:117–126.

Bramble, D. M. 1978. Origins of mammalian feeding complex: Models and mechanisms. *Paleobiology* 4:271–301.

Buckland-Wright, J. C. 1971. The distribution of biting forces in the skulls of dogs and cats. *J. Dent. Res.* 50:1168–1169.

Buckland-Wright, J. C. 1972. The shock-absorbing effect of cranial suture in certain mammals. *J. Dent. Res.*, Supp., 5:1241.

Buckland-Wright, J. C. 1978. Bone structure and the patterns of force transmission in the cat skull (*Felis catus*). *J. Morph.* 155:35–62.

Butler, P. M. 1946. The evolution of carnassial dentition in the Mammalia. *Proc. Zool. Soc. London* 116:198–220.

Caro, T. M. 1994. *Cheetahs of the Serengeti Plains.* Chicago: Univ. Chicago Press.

Covey, D. S. G., and Greaves, W. S. 1994. Jaw dimensions and torsion resistance during canine biting in the Carnivora. *Canadian J. Zool.* 72:1055–1060.

Crusafont-Pairo, M., and Truyols-Santonja, J. 1956. A biometric study of the evolution of fissiped carnivores. *Evolution* 10:314–332.

Daegling, D. J., Ravosa, M. J., Johnson, K. R., and Hylander, W. L. 1992. Influence of teeth, alveoli, and periodontal ligaments on torsional rigidity in human mandibles. *Amer. J. Phys. Anthropol.* 89:59–72.

Demes, B. 1982. The resistance of primate skulls against mechanical stresses. *J. Human Evol.* 11:687–691.

Dessem, D. 1985. The transmission of muscle force across the unfused symphysis in mammalian carnivores. In: H.-R. Duncker and G. Fleischer, eds. *Vertebrate Morphology*, pp. 289–291. New York: Gustav Fischer Verlag.

Dessem, D. 1989. Interactions between jaw-muscle recruitment and jaw-joint forces in *Canis familiaris. J. Anat.* 164:101–121.

Dessem, D., and Druzinsky, R. E. 1992. Jaw-muscle activity in ferrets, *Mustela putorius furo. J. Morphol.* 213:275–286.

Emerson, S. B., and Bramble, D. M. 1993. Scaling, allometry, and skull design. In: V. Hanken and B. K. Hall, eds. *The Skull*, Vol. 3, pp. 389–421. Chicago: Univ. Chicago Press.

Emerson, S. B., and Radinsky, L. 1980. Functional analysis of sabertooth cranial morphology. *Paleobiology* 6:295–312.

Estes, R. D., and Goddard, J. 1967. Prey selection and hunting behavior of the African wild dogs. *J. Wildl. Mgmt.* 31:52–69.

Ewer, R. F. 1973. *The Carnivores.* Ithaca, N.Y.: Cornell Univ. Press.

Flynn, J. J., Neff, N. A., and Tedford, R. H. 1988. Phylogeny of the Carnivora. In: M. J. Benton, ed. *The Phylogeny and Classification of the Tetrapods*, Vol. 2, pp. 73–116. Oxford: Clarendon Press.

Gans, C., Gorniak, G. C., and Morgan, W. K. 1990. Bite-to-bite variation of muscular activity in cats. *J. Exp. Biol.* 151:1–19.

Gaunt, W. A. 1959. The development of the deciduous cheek teeth of the cat. *Acta Anat.* 38:187–212.

Gingerich, P. D., and Winkler, D. A. 1979. Patterns of variation and correlation in the dentition of the red fox, *Vulpes vulpes*. J. Mamm. 60:691–704.

Gorniak, G. C. 1986. Correlation between histochemistry and muscle activity of jaw muscles in cats. *J. Applied Physiol.* 60:1393-1400.

Gorniak, G. C., and Gans, C. 1980. Quantitative assay of electromyograms during mastication in domestic cats (*Felis catus*). *J. Morphol.* 163:253–281.

Greaves, W. S. 1974. Functional implications of mammalian jaw joint position. *Forma et Functio* 74:363–376.

Greaves, W. S. 1983. A functional analysis of carnassial biting. *Biol. J. Linnean Soc.* 20:353–363.

Greaves, W. S. 1985. The generalized carnivore jaw. *Zool. J. Linnean Soc.* 85:267–274.

Greaves, W. S. 1988. The maximum average bite force for a given jaw length. *J. Zool. (London)* 184:271–285.

Heidt, G. A. 1972. Anatomical and behavioral aspects of killing and feeding by the least weasel, *Mustelis nivalis* L. *Proc. Arkansas Acad.* 26:53–54.

Hornocker, M. G. 1970. An analysis of mountain lion predation upon mule deer and elk in the Idaho Primitive Area. *Wildl. Monogr.* 21:1–39.

Hylander, W. L. 1981. Patterns of stress and strain in the macaque mandible. In: D. S. Carlson, ed. *Craniofacial Biology, Monograph #10, Craniofacial Growth Series, Center of Human Growth and Development*, pp. 1–35. Ann Arbor: Univ. Michigan.

Hylander, W. L. 1988. Implications of *in vivo* experiments for interpreting the functional significance of "robust" australopithecine jaws. In: F. E. Grine, ed. *Evolutionary History of the "Robust" Australopithecines*, pp. 55–83. New York: Aldine de Gruyter.

Hylander, W. L., and Johnson, K. R. 1992. Strain gradients in the craniofacial region of primates. In: Z. Davidovitch, ed. *The Biological Mechanisms of Tooth Movement and Craniofacial Adaptations*, pp. 559–569. Columbus: Ohio State Univ. College of Dentistry.

Kakudo, Y., and Amano, N. 1968. Strain in the rabbit and dog maxillomandibular bones during biting and mastication. *J. Dent. Res.* 47:496.

King, C. 1989. *The Natural History of Weasels and Stoats*. Ithaca, N.Y.: Cornell Univ. Press.

Kitchener, A. 1991. *The Natural History of the Wild Cats*. Ithaca, N.Y.: Cornell Univ. Press.

Koenigswald, W. von, and Pfretschner, H. V. 1991. Biomechanics in the enamel of mammalian teeth. In: N. Schmidt-Kittler and K. Vogel, eds. *Constructional Morphology and Evolution*, pp. 113–125. Berlin: Springer-Verlag.

Kruuk, H. 1972. *The Spotted Hyena*. Chicago: Univ. Chicago Press.

Kruuk, H., and Turner, M. 1967. Comparative notes on predation by lion, leopard, cheetah and wild dog in the Serengeti area, East Africa. *Mammalia* 31:1–27.

Kurtén, B. 1952. The Chinese *Hipparion* fauna. *Soc. Sci. Fennica, Commentationes Biol.* 13:1–82.

Kurtén, B., and Anderson, E. 1980. *Pleistocene Mammals of North America*. New York: Columbia Univ. Press.

Lanyon, L. E., and Rubin, C. T. 1985. Functional adaptation in skeletal structures. In: M. Hildebrand, D. M. Bramble, K. F. Liem, and D. B. Wake, eds. *Functional Vertebrate Morphology*, pp. 1–25. Cambridge, Mass.: Harvard Univ. Press.

Leyhausen, P. 1979. *Cat Behavior*. New York: Garland STPM Press.

Liebman, F. M., and Kussick, L. 1965. An electromyographic analysis of masticatory muscle imbalance with relation to skeletal growth in dogs. *J. Dent. Res.* 44:768–774.

Lucas, P. W., and Luke, D. A. 1984. Chewing it over: Basic principles of food breakdown. In: D. J. Chivers, B. A. Wood, and A. Bilsborough, eds. *Food Acquisition and Processing in Primates*, pp. 283–301. New York: Plenum Press.

Lund, J. P., Drew, T., and Rossignol, S. 1984. A study of jaw reflexes of the awake cat during mastication and locomotion. *Brain Behav. Evol.* 25:146–156.

MacKenna, B. R., and Türker, K. S. 1978. Twitch tension in the jaw muscles of the cat at various degrees of mouth opening. *Archs. Oral. Biol.* 23:917–920.

MacLeod, N., and Rose, K. D. 1993. Inferring locomotor behavior in Paleogene mammals via eigenshape analysis. *Amer. J. Sci.* 293A:300–355.

Marean, C. W., Spencer, L. M., Blumenschine, R. J., and Capaldo, S. D. 1992. Captive hyena bone choice and destruction, the schlepp effect, and Olduvai archaeofaunas. *J. Archaeol. Sci.* 19:101–121.

Matthew, W. D. 1910. The phylogeny of the Felidae. *Bull. Amer. Mus. Nat. Hist.* 26:289–316.

Mech, L. D. 1970. *The Wolf.* Minneapolis: Univ. Minnesota Press.

Mellett, J. S. 1981. Mammalian carnassial function and the "Every effect." *J. Mamm.* 62:164–166.

Merriam, J. C. 1912. The fauna of Rancho la Brea, II: Canidae. *Mem. Univ. Calif.* 1:215–272.

Merriam, J. C., and Stock, C. 1932. *The Felidae of Rancho La Brea.* Washington, D.C.: Publ. Carnegie Inst., 4.

Miller, G. J. 1969. A new hypothesis to explain the method of food ingestion used by *Smilodon californicus* Bovard. *Tebiwa* 12:9–19.

Miller, G. J. 1980. Some new evidence in support of the stabbing hypothesis for *Smilodon californicus* Bovard. *Carnivore* 3:8–26.

Mills, M. G. L. 1990. *Kalahari Hyaenas: Comparative Behavioural Ecology of Two Species.* Boston: Unwin Hyman.

Noble, H. W. 1979. Comparative functional anatomy of the temporomandibular joint. In: G. A. Zarb and G. E. Carlsson, eds. *Temporomandibular Joint Function and Dysfunction*, pp. 15–41. St. Louis: C. V. Mosby.

Packer, C. 1986. The ecology of sociality in felids. In: D. I. Rubenstein and R. W. Wrangham, eds. *Ecological Aspects of Social Evolution*, pp. 429–451. Princeton, N.J.: Princeton Univ. Press.

Pengilly, D. 1984. Developmental versus functional explanations for patterns of variability and correlation in the dentitions of foxes. *J. Mamm.* 65:34–43.

Plishka, J., and Bardack, D. 1992. Bending strength of upper canine tooth in domestic dogs. *Ann. Anat.* 174:321–326.

Radinsky, L. B. 1981a. Evolution of skull shape in carnivores, 1: Representative modern carnivores. *Biol. J. Linnean Soc.* 15:369–388.

Radinsky, L. B. 1981b. Evolution of skull shape in carnivores, 2: Additional modern carnivores. *Biol. J. Linnean Soc.* 16:337–355.

Rensberger, J. M. 1995. Determination of stresses in mammalian dental enamel and their relevance to the interpretation of feeding behaviors in extant taxa. In: J. J. Thomason, ed. *Functional Morphology in Vertebrate Paleontology*, pp. 151–172. Cambridge: Cambridge Univ. Press.

Riley, M. A. 1985. An analysis of masticatory form and function in three mustelids (*Martes americana, Lutra canadensis, Enhydra lutra*). *J. Mamm.* 66:519–528.

Roberts, D. 1979. Mechanical structure and function of the craniofacial skeleton of the domestic dog. *Acta Anat.* 103:422-433.

Rowe-Rowe, D. T. 1978. Comparative prey capture and food studies of South African mustelines. *Mammalia* 42:175–196.

Russell, A. P., and Thomason, J. J. 1993. Mechanical analysis of the mammalian head skeleton. In: J. Hanken and B. K. Hall, eds. *The Skull*, Vol. 3, pp. 345–383. Chicago: Univ. Chicago Press.

Scapino, R. 1965. The third joint of the canine jaw. *J. Morph.* 116:23–50.

Scapino, R. 1976. Function of the digastric muscle in carnivores. *J. Morph.* 150:843–860.

Scapino, R. 1981. Morphological investigation into functions of the jaw symphysis in carnivorans. *J. Morph.* 167:339–375.

Schaller, G. B. 1972. *The Serengeti Lion.* Chicago: Univ. Chicago Press.

Seidensticker, J. C. 1976. On the ecological separation between tigers and leopards. *Biotropica* 8:225–234.

Selbie, W. S., Thomson, D. B., and Richmond, F. J. R. 1993. Suboccipital muscle in the cat neck: Morphometry and histochemistry of the rectus captitis muscle complex. *J. Morphol.* 216:47–63.

Sherrington, C. S. 1917. Reflexes elicitable in the cat from pinna, vibrissae and jaws. *J. Physiol. (London)* 51:404–431.

Simpson, G. G. 1941. The function of saber-like canines in carnivorous mammals. *Amer. Mus. Novitates* 1130:1–12.

Smith, J. M., and Savage, R. J. G. 1959. The mechanics of mammalian jaws. *Sci. Rev.* 141:289–301.

Smith, K. K. 1993. The form of the feeding apparatus in terrestrial vertebrates: Studies of adaptation and constraint. In: J. Hanken and B. K. Hall, eds. *The Skull*, Vol. 3, pp. 150–196. Chicago: Univ. Chicago Press.

Solounias, N., Teaford, M., and Walker, A. 1988. Interpreting the diet of extinct ruminants: The case of a non-browsing giraffid. *Paleobiology* 14:287–300.

Spencer, M. A., and Demes, B. 1993. Biomechanical analysis of masticatory system configuration in Neandertals and Inuits. *Amer. J. Phys. Anthropol.* 91:1–20.

Spoor, C. F., and Badoux, D. M. 1986. Descriptive and functional morphology of the neck and forelimb of the striped hyena. *Anat. Anz., Jena* 161:375–387.

Stock, C., and Harris, J. M. 1992. Rancho La Brea: A record of Pleistocene life in California, 7th ed. *Nat. Hist. Mus. Los Angeles County, Science Series* 37:1–113. (Revision of work originally published as Merriam and Stock 1932; see entry above.)

Tamari, J. W., Tomey, G. F., Ibrahim, M. Z. M., Baraka, A., Jabbur, S. J., and Bahuth, N. 1973. Correlative study of the physiologic and morphologic characteristics of the temporal and masseter muscles of the cat. *J. Dent. Res.* 52:538–543.

Taylor, A., Cody, F. W. J., and Bosley, M. A. 1973. Histochemical and mechanical properties of the jaw muscles of the cat. *Exp. Neurol.* 38:99–109.

Teaford, M. F., and Oyen, O. J. 1989. *In vitro* and *in vivo* turnover in dental microwear. *Amer. J. Phys. Anthropol.* 80:447–460.

Thomason, J. J. 1991a. Comparing rostral and mandibular strength estimates for carnivorans: Implications for cranial design. *Amer. Zool.* 31:39A.

Thomason, J. J. 1991b. Cranial strength in relation to estimated biting forces in some mammals. *Canadian J. Zool.* 69:2326–2333.

Thomason, J. J., and Russell, A. P. 1986. Mechanical factors in the evolution of the mammalian secondary palate: A theoretical analysis. *J. Morph.* 189:199–213.

Tucker, R. 1954a. Studies in functional and analytical craniology, I: The elements of analysis. *Australian J. Zool.* 2:381–390.

Tucker, R. 1954b. Studies in functional and analytical craniology, II: The functional classification of mammalian skulls. *Australian J. Zool.* 2:391–398.

Tucker, R. 1954c. Studies in functional and analytical craniology, III: The brevicaudate skull and its analysis. *Australian J. Zool.* 2:399–411.

Tucker, R. 1954d. Studies in functional and analytical craniology, IV: More extreme forms of the brevicaudate skull. *Australian J. Zool.* 2:412–417.

Tucker, R. 1954e. Studies in functional and analytical craniology, V: The functional metamorphoses of the carnivorous skull. *Australian J. Zool.* 2:418–426.

Tucker, R. 1954f. Studies in functional and analytical craniology, VI: Strains and the direction of certain vectors in the brevicaudate skull. *Australian J. Zool.* 2:427–430.

Tucker, R. 1955. Studies in functional and analytical craniology, X: The density of the bony tissue in nodes and tracts. *Australian J. Zool.* 3:541–546.

Turnbull, W. D. 1970. Mammalian masticatory apparatus. *Fieldiana (Zool.)* 18:147–356.

Van Valen, L. 1969. Evolution of dental growth and adaptation in mammalian carnivores. *Evolution* 23:96–117.

Van Valkenburgh, B. 1985. Locomotor diversity within past and present guilds of large predatory mammals. *Paleobiology* 11:406–428.

Van Valkenburgh, B. 1988. Incidence of tooth breakage among large, predatory mammals. *Amer. Nat.* 131:291–302.

Van Valkenburgh, B. 1989. Carnivore dental adaptations and diet: A study of trophic diversity within guilds. In: J. L. Gittleman, ed. *Carnivore Behavior, Ecology, and Evolution*, pp. 410–436. Ithaca, N.Y.: Cornell Univ. Press.

Van Valkenburgh, B. 1991. Iterative evolution of hypercarnivory in canids (Mammalia: Carnivora): Evolutionary interactions among sympatric predators. *Paleobiology* 17:340–362.

Van Valkenburgh, B. In press. Feeding behavior in free-ranging, large African predators. *J. Mammal.*

Van Valkenburgh, B., and Hertel, F. 1993. Tough times at La Brea: Tooth breakage in large carnivores of the Late Pleistocene. *Science* 261:456–459.

Van Valkenburgh, B., and Koepfli, K.-P. 1993. Cranial and dental adaptations to predation in canids. *Symp. Zool. Soc. Lond.* 65:15–37.

Van Valkenburgh, B., and Ruff, C. B. 1987. Canine tooth strength and killing behavior in large carnivores. *J. Zool. (London)* 212:379–397.

Van Valkenburgh, B., Teaford, M. F., and Walker, A. 1990. Molar microwear and diet in large carnivores: Inferences concerning diet in the sabretooth cat, *Smilodon fatalis*. *J. Zool. (London)* 222:319–340.

Vitti, M. 1965. Estudo electromiográfico dos músculos mastigadores no cao. *Folia Clin. Biol.* 34:101–114.

Vogel, S. 1988. *Life's Devices*. Princeton, N.J.: Princeton Univ. Press.

Wainwright, S. A., Biggs, W. D., Currey, J. D., and Goslin, J. M. 1976. *Mechanical Design of Organisms*. Princeton, N.J.: Princeton Univ. Press.

Wayne, R. K. 1986. Cranial morphology of domestic and wild canids: The influence of development on morphological change. *Evolution* 40:243–261.

Werdelin, L. 1988. Correspondence analysis and the analysis of skull shape and structure. *Ossa* 13:207–215.

Werdelin, L. 1989. Constraint and adaptation in the bone-cracking canid *Osteoborus* (Mammalia: Canidae). *Paleobiology* 15:387–401.

Whitfield, P. 1978. *The Hunters*. New York: Simon and Schuster.

Zar, J. H. 1984. *Biostatistical Analysis*. Second ed. Englewood Cliffs, N.J.: Prentice-Hall.

CHAPTER 13

Genetic Relatedness among Individuals within Carnivore Societies

MATTHEW E. GOMPPER AND ROBERT K. WAYNE

Estimations of genetic kinship are a prerequisite for understanding the behavioral and ecological interactions occurring within animal societies. This chapter reviews molecular genetic studies that attempt to measure degree of kinship and to account for observed patterns of interactions among individuals within carnivore societies. We first discuss general patterns of carnivore social organization, within-group relatedness, and the problems inherent in estimating relatedness among individuals by means of behavioral approaches. We then briefly explain the molecular genetic techniques most commonly used to assess relatedness. Finally, we review the results and insights provided by within-population genetic studies of carnivores.

The common designation of a carnivore species as either group-living or solitary, social or asocial, monogamous or polygynous, often conceals a high level of intraspecific behavioral variability. For example, the extent of group-living and group size can vary dramatically within and between geographical regions. In populations of some basically group-living species, a substantial fraction of the population may live as solitary individuals (European badger, *Meles meles:* Kruuk 1989; white-nosed coati, *Nasua narica:* Kaufmann 1962). Similarly, species commonly perceived as solitary may show extensive spatial variability in sociality (Egyptian mongoose, *Herpestes ichneumon:* Ben-Yaacov and Yom-Tov 1983; Palomares and Delibes 1993; white-tailed mongoose, *Ichneumia albicauda:* Waser and Waser 1985; red fox, *Vulpes vulpes:* Macdonald 1981; Doncaster and Macdonald 1991; coyote, *Canis latrans:* Bekoff and Wells 1982), or may form ephemeral aggregations during periods of high resource availability (raccoon, *Procyon lotor:* Sharp and Sharp 1956; Mech and Turkowski 1966; polar bear, *Ursus maritimus:* Derocher and Stirling 1990). The development of social organization is often considered to reflect evolutionary pressures to maximize inclusive fitness in specific environments, and, consequently, recent carnivore research has emphasized the benefits of social interactions to the individual. To quantify the benefits of social interactions accurately, and to understand their temporal and spatial variability, the genetic relatedness of individuals within carnivore populations needs to be measured using population- and observer-independent techniques.

Behavioral studies have been careful to stress that within carnivore social groups both mean relatedness, r (the fraction of genes in two individuals that are identical by descent: Wright 1922; Hamilton 1964), and the variance in r may be high. For example, spotted hyenas (*Crocuta crocuta*) form clans that may contain over 80 individuals, including several long-lived breeding females (Kruuk 1972; Mills 1990; Frank, this volume). If close inbreeding is avoided, and both mothers and daughters are philopatric, relatedness among clan founders and individuals born in later generations rapidly diverges. In support of this prediction, relatedness estimated from behavioral observations of 22 individuals from a single clan varied from $r = 0.05$ to $r = 0.50$ (Mills 1990). Carnivores with a similar potential for high variance in within-group relatedness include white-nosed coatis (*Nasua narica*: Russell 1983; Gompper and Krinsley 1992; Gompper 1994), meerkats (*Suricata suricatta*: Macdonald 1992), banded mongooses (*Mungos mungo*: Rood 1975), dwarf mongooses (*Helogale parvula*: Rood 1980, 1990; Creel and Waser 1991), brown hyenas (*Hyaena brunnea*: Mills 1982, 1984, 1990; Owens and Owens 1984), feral domestic cats (*Felis catus*: Macdonald et al. 1987), and European badgers (*Meles meles*: Kruuk 1989; Evans et al. 1989; Woodroffe 1993; Woodroffe and Macdonald 1993). If inclusive-fitness considerations influence behavioral interactions within carnivore societies, then variability in relatedness is hypothesized to correlate with variability in behavioral interactions. In testing this hypothesis, however, we must take an approach that does not implicitly assume a relationship between relatedness and the degree of benefit afforded by social interactions. For example, provisioning of young by adults cannot be assumed to indicate a genetic relationship, since several examples of adults provisioning young other than their own have been described (Schaller 1972; Rood 1978, 1980; Ben-Yaacov and Yom-Tov 1983; Russell 1983; Macdonald et al. 1987; Rasa 1987; Knight et al. 1992; Packer et al. 1992; Woodroffe 1993).

Several other problems are inherent in efforts to infer relatedness from behavioral observations. Cuckoldry may occur (e.g., in the aardwolf, *Proteles cristatus*: Richardson 1987). Species with delayed implantation, second estrus periods, or induced ovulations may have multipaternity litters (Johansson and Venge 1951; Harrison and Neal 1956; Enders and Enders 1963; Ahnlund 1980). In several species, male mating coalitions occur that may vary greatly in size and relatedness (e.g., Caro and Collins 1986, 1987; Packer et al. 1991a). A common assumption, based on observations of copulation or female access, is that all coalition males breed, or breed equally (Bertram 1976; Bygott et al. 1979). Mating access, however, is not necessarily equivalent to reproductive success, and ample opportunity may exist for asymmetric reproduction. Finally, because the reasons for individuals to associate (i.e., sociality or group-living) are varied, and may be distinct from those underlying mate choice, social systems should not be confused with mating systems. For all these reasons, an independent, nonbehavioral approach is often necessary in attempting to confirm (or refute) the relatedness of individuals. As we discuss below, approaches provided by molecular genetic techniques may offer sufficient resolution to address these problems of relatedness in natural populations.

Molecular Genetic Techniques

Over the past two decades, the analysis of allozyme polymorphisms using gel electrophoresis has been used to improve our understanding of relatedness within animal populations (e.g., Schwartz and Armitage 1980; Lester and Selander 1981; Olivier et al. 1981; Wrege and Emlen 1987; Xia and Millar 1991). Allozyme loci, however, are frequently monomorphic (typically only 10–15% of loci are variable), and often only a few alternative alleles are found at a specific locus. Consequently, allozyme electrophoresis can generally be used to deduce parental relationships for only a fraction of the sampled individuals in a population, and cannot confidently be used to distinguish relatedness classes (but see McCracken and Bradbury 1981; Wilkinson 1984, 1985, 1992; Pope 1990; Hughes and Queller, 1993). For example, when allozyme variability is low or the genotypes of possible parents are similar, analysis of allozyme polymorphisms underestimates the actual frequencies of extra-pair matings (Wrege and Emlen 1987; Gelter and Tegelström 1992). Because many species of carnivores exhibit relatively low allozyme variability, allozyme electrophoresis may not often be a useful technique for deducing parentage or relatedness (Bonnell and Selander 1974; Allendorf et al. 1979; Simonsen 1982; Simonsen et al. 1982a, 1982b; Larsen et al. 1983; O'Brien et al. 1983; Kilpatrick et al. 1986).

Analysis of mitochondrial DNA (mtDNA) polymorphisms potentially provides a more detailed resolution of within-population relatedness (Hoelzel and Dover 1991; Lehman et al. 1992; Melnick and Hoelzer 1992; Wilkinson 1992; Edwards 1993). Because the mitochondrial genome is nonrecombining and maternally inherited, siblings will have a genotype identical to that of their mother, barring mutations. Generally, analysis of mtDNA polymorphisms is not used to deduce relatedness (r), but because the mitochondrial genome is often highly polymorphic in a population, matrilines can potentially be deduced and putative mothers excluded. Moreover, the evolutionary relationships of matrilines can be determined through phylogenetic analysis of restriction site or sequence data (see reviews in Brown 1985; Avise et al. 1987; Moritz et al. 1987; Harrison 1989). Most past studies of mtDNA sequence differences were based on restriction fragment length polymorphisms (RFLP). Recently, direct sequencing of the hypervariable mitochondrial control region has resulted in the finding of an increased number of polymorphisms in natural populations that can be readily analyzed using phylogenetic techniques (e.g., Thomas et al. 1990; Vigilant et al. 1991).

Problematic aspects of mtDNA analyses include interpretation of the role of males, biases caused by female philopatry, and the limited variability within isolated, recently colonized, or bottlenecked populations. For example, in group-living species with strong female philopatry, the within-population variability of mitochondrial DNA should be low, because the number of mitochondrial genotypes is a function of population size and the extent of genetic exchange (Lehman and Wayne 1991; Melnick and Hoelzer 1992; Wayne et al. 1992; Hoelzer et al. 1994). Moreover, individuals with different genotypes may appear genetically dissimilar but have the same or related fathers. Thus, analysis of mtDNA polymorphisms is most useful when variability of genotypes is high within populations, and when it is used in conjunction with biparentally inherited markers to determine the genetic contribution of males.

Some loci that have been described recently are hypervariable and extremely common in the nuclear genome. These loci consist of tandem repeats of short DNA sequences that can be assayed through hybridization with DNA probes or amplified by the polymerase chain reaction (Jeffreys et al. 1985a, 1985b; Vassart et al. 1987; Tautz et al. 1989; Rassmann et al. 1991). Tandem-repeat loci, often called variable number of tandem-repeat loci (VNTR), frequently have levels of heterozygosity exceeding 90% and are nearly always polymorphic. The first VNTR studies focused on multilocus "genetic fingerprints" that provided phenotypes which could be used to deduce paternity and, potentially, relatedness (Burke 1989; Lynch 1991; Packer et al. 1991a). More recent studies have used single-locus approaches that will likely allow a more precise quantification of relatedness, even in small, isolated populations (Schlötterer et al. 1991; Bruford and Wayne 1993; Queller et al. 1993; Gottelli et al. 1994). Moreover, because recent approaches utilize the polymerase chain reaction, minute amounts of DNA can be analyzed, even from old, poorly preserved samples (Hagelberg et al. 1991; Ellegren 1991; Kohn et al. 1995).

The molecular and statistical techniques used depend on the hypotheses being tested, background knowledge of the study population, and the fraction of the population sampled. Each molecular technique has its strengths and limitations. For example, multilocus fingerprints are well suited for the confirmation of extra-pair copulations if all putative parents and offspring are sampled in a study population. Measuring relatedness in populations is more difficult, however, especially if little is known in advance about the relatedness of individuals (Lynch 1988, 1990). Single-locus VNTR or microsatellite analyses are more powerful statistically, but may be taxon-specific and may involve more expense and effort. Allozyme and mtDNA restriction site analyses are potentially easier and less expensive to use, but levels of variability within populations may be low, and thus relatedness and paternity may be poorly resolved. Field researchers considering a molecular genetic analysis should carefully evaluate the effort and expense presented by the technique considered relative to the likelihood of answering specific behavioral hypotheses.

Case Studies

Canidae

Intensive examination of gray wolves (*Canis lupus*) from three well-studied and geographically distinct regions (Lehman et al. 1992) has provided new insights into the structure of wolf societies. By analyzing multilocus genetic fingerprints and mitochondrial DNA RFLPs and using as a baseline the calibration data obtained from two captive wolf colonies, Lehman and colleagues were able to classify dyads of free-living individuals as unrelated ($r < 0.1$), moderately related ($r > 0.25$), or highly related ($r \geq 0.5$). The genetic analyses showed, first, that wolf packs consisted primarily of individuals that are closely related genetically (Figure 13.1), a result supported by field studies. But if a

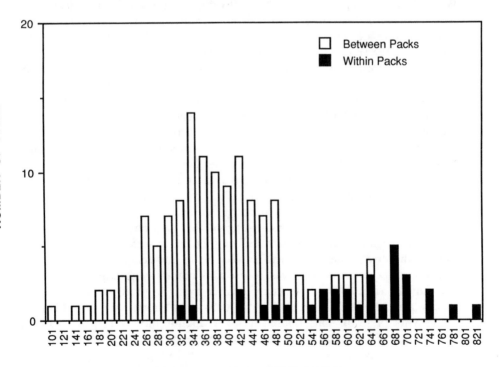

Figure 13.1. Histograms of multilocus DNA fingerprint-similarity values (times 1000) among 13 Minnesota gray wolf (*Canis lupus*) packs. Filled bars indicate similarity of individuals within packs. Open bars indicate similarity of individuals from different packs. For histograms of other pack populations, see Lehman et al. 1992, from which this figure is taken, by permission.

pack is composed of a dominant reproductive pair and its young (Mech 1970; Kleiman and Eisenberg 1973; Macdonald and Moehlman 1982), no more than two unrelated individuals (the breeding pair), as discerned by low DNA fingerprint similarity values, should be present. In 13 packs from Minnesota, three contained individuals unrelated to the breeding pair. The same result was found in two of 13 packs examined from Denali, Alaska, and in four of nine packs examined from Inuvik, Northwest Territories. In fact, because many packs were not completely sampled, these figures are probably an underestimation of the number of packs containing individuals unrelated to the breeding pair. Thus, although wolf packs are dominated by highly related individuals, they are not necessarily family units consisting solely of a mated pair and their offspring (Mech 1970).

A second result was the high frequency of dispersal to neighboring packs, or at least to packs within the same area (Figure 13.2). This conclusion was based on the observation that in all three regions, fingerprint similarity values among individuals from different packs were sometimes as high as that between siblings, or between parents and their offspring. This result is an independent confirmation of field observations demonstrating that dispersing individuals may commonly colonize areas near their natal pack (Mech 1970, 1987; Fritts and Mech 1981; Van Ballenberghe 1983; Gese and Mech 1991; Waser, this volume). Paradoxically, the genetic results suggest that such dispersal may occur from the natal pack to neighboring packs even where interpack aggression is so intense that mortality often results (Mech 1970, 1994; Fritts and Mech 1981; Van Ballenberghe 1983). The genetic results also suggest the possibility of local inbreeding between related pack founders derived from neighboring packs, which is consistent with computer simulations (Woolpy and Eckstrand 1979) and studies of captive (Packard et al. 1985; Laikre and Ryman 1992) and wild (Mech 1987; Wayne et al. 1991) populations.

The third finding of Lehman et al. (1992) was that in all three regions, the female-female mean fingerprint-similarity coefficients were higher than the male-male mean values. This result suggests that long-range dispersal is more common for males than for females, or that males suffer greater mortality during dispersal. The difference between male and female similarities was greatest in the Minnesota wolves, where pack home ranges exhibited the least overlap. In addition, high genetic-similarity dyads from different Minnesota packs ($n = 22$) were significantly more likely to be female pairs ($n = 11$) than male pairs ($n = 0$). This finding was perhaps the most surprising, since it contrasts with previous observations in Minnesota indicating no sex bias in dispersal distance (Mech 1987; Gese and Mech 1991).

The Isle Royale gray wolves, a small island population in Lake Superior adjacent to Northern Michigan, are probably descendants of a single pair of wolves that migrated to the island about 1950. These wolves have varied in population size from as much as 50 individuals to the present number of about 12 (Peterson 1979; Peterson and Page 1988; Wayne and Koepfli, this volume).

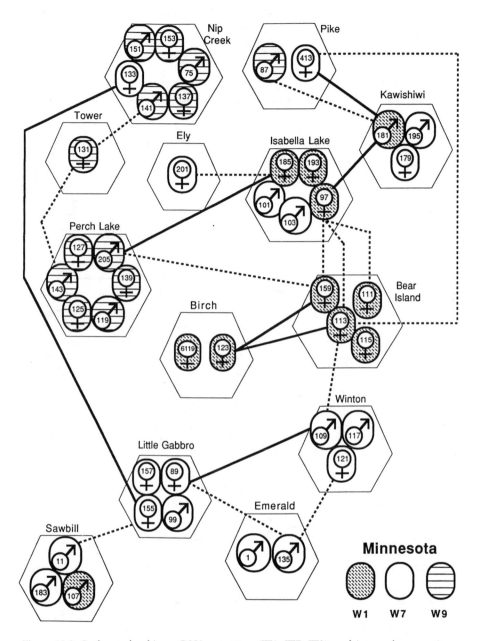

Figure 13.2. Pack membership, mtDNA genotypes (W1, W7, W9), and interpack connections suggested by high DNA fingerprint bandsharing values. Individuals identified by numbers have mtDNA genotypes indicated by the pattern of background shading behind each wolf symbol. High similarity values among wolves from different packs are indicated by lines: solid lines indicate similarity as high as that observed between parents and offspring and between siblings; dotted lines indicate a possible but more distant relationship. The placement of hexagons is intended to improve interpretations of connections and does not reflect the exact spatial relationships of packs. (From Lehman et al. 1992; reprinted by permission.)

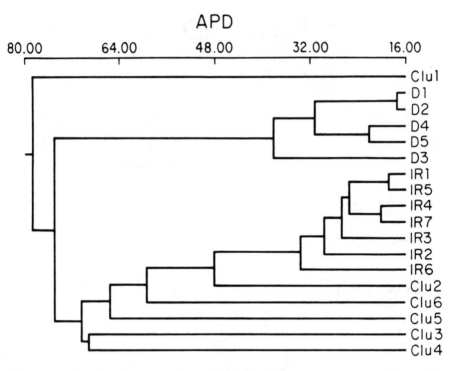

Figure 13.3. Clustering diagram based on dissimilarity (APD = average percent difference) in genetic fingerprints of unrelated mainland (Clu2–6) and Isle Royale (IR1–7) and captive (D1–5) gray wolves related as siblings. (From R. K. Wayne et al., "Conservation Genetics of the Endangered Isle Royale Gray Wolf," *Conservation Biology* 5 [March 1991]: 41–51. Reprinted by permission of Blackwell Sciences, Inc.)

Allozyme, mtDNA, and multilocus fingerprinting analyses have shown that this population has about half the genetic variability of mainland populations, and its members are genetically as similar as siblings (Figure 13.3; Wayne et al. 1991). Surprisingly, all individuals have a mtDNA genotype that is extremely rare on the mainland, providing genetic support for the theory that a single dispersing female founded the island population (Wayne et al. 1991). Inbreeding, combined with a recent population decline, may explain the recent failure of the Isle Royale wolves to breed (Peterson and Page 1988; Wayne et al. 1991). Initially, after several litters were born to the founding pair, dispersed siblings of vastly different ages did not recognize each other as close relatives and would form breeding pairs. In 1982, however, the population dropped to just 12 wolves. Wayne et al. (1991) hypothesize that these wolves may have been overlapping offspring from a single breeding pair and did not mate because they recognized each other as siblings.

The significance of multiple paternity in wild canid populations is uncertain. Recently, from the results of VNTR analysis, multiple paternity was discovered in a population of the Ethiopian wolf (*Canis simensis*). This highly endangered species, now numbering fewer than 1000 individuals, is restricted to six isolated areas in the Ethiopian highlands (Gottelli and Sillero-Zubiri 1992; Wayne and Koepfli, this volume). One population from the Bale Mountains National Park contained a pack wherein a female had a litter with at least two fathers and another pack wherein a female had a litter possibly having two fathers, one being a domestic dog or wolf-dog hybrid, the other an Ethiopian wolf (Sillero-Zubiri and Gottelli 1991; Gottelli et al. 1994). Field observations suggested that Ethiopian wolves appear to have a pack structure composed of a dominant pair and its offspring, as gray wolves do (Gottelli and Sillero-Zubiri 1992; Sillero-Zubiri 1994). A long-term study found a more open mating system, however: female wolves were observed to copulate with males from neighboring packs more often than with the alpha male in their own pack (Sillero-Zubiri and Gottelli 1991). This observation, supported by the genetic results, provides an important example of the need to distinguish mating from rearing systems when considering the genetic consequences of social behavior.

Mustelidae

Clans of the European badger (*Meles meles*) are territorial social units of 2–29 individuals of both sexes with a single breeding pair. Clan territories are marked by latrines and defended by aggressive behavior (Cheeseman et al. 1988; Kruuk 1989; da Silva et al. 1994). Two genetic studies have examined the relationships within and between badger clans. Using allozyme electrophoresis, Evans et al. (1989) studied 13 neighboring clans in Gloucestershire, England, and da Silva et al. (1994) studied 16 clans in Oxfordshire, England. Two of the 23 loci examined in the two studies were found to be both variable and informative. F-statistics based on these two variable loci revealed some genetic structuring among clusters of clans (termed "supergroups": Evans et al. 1989) within an area, and a deficiency of heterozygotes. Evans et al. (1989) attributed this genetic structure to stable clan membership and philopatric recruitment of clan offspring leading to greater differentiation among clans. Genetic heterogeneity among clans was reduced by occasional adult dispersal from one clan to another (Woodroffe et al. 1995) and by some clan young being sired by males from different clans. For example, da Silva et al. (1994) found that six of eight clans showed evidence of extra-clan paternity. Moreover, at least one clan was found to have cubs that were the offspring of several females. Because these results are based on only two loci, additional inter-clan matings and dispersal may have been missed.

Ursidae

Studies of black bear (*Ursus americanus*) ecology have shown males to be the dispersing sex, whereas females are philopatric (Alt 1978; Rogers 1987).

Wathen et al. (1985) examined allozyme variability among 49 black bears from three regions, separated by distances of less than 50 kilometers, in Great Smoky Mountain National Park, Tennessee. Five of 23 examined loci were polymorphic, of which two were sufficiently polymorphic to use in population genetic analyses. The observed heterogeneity among populations was due primarily to differences in the allele frequencies of subadult males. No heterogeneity was observed among adult females; homogeneity would be expected if females were strictly philopatric (Wathen et al. 1985). These data are very limited, however, because only two loci were polymorphic, and even a single aberrant individual could thus strongly bias the results.

Procyonidae

White-nosed coatis (*Nasua narica*) display a unique social organization of cohesive, female-bonded bands (made up of adult females and immature offspring) and solitary adult males (Kaufmann 1962; Gompper 1995). Observations of extensive social interactions within these bands, and occasional female migration between bands, led Russell (1983) to hypothesize reciprocal altruism as a primary evolutionary mechanism selecting for group-living in coatis. Gompper (1994) used multilocus fingerprinting to examine the relatedness of nine of the coati bands on Barro Colorado Island, Panama. In addition, 47 adult males whose home ranges overlapped those of the nine bands were analyzed to determine male-male relatedness and male-band relatedness. The relationship between fingerprint similarity and genetic relatedness was determined from studies of individuals of known relationship from wild and captive populations.

The analysis showed that each of the nine coati bands was composed primarily of related individuals. The within-band variance in relatedness, however, was high, and three of the bands included unrelated individuals (Figure 13.4). By comparing behavioral observations with the genetic data it was determined that none of the bands consisted of simple nuclear families; rather, coati bands appear to consist of extended families (excluding adult males) as well as occasional unrelated individuals. Severely complicating the picture is uncertain paternity. Coatis mate during a brief, synchronous breeding season, during which time a single male appears to monopolize access to all band females (Gompper 1995). Yet the genetic analysis showed that rarely was the monopolizing male wholly successful. Females often leave the band temporarily during the mating period and, therefore, although the male attempts to restrict access to females in the band to himself, the mating system may nonetheless be promiscuous.

Adult males captured in the vicinity of a band were generally closely related to band members and to one another. Gompper (1994) hypothesized that when maturing adult males shift from a social to a solitary existence, they

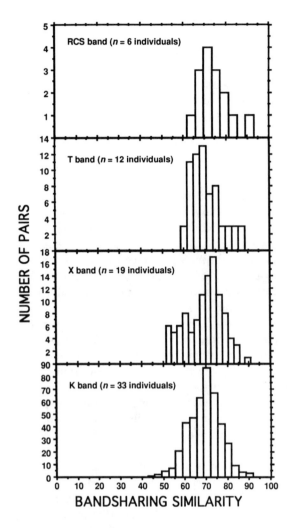

Figure 13.4. Histograms of multilocus DNA fingerprint bandsharing values from four bands of Barro Colorado Island white-nosed coatis (*Nasua narica*). Each pair represents the value of a different dyad of band individuals. Similarity values falling below 60.5 suggest the individuals involved are unrelated. (From Gompper 1994.)

occupy home ranges within that of their natal band and disperse beyond those ranges only for brief periods during the mating season. This behavior, seemingly unique among carnivores, is supported by observations of movements of adult and subadult males. A male maintaining a home range within that of his natal band would benefit from home range familiarity, and possibly be tolerated by the resident band because of their close kinship.

Forman (1985) measured allozyme variability in kinkajous (*Potos flavus*) from Puerto Bermudez, Peru. She analyzed 32 loci and found some of the highest levels of polymorphism (*P*) and heterozygosity (*H*) reported for any carnivore ($n = 27$, $P = 0.39$, $H = 0.13$). Although the study did not directly examine the relatedness of individuals, Forman did examine a single kinkajou social unit comprising just four individuals. Genetic and behavioral data revealed that this small group consisted of an adult female, her immature daughter, an immature male, and an adult male. The adult male appeared not to be the sire of the immature female, and the immature male was not the offspring of either adult. Thus, Forman's study suggests that the Puerto Bermudez kinkajous have a complex social organization and are not solitary as is suggested by other studies (Ford and Hoffman 1988; Julien-Laferriere 1993).

Phocidae

Using analysis of multilocus fingerprints and single-locus minisatellites, Amos et al. (1993, 1995) assigned paternity in colonies of gray seals (*Halichoerus grypus*) on the island of North Rona, Scotland. In 89% of the assigned paternity cases, the assigned father had been observed near the female during estrus. But in 36% of these cases the male predicted to be the parent on the basis of observed copulations and proximity to the female was excluded as the father. Thus, dominant males were reproductively more successful than were subordinate or peripheral males, but to a degree less than that expected on the basis of observations of copulation. This result may indicate either that presumed subordinate males are able to copulate with guarded females, or that some females mate before harems are established. Of 20 males who successfully sired offspring, four fathered more than one pup, confirming field observations of polygyny. In addition, in six of 100 cases the bandsharing value between a putative mother-pup dyad was very low, indicating the possibility that these females were nursing unrelated pups. Whether these events are genuine adoptions or accidental failures to reject the pups is unknown (Amos et al. 1993).

Long-term mate fidelity in gray seals was also studied by examining the distribution of full siblings and half-siblings born to particular females in different years (Amos et al. 1995). Thirty percent of a female's pups born in different years were identified as full siblings, and the authors were able to estimate that only 1 to 2.9 males father 59–100% of an average female's pups over the lifetime of the female. This result indicates a surprisingly high degree of mate fidelity for this basically polygynous species. The authors (Amos et al. 1995) suggest that this finding, coupled with the inability of male dominance and proximity to fully explain paternity (Amos et al. 1993), implies that gray seals recognize and select previous breeding partners.

The mating system of the hooded seal (*Cystophora cristata*), which breeds on pack ice, has been described as monogamous, since a male is often found in proximity to a female and her pup during the breeding season, and these trios are implied to be family groups (Olds 1950; Stirling 1975; Kleiman 1977). Behavioral data, however, have shown that males fight for access to females, that trio liaisons are short-lived, and that a male may gain access to several females (Boness et al. 1988; Kovacs 1990). McRae and Kovacs (1994) used multilocus fingerprinting to examine paternal relationships for 12 trios. In all cases the attending male was excluded from the paternity of the pup. Rather, McRae and Kovacs hypothesize that the males may be guarding access to the females during estrus to secure future paternity, although the effectiveness of this strategy has not been closely examined.

Felidae

Lion (*Panthera leo*) prides generally consist of several adult females, their dependent offspring, and a coalition of two to six adult males who are unrelated to the adult females (Schaller 1972; Bertram 1975; Hanby and Bygott 1987; Packer et al. 1988, 1990). Females are philopatric, and thus adult females in a pride are closely related. Males emigrate when mature, frequently in coalitions with other related males. Although mating in a given pride does occur primarily with a single pride male, paternity is often difficult to determine with certainty, because females also may mate with several coalition males during estrus. Maternity may also be difficult to discern, owing to synchronous breeding and communal rearing of young.

Gilbert et al. (1991) used genetic fingerprinting to establish unequivocally the parentage of 78 offspring from 11 lion prides. All fathers were found to be pride members, and multiple paternity was discovered for just one of 24 litters. Using the relationship between fingerprint-similarity values and the known coefficient of relatedness, Gilbert et al. determined that female pride members were more closely related than are second cousins (Figure 13.5). This result confirmed behavioral observations that females either are recruited into their natal pride or emigrate with other females to form a new pride nearby (Hanby and Bygott 1987; Pusey and Packer 1987; Packer et al. 1991a; Packer and Pusey 1993). Fingerprint-similarity values between resident males ($n = 44$ from 18 coalitions) and pride females ($n = 52$ from 15 prides) indicated that males were generally unrelated to the female pride members. One male coalition, however, joined distantly related females that had been born in the same pride at a different time (Packer et al. 1991a).

Fingerprint similarities between male coalition members showed either that coalitions included only individuals related as siblings or first cousins ($n = 6$) or that the coalitions were dominated by unrelated males ($n = 9$) (Figure 13.5A; Gilbert et al. 1991; Packer et al. 1991a). In general, large coalitions—three or more individuals—consisted of related males who each had different reproductive success. In contrast, small coalitions—two individuals—consisted of unrelated males; in these cases, reproductive success was more equitable. A few coalitions of three males consisted of one unrelated and two

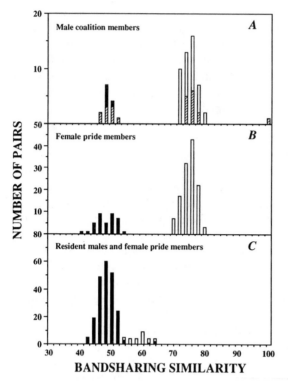

Figure 13.5. Histograms of multilocus DNA fingerprint bandsharing values from Serengeti lions (*Panthera leo*). *A*, male coalition members: dark area represents males born in different prides; unfilled area represents males born in the same pride; hatched area represents males of unknown origin. *B*, female pride members: dark portions represent females from different prides; unfilled portions represent females born in the same pride. *C*, pride resident adult males and pride females: dark area represents males with no known kinship links to females; unfilled area represents relationships in which coalition males and pride females are derived from the same natal pride. (From Packer et al. 1991a; reprinted by permission.)

closely related individuals, and in the one coalition for which paternity was known the unrelated male sired eight offspring, whereas the two mutually related individuals sired five and zero offspring, respectively (Packer et al. 1991a). For large coalitions, these results are consistent with inclusive fitness models; males are willing to sacrifice access to females only if the reproductive success of a close relative is increased. The direct fitness gains of unrelated males in smaller coalitions may be balanced against their decreased success in defending prides from larger coalitions of related males.

The fingerprint results suggest that the formation of a new pride involves the emigration of a group of pride females to an area within or adjacent to the natal pride's home range. Therefore, the initial level of relatedness between the two prides is high, as great as that between female members of a single pride (Figure 13.6). This high relatedness gradually decays, however, until after about 10 years, when (1) fingerprint similarity decreases sharply because older foundresses die and (2) their offspring, derived from matings with different male coalitions, attain reproductive age (Packer et al. 1991a). An interesting predicted consequence of this decay in relatedness is that ancestor and descendant prides may become increasingly intolerant of each other with time.

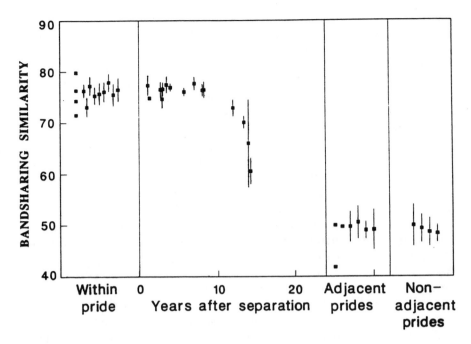

Figure 13.6. DNA fingerprint bandsharing within and between Serengeti lion (*Panthera leo*) prides. Relatedness is high within prides and low between prides. When a pride first splits, relatedness is as high within prides as between prides, but declines sharply after approximately 10 years. (From Packer et al. 1991a; reprinted by permission.)

Herpestidae

Dwarf mongooses (*Helogale parvula*) are communal breeders. Packs in the Serengeti have an average size of 9 ($\pm$ 4.0) individuals and consist of a dominant breeding pair, subordinate adults (helpers), and offspring (Rood 1980, 1987, 1990; Creel and Waser 1991). Field observations indicate that subordinate females may sometimes copulate, become pregnant, and reproduce, but they do so less often than the female of the dominant pair, owing to behavioral and hormonal reproductive suppression and infanticide (Rood 1980; Creel and Waser 1991). On the basis of pedigrees, and an assumption that dominant individuals parent all offspring, average pack relatedness was found to be high ($r = 0.33$: Keane et al. 1994), and subordinates were assumed to gain inclusive-fitness benefits primarily by helping with the rearing of the young (Creel and Waser 1991). Fingerprinting analyses of nine packs, however, showed that subordinates of both sexes were more successful in reproduction than previously thought: 24% of the offspring had subordinate fathers, and 15% had subordinate mothers (Keane et al. 1994). Twelve of 17 offspring from packs with a single pregnant female were fathered by the dominant pack male, and the remainder by subordinate males. In packs that had two or more pregnant females, maternity could be determined for eight young, and paternity for nine. Subordinate females reproduced in four instances and subordinate males in two. Multiple paternity occurred in four litters. Thus subordinate "helpers" may gain significant fitness benefits both directly (by mating) and indirectly (by helping) before they reach dominant status. Interestingly, subordinates that bred were generally higher-ranking and less closely related to the dominant individuals than were the nonbreeders. Presumably, the latter receive indirect fitness benefits through their relationship to breeding individuals, a situation reminiscent of male lion coalitions in which mating is asymmetric if the coalition is composed of close relatives (Packer et al. 1991a).

Most fingerprinting studies of carnivores have examined social species. In one of the few examinations of a solitary carnivore, Waser et al. (1994) examined a population of slender mongooses (*Herpestes sanguineus*) in the Serengeti. Although they have usually been observed as solitary individuals, field observations indicated that as many as four males may jointly use the same home range, and that as many as six females may also use that range. Male associations are stable for as long as 7 years and probably function as coalitions in securing mating access to females. Fingerprinting results, in agreement with field observations, indicated that no single association member monopolizes access to females (Rood 1989; Waser et al. 1994). In addition, genetic and behavioral data suggest that these associations may include both related and unrelated males. The fingerprint similarity between neighboring male associations was low, indicating that they are unrelated.

Implications

Within-population relatedness is often difficult to quantify in wild populations because of the limitations of behavioral data and the low levels of genetic polymorphism generally exhibited. The use of new molecular techniques in behaviorally well-studied species has nonetheless revealed intriguing and unexpected results. Although only a few carnivores have been analyzed using molecular techniques, four important points emerge from these studies.

First, behavioral observations may not always provide accurate assessments of paternity. Genetic evidence of extra-pair mating or subordinate breeding was observed in Ethiopian wolves, dwarf mongooses, gray seals, European badgers, and white-nosed coatis. Multiple paternity was observed in lions (1/24 litters: Gilbert et al. 1991), Ethiopian wolves (2/5 litters: Gottelli et al. 1994) and dwarf mongooses (4/21 litters: Keane et al. 1994). Female white-nosed coatis and gray seals are apparently able to choose alternative mates despite being guarded by a single male. These observations suggest the need for molecular genetic studies to confirm or refute behavioral observations of suspected paternity.

Second, although the genetic data are limited, dispersal patterns appear to be more complex than are suggested by comparative studies across species. The generality that males disperse to avoid inbreeding with philopatric females (Greenwood 1980) may be overly simplistic, for genetic studies suggest diverse dispersal strategies (see Waser, this volume). In gray wolves (Lehman et al. 1992), short-distance dispersal is less frequent as home range overlap increases. Males and females both emigrate, but appear to have different mean dispersal distances, thereby avoiding close inbreeding. Coati males apparently do not disperse, but close inbreeding is avoided through brief, periodic, breeding forays (Gompper 1994). Inbreeding does occur, however, in closed populations such as the Isle Royale wolves (Wayne et al. 1991) and the Ngorongoro lions (Packer et al. 1991b; Packer and Pusey 1993).

Third, solitary species are not asocial stereotypes, but rather may form breeding coalitions involving cooperation, much as their more social counterparts do. Although some species apparently conform to the traditional notion of a solitary lifestyle and are truly asocial (e.g., martens, *Martes:* Powell 1979; Balharry 1993), the slender mongoose study (Waser et al. 1994) indicates that even if individuals are usually found alone they may be social for the purposes of mating, or form part of a cooperating social unit. Male coalitions in species often considered solitary, such as the slender mongoose and the cheetah (*Acinonyx jubatus*), may include unrelated individuals that cooperate in territorial defense, in gaining mate access, and in hunting (Eaton 1970; Caro and Collins 1986, 1987; Rood 1990; Caro and Durant 1991; Waser et al. 1994). These results emphasize the need to distinguish the factors influencing social systems from those affecting mating systems.

Finally, the degree of genetic relatedness between cooperating individuals varies substantially. Although cooperating individuals are generally closely related, studies have often found a few individuals that were unrelated to other group or coalition members. Historically, carnivore social groups were often considered kin groups, and thus studies of cooperation in carnivores have used indirect, potentially circular evidence to support kin selection as the mechanism underpinning cooperative behavior (Bertram 1976; Rodman 1981; Macdonald and Moehlman 1982). Yet kin selection is only one of several nonexclusive mechanisms likely to be acting within carnivore societies. Molecular genetic studies of carnivores indicate that cooperation between unrelated individuals may be common, and, consequently, kin selection alone should perhaps be considered just one extreme of a spectrum encompassing reciprocity, mutualism, and group selection. Alternative hypotheses have been advanced for some cooperative behaviors (Macdonald and Moehlman 1982; Russell 1983; Knight et al. 1992). Future studies should focus more on the quantitative relationship of relatedness and cooperative behavior, and field studies should be designed to include individuals from a diversity of relatedness categories.

Past studies, because they implicitly assumed kin selection as the primary mechanism structuring carnivore social behavior, may have missed some of the most interesting behavioral and ecological aspects of sociality.

Acknowledgments

Thanks to J. Gittleman, A. Hoylman, K. Koepfli, D. Macdonald, C. Sillero, and P. Waser for criticisms on various versions of the manuscript. Support for MEG came from the Smithsonian Tropical Research Institute and the National Science Foundation.

References

Ahnlund, H. 1980. Sexual maturity and breeding system of the badger, *Meles meles*, in Sweden. *J. Zool. (London)* 190:77–95.

Allendorf, F. W., Christiansen, F. B., Dobson, T., Eanes, W. F., and Frydenberg, O. 1979. Electrophoretic variation in large mammals, I: The polar bear, *Thalarctos maritimus*. *Hereditas* 91:19–22.

Alt, G. L. 1978. Dispersal patterns of black bears in northeastern Pennsylvania: A preliminary report. In: R. D. Hugie, ed. *Proc. 4th Eastern Black Bear Workshop*, pp. 186–199. Orono: Maine Dept. Inland Fisheries and Wildlife.

Amos, B., Twiss, S., Pomeroy, P. P., and Anderson, S. S. 1993. Male mating success and paternity in the grey seal, *Halichoerus grypus*: A study using DNA fingerprinting. *Proc. Roy. Soc. London* (B)252:199–207.

Amos, B., Twiss, S., Pomeroy, P., and Anderson, S. 1995. Evidence for mate fidelity in the gray seal. *Science* 268:1897–1899.

Avise, J. C., Arnold, J., Ball, R. M., Bermingham, E., Lamb, T., Neigel, J. E., Reeb, C. A., and Saunders, N. C. 1987. Intraspecific phylogeography: The mitochondrial DNA bridge between population genetics and systematics. *Ann. Rev. Ecol. Syst.* 18:489–522.

Balharry, D. 1993. Social organization in martens: An inflexible system? *Symp. Zool. Soc. London* 65:321–345.

Bekoff, M., and Wells, M. C. 1982. Behavioral ecology of coyotes: Social organization, rearing patterns, space use, and resource defense. *Z. Tierpsychol.* 60:281–305.

Ben-Yaacov, R., and Yom-Tov, Y. 1983. On the biology of the Egyptian mongoose, *Herpestes ichneumon*, in Israel. *Z. Säugetier.* 48:34–45.

Bertram, B. C. R. 1975. Social factors influencing reproduction in wild lions. *J. Zool.* 177:463–482.

Bertram, B. C. R. 1976. Kin selection in lions and in evolution. In: P. P. G. Bateson and R. A. Hinde, eds. *Growing Points in Ethology*, pp. 281–301. Cambridge: Cambridge Univ. Press.

Boness, D. J., Bowen, W. D., and Oftedal, O. T. 1988. Evidence of polygyny from spatial patterns of hooded seals (*Cystophora cristata*). *Canadian J. Zool.* 66:703–706.

Bonnell, M. L., and Selander, R. K. 1974. Elephant seals: Genetic variation and near extinction. *Science* 184:908–909.

Brown, W. M. 1985. The mitochondrial genome of animals. In: R. J. MacIntyre, ed. *Molecular Evolutionary Genetics*, pp. 95–130. New York: Plenum Press.

Bruford, M. W., and Wayne, R. K. 1993. Microsatellites and their application to population genetic studies. *Current Opinion Genet. Developm.* 3:939–943.

Burke, T. 1989. DNA fingerprinting and other methods for the study of mating success. *Trends Ecol. Evol.* 4:139–144.

Bygott, J. D., Bertram, B. C. R., and Hanby, J. P. 1979. Male lions in large coalitions gain reproductive advantages. *Nature* 282:839–841.

Caro, T. M., and Collins, D. A. 1986. Male cheetahs of the Serengeti. *Natl. Geogr. Res.* 2:75–86.

Caro, T. M., and Collins, D. A. 1987. Male cheetah social organization and territoriality. *Ethology* 74:52–64.

Caro, T. M., and Durant, S. M. 1991. Use of quantitative analyses of pelage to reveal family resemblances in genetically monomorphic cheetahs. *J. Hered.* 82:8–14.

Cheeseman, C. L., Cresswell, W. J., Harris, S., and Mallinson, P. J. 1988. Comparison of dispersal and other movements in two badger (*Meles meles*) populations. *Mamm. Rev.* 18:51–59.

Creel, S. R., and Waser, P. M. 1991. Failures of reproductive suppression in dwarf mongooses (*Helogale parvula*): Accident or adaptation? *Behav. Ecol.* 2:7–15.

da Silva, J., Macdonald, D. W., and Evans, P. G. H. 1994. Net costs of group living in a solitary forager, the Eurasian badger. *Behav. Ecol.* 5:151–158.

Derocher, A. E., and Stirling, I. 1990. Observations of aggregating behavior in adult male polar bears (*Ursus maritimus*). *Canadian J. Zool.* 68:1390–1394.

Doncaster, C. P., and Macdonald, D. W. 1991. Drifting territoriality in the red fox, *Vulpes vulpes. J. Anim. Ecol.* 60:423–439.

Eaton, R. L. 1970. Hunting behavior of the cheetah. *J. Wildl. Mgmt.* 34:56–67.

Edwards, S. V. 1993. Long-distance gene flow in a cooperative breeder detected in genealogies of mitochondrial DNA sequences. *Proc. Roy. Soc. London* (B)252:177–185.

Ellegren, H. 1991. DNA typing of museum birds. *Nature* 354:113.

Enders, R. K., and Enders, A. C. 1963. Morphology of the female reproductive tract during delayed implantation in the mink. In: A. C. Enders, ed. *Delayed Implantation,* pp. 129–139. Chicago: Univ. Chicago Press.

Evans, P. G. H., Macdonald, D. W., and Cheeseman, C. L. 1989. Social structure of the Eurasian badger (*Meles meles*): Genetic evidence. *J. Zool. (London)* 218:587–595.

Ford, L. S., and Hoffmann, R. S. 1988. *Potos flavus. Mamm. Species* 321:1–9.

Forman, L. 1985. Genetic variation in two procyonids: Phylogenetic, ecological and social correlates. Ph.D. dissert., New York University, New York.

Fritts, S. H., and Mech, L. D. 1981. Dynamics, movements, and feeding ecology of a newly protected wolf population in northwestern Minnesota. *Wildl. Monogr.* 80:1–79.

Gelter, H. P., and Tegelström, H. 1992. High frequency of extra-pair paternity in Swedish pied flycatchers revealed by allozyme electrophoresis and DNA fingerprinting. *Behav. Ecol. Sociobiol.* 31:1–7.

Gese, E. M., and Mech, L. D. 1991. Dispersal of wolves (*Canis lupus*) in northeastern Minnesota, 1969–1989. *Canadian J. Zool.* 69:2946–2955.

Gilbert, D. A., Packer, C., Pusey, A. E., Stephens, J. C., and O'Brien, S. J. 1991. Analytical DNA fingerprinting in lions: Parentage, genetic diversity, and kinship. *J. Hered.* 82:378–386.

Gompper, M. E. 1994. The importance of ecology, behavior, and genetics in the maintenance of coati (*Nasua narica*) social structure. Ph.D. dissertation, University of Tennessee, Knoxville.

Gompper, M. E. 1995. *Nasua narica. Mamm. Species* 487:1–10.

Gompper, M. E., and Krinsley, J. S. 1992. Variation in social behavior of adult male coatis (*Nasua narica*) in Panama. *Biotropica* 24:216–219.

Gottelli, D., and Sillero-Zubiri, C. 1992. The Ethiopian wolf: An endangered endemic canid. *Oryx* 26:205–214.

Gottelli, D., Sillero-Zubiri, C., Applebaum, G. D., Roy, M. S., Girman, D. J., Garcia-Moreno, J., Ostrander, E. A., and Wayne, R. K. 1994. Molecular genetics of the most endangered canid: The Ethiopian wolf, *Canis simensis. Mol. Ecol.* 3:301–312.

Greenwood, P. J. 1980. Mating systems, philopatry and dispersal in birds and mammals. *Anim. Behav.* 28:1140–1162.

Hagelberg, E., Gray, I. C., and Jeffreys, A. J. 1991. Identification of the skeletal remains of a murder victim by DNA analysis. *Nature* 352:427–429.

Hamilton, W. D. 1964. The genetical evolution of social behaviour, I. *J. Theor. Biol.* 7:1–16.

Hanby, J. P., and Bygott, J. D. 1987. Emigration of subadult lions. *Anim Behav.* 35:161–169.

Harrison, R. G. 1989. Animal mitochondrial DNA as a genetic marker in population and evolutionary biology. *Trends Ecol. Evol.* 4:6–11.

Harrison, R. J., and Neal, E. G. 1956. Ovulation during delayed implantation and other reproductive phenomena in the badger (*Meles meles* L.). *Nature* 177:977–979.

Hoelzel, A. R., and Dover, G. A. 1991. Genetic differentiation between sympatric killer whale populations. *Heredity* 66:191–195.

Hoelzer, G. A., Dittus, W. P. J., Ashley, M. V., and Melnick, D. J. 1994. The local distribution of highly divergent mitochondrial DNA haplotypes in toque macaques *Macaca sinica* at Polonnaruwa, Sri Lanka. *Mol. Ecol.* 3:451–458.

Hughes, C. R., and Queller, D. C. 1993. Detection of highly polymorphic microsatellite loci in a species with little allozyme polymorphism. *Mol. Ecol.* 2:131–137.

Jeffreys, A. J., Wilson, V., and Thein, S. L. 1985a. Hypervariable 'minisatellite' regions in human DNA. *Nature* 314:67–73.

Jeffreys, A. J., Wilson, V., and Thein, S. L. 1985b. Individual-specific 'fingerprints' of human DNA. *Nature* 316:76–79.

Johansson, I., and Venge, O. 1951. Relation of the mating interval to the occurrence of superfetation in the mink. *Acta Zool. Stockholm* 32:255–258.

Julien-Laferriere, D. 1993. Radio-tracking observations on ranging and foraging patterns by kinkajous (*Potos flavus*) in French Guiana. *J. Trop. Ecol.* 9:19–32.

Kaufmann, J. H. 1962. Ecology and social behavior of the coati (*Nasua narica*) on Barro Colorado Island, Panama. *Univ. California Publ. Zool.* 60:95–222.

Keane, B., Waser, P. M., Creel, S. R., Creel, N. M., Elliot, L. F., and Minchella, D. J. 1994. Subordinate reproduction in dwarf mongooses. *Anim. Behav.* 47:65–75.

Kilpatrick, C. W., Forrest, S. C., and Clark, T. W. 1986. Estimating genetic variation in the black-footed ferret: A first attempt. *Great Basin Naturalist Mem.* 8:145–149.

Kleiman, D. G. 1977. Monogamy in mammals. *Quart. Rev. Biol.* 52:39–69.

Kleiman, D. G., and Eisenberg, J. 1973. Comparison of canid and felid social systems from an evolutionary perspective. *Anim. Behav.* 21:637–659.

Knight, M. H., Vanjaarsveld, A. S., and Mills, M. G. L. 1992. Allo-suckling in spotted hyaenas (*Crocuta crocuta*): An example of behavioral flexibility in carnivores. *African J. Ecol.* 30:245–251.

Kohn, M., Knaver, F., Stoffella, A., Schröder, W., and Pääbo, S. 1995. Conservation genetics of the European brown bear: A study using excremental PCR of nuclear and mitochondrial sequences. *Mol. Ecol.* 4:95–103.

Kovacs, K. M. 1990. Mating strategies in the hooded seal (*Cystophora cristata*). *Canadian J. Zool.* 68:2499–2502.

Kruuk, H. 1972. *The Spotted Hyena: A Study of Predation and Social Behavior.* Chicago: Univ. Chicago Press.

Kruuk, H. 1989. *The Social Badger: Ecology and Behavior of a Group-Living Carnivore (Meles meles).* Oxford: Oxford Univ. Press.

Laikre, L., and Ryman, N. 1992. Inbreeding depression in a captive wolf (*Canis lupus*) population. *Conserv. Biol.* 5:33–40.

Larsen, T., Tegelström, H., Kumar Juneia, R., and Taylor, M. K. 1983. Low protein variability and genetic similarity between populations of the polar bear (*Ursus maritimus*). *Polar Res.* 1:97–105.

Lehman, N., Clarkson, P., Mech, L. D., Meier, T. J., and Wayne, R. K. 1992. A study of the genetic relationships within and among wolf packs using DNA fingerprinting and mitochondrial DNA. *Behav. Ecol. Sociobiol.* 30:83–94.

Lehman, N., and Wayne, R. K. 1991. Analysis of coyote mitochondrial DNA genotype frequencies: Estimation of the effective number of alleles. *Genetics* 128:405–416.

Lester, L. J., and Selander, R. K. 1981. Genetic relatedness and the social organization of *Polistes* colonies. *Amer. Nat.* 117:147–166.

Lynch, M. 1988. Estimation of relatedness by DNA fingerprinting. *Mol. Biol. Evol.* 5:584–599.

Lynch, M. 1990. The similarity index and DNA fingerprinting. *Mol. Biol. Evol.* 7:478–484.

Lynch, M. 1991. Analysis of population genetic structure by DNA fingerprinting. In: T. Burke, G. Dolf, A. J. Jefferys, and R. Wolff, eds. *DNA Fingerprinting: Approaches and Applications*, pp. 113–126. Basel: Birkhouser Verlag.

Macdonald, D. W. 1981. Resource dispersion and the social organisation of the red fox, *Vulpes vulpes*. In: J. Chapman and D. Pursley, eds. *Proceedings of the World-wide Furbearer Conference*, Vol. 2, pp. 918–949. Frostburg: Univ. Maryland Press.

Macdonald, D. W. 1992. *The Velvet Claw.* London: BBC Books.

Macdonald, D. W., Apps, P. J., Carr, G. M., and Kerby, G. 1987. Social dynamics, nursing coalitions and infanticide among farm cats, *Felis catus*. Berlin: Paul Parey.

Macdonald, D. W., and Moehlman, P. D. 1982. Cooperation, altruism, and restraint in the reproduction of carnivores. In: P. P. G. Bateson and P. H. Klopfer, eds. *Perspectives in Ethology*, vol. 5, pp. 433–467. New York: Plenum Press.

McCracken, G. F., and Bradbury, J. W. 1981. Social organization and kinship in the polygynous bat, *Phyllostomus hastatus*. *Behav. Ecol. Sociobiol.* 8:11–34.

McRae, S. B., and Kovacs, K. M. 1994. Paternity exclusion by DNA fingerprinting and mate guarding in the hooded seal *Cystophora cristata*. *Mol. Ecol.* 3:101–107.

Mech, L. D. 1970. *The Wolf: The Ecology and Behavior of an Endangered Species.* Garden City, N.Y.: Natural History Press.

Mech, L. D. 1987. Age, season, distance, direction, and social aspects of wolf dispersal from a Minnesota pack. In: B. D. Chepko-Sade and Z. T. Halpin, eds. *Mammalian Dispersal Patterns*, pp. 55–74. Chicago: Univ. Chicago Press.

Mech, L. D. 1994. Buffer zones of territories of gray wolves as regions of intraspecific strife. *J. Mamm.* 75:199–202.

Mech, L. D., and Turkowski, F. J. 1966. Twenty-three raccoons in one winter den. *J. Mamm.* 47:529–530.

Melnick, D. J., and Hoelzer, G. A. 1992. Differences in male and female macaque dispersal lead to contrasting distributions of nuclear and mitochondrial DNA variation. *Intl. J. Primatol.* 13:379–393.

Mills, M. G. L. 1982. The mating system of the brown hyaena, *Hyaena brunnea*, in the southern Kalahari. *Behav. Ecol. Sociobiol.* 10:131–136.

Mills, M. G. L. 1984. The comparative behavioral ecology of the brown hyaena *Hy-*

aena brunnea and the spotted hyaena *Crocuta crocuta* in the southern Kalahari. *Koedoe* (supplement):237–247.

Mills, M. G. L. 1990. *Kalahari Hyaenas: Comparative Behavioural Ecology of Two Species*. London: Unwin Hyman.

Moritz, C., Dowling, T. E., and Brown, W. M. 1987. Evolution of animal mitochondrial DNA: Relevance for population biology and systematics. *Ann. Rev. Ecol. Syst.* 18:269–292.

O'Brien, S. J., Wildt, D. E., Goldman, D., Merril, C. R., and Bush, M. 1983. The cheetah is depauperate in biochemical genetic variation. *Science* 221:459–462.

Olds, J. M. 1950. Notes on the hooded seal, *Cystophora cristata. J. Mamm.* 31:450–452.

Olivier, T. J., Ober, C., Buettner-Janusch, J., and Sade, D. S. 1981. Genetic differentiation among matrilines in social groups of rhesus monkeys. *Behav. Ecol. Sociobiol.* 8:279–285.

Owens, D. D., and Owens, M. J. 1984. Helping behaviour in brown hyenas. *Nature* 308:843–845.

Packard, J. M., Seal, U. S., and Mech, L. D. 1985. Causes of reproductive failure in two family groups of wolves (*Canis lupus*). *Z. Tierpsych.* 68:24–40.

Packer, C., Gilbert, D. A., Pusey, A. E., and O'Brien, S. J. 1991a. A molecular genetic analysis of kinship and cooperation in African lions. *Nature* 351:562–565.

Packer, C., Herbst, L., Pusey, A. E., Bygott, D., Hanby, J. P., Cairns, S. J., and Borgerhoff Mulder, M. 1988. Reproductive success of lions. In: T. H. Clutton-Brock, ed. *Reproductive Success: Studies of Individual Variation in Contrasting Breeding Systems*, pp. 363–383. Chicago: Univ. Chicago Press.

Packer, C., Lewis, S., and Pusey, A. E. 1992. A comparative analysis of non-offspring nursing. *Anim. Behav.* 43:265–281.

Packer, C., and Pusey, A. E. 1993. Dispersal, kinship, and inbreeding in African lions. In: N. W. Thornhill, ed. *The Natural History of Inbreeding and Outbreeding*, pp. 375–391. Chicago: Univ. Chicago Press.

Packer, C., Pusey, A. E., Rowley, H., Gilbert, D. A., Martenson, J., and O'Brien, S. J. 1991b. A study of a population bottleneck: Lions of the Ngorongoro Crater. *Conserv. Biol.* 5:219–230.

Packer, C., Scheel, D., and Pusey, A. E. 1990. Why lions form groups: Food is not enough. *Amer. Nat.* 136:1–19.

Palomares, F., and Delibes, M. 1993. Social organization in the Egyptian mongoose: Group size, spatial behaviour and inter-individual contacts in adults. *Anim. Behav.* 45:917–925.

Peterson, R. O. 1979. The wolves of Isle Royale: New developments. In: E. Klinghammer, ed. *The Behavior and Ecology of Wolves*, pp. 3–18. New York: Garland STPM Press.

Peterson, R. O., and Page, R. E. 1988. The rise and fall of Isle Royale wolves, 1975–1986. *J. Mamm.* 69:89–99.

Pope, T. R. 1990. The reproductive consequences of male cooperation in the red howler monkey: Paternity exclusion in multi-male and single-male troops using genetic markers. *Behav. Ecol. Sociobiol.* 27:439–446.

Powell, R. A. 1979. Mustelid spacing patterns: Variations on a theme by *Mustela. Z. Tierpsychol.* 50:153–165.

Pusey, A. E., and Packer, C. 1987. The evolution of sex-biased dispersal in lions. *Behaviour* 101:275–310.

Queller, D. C., Strassmann, J. E., and Hughes, C. R. 1993. Microsatellites and kinship. *Trends Ecol. Evol.* 8:285–288.

Rasa, O. A. E. 1987. The dwarf mongoose: A study of behavior and social structure in relation to ecology in a small, social carnivore. *Advances in the Study of Behav.* 17:121–163.

Rassmann, K., Schlötterer, C., and Tautz, D. 1991. Isolation of simple-sequence loci for use in polymerase chain reaction-based DNA fingerprinting. *Electrophoresis* 12: 113–118.

Richardson, P. R. K. 1987. Aardwolf mating systems: Overt cuckoldry in an apparently monogamous mammal. *South African J. Sci.* 83:405–410.

Rodman, P. S. 1981. Inclusive fitness and group size with a reconsideration of group sizes in lions and wolves. *Amer. Nat.* 118:275–283.

Rogers, L. L. 1987. Factors influencing dispersal in the black bear. In: B. D. Chepko-Sade and Z. T. Halpin, eds. *Mammalian Dispersal Patterns*, pp. 75–84. Chicago: Univ. Chicago Press.

Rood, J. P. 1975. Population dynamics and food habits of the banded mongoose. *East African Wildl. J.* 13:89-111.

Rood, J. P. 1978. Dwarf mongoose helpers at the den. *Z. Tierpsychol.* 48:277–287.

Rood, J. P. 1980. Mating relationships and breeding suppression in the dwarf mongoose. *Anim. Behav.* 28:143–150.

Rood, J. P. 1987. Dispersal and intergroup transfer in the dwarf mongoose. In: B. D. Chepko-Sade and Z. T. Halpin, eds. *Mammalian Dispersal Patterns*, pp. 85–103. Chicago: Univ. Chicago Press.

Rood, J. P. 1989. Male associations in a solitary mongoose. *Anim. Behav.* 38:725–727.

Rood, J. P. 1990. Group size, survival, reproduction, and routes to breeding in dwarf mongooses. *Anim. Behav.* 39:566–572.

Russell, J. K. 1983. Altruism in coati bands: Nepotism or reciprocity? In: S. K. Wasser, ed. *Social Behavior of Female Vertebrates*, pp. 263–290. New York: Academic Press.

Schaller, G. B. 1972. *The Serengeti Lion.* Chicago: Univ. Chicago Press.

Schlötterer, C., Amos, B., and Tautz, D. 1991. Conservation of polymorphic simple sequence loci in cetacean species. *Nature* 354:63–65.

Schwartz, O. A., and Armitage, K. B. 1980. Genetic variation in social mammals: The marmot model. *Science* 207:665-667.

Sharp, W. M., and Sharp, L. H. 1956. Nocturnal movements and behavior of wild raccoons at a winter feeding station. *J. Mamm.* 37:170–177.

Sillero-Zubiri, C. 1994. Behavioural ecology of the Ethopian wolf (*Canis simensis*). D. Phil. dissert., Oxford University.

Sillero-Zubiri, C., and Gottelli, D. 1991. Ethiopia: Domestic dogs may doom endangered jackal. *Wildl. Conserv.* 42:15.

Simonsen, V. 1982. Electrophoretic variation in large mammals, II: The red fox, *Vulpes vulpes*, the stoat, *Mustela erminea*, the weasel, *Mustela nivalis*, the polecat, *Mustela putorius*, the pine marten, *Martes martes*, the beech marten, *Martes foina*, and the badger, *Meles meles*. *Hereditas* 96:299–305.

Simonsen, V., Allendorf, F. W., Eanes, W. F., and Kapel, F. O. 1982a. Electrophoretic variation in large mammals, III: The ringed seal, *Pusa hispida*, the harp seal, *Pagophilus groenlandicus*, and the hooded seal, *Cystophora cristata*. *Hereditas* 97:87–90.

Simonsen, V., Born, E. W., and Kristensen, T. 1982b. Electrophoretic variation in large mammals, IV: The Atlantic walrus, *Odobenus rosmarus rosmarus* (L). *Hereditas* 97:91–94.

Stirling, I. 1975. Factors affecting the evolution of social behavior in the Pinnipedia. *Rapport P.-V. Réunion du Conseil International de l'Exploration de la Mer.* 169:205–212.

Tautz, D., Trick, M., and Dover, G. A. 1989. Cryptic simplicity in DNA as a major source of genetic variation. *Nature* 322:652–656.

Thomas, W. K., Pääbo, S., Villablanca, F. X., and Wilson, A. C. 1990. Spatial and temporal continuity of kangaroo rat populations shown by sequencing mitochondrial DNA from museum specimens. *J. Mol. Evol.* 31:101–112.

Van Ballenberghe, V. 1983. Extraterritorial movements and dispersal of wolves in south central Alaska. *J. Mamm.* 64:168–171.

Vassart, G., Georges, M., Monsieur, R., Brocas, H., Lequarre, A. S., and Christophe, D. 1987. A sequence in M13 phage detects hypervariable minisatellites in human and animal DNA. *Science* 235:683–684.

Vigilant, L., Pennington, R., Harpending, H., Kocher, T. D., and Wilson, A. C. 1991. African populations and the evolution of human mitochondrial DNA. *Science* 239:1263–1268.

Waser, P. M., Keane, B., Creel, S. R., Elliot, L. F., and Minchella, D. J. 1994. Possible male coalitions in a solitary mongoose. *Anim. Behav.* 47:289–294.

Waser, P. M., and Waser, M. S. 1985. *Ichneumia albicauda* and the evolution of viverrid gregariousness. *Z. Tierpsychol.* 68:137–151.

Wathen, W. G., McCracken, G. F., and Pelton, M. R. 1985. Genetic variation in black bears from the Great Smoky Mountains National Park. *J. Mamm.* 66:564–567.

Wayne, R. K., Gilbert, D. A., Lehman, N., Hansen, K., Eisenhawer, A., Girman, D., Peterson, R. O., Mech, L. D., Gogan, P. J. P., Seal, U. S., and Krumenaker, R. J. 1991. Conservation genetics of the endangered Isle Royale gray wolf. *Conserv. Biol.* 5: 41–51.

Wayne, R. K., Lehman, N., Allard, M. W., and Honeycutt, R. L. 1992. Mitochondrial DNA variability of the gray wolf: Genetic consequences of population decline and habitat fragmentation. *Conserv. Biol.* 6:559–569.

Wilkinson, G. S. 1984. Reciprocal food sharing in vampire bats. *Nature* 308:181–184.

Wilkinson, G. S. 1985. The social organization of the common vampire bat. II: Mating system, genetic structure, and relatedness. *Behav. Ecol. Sociobiol.* 17:123–134.

Wilkinson, G. S. 1992. Communal nursing in the evening bat, *Nycticeius humeralis*. *Behav. Ecol. Sociobiol.* 31:225–235.

Woodroffe, R. 1993. Alloparental behaviour in the European badger. *Anim. Behav.* 46:413–415.

Woodroffe, R., and Macdonald, D. W. 1993. Badger sociality: Models of spatial grouping. *Symp. Zool. Soc. London* 65:145–169.

Woodroffe, R., Macdonald, D. W., and da Silva, J. 1995. Dispersal and philopatry in the European badger, *Meles meles. J. Zool. (London)* 237:227–239.

Woolpy, J. H., and Eckstrand, I. 1979. Wolf pack genetics: A computer simulation with theory. In: E. Klinghammer, ed. *The Behavior and Ecology of Wolves*, pp. 206–224. New York: Garland STPM Press.

Wrege, P. H., and Emlen, S. T. 1987. Biochemical determination of parental uncertainty in white-fronted bee-eaters. *Behav. Ecol. Sociobiol.* 20:153–160.

Wright, S. 1922. Coefficients of inbreeding and relationship. *Amer. Nat.* 56:330–338.

Xia, X., and Millar, J. S. 1991. Genetic evidence of promiscuity in *Peromyscus leucopus. Behav. Ecol. Sociobiol.* 28:171–178.

CHAPTER 14

Demographic and Historical Effects on Genetic Variation of Carnivores

ROBERT K. WAYNE AND KLAUS-PETER KOEPFLI

The rate and direction of evolutionary change in carnivore populations is determined basically by the dynamic interactions of gene flow, genetic drift, mutation, and selection. Carnivores with high rates of dispersal and large dispersal distances potentially have high rates of gene flow among populations if the individuals that disperse are able to reproduce. Genetic differences among populations in highly mobile species will therefore tend to be small, especially if selection is weak. Conversely, in less mobile species with low dispersal rates, genetic drift in small, highly isolated populations will lead to the accumulation of genetic differences among populations. But even in highly mobile species, significant levels of genetic substructure may exist if their distribution has been fragmented by recent habitat alterations or prehistoric barriers to gene flow (O'Brien et al. 1987a; Wayne et al. 1992; Roelke et al. 1993; Taberlet and Bouvet 1994; Roy et al. 1994a). The demographic history of a population can also have a substantial influence on its current levels of genetic variability and differentiation. The harmonic mean of population size over many generations may often be much smaller than population figures based on recent census estimates, and, consequently, genetic variation may be much lower than would be predicted by current population sizes (e.g., Avise et al. 1984; Avise 1992). Moreover, population bottlenecks that may have occurred in the past may have left their imprint on current levels of genetic variation (Bonnell and Selander 1974; O'Brien et al. 1987a).

In what follows, we discuss molecular genetic studies of carnivore populations that have revealed important aspects of their history, population size, and evolutionary relationships. These studies, using several molecular genetic tools, have focused primarily on levels of genetic variation within populations and the degree of genetic differentiation among populations. We also discuss the influence of interspecific hybridization on the genetic integrity of populations, and the conservation implications of molecular genetic studies. We begin with a short review of the molecular genetic techniques that have been used to

Table 14.1. Carnivore species, numbers of populations, and numbers of individuals surveyed for genetic variation

Family and species	Common name	Populations (no.)	Individuals (no.)	Reference(s)
CANIDAE				
Canis latrans	Coyote	6	327	Lehman and Wayne 1991
		1	6	Fisher et al. 1976
Canis lupus	Gray wolf	5	276	Wayne et al. 1991a, 1992
		1	12	Fisher et al. 1976
		9	188	Kennedy et al. 1991
		1	34	Randi 1993
Canis simensis	Ethiopian wolf	2	4	Gottelli et al. 1994
Canis adustus	Side-striped jackal	1	7	Wayne et al. 1990a, 1990b
Canis aureus	Golden jackal	1	20	Wayne et al. 1990a, 1990b
Canis mesomelas	Black-backed jackal	1	64	Wayne et al. 1990a, 1990b
Lycaon pictus	African wild dog	3	104	Girman et al. 1993
Vulpes macrotis/V. velox	Kit and swift fox	10	256	Mercure et al. 1993
Vulpes vulpes	Red fox	5	103	Dragoo et al. 1990
		1	282	Simonsen 1982
Urocyon littoralis	Island gray fox	6	150	Wayne et al. 1991b
FELIDAE				
Felis catus	Domestic cat	1	56	O'Brien 1980
Felis silvestris	Wild cat	4	27	Randi and Ragni 1991
Felis concolor	Puma	12	56	O'Brien et al. 1990; Roelke et al. 1993
Leopardus pardalis	Ocelot	3	6	Newman et al. 1985
Leopardus wiedi	Margay	3	11	Newman et al. 1985
Caracal caracal	Caracal	2	16	Newman et al. 1985
Leptailurus serval	Serval	3	16	Newman et al. 1985
Panthera pardus	Leopard	5	18	Newman et al. 1985
Panthera leo	Lion	6	42	Newman et al. 1985
		5	104	O'Brien et al. 1987a
Panthera tigris	Tiger	5	40	Newman et al. 1985
Neofelis nebulosa	Clouded leopard	5	25	Newman et al. 1985
Acinonyx jubatus	Cheetah	9	85	O'Brien et al. 1985, 1987c

MUSTELIDAE				
Mustela nigripes	Black-footed ferret	1	12	O'Brien et al. 1989
Mustela erminea	Stoat or ermine	1	39	Simonsen 1982
Mustela nivalis	Common weasel	1	13	Simonsen 1982
Mustela putorius	Polecat	1	24	Simonsen 1982
Martes martes	Pine marten	1	2	Simonsen 1982
Martes foina	Beech marten	1	121	Simonsen 1982
Meles meles	Eurasian badger	1	5	Simonsen 1982
			170	Evans et al. 1989
PINNIPEDIA				
Arctocephalus forsteri	New Zealand fur seal	7	16	Lento et al. 1994
Eumetopias jubatus	Steller's sea lion	5	157	Lidicker et al. 1981
Mirounga angustirostris	Northern elephant seal	5	159	Bonnell and Selander 1974
		4	107	Hoelzel et al. 1993
		2	101	Lehman et al. 1993
Mirounga leonina	Southern elephant seal	2	48	Hoelzel et al. 1993
			24	Slade 1992
Odobenus rosmarus	Walrus	5	87	Cronin et al. 1994
Phoca vitulina	Harbor seal	5	79	Lehman et al. 1993
Zalophus californianus	California sea lion	4	49	Maldonado et al. 1995
PROCYONIDAE				
Procyon lotor	Raccoon	23	526	Beck and Kennedy 1980
		5	298	Dew and Kennedy 1980
		1	75	Forman 1985
Potos flavus	Kinkajou	1	27	Forman 1985
URSIDAE				
Ursus americanus	American black bear	6	233	Manlove et al. 1980
		4	41	Cronin et al. 1991
		3	86	Paetkau and Strobeck 1994
		1	46	Wathen et al. 1985
Ursus arctos	Brown bear (Eurasia)	17	60	Taberlet and Bouvet 1994
		2	12	Randi 1993
	Brown bear (N. America)	3	59	Cronin et al. 1991
Ursus maritimus	Polar bear	2	52	Allendorf 1979
		2	40	Cronin et al. 1991

analyze genetic variation in carnivores (here specifically the Carnivora) and then follow with a discussion of studies that measure genetic variation within populations, the relationships and levels of gene flow among populations, and interspecific hybridization. Most population-genetic studies on carnivores have concentrated on the felids, canids, ursids, and pinnipeds. We therefore organize the discussion under these main groupings, with some limited comments on findings in other carnivore species (Table 14.1). Throughout, we will emphasize the historical characteristics that may explain the observed patterns of genetic variation.

Molecular Genetic Techniques

One of the first molecular genetic techniques to be used on carnivores involved electrophoretic separation of protein variants on starch gels. Often called allozyme or protein electrophoresis, this technique allows for the identification of alternative alleles at many loci for large population samples, and requires minimal expense. Only a fraction of the loci, however, often less than 20%, are generally polymorphic, and heterozygosity levels are low, frequently less than 6%, especially in carnivores (e.g., Dew and Kennedy 1980; Ferrell et al. 1978; Newman et al. 1985; O'Brien et al. 1983; Wayne et al. 1991a, b). A large number of loci, therefore, need to be assessed in order to provide high-resolution estimates of genetic variation and gene flow.

More recently, hypervariable nuclear loci have been assayed in populations of carnivores and found to provide high-resolution estimates of genetic variation that correspond with differences in population size or history (Gilbert et al. 1990; Wayne et al. 1991a, b; Packer et al. 1991). These loci consist of variable numbers of repeats of a core sequence, and hence can be separated by size on agarose gels and visualized by hybridization with radio-labeled probes. Most commonly, a multilocus probe is used, such that the resulting "genetic fingerprint" shows variation at many loci simultaneously (Jeffreys et al. 1985). Single-locus probes reveal variation that is more interpretable in terms of Mendelian genetics but can often be applied only to a limited range of related species. More recently, short tandem repeats, two to five base pairs in length, have been identified; these repeats can be amplified by the polymerase chain reaction and separated on acrylamide gels. This procedure allows an assay of individual loci that are highly polymorphic in large population samples, and permits the use of such degraded material as bones, hair, and fecal material (Hagelberg et al. 1991; Hoss et al. 1992; Roy et al. 1994b). A panel of only a dozen or fewer microsatellite loci may be sufficient to quantify components of variation within and among populations, and to study individual relatedness within social groups (Amos et al. 1993; Gottelli et al. 1994; Taylor et al. 1994; Roy et al. 1994a; see also Gompper and Wayne, this volume).

Both of these approaches, however, assess variation in nuclear genes and do not readily produce a precise reconstruction of evolutionary events among populations. Several unique properties of the mitochondrial DNA (mtDNA) genome, an extranuclear sequence housed in the mitochondria and containing about 16 to 18 thousand base pairs in mammals, make it ideal for the analysis of population variability and relatedness. The mitochondrial genome is inherited maternally, in a clonal fashion, without recombination. Consequently, analysis of mtDNA sequences in populations provides a history of maternal lineages that avoids the reticulation caused by recombination, and may allow for a precise reconstruction of colonization events, gene flow, and hybridization (Avise et al. 1987; Slatkin and Madisson 1989; Avise 1992). In many mammals, moreover, mtDNA exhibits a rapid rate of substitution, about five to ten times faster than a typical single-copy nuclear gene, making it an ideal marker for studying recent divergences among populations or closely related species (Brown 1986). In the past, mtDNA sequence differences were commonly assessed indirectly by restriction site analysis, but with the advent of the polymerase chain reaction and the identification of universal primers, researchers are beginning to compare hypervariable mitochondrial sequences directly, in large population samples (Kocher et al. 1989; Hoelzel and Dover 1991; Wenink et al. 1993; Taberlet and Bouvet 1994).

Felidae

The genetic variability of many species of felids has been studied in detail by Stephen O'Brien and colleagues. These species range from lions to domestic cats and include several taxa that have suffered dramatic range and population declines (see O'Brien 1994). The studies provide valuable general lessons for conservation genetics and for the significance of variation in natural populations.

The Lion (*Panthera leo*)

The lion formerly had a geographic range including most of sub-Saharan Africa, parts of the Middle East, and southwestern Asia (O'Brien et al. 1987a). Because of habitat destruction and predator-control measures, the lion is now found outside of Africa only in the Gir Forest of India, where there is a population of less than 250 individuals (O'Brien et al. 1987a). Allozyme electrophoresis of 46 to 50 loci found no polymorphisms in the Gir Forest lions, which contrasts sharply with levels of polymorphism of nearly 10% in most African lion populations. The African populations were genetically very similar to Gir Forest lions, and were estimated to have diverged from them for about 200,000 years. But the African lions contained three unique alleles not found in Asiatic lions, and when O'Brien and colleagues (1987b) surveyed zoo lions that had been identified as Asiatic, they found that they were in fact polymorphic for African lion alleles. They concluded that two of the five founders of the Asiatic lion breeding program likely were descendants of African subspecies. These results clearly showed that genetic variation can be lost rapidly in small isolated populations of large carnivores, and emphasized the need for careful genetic analysis of founders for captive breeding programs.

Similar results were obtained on a smaller geographic scale for lions of the Serengeti and the Ngorongoro Crater (Packer et al. 1991; Gilbert et al. 1991). In 1962, the population of about 60–75 lions in the Ngorongoro Crater plummeted, owing to an epidemic of biting flies, and was reduced to one male and nine females. The Crater population recovered through the breeding of the survivors and a limited immigration of no more than seven males from outside the area. Packer et al. (1991) sought to determine how much allelic variation and heterozygosity had been lost through founder effect and inbreeding. They found convincing evidence of reduced levels of genetic variation. Four alleles found in Serengeti lions were missing from lions in the Crater, and genetic heterozygosity of allozyme loci in the Crater population was about 2.2%, significantly lower than the 3.3% found in the Serengeti lions. Detailed pedigree analysis suggested that inbreeding had been occurring even before the 1962 bottleneck.

The conclusions of these allozyme studies were later confirmed by a DNA fingerprint analysis. Serengeti lions had fingerprint heterozygosity values of 48%, Crater lions, 43.5%, and Gir Forest lions, 2.9%. This pattern of variability reflected the relative importance of long-term isolation and small population size on levels of variability. The Gir Forest lions, which have been completely isolated for several hundred years, during which they have consisted of a few hundred individuals, have just 5% of the heterozygosity of the partially isolated Ngorongoro Crater population that has recently gone through a one-generation bottleneck and recovered rapidly.

The Cheetah (*Acinonyx jubatus*)

The study of genetic variation in the African cheetah is an evolving story that demonstrates both the utility and limitations of molecular-genetic techniques, as well as the effect of history on population structure and variability. The first published account by O'Brien and colleagues showed dramatically that cheetahs had no detectable levels of genetic variation in 47 allozyme loci and barely perceptible levels of variation in 155 protein loci detected by two-dimensional gel electrophoresis (O'Brien et al. 1983). This surprising result, genetic monomorphism in a widely abundant species, was followed by the more unexpected demonstration that reciprocal skin grafts between unrelated animals were not rejected. Because skin-graft acceptance is presumably due to monomorphism at the major histocompatibility complex (MHC), which includes some of the most polymorphic loci in mammals, graft acceptance might be expected only between siblings or between individuals from the same completely inbred laboratory strain. Shortly thereafter, a devastating epizootic outbreak in an Oregon colony of cheetahs suggested that genetic monomorphism may increase susceptibility to epizootics, since pathogens adapt to and are spread readily among genetically similar populations (O'Brien et al. 1985; O'Brien et al. 1987c; O'Brien and Evermann 1988). Associated, moreover, with the lower levels of genetic variation were decreased sperm viability, increased juvenile mortality, and lower overall fecundity, especially in captive populations (O'Brien et al. 1985; Wildt et al. 1987). O'Brien suggested that an ancient bottleneck, perhaps 1000 generations ago, was responsible for the genetic monomorphism of cheetahs. Other scenarios, however, not involving a single population contraction, are also conceivable (Pimm et al. 1989).

On the basis of a field study of cheetah reproduction in the Serengeti Plain of Tanzania, Caro and Laurenson (1994) have questioned the consequences of genetic monomorphism in cheetahs. Predation by lions and spotted hyenas was observed to be the primary cause of high mortality in cheetah cubs, and high densities of these predators may explain the low population density of wild cheetahs in general. The authors stressed also that human-induced demographic problems (i.e., habitat loss and poaching) constitute a more immediate concern for the survival of wild cheetahs. In any event, the effect of low genetic variability on the long-term survival of cheetahs or on their susceptibility to epizootics cannot easily be assessed from such short-term data.

The single-bottleneck hypothesis has been modified to accommodate the results of new genetic analyses (Menotti-Raymond and O'Brien 1993). South and East African cheetahs were found to be genetically very similar, but a small amount of genetic variation was discovered in the East African population (O'Brien et al. 1987c). The presence of variation in the East African subspecies (but not the South African) suggested the possibility of *two* bottlenecks, one occurring about 10,000 years ago in both populations and the other occurring only in the South African population, within the last century, causing genetic uniformity in that population. Recently, Menotti-Raymond and O'Brien (1993) assessed the genetic variation of cheetahs in the rapidly evolving mitochondrial DNA genome and in hypervariable nuclear genetic fingerprints. Levels of variation were moderate, and only slightly lower than those found in outbred populations of other mammal species. Their results, which led to a better calibration of the earlier bottleneck at 10,000 years before present, make an important point that different loci have different rates of evolution. Some loci, such as the MHC, may have high levels of polymorphism that are maintained by selection, but if these levels are reduced, owing to a severe bottleneck, they are slow to recover (Yuhki and O'Brien 1990). The cheetah results suggest that conclusions about population history should be based on the comparison of multiple independent measures of genetic variation.

The Puma, or Panther (*Felis concolor*)

The puma has one of the largest geographic distributions of any mammal, as its many common names can testify, but in many areas of North America its populations are fragmented and highly isolated. The most extreme example is in southern Florida, where the only extant puma population east of the Mississippi clings to existence. The Florida panther—that is the name it is known by there—has been isolated for probably more than 150 years, and now numbers fewer than 50 individuals distributed in two distinct southern Florida populations, one in the Big Cypress Swamp and the other in the Everglades National Park. The Florida panther is considered a distinct subspecies, because of diagnostic characteristics such as a kinked tail and a cowlick, and was placed on the federal endangered species list in 1967. Genetic analysis of multiple independent loci demonstrated convincingly that the two Florida populations were greatly reduced in genetic variation relative to other populations of panthers (Roelke et al. 1993). The Florida panther had less than half the typical heterozygosity of outbred populations, and 85% less variation in genetic-fingerprint loci than did an outbred population of panthers. Only two mitochondrial DNA genotypes were found. Lower heterozygosity was associated with a suite of physiological problems: sperm viability was 18–38 times lower than that of other panther subspecies, 56% of males were cryptorchids, and cardiac defects and disease incidence were high (Roelke et al. 1993). The Florida panther is the clearest example yet of a natural carnivore population suffering from inbreeding depression.

But this story has a curious twist. The two mtDNA genotypes that are found in the Florida panther are very divergent in sequence and are largely population-specific. The genotype found in the Everglades population is related to genotypes found in South American puma stocks that probably became established following the release by the National Park Service of genetically mixed captive-bred panthers between 1956 and 1966 (O'Brien et al. 1990). The genotype found in the Big Cypress Swamp is related to North American puma genotypes, and likely represents that of the authentic Florida subspecies. These conclusions were supported by an allozyme analysis that found that two alleles common in one subspecies of South American pumas were also found in individuals from the Everglades population. Should individuals from the two populations be allowed to interbreed? The number of Florida panthers is fewer than 50 individuals today, and a captive-breeding program probably would not be viable with so limited a founding stock. But as might be expected, inbreeding depression is less severe in the genetically mixed Everglades population, and the influx of genes from outside Florida may therefore have increased the probability of survival of the entire population. Additional genetic augmentation plans for the Florida panther, using individuals from elsewhere in the United States, are being considered. Such managed introgression may be the only way to reverse the consequences of genetic depletion in the Florida panther (Roelke et al. 1993).

Other Cats

Levels of allozyme variation have been surveyed in wild cats, *Felis silvestris* (Randi and Ragni 1991), domestic cats, *Felis catus* (O'Brien 1980), and eight endangered species of cats (Newman et al. 1985; Table 14.1). Generally, variation in wild cats and domestic cats was high, and variation in the eight endangered species was moderate to low. The clouded leopard (*Neofelis nebulosa*) had the lowest level of variation, approaching that of the cheetah. The wild cat study, which utilized samples from throughout the species' geographic range, showed that domestic cats probably derive from African wild cats. African and European wild cats, however, belong to the same polytypic species, and wild cats have high levels of variation, despite the fact that their various populations are often highly isolated and have been established in many cases from just a few founding individuals. In some areas, variation in wild cats may be augmented by interbreeding with feral populations of the domestic cat. A morphological and limited genetic study of wild cats in Scotland suggested that hybridization may be common, and may be influencing the phenotypic characteristics of wild cats (French et al. 1988; Hubbard et al. 1992).

Canidae

The Wolflike Canids
(*Canis lupus, C. rufus,* and *C. latrans*)

The gray wolf, red wolf, and coyote are the three species of wolflike canids found in North America. The geographic range of the gray wolf also includes parts of Asia and Europe, but in Europe the distribution of gray wolves is highly fragmented, with many populations persisting as only a few hundred individuals or less. The degree of genetic subdivision among populations of wolflike canids is an interesting issue, because individuals may travel a hundred kilometers before reproducing, and, consequently, genetic differentiation between populations may be suppressed by gene flow. High rates of dispersal also have implications for interspecific hybridization. For example, theoretical calculations predict that in the absence of selection, hybrid-zone width may be 50 times the variance in dispersal distance (Barton and Hewitt 1989). Because the variance in dispersal of wolflike canids may be a hundred kilometers or more (Mech 1987), hybrid zones over a thousand kilometers in width are conceivable. The presence of such broad hybrid zones may greatly alter the

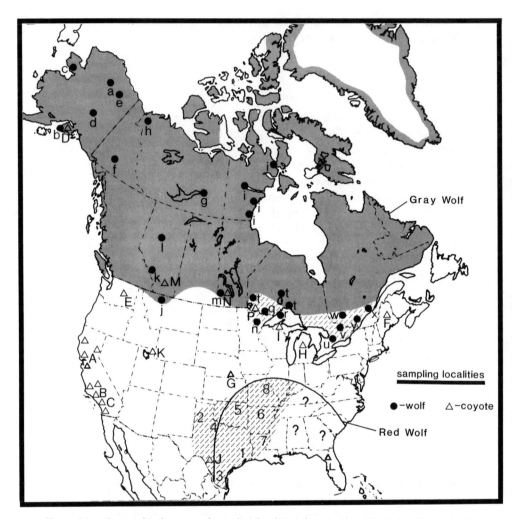

Figure 14.1. Geographical range and sampling localities of gray wolves (*Canis lupus*), coyotes (*C. latrans*), and red wolves (*C. rufus*) (Lehman et al. 1991; Wayne and Jenks 1991). The darkly shaded area indicates the present North American geographical range of the gray wolf. The stippled areas in the Great Lakes region and the south-central United States are hybrid zones between gray wolves, coyotes, and/or red wolves, based on mtDNA analysis. The past geographical range of the red wolf, circa 1700, is also shown (semicircle). Lowercase letters identify sampling localities of gray wolves; uppercase letters identify sampling localities of coyotes; and numbers identify sampling localities of red wolves from the south-central United States. Question marks indicate that no samples have been obtained from these areas. (For sample-locality locations, see table 1 in Wayne and Lehman 1992, from which this art is taken, by permission.)

genetic landscape of a species and affect its long-term evolutionary response to selection (Wayne 1993; Dowling and DeMarais 1993).

The three species of wolflike canids differ in population substructure. Coyotes, which can live in a wide variety of environments, flourish in the United States and in areas of southeastern Canada (Figure 14.1). With the extirpation of gray wolves over the past few hundred years, coyotes have rapidly expanded their geographic range to include most of North America. Restriction fragment length polymorphism (RFLP) analysis of mtDNA found that populations as distantly separated as California and Maine shared genotypes and had similar frequencies of hypervariable microsatellite alleles (Lehman and Wayne 1991; Roy et al. 1994a). In contrast, North American gray wolves showed significant genetic subdivision in terms of mitochondrial DNA variability and microsatellite alleles. In North America, gray wolves showed some west-to-east subdivisions, the Alaskan wolves missing a genotype found in eastern populations. Subdivision was even more pronounced in Europe, where, with one exception, each of eight localities examined had a single unique genotype. In contrast, in North America, many genotypes were widespread, and each locality had on average 3.2 genotypes. These patterns of genetic subdivision have different causes. The east-to-west clinal pattern in North America may reflect differentiation in past Ice Age refugia, followed by recent intergradation. The more extreme partitioning in Europe reflects habitat fragmentation imposed historically by the extensive agricultural development pursued by humans. In contrast, among newly founded coyote populations, there has been insufficient time for drift to cause differentiation, and rates of gene flow may in any case be too large for differentiation to develop.

An important confirmation of the effects of small population size and isolation is presented by the wolves found on Isle Royale in Lake Superior. The population was founded probably by one mated pair, around 1950, but expanded rapidly to more than 50 individuals (Peterson and Page 1988; Gompper and Wayne, this volume). Since 1980, however, the population has fallen to about ten individuals. Reduced food or increased disease is likely to be the primary cause of the persistently small population size (Mech 1966; Mech et al. 1986; Peterson and Page 1988), but analysis of mtDNA and hypervariable nuclear fingerprints showed that the entire population was very closely related and may be suffering inbreeding depression (Wayne et al. 1991a; Laikre and Ryman 1991). Reproductive-age individuals are as closely related as siblings, and if they recognize each other as such they may not form mating bonds (Wayne et al. 1991a).

Mitochondrial DNA analysis of gray wolves and coyotes from eastern Canada and Minnesota found that some gray wolves had mtDNA genotypes identical to or very similar to those found in coyotes, suggesting that transfer of mtDNA genotypes had occurred through interspecific hybridization (Lehman et al. 1991). The genetic influence of interspecific hybridization was quantified later using nuclear microsatellite loci (Roy et al. 1994a). As few as two or three hybridization events per generation could cause the observed similarity of allele frequencies in the coyotes and wolves of Minnesota and eastern Canada. Both approaches also suggested that the most reproductively successful hybridization events occurred between male gray wolves and female coyotes whose offspring backcrossed to gray wolves. This sort of asymmetric hybridization is consistent with the expectation that the larger male gray wolf can dominate matings with the smaller female coyote, but the smaller male coyote cannot dominate matings with the larger female gray wolf.

An older hybrid zone between gray wolves, red wolves, and coyotes may exist in the south-central United States (Mech 1970; Wayne and Jenks 1991). Mitochondrial DNA analysis of red wolves has shown them to contain genotypes identical to or very similar to those found in gray wolves and coyotes. No distinct mtDNA genotypes were found, as might be expected if the divergence of red wolves predated the divergence of coyotes and gray wolves. The result is consistent with the hypothesis that the origin of the red wolf was caused by repeated hybridization events between gray wolves and coyotes in historic times (Wayne and Jenks 1991; Jenks and Wayne 1992). Recently, nuclear DNA analysis of ten hypervariable microsatellite loci has shown that red wolves have no unique alleles relative to coyotes, whereas computer-simulation tests show that they should have at least four or five unique alleles (Roy et al. 1994). Moreover, allele-frequency similarities placed the red wolf closer to coyotes and hybridizing populations of gray wolves in eastern Canada than to any other canid population elsewhere. These data thus confirmed the hypothesis that the red wolf was a form derived from hybridization of gray wolves and coyotes in historic times or earlier. The red wolf hybrid zone may represent a more advanced stage than the zone existing in eastern Canada. Even in eastern Canada, however, intermediate-sized animals are frequently reported, and when more time passes and if the populations of wolves continue to decline, we may see a more extensive phenotype shift. These results demonstrate the potential of hybridization to produce new phenotypic forms possibly adapted to different prey size distributions (Thurber and Peterson 1991).

The Simien Jackal, or Ethiopian Wolf (*Canis simensis*)

The Simien jackal, a highly endangered canid, lives in the Ethiopian highlands above 3000 meters. Phylogenetic analysis of mitochondrial DNA sequence data has shown that the closest living relatives of Simien jackals are gray wolves and coyotes (Gottelli et al. 1994). Perhaps a more appropriate name for this species, then, is the less commonly used "Ethiopian wolf." A possible evolutionary scenario is that the Ethiopian wolf is a descendant of a past invasion of wolflike canids into the African highlands, which were much more extensive in area during the late Pleistocene (Yalden 1983; Kingdon 1990; Gottelli et al. 1994). Since the last Ice Age, the highlands have contracted about 95% (Yalden 1983) and, presumably, populations of Ethiopian wolves have been vastly reduced. More recently, the Ethiopian wolf has become restricted to the few habitat fragments of Ethiopian alpine moorland that remain relatively undisturbed by human activities. Fewer than 500 individuals are likely to exist today.

Analysis of mtDNA and microsatellite polymorphisms demonstrated convincingly that Ethiopian wolves have significantly lower levels of genetic variation than other canids, which is consistent with an equilibrium population of about a thousand individuals (Gottelli et al. 1994). No mtDNA polymorphisms were found in 50 individuals, and heterozygosity was less than 40% of that found in outbred wolflike canids (Table 14.2). The microsatellite data showed that in one area of the Bale Mountains, feral domestic dogs are hybridizing with the Ethiopian wolves. Genetic evidence of domestic dog hybridization was found in about 15% of the wolves there, which accords with phenotypic identification of them as hybrids (Gottelli et al. 1994). In one case, a single litter was shown to have at least two fathers, one of which may have been a domestic dog. The molecular results demonstrated the profound genetic effects of long-term population reduction compounded by recent habitat fragmentation. Moreover, hybridization has changed the genetic characteristics of at least one population of Ethiopian wolves.

The African Wild Dog (*Lycaon pictus*)

The African wild dog once ranged over most of Africa south of the Sahara Desert, inhabiting areas of dry woodland and savannah. Owing to habitat disturbance and disease, however, many populations have vanished or have been severely reduced in number. The present populations are highly fragmented and total no more than several thousand individuals (Ginsberg and Macdonald 1990). The most severe losses have occurred in western and eastern Africa; populations in southern Africa are currently stabilized in protected areas.

Over 100 African wild dogs from eastern populations (Kenya and Tanzania) and southern populations (Zimbabwe and the Republic of South Africa) were examined for mtDNA restriction fragment length polymorphisms and variation in cytochrome *b* sequence (Girman et al. 1993). The results showed that southern populations differed from eastern populations by about 1% of the cytochrome *b* sequence. Each population had three genotypes defining two distinct monophyletic groups. Consequently, the two populations have independent evolutionary histories, and probably diverged about half a million years ago. The barrier to dispersal may have been the Rift Valley lake system (Girman et al. 1993).

To determine whether barriers to dispersal existed between East and West African populations, DNA was extracted from museum skins, and a 130–250

Table 14.2. Measures of genetic variation in mitochondrial DNA and microsatellite loci for ten species of canids

Species	Common name	mtDNA			Microsatellites		
		No. of individuals	No. of genotypes	% sequence divergence	Sample size per locus	Allelic diversity	Mean heterozygosity
Canis latrans	Coyote	327	32	2.5	17.0 (3.0)	5.9 (0.7)	0.675 (0.035)
Canis lupus	Gray wolf	276	9	0.8	17.7 (2.8)	4.5 (1.1)	0.620 (0.070)
Canis simensis	Ethiopian wolf	54	1	0	19.6 (0.4)	2.4 (0.3)	0.241 (0.063)
Canis adustus	Side-striped jackal	7	2	0.2			
Canis aureus	Golden jackal	20	2	0.1	16.4 (0.7)	4.8 (0.8)	0.520 (0.103)
Canis mesomelas	Black-backed jackal	64	4	8.4	13.7 (1.7)	5.0 (0.4)	0.674 (0.063)
Lycaon pictus	African wild dog	104	6	0.9			
Vulpes macrotis	Kit fox	256	24	1.5			
Vulpes vulpes	Red fox	3	3	1.2			
Urocyon littoralis	Island fox	150	5	1.8			

base-pair segment of the hypervariable mtDNA control region was sequenced (Roy et al. 1994b). A museum skin collected from Nigeria had a genotype very distinct from those in East and South African populations. Hence, population subdivision is evident even in this large, presumably highly mobile canid, which may disperse over several hundred kilometers (Fuller et al. 1993). The wild dog may be more habitat-specific than North American wolflike canids, and its dispersal may be more limited by the distribution of dry savannah and woodland habitats (Girman et al. 1993).

Small Canids (*Urocyon littoralis*, *Vulpes macrotis*, and *V. velox*)

The island fox (*U. littoralis*), found only on the six Channel Islands off the coast of Southern California, is an example of insular dwarfism. Island foxes are about two-thirds the size of their mainland progenitor, the gray fox, *U. cinereoargenteus* (Collins 1991; Wayne et al. 1991b). The colonization of the islands likely occurred in stages, first to the three northern islands about 15,000 years ago, when the three were a single, connected landmass and the channel separating them from the mainland was only a few kilometers wide (Figure 14.2). As sea level rose 9,500 to 11,500 years ago, the landmass was reduced to the three separate islands. About 4000 years ago foxes were transported to the southern islands, probably by Native Americans. An understanding of this series of isolation and colonization events on the islands, combined with knowledge of effective population sizes, allowed predictions of levels of genetic variation. The smallest islands, such as San Miguel and San Nicolas, having only a few hundred foxes each, should have much lower levels of variation than do the large islands, such as Santa Rosa, which supports over a thousand foxes. Moreover, a historic imprint should be apparent in the distribution of genetic variation, since the islands colonized last should have lower levels of genetic variation, owing to prior founding events.

In general, these expectations are confirmed. The smallest islands, San Nicolas and San Miguel, were found to have low levels of morphologic and genetic variation (Table 14.3). In fact, the San Nicolas fox population had invariant genetic fingerprints (Figure 14.3), a result described otherwise only in inbred mice or eusocial naked mole rats (Reeve et al. 1990). The patterns of within-island genetic variability, however, did not exactly correspond with the hypothesized colonization history. Allozyme loci showed the expected pattern: the last-colonized islands, such as San Clemente and Santa Catalina, show almost no genetic variation, even though their populations are large. Fingerprint variation, by contrast, did not show this pattern (Table 14.3). Because genetic variation at a locus is restored at a rate that is a function of mutation rate and population size (Nei 1987), the much lower mutation rate of allozyme loci, relative to loci revealed by genetic fingerprinting, resulted in a greater

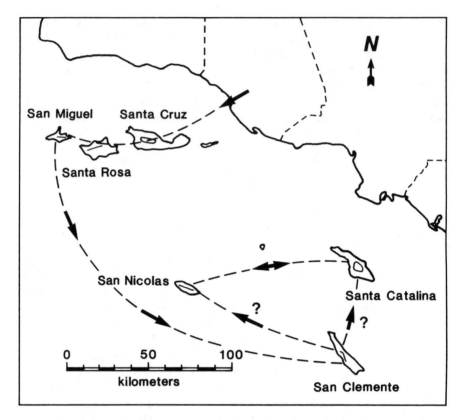

Figure 14.2. Size (from km scale) and location of the six Channel Islands of Southern California, where island foxes (*Urocyon littoralis*) are found. Solid line on each island indicates transect line along which fox samples were obtained. Dotted line across water indicates hypothesized colonization route of island foxes over the past 3–15,000 years. (From Gilbert et al. 1990; reprinted by permission.)

historical influence on allozyme variability. A general lesson, also indicated by the cheetah genetic results, is that evolutionary biologists need to frame their interpretations carefully in the light of historical information and dispersal abilities as well as in differences in the mutation rate of genes.

These principles are exemplified by a genetic study of kit foxes (*V. macrotis*) and swift foxes (*V. velox*), the smallest of the North American canids (Mercure et al. 1993). In large, wolflike canids, population substructure is generally weak, exhibiting perhaps only a hint of regional differentiation (see above). The kit and swift foxes, however, have much lower dispersal abilities than do the large wolflike canids, dispersing on average only 11 km (O'Farrell 1987). Consequently, we might expect a more distinct pattern of genetic differentiation over distance than is seen in large canids (Slatkin 1993). Moreover, the

Table 14.3. Relative ranks of island foxes on individual islands of the Channel Islands, California, by effective population size, and variability in morphology, fingerprinting profiles, allozymes, and mtDNA. Actual values in parentheses. (For details, see Wayne et al. 1991b.)

Island of population	Effective population size	Morphology (log of measurement variance)	Fingerprintings (average percent difference)	Allozymes (expected heterozygosity)	mtDNA (number of genotypes)
San Miguel	1 (157)	1 (0.012)	2 (4.7)	2 (0.008)	2 (2)
San Nicolas	2 (247)	2 (0.013)	1 0.0	1 0.0	1 (1)
San Clemente	3 (551)	5 (0.017)	3 (8.5)	3 (0.013)	1 (1)
Santa Rosa	4 (922)	4 (0.016)	5 (23.7)	5 (0.055)	2 (2)
Santa Catalina	5 (979)	5 (0.017)	6 (25.3)	1 0.0	3 (3)
Santa Cruz	6 (984)	3 (0.015)	4 (10.0)	4 (0.041)	2 (2)

geographic ranges of the kit and swift foxes are divided by the Rocky Mountains of the American West, and such a pronounced barrier may have the effect of completely isolating populations on the east and west sides. Past studies, however, have considered swift foxes living on the eastern side of the Rocky Mountains as conspecific with kit foxes found on the western side (Packard and Bowers 1970; Rohwer and Kilgore 1973). It may be that interbreeding between the two foxes can occur only in the southern reaches of the Rockies, where the mountains are lower.

These expectations were well supported by mtDNA analysis (Figure 14.4) (Mercure et al. 1993). The kit and swift fox populations are genetically distinct, differing as much from each other as either does from the arctic fox, a species often placed in a distinct genus (*Alopex lagopus*). A hybrid zone was delineated between these two distinct groupings in northwest Texas, where morphologic intermediates have been described (Packard and Bowers 1970) and mitochondrial genotypes of both clades were found (Mercure et al. 1993). Differentiation was also apparent between the populations on the two sides of the Rocky Mountains. San Joaquin kit foxes, isolated in the Central Valley of California, appeared to form a unique clade (Figure 14.4). Similarly, Arizona populations are weakly isolated from neighboring populations by barriers such as the Colorado River. Still, past morphologic and allozyme analyses of kit foxes (Dragoo et al. 1990) do not reveal these divisions, probably because selection and mutation rates impose different constraints on the rates of allozyme and morphologic divergence. Thus, two aspects of this study have more general implications. First, different analyses may not provide the same degree of resolution with regard to population history and genetic exchange,

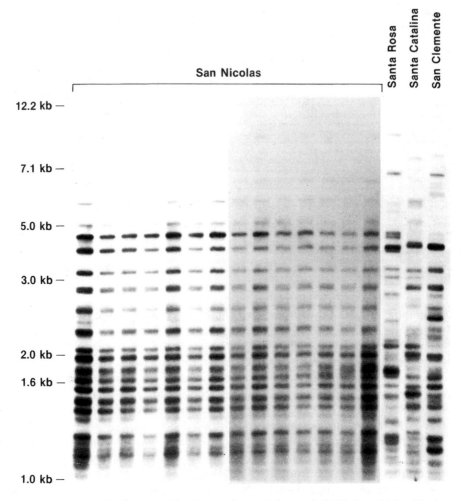

Figure 14.3. Example of an intra-island autoradiogram of genomic DNA from 14 San Nicolas Island foxes, showing their invariant fingerprints, contrasted with those of three foxes from other islands where variability is evident. (For methods, see Gilbert et al. 1990.)

and, second, size and behavioral attributes of species, even species that are similar in body design and life history, may impose different levels of population structure. These factors need to be considered in the formation of conservation plans aimed at the preservation of genetic diversity.

Ursidae

Because bears are large, highly mobile animals, some species migrating over long distances, their dispersal distances may be substantial. Thus, low levels of

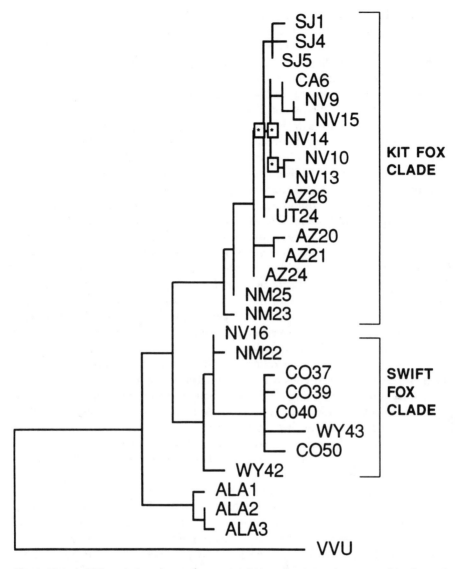

Figure 14.4. A 50% majority-rule consensus tree of 30 most-parsimonious trees of kit fox and swift fox genotypes, based on analysis of mtDNA restriction-site data. Nodes indicated by stars are supported in only 60% of the trees; other nodes are found in all 30 most-parsimonious trees. Genotype designations: ALA, Arctic fox (*Alopex lagopus*); AZ, Arizona; CA, California; CO, Colorado; NM, New Mexico; NV, Nevada; SJ, San Joaquin Valley (California); UT, Utah; WY, Wyoming. The tree was rooted with red fox (VVU) restriction-site data. (For methods and genotype codes, see Mercure et al. 1993; from which this art is taken, by permission.)

genetic differentiation among populations might be predicted. Early studies of electrophoretic variation at allozyme loci in several bear species seemed to confirm this prediction (Allendorf 1979; Manlove 1980; but see Wathen et al. 1985), but recent studies of mtDNA sequence polymorphisms in the hypervariable control region of brown bears (*Ursus arctos*) showed that such predictions do not consider the importance of historical effects (Taberlet and Bouvet 1994; Randi et al. 1994). Taberlet and Bouvet (1994), in a study involving the use of blood and hair samples from about 60 European brown bears, showed a striking pattern of phylogenetic subdivision across their geographic range (Figure 14.5). Brown bears from Scandinavia, France, and Italy formed one distinct clade, those from Greece and the Balkans another closely related clade, and those from Russia and Romania a third very distinct grouping. The existence of distinct western and eastern European clades is probably due to the isolation of brown bears in different Ice Age refugia, and populations farther north and east may have had a longer history of genetic isolation. In fact, the extent of genetic divergence between Russian and Romanian brown bears, as a group, and other Old World brown bears, as a group, is greater than that between polar bears (*Thalarctos maritimus*) and North American brown bears. (The latter two species likely diverged only in the late Pleistocene: Shields and Kocher 1991.) The nuclear genes of European brown bears should be examined carefully, however, because these divisions may reflect an absence of female migration. More precise data are thus needed on the dispersal of brown bears. Moreover, reintroduction and augmentation plans for the brown bear in Europe need to consider the population origin of the individuals selected, so as to avoid mixing genetically distinct populations with a long history of independent evolution (Avise 1992; Avise and Nelson 1989). An important technical advance in this regard is that both mitochondrial and nuclear genes can be characterized by noninvasive sampling of hair, feces, or bone (Hoss et al. 1992; Taberlet and Bouvet 1994; Randi et al. 1994; Hanni et al. 1994; Kohn et al. 1995).

A mitochondrial DNA analysis of American black bears (*U. americanus*) has also found genetically divergent mtDNA genotypes, but these genotypes were found in the same population (Cronin et al. 1991). Because the geographic distribution of samples in this study was limited, it is difficult to reconstruct the origin of these distinct genotypes. Their presence in a single population, however, indicates that the black bear may have a complex phylogeographic history that could be reconstructed through a careful sampling design and by applying several molecular genetic techniques and data from the fossil record. Paetkau and Strobeck (1994) used microsatellites to examine differences in patterns of variability in three subspecies of black bear from three Canadian National Parks. Analysis of four microsatellite loci revealed that the population from the island of Newfoundland was well differentiated, had significantly fewer alleles, and had less than half the heterozygosity (36%, compared with 80%) of the two continental populations. The two continental popula-

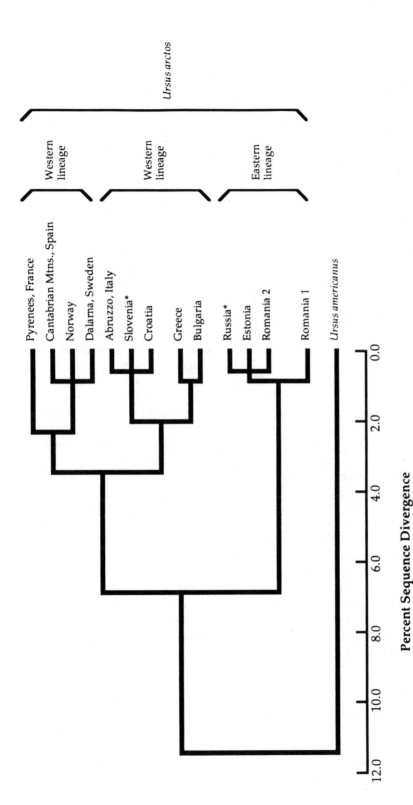

Figure 14.5. UPGMA phenogram summarizing the mtDNA relationships among 13 haplotypes of European brown bear (*Ursus arctos*), rooted with one haplotype from North American black bear (*Ursus americanus*). Slovenia* is a composite of four haplotypes of brown bears from the following localities: Trentino, Italy; Slovenia; Croatia; and Bosnia. Russia* is a composite of five haplotypes of brown bears from following localities: Slovakia; Estonia; Lapland (Sweden); Finland; and Karelia (Russia). (Adapted from Taberlet and Bouvet 1994.)

tions, separated from each other by about 2700 km, had significant differences in allele frequencies. Black bears on Newfoundland may have been isolated in a glacial refugium, and the authors speculated that prehistoric population bottlenecks combined with genetic drift may account for the reduced variability observed in this population. The reduction has not adversely affected the population as a whole, however, since 3000–10,000 bears currently live on Newfoundland (Paetkau and Strobeck 1994).

Pinnipedia

The seals, sea lions, and walruses are arctoid carnivores that are sometimes placed in a separate suborder (Wayne et al. 1989; Zhang and Ryder 1993; Vrana et al. 1994). In the late 1880s, the northern elephant seal (*Mirounga angustirostris*) suffered a severe population bottleneck, falling to about 20 individuals, but increased progressively to a current population of more than 100,000 individuals, owing to the relaxation of hunting pressure attributable to legislative protection (Bonnell and Selander 1974). Analysis of 24 allozyme loci clearly established that the species was genetically monomorphic. In contrast, the southern elephant seal (*M. leonina*), which suffered a less severe population contraction, had levels of heterozygosity of 2.8%, and five of its 18 allozyme loci were polymorphic. More recent molecular genetic analyses of elephant seals have confirmed this initial finding. An analysis of hypervariable minisatellite nuclear loci in northern elephant seals found high levels of genetic similarity: 90% of fingerprint bands were shared among sampled individuals. By contrast, the widely abundant harbor seal (*Phoca vitulina*) had low levels of similarity; on average, two individuals shared just 57% of fingerprint bands (Lehman et al. 1993). Hoelzel et al. (1993) examined sequence diversity in the mtDNA control region and found only two genotypes in the northern elephant seal, compared to 23 in the southern species. Moreover, an extended analysis of allozyme diversity involving 43 loci confirmed that northern elephant seals lack genetic variability. These data and computer-simulation studies, which were used to estimate that there were less than 20 breeding individuals during the bottleneck, support the theory that all extant northern elephant seals are descendants of the last observed colony of 20–100 individuals on Guadalupe Island, 240 km west of Baja California (Hoelzel et al. 1993).

Genetic studies of marine mammals have shown them to possess lower levels of genetic variation than do their terrestrial relatives (Slade 1992). An analysis of 32 allozyme loci in 157 Steller's sea lions (*Eumetopias jubatus*) from five localities in the Gulf of Alaska found only one polymorphic locus (Lidicker et al. 1981). Additional areas need to be sampled, but a prehistoric population bottleneck was suggested by these authors as a possible cause for the low variability. Low levels of polymorphism in the MHC were found in two populations of the southern elephant seal and in two whale species (Slade 1992). On the basis of these results, it has been tentatively hypothesized that selection

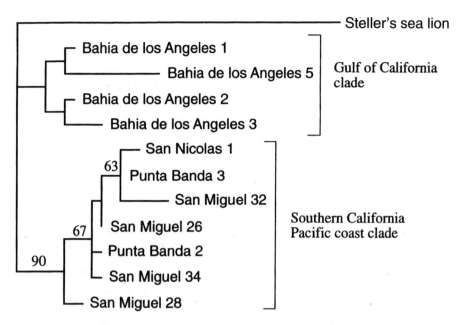

Figure 14.6. Maximum-parsimony tree of 11 different genotypes, based on a 398 base pair fragment of the mitochondrial control region from 49 California sea lions (*Zalophus californianus*) rooted with homologous sequences from the Steller's sea lion (*Eumetopias jubatus*). Groupings supported in over 50% of 1000 bootstrap replications are shown at phylogenetic nodes, as percentages. (Modified from Maldonado et al. 1995.)

pressure to maintain genetic diversity at immune-response loci is reduced because of the lower diversity and density of pathogens in the marine environment (Slade 1992).

The extent of genetic subdivisions among populations and the physical barriers that restrict gene flow are poorly understood in marine mammals (Hoelzel 1993). Mitochondrial DNA variability was studied on California sea lions (*Zalophus californianus*) from populations in the Channel Islands of Southern California and in the Gulf of California (Maldonado et al. 1995). The two areas shared no genotypes, and genotypes in each population defined distinct monophyletic groups (Figure 14.6). These results suggest a long history of isolation potentially caused by female philopatry. A population on the west coast of Baja California, halfway between the southern tip and Southern California, had genotypes identical to those in the Channel Islands, suggesting extensive genetic mixing between the two populations. The isolation of the Gulf population may be the result of upwelling, which provides a continuous source of food, allowing that population to be resident year-round. In contrast, the coastal populations migrate annually to areas as far north as British Columbia. Thus, differences in migratory behavior could have led to the observed differentiation between the two populations, but this hypothesis should be tested with direct radiotelemetry observations.

Similar results on intraspecific genetic differentiation have been reported for other pinniped species. Walruses (*Odobenus rosmarus*) are currently distributed as disjunct subspecies in the Arctic regions of the Atlantic and Pacific Oceans. To examine genetic differentiation within walruses, Cronin et al. (1994) used RFLP analysis of DNA, amplified by the polymerase chain reaction, from three regions of the mitochondrial genome. The distribution of haplotypes corresponded with geographic locality, and phylogenetic analysis showed significant geographic differentiation between Atlantic and Pacific walruses. Moreover, populations from the east and west coasts of Greenland were also distinct, suggesting that negligible, female-mediated gene flow had occurred between the two populations (Cronin et al. 1994). Lento et al. (1994) sampled New Zealand fur seals (*Arctocephalus forsteri*) from rookeries in Western Australia and the east and west coasts of New Zealand and examined them for variability in mitochondrial cytochrome *b* sequences. Analysis of five observed haplotypes from 16 fur seals showed no fixed differences among the rookeries, but the haplotype frequencies of the Western Australia and New Zealand populations differed, suggesting restricted gene flow between the rookeries of these two regions. As suggested by Lento et al., the genetic differences between regions and rookeries may be more fully resolved by analysis of the more variable mitochondrial control region.

Other Carnivores: Procyonids and Mustelids

Population-genetic studies have been published on raccoons (*Procyon lotor*), on Eurasian badgers (*Meles meles*), and on black-footed ferrets (*Mustela nigripes*) and their relatives (Beck and Kennedy 1980; Dew and Kennedy 1980; Simonsen 1982; Forman 1985; O'Brien et al. 1989; Evans et al. 1989). The raccoon study involved analysis of 24 allozyme loci, six of which were polymorphic, in populations of the U.S. Southeast. Levels of heterozygosity were low, about 2.8%, and populations were not well differentiated, suggesting high rates of gene flow among populations (Dew and Kennedy 1980).

The Eurasian badger (*Meles meles*) lives in highly stable, highly territorial social groups allowing only restricted juvenile dispersal (Kruuk and Parish 1982; see Waser, this volume). A sample of 170 badgers from 13 social groups organized into four primary supergroups was examined for allozyme variation in 23 loci (Evans et al. 1989). The heterogeneity of allele frequencies at the single polymorphic locus revealed genetic structuring among the supergroups. The small number of breeding adults, combined with limited juvenile dispersal from their natal social groups, was hypothesized as the cause of the genetic substructure. Frequent migration, however, especially of males, between different supergroups decreased the level of differentiation (Evans et al. 1989; see also Gompper and Wayne, this volume).

The black-footed ferret is an endangered species that was once abundant in the American West. Programs to control the prairie dog had a more dramatic effect on their primary predator, the black-footed ferret, than on the prairie dogs. In 1986, the known remaining ferrets were captured for a captive-breeding program. Allozyme variation appeared to be low in this species: heterozygosity was 0.8%, and only two of 46 loci were polymorphic. But such low variation needs to be evaluated in a phylogenetic context. Allozyme studies of six other mustelid species, including close relatives of the black-footed ferret—the stoat (*M. erminea*), polecat (*M. putorius*), and least weasel (*M. nivalis*)—and even distant relatives such as the Eurasian badger (O'Brien et al. 1989), showed all these species to be monomorphic in 21 allozyme loci. Thus, mustelids may have characteristically low levels of allozyme variation. Until additional genetic studies are done, it is premature to consider the low level of allozyme variation in black-footed ferrets as due to population contraction alone.

Conclusions and Prospects

Interpretations of molecular genetic results are limited by known observational data on behavior and demography. Molecular genetic studies of large carnivores suggest that although dispersal limits genetic differentiation in some species, differences in population size and in the past degree of fragmentation of populations have greatly influenced population substructure and levels of within-population variation. Population bottlenecks, both recent and ancient, may have left their effects on current levels of within-population variation.

The population bottleneck that may have occurred in the cheetah is evident in allozyme and MHC loci but not to the same degree at loci with higher mutation rates. Presumably, the persistence of the cheetah may be reduced relative to outbred species as a consequence of monomorphism at MHC loci and other slowly evolving genes of high fitness value. But in applying this reasoning, we need to exercise caution, because some populations, such as that of the several hundred foxes on San Nicolas Island, may have survived for thousands of years despite being small and genetically monomorphic. Random demographic factors may prove more important in such small, endangered populations (Lande 1988). Small populations may face other genetic threats, however, such as interspecific hybridization, as exemplified by the Ethiopian wolves in Africa and the red wolves in North America.

The extent of population substructure in carnivores also reflects past isolation events and habitat changes. The Ice Ages may have had a dramatic effect on some species, such as brown bears, but more modest effects on such species as gray wolves and coyotes, which commonly disperse over great distances. Topographic obstacles and distance may impose different degrees of genetic subdivision on a species, reflecting differences in habitat specificity, body size, physiological tolerance, behavioral characteristics, and other species-specific characteristics. Large carnivores such as lions, cheetahs, pumas, and wild dogs show slight regional divisions likely reflecting differentiation by distance and geographic barriers. Small carnivores, such as kit foxes, show much more subdivision due to topographic obstacles, as well as patterns of differentiation by distance. Indeed, carnivores provide a fertile ground for understanding the causes of population differentiation, because they exist in a wide diversity of habitats punctuated by an equally diverse array of physical and environmental barriers. Moreover, within a habitat, differences among carnivores in their life-history patterns have clear implications for their relative degrees of genetic differentiation.

Future molecular genetic studies can be expected to provide new insights into the biology of carnivores. Recently, population-genetic studies have been greatly advanced by the discovery of the polymerase chain reaction, which permits the amplification of degraded DNA from hair, old tissue samples, fecal material, and bone (Hagelberg et al. 1991; Hoss et al. 1992; Taberlet and Bouvet 1992). It may thus become possible to measure the genetic variation of ancient populations, which would offer a direct historical framework for the study of extant populations (Thomas et al. 1990; Cooper et al. 1992; Roy et al. 1994b). The use of the polymerase chain reaction to amplify mtDNA and hypervariable nuclear sequence promises to offer a finer resolution of population-genetic parameters than has heretofore been available. Molecular geneticists need to match these new approaches to fundamental questions in the population biology of carnivores.

Acknowledgments

We thank D. Girman, J. Maldonado, and B. Van Valkenburgh for helpful comments. This work was supported in part by the U.S. Fish and Wildlife Service (Region 2) and a National Science Foundation grant (BSR 9020282).

References

Allendorf, F. W. 1979. Electrophoretic variation in large mammals, I: The polar bear, *Thalarctos maritimus*. *Hereditas* 91:19–22.

Amos, B., Schlotterer, C., and Tautz, D. 1993. Social structure of pilot whales revealed by analytical DNA profiling. *Science* 260:670–672.

Avise, J. C. 1992. Molecular population structure and the biogeographic history of a regional fauna: A case history with lessons for conservation biology. *Oikos* 63:62–76.

Avise, J. C., Arnold, J., Ball, R. M., Bermingham, E., Lamb, T., Neigel, J. E., Reeb, C. A., and Saunders, N. C. 1987. Intraspecific phylogeography: The mitochondrial DNA bridge between population genetics and systematics. *Ann. Rev. Ecol. Syst.* 18:489–522.

Avise, J. C., Neigel, J. E., and Arnold, J. 1984. Demographic influences on mtDNA lineage survivorship in animal populations. *J. Mol. Evol.* 20:99–105.

Avise, J. C., and Nelson, W. S. 1989. Molecular genetic relationships of the extinct Dusky seaside sparrow. *Science* 243:646–648.

Barton, N. H., and Hewitt, G. M. 1989. Adaptation, speciation, and hybrid zones. *Nature* 341:497–503.

Beck, M. L., and Kennedy, M. L. 1980. Biochemical genetics of the raccoon, *Procyon lotor*. *Genetica* 54:127–132.

Bonnell, M. L., and Selander, R. K. 1974. Elephant seals: Genetic variation and near extinction. *Science* 184:908–909.

Brown, W. M. 1986. The mitochondrial genome of animals. In: R. MacIntrye, ed. *Molecular Evolutionary Biology*, pp. 95–128. Ithaca, N.Y.: Cornell Univ. Press.

Caro, T. M., and Laurenson, M. K. 1994. Ecological and genetic factors in conservation: A cautionary tale. *Science* 263:485–486.

Collins, P. W. 1991. Origin and differentiation of the island fox: A study of evolution in insular populations. In: F. G. Hochberg, ed. *Recent Advances in California Islands Research*. Santa Barbara, Calif.: Santa Barbara Mus. Nat. Hist.

Cooper, A., Mourer-Chauvire, C., Chambers, G. K., von Haeseler, A., Wilson, A. C., and Paabo, S. 1992. Independent origins of New Zealand moas and kiwis. *Proc. Natl. Acad. Sci.* 89:8741–8744.

Cronin, M. A., Amstrup, S. C., Garner, G. W., and Vyse, E. R. 1991. Interspecific and intraspecific mitochondrial DNA variation in North American bears (*Ursus*). *Canadian J. Zool.* 69:2985–2992.

Cronin, M. A., Hills, S., Born, E. W., and Patton, J. C. 1994. Mitochondrial DNA variation in Atlantic and Pacific walruses. *Canadian J. Zool.* 72:1035–1043.

Dew, R. D., and Kennedy, M. L. 1980. Genic variation in raccoons, *Procyon lotor*. *J. Mamm.* 61:697–702.

Dowling, T. E., and DeMarais, B. D. 1993. Evolutionary significance of introgressive hybridization in cyprinid fishes. *Nature* 362:444–446.

Dragoo, J. W., Choate, J. R., Yates, T. L., and O'Farrell, T. P. 1990. Evolutionary and taxonomic relationships among North American arid land foxes. *J. Mamm.* 71:318–332.

Evans, P. G. H., Macdonald, D. W., and Cheeseman, C. L. 1989. Social structure of the Eurasian badger (*Meles meles*): Genetic evidence. *J. Zool. (London)* 218:587–595.

Ferrell, R. E., Morizot, D. C., Horn, J., and Carley, C. J. 1978. Biochemical markers in species endangered by introgression: The red wolf. *Biochem. Genet.* 18:39–49.

Fisher, R. A., Putt, W., and Hackel, E. 1976. An investigation of the products of 53 gene loci in three species of wild Canidae: *Canis lupus, Canis latrans,* and *Canis familiaris*. *Biochem. Genet.* 14:963–974.

Forman, L. 1985. Genetic variation in two procyonids: Phylogenetic, ecological, and social correlates. Ph.D. dissert., New York Univ.

French, D. D., Corbett, L. K., and Easterbee, N. 1988. Morphological discriminants of Scottish wildcats (*Felis silvestris*), domestic cats (*F. catus*) and their hybrids. *J. Zool. (London)* 214:235–259.

Fuller, T. K., Mills, M. G. L., Borner, M., Laurenson, K., Lelo, S., Burrows, R., and Kat, P. W. 1992. Long distance dispersal by African wild dogs in East and South Africa. *J. African Zool.* 106:535–537.

Gilbert, D. A., Lehman, N., O'Brien, S. J., and Wayne, R. K. 1990. Genetic fingerprinting reflects population differentiation in the Channel Island Fox. *Nature* 344:764–767.

Gilbert, D. A., Packer, C., Pusey, A. E., Stephens, J. C., and O'Brien, S. J. 1991. Analytical DNA fingerprinting in lions: Parentage, genetic Diversity, and kinship. *J. Hered.* 82:378–386.

Ginsberg, J. R., and Macdonald, D. W. 1990. *Foxes, Wolves, Jackals and Dogs: An Action Plan for the Conservation of Canids*. Gland, Switzerland: International Union for Conservation of Nature and Natural Resources.

Girman, D. J., Kat, P.W., Mills, M. G. L., Ginsberg, J. R., Borner, M., Wilson, V., Fanshawe, J. H., Fitzgibbon, C., Lau, L. M., and Wayne, R. K. 1993. Molecular genetic and morphological analyses of the African wild dog (*Lycaon pictus*). *J. Hered.* 84:450–459.

Gottelli, D., Sillero-Zubiri, C., Applebaum, G. D., Roy, M. S., Girman, D. J., Garcia-Moreno, J., Ostrander, E. A., and Wayne, R. K. 1994. Molecular Genetics of the most endangered canid: The Ethiopian wolf, *Canis simenesis*. *Mol. Ecol.* 3:301–312.

Hagelberg, E., Gray, I. C., and Jeffreys, A. J. 1991. Identification of the skeletal remains of a murder victim by DNA analysis. *Nature* 352:427–429.

Hanni, C., Laudet, V., Stehelin, D., and Taberlet, P. 1994. Tracking the origins of the cave bear (*Ursus spelaeus*) by mitochondrial DNA sequencing. *Proc. Natl. Acad. Sci.* 91:12336–12340.

Hoelzel, A. R. 1993. Genetic ecology of marine mammals. *Symp. Zool. Soc. Lond.* 66:15–32.

Hoelzel, A. R., and Dover, G. A. 1991. Genetic differentiation between sympatric populations of killer whale populations. *Heredity* 66:191–195.

Hoelzel, A. R., Halley, J., O'Brien, S. J., Campagna, C., Arnbom, T., Le Boeuf, B., Ralls, K., and Dover, G. A. 1993. Elephant seal genetic variation and the use of simulation models to investigate historical population bottlenecks. *J. Hered.* 84:443–449.

Hoss, M., Kohn, M., Paabo, S., Knauer, F., and Schroder, W. 1992. Excremental analysis by PCR. *Nature* 359:199.

Hubbard, A. L., McOrist, S., Jones, T. W., Boid, R., Scott, R., and Easterbee, N. 1992. Is survival of European wildcats *Felis silvestris* in Britain threatened by interbreeding with domestic cats? *Biol. Conserv.* 61:203–208.

Jeffreys, A. J., Wilson, V., and Thein, S. L. 1985. Hypervariable minisatellite regions in human DNA. *Nature* 327:139–144.

Jenks, S. M., and Wayne, R. K. 1992. Problems and policy for species threatened by hybridization: The red wolf as a case study. In: D. R. McCullough and R. H. Barrett, eds. *Wildlife 2001: Populations*, pp. 237–251. London: Elsevier.

Kennedy, P. K., Kennedy, M. L., Clarkson, P. L., and Liepins, I. S. 1991. Genetic variability in natural populations of the gray wolf, *Canis lupus. Canadian J. Zool.* 69:1183–1188.

Kingdon, J. 1990. *Island Africa*. London: Collins Press.

Kocher, T. D., Thomas, W. K., Meyer, A., Edwards, S. V., Paabo, S., Villablanca, F. X., and Wilson, A. C. 1989. Dynamics of mitochondrial DNA evolution in animals: Amplification and sequencing with conserved primers. *Proc. Natl. Acad. Sci.* 86:6196–6200.

Kohn, M., Knauer, F., Stoffella, A., Schröder, W., and Pääbo, S. 1995. Conservation genetics of the European brown bear: A study using excremental PCR of nuclear and mitochondrial sequences. *Mol. Ecol.* 4:95–103.

Kruuk, H., and Parish, T. 1982. Factors affecting population density, group size and territory size of the European badger, *Meles meles. J. Zool. (London)* 196:31–39.

Laikre, L., and Ryman, N. 1991. Inbreeding depression in a captive wolf (*Canis lupus*) population. *Conserv. Biol.* 5:33–40.

Lande, R. 1988. Genetics and demography in biological conservation. *Science* 241:1455–1460.

Lehman, N., Eisenhawer, A., Hansen, K., Mech, D. L., Peterson, R. O., Gogan, P. J. P., and Wayne, R. K. 1991. Introgression of coyote mitochondrial DNA into sympatric North American gray wolf populations. *Evolution* 45:104–119.

Lehman, N., and Wayne, R. K. 1991. Analysis of coyote mitochondrial DNA genotype frequencies: Estimation of effective number of alleles. *Genetics* 128:405–416.

Lehman, N., Wayne, R. K. and Stewart, B. S. 1993. Comparative levels of genetic variability in harbour seals and northern elephant seals as determined by genetic fingerprinting. *Symp. Zool. Soc. London* 66:49–60.

Lento, G. M., Mattlin, R. H., Chambers, G. K., and Baker, C. S. 1994. Geographic distribution of mitochondrial cytochrome *b* DNA haplotypes in New Zealand fur seals (*Arctocephalus forsteri*). *Canadian J. Zool.* 72:293–299.

Lidicker, W. Z., Sage, R. D., and Calkins, D. G. 1981. Biochemical variation in northern sea lions from Alaska. In: M. H. Smith and J. Joule, eds., *Mammalian Population Genetics*, pp. 231–241. Athens: Univ. Georgia Press.

Maldonado, J. E., Geffen, E., Wayne, R. K., Orta-Davila, F., and Stewart, B. S. 1995. Interspecific differentiation in two populations of California sea lions (*Zalophus californicus*) determined by analysis of hypervariable-control-region sequences. *Mar. Mamm. Sci.* 11:46–58.

Manlove, M. N., Baccus, R., Pelton, M. R., Smith, M. H., and Graber, D. 1980. Biochemical variation in the black bear. In: C. J. Martinka and K. L. McArthur, eds. *Bears: Their Biology and Management*, pp. 37–41. Bear Biol. Assoc. Conf. Ser., Publ. 3.

Mech, L. D. 1966. The wolves of Isle Royale. *Fauna Series No. 7* Washington, D.C.: U.S. National Park Service.

Mech, L. D. 1970. *The Wolf: The Ecology and Behavior of an Endangered Species.* Minneapolis: Univ. Minnesota Press.

Mech, L. D. 1987. Age, season, distance, direction, and social aspects of wolf dispersal from a Minnesota pack. In: B. D. Chepko-Sade and Z. Tang Halpin, eds. *Mammalian Dispersal Patterns*, pp. 55–74. Chicago: Univ. Chicago Press.

Mech, L. D., Goyal, S. M., and Bota, C. N. 1986. Canine parvovirus infection in wolves (*Canis lupus*) from Minnesota. *J. Wildl. Disease* 22:104–106.

Menotti-Raymond, M., and O'Brien, S. J. 1993. Dating the genetic bottleneck of the African cheetah. *Proc. Natl. Acad. Sci.* 90:3172–3176.

Mercure, A., Ralls, K., Koepfli, K.-P., and Wayne, R. K. 1993. Genetic subdivisions among small canids: Mitochondrial DNA differentiation of swift, kit, and Arctic foxes. *Evolution* 47:1313–1328.

Nei, M. 1987. *Molecular Evolutionary Genetics.* New York: Columbia Univ. Press.

Newman, A., Bush, M., Wildt, D. E., Van Dam, D., Frankenhuis, M. T., Simmons, L., Phillips, L., and O'Brien, S. J. 1985. Biochemical genetic variation in eight endangered or threatened felid species. *J. Mamm.* 66:256–267.

O'Brien, S. J. 1980. The extent and character of biochemical genetic variation in the domestic cat. *J. Hered.* 71:2–8.

O'Brien, S. J. 1994. Genetic and phylogenetic analyses of endangered species. *Ann. Rev. Genet.* 28:467–489.

O'Brien, S. J., and Evermann, J. F. 1988. Interactive influence of infectious disease and genetic diversity of natural populations. *Trends Ecol. Evol.* 3:254–259.

O'Brien, S. J., Joslin, P., Smith, G. L., Wolfe, R., Heath, E., Ott-Joslin, J., Rawal, P. P., Bhattacherjee, K. K., and Martenson, J. S. 1987b. Evidence for African origin of founders of the Asiatic lion species survival plan. *Zoo Biology* 6:99–116.

O'Brien, S. J., Martenson, J. S., Eichelberger, M. A., Thorne, E. T., and Wright, F. 1989. Genetic variation and molecular systematics of the black-footed ferret. In: U.S. Seal, E. T. Thorne, M. A. Bogan, and S. H. Anderson, eds. *Conservation Biology and the Black-Footed Ferret*, pp. 21–33. New Haven, Conn.: Yale Univ. Press.

O'Brien, S. J., Martenson, J. S., Packer, C., Herbst, L., de Vos, V., Joslin, P., Ott-Joslin, J., Wildt, D. E., and Bush, M. 1987a. Biochemical genetic variation in geographic isolates of African and Asian lions. *Natl. Geogr. Res.* 3:114–124.

O'Brien, S. J., Roelke, M. E., Marker, L., Newman, A., Winkler, C. A., Meltzer, D.,

Colly, L., Evermann, J. F., Bush, M., and Wildt, D. E. 1985. Genetic basis for species vulnerability in the cheetah. *Science* 227:1428–1434.

O'Brien, S. J., Roelke, M. E., Yuhki, N., Richards, K. W., Johnson, W. E., Franklin, W. L., Anderson, A. E., Bass, O. L., Belden, R. C., and Martenson, J. S. 1990. Genetic introgression within the Florida panther *Felis concolor coryi*. *Natl. Geogr. Res.* 6:485–494.

O'Brien, S. J., Wildt, D. E., Bush, M., Caro, T. M., FitzGibbon, C., Aggundey, I., and Leaky, R. E. 1987c. East African cheetahs: Evidence for two population bottlenecks? *Proc. Natl. Acad. Sci.* 84:508–511.

O'Brien, S. J., Wildt, D. E., Goldman, D., Merril, C. R., and Bush, M. 1983. The cheetah is depauperate in genetic variation. *Science* 221:459–462.

O'Farrell, T. P. 1987. Kit Fox. In: M. Novak. J. A. Baker, M. E. Obbard, and B. Malloch, eds. *Wild Furbearer Management and Conservation in North America*, pp. 423–431. Toronto: Ontario Ministry of Natural Resources.

Paabo, S. 1989. Ancient DNA: Extraction, characterization, molecular cloning, and enzymatic amplification. *Proc. Natl. Acad. Sci.* 86:1939–1943.

Packard, R. L., and Bowers, J. H. 1970. Distributional notes on some foxes from western Texas and eastern New Mexico. *Southwestern Naturalist* 14:450–451.

Packer, C., Pusey, A. E., Rowley, H., Gilbert, D. A., Martenson, J., and O'Brien, S. J. 1991. Case study of a population bottleneck: Lions of the Ngorongoro Crater. *Conserv. Biol.* 5:219–230.

Paetkau, D., and Strobeck, C. 1994. Microsatellite analysis of genetic variation in black bear populations. *Mol. Ecol.* 3:489–495.

Peterson, R. O., and Page, R. E. 1988. The rise and fall of Isle Royale wolves, 1975–1986. *J. Mamm.* 69:89–99.

Pimm, S. L., Gittleman, J. L., McCracken, G. F., and Gilpin, M. 1989. Plausible alternatives to bottlenecks to explain reduced genetic diversity. *Trends Ecol. Evol.* 4:176–178.

Randi, E. 1993. Effects of fragmentation and isolation on genetic variability of the Italian populations of wolf *Canis lupus* and brown bear *Ursus arctos*. *Acta Theriol.* 38:113–120.

Randi, E., Gentile, L., Boscagli, G., Huber, D., and Roth, H. U. 1994. Mitochondrial DNA sequence divergence among some west European brown bear (*Ursus arctos* L.) populations: Lessons for conservation. *Heredity* 73:480–489.

Randi, E., and Ragni, B. 1991. Genetic variability and biochemical systematics of domestic cat and wild cat populations. *J. Mamm.* 72:79–88.

Reeve, H. K., Westneat, D. F., Noon, W. A., Sherman, P. W., and Aquadro, C. F. 1990. DNA fingerprinting reveals high levels of inbreeding in colonies of the eusocial naked mole rat. *Proc. Natl. Acad. Sci.* 87:2496–2499.

Roelke, M. E., Martenson, J. S., and O'Brien, S. J. 1993. The consequences of demographic reduction and genetic depletion in the endangered Florida panther. *Curr. Biol.* 3:340–350.

Rohwer, S. A., and Kilgore, D. L., Jr. 1973. Interbreeding in the aridland foxes, *Vulpes velox* and *V. macrotis*. *Syst. Zool.* 22:157–165.

Roy, M.S., Geffen, E., Smith, D., Ostrander, E., and Wayne, R. K. 1994a. Patterns of differentiation and hybridization in North American wolf-like canids revealed by analysis of microsatellite loci. *Mol. Biol. Evol.* 11:553–570.

Roy, M. S., Girman, D. J., and Wayne, R. K. 1994b. The use of museum specimens to reconstruct the genetic variability and relationships of extinct populations. *Experientia* 50:551–557.

Shields, G. F., and Kocher, T. D. 1991. Phylogenetic relationships of North American ursids based on analysis of mitochondrial DNA. *Evolution* 45:218–221.

Simonsen, V. 1982. Electrophoretic variation in large mammals, II: The red fox, *Vulpes vulpes*, the stoat, *Mustela erminea*, the weasel, *Mustela nivalis*, the polecat, *Mustela putorius*, the pine marten, *Martes martes*, the beech marten, *Martes foina*, and the badger, *Meles meles. Hereditas* 96:299–305.

Slade, R. W. 1992. Limited MHC polymorphism in the southern elephant seal: Implications for MHC evolution and marine mammal population biology. *Proc. Roy. Soc. London* B249:163–171.

Slatkin, M. 1993. Isolation by distance in equilibrium and non-equilibrium populations. *Evolution* 47:264–279.

Slatkin, M., and Maddison, W. P. 1989. A cladistic measure of gene flow inferred from the phylogeny of genes. *Genetics* 123:603–613.

Taberlet, P., and Bouvet, J. 1992. Bear conservation genetics. *Nature* 358:197.

Taberlet, P., and Bouvet, J. 1994. Mitochondrial DNA polymorphism, phylogeography, and conservation genetics of the brown bear *Ursus arctos* in Europe. *Proc. Roy. Soc. London* B255:195–200.

Taylor, A. C., Sherwin, W. B., and Wayne, R. K. 1994. Genetic variation of simple sequence loci in bottlenecked species: The decline of the northern hairy-nosed wombat (*Lasiorhinus krefftii*). *Mol. Ecol.* 3:277–290.

Thomas, W. K., Paabo, S., Villablanca, F. X., and Wilson, A. C. 1990. Spatial and temporal continuity of kangaroo rat populations shown by sequencing mitochondrial DNA from museum specimens. *J. Mol. Evol.* 31:101–112.

Thurber, J. M., and Peterson, R. O. 1991. Changes in body size associated with range expansion in the coyote (*Canis latrans*). *J. Mamm.* 72:750–755.

Vrana, P. B., Milinkovitch, M. C., Powell, J. R., and Wheeler, W. C. 1994. Higher level relationships of the arctoid Carnivora based on sequence data and "total evidence." *Mol. Phyl. Evol.* 3:47–58.

Wathen, W. G., McCracken, G. F., and Pelton, M. R. 1985. Genetic variation in black bears from the Great Smoky Mountains National Park. *J. Mamm.* 66:564–567.

Wayne, R. K. 1993. Molecular evolution of the dog family. *Trends Genet.* 9:218–224.

Wayne, R. K., Benveniste, R. E., and O'Brien, S. J. 1989. Molecular and biochemical evolution of the Carnivora. In: J. L. Gittleman, ed. *Carnivore Behavior, Ecology and Evolution*, pp. 465–494. Ithaca, N.Y.: Cornell Univ. Press.

Wayne, R. K., George, S., Gilbert, D., Collins, P., Kovach, S., Girman, D., and Lehman, N. 1991b. A morphologic and genetic study of the island fox, *Urocyon littoralis. Evolution* 45:1849–1868.

Wayne, R. K., Gilbert, D. A., Eisenhawer, A., Lehman, N., Hansen, K., Girman, D., Peterson, R. O., Mech, L. D., Gogan, P. J. P., Seal, U. S., and Krumenaker, R. J. 1991a. Conservation genetics of the endangered Isle Royale gray wolf. *Conserv. Biol.* 5:41–51.

Wayne, R. K., and Jenks, S. M. 1991. Mitochondrial DNA analysis supports extensive hybridization of the endangered red wolf (*Canis rufus*). *Nature* 351:565–568.

Wayne, R. K., and Lehman, N. 1992. Mitochondrial DNA analysis of the Eastern coyote: Origins and hybridization. In: A. H. Boer, ed., *Ecology and Management of the Eastern Coyote*, pp. 9–22. Fredericton, New Brunswick: Wildl. Res. Unit.

Wayne, R. K., Lehman, N., Allard, M. W., and Honeycutt, R. L. 1992. Mitochondrial DNA variability of the gray wolf: Genetic consequences of population decline and habitat fragmentation. *Conserv. Biol.* 6:559–569.

Wayne, R. K., Meyer, A., Lehman, N., Van Valkenburgh, B., Kat, P. W., Fuller, T. K., Girman, D., and O'Brien, S. J. 1990b. Large sequence divergence among mitochondrial DNA genotypes within populations of East African black-backed jackals. *Proc. Natl. Acad. Sci.* 87:1772–1776.

Wenink, P. W., Baker, A. J., and Tilanus, M. G. L. 1993. Hypervariable-control-region sequences reveal global population structuring in a long-distance migrant shorebird, the dunlin (*Calidris alpina*). *Proc. Natl. Acad. Sci.* 90:94–98.

Wildt, D. E., Bush, M., Goodrowe, K. L., Packer, C., Pusey, A. E., Brown, J. L., Joslin, P., and O'Brien, S. J. 1987. Reproductive and genetic consequences of founding isolated lion populations. *Nature* 329:328–331.

Yalden, D. W. 1983. The extent of high ground in Ethiopia compared to the rest of Africa. *Ethiopian J. Sci.* 6:35–38.

Yuhki, N., and O'Brien, S. J. 1990. DNA variation of the mammalian major histocompatibility complex reflects genomic diversity and population history. *Proc. Natl. Acad. Sci.* 87:836–840.

Zhang, Y.-P., and Ryder, O. A. 1993. Mitochondrial DNA sequence evolution in the Arctoidea. *Proc. Natl. Acad. Sci.* 90:9557–9561.

CHAPTER 15

Biogeography of
the Order Carnivora

ROBERT M. HUNT, JR.

By definition, a carnivore is a flesh-eater (L., *carnis*, flesh; *vorare*, to devour). In the widest sense, the term refers to organisms that prey upon animals, and thus may include vertebrate and invertebrate predators and even insectivorous plants. Insects constitute an enormous source of animal protein, and the many species of primarily insectivorous mammals that have exploited this resource during the Cenozoic could be termed carnivores. In this chapter, however, I restrict the term "carnivore" to the three great global radiations of Cenozoic therian mammals that have exploited predation on, and/or scavenging of, vertebrates as a principal feeding mode: (1) predaceous marsupials; (2) among the placentals, the creodonts, and (3) the true Carnivora (here termed *carnivorans,* following Van Valen [1969], to distinguish them from carnivorous mammals in general). These three groups have been formally classified at various times as mammalian orders of coordinate rank: (1) Marsupicarnivora; (2) Creodonta; (3) Carnivora. I designate the three simply as marsupial carnivores, creodonts, and carnivorans. My primary focus will be the biogeography of placental carnivorans (Order Carnivora); more detailed treatment can be found, for marsupial carnivores, in Archer (1976) and Marshall (1976, 1978a), and, for creodonts, in Barry (1988), Denison (1938), Gingerich and Deutsch (1989), Ginsburg (1980), Gunnell and Gingerich (1991), Lange-Badré (1979), Lange-Badré and Haubold (1990), Mellett (1977), and Springhorn (1980).

During the Cenozoic, the evolution of terrestrial carnivores is manifested in a number of adaptive radiations geographically confined to northern and southern continental areas: (1) North America–Eurasia; (2) South America–Antarctica–Australia; (3) Africa. An excellent Cenozoic record of fossil carnivores, coupled with the geographic isolation of these radiations, allows us to observe multiple natural experiments in the evolution of terrestrial mammalian predators. Marsupial carnivores dominate the Cenozoic faunas of Australia and South America, and may have flourished in Antarctica. A radiation of marsupial carnivores (Borhyaenidae, 33 genera including Thylacosmilinae) and large phororhacoid birds developed in South America from the Paleocene until the Pliocene, when placental carnivorans entered from the north (Marshall 1978a, b, 1985). In Australia, marsupial carnivores of the families Dasyuridae and Thylacinidae attained a diversity of about 19 genera (Marshall 1981), none of large body size, the niche for large carnivores having been filled in part by giant varanid lizards, crocodilians, and boid snakes.

Placental carnivores radiated first in the early Cenozoic as creodonts, and then in the early to late Cenozoic as carnivorans, both groups confined to the northern continents (Eurasia, North America) and Africa. Africa's early Cenozoic fossil record is poorly known; the oldest African carnivores are represented only by rare, indeterminate teeth from the Paleocene. The first well-sampled African carnivore fauna, of Oligocene age, includes only creodonts; the earliest carnivoran diversity appears in the early Miocene rift sediments of East Africa, and these carnivorans are still accompanied by creodonts, in some numbers. But from at least the beginning of the Pliocene to the present, the carnivorans have been the dominant African predators. However, the endemic carnivore fauna of Africa remains in doubt because of the limited number of early Cenozoic sites and some uncertainty in distinguishing Eurasian immigrants from native lineages.

Late in the Cenozoic, the marsupial sanctuary in the austral salient is finally violated by the immigration of placental carnivorans into South America, by human introduction of placental carnivorans into Australia, and by probable local extinctions in Antarctica caused by the development of the continental ice cap.

Biogeography of the Order

Recent progress in our understanding of carnivoran phylogeny has made possible a marked clarification in the biogeography of the order. Studies utilizing the basicranium, teeth, and feet of extinct and living Carnivora have supplied reliable taxonomic units with clear distributional patterns. Since 1965 the sampling of middle to late Cenozoic fossil Carnivora has been enriched by access to the American Museum's Childs Frick collection (New York), accumulated in North America primarily during the late 1920s to 1965; by energetic collecting and continual refinement of the European fossil mammal record; and by the inauguration of intensive Cenozoic explorations in western and southern Asia, China, and Africa. In broad outline, the biogeographic history of the major groups of carnivorans is now evident, and merits review. The information thus marshaled should be useful in the study of other mammalian groups, whose distribution and ecological relationships were frequently influenced by the biogeographic evolution of carnivorous mammals.

I discuss carnivoran biogeography here in terms of the major divisions of the order first outlined by the English zoologist W. H. Flower (1869). Today, these three divisions are usually accorded subordinal rank, and include the various living and †extinct families: Arctoidea (Ursidae, †Amphicyonidae, Otariidae, Phocidae, Odobenidae, Mustelidae, Procyonidae); Cynoidea (Canidae), Aeluroidea (Felidae, Viverridae, Herpestidae, ?†Nimravidae). Current informed opinion recognizes the arctoids and cynoids as sister-taxa, together constituting the Caniformia, and the aeluroids constitute the Feliformia.

The biogeography of arctoids, cynoids, and aeluroids can be confidently set forth for the middle and late Cenozoic, that is, for the Oligocene to Recent. Some of these lineages can also be recognized in the later Eocene, but in general the tracking of middle and late Cenozoic lineages into the early Cenozoic (Paleocene and Eocene) has been a frustrating and often unsatisfactory process. A principal reason for this difficulty has been the scarcity of good cranial material of Paleo-Eocene Carnivora (only a single skull with basicranium of a Paleocene carnivoran is known). Despite these problems, however, two major groups of early Cenozoic carnivorans, the Miacidae and Viverravidae, can be identified, primarily on the basis of dental traits. As the only carnivorans currently known from Paleocene rocks, viverravids claim the place of oldest members of the order. Miacids, which first appear in the earliest Eocene, are the most abundant of the Eocene Carnivora despite coexistence with viverravids during most of the epoch. The geographic distribution of miacids and viverravids is not particularly revealing: both groups are found on the northern continents, but the record is so poor on the southern landmasses that little can be concluded about their early distribution or areas of origin. Most miacids and viverravids occur in three geographic regions: fossils from (1) western North America are more abundant and complete than the sparse record of (2) western Europe and (3) China, where both families are represented largely by fragmentary dental remains.

Thus it is the chronicle of the Oligocene to Recent Carnivora that will be set forth in this chapter in an attempt to clarify the distribution and migrations of major carnivoran families during the middle and late Cenozoic. The biogeographic history and distributions of these carnivoran families, presented in Figures 15.3–9 and 15.11–16, emphasize the Neogene record.

Biogeography of the Arctoids

Arctoids, the carnivorans of the northern continents (the Holarctic region), dominate the faunas of Eurasia and North America from the late Eocene to Recent. Their history is one of Holarctic regional evolution *in situ* coupled with migration events southward into Africa and South America. Members of the Arctoidea attain the largest body size in the order (the Pleistocene ursid *Arctodus*) and the smallest (the living mustelid *Mustela nivalis*, the least weasel).

Their early diversity is best recorded in the fossil faunas of the Phosphorites du Quercy, France, where numerous lineages of small to mid-sized arctoids were preserved in late Eocene to Oligocene fissure fillings (Remy et al. 1987; Teilhard 1915). Early arctoids, for example certain White River Group amphicyonids in the midcontinent (Hunt, in press, c) are in some cases abundant in the late Eocene to Oligocene of North America, but the Nearctic record lacks the diversity of lineages seen in Europe at this time (Bryant 1993; Emry 1992; Gustafson 1986). The early Asian arctoid record is poorer than that in North America, and very few specimens are yet known (Russell and Zhai 1987). The richest sample in Asia derives from the early Oligocene of Hsanda Gol in Mongolia. Hsanda Gol's few arctoids are comparable to lineages from Quercy in Europe, suggesting a Palaearctic commonality.

Four arctoid families can be recognized: Ursidae, Amphicyonidae, Mustelidae, and Procyonidae. Pinnipeds, also derived from an arctoid stock (Hunt and Barnes 1994; Mitchell and Tedford 1973), comprise three families: Otariidae, Odobenidae, and Phocidae.

Ursidae

Ursids living today are widely distributed across Eurasia and North America, and are also present in northwestern South America. The living ursids are a highly uniform group in terms of their general postcranial anatomy, basicranial and auditory structure, and dental features. In the past, however, ursid diversity was much richer in species and more varied in postcranial adaptations, body size, and teeth.

The living bears are all placed in the Ursinae (Tribe Ursini), with the exception of the spectacled bear, *Tremarctos ornatus,* of South America (Ursinae, Tribe Tremarctini) and the giant panda, *Ailuropoda melanoleuca,* of southeast Asia (Ailuropodinae). Although an ursid, the giant panda has had a long, separate history of evolution relative to the ursine bears. It descends as a distinct lineage from the late Miocene protopanda *Ailurarctos* (7–8 Ma, Turolian, Lufeng, China), whose teeth already display the beginnings of the elaborate cuspation so spectacularly developed in the living giant panda (Qiu and Qi 1989). The tribes Ursini and Tremarctini are related, probably sharing a common ancestor in the small early and middle Miocene ursid *Ursavus* from Eurasia and North America.

G. G. Simpson's concept of the Ursidae (Simpson 1945) included only the living ursine bears and a few extinct relatives, all characterized by morphological uniformity in the postcranial skeleton, cranium, and teeth, and by their ambulatory plantigrade style of locomotion. Today, ursids are seen in a broader and more inclusive perspective derived from new paleontological evidence: the family comprises a series of three middle to late Cenozoic radiations (Amphicynodontinae, Hemicyoninae, and Ursinae) that more accurately reveal the Ursidae as an arctoid group experimenting with both cranial and postcranial functional strategies throughout their 40-million-year history (Hunt, in press, a).

Amphicynodontines and hemicyonines include a varied array of extinct ursids that dominated the arctoid fauna of the northern continents in the Oligocene and Miocene. Simpson (1945), who listed 14 ursid genera in his classification of the family, included primarily the living ursines, and omitted the amphicynodonts and hemicyonines altogether. Moreover, he classified amphicynodonts and hemicyonines as canids, not ursids, owing to the mistaken practice at that time of using teeth as the principal index of relationships.

Simpson failed to recognize that ursids are united by a common basicranial pattern different from that found in true canids (Hunt 1974a, b, 1977). The auditory bulla pattern in living ursids is Type A (Hunt 1974a), believed to be representative of the plesiomorphic arctoid state (Figure 15.1). The Type A bulla is formed primarily by an ectotympanic. A rostral entotympanic element occupies the anterointernal corner of the bulla (enclosing the internal carotid artery), and a small, elongate, elliptical caudal entotympanic (E1) is attached to the medial rim of the ectotympanic. In addition, the posterointernal corner of the bulla is closed by a very small, cup-shaped caudal entotympanic (E2). In the adult the ectotympanic and the three entotympanic elements fuse together into a unit structure in which the sutures joining the elements may or may not be distinguishable. Most important, however, is that the relative contribution of each of these elements to the bulla does not change significantly during ontogeny. This situation is in marked contrast to that in true canids, in which the caudal entotympanic is a single element that strongly inflates during ontogenetic development to form the highly distinctive and recognizable bulla shape common to all living and extinct Canidae (Hunt 1974a, b).

In addition to the Type A auditory bulla, ursids also share a deeply embayed basioccipital bone that in living species contains a loop of the internal carotid artery nested within the inferior petrosal venous sinus (Figure 15.2). This artery-vein complex is believed to be a countercurrent heat-exchange device cooling arterial blood flowing to the brain (Hunt 1977, 1990; Hunt and Barnes 1994). Extinct ursids possess embayed basioccipital bones, presumably containing an expanded venous sinus that could have housed at first a rudimentary, later a more developed, loop of the internal carotid. Living and extinct canids do not show embayed basioccipitals, nor do living species have a carotid loop.

Finally, most ursids (a few amphicynodonts with primitive molars are exceptions) have upper molars in which the crown is quadrate (subquadrate in some amphicynodonts and some species of *Cephalogale*), with four principal cusps (protocone, paracone, metacone, metaconule). The metaconule is enlarged and connected to the protocone by an anteroposterior ridge. The protocone basin opens to the rear and is not closed by a postprotocrista (except in a few dentally plesiomorphic amphicynodonts and species of *Cephalogale* in which the metaconule is not enlarged and a normal postprotocrista closes the proto-

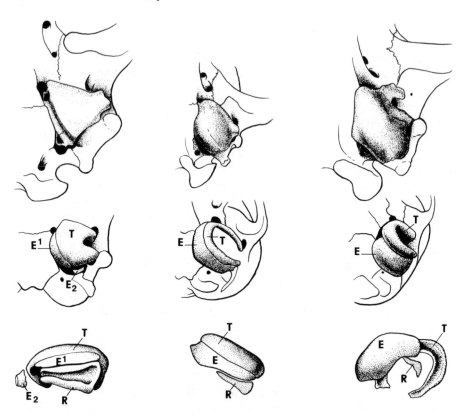

Figure 15.1. Ontogenetic elements forming the auditory bulla in (*left to right*) arctoid (ursid), cynoid (canid), and aeluroid (felid) carnivorans. *Top row:* adult basicrania with left auditory bulla in ventral view; *middle row:* neonatal stage, same view showing ectotympanic (T) and caudal entotympanic (E) elements that unite to form the adult bulla; *bottom row:* neonatal bullae in medial view (anterior to right), removed from skull to show relationship of rostral entotympanic (R, not visible in ventral view) to ectotympanic and caudal entotympanic elements. Two caudal entotympanics (E1, E2) are present in ursids. (After Hunt and Tedford 1993.)

cone basin). This molar pattern accompanies a retracted protocone on the upper carnassial and often low simple premolars surrounded by distinct cingula. The migration of the metaconule to the posterointernal corner of the tooth (Beaumont 1982) and the open protocone basin are useful diagnostic traits of ursid molars. In ursine ursids, the posterior border of M2 enlarges (rudimentary in some *Ursavus* and *Agriotherium*) and elongates to form a prominent talon not seen in amphicynodonts and hemicyonines. Ursids never evolve greatly enlarged crushing premolars, nor do they develop the hypersectorial carnassial blades generally seen in felids and hyaenids. Tooth-wear facets on the carnassials and molars of large hemicyonines and living ursines (Ursini) indicate that the former retained some shear in their bite, but the latter use the teeth primarily for crushing, an interpretation supported by the strong reduction in size of the carnassial teeth in Ursini.

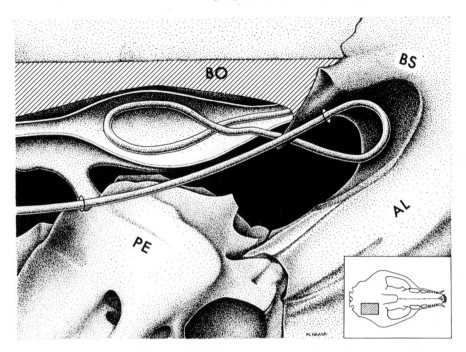

Figure 15.2. Auditory region of a living ursid, demonstrating the looped internal carotid artery nested within the inferior petrosal venous sinus. The sinus is emplaced in the lateral margin of the basioccipital bone. In the illustration the left side of the basioccipital has been removed to show the artery-vein complex, and the venous sinus has been opened to show the arterial loop within. Small circles around the artery show the locations of the anterior and posterior carotid foramina: the segment of the artery between the circles lies within the medial wall of the auditory bulla. Abbreviations: BO, basioccipital; BS, basisphenoid; AL, alisphenoid; PE, petrosal. (After Hunt 1977.)

Amphicynodonts are a paraphyletic late Eocene-Oligocene stem group from which evolved all other ursids (the last amphicynodonts are extinct by the end of the early Miocene). They are small in body size (<15 kg except for the large North American aquatic amphicynodont *Kolponomos*), but the basicranium and plesiomorphic dental pattern of the Ursidae are already evident. The earliest amphicynodonts are more diverse and better represented in Europe than in Asia or North America (Figure 15.3). The core genera in Europe are *Amphicynodon* and *Pachycynodon,* both well represented in the Quercy fissures (Teilhard 1915; Cirot and Bonis 1992). North American amphicynodonts have been placed in *Parictis* and are not yet identified in Eurasia; they occur only in the White River Group of the Great Plains and northern Rocky Mountains, and in the John Day beds of Oregon (Clark and Guensburg 1972). In Asia the only amphicynodonts belong to *Amphicynodon* and the small but derived *Amphicticeps,* both from the Oligocene Hsanda Gol Formation of Mongolia; the latter genus seems to be a plausible ancestor for the highly specialized North American amphicynodonts *Allocyon* and *Kolponomos* of the Pacific Northwest, *Allocyon* known from a single individual (Merriam

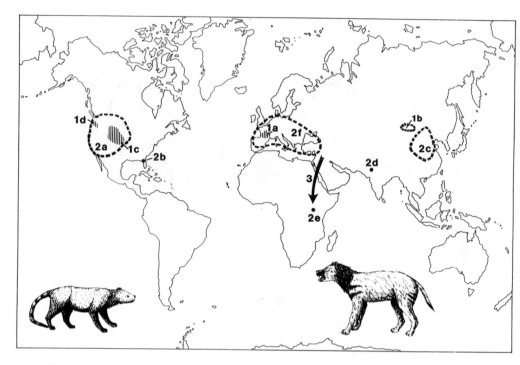

Figure 15.3. Geographic distribution of Late Eocene, Oligocene, and Miocene fossil ursids (Amphicynodontinae, Hemicyoninae). 1, amphicynodontines: 1a, western Europe (*Amphicynodon, Pachycynodon*); 1b, Mongolia (*Amphicynodon, Amphicticeps*); 1c, White River Group of the Great Plains and Rocky Mountains, North America (*Parictis*); 1d, John Day beds, Oregon (*Parictis, Allocyon*); Washington (*Kolponomos*). 2, hemicyonines: 2a, western United States (*Plithocyon, Cephalogale*, undescribed genus); 2b, Florida (*Phoberocyon*); 2c, eastern China (*Hemicyon, Phoberocyon*); 2d, Dera Bugti, Pakistan (*Cephalogale*); 2e, Rusinga, Kenya (hemicyonine); 2f, Europe and Asia Minor (*Hemicyon, Plithocyon, Phoberocyon, Dinocyon, Cephalogale*). 3, early Miocene migration of hemicyonines into east Africa about 18 Ma.

1930), *Kolponomos* from only three (Stirton 1960; Tedford et al. 1991). Since there are no records of amphicynodonts on the southern continents, at present they remain a Holarctic group.

Hemicyonines are mid-sized to large carnivorous ursids (5 to >200 kg), but in contrast to ursines and amphicynodonts they show pronounced cursorial adaptations of the postcranial skeleton. The principal derivative feature of the group is a digitigrade stance in which both the metatarsals and metacarpals are elongate; because the major weight-bearing axis passes between digits 3 and 4, the foot is functionally paraxonic.

Hemicyonines were an Oligocene-Miocene group, becoming extinct by the end of the late Miocene (Figure 15.3). The Miocene hemicyonine radiation (*Hemicyon, Plithocyon, Phoberocyon, Dinocyon, Cephalogale*) includes the first ursids to attain large body size. These bears were long thought to be members of the Canidae, but their basicrania and teeth demonstrate unequivocally that they are long-footed ursids with well-developed running ability and carnivorous dentitions. Cursorially adapted Old and New World hemicyonines paralleled the cursorial predatory canids evolving in the New World during the Miocene.

Hemicyonines were found only in Eurasia and North America, until a single tooth was reported from the early Miocene of East Africa (Schmidt-Kittler 1987). These, then, are evidently the first ursids to invade the southern continents, and probably a richer record of the group will emerge in future years from Africa.

The Ursinae are here divided into three tribes: Ursavini, Ursini, and Tremarctini. The Tremarctini and Ursini include all living ursids (except the giant panda), and are the product of relatively recent Plio-Pleistocene ursid radiations. A separate tribe, the Ursavini, is suggested for the Holarctic Miocene ursines (Figure 15.4): the large *Agriotherium* and *Indarctos* and the small, ancestral ursine *Ursavus*.

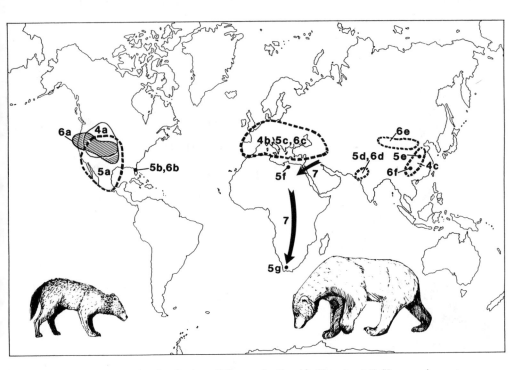

Figure 15.4. Geographic distribution of Miocene fossil ursids (Ursavinae). 4, *Ursavus;* 4a, western United States and Canada; 4b, Europe and western Asia; 4c, eastern China (Shanwang, Lufeng). 5, *Agriotherium:* 5a, western United States and Mexico; 5b, Florida (Bone Valley); 5c, Europe and western Asia (to Iran); 5d, Siwalik Group; 5e, Xiaoxian fissure, China; 5f, Libya (Sahabi); 5g, south Africa (Langebaanweg). 6, *Indarctos:* 6a, western United States; 6b, Florida; 6c, Europe and western Asia; 6d, Siwalik Group; 6e, north China; 6f, south China (Lufeng). 7, latest Miocene to early Pliocene migration of *Agriotherium* into Africa.

The theater of ursid evolution has been the northern continents. Only the ursine bears have effectively migrated into Africa and South America, and these migrants are commonly large, predaceous ursids. But despite their penetration into the southern continents, they achieved only limited species diversity. *Arctodus* is the first ursid to reach South America, appearing in Argentinian deposits judged to be younger than 2 Ma and containing other North American immigrant mammals (Marshall 1985; Kurtén 1967). In Africa, the large agriotheres (*Agriotherium*), which range from Libya (Boaz et al. 1979) to South Africa (Hendey 1980) in the earliest Pliocene, are the only ursids known to have spread over the African continent. *Ursus* enters North Africa in the Pleistocene (Hopwood and Hollyfield 1954), but there is no evidence that the genus spread beyond this limited area: *Ursus arctos* is said to have disappeared from northwest Africa during the mid-nineteenth century (Nowak 1991:1091).

Tremarctini includes the ancestral *Plionarctos,* which is confined to western North America, and its probable descendants *Tremarctos* and *Arctodus* (Figure 15.5). *Plionarctos* first appears ~7 Ma in North America in the late

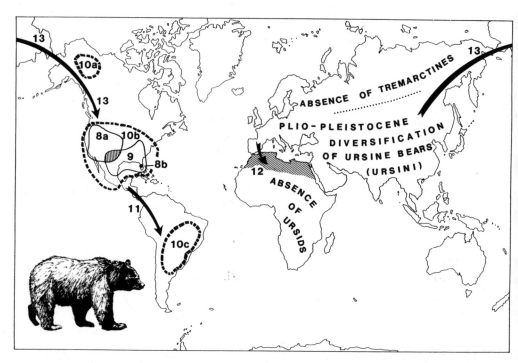

Figure 15.5. Geographic distribution of late Miocene, Pliocene, and Pleistocene ursids (Ursinae: Tremarctini, Ursini). 8, late Miocene and Pliocene *Plionarctos* and *Tremarctos:* 8a, western United States; 8b, Florida (*Plionarctos* only). 9, Pleistocene *Tremarctos,* southeastern United States and Mexico. 10, Pleistocene *Arctodus:* 10a, Alaskan *Arctodus;* 10b, North America (*Arctodus simuspristinus*); 10c, South America (*Arctodus bonariensis-brasiliensis*). 11, migration of *Arctodus* to South America about 2 Ma. 12, Pleistocene migration of *Ursus* to north Africa. 13, Pliocene migration of *Ursus* to North America about 4.3 Ma.

Miocene Rattlesnake fauna of Oregon. No postcranials of the rare *Plionarctos* are known, but Pleistocene discoveries of short-footed *Tremarctos* and long-footed *Arctodus* demonstrate that these are plantigrade bears (Kurtén 1966, 1967). Kurtén believed that the Tremarctini never evolved the turning-in of the forefeet seen in the gait of the living Ursini.

In North America *Tremarctos* is first known at ~2.5 Ma (Opdyke et al. 1977) in the Vallecito (California) and Grand View (Idaho) faunas, and is last recorded at 7000–8000 B.P. in the Holocene Devil's Den fauna, Florida (Kurtén and Anderson 1980).

Arctodus and *Tremarctos* are the only ursids to reach South America. Whereas *Arctodus* possesses a fossil record derived from the Argentine pampas and Brazilian caves, *Tremarctos* is known only from the presence of the living spectacled bear *T. ornatus* in the mountainous regions of northwest South America, extending once perhaps as far south as northern Argentina (Nowak 1991).

Bears of the genus *Arctodus*, at 250–630 kg the largest land carnivorans that ever lived (Kurtén 1967), do not appear in North America until Pleistocene (Irvingtonian) time, and are last known at 12,770 ± 900 B.P. (Wyoming) and 12,650 ± 350 B.P. (Texas) (Kurtén and Anderson 1980). The great arctodonts ranged over all of North America, eventually migrating through Central America to reach South America no earlier than 1.9–2.0 Ma (Marshall 1985:77). The South American arctodonts are considered to belong to at least two species, on the basis of dental traits and a marked size difference between the Brazilian and Argentine samples (Kurtén 1967): the largest form, *A. bonariensis*, rivals in size the large North American *Arctodus simus*, but *A. brasiliensis* from Brazil, only somewhat larger than the living spectacled bear, is the smallest known arctodont. Why the tremarctine bears were able to penetrate South America but the species of *Ursus* failed to do so remains an enigma.

Procyonidae

The Procyonidae are usually defined as a New World family of five to six genera (*Bassariscus, Bassaricyon, Nasua, Potos, Procyon*) of living arctoid carnivorans that favor forested environments, with all members capable of or strongly inclined toward an arboreal life style. Body size rarely exceeds 15 kg, and the living species (except *Nasua*) are chiefly nocturnal. Most procyonids live in southern North America or in warmer Central and South American subtropical to tropical habitats, with only one genus, *Procyon* (the raccoons), occurring in more northern temperate regions. Despite diverse dentitions and varied appearance, karyotypic uniformity of the living New World procyonids argues that they are a closely related arctoid group (Wurster and Benirschke 1968).

Following Hough's (1948) and Beaumont's (1968) identification of pro-

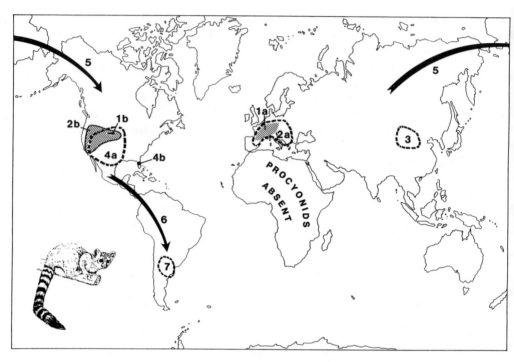

Figure 15.6. Geographic distribution of Oligocene to Pliocene fossil procyonids. 1, Oligocene to early Miocene procyonids: 1a, western Europe (*Amphictis, Broiliana, Stromeriella, Plesictis*); 1b, North America, early Miocene only (*Zodiolestes, Stromeriella*). 2, Simocyonines: 2a, Europe, middle to late Miocene (*Alopecocyon, Simocyon*); 2b, North America, late Miocene (*Alopecocyon, Simocyon*). 3, Simocyonines, middle to late Miocene, north China and Mongolia. 4, early to late Miocene and Pliocene procyonine and bassariscine procyonids: 4a, western United States; 4b, Florida (procyonines only). 5, migration of stem procyonids in early Miocene and simocyonines in middle or late Miocene to North America. 6, late Miocene migration of *Cyonasua*-group procyonids to South America about 7 Ma. 7, late Miocene and Pliocene procyonids, Argentina (*Cyonasua, Chapalmalania*).

cyonid basicranial structure in several genera of small arctoids from the latest Oligocene and early Miocene of Europe, it was realized that the family might have been originally Holarctic in distribution (Figure 15.6). Hough and Beaumont recognized three principal European genera with procyonid auditory regions: *Plesictis*, latest Oligocene, Peu Blanc, France, including the genoholotypic species, *P. genettoides*, and *Broiliana* and *Stromeriella*, early Miocene, Wintershof-West, Bavaria. But it has been difficult to segregate these genera morphologically from a plethora of other small European arctoids that radiated in the Oligocene and Miocene (Schmidt-Kittler 1981). Hough's concept of the procyonid auditory region was based upon the living New World procyonids, in which a suprameatal fossa is conspicuously developed. This fossa was central to Hough's (1944, 1948) procyonid concept. In addition, the auditory bulla was a slightly to strongly inflated flask-shaped structure with a laterally directed bony meatus covering a broad, low petrosal promontorium (in many living mustelids the bony meatus is directed anterolaterally and the promontorium is small, narrow, and constricted by the encroaching bulla margins). The bulla of living procyonids is single-chambered and built from an ectotympanic, a single medially placed caudal entotympanic, and a small rostral entotympanic (Hunt 1974a). When Hough (1948) first recognized a procyonid auditory region in a European carnivoran, the small *Plesictis* from Peu Blanc, it was the suprameatal fossa, the shape and orientation of the bulla and its bony external meatus, the form of the promontorium, carotid canal, and tensor tympani fossa, and the path of the facial nerve that influenced her identification.

In living procyonids the suprameatal fossa lies in the squamosal bone and penetrates both upward into the roof of the external auditory meatus and backward into the mastoid process. It is broadly open ventrally, and among living carnivorans it is developed in this way only in procyonids. The fossa went unnoticed in early studies of the carnivoran auditory region (Turner 1848; Flower 1869) and was not mentioned in Pocock's (1929) investigation of the procyonid auditory region. It first received attention through anatomical dissections of living and fossil carnivorans carried out by Segall (1943) and Hough (1944, 1948) at the Field Museum, Chicago. A hint of the complexity of the fossa was present in Segall's initial work and in later observations by Lavocat (1951, 1952), who observed it not only in living procyonids but also in certain mustelids. Segall explicitly regarded the fossa of procyonids and a similar cavity in mustelids as homologues. Most recently, Schmidt-Kittler (1981) has shown that a suprameatal fossa is widely distributed among arctoid carnivorans: it is rudimentary in many arctoid lineages, never developing beyond a preliminary stage in ursids and ailurids but evolving into a complex feature in procyonids and mustelids.

A ventrally open suprameatal fossa, considered the primitive state by Schmidt-Kittler, occurs not only in a number of Oligocene and Miocene European arctoids but in living procyonids as well. It was presumably floored in these fossil arctoids by a thin connective-tissue membrane like that found in the living *Procyon*. The fossa in raccoons dissected by myself and Matt Joeckel is an empty, air-filled space lined by a thin periosteal covering. The membrane, which is in continuity with the tympanum, is flaccid and distensible. Injection of air or fluid into the fossa distends the membrane, demonstrating that it can protrude freely into the meatus. Its function remains an enigma.

The suprameatal fossa of the living mustelids that possess it is differently configured from that of procyonids. The mustelid fossa is floored by bone rather than membrane, hence is usually not visible when one looks into the auditory meatus. Dissection is required to fully reveal the fossa in living mustelids, and Schmidt-Kittler believes it was reduced or lost, or incorporated into surrounding cavities, in many mustelid species. Despite some uncertainty concerning the homology of the fossa in procyonids and mustelids, it seems reasonable to interpret the procyonid type as a primitive state that in some mustelid lineages evolves to a closed chamber in which the cavity is floored by bone.

The most important implication of this interpretation is that procyonids are simply plesiomorphic in this feature, and may share it with other early arctoids. Furthermore, primitive mustelids might possess such a ventrally open fossa that then in some mustelid lineages transforms to the closed state. Thus the Mustelidae could include species with both ventrally open and ventrally closed suprameatal fossae. I suggest here that this is a plausible explanation of the contradiction in dental and auditory structure of the small arctoid *Plesictis* that figured so significantly in Hough's early studies, and I reinterpret the biogeographic patterns on the basis of this viewpoint.

Plesictis remains problematical in its affinities because of its "procyonid" auditory structure coupled with a dentition lacking the posterior molars. Living procyonids always retain M1-2/m1-2, with m2 elongate and m3 absent, and so are unlikely to have evolved from *Plesictis*. The oldest arctoid crania in Europe with well-developed suprameatal fossae belong to *Plesictis* from Peu Blanc and Coderet (MP30, latest Oligocene, ~24 Ma). Because these *Plesictis* species all lack M2 and do not have an elongate m2, they are improbable ancestors for all later procyonids. At nearly the same time (latest Oligocene, MP28, Quercy), the oldest procyonid, *Amphictis,* has only a rudimentary fossa but retains M2/m2. Since the procyonid auditory region may in fact be simply a primitive musteloid feature, these early *Plesictis* might reasonably be regarded as early musteloids with plesiomorphic auditory regions. From a plethora of species of *Plesictis* and *Mustelictis,* the diversity of Miocene and younger mustelids may have arisen (on mustelids, see below), but at present we should not rule out the possibility that some true procyonids may have been included in *Plesictis,* on the basis of Beaumont's (1968) basicranial comparisons.

Beaumont (1968) evaluated specimens of early Miocene *Plesictis* from St.-Gerand, France, and skulls of early Miocene *Broiliana* and *Stromeriella* from Wintershof-West, concluding that they were very similar to *Bassariscus* and the living procyonids in auditory structure. Early Miocene fissure fillings at Wintershof-West, Bavaria (~18 Ma), have produced the best samples of a diverse procyonid ancestral stock, placed by Dehm (1950) in three genera: *Broiliana, Stromeriella,* and *Amphictis.* Other, less well-sampled arctoids in the fissures may also belong to this stem group.

The New World procyonids are most likely derived from these Wintershof genera and their earlier European ancestry. All possess an elongate, similarly configured m2, retain M2, and have lost m3, the key dental traits of New World procyonids. *Broiliana* and *Stromeriella* range in the early Miocene from ~18 Ma at Wintershof-West (MN3b) to ~20.5 Ma at Laugnac (MN2b: Bonis and Guinot 1987). They appear to evolve from *Amphictis,* a plausible stem procyonid, which is known from the earliest Miocene to late Oligocene. The cranium of *Amphictis* is known from a single specimen from Quercy (MP28, Pech du Fraysse: Cirot and Bonis 1993), and its auditory region displays many procyonid features. Because this *Amphictis* skull shows a rudimentary supra-meatal fossa and the Wintershof *Broiliana* and *Stromeriella* have well-developed ones, the fossa appears to have developed in the early Miocene *Amphictis*.

The New World procyonid radiation probably stems from one or at most a very few species of small Eurasian immigrant procyonids that reached North America in the early Miocene. These immigrants retained the ventrally open suprameatal fossa, elongate m2, and M2. Derivation from one or two immi-grant species would explain the strong karyotypic identity of the New World forms.

Paleontological evidence from North American fossils provides additional perspective on the origin of the New World procyonids. First, the living and extinct North and South American procyonines *Nasua, Procyon, Edaphocyon, Arctonasua, Paranasua,* and *Cyonasua* are the product of a New World radia-tion first recognized in the early Miocene and identified by derived dental traits emphasizing a hypocarnivorous feeding mode like that found in the living raccoon, *Procyon lotor,* and coati, *Nasua nasua* (Baskin 1982, 1989). Second, the New World ring-tailed cat, *Bassariscus astutus,* retains a more ples-iomorphic hypercarnivorous procyonid dentition than do these procyonines, and its teeth and auditory regions are derivable from Oligocene or early Mi-ocene Old World procyonids such as *Amphictis* or *Broiliana.* Bassariscine procyonids (*Bassariscus, Probassariscus*) stand apart from the hypocar-nivorous procyonines as a distinct lineage first identified in North America in the middle and late Miocene (Baskin 1989).

The procyonines and bassariscines first appear in North America at ~16–18 Ma and are most likely derived from immigrants belonging to the *Broiliana-Amphictis-Stromeriella* group that existed in Europe ~18–26 Ma. The only record of these genera in the New World is a mandible of *Stromeriella* (or a closely allied taxon) from the early Miocene Runningwater Formation of west-ern Nebraska. The oldest New World procyonine, *Edaphocyon,* also comes from the Runningwater Formation; hence the presence of at least two pro-cyonid lineages in the early Miocene of North America at ~18 Ma provides the earliest evidence of procyonid diversity after the arrival of the family in the New World.

From the New World procyonines, possibly from the Nearctic *Arctonasua* or Holarctic *Stromeriella,* evolve the early procyonid immigrants into South America (Baskin 1982), the *Cyonasua* group (*Cyonasua, Chapalmalania*), which first appears on that continent ~7 Ma and extends to ~2.5 Ma (Mar-shall 1985).

The principal Eurasian group that evolved from these early procyonids is the Simocyoninae (*Alopecocyon, Simocyon*). Originally believed by Simpson (1945) to be canids, they are clearly identifiable, on the basis of both dental and basicranial structure, as arctoids, and were claimed as procyonids by Beaumont (1968). The two genera, *Alopecocyon* and *Simocyon,* are Holarctic parallels to canids in their dentition. Both are present in Europe, Asia, and

North America, never diversifying, and become extinct by the end of the
Miocene. Simocyonines have an elongate m2, but the hypoconulid is not developed; M2 is not lost or even strongly reduced; the P4 protocone is reduced,
retracted; and metaconules are present on the caniform upper molars, but not
emphasized. The small arctoid *Amphictis*, with a rudimentary suprameatal
fossa, seems likely to be the ancestral simocyonine (the fossa is not strongly
developed in the subfamily). Simocyonines and procyonines share the loss of
m3 coupled with an elongate m2 talonid and the retention of M2 and a
persistent metaconule on M1.

Current consensus favors a relationship between procyonids and mustelids,
as either Mustelida (Tedford 1976) or Musteloidea (Schmidt-Kittler 1981).
But other than the loss of m3 in both groups and the tendency to develop an
elaborate suprameatal fossa in the auditory region (characters that could have
developed in parallel), few reliable derived traits unite them relative to an
ursid-procyonid dichotomy. The pronounced similarity of the basicranium,
bulla, teeth, and skull form shared by early procyonids (*Broiliana* and *Amphictis*) and amphicynodont ursids (*Amphicynodon*) supports an ursid-procyonid
relationship (see Wyss and Flynn 1993).

The migration of procyonids from Holarctica into the southern continents is
a very limited affair. The Old World groups herein attributed to Procyonidae
have no record of migration into Africa or southeastern Asia and are restricted
to the Eurasian mainland. In South America, however, there are two separate
migration events: (1) the invasion of the *Cyonasua* group about 7 Ma (*Cyonasua* is the first placental carnivoran to reach South America: Marshall 1985;
Berta 1988); and (2) the later invasion of modern procyonine genera (*Procyon,
Nasua, Potos, Bassaricyon*). These living genera are occupants of woodland
and forested environments, and it is no surprise that a fossil record is lacking;
dating their movement into South America is problematic. Dispersal may have
occurred from Central America, beginning at ~2.4 Ma upon completion of the
Panamanian land bridge (Marshall and Sempere 1993), and plausibly continued to the present; Pleistocene dispersal is often regarded as probable (Marshall 1985).

Amphicyonidae

The amphicyonids, or beardogs, are first recognized in the late Eocene of
North America, Europe, and Asia. One of the earliest arctoid groups, they are
Holarctic in distribution from the time of their appearance in the fossil record
(Figure 15.7). They are the first arctoids to reach large body size, and may in
fact be the sister group to all other caniform carnivorans.

Alone among arctoids, the beardogs primitively retain the full eutherian
dentition, 3-1-4-3/3-1-4-3. Although the upper molars are reduced in number
to two in several lineages (temnocyonines, *Ischyrocyon*, and terminal species

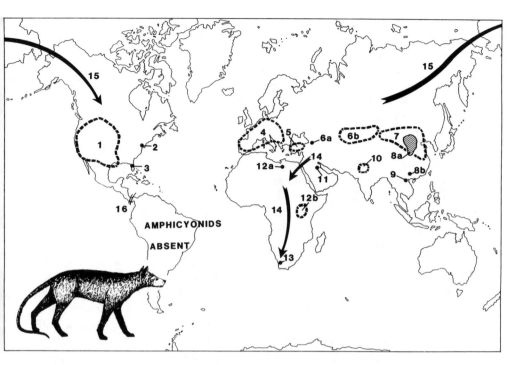

Figure 15.7. Geographic distribution of late Eocene to late Miocene fossil amphicyonids. 1, western North America, late Eocene to late Miocene. 2, eastern United States (Calvert Formation), middle Miocene. 3, Florida, Oligocene and Miocene. 4, Europe, late Eocene to late Miocene. 5, Turkey, middle Miocene. 6, Oligocene amphicyonids: 6a, Georgia; 6b, Kazakhstan. 7, northeastern China, middle Miocene. 8, Eocene and Oligocene amphicyonids: 8a, north China; 8b, south China (Guangxi). 9, North Vietnam, early Miocene (*Amphicyon*). 10, Siwalik Group, middle to late Miocene. 11, Saudi Arabia, early Miocene. 12, Africa, early Miocene amphicyonids: 12a, Libya (Gebel Zelten); 12b, Kenya. 13, Arrisdrift, Namibia, middle Miocene (including *Amphicyon*). 14, early Miocene migration of amphicyonids into Africa. 15, early Miocene migration of *Cynelos-Ysengrinia* to North America about 19–20 Ma and *Amphicyon* at about 18 Ma. 16, probable southern limit of amphicyonids in the Americas, based on the early Miocene Cucuracha fauna, Panama.

of *Daphoenodon*), the plesiomorphic condition for the family (three upper molars) is preserved in many genera (*Amphicyon, Cynelos, Daphoenus,* early *Daphoenodon, Ysengrinia*) into the early and middle Miocene.

Early beardogs also display the most plesiomorphic condition of the arctoid auditory bulla: the ectotympanic is an unspecialized, unexpanded, simple bony crescent in late Eocene and Oligocene *Daphoenus* (North America) and *Cynodictis* (Europe). No ossified entotympanic(s) occur, but a space between the petrosal and the inner rim of ectotympanic is almost certainly the locus for an entotympanic element (possibly hyaline cartilage or a fibrous connective tissue). Some late Oligocene beardogs have expanded the ectotympanic into a wide, shallow, bony crescent, still quite rudimentary in construction, lacking any ossified entotympanic contribution. By the early Miocene, several beardog lineages in Europe (*Cynelos*) and North America (*Daphoenodon*) have developed an auditory bulla that completely encloses the middle-ear cavity, comprising fully ossified and fused ectotympanic and entotympanic elements.

Despite the fact that the bulla of Miocene amphicyonids is nearly identical in form and construction to that of ursids, it must have evolved independently in the two families, since the earliest members of both display bullae at different stages of evolution.

In North America the fossil record of beardogs is unexcelled. Complete skeletons and crania of many genera record a rich diversity that also existed in Europe and, presumably, in Asia. Unfortunately, the European and Asian fossils are often limited to isolated teeth, jaw fragments, and scattered postcranials. But even this sparse evidence demonstrates that many lineages were anatomically similar in dental and postcranial features to North American taxa, indicating a broad Holarctic continuity during the Miocene, the time of greatest diversity for the family.

Beardogs do not survive the end of the Miocene in Holarctic faunas. They are extinct in North America by the end of the Clarendonian land-mammal age (~9 Ma). In Europe the last known amphicyonid comes from Kohfidisch, Austria, MN11, basal Turolian (Fejfar 1989) or MN10, latest Vallesian (Mein 1989), and is also dated at ~9 Ma (Beaumont 1984). Beardogs persist until ~7.4 Ma in southern Asia as relicts in Siwalik faunas of Pakistan (Barry 1985; Barry et al. 1982:112; Barry and Flynn 1989), and in the Miocene they penetrate into Southeast Asia as far as northern Vietnam (Hangmon site: Ginsburg et al. 1992). They are unknown in China after the Tunggurian (no amphicyonids younger than 13 Ma are known in eastern Asia).

Beardogs first appear on the African continent in the early Miocene of the Kenyan rift (*Cynelos*: Savage 1965; Schmidt-Kittler 1987), in Libya (Gebel Zelten: Tchernov et al. 1987), and in Saudi Arabia (Al Sarrar: Thomas et al. 1982; Whybrow 1987), and are among the first placental carnivorans to enter Africa from Eurasia. The youngest African occurrence of an amphicyonid, represented by a large, robust rostrum, is from about the middle of the Ngorora Formation, Kenya, and is estimated to be ~11.2 Ma in age (early late Miocene: Barry 1987, pers. comm.; Hill et al. 1985, table 3, *Pseudocyon*[?]). Several amphicyonid lineages are present in Africa during the Miocene, and at least two genera reached South Africa (Namibia), where the great bearlike *Amphicyon* itself has been found in the Arrisdrift fluvial gravels together with middle Miocene ungulates, proboscideans, hyracoids, rodents, and insectivores (Corvinus and Hendey 1978; Hendey 1978).

The biogeography of Holarctic amphicyonids reflects a degree of endemism in the late Eocene and Oligocene followed by evidence of European-North American faunal connections in the Miocene. In the former instance, the late Eocene and Oligocene genera of amphicyonids from the White River beds of North America and from the Quercy fissures of France are distinct lineages, although a familial relationship is evident. Diversity seems greater at Quercy, and several lineages present there continue into the European Miocene. Some of these European genera suddenly appear as immigrants in North America during the early Miocene, suggesting that European amphicyonids spread rapidly across Eurasia, becoming a key element in the Holarctic carnivoran fauna of the early and middle Miocene.

No amphicyonids ever reached South America, probably due to the absence of a land connection in the Miocene when they, together with the hemicyonine ursids and borophagine canids, were the dominant, large terrestrial carnivorans of North America. Today, the southernmost occurrences of New World amphicyonids are found in the southwestern United States and in Florida. An amphicyonid was recorded in Miocene rocks from Honduras by Olson and McGrew (1941; also Patterson and Pascual 1972: table 2), but later study of these fossils (an essentially edentulous mandible, a partial upper molar, and postcranials) has shown that they belong to the canid *Osteoborus* (Webb and Perrigo 1984). However, a Miocene terrestrial mammal fauna of undoubted northern aspect occurs as far south as Panama in the Cucaracha Formation (Whitmore and Stewart 1965). The scarcity of fossil mammals in the Cucaracha assemblage, together with the fact that only ungulates have been found so far, leaves us ignorant of the carnivore component of the fauna. In North America, however, such faunas are commonly accompanied by diverse amphicyonids. Thus the eventual discovery of beardogs in Central America is plausible and, on the basis of the ungulate component of the Cucaracha fauna, to be expected. The Cucaracha Formation mammals of Panama would seem to indicate the approximate southern limit of penetration for large North American land carnivorans before the completion of the Panamanian land bridge in the late Cenozoic.

Mustelidae

Mustelids rank as the most successful of the small predatory Carnivora. Arising in Holarctica as very small animals (~1–2 kg), the mustelids have never evolved species of large body size (>100 kg), nor do they ever evolve as cursorial open-country predators. Yet they dominate the more densely vegetated habitats, where they occur in both terrestrial and arboreal settings. Today, 23 genera and about 65 species occur throughout the Holarctic, Neotropical, and African regions.

The first mustelids appear in the late Eocene to Oligocene of Europe (*Mustelictis*, Quercy) and North America (*Mustelavus*, South Dakota), but little species diversity is in evidence (Figure 15.8). Diverse mustelids are evident for the first time in the early Miocene of Europe and North America. European (Aquitanian) and North American (late Arikareean) deposits contain multiple lineages that are highly endemic yet broadly comparable in stage of cranial and skeletal evolution. In North America the stem mustelids (*paleomustelids*) of the early Miocene (*Promartes, Oligobunis, Megalictis, Brachypsalis*) represent the first New World mustelid radiation, ranging from small forms less than

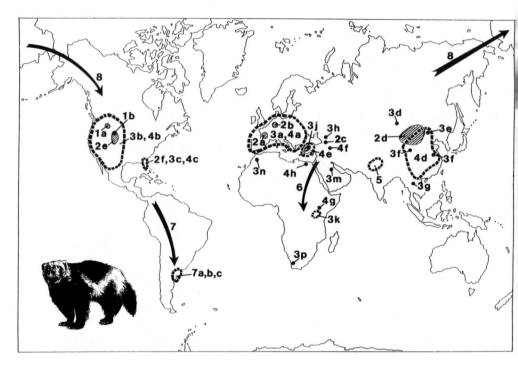

Figure 15.8. Geographic distribution of late Eocene, Oligocene, and Miocene fossil mustelids. 1, late Eocene: 1a, southwestern Montana (Chadronian, *Palaeogale*); 1b, southwestern South Dakota (Chadronian, *Mustelavus*). 2, Oligocene: 2a, western Europe (Quercy: *Mustelictis, Plesictis, Palaeogale, ?Stenogale*); 2b, western Europe (Mainz Basin: cf. *Plesictis*, mustelids); 2c, Georgia (Benara: *Plesictis*); 2d, Mongolia and north China (*Plesictis, Palaeogale*); 2e, western North America (White River and Arikaree Groups: *Mustelavus, Palaeogale;* 2f, north Florida (I-75 fauna: *?Palaeogale*). 3, early and middle Miocene: 3a, Europe (early Miocene, MN1–3: *Plesictis, Plesiogale, Paragale, Palaeogale, Martes, Dehmictis, Miomephitis;* MN4: *Martes, Palaeogale, Mionictis, Trochictis, Iberictis, Ischyrictis, Hoplictis*); (middle Miocene, MN5–8: *Mionictis, Trochictis, Ischyrictis, Hoplictis, Martes, Taxodon, Trocharion, Trochotherium, Proputorius, Paralutra, Palaeomeles, ?Heterictis*); 3b, western North America (early Miocene: *Promartes, Oligobunis, Megalictis, Brachypsalis, Palaeogale, Leptarctus, Craterogale*, musteline *n. gen.*; middle Miocene: *Mionictis, Brachypsalis, Martes, Miomustela, Leptarctus, Sthenictis, Plionictis, Dinogale;* 3c, Florida (earliest Miocene or late Oligocene: *Megalictis frazieri*, SB-1A; indet. Mustelidae, Buda; early Miocene: *Leptarctus, ?Oligobunis;* middle Miocene: *Leptarctus;* 3d, northwest China (early Miocene: Suosuoquan, *Palaeogale*); 3e, northeast China (middle Miocene: Tung-Gur, *Martes, Mionictis, Melodon, Leptarctus*); 3f, north China (early Miocene: Xie-Jia, Sihong, mustelids); 3g, Thailand (?middle Miocene: *Siamogale*); 3h, Georgia (middle Miocene: Belometcheskaya, mellivorine, eumustelid); 3j, Turkey (middle Miocene: *Proputorius, Trochictis, Hoplictis, Anatolictis, Plesiogulo*, lutrine); 3k, Kenya (early Miocene: *Luogale, Kenyalutra*); 3m, Saudi Arabia (early Miocene: *Mionictis*, cf. *Martes*); 3n, Morocco (middle Miocene: Beni Mellal, *Martes, Mellalictis*); 3p, Namibia (middle Miocene: Arrisdrift, cf. *Ischyrictis*). 4, late Miocene: 4a, Europe (*Martes, Ischyrictis, Sabadellictis, Circamustela, Marcetia, Mesomephitis, Promephitis, Trochictis, Trocharion, Baranogale, Paralutra, Limnonyx, Sivaonyx, Enhydriodon, Sinictis, Promeles, Plesiomeles, Taxodon, Eomellivora, Plesiogulo;* 4b, western United States and northern Mexico (*Pliogale, Martinogale, Buisnictis, Mionictis, Pliotaxidea, Martes, Plionictis, Sthenictis, Ischyrictis, Eomellivora, Plesiogulo, Lutravus, Enhydritherium, Cernictis, Sminthosinus*); 4c, Florida (*Ischyrictis, Plionictis, Sthenictis, Leptarctus, Enhydritherium*); 4d, China (*Proputorius, Promephitis, Sinictis, Mustela, Martes, Meles, Melodon, Parataxidea, Plesiogulo, Eomellivora, Lutra, Sivaonyx*); 4e, Turkey and Samos (*Promephitis, Promeles, Eomellivora, Parataxidea*); 4f, Iran

5 kg to the largest terrestrial mustelids that ever lived (*Megalictis*, 25–65 kg: Hunt and Skolnick, in press). These genera possess mustelid auditory regions but lack important derived features of the upper carnassial (slitlike notch in shearing blade) and M1 (lingual expansion) that distinguish the living species. North American paleomustelids lack a developed suprameatal fossa. The early Miocene European genera (*Plesiogale, Paragale*) with mustelid auditory regions also retain the plesiomorphic M1 seen in the North American radiation, but precociously develop the modern upper carnassial (Petter 1967; Schmidt-Kittler 1981), the earliest mustelids to do so.

Representatives of the living subfamilies, here termed *eumustelids*, first appear in the fossil record in the early Miocene. By that time these small predatory arctoids had already attained considerable diversity. Members of the eumustelid radiation are currently recognized by two key anatomical features (Schmidt-Kittler 1981): a suprameatal fossa in the auditory region, ventrally floored by bone (the plesiomorphic arctoid suprameatal fossa lacks a bony floor), and absence of the slitlike notch in the upper carnassial tooth. Some living mustelids lack a developed suprameatal fossa, but in these forms it is presumed to be secondarily reduced or incorporated into adjacent auditory cavities (Schmidt-Kittler 1981).

The rich carnivoran assemblage from the fissure fills at Wintershof-West, Bavaria, dated at ~18 Ma (MN3b: Dehm 1950; Bonis 1981), is the first in the Old World to include eumustelid auditory regions, upper carnassials, and even the lingually expanded M1 so characteristic of the living groups. (W. D. Matthew, as early as 1924, accurately observed that "in the Miocene Mustelidae the tendency to broaden the inner half of M1, so characteristic of the later genera, becomes progressively apparent.") Species of small Wintershof eumustelids attributed by Dehm (1950:65–71) to *Martes* and *Laphyctis*? (=*Dehmictis*: Ginsburg and Morales 1992) show upper carnassial and M1 form like that in living mustelids. Their upper carnassials lack the deep slit between paracone and metastylar blade, and their first upper molars have expanded lingual portions without metaconules or postprotocristae (the well-known "wasp-waisted" M1 of many living mustelids). At least two other early Miocene mustelid lineages (MN4: *Trochictis, Iberictis*) also possess lingually expanded first upper molars (Ginsburg and Morales 1992).

Somewhat prior to this, in the early Miocene Aquitanian deposits at Montaigu-le-Blin, France (MN2a, 22–23 Ma), crania of *Plesiogale* and *Paragale* display eumustelid upper carnassials and auditory bullae configured as in

---

(*Martes, Promeles, Parataxidea*); 4g, Kenya (mellivorine, *Vishnuonyx, Enhydriodon*); 4h, Wadi Natrun (*Enhydriodon, Aonyx, ?Topolutra*). 5, Siwalik Group (middle Miocene: *Martes, Hoplictis, Vishnuonyx;* late Miocene: *Eomellivora, Sivaonyx*, cf. *Ischyrictis*, mustelid indet.). 6, early Miocene migration of mustelids to Africa (earliest records at Rusinga and Songhor in Kenya, and As-Sarar, Saudi Arabia). 7, late Pliocene and early Pleistocene migrations of mustelids to South America: 7a, Chapadmalal Fm., ~2.4–2.5 Ma (*Conepatus*); 7b, Vorohue and San Andres Fms., ~1.9–2.4 Ma (*Galictis, Stipanicicia*); 7c, Ensenadan, ~1.4 Ma (*Lutra, Lyncodon*). 8, periodic Miocene migrations of mustelids from Eurasia to North America (*Plesictis*-group, leptarctines, *Martes, Mionictis, Ischyrictis, Eomellivora, Plesiogulo*, and others).

living mustelids (Viret 1929: pl. 13, figs. 4, 5; Beaumont 1968: pl. 1; Petter 1967, pl. 1, fig. 1), but the suprameatal fossa is not yet fully floored ventrally by bone and M1 shows only rudimentary lingual expansion. Thus, in both morphology and placement in time, *Plesiogale* and *Paragale* can be considered as transitional forms between the archaic paleomustelids and the eumustelid radiation. Schmidt-Kittler (1981) designated the Aquitanian species as a *mustelides Fruhstadium*, or early stage, visualizing the living eumustelids as the late stage, or *Hauptstadium*.

In North America the oldest eumustelid is represented by a complete skull with basicranium from the Kimberly Member of the John Day Formation, Oregon (~23 Ma), of earliest Miocene age. The same lineage appears in the late early Miocene Runningwater Formation at Cottonwood Creek Quarry, Dawes County, Nebraska (~18 Ma), where it is represented by a posterior cranium with well-preserved auditory region. This Runningwater cranium is an excellent morphological antecedent stage to a nearly complete skull with basicranium of a small musteline from Observation Quarry, western Nebraska (~16–16.5 Ma). These three crania document an evolving lineage of small mustelines, closely related to *Mustela*, over an interval of about 7 million years in the early Miocene. The fossils demonstrate that by the beginning of the Miocene in North America this eumustelid line had already evolved the enlarged braincase and small rostrum of living *Mustela*. As shown by the basicrania, the specimens also record a remarkable transition from a simple flask-shaped auditory bulla to the elongate and enlarged bulla of living *Mustela* species.

The skull form and basicranial structure of these earliest North American eumustelids are similar to those of the European early Miocene *Plesictis julieni* from Aquitanian deposits at St.-Gerand, France (Viret 1929: pl. XV, fig. 5). Although not identical in cranial morphology, the European and North American fossils are plausibly related, sharing a highly similar configuration of the auditory bulla and surrounding cranial features. The simple flask-shaped bulla of the European *P. julieni* is found in the two earliest eumustelid crania from North America, and it is this type of bulla that becomes progressively modified in the North American fossils during the later early Miocene into the inflated, elongated bulla of *Mustela* seen at Observation Quarry in Nebraska.

In contrast to the European fossil record, the North American record lacks transitional forms between the paleomustelids of the late Arikareean and the first eumustelids of the medial Arikareean–early Hemingfordian age, suggesting that the modern eumustelid subfamilies originated in Eurasia and entered North America as immigrant lineages, beginning in the earliest Miocene and continuing into the middle Miocene. Immigrant eumustelids replaced the North American paleomustelids, which largely became extinct by the end of the early Miocene. The morphological similarity of the oldest skulls of New World eumustelids to skulls of European mustelids like *Plesictis julieni* strongly supports the argument for a common Old World ancestry for these North American eumustelid groups.

Mustelines (weasels, martens, wolverines, badgers), mellivorines (ratels), mephitines (skunks), and lutrines (otters) are all represented as distinct lineages in the Holarctic late Miocene.

A particularly striking feature of all living mustelids is the loss of their posterior molars; they retain only the first molar in the maxilla and the first and second molars in the mandible. The elongate m2 of procyonids does not occur. In fact, the early Miocene European fossils from Montaigu and Wintershof-West have already attained this degree of molar reduction. The loss of M2 in these earliest predecessors of the eumustelid radiation explains the elaborate talon appended to the M1 in many lineages of living mustelids: in order for the m2 in the mandible to maintain contact with a tooth in the maxilla, the M1 must grow backward by developing an elaborate talon, which then occludes with m2 (commonly only the m2 trigonid contacts M1). Thus the occlusion of the M1 talon with m2 is a neomorphic relationship, a derived trait appearing independently in many eumustelid lineages (some living eumustelids lack M1/m2 occlusion or develop only a rudimentary contact, as for example in *Mustela*). The presence of an M1 talon therefore demonstrates beyond any doubt that at the time of origin of the lineage the second upper molars had already been lost (which implies loss of m3 as well).

In Asia, mustelids of Oligocene and early Miocene age are few and largely undescribed. Accordingly, we know very little about the presence of stem mustelids or eumustelids relative to those of Europe and North America, where the best fossil record occurs. In the most recent summaries of Chinese Neogene mammal faunas, Li, Wu, and Qiu (1984) and Qiu and Qiu (1990) report undetermined mustelids, "Mustela," and *Proputorius* from the early Miocene (Xiejia and Sihong faunas); and *Leptarctus, Martes, Mionictis,* and *Melodon* from the middle Miocene (Tung Gur fauna).

Migration to the southern continents occurred first in Africa in the early Miocene, much later in South America in the Quaternary. The immigration of mustelids out of Holarctica and into Africa is recorded by the presence of a probable lutrine and a stem mustelid (near the European *Paragale*) in the early Miocene of east Africa (Schmidt-Kittler 1987), and also by *Mionictis* and cf. *Martes* from the early Miocene of Saudi Arabia (Thomas et al. 1982). Surprisingly, the oldest African mustelid is reported to be the otter *Kenyalutra* from Songhor (European zone MN3, ~19.6 Ma: Schmidt-Kittler 1987), known only from a specialized lower carnassial and a tooth fragment. This occurrence is followed at Rusinga (MN4, ~17.8 Ma) by a second African mustelid, *Luogale,* known from two maxillary fragments and a partial mandible (Schmidt-Kittler 1987). As in the eumustelids, the P4 of *Luogale* appears to lack the carnassial notch. *Luogale* is more generalized, and appears to be derivable from early Miocene *Paragale-Plesiogale* from the Aquitanian of France, which are much like martens and are the oldest European fossils with the eumustelid P4.

It is difficult to determine if in fact mustelids were as diverse in the African early Miocene as they were in Europe. The occurrence of mustelids in the early

Miocene of Saudi Arabia at the Al Sarrar site (Thomas et al. 1982) suggests that the family may have been widely distributed in northeast and east Africa by this time. Penetration of the family into southern Africa by the middle Miocene (about 15–17 Ma) is evidenced by a carnivorous mellivorine in the Arrisdrift fauna of Namibia (Hendey 1978). A mellivorine is also present in the late Miocene of north Africa (Ginsburg 1977).

Mephitines, mustelines, and lutrines invaded South America from the north, beginning in the Pliocene, ~2.4–2.5 Ma (Chapadmalalan land-mammal age: Marshall et al. 1983; Marshall and Sempere 1993), and continuing into the Pleistocene, and are today represented by 14 species distributed widely over the continent in varied habitats. Eleven of the 14 have been considered endemic (Hershkovitz 1972). Mustelids are among the first carnivoran families of North American origin to enter South America via the Panamanian land bridge: the earliest recorded immigrant is the mephitine *Conepatus* in the Chapadmalalan Formation of Buenos Aires Province, Argentina, where it occurs with the endemic procyonids *Cyonasua* and *Chapalmalania* (Reig 1952; Marshall et al. 1983). Species of *Conepatus* are the only mephitine skunks to reach South America; three living and multiple fossil species are recognized (seven fossil species are known from Argentina, Brazil, Bolivia, and Peru: Berta 1988). In the overlying Vorohue and San Andres formations (Marshall et al. 1984), the endemic weasels *Galictis* and *Stipanicicia* appear, followed in the Ensenadan age by the endemic Patagonian weasel *Lyncodon* and the first otters (*Lutra*). In the Lujanian, *Mustela* makes its appearance in South America, where today it is known from three species confined to the northern part of the continent. The martenlike *Eira* and the giant Brazilian otter *Pteronura*, each known from a single species endemic to South America, have no fossil record certainly older than Holocene.

South American mustelid genera with present-day northern congeners include nine species of skunks, otters, and weasels (*Conepatus, Lutra, Mustela*); southern endemics are represented by three species of weasels (*Galictis, Lyncodon*), the martenlike *Eira,* and the giant Brazilian otter *Pteronura*. On the basis of the temporally staggered appearance of various mustelid lineages in the South American fossil record, we can state that mustelid immigration into the continent probably involved a series of pulses distributed over the last 2.5 million years, rather than a single immigration event.

Ailuridae

The fossil record of the red panda (*Ailurus fulgens*) is limited to the late Ruscinian-Early Villafranchian (middle Pliocene, 3–4 Ma) of Europe (Tedford and Gustafson 1977 for localities) and to Pleistocene cave sites in China (Bien and Chia 1938). In North America, a single upper molar has been discovered in early Blancan fluvial sediments of Washington. Dental remains from Europe

(Schmidt-Kittler 1983) and from the Siwaliks of southern Asia (Gingerich and Sahni 1979), referred to *Sivanasua* and once believed to be ancestral ailurids, are now known to belong to feliform Carnivora and primates, respectively. Thus the oldest ailurids are Pliocene.

Ailurids were once Holarctic in their distribution, but today are restricted to mountain forests in Nepal, north Burma, and south China. Known fossils are limited to cranial and dental remains very similar to those features in the living animal; no postcranials are known, but the primitive, arboreally adapted skeleton of *Ailurus* with plantigrade feet has probably not changed since at least the Pliocene.

The red panda (*Ailurus, Parailurus*) and giant panda (*Ailuropoda*) have been variously considered as most closely related to procyonids or ursids, or placed in their own families, Ailuridae Flower 1869 and Ailuropodidae Pocock 1921. Basicranial structure and the classic monograph by Davis (1964) establish that the giant panda is an ursid. The nature of the red panda has been more difficult to solve. Attempts to address this question have usually proceeded from a strong focus on, and comparison with, living forms, most often ursids and procyonids. This perspective should be modified, however, in view of the much better fossil record of arctoid Carnivora available since the question of panda origins was first addressed in the nineteenth century. From a paleontological perspective, the red panda is a Eurasian derivative of the early arctoid radiation, and hence has relationships with both ursids (Ginsburg 1982) and procyonids (Gregory 1936; Roberts and Gittleman 1984). Its basicranial anatomy is very similar to that of middle Cenozoic ursids such as *Cephalogale,* but its molar teeth lack the characteristic ursid occlusal pattern, in which the metaconule shifts to the posterointernal corner of the tooth, creating a quadrate molar. *Ailurus* most likely evolved from a dentally plesiomorphic species of amphicynodont or *Cephalogale* that subsequently developed the elaborate cuspation of the cheek teeth characteristic of the living species, *A. fulgens.*

The red panda serves as a particularly instructive example of the value of the fossil record, by emphasizing the fact that present-day family-level groupings of living arctoids (ursids, procyonids, mustelids) do not adequately communicate the known paleontological diversity that characterized the arctoid radiation of the middle to late Cenozoic.

Otariidae-Phocidae-Odobenidae

The pinniped carnivorans (sea lions, seals, walruses) are aquatically adapted arctoids that underwent a major Neogene radiation in the world's oceans. Current evidence suggests that the oldest pinnipeds, the enaliarctines, are derived from terrestrial ursids (Hunt and Barnes 1994; Mitchell and Tedford 1973), and that pinnipeds may be monophyletic (Fay et al. 1967; Wyss 1987). Their biogeography, a subject in its own right, has recently been reviewed by

Repenning et al. (1979). The striking biogeographic separation of otariids in the Pacific and phocids in the Atlantic and Tethyan realms argues for distinct and early origins for the two major groups, plausibly from amphicynodont ursids. Unfortunately, the paucity of early phocid skulls with plesiomorphic basicrania hinders our comprehension of phocid origins within the arctoid group. Additional studies presenting relevant biogeographic information include Repenning (1976), Barnes et al. (1985), Mitchell (1975), and Repenning and Tedford (1977).

Biogeography of the Cynoids

Basicranial structure and dental traits indicate that the canids, the only cynoid family, are more closely related to the arctoids than to aeluroids. All living and extinct canids are united by the form and structure of the auditory bulla, its ontogeny (including the development of an *entoseptum* within the bulla), and the path of the internal carotid artery in the auditory region (Hunt 1974a, 1987). In the early ontogeny of living canids the bulla is made up of three separate bony elements: ectotympanic, rostral entotympanic, and caudal entotympanic (Figure 15.1). Initially, the bulla is of little volume, lacking any degree of internal expansion. During ontogeny, however, the caudal entotympanic enlarges relative to the other elements, producing an inflated middle-ear cavity of increased volume in the adult. During this growth process a shelflike bony septum develops within the bulla, largely or exclusively from the inflected margin of the caudal entotympanic (hence the term *entoseptum*). In early canids, the septum appears to have been contributed by the full circumference of the inflected caudal element, whereas in living Caninae the septum is restricted to the anterior part of the bulla (Wang, pers. comm.).

Furthermore, in canids the internal carotid artery enters the basicranium at the posterior lacerate foramen, and travels in a common passageway with the inferior petrosal venous sinus, rather than, as in arctoids, in a separate bony tube (carotid canal) in the medial bulla wall. In the anterointernal corner of the auditory region the artery runs in a short bony tube formed by the rostral entotympanic before entering the cranial cavity (Hunt 1974a). The extrabullar location of the internal carotid relative to the caudal entotympanic, found in all but the earliest canids, is an important derived trait uniting the family.

The canid dental formula is generally 3-1-4-2/3-1-4-3 (except in the African *Otocyon*, where it is M3/4).

In the living Caninae we find evidence of cursorial adaptation, in the lengthening of the lower limb segments and the restriction of limb movement to favor fore-aft motion, with lessened ability to pronate-supinate the forelimbs. The Caninae achieve this high degree of limb specialization by restriction of joint motion, appression of the metapodials and digits, and lengthening of distal limb elements, including tarsus and carpus.

Canids are originally endemic to North America, a fact still unappreciated by some students of carnivorans, who mistakenly place several Old World canid parallels, particularly certain amphicyonids and hemicyonine ursids, with the true canids. Canids first migrate to the Old World at ~9 Ma, when the first true fossil canid appears in the late Miocene of Spain (Crusafont-Pairo 1950). Shortly thereafter, *Canis, Vulpes,* and *Nyctereutes* all occur in the Pliocene of Eurasia.

Canidae

Canids are represented by three successive radiations (hesperocyonine, borophagine, canine) that take place in the Nearctic region during the middle to late Cenozoic (Figure 15.9). From their first appearance in the late Eocene until the late Miocene, the family is exclusive to North America, filling a broad spectrum of omnivorous and carnivorous niches for predators of small to medium body size (<100 kg). Beginning in the late Miocene and continuing into the Plio-Pleistocene, the third radiation (Caninae) spreads over Eurasia (9 Ma), Africa (4–5 Ma), and South America (2 Ma).

The first canid radiation during the late Eocene to early Miocene is distinguished by a diverse array of small canids of the subfamily Hesperocyoninae. The largest species are early Miocene end members of the *Mesocyon* and *Enhydrocyon* lineages, which have body weights of less than 20 kg. Both lineages evolved directly from the stem canid *Hesperocyon* (<3 kg), which is the first and only representative of the family in the late Eocene of North America. Hesperocyonines retain a plesiomorphic triangular M1, bordered by a strong lingual cingulum, that bears conules but lacks the enlarged metaconule of ursids and many procyonids. Most species retain primitive canid dentitions, with the exception of the *Enhydrocyon* line, in which the premolars become robust crushing teeth and the rear molars are reduced, resulting in the first canid parallel to durophagous Old World hyaenids. Only *Hesperocyon* occurs in the late Eocene, but it is joined in the early Oligocene (Brule Formation of the Great Plains) by the somewhat larger *Mesocyon*. During the late Oligocene and early Miocene, hesperocyonine diversity reaches its peak: rocks of that age from the Great Plains and Rocky Mountain basins have produced multiple species of small hesperocyonines.

Borophagine canids, the second canid radiation, evolved from the hesperocyonines in the late Oligocene, stemming from small, fox-sized species of *Tomarctus* and "Nothocyon." During the Miocene, borophagines increased steadily in size and filled a variety of niches paralleling those of small foxes, coyotes, wolves, omnivorous procyonids, and bone-crushing hyaenids. Among them are large cursorial wolflike (*Aelurodon*) and hyena-like (*Osteoborus, Borophagus*) canids that evolved in response to open savanna and grassland environments in the late Miocene and Pliocene. The M1 in borophagines is initially more quadrate than in hesperocyonines, largely as the result of a developed metaconule which, with the lingual cingulum, contributes to the

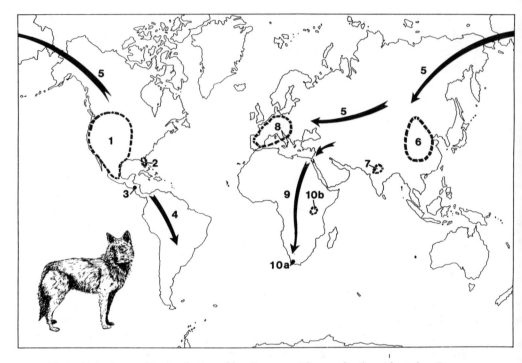

Figure 15.9. Geographic distribution of late Eocene to Pliocene fossil canids. 1, late Eocene to Pliocene, North America (Hesperocyoninae, Borophaginae, Caninae). 2, Oligocene to Pliocene, Florida. 3, southern limit of canids in the New World during the Oligocene and Miocene (Honduras, *Osteoborus*). 4, late Pliocene and Pleistocene migrations of canids (Caninae only) to South America, 1.4 Ma and 2.0–2.4 Ma. 5, late Miocene and Pliocene migrations of canids (Caninae only) to Eurasia, as early as 7 Ma. 6, Pliocene canids, eastern Asia (*Nycterreutes*-"Canis," first appearance about 4–5 Ma; *Vulpes-Canis*, about 3 Ma; *Cuon*, about ?2–3 Ma). 7, Pliocene canids, Siwalik Group, first appearing in *Elephas planifrons* interval-zone, about 1.5–2.9 Ma. 8, Europe, late Miocene ("Canis" cipio, Nycterreutes) and Pliocene (*Canis, Vulpes, Nycterreutes*) canids. 9, early Pliocene migration of canids into Africa about 4–5 Ma. 10, Pliocene canids, Africa: 10a, Langebaanweg (*Vulpes*); 10b, Laetoli (*Cerdocyon*-group and other canids).

quadrate outline of the molar. The radiation terminates with the hyena-parallel *Borophagus* in the Pliocene of North America; no borophagines survive into the Quaternary.

The rich diversity of living canids on all continents reflects the success of the final radiation, the subfamily Caninae. Canines evolve from a single lineage of early cynoids, the small, fox-sized *Leptocyon* of the late Oligocene and early to middle Miocene of North America. The dental pattern in living Caninae does not differ to any significant extent from that of this small carnivore except in tooth size, indicating the remarkably conservative nature of the canine dentition. Diversification of canine canids does not occur until the late Miocene, when *Canis* and *Vulpes* first appear, followed by the principal canine radiation in the Plio-Pleistocene.

Canines are the first canids to enter the Old World (the borophagines and hesperocyonines are restricted to North America), appearing in Europe in the late Miocene (Crusafont-Pairo 1950), and in Africa in the early Pliocene (Langebaanweg: Hendey 1974; Laetoli: Petter 1987; Barry 1987), where at the Tanzanian locality they are already diverse. Canids enter South America in the late Pliocene at ~2.0–2.4 Ma, with the appearance of *Protocyon* in Argentina (Vorohue Formation, Buenos Aires Province). Three other genera of large canids (*Theriodictis, Chrysocyon,* and *Canis*) and the small *Cerdocyon* appear in South America in association with an early Pleistocene dispersal event at ~1.4 Ma (Marshall 1985; Berta 1988). All of the large canids, with the exception of *Chrysocyon,* become extinct by the Holocene. In addition to a single species of *Chrysocyon* (the maned wolf, *C. brachyurus*) that survives in central South America, six other genera of living canids are found today in South America (*Urocyon, Speothos, Atelocynus,* all from northern South America; *Cerdocyon,* northern and central South America; *Pseudalopex,* broadly distributed in South America; *Dusicyon australis,* the Falkland Island fox, extinct about 1876). The South American canids very probably evolved from a few northern immigrant forms of the Caninae that entered the continent soon after land connections were established, and then evolved endemic species adapted to local conditions.

In Africa, *Canis* is represented by three species of jackals with a sub-Saharan distribution, and the gray wolf (*C. lupus*), which penetrated only into northeast Africa. Though it is not found in South America, the fox (*Vulpes*) occurs throughout Africa, and the smallest living canid, *Fennecus zerda,* the fennec fox, occupies northern Africa. The African wild dog (*Lycaon pictus*) and bat-eared fox (*Otocyon megalotis*) are endemic and confined to sub-Saharan Africa. This degree of canine diversity is already evident by ~3.5 Ma at Laetoli, in Tanzania.

In the Oligocene and Miocene, canids occupy many niches in North America that are the province of arctoids in the Old World. Eurasia clearly emerges as the staging ground of arctoid evolution, certain lineages spilling over by immigration into North America and Africa. As the sister group to the arctoids, cynoids flourish in North America (and eventually South America as well), occupying arctoid roles for small to mid-sized carnivores and omnivores, but relinquishing the large carnivoran niche (>100 kg) to amphicyonids, ursids, and felids.

Biogeography of the Aeluroids

Aeluroid carnivorans include, as a core group, the families Viverridae, Herpestidae, Felidae, and Hyaenidae (Flower 1869; Hunt and Tedford 1993). The extinct catlike Nimravidae and the Recent Malagasy aeluroids have been the focus of recent research and are also discussed here. Aeluroids are essentially Old World carnivorans: the viverrids, herpestids, and hyaenids are restricted almost entirely to Eurasia and Africa, the only significant exception being the immigration into North America in the Plio-Pleistocene of a single hyaenid lineage (*Chasmaporthetes*), known only from the southern United States and Mexico. Felids are the only aeluroid family that successfully attains a degree of species diversity in the New World, reaching all of North America and penetrating into South America during the glacial maxima at ~1.9–2.0 Ma, when *Felis* and *Smilodon* appear for the first time in company with other carnivorans such as large arctodont bears (Marshall 1985).

Just as for arctoids, the early diversity of aeluroids is first recorded in the Quercy fissure fills of France (late Eocene to Oligocene), where several aeluroid lineages are interpreted to be ancestral stocks of felids and viverrids. The evidence in Asia at this time is very poor, but the principal site yielding aeluroids, Hsanda Gol (Mongolia), provides evidence of lineages similar to, if not the same as, those at Quercy, indicating a probable Palaearctic distribution for these early aeluroids. No aeluroids are known from Africa until ~20 Ma in the early Miocene (Schmidt-Kittler 1987; Hunt and Tedford 1993), when multiple lineages of small aeluroid carnivorans occur in the rift sediments of Kenya.

The migration of a single aeluroid family into the New World is an early middle Miocene event: true felids enter North America via Beringia and make their first appearance at ~16–17 Ma in the Great Plains (Hunt 1987:3).

Aeluroids are most reliably identified by the uniquely derived morphology of the auditory region (Figure 15.1), particularly the construction of the auditory bulla and the form of the petrosal (Hunt 1987, 1989, 1991; Hunt and Solounias 1991; Hunt and Tedford 1993 and references therein). Important features include the relationship of the caudal entotympanic element of the bulla to the ectotympanic bone, and the development of a ventral promontorial process of the petrosal. Moreover, in contrast to arctoids, each of the aeluroid families is recognized by a distinctive auditory bulla morph and a pattern of ontogenetic development that remains stable over a significant time interval in the middle and late Cenozoic (Figure 15.10).

Among the living Carnivora, the aeluroids alone have developed, in some lineages, elaborate carotid retia in the orbital region for cooling warm arterial blood flowing to the brain (Baker 1979, 1980, 1982; Paule and Baker 1978; Frackowiak 1989; Davis and Story 1943). The aeluroids that attain large body size, the felids and hyaenids, have this countercurrent heat-exchange mechanism. This rete may be a significant factor in the dissipation of body heat in large, exercising carnivorans and in the adaptation of large aeluroids to warm, open environments.

Viverridae

Living viverrids, small to mid-sized (to 20 kg) carnivorans of the Old World tropics, include about 19 genera of chiefly nocturnal (but some diurnal) predators, omnivores, and frugivores, occupying Africa, southern Asia, the East

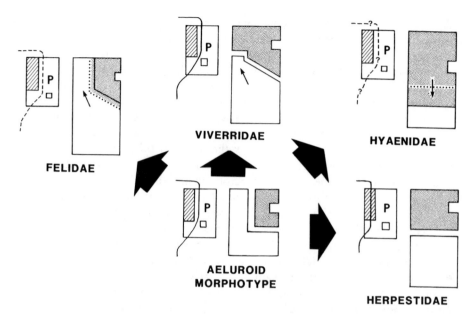

Figure 15.10. Evolution of the auditory bulla in aeluroid carnivorans. The diagrams, which represent the left auditory region in ventral view, anterior to top, illustrate a progression from a morphotype to the derived character states of the living families. The left side of each diagram indicates the relationship of petrosal (P), rostral entotympanic (hatched), and internal carotid artery (solid line, artery present; dashed line, artery reduced or lost). The right side of each diagram illustrates the ectotympanic (stippled) and caudal entotympanic (open), which lie ventral to the structures on the left. Small arrows indicate direction of ontogenetic growth in felids, viverrids, and hyaenids. The morphotype is most closely approximated by the living aeluroid *Nandinia* of Africa. (After Hunt and Tedford 1993.)

Indies, and Madagascar, with *Genetta* alone surviving in southern Europe. They inhabit well-vegetated environments (forest, brushland, grassland), where they exploit arboreal, terrestrial, and aquatic habitats. Viverrines and hemigalines, although often good climbers, are predominantly terrestrial, whereas the paradoxures are arboreal. Arboreal species are commonly plantigrade; terrestrial types tend to digitigrady. In their lack of cursorial specializations, which is reflected in a generalized postcranial anatomy, small body size, and (in many) unspecialized dentition, the viverrids have been regarded as skeletally primitive representatives of early Tertiary carnivorans.

In contrast to their present diversity, the Cenozoic fossil record of viverrids is poor. Moreover, because of the often fragmentary state of these fossils, the affinities of many taxa are problematical, and their biogeography remains less clear relative to other carnivoran families (Figure 15.11).

True fossil viverrids can be identified from basicranial and dental features. No living or fossil viverrid has more than two upper or lower molars, and premolars are not reduced in number; hence the dental formula is usually I3/3, C1/1, P4/4, M2/2. The upper molars generally lack strong internal cingula or hypocones, and often retain a developed parastyle. The lower carnassial usually maintains a primitive arrangement of the trigonid cusps: the paraconid-protoconid shearing blade is not developed to the extent seen in felids or most hyaenids. Similarly, the elongate upper carnassial and strong molar reduction of hyaenids and felids is absent. The basined heel of the lower carnassial usually displays a characteristic arrangement of three sharp cusps, including a hypoconulid, placed toward the rear of the talonid.

In living viverrids the ontogenetic growth pattern of the auditory bulla is diagnostic of the family (Figure 15.10). In early ontogeny, ectotympanic-rostral entotympanic contact produces the aeluroid thictic condition of the bulla (Hunt 1987). During ontogeny, the caudal entotympanic enlarges and grows forward over the conjoined ectotympanic-rostral entotympanic elements that form the anterior bulla chamber. The uniformity of this bulla pattern in living viverrids is particularly striking: it first appears in the early Miocene viverrid *Herpestides* from lacustrine deposits of the St.-Gerand sub-basin, France (Aquitanian, zone MN2a, 22–23 Ma), in which the bulla pattern and petrosal morph are fully modern in aspect (Hunt 1991). Notwithstanding the existence of a magnificent sample of early Miocene *Herpestides* crania, middle and late Miocene viverrids are rarely represented by skulls in Old World deposits. Before the Plio-Pleistocene, fossil viverrid skulls with well-preserved basicrania are poorly known, yet many lineages with fully modern basicrania must have been in existence, given the advanced state of the *Herpestides* auditory region in the early Miocene.

Fossil viverrids can be grouped into five temporal intervals:

1. 30(?) to 34 Ma, early Oligocene: protoviverrids (*Palaeoprionodon*) known from several skulls and a large number of mandibles from fissure fills at Quercy, France (see Hunt 1991:32), as well as from fragmentary dental material of *Palaeoprionodon* from Hsanda Gol, Mongolia.

2. 21 to 23 Ma, MN zone 2: a rich sample of crania, mandibles, and postcrania of *Herpestides,* from the Aquitanian of France, the oldest known viverrid with a modern auditory region.

3. 13 to 20 Ma, MN zones 3 to 8: the first diverse viverrids are identified from teeth in jaw fragments using dental traits common to living genera, and are accompanied by rare postcranial bones that suggest skeletons much like modern forms; no skulls with intact basicrania are known.

Viverra (subgenus *Viverrictis*), *Jourdanictis*, and *Semigenetta* are known from Europe; recently, *Semigenetta* has been found in China. In contrast to these rare (chiefly dental) remains is a completely articulated viverrid skeleton from freshwater limestone at Oeningen (MN8, ~13 Ma) in the British Museum, which represents the most complete European viverrid yet discovered; unfortunately, the skull is crushed and the basicranium is obscured. This large (15 kg) civetlike species demonstrates that by the middle Miocene some viverrids had attained a size equal to the larger living terrestrial civets (such as *Civettictis*).

The first African viverrids are known from this interval: these are represented primarily by dental remains from the early Miocene African rift sites of Rusinga and Songhor and nearby localities (Schmidt-Kittler 1987). Sites (such as Songhor) associated with the Tinderet volcanic complex (19.6 Ma) have yielded fossils attributed to *Stenoplesictis*, but these may be closer to the viverrid *Semigenetta;* sites (such as Rusinga) tied to the Rangwa volcano (17.8 Ma) have produced similar material but also dental remains referred to *Herpestides*, though the latter differ somewhat in details from the Aquitanian *Herpestides* of France. In the rift deposits these viverrids are associated with herpestids, and they indicate the oldest viverrid-herpestid differentiation known in the Old World, one that is based on dental evidence. When better fossils are found, the nature of the basicrania of these viverrids and herpestids will be of considerable interest.

4. 6 to 13 Ma, MN zones 9–13: although this is an interval with very few described viverrid remains, future discoveries in Asia may fill this gap. Few European and no African viverrids are certainly known. In Asia, Chinese faunas from Ertemte (MN13: Fahlbusch et al. 1983) and Lufeng (8 Ma, MN12) have produced good viverrid samples that are, however, yet to be described. In the Siwaliks, the "U-level" (~8 Ma) yielded a large viverrine and *Viverra chinjiensis;* the type of the latter, from the Chinji Siwaliks, probably came from the MN8–12 interval. In Europe a small viverrid attributed to *Semigenetta* (Petter 1976) is known at Can Llobateres (Vallesian, MN9).

5. 0 to 5 Ma, MN zones 14–17, MQ zones 1–2: Plio-Pleistocene discoveries of true viverrids in recent years, particularly in Africa, have greatly increased our knowledge of the family. Many of these species are large animals, equaling and even exceeding the living *Civettictis* (to 20 kg) and *Viverra* (to 12 kg) in body size (*Civettictis*, *Viverra*, and *Genetta* all appear for the first time in the African Pliocene). Most fossils not certainly attributable to these three genera belong to a complex of Eurasian and African species referred to "*Viverra*" (*leakeyi*, *peii*, *bakeri*, *chinjiensis*, *pepratxi*), "*Vishnuictis*" (*durandi*, *salmontanus*), and "*Megaviverra*" (*carpathorum*, *apennina*). These fossils indicate that large viverrids existed in Europe as far north as central Czechoslovakia and in Asia to the latitude of Zhoukoudian, near Beijing. A number of these species belong to lineages in which large body size is coupled with a primitive shearing dentition and, in rare instances when the basicranium is present, a viverrid bulla. During this time interval and from this group evolve the largest viverrids that ever lived: the African endemic *Pseudocivetta* (whose skull is undescribed) and the south Asian *Vishnuictis durandi* (known from two crania), the former a dental omnivore with specialized crushing teeth, the latter a huge predatory wolflike viverrid with shearing teeth and an auditory bulla unquestionably of the viverrid type. *Vishnuictis durandi*, from the Siwaliks, also seems to occur in north China, where it has been called "*Viverra*" *peii*. Its skull form closely parallels the living *Civettictis*, which may be a close relative.

Recently, a skull and a large number of mandibles of a large predatory viverrid have been found at Langebaanweg, in South Africa (Hendey 1974):

the teeth correspond to those of the East African *Viverra leakeyi,* and the skull form of the Langebaanweg carnivoran is indeed similar to that of living *Viverra.* The Langebaanweg skull, when compared to a nearly complete Siwalik skull of *Vishnuictis durandi,* demonstrates that at least two large hypercarnivorous viverrids of very different skull form existed in the Plio-Pleistocene (Hunt, in press, b): (1) *durandi,* with a *Civettictis*-like cranium, a short snout, and an inflated frontal region; and (2) *leakeyi,* with a *Viverra*-like cranium, a long snout, and a normal frontal region. The genus *Vishnuictis* can be used for the former, and *Viverra* for the latter. Because the specialized crushing teeth of the African *Pseudocivetta* establish this genus as a unique hypocarnivorous lineage, the Plio-Pleistocene of the Old World now documents three lineages of large viverrids (*Viverra, Vishnuictis,* and *Pseudocivetta*).

The origins and fossil record of many of the living viverrids, such as the linsangs, paradoxures, hemigalines, and Malagasy *Eupleres, Fossa,* and *Cryptoprocta,* are largely unknown: at best, only Holocene records exist.

The Malagasy viverrids provide an important corroboration of the usefulness of the auditory bulla and associated basicranium in determining carnivoran affinities. On Madagascar the resident carnivorans are believed by several workers to have radiated from a single stock that reached the island at some time in the middle to late Cenozoic. Seven genera comprising eight species occur today on the island. The basicrania of the seven genera can be grouped in two categories: a viverrid bulla morph and a herpestid bulla morph. The genera *Fossa, Eupleres,* and *Cryptoprocta* have viverrid auditory regions, are larger animals ranging upwards from 1.5 to 12 kg, and are generally nocturnal to crepuscular in habit. In contrast, the genera *Galidia, Galidictis, Mungotictis,* and *Salanoia* have herpestid auditory regions, are smaller in size (usually <1.5–2 kg), and are chiefly diurnal (a species of *Galidictis,* however, is nocturnal). The larger animals are viverrids, the smaller ones herpestids. The island must have been invaded at least twice (by separate viverrid and herpestid stocks) at some time in the Cenozoic when the bulla morphs were already established. Because the viverrid morph had appeared by the early Miocene (Aquitanian, MN2a) in Europe, and on present evidence is unknown in the Oligocene, the colonization of Madagascar by viverrids would appear to be a post-Oligocene (probably early Neogene) event.

The first appearance of viverrids in Africa occurs in the early Miocene of Kenya and Saudi Arabia. But the time of migration of viverrids *sensu stricto* into Africa is less certain, for the reason that the diverse aeluroids found at Songhor, Rusinga, and other early Miocene sites (Schmidt-Kittler 1987) suggest some degree of *in situ* evolution, possibly a period of endemic development within the African continent. Demonstration of viverrid (or aeluroid) endemism in Africa or elsewhere must await a fuller knowledge of Oligocene faunas on that continent and in Eurasia. Consequently, the place of origin of viverrids in the Old World remains an open question.

Herpestidae

The living mongooses comprise about 17 genera of small (0.2–6 kg), chiefly terrestrial, digitigrade carnivorans, found mainly in Africa (*Herpestes* occurs in Africa, Asia Minor, southern Asia, and the East Indies). A Cenozoic radiation of mongooses in Africa is credited with the production of diverse species (Hendey 1974; Petter 1969) tending to predatory diurnal ground-dwellers, constituting at least seven gregarious genera (although some are nocturnal and solitary and can climb). Their fossil record is poor, with best representation in the Plio-Pleistocene of Africa, where about eight genera are known (Figure 15.11). As they have for viverrids, the sediments of the east African rift

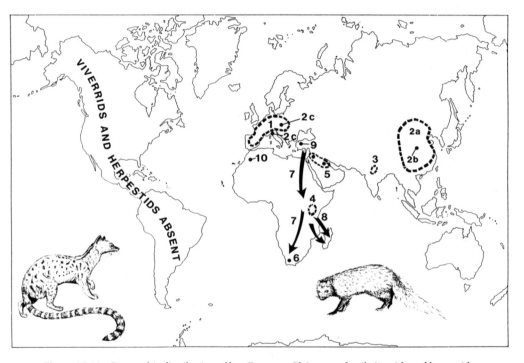

Figure 15.11. Geographic distribution of late Eocene to Pleistocene fossil viverrids and herpestids. 1, Europe, Oligocene and Miocene viverrids and herpestids (*Palaeoprionodon, Herpestides, Semigenetta, Viverrictis, Jourdanictis, Leptoplesictis*). 2, Oligocene to Pliocene viverrids and herpestids: 2a, China and Mongolia (*Palaeoprionodon, Semigenetta, Viverra,* undescribed genera from Lufeng, south China); 2b, Bahe, north China (late Miocene herpestid); 2c, Europe, large late Pliocene viverrids (Czechoslovakia: Ivanovce, upper MN15; Italy: Villafranca d'Asti, lower MN16). 3, middle Miocene to Plio-Pleistocene viverrids and herpestids, Siwalik Group. 4, early Miocene to Plio-Pleistocene viverrids and herpestids, Kenya and Tanzania. 5, early Miocene viverrids, Israel and Saudi Arabia. 6, Pliocene viverrids and herpestids, Langebaanweg, south Africa (*Viverra, Genetta, Herpestes*). 7, early Miocene migration of viverrids and herpestids into Africa. 8, Neogene migration of separate viverrid and herpestid stocks to Madagascar. 9, middle Miocene viverrids, Turkey. 10, middle Miocene viverrid, north Africa (Beni Mellal, Morocco).

at Olduvai, Laetoli, and Omo have produced most of these carnivorans (*Helogale, Atilax, Ichneumia, Mungos,* and *Herpestes*), in association with early hominids and other mammals. South African Plio-Pleistocene sites have yielded *Herpestes, Cynictis, Crossarchus, Atilax,* and *Suricata,* only the first two of which extend into the Pliocene (Cooke 1964).

Early Miocene rift sediments in Kenya and Uganda, especially the Rusinga and Songhor localities, have produced the oldest African carnivoran assemblage, including six genera of small aeluroids. Schmidt-Kittler (1987) believes this is an immigrant assemblage, one that entered Africa during the later Oligocene or early Miocene. The oldest aeluroids, which occur at Songhor (~19.6 Ma), are already diverse; they include two hypercarnivores (*Mioprionodon* and a ?stenoplesictine) and two hypocarnivores that may be early herpestids (*Legetetia, Kichechia*). Basicrania of all these genera, however, are unknown. Shortly thereafter, at the Rusinga sites (~17.8 Ma), a definite herpestid (*Leptoplesictis*) and a true viverrid (*Herpestides*) appear. This early Miocene diversity suggests a middle Cenozoic radiation of small aeluroids occurring earlier than previously believed. Because these small aeluroids occur in the faunas at Rusinga and Songhor with other definitive immigrants, such as amphicyonids and mustelids, they may well be immigrants themselves, but their diversity implies a degree of endemism or an arrival earlier than 19.6 Ma, in order to accommodate the differentiation of species observed at these sites.

Dental remains of *Leptoplesictis* in middle Miocene rocks (MN4) represent the oldest European herpestids. Because no complete skulls with basicrania have been found, the auditory region is unknown. These herpestids are small hypercarnivores, little different in tooth form and size from some of the living species of *Herpestes*. Both European and African *Leptoplesictis* are believed to be immigrants derived from southern Asia (Schmidt-Kittler 1987). Herpestids, however, are unknown before the late Miocene in Asia, where they occur in the Siwalik beds of Pakistan (*Herpestes:* Barry et al. 1980; Barry 1985) and in the Bahe fauna of north China (9–10 Ma, MN 10: Qiu 1989); these are the only Asian fossil herpestids, and they are as yet undescribed. Because postcranial remains of pre-Pliocene herpestids are extremely rare, almost nothing is known about their early skeletal structure.

The living herpestids share a number of derived anatomical traits, suggesting their origin from a common ancestral stock: (1) auditory bulla (Stage 3: Hunt 1987) with perbullar path of the internal carotid artery; (2) absence of marker chromosome; (3) specialized ear cartilages to close auditory meatus; (4) anal glands and pouches; and (5) a unique anastomosis of the internal carotid artery in the anterointernal corner of the auditory region (anastomosis Y of Bugge 1978). This complex of features, shared by all living mongooses, a morphologically uniform group, indicates that the African herpestid radiation stems from an ancestor possessing these traits, but as yet we have no fossil older than Pliocene displaying the single osteological feature (the auditory bulla) that might signal the presence of this character complex.

Hyaenidae

The four species of hyaenids living today are confined essentially to Africa, with only the striped hyena ranging to southern Asia. Striped and brown hyenas (*Hyaena*) weigh generally less than 60 kg, and the aardwolf (*Proteles cristatus*) less than 15 kg, but the spotted hyena (*Crocuta crocuta*) may reach a weight of nearly 90 kg. Recent field studies (Kruuk 1972; Frank, this volume) have clarified the role of spotted hyenas as both active predators and scavengers, and it is clear from the diverse fossil record that the family has consistently included a wide range of feeding types.

Living hyaenids (*Hyaena, Crocuta*) can be identified by their robust crushing premolars, elongate upper carnassial (with prominent parastyle), bladelike lower carnassial (metaconid lacking in *Crocuta,* but present in *Hyaena*), and loss of M2/m2. M1/ is greatly reduced, serving only as a stop for the lower carnassial. These dental traits are obscured (if they ever existed) in the highly reduced dentition of the aardwolf, as a result of its myrmecophagy.

The hyaenid basicranium, particularly the distinctive auditory bulla, has remained anatomically conservative since the middle Miocene (Hunt and Solounias 1991). Although first believed to contain only a single chamber (Flower 1869), the unusual hyaenid bulla (Hunt 1974a, 1987; Hunt and Solounias 1991) was discovered by Pocock (1916) to be double-chambered. In *Crocuta* and *Hyaena,* the bulla displays a unique ontogenetic growth pattern (Figure 15.10) in which the anterior (ectotympanic) chamber grows backward beneath a much smaller posterior chamber (*Proteles,* however, does not show the backward growth of the ectotympanic chamber, instead maintaining the anterior-posterior alignment of the chambers without any overgrowth of one chamber by the other). Almost all lineages of fossil hyaenids include at least one skull demonstrating the association of dentition with the unique hyaenid bulla pattern; hence the broad adaptive radiation of the family in the Miocene and Plio-Pleistocene is well documented by both basicranial and dental evidence.

An abundant fossil record of hyaenids stands in marked contrast to their present low diversity (Werdelin and Solounias 1991). Two major hyaenid assemblages can be recognized in time: (1) 5 to ~18 Ma, MN4b–13, early, middle, and late Miocene hyaenids (Figures 15.12, 15.13), a highly diverse radiation of Old World species (with good fossil samples in the Eurasian Vallesian-Turolian interval), which occupied bone-crushing, predatory, and generalized niches, paralleling in range of body size, dentition, and skull shape the New World canids of the same time interval; and (2) 0 to 5 Ma, MN14–17, MQ1–2, Plio-Pleistocene hyaenids (Figure 15.14), in essence, the living genera (extending back to about 3 Ma), which coexist with archaic bone-crushing (*Pachycrocuta*) and hunting hyaenids (the *Euryboas-Chasmaporthetes* group). Basicranial structure is important to the demonstration of parallel hyaenid-canid radiations in the Old and New World.

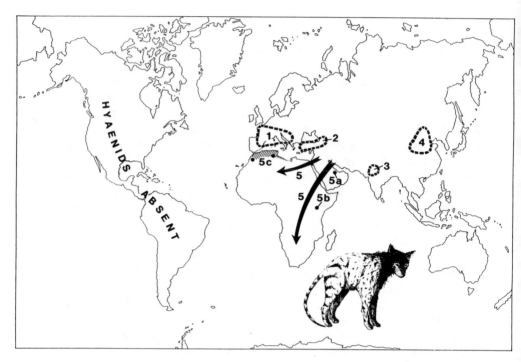

Figure 15.12. Geographic distribution of early and middle Miocene fossil hyaenids. 1, Europe (*Protictitherium, Percrocuta, Miohyaena*). 2, Turkey and Georgia (*Protictitherium, Percrocuta, Miohyaena*). 3, Siwalik Group (*Percrocuta*). 4, northeast China (*Protictitherium, Percrocuta, Miohyaena*). 5, middle Miocene migration of hyaenids into Africa about 14 Ma (MN6/MN7 boundary): 5a, Saudi Arabia (*Percrocuta*); 5b, Ft. Ternan, Kenya (*Percrocuta, Protictitherium*); 5c, Morocco and Tunisia (*Protictitherium* and large hyaenid). The oldest African fossil hyaenids are from Ft. Ternan (Kenya), Al Jadidah (Saudi Arabia), and Beni Mellal (Morocco); these faunas correlate with the MN6/MN7 boundary or within European zone MN7.

Prior to 15 Ma, hyaenid fossils are rare, and the family appears to have been represented by two main branches (the diphyly of Chen and Schmidt-Kittler 1983, demonstrated on milk teeth): the *Percrocuta* species group and the central hyaenid stock (*Protictitherium*). The *Percrocuta* group is thought to have arisen from stenoplesictine aeluroids (Chen and Schmidt-Kittler 1983), whereas the origin of the protictitheres is uncertain. The structure and ontogenetic growth pattern of the auditory bulla (Figure 15.10) unites the *Percrocuta* group and the protictitheres in a monophyletic Hyaenidae (for a different view, see Werdelin and Solounias 1991).

Before ~18 Ma, no hyaenids can be identified in the fossil record. Hyaenids are found first in Europe, as rare dental remains from zones MN4b and MN5 in France (Ginsburg and Bulot 1982; Mein 1958; Stehlin 1925). Although not abundant, hyaenids occur in European Miocene to Pleistocene deposits: five different lineages are present in the Pleistocene (*Crocuta, Hyaena, Pachycrocuta* [two spp.], and the *Euryboas-Chasmaporthetes* group).

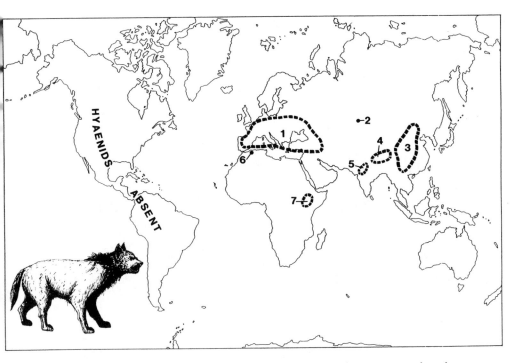

Figure 15.13. Geographic distribution of late Miocene fossil hyaenids. 1, Europe and southwest Asia (*Ictitherium, Thalassictis, Adcrocuta, Percrocuta, Miohyaena, Protictitherium, Plioviverrops, Lycyaena*). 2, central Asia (Pavlodar). 3, eastern China (*Ictitherium, Thalassictis, Adcrocuta, Palinhyaena*). 4, Tibet (Gyirong, Bulong). 5, Siwalik Group (*Percrocuta, Adcrocuta, Lycyaena*). 6, Algeria (Bou Hanifia). 7, Kenya (Lothagam, Baringo).

The first record of African hyaenids occurs at Fort Ternan in Kenya, where both *Protictitherium* and *Percrocuta* are dated at ~14 Ma (MN6/MN7 boundary: Schmidt-Kittler 1987). *Percrocuta* is also reported from the Al Jadidah fauna of Saudi Arabia, now placed at about the MN6/MN7 boundary (Savage 1989). Both *Percrocuta* and protictitheres occur from ~11–14 Ma in northern and eastern Africa at several sites (Beni Mellal, Beglia, Bou Hanifia, Fort Ternan, and Kabarsero). In Africa, diverse hyaenids appear for the first time at ~5 Ma, at or near the beginning of the Pliocene (Langebaanweg, Sahabi, Klein Zee, and Kanapoi).

The earliest record of an Asian hyaenid was believed to be a maxilla of *Protictitherium* from eastern China, probably MN4 or MN5 (Li et al. 1983), but Qiu (1989) reports that this record is in error and the maxilla should be referred to the viverrid *Semigenetta*. In Asia, the Turkish sites at Pasalar (early MN6) and Candir (late MN6) produced early records of both *Percrocuta* and protictitheres (14–15.5 Ma), and in north-central China the same two groups occur with *Miohyaena* in the Tongxin basin (MN6: Guan 1988; Harrison et al. 1991). In the Siwaliks, rare hyaenids from the Chinji levels could be as old

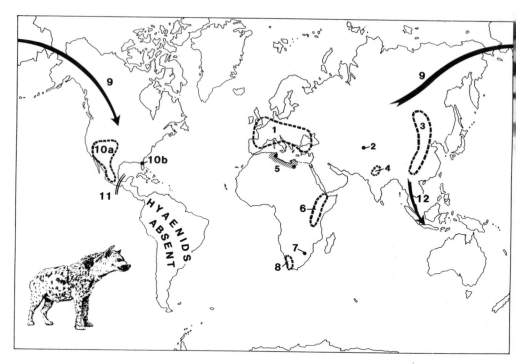

Figure 15.14. Geographic distribution of Pliocene fossil hyaenids. 1, Europe and southwest Asia (*Chasmaporthetes, Pachycrocuta*). 2, Tadjikistan (Kuruksay). 3, eastern Asia (China and Baikal: *Chasmaporthetes, Pachycrocuta*). 4, Siwalik Group. 5, north Africa (Ain Brimba, Sahabi: *Chasmaporthetes, Pachycrocuta*). 6, Ethiopia, Kenya, Tanzania (*Pachycrocuta, Crocuta, Hyaena*). 7, south Africa (Sterkfontein 4, Makapansgat 3: *Chasmaporthetes, Pachycrocuta, Crocuta, Hyaena*). 8, south Africa (Kleinzee, Langebaanweg: *Hyaena, Chasmaporthetes,* "Hyaena" *namaquense*). 9, Pliocene migration of *Chasmaporthetes* to North America about 3.5 Ma. 10, North America (*Chasmaporthetes*): 10a, southwestern United States and Mexico (middle and late Pliocene, early Pleistocene); 10b, Florida (late Pliocene, early Pleistocene). 11, southern limit of fossil hyaenids in the New World. 12, Pleistocene migration of hyaenids (*Pachycrocuta*) to Java (Djetis, Sangiran).

as 13–14 Ma (13.2 Ma, the oldest Siwalik hyaenids: Barry and Flynn 1989). Just one basicranium is associated with any of these early occurrences: the holotype of *Percrocuta primordialis,* a skull with hyaenid auditory structure from Yinziling in the Tongxin basin (MN6: Qiu et al. 1988b). This specimen has recently been placed in a new genus, *Tongxinictis,* by Werdelin and Solounias (1991).

Collectively, these records demonstrate that the early hyaenids were small carnivorans widely distributed in Eurasia and Africa by 14–15 Ma, in the middle Miocene (MN6).

Three genera of hyaenids are at present recognized in Eurasia in the MN4–MN8 interval: *Protictitherium, Percrocuta,* and *Miohyaena.* The dentition of *Protictitherium* is the least specialized of these middle Miocene hyaenids: most fossils of the genus are distributed from western Europe to Asia Minor. In contrast, the teeth of *Percrocuta* and *Miohyaena,* with their enlarged premolars and reduced molars, are already specialized for durophagous feeding. By MN6 these two genera exist across Eurasia. By the later middle Miocene (MN8), Eurasian hyaenids were remarkably diverse, as evidenced by the association of small, foxlike *Tungurictis* and large, bone-crushing *Percrocuta* species in the Tung Gur Formation of Inner Mongolia and in deposits of equivalent age elsewhere in Eurasia.

Diversity of body size was much greater in extinct hyaenids than it is in the living species. *Plioviverrops* from Samos and Pikermi was the size of a small fox, weighing ~2 kg; giant predaceous hyaenids (to 150 kg) evolved in the Plio-Pleistocene (*Pachycrocuta brevirostris*) and late Miocene of Eurasia (*Dinocrocuta gigantea*), the latter known to have had an upper carnassial length of over 5 cm and a basal skull length of 32 cm (Qiu, Xie, and Yan 1988a)! Dental diversity paralleled this range in body size. In early hyaenids, the two upper and two lower molars are present, the carnassials are not as exaggerated in either length or shearing structure, and the premolars are less robust, less inflated. From such a generalized dental pattern, bone-crushing hyaenids repeatedly evolve in the middle to late Cenozoic (*Percrocuta, Adcrocuta, Pachycrocuta, Dinocrocuta, Crocuta*), distinguished by the same suite of dental traits: robust crushing premolars, enormously elongated carnassials specialized for shear, and reduction of the rear molars (M2/m2 lost). As seen in living *Crocuta,* the bone-crushers represent the repeated evolution of an opportunistic feeding strategy: actively hunting prey and/or scavenging carcasses, breaking bones with their greatly enlarged premolars to exploit a nutrient source largely denied to other carnivorans. The evolution of similar canid ecomorphs (*Enhydrocyon, Osteoborus, Borophagus, Epicyon*) in North America lends support to this view. Alongside the bone-crushing hyaenids were generalized (*Ictitherium, Thalassictis*) and more predatory species (the *Euryboas-Chasmaporthetes* group), all of them today extinct and none represented by ecological equivalents among living hyaenids.

The hunting hyenas (*Chasmaporthetes*) are the only hyaenids to invade the New World, reaching North America (Fox Canyon, Kansas) ~3.5 Ma. These are Asian immigrants—the genus is known in China in the late Pliocene and early Pleistocene (Galiano and Frailey 1977), and an ancestral species occurs in the early Pliocene of Moldavia. No hyaenids have ever reached South America, notwithstanding the southernmost records of *Chasmaporthetes* in Mexico (Berta 1981).

Although hyaenid postcranials are not plentiful, they are much better represented than are those of viverrids and herpestids, and demonstrate that many lineages exhibit normally proportioned limbs (Gaudry 1862–63; Hunt and Solounias 1991); many species did not evolve the robust, exaggerated forequarters and low hindquarters that distinguish living *Crocuta* and *Hyaena.* The suggested ancestors of the striped hyena (*Hyaena hyaena*), found at Makapansgat (3 Ma) and Langebaanweg (5 Ma) in South Africa, retain normal limb proportions (Hendey 1974).

Postcranial skeletons of four genera (*Protictitherium, Ictitherium, Adcrocuta,* and *Plioviverrops*) of middle and late Miocene hyaenids demonstrate that digitigrady and a cursorial gait were common attributes of Miocene hyaenids, and that a digitigrade stance must have been achieved early in the history of the family (Hunt and Solounias 1991).

Felidae-Nimravidae

Felids are commonly considered the true cats of the Cenozoic, whereas nimravids are seen as catlike ecomorphs that parallel the felids in dental and cranial/postcranial osteology to a striking degree. The nimravids, however, can be shown from basicranial structure to be only distantly related to the felids (Neff 1983; Hunt 1987).

All fossil and living felids are short-faced predators with prominent biting canines and a tendency toward reduction and loss of teeth in front of and behind the carnassial pair, often with modification of the carnassials into simple shearing blades. It is this dental/cranial morph that is usually central to the popular concept of a "cat." In the postcranial skeleton, retractile claws accompany digitigrade feet, and the forelimbs retain a strong ability for pronation-supination as an aid in grasping prey. The felid basicranial and bulla structure is highly diagnostic in having a double-chambered aeluroid bulla with *septum bullae,* strong ontogenetic overgrowth of the anterior chamber of the bulla by the posterior (caudal entotympanic) chamber (Figure 15.10), and the bradynothictic pattern of ectotympanic-rostral entotympanic development during bulla ontogeny (Hunt 1987). Felids strongly reduce the internal carotid and basilar arterial blood supply to the brain, and utilize instead the external carotid route, commonly developing arterial retia along the orbital course of this artery to cool blood flowing to the brain.

In a superb survey of the living large felids, Neff summarizes the evolution of felid and nimravid carnivorans, setting her account against the artwork of Guy Coheleach (Neff and Coheleach 1982). Among the placental carnivorans, the nimravids and felids represent two entirely independent yet parallel Cenozoic radiations of catlike mammals (Figures 15.15–17). Nimravids, confined to the late Eocene to late Miocene, precede the felids in the development of their radiation, whereas felids diversify in the late Miocene to Pleistocene, an event that continues to be reflected today in their broad geographic distribution and rich species diversity. At first appearance, the nimravids are Holarctic in distribution; the earliest felids are confined to Eurasia.

Catlike parallels among the mammals evolve independently a number of times among both the placental (creodonts, carnivorans) and the marsupial (borhyaenids) carnivores. The earliest catlike mammals are Eocene hyaenodont creodonts (*Apataelurus:* Scott 1938; *Machaeroides:* Matthew 1909; Dawson et al. 1986) from western North America. Although known primarily

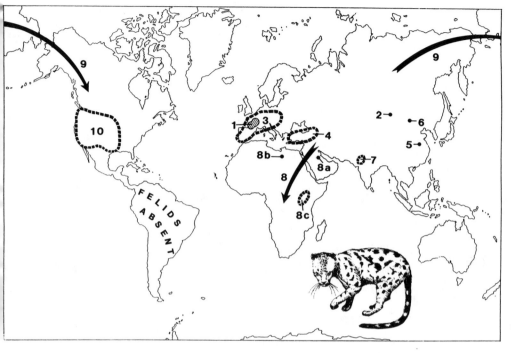

Figure 15.15. Geographic distribution of late Eocene to middle Miocene felids. 1, late Eocene to early Miocene, France (*Haplogale, Proailurus*). 2, early Oligocene, Hsanda Gol, Mongolia (proailurine). 3, early and middle Miocene, Europe (*Pseudaelurus*). 4, middle Miocene, Turkey and Georgia (Belometcheskaya, *Pseudaelurus*). 5, early Miocene, China (Xiacaowan, *Pseudaelurus*). 6, middle Miocene, Inner Mongolia (Tung Gur, *Metailurus*, ?machairodont). 7, middle Miocene, Siwalik Group (*Sivasmilus, Sivaelurus, Vishnufelis,* "Vinayakia"). 8, early Miocene migration of felids into Africa about 19.6 Ma: 8a, early Miocene, Saudi Arabia (*Pseudaelurus*); 8b, early Miocene, Libya (felid); 8c, early Miocene, Kenya (*Afrosmilus*). 9, Miocene migration of felids to North America about 16–17 Ma. 10, middle Miocene, western North America (*Pseudaelurus*).

from only a few jaws, the group is best represented by a skull, mandibles, and associated postcrania from the Bridger Eocene of Wyoming (Gazin 1946).

These creodonts are succeeded in time by the Oligocene and Miocene radiation of nimravid cats of the northern continents (Cope 1880; Neff 1983). Skeletons of nimravids, many essentially complete, are often found preserved in the White River Group of the North American midcontinent and in the John Day beds of Oregon. Nimravids are largely extinct by the end of the early Miocene, but a single subfamily, the Barbourofelinae, persists in Eurasia (*Sansanosmilus*) and North America (*Barbourofelis*) until the late Miocene. Only a single nimravid (*Syrtosmilus*) reached Africa (Ginsburg 1978), the only record of the family on the southern continents. *Syrtosmilus* occurs in the Gebel Zelten fauna, Libya, in the late early Miocene, MN3b or MN4a (Savage 1989). There is no evidence that nimravids penetrated farther south into cen-

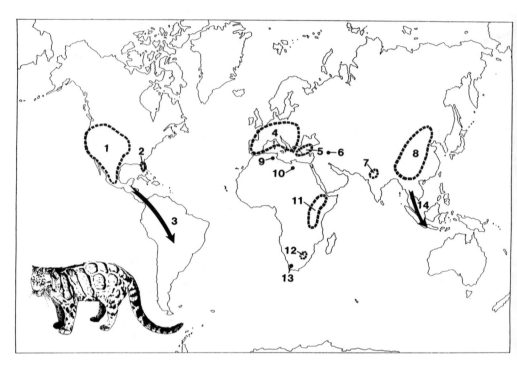

Figure 15.16. Geographic distribution of late Miocene and Pliocene felids. 1, western North America (*Homotherium, Machairodus, Megantereon, Nimravides, Dinofelis, Panthera, Felis, Lynx,* "*Acinonyx*"). 2, Florida (*Homotherium, Machairodus, Megantereon, Nimravides, Smilodon, Lynx*). 3, late Pliocene migration of felids to South America about 1.9–2.0 Ma (*Felis, Smilodon*). 4, Europe (*Homotherium, Machairodus, Megantereon, Dinofelis, Panthera, Felis, Lynx, Acinonyx, Metailurus, Pontosmilus* (=*Paramachairodus*). 5, Turkey (*Megantereon, Machairodus, Miomachairodus, Metailurus, Felis*). 6, Iran (Maragheh: *Machairodus, Pontosmilus, Metailurus, Felis*). 7, Siwalik Group (*Machairodus, Megantereon, ?Homotherium, Panthera, Felis,* "Propontosmilus," "Sivafelis," "Aeluropsis"). 8, eastern China and Tibet (*Homotherium, Machairodus, Miomachairodus, Megantereon, Dinofelis, Felis, Lynx, Acinonyx, Metailurus, Pontosmilus*). 9, Tunisia (*Machairodus*). 10, Libya (*Machairodus*). 11, East Africa (*Homotherium, Machairodus, Megantereon, Dinofelis, Panthera, Felis, Acinonyx*). 12, south Africa (Sterkfontein 4 and Makapansgat 3: *Homotherium, Megantereon, Dinofelis, Panthera, Felis, Acinonyx*). 13, south Africa (Langebaanweg: *Homotherium, Machairodus, Dinofelis, Felis*). 14, Pleistocene migration of *Homotherium* and *Megantereon* to Java.

tral or southern Africa, and nimravids are extinct worldwide by the beginning of the Pliocene.

Although the earliest true felids (*Proailurus, Haplogale:* Ginsburg 1983) exist in Europe and Asia alongside the nimravids during the Oligocene and early Miocene, they are not diverse, and it is not until the middle Miocene decline in nimravid diversity that felids begin their radiation, culminating in the great Plio-Pleistocene sabertooths (*Machairodus, Homotherium, Megantereon, Smilodon*) and other true cats with normal canines (*Felis, Panthera, Metailurus, Acinonyx*).

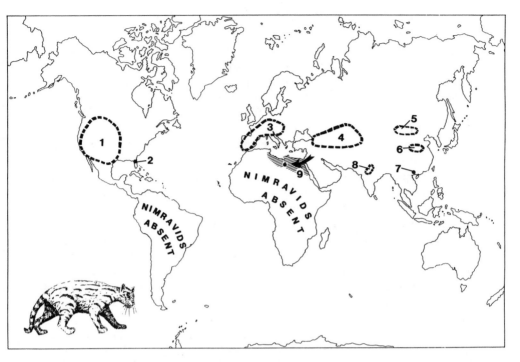

Figure 15.17. Geographic distribution of late Eocene to late Miocene fossil nimravids. 1, western North America, late Eocene to late Miocene (*Dinictis, Hoplophoneus, Nimravus, Eusmilus, Dinaelurus, Barbourofelis*). 2, Florida, late Miocene (*Barbourofelis*). 3, Europe, late Eocene to late Miocene (late Eocene to Oligocene: *Nimravus, Eusmilus, Dinailuristis, Quercylurus, Eofelis*; Miocene, 10.6–18 Ma: *Prosansanosmilus, Sansanosmilus*). 4, southwest Asia, Oligocene nimravids (Benara, Aral Sea, Askazansor). 5, Mongolia, Oligocene (Hsanda Gol, Khoer-dzan: *Nimravus*). 6, north China, middle Miocene (*Sansanosmilus*). 7, south China, ?Eocene (Bose Basin). 8, Siwalik Group, middle to late Miocene (7.4 to >15.1 Ma). 9, early Miocene migration of nimravids to north Africa (Libya: *Syrtosmilus*).

True felids are not found in the Oligocene of North America, but first appear in the Oligocene of Europe (Quercy) and Asia (Hsanda Gol). These earliest felids, the proailurines, are small, none larger than a lynx. They spread to North America ~16–17 Ma, in the late early Miocene or early middle Miocene, when a proailurine first appears at Ginn Quarry in western Nebraska. This is the earliest record of a New World felid (Hunt 1987). In the middle Miocene of North America, the descendants of the proailurines, the pseudaelurine felids (*Pseudaelurus*), do not exceed the size of living lynx and leopards and are not diverse. By the late Miocene, North American and Eurasian felids display a wider range in size and diversity, heralding the Plio-Pleistocene radiation of the family. Felids of large size do not appear and diversify until the extinction of the great amphicyonids and hemicyonine ursids that dominate the middle Miocene and early late Miocene faunas of Holarctica.

True felids migrate to Africa in the early Miocene. Sediments of the East

African rift system yield the oldest African felids (*Afrosmilus*), at Songhor (MN3, ~19.6 Ma) and Rusinga (MN4, ~17.8 Ma in Schmidt-Kittler 1987; 17.5–18.0 Ma in Savage 1989; Drake et al. 1988). Because of its incipient sabertooth features, *Afrosmilus* differs sufficiently from the typical Holarctic Miocene felid *Pseudaelurus* to suggest a separate lineage. A species of *Pseudaelurus* is also recorded in the Dam Formation of Saudi Arabia, at the As-Sarrar locality (Thomas et al. 1982; Savage 1989) at about MN4a. These two felid genera at MN4 sites in Africa and Saudi Arabia indicate that at least two lineages are present at the earliest appearance of the family on that continent in the early Miocene.

The earliest record of felids in South America occurs in the Vorohue Formation of Argentina, where fossils of two genera (*Felis vorohuensis, Smilodon*) appear as part of a dispersal event in the late Pliocene, ~1.9–2.4 Ma (Marshall 1985; Marshall et al. 1984; Marshall and Sempere 1993). *Felis vorohuensis,* a small cat represented by upper and lower jaw fragments and a partial post-cranial skeleton, is regarded as the oldest certainly determined felid in South America (Berta 1983). This fossil species seems to be much like the living small Pampas cat, *Felis colocolo.* The reported occurrence of *Smilodon* in the Vorohue Formation is disputed (Berta 1985:2); the material in question seems to have come from younger Ensenadan-age sediments.

By the Ensenadan mammal age, the puma (*Felis concolor*), jaguar (*F. onca*), and sabercat (*Smilodon*) are all present in South America (Marshall et al. 1984:20). Among these large cats, only *Smilodon* becomes extinct prior to the Holocene. The extinction of the marsupial borhyaenid carnivores by the end of the Chapadmalalan age, succeeded in the immediately overlying deposits by the arrival of true felids, is considered evidence of ecologic replacement of marsupial by placental carnivores (Marshall 1978a; Berta 1988). The absence of felids in South America for most of the Cenozoic led to the evolution among the carnivorous borhyaenids of a marsupial sabertooth cat, *Thylacosmilus* (Huayquerian and Montehermosan mammal ages), closely paralleling in cranial anatomy the sabertooths of the northern continents (Riggs 1934; Marshall 1976; Turnbull and Segall 1984).

The evolutionary radiation of felids has taken place primarily over the last 10 million years. During that interval (late Miocene to Recent), felid cats with saber canines evolved, paralleling the earlier nimravid sabertooth cats and sabertooth marsupials. Sabertooth cats, whether marsupial or placental, share in a number of osteological traits that suggest the different lineages are ecomorphs: all possess elongate, bladelike canines, shearing cheek teeth, short cranium with developed mastoid and occipital regions, massive neck and fore-quarters, and powerful forelimbs with good ability for rotation (supination-pronation) of the lower forelimb. Convergence in these skeletal traits indicates a common sabertooth feeding behavior, one in which the sabers are driven into the prey using particularly strong head and neck musculature, and in which prey capture and manipulation are accomplished through powerful forelimbs and clawed feet. Postcranial skeletons demonstrate that no sabertooth is a cursor; hence these animals must have attacked quickly from cover, and could run for only short distances. Sabertooths persisted nearly to the present: *Homotherium* is known from the late Pleistocene Friesenhahn Cave in Texas (Rawn 1981; Meade 1961), and *Smilodon* was probably living into the early Holocene (Kurtén and Anderson 1980).

Summary

Once the major lineages of the Carnivora are identified and traced through Cenozoic time, a clearer picture of the geographic distribution and evolution of carnivorous mammals emerges. Important to this new biogeography is the sharper definition of the doglike arctoid and cynoid carnivorans: the canids, amphicyonids, and ursids. In earlier mammalian classifications, particularly that of G. G. Simpson in 1945, these lineages were seriously confused, resulting in the mistaken impression that "canids," a waste-basket taxon for the doglike Carnivora, were broadly distributed on the northern continents. Endemism went largely unrecognized and was restricted to the level of genera and species. Through the efforts of French and American paleontologists in the 1960s, the defining features of true canids (primarily basicranial) were recognized for the first time, and it then became apparent that the family had been geographically restricted to the New World until the late Miocene, immigrating to Eurasia only about 9 million years ago. The numerous species of doglike carnivorans of the Old World were in fact not canids, but amphicyonids and ursids. The ursids and amphicyonids, together with the durophagous and carnivorous hyaenids, filled the niches in the Old World so thoroughly exploited by canids in the Americas. Indeed, the endemism of doglike carnivorans at the level of the family was a key aspect of Holarctic carnivoran biogeography unappreciated by Simpson and his contemporaries.

Similarly, the study of basicranial structure has resolved and better defined the limits of the aeluroid families, thus contributing to clarification of aeluroid biogeography. In contrast to arctoids, the aeluroids remain largely confined to the Old World throughout the Cenozoic. Only felids successfully colonize the New World, entering North America ~16–17 Ma and migrating to South America within the last 2 million years. Despite their marked diversity and broad distribution in the Old World, only a single hyaenid species penetrated the Americas, and is limited to the southern United States and northern Mexico during the Pliocene and early Pleistocene. The small aeluroids, the herpestids and viverrids, are confined to Eurasia and Africa throughout their history and failed to invade the New World.

The dominance of placental creodonts and absence of carnivorans in the Oligocene of Africa, and the sudden appearance there of Carnivora of several families (both arctoids and aeluroids) in the early Miocene, suggest to most biogeographers that the Carnivora entered Africa as migrants in the earliest Miocene, during the initiation of major faunal exchange between Eurasia and Africa. By the end of the Miocene, creodonts were extinct in Africa. The placental carnivorans radiated rapidly during the Miocene, becoming dominant and widely distributed in the Plio-Pleistocene. The modern African carnivoran fauna is a relict of this late Cenozoic radiation.

In South America during the Cenozoic, the borhyaenid marsupials and phororhacoid birds occupied the principal niches for carnivores, becoming extinct during the 2.0–2.5 Ma interval. Placental carnivores of the arctoid, cynoid, and aeluroid groups entered South America from the north during the great Plio-Pleistocene faunal interchange. Rare arctoids of the *Cyonasua* group reach South America ~7 Ma, followed by invading mustelids, procyonids, canids, felids, and ursids at ~2–2.5 Ma. Hence the modern carnivore fauna of South America is, as in Africa, the product of late Cenozoic tectonic and climatic events.

The Australasian region throughout Cenozoic time was the province of marsupial carnivores, and on present evidence never felt the impact of placental carnivorans until the advent of man.

Carnivoran biogeography reflects the isolation of the continents in the earlier Cenozoic, an isolation created not only by the spatial separation of the continental landmasses, but also by barriers to migration resulting from the invasion of the land by epeiric seas. As continental connections were established in the middle and late Cenozoic (and as epeiric seas withdrew), the migration of placental Carnivora into the southern continents apparently replaced the endemic creodonts of Africa and borhyaenids of South America. These placentals belonged to established and well-defined Holarctic groups: the arctoids, aeluroids, and cynoids.

Less well known is the succession of placental carnivoran faunas on the two northern continents: (1) late Eocene-Oligocene (~24–40 Ma): in Eurasia a hyaenodont-nimravid-arctoid fauna (with sparse aeluroids); in North America a hyaenodont-nimravid-canid fauna (with few arctoids); (2) early Miocene (~18–24 Ma): in Eurasia an arctoid (amphicyonids, small species of ursids, mustelids, and procyonids) and aeluroid (all four of the living families present for the first time) fauna; in North America an arctoid (chiefly an amphicyonid-mustelid assemblage with rare ursid and procyonid species) and canid fauna; (3) middle Miocene (~ 9–18 Ma): in Eurasia an arctoid (large forms are amphicyonids and hemicyonine ursids, small species include mustelids, procyonids, other ursids) and aeluroid fauna; in North America an arctoid (large amphicyonids and hemicyonines, smaller mustelids, procyonids, and rare ursine ursids) and canid fauna (with rare felids); and (4) late Miocene to Recent (~9 Ma to the present): in Eurasia, arctoids and aeluroids, with introduction of canids at 9 Ma; in North America, canids and arctoids (ursids, mustelids, and procyonids), with diverse Plio-Pleistocene felids and a single hyaenid. These faunas of Holarctic carnivorans opportunistically migrated to the southern continents when dispersal was possible: in Africa these successive carnivoran faunas can be recognized in Miocene to Recent assemblages; in South America only the most recent fauna reached that continent because of the absence of an effective land connection prior to the middle Pliocene.

Our perceptions of paleobiogeographic patterns are strongly influenced by the known geographic distribution of fossils (Figures 15.3–9, 15.11–17). The rich carnivoran samples of the Holarctic continents have in recent years provided a reliable framework for new biogeographic inferences concerning middle to late Cenozoic carnivoran distributions. At the higher taxonomic levels these patterns are now confirmed on a global basis. New discoveries from Asia, however, are necessary to complete our knowledge of the Palearctic region, especially with regard to Asia's role as a place of origin for a number of supposedly immigrant groups appearing abruptly in Europe and North America during the late Paleogene and early Neogene.

Emerging evidence from Africa and South America has begun to clarify the timing of faunal transformations on the southern continents during the Cenozoic. In South America, future work can focus profitably on a more detailed resolution of the Plio-Pleistocene chronology of carnivoran immigration during the last 2.5 million years. In Africa, changes in the carnivore component of the faunas from the Paleogene into the Neogene are important and still obscure, and will only be comprehended through new discoveries in the field. The nature of African and Asian carnivore faunas and evolution is perhaps the principal challenge within our newly emergent concept of carnivoran paleobiogeography.

Acknowledgments

This work owes a particular debt to the curators who provided access to collections and made fossils available for this study: R. H. Tedford and M. C. McKenna, American Museum, New York; B. MacFadden and S. D. Webb, Florida State Museum, Gainesville; F. A. Jenkins, Jr., Museum of Comparative Zoology, Cambridge; R. Emry and C. Ray, Smithsonian Institution, Washington; L. G. Barnes and D. Whistler, Los Angeles County Museum, Los Angeles; M. Dawson, Carnegie Museum, Pittsburgh; P. Freeman, University of Nebraska State Museum, Lincoln; A. Gentry, British Museum (Natural History), London; L. Ginsburg and M. Pickford, Muséum National d'Histoire Naturelle, Paris; M. Hugueney and P. Mein, Université de Lyon, Lyon; M. Philippe, Musée Guimet d'Histoire Naturelle, Lyon; J. Hürzeler and B. Engesser, Naturhistorisches Museum, Basel; M. Leakey, National Museum of Kenya, Nairobi; Qiu X.-Z. and Li C.-K., Institute of Vertebrate Paleontology and Paleoanthropology, Beijing.

My thanks to Polly Denham, University of Nebraska, for the restorations of carnivorans in Figures 15.3–9, 15.11–17; to Ted Browne, Bozeman, Montana, for the restoration of the hemicyonine in Figure 15.3; and to Marti Haack for Figure 15.2.

References

Archer, M. 1976. The basicranial region of marsupicarnivores (Marsupialia), interrelationships of carnivorous marsupials, and affinities of the insectivorous marsupial peramelids. *J. Linn. Soc. London (Zool.)* 59(3):217–322.

Baker, M. A. 1979. A brain-cooling system in mammals. *Sci. Amer.* 240(5):130–139.

Baker, M. A. 1980. Anatomical and physiological adaptations of panting mammals to heat and exercise. In: S. M. Horvath and M. K. Yousef, eds. *Environmental Physiology: Aging, Heat and Altitude,* pp. 121–146. New York: Elsevier North Holland.

Baker, M. A. 1982. Brain cooling in endotherms in heat and exercise. *Ann. Rev. Physiol.* 44:85–96.

Barnes, L. G., Domning, D. P., and Ray, C. E. 1985. Status of studies on fossil marine mammals. *Marine Mamm. Sci.* 1(1):15–53.

Barry, J. C. 1985. Les variations des faunes du Miocène moyen et supérieur des formations des Siwaliks au Pakistan. *L'Anthropologie (Paris)* 89(3):268.

Barry, J. C. 1987. Large carnivores (Canidae, Hyaenidae, Felidae) from Laetoli. In: M. D. Leakey and J. M. Harris, eds. *Laetoli: A Pliocene Site in Northern Tanzania,* pp. 235–258. Oxford: Clarendon Press.

Barry, J. C. 1988. *Dissopsalis,* a middle and late Miocene proviverrine creodont (Mammalia) from Pakistan and Kenya. *J. Vert. Paleontol.* 8(1):25–45.

Barry, J. C., Behrensmeyer, A. K., and Monaghan, M. 1980. A geologic and biostratigraphic framework for Miocene sediments near Khaur village, northern Pakistan. *Yale Peabody Mus. Postilla* 183(1–2):1–19.

Barry, J. C., and Flynn, L. J. 1989. Key biostratigraphic events in the Siwalik sequence. In: E. H. Lindsay, V. Fahlbusch, and P. Mein, eds. *European Neogene Mammal Chronology,* pp. 557–571. New York: Plenum Press.

Barry, J. C., Lindsay, E. H., and Jacobs, L. L. 1982. A biostratigraphic zonation of the Middle and Upper Siwaliks of the Potwar Plateau of northern Pakistan. *Palaeogeogr., Palaeoclimatol., Palaeoecol.* 37:112.

Baskin, J. 1982. Tertiary Procyoninae (Mammalia, Carnivora) of North America. *J. Vert. Paleontol.* 2(1):71–93.

Baskin, J. 1989. Comments on New World Tertiary Procyonidae (Mammalia, Carnivora). *J. Vert. Paleontol.* 9(1):110–117.

Beaumont, G. de. 1968. Note sur la région auditive de quelques carnivores. *Archiv. des Sci., Genève* 21(2):213–223.

Beaumont, G. de. 1982. Brèves remarques sur la dentition de certains ursidés (mammifères). *Archiv. des Sci., Genève* 35(2):153–156.

Beaumont, G. de. 1984. Des dents d'*Amphicyon* (mammifère, carnivore, Ursidé) du Turolien basal de Kohfidisch, Burgenland, Autriche. *Archiv. des Sci., Genève* 37(1):75–83.

Berta, A. 1981. The Plio-Pleistocene hyaena *Chasmaporthetes ossifragus* from Florida. *J. Vert. Paleontol.* 1(3–4):341–356.

Berta, A. 1983. A new species of small cat (Felidae), from the late Pliocene–early Pleistocene (Uquian) of Argentina. *J. Mammal.* 64(4):720–725.

Berta, A. 1985. The status of *Smilodon* in North and South America. *Contrib. Sci., Nat. Hist. Mus., Los Angeles Co.* 370:1–15.

Berta, A. 1988. Quaternary evolution and biogeography of the large South American Canidae (Mammalia: Carnivora). *Univ. California Publ. Geol. Sci.* 132:1–149.

Bien, M., and Chia, L. 1938. Cave and rock-shelter deposits in Yunnan. *Bull. Geol. Soc. China* 18(4):325–347.

Boaz, N., Gaziry, A., and El-Arnauti, A. 1979. New fossil finds from the Libyan Upper Neogene site of Sahabi. *Nature* 280:137–140.

Bonis, L. de. 1981. Contribution à l'étude du genre *Palaeogale* Meyer (Mammalia, Carnivora). *Ann. Paléontol.* 67(1):37–56.

Bonis, L. de, and Guinot, Y. 1987. Le gisement de vertébrés de Thézels (Lot) et la limite Oligo-Miocène dans les formations continentales du bassin d'Aquitaine. *München Geowiss. Abhandl.* 10:49–58.

Bryant, H. N. 1993. Carnivora and Creodonta of the Calf Creek Local Fauna (late Eocene, Chadronian), Cypress Hills Formation, Saskatchewan. *J. Paleontol.* 67(6):1032–1046.

Bugge, J. 1978. The cephalic arterial system in carnivores, with special reference to the systematic classification. *Acta Anat.* 101:45–61.

Chen, G., and Schmidt-Kittler, N. 1983. The deciduous dentition of *Percrocuta* Kretzoi and the diphyletic origin of the hyaenas (Carnivora, Mammalia). *Paläont. Zeitschr.* 57:159–169.

Cirot, E., and Bonis, L. de. 1992. Revision du genre *Amphicynodon,* carnivore de l'Oligocène. *Palaeontographica* (Abt. A) 220:103–130.

Cirot, E., and Bonis, L. de. 1993. Le crâne d'*Amphictis ambiguus* (Carnivora, Mammalia): Son importance pour la compréhension de la phylogénie des mustéloïdes. *C. R. Acad. Sci. Paris* (ser. II) 316:1327–1333.

Clark, J., and Guensburg, T. 1972. Arctoid genetic characters as related to the genus *Parictis. Fieldiana: Geol.* 26(1):1–71.

Cooke, H. B. S. 1964. The Pleistocene environment in southern Africa. In: D. H. S. Davis, ed. *Ecological Studies in Southern Africa,* pp. 1–23. The Hague: Junk.

Cope, E. D. 1880. On the extinct cats of America. *Amer. Nat.* 14(12):833–858.

Corvinus, G., and Hendey, Q. B. 1978. A new Miocene vertebrate locality at Arrisdrift in South West Africa (Namibia). *Neue Jahrb. Geol. Paläont. Monatshefte,* Heft 4:193–205.

Crusafont-Pairo, M. 1950. El primer representante del genera "Canis" en el pontiense Eurasiatico ("*Canis*" *cipio* nova sp.). *Bolet. Real Soc. Espan. Hist. Natur.* 48(1):43–51.

Davis, D. D. 1964. The Giant Panda: A morphological study of evolutionary mechanisms. *Fieldiana: Zool., Mem.* 3:1–339.

Davis, D. D., and Story, H. E. 1943. The carotid circulation in the domestic cat. *Zool. Ser., Field Mus. Nat. Hist.* 28:1–47.

Dawson, M. R., Stucky, R. K., and Krishtalka, L. 1986. *Machaeroides simpsoni,* new species, oldest known sabertooth creodont (Mammalia), of the Lost Cabin Eocene. *Contrib. Geol., Univ. Wyoming,* Spec. Paper 3:177–182.

Dehm, R. 1950. Die Raubtiere aus dem Mittel-Miocän (Burdigalium) von Wintershof-West bei Eichstätt in Bayern. *Abhandl. Bayerisch. Akad. d. Wiss. (Math.-naturwiss. Klasse),* N.F., Heft 58:1–141.

Denison, R. H. 1938. The broad-skulled Pseudocreodi. *Ann. New York Acad. Sci.* 37:163–257.

Drake, R., Van Couvering, J., Pickford, M., Curtis, G., and Harris, J. 1988. New chronology for the Early Miocene mammalian faunas of Kisingiri, Western Kenya. *J. Geol. Soc. London* 145:479–491.

Emry, R. J. 1992. Mammalian range zones in the Chadronian White River Formation at Flagstaff Rim, Wyoming. In: D. Prothero and W. A. Berggren, eds. *Eocene-Oligocene Climatic and Biotic Evolution,* pp. 106–115. Princeton, N.J.: Princeton Univ. Press.

Fahlbusch, V., Qiu Z., and Storch, G. 1983. Neogene mammalian faunas of Ertemte and Harr Obo in Nei Mongol, China. *Scientia Sinica* (Ser. B) 26(2):205–224.

Fay, F. H., Rausch, V. R., and Feltz, E. T. 1967. Cytogenetic comparison of some pinnipeds (Mammalia: Eutheria). *Canadian J. Zool.* 45:773–778.

Fejfar, O. 1989. The Neogene VP sites of Czechoslovakia. In: E. H. Lindsay, V. Fahlbusch, and P. Mein, eds. *European Neogene Mammal Chronology,* pp. 211–236. New York: Plenum Press.

Flower, W. H. 1869. On the value of the characters of the base of the cranium in the classification of the order Carnivora. *Proc. Zool. Soc. London* 1869:4–37.

Frackowiak, H. 1989. Das Rete mirabile der Arteria maxillaris des Löwen (*Panthera leo* L. 1758). *Anat. Histol. Embryol.* 18:342–348.

Galiano, H., and Frailey, D. 1977. *Chasmaporthetes kani,* new species from China, with remarks on phylogenetic relationships of genera within the Hyaenidae (Mammalia, Carnivora). *Amer. Mus. Novit.* 2632:1–16.

Gaudry, A. 1862–63. *Animaux fossiles et géologie de l'Attique.* Carnivores, pp. 37–120. Paris.

Gazin, C. L. 1946. *Machaeroides eothen* Matthew, the sabertooth creodont of the Bridger Eocene. *Proc. U.S. Natl. Mus.* 96:335–347.

Gingerich, P., and Deutsch, H. 1989. Systematics and evolution of early Eocene Hyaenodontidae (Mammalia, Creodonta) in the Clark's Fork Basin, Wyoming. *Contrib. Mus. Paleontol., Univ. Michigan* 27(13):327–391.

Gingerich, P., and Sahni, A. 1979. *Indraloris* and *Sivaladapis:* Miocene adapid primates from the Siwaliks of India and Pakistan. *Nature* 279(5712):415–416.

Ginsburg, L. 1977. Les carnivores du Miocène de Beni Mellal (Maroc). *Géol. Mediterr.* 4(3):225–240.

Ginsburg, L. 1978. *Syrtosmilus syrtensis,* n. gen., n. sp., Félidé machairodontiforme du Burdigalien de Libye. *C. R. somm. Soc. Géol. France,* 1978, 2:73–74.

Ginsburg, L. 1980. *Hyainailouros sulzeri,* Mammifère créodonte du Miocène d'Europe. *Ann. Paléontol. (Vert.)* 66(1):19–73.

Ginsburg, L. 1982. Sur la position systématique du petit panda, *Ailurus fulgens* (Carnivora, Mammalia). *Géobios, Spéc. Mém.* 6:247–258.

Ginsburg, L. 1983. Sur les modalités d'évolution du genre Néogène *Pseudaelurus* Gervais (Felidae, Carnivora, Mammalia). In: J. Chaline, ed. *Modalités, rythmes, mécanismes de l'évolution biologique: Gradualisme phylétique ou équilibres ponctués? CNRS Colloq.* 330:131–136.

Ginsburg, L., and Bulot, C. 1982. Les carnivores du Miocène de Bézian pres de la Romieu (Gers, France). *Proc. K. Nederland Akad. Wet.* (ser. B) 85(1):53–76.

Ginsburg, L., and Morales, J. 1992. Contribution à la connaissance des Mustélidés (Carnivora, Mammalia) du Miocène d'Europe *Trochictis* et *Ischyrictis,* genres affines et genres nouveaux. *C. R. Acad. Sci. Paris* (ser. II) 315:111–116.

Ginsburg, L., Le van Minh, Kieu Qui Nam, and Din van Thuan. 1992. Premières découvertes des vertébrés continentaux dans le Néogène du Nord du Vietnam. *C. R. Acad. Sci. Paris* (ser. II) 314:627–630.

Gregory, W. K. 1936. On the phylogenetic relationships of the giant panda (*Ailuropoda*) to other arctoid Carnivora. *Amer. Mus. Novit.* 878:1–29.

Guan, J. 1988. The Miocene strata and mammals from Tongxin, Ningxia and Guanghe, Gansu. *Mem. Beijing Nat. Hist. Mus.* 42:1–21.

Gunnell, G., and Gingerich, P. 1991. Systematics and evolution of late Paleocene and early Eocene Oxyaenidae (Mammalia, Creodonta) in the Clark's Fork Basin, Wyoming. *Contrib. Mus. Paleontol., Univ. Michigan* 28(7):141–180.

Gustafson, E. P. 1986. Carnivorous mammals of the Late Eocene and Early Oligocene of Trans-Pecos Texas. *Texas Memorial Mus. Bull.* 33:1–66.

Harrison, T., Delson, E., and Guan, J. 1991. A new species of *Pliopithecus* from the middle Miocene of China and its implications for early catarrhine zoogeography. *J. Human Evol.* 21:329–361.

Hendey, Q. B. 1974. The late Cenozoic Carnivora of the South-Western Cape Province. *Ann. South African Mus.* 63:1–369.

Hendey, Q. B. 1978. Preliminary report on the Miocene vertebrates from Arrisdrift, South West Africa. *Ann. South African Mus.* 76(1):1–41.

Hendey, Q. B. 1980. *Agriotherium* (Mammalia, Ursidae) from Langebaanweg, South Africa, and relationships of the genus. *Ann. South African Mus.* 81(1):1–109.

Hershkovitz, P. 1972. The Recent mammals of the Neotropical region: A zoogeographic and ecological review. In: A. Keast, F. Erk, and B. Glass, eds. *Evolution, Mammals, and Southern Continents,* pp. 311–431. Albany: State Univ. New York Press.

Hill, A., Drake, R., Tauxe, L., Monaghan, M., Barry, J. C., Behrensmeyer, A. K., Curtis, G., Jacobs, B. F., Jacobs, L., Johnson, N., and Pilbeam, D. 1985. Neogene paleontology and geochronology of the Baringo Basin, Kenya. *J. Human Evolution* 14:768.

Hopwood, A., and Hollyfield, J. 1954. An annotated bibliography of the fossil mammals of Africa (1742–1950). *Fossil Mammals of Africa No. 8, British Mus. Nat. Hist.,* p. 136.

Hough, J. R. 1944. The auditory region in some Miocene carnivores. *J. Paleontol.* 18(5):470–479.

Hough, J. R. 1948. The auditory region in some members of the Procyonidae, Canidae, and Ursidae. *Bull. Amer. Mus. Nat. Hist.* 92(2):67–118.

Hunt, R. M., Jr. 1974a. The auditory bulla in Carnivora: An anatomical basis for reappraisal of carnivore evolution. *J. Morphol.* 143(1):21–76.

Hunt, R. M., Jr. 1974b. *Daphoenictis,* a cat-like carnivore (Mammalia, Amphicyonidae) from the Oligocene of North America. *J. Paleontol.* 48(5):1030–1047.

Hunt, R. M., Jr. 1977. Basicranial anatomy of *Cynelos* Jourdan (Mammalia, Carnivora), an Aquitanian amphicyonid from the Allier Basin, France. *J. Paleontol.* 51(4):826–843.

Hunt, R. M., Jr. 1987. Evolution of the aeluroid Carnivora: Significance of auditory structure in the nimravid cat *Dinictis. Amer. Mus. Novit.* 2886:1–74.

Hunt, R. M., Jr. 1989. Evolution of the aeluroid Carnivora: Significance of the ventral promontorial process of the petrosal, and the origin of basicranial patterns in the living families. *Amer. Mus. Novit.* 2930:1–32.

Hunt, R. M., Jr. 1990. Vascular countercurrent cooling mechanisms in Carnivora. *J. Vert. Paleontol.* 10(3, suppl.):28A.

Hunt, R. M., Jr. 1991. Evolution of the aeluroid Carnivora: Viverrid affinities of the Miocene carnivoran *Herpestides. Amer. Mus. Novit.* 3023:1–34.

Hunt, R. M., Jr. In press, a. North American Tertiary Ursidae. In: C. Janis, K. Scott, and L. Jacobs, eds. *The Tertiary Mammals of North America.* Cambridge: Cambridge Univ. Press.

Hunt, R. M., Jr. In press, b. Basicranial anatomy of the giant viverrid from 'E' Quarry, Langebaanweg, South Africa. In: K. Stewart and K. Seymour, eds. *Paleoecology and Palaeoenvironments of Late Cenozoic Mammals: Tributes to the Career of C. S. Churcher.* Toronto: Univ. Toronto Press.

Hunt, R. M., Jr. In press, c. North American Eocene-Oligocene Amphicyonidae. In: D. Prothero and R. J. Emry, eds. *The Terrestrial Eocene-Oligocene Transition in North America.* Cambridge: Cambridge Univ. Press.

Hunt, R. M., Jr., and Barnes, L. 1994. Basicranial evidence for ursid affinity of the oldest pinnipeds. *Proc. San Diego Soc. Nat. Hist.* 29:57–67.

Hunt, R. M., Jr., and Skolnick, R. I. In press. The giant mustelid *Megalictis* from the early Miocene carnivore dens at Agate Fossil Beds National Monument, Nebraska: Earliest evidence of dimorphism in New World Mustelidae (Carnivora, Mammalia). *Univ. Wyoming Contrib. Geol.*

Hunt, R. M., Jr., and Solounias, N. 1991. Evolution of the aeluroid Carnivora: Hyaenid affinities of the Miocene carnivoran *Tungurictis spocki* from Inner Mongolia. *Amer. Mus. Novit.* 3030:1–25.

Hunt, R. M., Jr., and Tedford, R. H. 1993. Phylogenetic relationships within the aeluroid Carnivora and implications of their temporal and geographic distribution. In: F. S. Szalay, M. J. Novacek, and M. C. McKenna, eds. *Mammal Phylogeny (Placentals)*, pp. 53–73. New York: Springer-Verlag.

Kruuk, H. 1972. *The Spotted Hyena.* Chicago: Univ. Chicago Press.

Kurtén, B. 1966. Pleistocene bears of North America, 1: Genus *Tremarctos,* spectacled bears. *Acta Zool. Fennica* 115:1–120.

Kurtén, B. 1967. Pleistocene bears of North America, 2: *Arctodus,* short-faced bears. *Acta Zool. Fennica* 117:1–60.

Kurtén, B., and Anderson, E. 1980. *Pleistocene mammals of North America.* New York: Columbia Univ. Press.

Lange-Badré, B. 1979. Les créodontes (Mammalia) d'Europe occidentale de l'Éocène supérieur a l'Oligocène supérieur. *Mem. Mus. nat. Hist. natl., Paris* (ser. C) 42:1–249.

Lange-Badré, B., and Haubold, H. 1990. Les créodontes (mammifères) du gisement du Geiseltal (Éocène Moyen, RDA). *Géobios* 23(5):607–637.

Lavocat, R. 1951. Sur les affinités de quelques carnassiers de l'Oligocène d'Europe, notamment du genre *Plesictis* et du genre *Proailurus. C. R. Soc. Paléont. Suisse, Eclog. Géol. Helv.* 44(2).

Lavocat, R. 1952. Sur les affinités de quelques carnassiers de l'Oligocène d'Europe, notamment du genre *Plesictis* Pomel et du genre *Proailurus* Filhol. *Mammalia* 16:62–72.

Li C., Lin Y., Gu Y., Hou L., Wu W., and Qiu Z. 1983. The Aragonian vertebrate fauna of Xiacaowan, Jiangsu. *Vert. Palasiatica* 21(4):313–327.

Li C., Wu W., and Qiu Z. 1984. Chinese Neogene: Subdivision and Correlation. *Vert. Palasiatica* 22(3):163–178.

Marshall, L. G. 1976. Evolution of the family Thylacosmilidae, fossil marsupial sabertooths of South America. *Paleobios* 23:1–20.

Marshall, L. G. 1978a. Evolution of the Borhyaenidae, extinct South American predaceous marsupials. *Univ. California Publ. Geol. Sci.* 117:1–89.

Marshall, L. G. 1978b. The Terror Bird. *Field Mus. Nat. Hist. Bull.* 49(9):6–15.

Marshall, L. G. 1981. The families and genera of Marsupialia. *Fieldiana: Geol.* (n.s.) 8:1–65.

Marshall, L. G. 1985. Geochronology and land mammal biochronology of the transamerican faunal interchange. In: F. Stehli and S. D. Webb, eds. *The Great American Biotic Interchange,* pp. 49–85. New York: Plenum Press.

Marshall, L., Berta, A., Hoffstetter, R., Pascual, R., Reig, O., Bombin, M., and Mones, A. 1984. Mammals and stratigraphy: Geochronology of the continental mammal-bearing Quaternary of South America. *Palaeovertebrata: Mém. Extraord.* 1984:1–76.

Marshall, L., Hoffstetter, R., and Pascual, R. 1983. Mammals and stratigraphy: Geochronology of the continental mammal-bearing Tertiary of South America. *Palaeovertebrata: Mém. Extraord.* 1983:1–93.

Marshall, L., and Sempere, T. 1993. Evolution of the Neotropic Cenozoic land mammal fauna in its geochronologic, stratigraphic, and tectonic context. In: P. Goldblatt, ed. *Biological Relationships between Africa and South America,* pp. 329–392. New Haven, Conn.: Yale Univ. Press.

Matthew, W. D. 1909. The Carnivora and Insectivora of the Bridger Basin, middle Eocene. *Amer. Mus. Nat. Hist. Memoir* 9:289–567.

Matthew, W. D. 1924. Third contribution to the Snake Creek fauna. *Bull. Amer. Mus. Nat. Hist.* 50(2):59–210.

Meade, G. E. 1961. The sabertoothed cat *Dinobastis serus. Bull. Texas Mem. Mus.* 2(2):23–60.

Mein, P. 1958. Les mammifères de la faune sidérolithique de Vieux-Collonges. *Nouv. Arch. Mus. Hist. Nat., Lyon* 5:1–122.

Mein, P. 1989. Updating of MN zones. In: E. H. Lindsay, V. Fahlbusch, and P. Mein, eds. *European Neogene Mammal Chronology,* pp. 73–90. New York: Plenum Press.

Mellett, J. S. 1977. Paleobiology of North American *Hyaenodon* (Mammalia, Creodonta). *Contrib. Vert. Evol.* 1:1–134.

Merriam, C. 1930. *Allocyon,* a new canid genus from the John Day beds of Oregon. *Univ. California Publ. Geol. Sci.* 19:229–244.

Mitchell, E. 1975. Parallelism and convergence in the evolution of the Otariidae and Phocidae. *Rapp. P.-v. Réun. Cons. Intl. Explor. Mer.* 169:12–26.

Mitchell, E., and Tedford, R. H. 1973. The Enaliarctinae: A new group of extinct aquatic Carnivora and a consideration of the origin of the Otariidae. *Bull. Amer. Mus. Nat. Hist.* 151(3):201–284.

Neff, N. 1983. The basicranial anatomy of the Nimravidae (Mammalia: Carnivora): Character analyses and phylogenetic inferences. Ph.D. dissert., City University of New York.

Neff, N., and Coheleach, G. 1982. *The Big Cats (The Paintings of Guy Coheleach).* New York: Harry Abrams.

Nowak, R. M. 1991. *Walker's Mammals of the World,* 5th Ed. Baltimore: Johns Hopkins Univ. Press, Vol. 2 (pp. 643–1629).

Olson, E. C., and McGrew, P. O. 1941. Mammalian fauna from the Pliocene of Honduras. *Bull. Geol. Soc. Amer.* 52:1219–1244.

Opdyke, N. D., Lindsay, E. H., Johnson, N. M., and Downs, T. 1977. The paleomagnetism and magnetic polarity stratigraphy of the mammal-bearing section of Anza-Borrego State Park, California. *Quaternary Res.* 7:316–329.

Patterson, B., and Pascual, R. 1972. The fossil mammal fauna of South America. In: A. Keast, F. Erk, and B. Glass, eds. *Evolution, Mammals, and Southern Continents,* pp. 247–309. Albany: State Univ. New York Press.

Paule, W. J., and Baker, M. A. 1978. Structural basis for heat-exchanger function of the carotid rete in the cat. *Anat. Rec.* 190:505.

Petter, G. 1967. *Paragale hürzeleri* nov. gen., nov. sp., mustélidé nouveau de l'Aquitanien de l'Allier. *Bull. Soc. géol. France* (7e ser.) 9:19–23.

Petter, G. 1969. Interprétation évolutive des caractères de la dentures des viverridés africains. *Mammalia* 33:607–625.

Petter, G. 1976. Étude d'un nouvel ensemble de petits carnivores du Miocène d'Espagne. *Géol. Mediterr.* 3(2):135–154.

Petter, G. 1987. Small carnivores (Viverridae, Mustelidae, Canidae) from Laetoli. In: M. D. Leakey and J. M. Harris, eds. *Laetoli: A Pliocene Site in Northern Tanzania,* pp. 194–234. Oxford: Clarendon Press.

Pocock, R. I. 1916. The tympanic bulla in hyaenas. *Proc. Zool. Soc. London* 1916:303–307.

Pocock, R. I. 1921. The external characters and classification of the Procyonidae. *Proc. Zool. Soc. London* 1921:389–422.

Pocock, R. I. 1929. The structure of the auditory bulla in the Procyonidae and the

Ursidae, with a note on the bulla of *Hyaena. Proc. Zool. Soc. London* 1929:963–974.

Qiu Z.-X. 1989. The Chinese Neogene mammalian biochronology: Its correlation with the European Neogene mammalian zonation. In: E. H. Lindsay, V. Fahlbusch, and P. Mein, eds. *European Neogene Mammal Chronology,* pp. 527–556. New York: Plenum Press.

Qiu Z.-X., and Qi G. 1989. Ailuropod found from the late Miocene deposits in Lufeng, Yunnan. *Vert. Palasiatica* 27:153–169.

Qiu Z.-X. and Qiu Z.-D. 1990. Late Tertiary mammalian local faunas of China. *J. Stratigraphy* 14(4):241–260.

Qiu Z.-X., Xie J., and Yan D. 1988a. Discovery of the skull of *Dinocrocuta gigantea. Vert. Palasiatica* 26:128–138.

Qiu Z.-X., Ye J., and Cao J. 1988b. A new species of *Percrocuta* from Tongxin, Ningxia. *Vert. Palasiatica* 26:116–127.

Rawn, V. M. 1981. Scimitar cats, *Homotherium serum* Cope, from Gassaway fissure, Cannon County, Tennessee, and the North American distribution of *Homotherium. J. Tennessee Acad. Sci.* 56(1):15–19.

Reig, O. 1952. Sobre la presencia de mustelidos mefitinos en la formacion de Chapadmalal. *Rev. Mus. Mun. Cien. Nat. Trad. Mar del Plata* 1:45–51.

Remy, J. A., et al. (nine others). 1987. Biochronologie des Phosphorites du Quercy: Mise à jour des listes fauniques et nouveaux gisements de mammifères fossiles. *München Geowiss. Abhandl.* 10:169–188.

Repenning, C. A. 1976. Adaptive evolution of sea lions and walruses. *Syst. Zool.* 25(4):375–390.

Repenning, C. A., Ray, C. E., and Grigorescu, D. 1979. Pinniped biogeography. In: J. Gray and A. Boucot, eds. *Historical Biogeography, Plate Tectonics, and the Changing Environment,* pp. 357–369. Corvallis: Oregon State Univ. Press.

Repenning, C. A., and Tedford, R. H. 1977. Otarioid Seals of the Neogene. *U.S. Geol. Surv. Prof. Paper* 992:1–93.

Riggs, E. S. 1934. A new marsupial sabertooth from the Pliocene of Argentina and its relationships to other South American predaceous marsupials. *Trans. Amer. Phil. Soc.* (n.s.) 24:1–31.

Roberts, M., and Gittleman, J. 1984. *Ailurus fulgens. Mammal. Species* 222:1–8.

Russell, D., and Zhai R.-J. 1987. The Paleogene of Asia: Mammals and stratigraphy. *Mem. Mus. Natl. d'Hist. Natur., Sci. de la Terre* (ser. C) 52:1–488.

Savage, R. J. G. 1965. Fossil mammals of Africa. 19: The Miocene Carnivora of East Africa. *Bull. British Mus. (Nat. Hist.), Geol.,* 10(8):239–316.

Savage, R. J. G. 1989. The African dimension in European early Miocene mammal faunas. In: E. H. Lindsay et al., eds. *European Neogene Mammal Chronology,* pp. 587–599. New York: Plenum Press.

Schmidt-Kittler, N. 1981. Zur Stammesgeschichte der marderverwandten Raubtiergruppen (Musteloidea, Carnivora). *Ecol. Geol. Helvet.* 74(3):753–801.

Schmidt-Kittler, N. 1983. On a new species of *Sivanasua* Pilgrim 1931 (Feliformia, Carnivora) and the phylogenetic position of this genus. *Proc. Konin. Nederl. Akad. Wetensch.* (ser. B) 86(3):301–318.

Schmidt-Kittler, N. 1987. The Carnivora (Fissipedia) from the Lower Miocene of East Africa. *Palaeontographica* (Abt. A) 197:85–126.

Scott, W. B. 1938. A problematical cat-like mandible from the Uinta Eocene, *Apataelurus kayi* Scott. *Annals Carnegie Mus.* 27:113–120.

Segall, W. 1943. The auditory region of the arctoid carnivores. *Fieldiana: Zool.* 29(3):33–59.

Simpson, G. G. 1945. The principles of classification and a classification of mammals. *Bull. Amer. Mus. Nat. Hist.* 85:1–350.

Springhorn, R. 1980. Radiation der Hyaenodonta (Mammalia, Deltatheridia) unter gebiss-strukturellen Gesichtspunkten und ein Vergleich mit fissipeden Carnivora. *Ber. Naturf. Ges. Freiburg i. Br.* 70:97–109.

Stehlin, H. G. 1925. Catalogue des ossements de mammifères tertiaires de la collection Bourgeois à l'École de Pont-Levoy (Loir-et-Cher). *Bull. Soc. Hist. Nat. Anthropol. Loir-et-Cher* 18:77–277.

Stirton, R. A. 1960. A marine carnivore from the Clallam Miocene Formation, Washington. *Univ. California Publ. Geol. Sci.* 36(7):345–368.

Tchernov, E., Ginsburg, L., Tassy, P., and Goldsmith, N. 1987. Miocene mammals from the Negev (Israel). *J. Vert. Paleontol.* 7(3):284–310.

Tedford, R. H. 1976. Relationships of pinnipeds to other carnivores (Mammalia). *Syst. Zool.* 25(4):363–374.

Tedford, R. H., Barnes, L. G., and Ray, C. E. 1991. Earliest Miocene littoral arctoid *Kolponomos. J. Vert. Paleontol.* 11(3):57A.

Tedford, R. H., and Gustafson, E. 1977. First North American record of the extinct panda *Parailurus. Nature* 265(5595):621–623.

Teilhard de Chardin, P. 1915. Les carnassiers des Phosphorites du Quercy. *Ann. Paléontol.* 9(3–4):103–192.

Thomas, H., Sen, S., Khan, M., Battail, B., and Ligabue, G. 1982. The lower Miocene fauna of Al-Sarrar (Eastern Province, Saudi Arabia). *ATLAL, J. Saudi Arabian Archaeol. Jeddah* 5:109–136.

Turnbull, W. D., and Segall, W. 1984. The ear region of the marsupial sabertooth, *Thylacosmilus:* Influence of the sabertooth lifestyle upon it, and convergence with placental sabertooths. *J. Morphol.* 181:239–270.

Turner, H. N. 1848. Observations relating to some of the foramina at the base of the skull in Mammalia, and on the classification of the Order Carnivora. *Proc. Zool. Soc. London* 1848:63–88.

Van Valen, L. 1969. The multiple origins of the placental carnivores. *Evolution* 23(1):118–130.

Viret, J. 1929. Les faunes des mammifères de l'Oligocène supérieur de la Limagne Bourbonnaise. *Ann. l'Univ. Lyon* (n.s.) 47:1–328.

Webb, S. D., and Perrigo, S. C. 1984. Late Cenozoic vertebrates from Honduras and El Salvador. *J. Vert. Paleontol.* 4(2):237–254.

Werdelin, L., and Solounias, N. 1991. The Hyaenidae: Taxonomy, systematics and evolution. *Fossils and Strata* 30:1–104.

Whitmore, F. C., Jr., and Stewart, R. H. 1965. Miocene mammals and Central American seaways. *Science* 148:180–185.

Whybrow, P. J. 1987. Miocene geology and paleontology of Ad Dabtiyah, Saudi Arabia: Summary. *Bull. British Mus. (Nat. Hist.)* 41(4):367–369.

Wurster, D., and Benirschke, K. 1968. Comparative cytogenetic studies in the Order Carnivora. *Chromosoma* 24:336–382.

Wyss, A. R. 1987. The walrus auditory region and the monophyly of pinnipeds. *Amer. Mus. Novit.* 2871:1–31.

Wyss, A. R., and Flynn, J. J. 1993. A phylogenetic analysis and definition of the Carnivora. In: F. S. Szalay, M. J. Novacek, and M. C. McKenna, eds. *Mammal Phylogeny (Placentals),* pp. 32–52. New York: Springer-Verlag.

CHAPTER 16

Carnivoran Phylogeny and Rates of Evolution: Morphological, Taxic, and Molecular

JOHN J. FLYNN

Several recent books have been devoted to the biology and systematics of the Carnivora (e.g., Ewer 1973; Gittleman 1989), and data from carnivoran phylogeny have been, and continue to be, used to evaluate the tempo and mode of taxic, morphologic, and molecular evolution (e.g., Simpson 1944, 1953, 1971; Werdelin and Solounias 1991; Wayne et al. 1991). Few of these evolutionary-rate studies, however, have considered explicitly the roles of fossil taxa and rigorously tested phylogenies. This chapter summarizes recent work on the higher-level phylogeny of the Carnivora, discusses the role of fossils in generating reliable phylogenies and estimating rates of evolution accurately, and calculates rates of evolution within the Carnivora, focusing on taxonomic diversification and molecular divergence.

To most of us, the mammalian Order Carnivora seems clearly recognizable. Yet the coherence of this group has often been supported by only a few distinctive features, primarily *carnivory* and the associated dental modifications, such as carnassial shear that is restricted to the last upper premolar against the first lower molar (P4/m1 shear). Carnivorans, however, retain many primitive features in their anatomy, making it difficult to diagnose the order, determine the relationships of the order to other Eutheria, and discern higher-level relationships within the order. Because of these difficulties, there has been much emphasis on finding a "holy grail" character system that is subject to little homoplasy, and that will yield unequivocal, or at least very reliable, phylogenetic information. Among the systems used at one time or another have been the dentition, auditory bulla, basicranial anatomy, pattern of cranial arteries, auditory ossicles, occasionally some postcranial features, and most recently genetic information, such as DNA-DNA hybridization, or protein or DNA sequences. Many of these systems have been favored because of a belief that they are not subject to functional or adaptive pressure leading to convergence (which would confound the process of phylogenetic reconstruction). For example, the nineteenth-century anatomist Flower (e.g., 1869) especially featured the basicranium, to avoid features related to feeding and other special adaptations of the skeleton upon which previous workers had relied. Detractors, however, have claimed that *all* of these systems have been strongly affected by "adaptive or functional selection" and thus would not or may not be useful indicators of phylogenetic history. It is clear that no single character system will fulfill these expectations. I believe strongly that reliance on a single, infallible character system to reconstruct phylogeny, be it features of the skull or DNA sequences, is naive and misplaced. Only careful analysis of characters (Neff 1986; Bryant 1989) drawn from a broad spectrum of character systems will permit the reconstruction of both the phylogeny of the group and inferences about evolutionary, ecological, and biogeographic patterns and processes.

Morphological Phylogenies of the Carnivora

Most prior investigations of the higher-level phylogeny of the Carnivora and of the relationships of the order to other Eutheria have focused on morphological data. More recently, these have been complemented by molecular-based phylogenies as well as studies integrating both molecular and morphological data. These studies, summarized below, have clarified many of the patterns of interrelationships of carnivoran clades, but have left several fascinating problems unresolved.

Relationships of Carnivora to Other Mammalia

Among the Eutheria, the entirely extinct Creodonta appear to be the closest sister-group of the Carnivora (the two together forming the "Ferae" of Simpson 1945; see also Flynn et al. 1988; Wyss and Flynn 1993). Hypothesized relationships between Carnivora and other "early," "primitive" eutherian taxa (e.g., "palaeoryctoids," "pantolestoids," Cretaceous "insectivores," etc.) are weakly supported, at best.

The relationships of the Carnivora to other extant orders of Eutheria are poorly understood; neither morphological nor molecular-phylogenetic studies have consistently and confidently identified the closest relatives of Carnivora, whether on morphological grounds (McKenna 1975; Novacek and Wyss 1986; Novacek et al. 1988; Novacek 1989, 1990), molecular grounds (Goodman et al. 1985; Miyamoto and Goodman 1986; McKenna 1987; Wyss et al. 1987; Czelusniak et al. 1990), or by integrating morphological and molecular data (Shoshani 1986). These studies proposed relationships of carnivorans to other eutherians ranging from unresolved ("bushes" or multichotomies with most or all other eutherians), to reasonable (sister-group to lipotyphlan insectivores), to unlikely (sister-group to Primates, or to only ochotonid

Lagomorpha, or to Insectivora + Chiroptera), to extremely unlikely (non-monophyletic order, with pangolins inside "Carnivora").

Wyss and Flynn (1993) suggested a number of anatomical features that may be synapomorphies allying the Carnivora and Lipotyphla (a restricted subset of the classically paraphyletic order "Insectivora"). Interestingly, Faith and Cranston (1991) used permutation tail probability (PTP) tests to reach a similar conclusion, by analyzing a combined morphological data set (highly significant PTP) and four sets (three of four marginal or significant PTP) of amino-acid sequence data (from Wyss et al. 1987). Novacek (1992, using morphological data updated from Wyss et al. 1987) also concluded that Carnivora and Lipotyphla were closely related, but only when data from *fossil* taxa were included in the analysis; analysis of living taxa alone had poorly resolved the basal relationships among groups of eutherians.

Interrelationships within the Carnivora

The living members of the order Carnivora clearly belong to a monophyletic group within the Eutheria, even though many authors (as reflected in this volume) have chosen to classify most marine Carnivora as an order (Pinnipedia: seals, sea lions, and walrus) separate from the terrestrial carnivoran taxa ("Fissipeda," or Order Carnivora in a restrictive sense) to emphasize their "distinctive specializations."

Morphologically based phylogenetic analyses (Flower 1869; Mivart 1882a, b; Flower and Lydekker 1891; Matthew 1909; Simpson 1945; Tedford 1976; Flynn and Galiano 1982; Flynn et al. 1988; Wyss and Flynn 1993; see Figure 16.1) indicate clearly that (1) the Carnivora are monophyletic, and (2) there are two major clades within the Carnivora, the Feliformia (cats, hyenas, civets, mongooses) and the Caniformia (dogs, bears, weasels and skunks, and raccoons). There are several other well-supported clades within the extant Carnivora, including (1) a dichotomy within the Caniformia, separating the Canoidea from the Arctoidea; (2) the monophyly of the Feloidea (Herpestidae, Hyaenidae, Felidae, Viverridae); and (3) the monophyly of the Pinnipedia, within Arctoidea (strong support, but controversial).

But attempts to produce finer resolution of the phylogenetic relationships of extant taxa within these clades have been controversial (see Wyss and Flynn 1993). Particularly problematic has been resolving the interrelationships of (1) families within Feloidea, (2) *Nandinia* (traditionally placed in Viverridae, but likely to be the sister-taxon to all other Feloidea), (3) families within Arctoidea, (4) *Ailurus* (in Procyonidae, or sister-taxon to Procyonidae, or sister-taxon to Ursidae + Pinnipedia), (5) Pinnipedia (monophyly or diphyly; if monophyletic, sister-group to Ursidae or Mustelidae), and (6) families within Pinnipedia (Odobenidae, sister-group to either Otariidae or Phocidae).

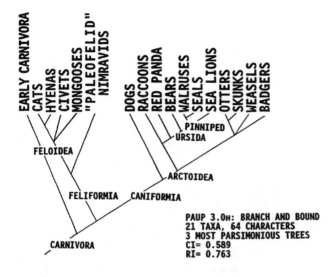

Figure 16.1. Higher-level phylogeny of the Carnivora (Wyss and Flynn 1993). This morphologically based phylogeny, with the slight modifications discussed in the text, is the base phylogeny used to calculate rates of molecular evolution of the mitochondrial cytochrome *b* gene.

FELIFORMIA, FELOIDEA

Flynn et al. (1988), in discussing the discrepancies between past hypotheses of relationships between Feloidea families, preferred to treat the Feloidea as an unresolved multichotomy. The analysis by Wyss and Flynn (1993) (Figure 16.1) was able to resolve only the Felidae/Hyaenidae as a monophyletic subclade within the Feloidea, although the data on fossil Hyaenidae presented by Werdelin and Solounias (1991) indicate that some of the putative synapomorphies linking Felidae and Hyaenidae may be convergent. My own preliminary molecular analyses (mtDNA sequence data for the cytochrome *b*, COIII, ND2, and ND3 genes) suggest a possible sister-group relation between Viverridae and Herpestidae, and indicate that the presumed viverrid *Nandinia binotata* (African palm civet) may actually be a sister-taxon to all other Feloidea/ Feliformia (Hunt [1987, 1989] had previously suggested this possibility, using morphological data).

CANIFORMIA, ARCTOIDEA

Within the Caniformia, the dichotomy between Canoidea and Arctoidea is well supported (Flynn et al. 1988; Wyss and Flynn 1993). The geometry of relationships of taxa *within* the Arctoidea, however, is more problematic. In particular, the relationships of the red panda *Ailurus* have been poorly resolved in morphologic, karyologic, and molecular (DNA-DNA hybridization, allozyme electrophoresis, albumin immunological distance) studies (Wyss and Flynn 1993). Moreover, varied geometries have been proposed for the interrelationships between the Ursidae, Procyonidae, and Mustelidae (Flynn et al. 1988, morphology; De Jong 1986, lens alpha crystallin A protein sequence;

Wyss and Flynn 1993). Additional analyses clearly are required to resolve this problem.

CANIFORMIA, PINNIPEDIA

Of particular interest is the phylogeny of the aquatic carnivorans, the Pinnipedia. The relationships of this group have been highly controversial (see Flynn et al. 1988): some workers (e.g., Simpson 1945) classified the group separately from the terrestrial Carnivora ("Fissipeda") in recognition of their anatomical/ecological specialization (but without explicit discussion of whether this should be considered a monophyletic grouping); some (e.g., Wozencraft 1989) consider the group to be of diphyletic origin within the Carnivora; and others (e.g., Wyss 1987; Flynn et al. 1988; Berta et al. 1989; Wyss and Flynn 1993) consider the pinnipeds a monophyletic group within the arctoid Carnivora. Most workers currently accept morphologically based phylogenetic analyses suggesting pinniped monophyly within the arctoid Carnivora (Wyss 1987; Flynn et al. 1988; Wyss and Flynn 1993; Berta et al. 1989) (Figure 16.1). If the pinnipeds are monophyletic, then one must determine their closest relatives within the Arctoidea. Wyss and Flynn (1993) considered the sister-group of the Pinnipedia to be the Ursidae, but Arnason and Widegren (1986) and Arnason and Ledje (1993) considered it to be the Mustelidae, with the Procyonidae as sister-group to the Mustelidae + Pinnipedia clade. It is important to note that the pinniped [Weddell seal, *Leptonychotes weddelli*] highly repetitive DNA probes used in the Arnason studies did not hybridize to the mephitine mustelid sample, implying either that the mustelines lie outside a Procyonidae + (Mustelidae + Pinnipedia) clade or that the phylogenetic implications of these hybridization studies are not straightforward (e.g., that the small number of taxa sampled were inadequate to reflect accurately the true phylogeny, the possible secondary loss of the highly repetitive DNA in mustelines, an originally broader distribution of the repetitive DNA [for instance, in all Arctoidea, and thus not a synapomorphy for pinnipeds, mustelids, and procyonids] and its subsequent independent loss in ursids and mephitine mustelids, or its convergent evolution in Pinnipedia, Procyonidae, and some Mustelidae). Wyss (1987) further proposed the controversial hypothesis that odobenids (walruses) are most closely related to phocids (true seals), rather than to otariids (fur seals). Resolution of pinniped diphyly versus monophyly will have particular relevance for understanding the relation between phylogeny and function in generating anatomical form in aquatic carnivorans.

Relationships between Fossil and Extant Carnivora

There has been substantial disagreement about the phylogenetic relationships of many fossil carnivorans, including the Nimravidae (sabertoothed "paleofelids"; see Flynn and Galiano 1982; Neff 1983; Flynn et al. 1988; Bryant 1988), early Cenozoic "Miacoidea" (the supposed stem-group for the later Carnivora; see Flynn and Galiano 1982; Wyss and Flynn 1993; Hunt and Tedford 1993; Flynn, in press), and many possible fossil members of extant families (particularly feloids). Nonetheless, the incorporation of fossil taxa into the phylogenetic analyses has provided additional insight into, and often resolution of, some of the preexisting areas of ambiguity and controversy in the phylogeny of the major clades of Carnivora (Wyss and Flynn 1993). The early Cenozoic "Miacoidea," which represent taxa in the basal radiation of the Carnivora, are of critical importance for understanding the phylogeny of the Carnivora.

The early Cenozoic Carnivora have generally been considered a basal stock of primitive Carnivora that were ancestral to the later Cenozoic Carnivora (which are sometimes considered to be a monophyletic group termed the "Neocarnivora"). The phylogenetic placement of these early Cenozoic "miacoid" carnivorans has been controversial; they have even been placed within the extinct order Creodonta because of their "primitiveness" or stem-group position. As a group within Carnivora, they were assigned to "primitive" taxa termed the "Miacidae" (Cope 1880), "Viverravidae" (Wortman and Matthew 1899), or "Miacoidea" (Teilhard de Chardin 1915). Animals within these groups have been allied primarily by their lack of features diagnostic of the later, more derived Carnivora (Cope 1880; Trouessart 1885; Matthew 1909).

Simpson (1945) was the first to place the early Cenozoic carnivorans within the order Carnivora, but he continued to lump them within a probably paraphyletic "stem-group," the "Miacoidea." More recently, Flynn and Galiano (1982) and Flynn (in press) challenged the assumption that the "Miacoidea" were a closely related "primitive" group that formed, in its entirety, the ancestral stock for all later, more "advanced" Carnivora. Their alternative proposition is that a monophyletic Carnivora split into two distinct lineages (Caniformia and Feliformia) early in its evolutionary history, and that these two lineages can be recognized both in living carnivorans and in their early Cenozoic relatives (the latter being "miacoid" taxa classically considered "Viverravidae" within the Feliformia and "Miacidae" within the Caniformia), thereby formalizing phylogenetic relationships first suggested by Matthew (1909) but rejected in his taxonomy. Although there was evidence supporting the assignment of various "miacoid" taxa to either the Caniformia or the Feliformia, Flynn et al. (1988) preferred to consider the "miacoids" as part of an unresolved multichotomy at the base of the Carnivora, so as to emphasize the need for further rigorous phylogenetic analyses of the entire Carnivora. A phylogenetic parsimony analysis of morphologic characters by Wyss and Flynn (1993, the first such study to include both fossil and living taxa) indicated that the possibly paraphyletic "Miacidae" and "Viverravidae" were sequentially closest sister-taxa to the remaining Carnivora, but including both taxa *within* the restricted Carnivora ("Miacidae" as Caniformia, and "Viverravidae" as

Feliformia) requires only five additional steps over the most parsimonious arrangement.

Hunt and Tedford (1993) even suggested the possibility of independent, diphyletic origins of the "Miacidae" and "Viverravidae" from within the Paleocene and Cretaceous members of the "*Cimolestes* group." They noted that this would suggest a diphyletic origin of Carnivora, for they accept the view that "Miacidae" may be most closely related to Caniformia and "Viverravidae" may be most closely related to Feliformia. Monophyly of the Carnivora, and an even more ancient splitting into two major clades, would be maintained if one considers "*Cimolestes*" paraphyletic and if one includes the species most closely related to the "Miacidae" and "Viverravidae" within the Carnivora.

It is important to note, however, that the morphologic evidence for the "Miacoidea" is incomplete, and it is clear that a more detailed study of all early Cenozoic carnivoran taxa represented by adequate morphologic material (Flynn, in press) is essential to clarifying the monophyly or paraphyly of the "Miacidae," "Viverravidae," or "Miacoidea," their phylogenetic relationships to other Carnivora, and the patterns and tempo of evolution within the Carnivora.

Biochemical Phylogenies of the Carnivora

A number of phenetic analyses of biochemical data (protein electrophoresis, serum immunology, karyotype patterns, DNA-DNA hybridization, etc.) have been applied to problems of higher-level, within-Carnivora systematics—analyses, for example, of immunology (Leone and Wiens 1956; Pauly and Wolfe 1957; Sarich 1969a, b; Farris 1972), of electrophoresis (Seal 1969; Feng et al. 1985), of karyotype (Fay et al. 1967; Wurster and Benirschke 1968; Arnason 1974; Couturier and Dutrillaux 1986; O'Brien et al. 1985; Arnason and Widegren 1986), of hybridization of highly repetitive DNA (Arnason and Widegren 1986; Arnason and Ledje 1993), and of multiple techniques (Wayne et al. 1989). I will focus here only on studies that include only discrete character analyses, since these are the only data sets for which it is possible to distinguish derived and primitive resemblances, and the only ones that currently can be incorporated into cladistic parsimony analyses. Some karyotype studies (e.g., Arnason 1974; Wurster-Hill and Gray 1975; Wurster-Hill and Bush 1980; Wurster-Hill and Canterwall 1982; O'Brien et al. 1985; Wurster-Hill et al. 1988) have identified discrete characters useful in cladistic analyses within the Carnivora. Protein and DNA sequencing approaches and restriction fragment/site analyses (e.g., Romero-Herrera et al. 1978; de Jong 1986; de Jong and Goodman 1982; Goodman et al. 1982; Miyamoto and Goodman 1986; Czelusniak et al. 1990; Geffen et al. 1992; Arnason and Johnsson 1992) have already yielded some cladistically analyzed results that are concordant with those generated by morphological studies, particularly in supporting the monophyly of Caniformia, Feliformia, and Pinnipedia. But there are also areas of discordance: Miyamoto and Goodman (1986) indicate that Canidae and Ursidae are sister-taxa (Ursidae not within Arctoidea), and Pinnipedia are the sister-group to a monophyletic "Musteloidea" (Mustelidae + Procyonidae); Czelusniak et al. (1990) indicate that Pinnipedia are the sister-group to all other Caniformia (rather than lying within Arctoidea); and Arnason and Johnsson (1992) analyzed only a single carnivoran, the harbor seal (*Phoca vitulina*), suggesting interordinal relationships between the Carnivora and "ungulates" (artiodactyls and cetaceans).

Vrana et al. (1994) recently completed a molecular DNA sequence analysis of the cytochrome *b* (307 b.p.; 123 potentially informative sites) and small ribosomal subunit 12S (394 b.p.; 125 potentially informative sites) mitochondrial genes in a variety of Carnivora, although emphasizing the Arctoidea. They analyzed the molecular data set (cyt *b* and 12S combined) alone, both unweighted and differentially weighting the transversions versus the transitions, as well as a combined molecular data set plus the morphological data set of Wyss and Flynn (1993, coded for their terminal taxa) in a "total-evidence" (Kluge 1989) analysis. Phylogenetic analysis of the molecular data alone yielded results that are largely in accordance with the morphological phylogeny proposed by Wyss and Flynn (1993, fig. 3.1), generally differing only in differential resolution of unresolved nodes rather than in conflicting geometries. Areas of similarity include: (1) two major clades within the Carnivora, consisting of a monophyletic Caniformia and a monophyletic Feliformia, (2) monophyly of Arctoidea, (3) Canidae as sister-group to the Arctoidea, (4) monophyly (individually) of the families Ursidae, Phocidae, and Canidae, (5) monophyly of the Pinnipedia, (6) a clade of the red panda (*Ailurus*) as sister-group to the Ursidae + Pinnipedia clade, and (7) raccoons as a possible sister-taxon to the clade of *Ailurus* (Ursidae, Pinnipedia) (resolved only in the differentially weighted analysis). The only major conflict between the two is that sea lions (Otariidae) are the sister-group to the walrus (Odobenidae) in Vrana et al.'s (1994) molecular analysis, whereas the walrus is sister-group to seals (Phocidae) in Wyss and Flynn's (1993) morphological analysis. The placement of some genera in Vrana et al.'s (1994) molecular analysis also suggests paraphyly (nonmonophyly) of several families (Mustelidae, Procyonidae, and Felidae) that have generally been considered to be monophyletic. The combined data set, total-evidence analysis (the molecular data of Vrana et al. plus the morphological data of Wyss and Flynn 1993), yielded results largely concordant with those summarized above (e.g., points 1–7 replicated) for both the molecular-alone and morphological-alone data sets. This result would be expected, given the high degree of concordance between the independent data sets, already noted. Of interest, in the total-evidence analysis, the walrus is sister-group to the seals (rather than sea lions, as suggested in the molecular-alone analyses), and only the Mustelidae and Procyonidae remain paraphyletic.

Integration and Resolution?

It is becoming increasingly clear that the richest and deepest understanding of evolution will emerge only from the integration of all available sources of evidence, and that integration will yield the most accurate reconstruction of the single evolutionary history of life on earth. Some of the most exciting questions in contemporary systematics are whether morphological and molecular analyses will yield concordant phylogenetic results, and what the role of fossils will be in interpreting evolutionary history. Although relevant comparative studies are few, it seems that molecular and morphological phylogenies are largely congruent (Wyss and Flynn 1993; Vrana et al. 1994), at least for higher-level relationships within the Carnivora. The next phase of research in the phylogeny of the Carnivora should focus on: (1) increasing our data sets, both in taxic breadth and numbers of characters, (2) integrating our data sets, via both concordance analyses of separate data sets and direct analysis of combined data sets, and (3) extending the search for congruence between molecular and morphological approaches and resolving the controversial parts of the phylogeny, with respect to both within-Carnivora relationships and the placement of Carnivora relative to other Eutheria. The role of fossils in all this will be elaborated on in the following sections.

Rates of Evolution

Once one has generated a well-supported phylogeny of the Carnivora, many other evolutionary problems can be addressed. The concept of *rate of evolution* has been applied to a variety of indicators of change through time, including morphology, diversity of taxa, and molecular or genetic change, among others. But it is critical that we recognize that all calculations of rates of evolutionary change require accurate temporal calibration, which is ultimately provided by the fossil record. Even in studies of molecular evolution that assume a "molecular clock," the rate at which the clock ticks must at some point be calibrated empirically by fossil data.

The Significance of Fossils and Reliable Phylogenies

The inclusion of fossils in any morphologic-phylogeny analysis is critical, wherever they are available. Their inclusion can dramatically affect interpretations of character homology, polarity, and transformation; the hierarchical level at which characters are synapomorphic (thus establishing the reconstructed ancestral morphotype for a particular clade); and even the topology of the phylogeny (Gauthier et al. 1988; Novacek 1992; Wyss and Flynn 1993). Prior studies (Flynn 1992; Wyss and Flynn 1993) indicate that this is clearly true for fossil taxa placed within living carnivoran families (e.g., canids, hyaenids; see also Werdelin, this volume) and within more inclusive clades (Feloidea, Caniformia/Feliformia), and it may be a general pattern in many groups, but it has been documented for only a few—for example, tetrapods (Gauthier et al. 1988), seed plants (Donoghue et al. 1989), and eutherian mammals and ungulates (Novacek 1991, 1992).

Further, a rigorous phylogeny should be used to calculate minimum ages of divergence (estimates calibrated from the fossil record are always minima, since ranges can become older with the discovery of new fossils) for *many* nodes spanning a *broad range* of divergence times. Wide variations in the phylogenetic alignment of taxa will yield correspondingly radical changes in the minimum-age estimates for the clade branch points (Marshall 1990; Norell 1992). The Carnivora are an excellent group for studies requiring reliable estimates for clade ages, since they have a good fossil record, have been intensively studied, and typically have broad geographic and environmental ranges. Using the techniques outlined in Norell (1992), I will determine divergence dates for all of the nodes within the phylogenies used in the following case studies.

Rates of Taxic Evolution

The Carnivora have long served as a focus for the evaluation of evolutionary rates. Simpson (1944), in his book *Tempo and Mode in Evolution,* used the Carnivora in a comparison of rates of taxic evolution between mammals and "invertebrates." He observed both higher taxic-origination rates and faster taxonomic turnover in carnivorans than in pelecypod (bivalve) mollusks, which led him to suggest fundamentally different patterns of evolution in mammals and mollusks. I will consider several other kinds of taxic-rate problems that can be evaluated by integrating information from fossils and phylogeny. Combining reliable phylogenies with temporal data derived from fossil occurrences often results in dramatically different interpretations of both the divergence age of groups and the patterns of taxic diversity through time.

Phylogeny and Fossils: Hyaenidae and "Ghost Lineages"

A study of the Hyaenidae (based on data in Werdelin and Solounias 1991) indicates that combining phylogenetic pattern with fossil-record temporal data for lineages yields estimates for taxic-diversity patterns over time that differ markedly from those derived from the use of fossil occurrences alone. The practice of combining temporal range information with phylogenetic pattern is complex and has seen a long history of debate. Many paleontologists prefer a literal reading of the fossil record to determine lineage temporal histories, although it is well understood that various sampling and preservational biases may greatly hamper accurate reconstruction of temporal ranges. As noted by Hennig (1965) and Norell (1992), however, there is a distinction between time of origin of a lineage (time of cladogenetic splitting) and time of origin of the diagnostic characteristics of a group (time of acquisition of apomorphies). The age of acquisition of apomorphy in a group is not necessarily concordant with the time this lineage split from its sister-taxon, and it is critically important to realize that the minimum age of a lineage is determined by recognition of either the oldest member of the group (the taxon possessing the diagnostic apomorphies) or the oldest member of the group's immediate sister-group. Thus the presence of an older sister-group provides evidence that its sister-lineage had already diverged, even if no direct evidence of this is yet evident in the fossil record. As a practical matter, if the fossil record is not ideal (e.g., does not accurately portray total taxon ranges and order of appearances), the integration of phylogenetic information with stratigraphic information generally (but not always) will yield older diversification ages for most groups and higher (and more accurate, if the phylogeny is reliable) taxic diversities through time. This approach also permits the incorporation of inferences about "ancestral lineages" that are not directly revealed in the fossil record (the "ghost taxa" of Norell 1992), as well as the temporal ranges of terminal taxa. As a corollary, because this method is sensitive to the implications of phylogenetic pattern, if the phylogeny is inaccurate, the calculations of taxic diversity through time may be incorrect. I would argue, however, that a diversity analysis that incorporates phylogenetic information, when a rigorously analyzed and robustly supported phylogeny is available, is preferable to an analysis using stratigraphic range data alone.

As an illustration, I use the Hyaenidae distributional and phylogenetic data set of Werdelin and Solounias (1991). Their phylogenetic and biogeographic analysis, which includes all known fossil and recent species, is one of the most detailed and complete available for a mammalian family. In discussing hyaenid taxic diversity, Werdelin and Solounias (1991) compared calculations based only on a literal reading of the fossil record with those based on integrating phylogenetic and biostratigraphic data, as advocated by Norell (1988, a Ph.D. dissertation; see Norell 1992 for the approach presented in Norell 1988). Their analysis used all "reasonably well known" hyaenids (61 taxa) to calculate taxic diversity, using direct calculations from stratigraphic occurrences and integrating phylogenetic information (using "ghosts," following the approach advocated by Norell 1988). Comparisons of the two methods, using the entire hyaenid taxic data set, are shown in their Fig. 55, but are not discussed in the text. That graphic comparison showed generally similar trends of diversity rise from the Orleanian through the Vallesian or Turolian, followed by general diversity decline to the present.

The analysis including phylogenetic information ("with ghosts"), however, differed from the information counted directly from the fossil record in having: (1) significantly higher taxic diversities in many temporal intervals (ranging from less than 10% higher in the Pleistocene to almost four times higher in the Vallesian and to seven to eight times higher in the Orleanian); (2) a diversity peak one land-mammal age earlier; and (3) a steady diversity decline from the Vallesian to the Recent, rather than an interval of slight diversity rise from the Ruscinian to the Pleistocene. Werdelin and Solounias (1991) preferred a literal reading of the fossil record, because they felt: (1) their stratigraphic resolution was too coarse for some taxa (possibly biasing ranges of ghosts to older intervals) and (2) the placement of a single early fossil taxon, *Adcrocuta eximia,* as sister-taxon to the living *Crocuta crocuta* nested well up within the crown of the hyaenid phylogeny would artificially inflate the age of origin of many lineages, and thus also total diversity through time.

I would argue that the influence of this old occurrence of a taxon nested high within the phylogeny is precisely why phylogeny must be incorporated into calculations of diversity. Werdelin and Solounias (1990, 1991:25) argued persuasively that *A. eximia* clearly is the sister-taxon to the living *C. crocuta.* Although the oldest occurrence (Vallesian MN10) of *A. eximia* was based on a single specimen (and therefore might inaccurately extend the age range for the species *and* its sister-lineages), good samples of *A. eximia* are known from the next younger temporal interval (Turolian MN11). Therefore, the clearly documented presence of *A. eximia* within MN11 would simply shift the high diversity noted in MN10 slightly younger (to MN11). Interestingly, new data (Werdelin, pers. comm.) definitely establishes the presence of *A. eximia* by MN10. In any case, the integration of the temporal information conveyed by the phylogenetic position of *A. eximia* (which seems secure) yields a dramatically different diversity pattern than would a literal reading of the fossil record.

I have attempted to address the other concern of Werdelin and Solounias (1991), regarding the coarseness of the temporal divisions and taxon ranges (to land-mammal ages rather than the more refined MN units), as well as the obvious problems that arise when dealing with taxa whose phylogenetic relationships are not robustly supported (because of homoplasy or missing data arising from the incompleteness of the fossils), by using the phylogeny and MN ranges of only the 18 best-known hyaenid species of Werdelin and Solounias (1991; their "core taxa" and "core tree"). These 18 taxa are not evenly spread across the largely "Hennigian-comb" phylogeny for all hyaenid taxa, which will result in diversity differences between the analyses of the two data sets. But at least the pattern of diversity changes over time, if not the actual magnitude, should be better reflected by restricting the analysis to only the better-known taxa and more robustly supported phylogeny ("core tree"). Figure 16.2 maps the stratigraphic range data for the 18 individual core species onto the phy-

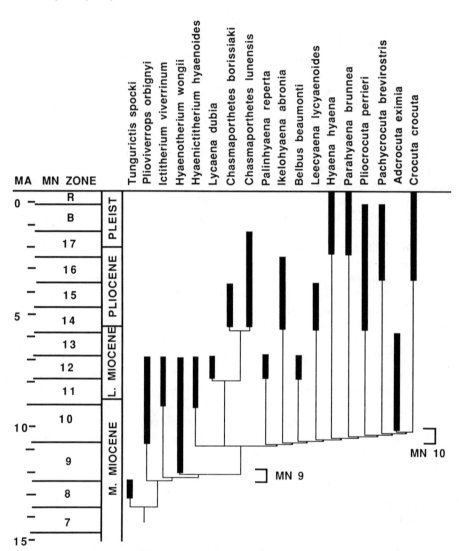

Figure 16.2. The phylogeny and stratigraphic ranges of the Hyaenidae. This diagram is based on the robust phylogeny of the "core" hyaenid taxa of Werdelin and Solounias (1991), with stratigraphic range data (numerical geologic-time scale, Ma; and mammalian biochrons, MN units 8–17, B [Biharian] and R [Recent]) derived from their text, and minimum divergence ages estimated from the phylogeny and taxic ranges using the method of Norell (1992). Abbreviations: Ma, Megannum; M. Miocene, Middle Miocene; L. Miocene, Late Miocene; Pleist, Pleistocene.

logeny. Even this simple graphic suggests that MN9 (four lineages present, versus only one taxon directly preserved) and MN10 (15 lineages present versus only three preserved in the fossil record) were intervals of rapid cladogenesis and high hyaenid diversity. Diversity through time for this group of hyaenid "core taxa" is shown in Figure 16.3. The left graph on the figure

TAXA OBSERVED: FOSSILS

TAXA INFERRED: FOSSILS & PHYLOGENY

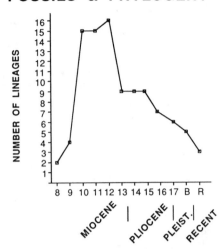

Figure 16.3. Comparison of taxic diversity of the hyaenid core taxa (Werdelin and Solounias 1991) from the Miocene to the present. *Left:* total diversity per unit time (mammalian biochrons; MN units 8–17, B [Biharian], and R [Recent]) observed directly in the stratigraphic record. *Right:* estimate of total lineage diversity, as inferred from the phylogeny plus taxic ranges observed in the fossil record (using the method of Norell 1992).

indicates the number of taxa (species) directly observed in the stratigraphic record through time (plotted using MN units, the finest temporal resolution possible from the data in Werdelin and Solounias 1991). The right graph presents inferred numbers of lineages (species directly observed plus minimum numbers of lineages inferred to have been present using the phylogeny and minimum age of sister taxa to determine minimum lineage ages). In contrast to Werdelin and Solounias' (1991: fig. 55) compilations for the entire suite of hyaenid taxa, the different methods of analyzing diversity using only the stratigraphic records and the more robust phylogeny of the "core taxa" showed sharply different diversity trends. Although both methods indicate a peak diversity in MN12, total diversity is twice as high in the analysis incorporating phylogeny, as might be expected. Further, the fossil record indicates a steady rise in diversity from MN9 to MN12 (from one to eight taxa), whereas diversity inferred using phylogeny indicates a much earlier (between MN9 and MN10) and sharper rise (almost quadrupling, from four lineages in MN9 to 15 in MN10) in diversity, with the same high diversity (15 taxa) maintained in MN11 and an increase of only one lineage to reach the peak diversity in MN12. As mentioned above, the higher total diversity, and earlier rise in diversity, are due in part to the early stratigraphic occurrence of *A. eximia,* a taxon nested well up in the termini of the phylogeny, emphasizing the importance of incorporating phylogenetic information into diversity analyses. Both methods indicate a diversity drop from MN12 to MN13 (about 6 million years ago, the time of the dramatic global environmental changes associated with the "Messinian Salinity Crisis"). The fossil record, however, suggests a catastrophic extinction event and only one surviving taxon, whereas the phylogenetic analysis indicates that there was indeed a significant diversity drop (about 45%) at this point, but that it was not nearly so drastic as might be inferred from the preserved fossil record. If the phylogenetic analysis better reflects the true diversity pattern, it suggests that many lineages were not preserved during this time because of the significant changes in depositional environments and reduction in fossiliferous localities of this age, and possibly because of the restriction of taxic geographic ranges during the episode of climatic change and habitat fragmentation. The phylogenetic diversity trend shows steady and moderate diversity from MN13 to MN15, then gradual decline from MN15 to the Recent, possibly tracking steady global cooling and climate/environmental change, whereas the fossil record indicates a dramatic diversity increase from MN13 to MN14 (one taxon to five), or apparent taxonomic radiation following the major extinction event, steady diversity from MN14 through MN16, a slight diversity increase in MN17, and a steady drop from MN17 to the Recent (in fact, the two methods indicate total diversities to be the same between MN17 and the Recent).

"Ancestors," or branches reflecting early phylogenetic events, should occur earlier in the stratigraphic record than do descendants or lineages reflecting later cladogenetic events, yet observed stratigraphic occurrences of taxa often conflict with this pattern. Norell and Novacek (1992) have shown that for tetrapod groups with "good" fossil records (high likelihood of preservation because of anatomy and habitus; intense sampling efforts across broad temporal and geographic spans; etc.), the patterns of phylogenetic branching sequence and taxic stratigraphic occurrences are generally concordant (e.g., lineages that diverge earlier in phylogenesis occur earlier in the fossil record).

Marshall (1990) attempted to use knowledge of sampling intensity and known patterns of occurrence within sampled sequences to estimate "error bars" on taxic ranges. Fisher (1992) has attempted to integrate stratigraphic occurrences of taxa, estimates of taphonomic sampling biases, and taxon character information to determine phylogeny. Both techniques are relevant to assessing the likelihood that anomalously old or young occurrences of taxa (relative to their phylogenetic position) may be due to stratigraphic sampling problems, rather than to an incorrect phylogeny. Although these techniques will provide deeper insights into the possible causes of such conflicts, they have serious limitations. Some of these limitations include: difficulties in assessing or assigning probabilities to taphonomic biases, uncertainties concerning the scope of applicability of the methods because they generally have been applied only to intra-basinal or regional studies, and methodological difficulties in combining stratigraphic and character information in phylogenetic analysis. The early occurrence of a late cladogenetic lineage (*Adcrocuta*) creates a strong conflict between diversity patterns derived from stratigraphic occurrences alone versus integrated analysis of a stratigraphically calibrated phylogeny. It would be much less parsimonious (more than five steps, or >10%: Werdelin, pers. comm.), however, to place *Adcrocuta* as an early hyaenid branch (which would be necessary to bring the results of the two methods into closer accord). It would be interesting to apply the techniques of Norell and Novacek (1992), Marshall (1990), and Fisher (1992) to an evaluation of the likelihood that all the lineages predicted (but not observed) by the early occurrence of *Adcrocuta* might have been present but not stratigraphically sampled.

Clearly, there can be dramatically different interpretations of both evolutionary patterns and processes using direct reading of the fossil record versus diversities inferred by combining phylogeny and stratigraphic information. If a robust phylogeny is available for the group under study, I would recommend analyzing diversity using both stratigraphic information and phylogenetic inferences of lineage existence. Comparisons between these results and those using only the fossil record may reveal new insights into biogeography, the timing and patterns of evolutionary extinction and diversification, and the effects of environmental and taphonomic changes on taxic preservation.

Phylogeny and Fossils: "Miacoidea"

The integration of fossils, stratigraphic ranges, and phylogeny is of significance not only in the analysis of patterns of taxic diversity over time. It is also of premier importance in calibrating lineage-divergence ages, and thus in calibrating rates of evolution between taxa across those lineages. Of particular interest are the implications of the phylogenetic placement of the early Cenozoic Carnivora ("Miacoidea"), and the problematic fossil Feloidea, for estimating the age of cladogenetic divergence for living clades. The "Miacoidea" have variously been considered to be the "stem-group" for all later carnivorans, or component taxa of the "Miacoidea" have been nested within Feliformia and Caniformia, the two major lineages of living carnivorans (as discussed above).

If one accepts the traditional placement of "miacoids" as a monophyletic sister-group, or paraphyletic "stem-group," for the remainder of the Carnivora ("Neocarnivora" of some), then the minimum divergence age of the Caniformia and Feliformia, the two major clades of living Carnivora, would be dated by the oldest taxon assignable to either clade (Figure 16.4, "Miacoids 1"). This phylogenetic hypothesis yields a Caniformia/Feliformia minimum divergence age as young as 36–45 million years ago (middle to late Eocene). Alternatively, if we accept the more recent (Flynn and Galiano 1982) phylogenetic hypothesis that various fossil "miacoids" are actually basal members of both the Feliformia and Caniformia lineages, the disposition I would currently favor, then the Caniformia/Feliformia divergence is at least as old as the oldest "miacoid"

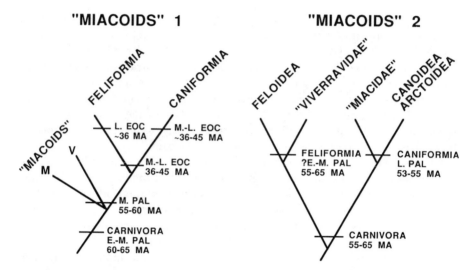

Figure 16.4. Estimates of minimum divergence ages for the Carnivora, Feliformia, and Canifor-mia. *Left:* "Miacoids 1," assumes the early Cenozoic "miacoids" to be sister-taxa (of unresolved monophyly or paraphyly) to the later Carnivora, which is consistent with the classical view of the phylogeny of the Carnivora. *Right:* "Miacoids 2," assumes the early Cenozoic "miacoids" to be actually members of the two major lineages of the Carnivora ("Viverravidae" as Feliformia, "Miacidae" as Caniformia), which is consistent with the phylogeny of Flynn and Galiano (1982). Age estimates are derived using the method of Norell (1992). Abbreviations: V, "Viverravidae"; M, "Miacidae"; E, Early; M, Middle; L, Late; MA, Megannum; PAL, Paleocene; EOC, Eocene.

(55–65 Ma; Figure 16.4, "Miacoids 2"). Although we are using the same fossils in both cases, the different phylogenetic placements result in a divergence-age estimate that is 50% greater, and comparison of rates of evolution across the feliform/caniform dichotomy would be 50% slower, in the preferred interpretation. Any study of rates of change (molecular, taxic, morphologic, etc.) between modern caniform and feliform Carnivora will be profoundly affected (rates could differ by 50%) by the divergence age that is accepted. Clearly, the generation of a stable and reliable phylogeny, including accurate placement of fossil taxa relative to modern taxa and lineages, is essential for accurately calibrating divergence estimates and calculating rates of evolution among taxa.

Phylogeny and Fossils: Fossil Taxa Assigned to Recent Families

We can similarly evaluate the role of fossils and phylogeny in calibrating estimates of divergence age for the families of living Carnivora. One might suspect that there would be fewer problems in dealing with taxa (families) having living representatives, on the basis of generally accepted hypotheses of monophyly, than there would be with the older, entirely fossil "miacoids." But even the divergence ages of the living families are profoundly affected by the phylogenetic placement of critical fossil taxa. There are many middle Cenozoic fossil carnivoran taxa that can be assigned confidently to living families, on the strength of clear synapomorphies. Others, however, are more problematic. These can usually be assigned to a major clade (e.g., Feloidea, Arctoidea, etc.) very easily, but they are primitive enough that it is difficult to assign them clearly to any family within the clade. They are often placed within living families for taxonomic convenience, and because they have morphologies that "could be ancestral to" (= primitive) or "are trending toward" those found in other members of that family (whatever that means!), even though they lack diagnostic synapomorphies. Figure 16.5 illustrates a case for the families of living feloid feliforms. This figure, which uses the phylogenetic results of Wyss and Flynn (1993), indicates that hyaenids and felids are sister-taxa (but see Werdelin, this volume) and that herpestids and viverrids are sister-taxa (I illustrate one of the three most parsimonious arrangements of Wyss and Flynn 1993; the other two place the herpestids or the viverrids, alternatively, as sister-taxa to the felid/hyaenid clade). In addition, I consider *Nandinia* likely

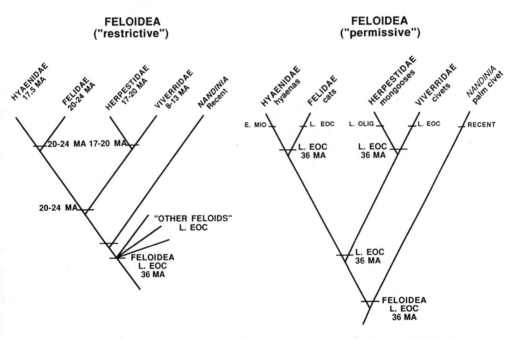

Figure 16.5. "Feloidea." Estimates of minimum divergence ages for the clades of Feloidea, using the method of Norell (1992) and the base phylogeny presented in the text (modified from Wyss and Flynn 1993). *Left:* "restrictive" takes a restrictive phylogenetic approach and accepts only fossil taxa that can be unambiguously assigned to the living families of Feloidea, while considering other taxa to be Feloidea *incertae sedis*. *Right:* "permissive" places the problematic fossil taxa within families that they have reasonably been assigned to by other workers. Abbreviations as in Figure 16.4, with the addition of OLIG = Oligocene.

to be the sister-taxon to all remaining feloids (following Hunt 1987, 1989; see also Wyss and Flynn 1993). Although different phylogenetic placement of the families will slightly alter the divergence-age estimates (the placement of *Nandinia* is irrelevant, because it is known only from the Recent and thus will not affect divergence-age estimates), they are trivial when compared to the effects of accurate placement of problematic fossil taxa.

If one uses "restrictive" concepts of the living feloid families (e.g., includes only taxa that unambiguously possess the diagnostic apomorphies of all the living taxa assigned to the family), then the earliest fossil taxa are 17.5 Ma for the Hyaenidae, 20–24 Ma for the Felidae, 17–20 Ma for the Herpestidae, and 8–13 Ma for the Viverridae (Figure 16.5, *left,* "Restrictive Feloidea"). These earliest lineage occurrences, and the geometry of the phylogenetic relationships, would suggest minimum divergence ages for the Hyaenidae/Felidae of 20–24 Ma, for the Herpestidae/Viverridae of 17–20 Ma, and for the clade of all the living Feloidea of 20–24 Ma.

Some authors have considered a variety of middle Cenozoic fossil feliform taxa (e.g., "stenoplesictines," "proailurines," *Herpestides,* etc.) to be assignable to the living feloid families Felidae, Herpestidae, and Viverridae (see discussion in Hunt 1987, 1989, 1991; Hunt and Tedford 1993; Wyss and Flynn 1993). But although these taxa clearly are closely related to the Feloidea, many of them appear to lack the synapomorphies diagnostic of the clades of the modern feloid families, and their inclusion within particular feloid families would require less-parsimonious convergent acquisition of various Feloidea "apomorphies" among the later members of those families. Thus they might be more appropriately considered "basal" Feloidea *incertae sedis.* I do not consider in detail here the simple nomenclatural issue of whether to restrict higher taxon names to the last common ancestor of relevant living taxa, and all of its descendants (see discussion in Wyss and Flynn 1993), for if one were to follow this convention, these problematic feliform taxa would not be classified within the Feloidea, since that taxon rank would consist of the last common ancestor of *Nandinia* and the Viverridae, Herpestidae, Felidae, and Hyaenidae. If we place them within the relevant feloid families, they greatly increase the age of the oldest fossil taxa within those families, and thereby increase the minimum divergence age for both the family and the relevant sister-taxa (Figure 16.5, *right,* "Permissive Feloidea"). For example, minimum divergence ages for the Felidae/Hyaenidae would be 36 Ma (Late Eocene) rather than 20–24 Ma, for the Herpestidae/Viverridae would be 36 Ma (Late Eocene) rather than 17–20 Ma, and for the clade of all living Feloidea would be 36 Ma (Late Eocene) rather than 20–24 Ma. Thus including these problematic "feloids" within the modern families would increase estimates of minimum divergence age by more than 50%. I believe that, in contrast to the preceding "miacoid" example, these problematic taxa currently are best excluded from the modern feloid families (see also Werdelin and Solounias 1991:78, 96), and thus the best divergence estimates are 50% younger, and rates of evolution would be

50% faster in comparisons across these cladogenetic events, than some that have been proposed recently.

Rates of Molecular Evolution

Many studies of rates of molecular change either have simply assumed a molecular clock or have raised questions about the linearity of the clock rate or the scale (both temporal and physico-molecular) over which changes may occur in a clocklike fashion. Some of these questions include whether there are constant rates of change (1) within individual genes or proteins (or particular regions within them, such as different nucleotide sites, membrane regions, stems vs. loops, etc.), (2) within taxa or lineages over all time spans of the lineage history, or (3) between different lineages (many studies simply extrapolate known or assumed rates of change for one lineage to studies of another lineage). Several studies (e.g., phenetic-distance studies including protein electrophoresis, immunological, and DNA hybridization data: Sheldon 1987; Krajewski 1989; Wayne et al. 1991; Marshall and Swift 1992; and temporally/phylogenetically calibrated RNA/DNA sequence analyses: Smith et al. 1992; Smith and Littlewood 1994) suggest significant differences in rates of molecular evolution across lineages. Given our state of knowledge, therefore, it would be important to test whether various molecular changes occur in a regular or "clocklike" fashion. The only adequate way to do this is by using robust phylogenies for which the age of branching points can be accurately calibrated by a good fossil record (see also Smith and Littlewood 1994). The Carnivora provide an excellent clade for examining these problems.

A Case Study: Rates of Evolution of the Mitochondrial Cytochrome *b*

To address questions of rates of molecular evolution we can, in essence, turn the typical phylogenetic usage of molecular-sequence data around, and use an existing, well-supported phylogeny for this group, then map molecular gene-sequence changes onto the phylogeny. In this context, I use preliminary mitochondrial cytochrome *b* gene-sequence data (399 base pairs for 13 carnivorans, representing almost all families, and a number of outgroups; generated at the Field Museum Biochemistry Laboratory). These data are analyzed both as relative-rate reciprocal comparisons, without temporal constraint, and as absolute, temporally calibrated rates of change. It must be emphasized that this analysis is a work in progress. These preliminary results are intended only to illustrate the kinds of inferences and tests of patterns and rates of molecular evolution that are possible if one integrates phylogeny and temporal calibration (from the fossil record) of cladogenesis and lineage durations.

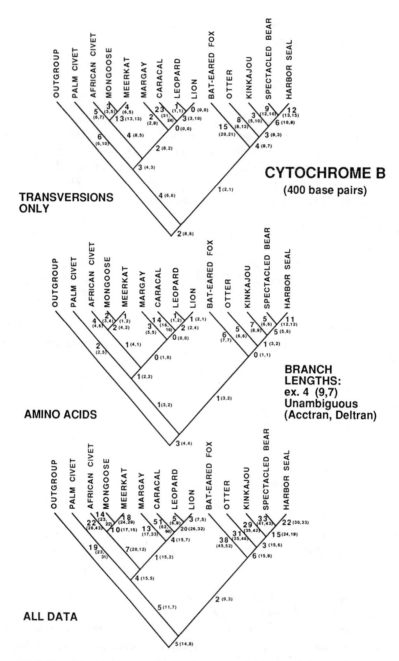

Figure 16.6. Branch lengths for unambiguously optimized character changes in 399 base pairs of the mitochondrial cytochrome *b* gene sequence, mapped onto the base phylogeny (see text). Numbers shown in parentheses for each branch are the number of changes assigned to that branch using the ACCTRAN and DELTRAN options of PAUP (these are algorithmically optimized, and include characters ambiguously placed at nodes). The data are partitioned to include all base-pair

The Method: A Base Phylogeny

The molecular-sequence data are mapped onto the consensus morphologically based phylogeny of Wyss and Flynn (1993) (Figure 16.1), with the two refinements discussed above: *Nandinia* is considered the sister-group to all other Feloidea, and the Viverridae and Herpestidae are considered sister-taxa, as suggested in one of our three most parsimonious solutions. I have chosen to resolve the Viverridae, Herpestidae, Felidae/Hyaenidae trichotomy of Wyss and Flynn (1993) in this way because there is good character evidence supporting it (but see Werdelin, this volume), and because rate calculations are sensitive to "artificial" or "methodological" multichotomies that are unresolved because of conflicting data (rather than those that represent true multichotomous phylogenetic branching), which will introduce artificial rate changes along branches by collapsing changes to lower node points. The phylogeny of Wyss and Flynn (1993) is concordant with almost all parts of the molecular-genetics-based phylogeny of Vrana et al. (1994), thus supporting its use as a reliable base phylogeny.

The Method: Calculating Branch Lengths

Variable characters from the 399 base-pair sequence of the mitochondrial cytochrome *b* gene are mapped onto the base phylogeny, using parsimony analysis (PAUP 3.0 for Macintosh: Swofford 1990) with both DELTRAN and ACCTRAN optimizations. Three different partitions of the original data are used: all the variable nucleotides ("all data"), transversions only (purine [A, G] to pyrimidine [T, C], or vice versa, to recognize that this mutation is less likely than transitions of A-G or T-C), and amino acids (Figure 16.6). These partitions were chosen because other workers have hypothesized that varying "selective constraints" may yield different rates of evolution within molecules/genes (e.g., at different nucleotide sites, transversions vs. transitions, particular codon positions vs. amino acids, etc.).

In determining branch lengths, I have chosen to use only variable characters distributed unambiguously on the base phylogeny (including autapomorphies for terminal taxa), since the process of assigning branch lengths in either the ACCTRAN or DELTRAN algorithms of parsimony analyses makes assumptions about optimization and tree topology. These assumptions can lead to a wide array (within bounds/limits) of possible branch-length assignments, as

---

changes (*bottom*), amino acid changes (*middle*) and transversions only (*top*). Taxa analyzed are the African palm civet, *Nandinia binotata;* African civet, *Viverra civetta;* slender mongoose, *Herpestes sanguineus;* meerkat, *Suricata suricatta;* margay, or tree ocelot, *Felis wiedii;* caracal, *Felis (Lynx) caracal;* leopard, *Panthera pardus;* lion, *Panthera leo;* bat-eared fox, *Otocyon megalotis;* otter, *Lutra longicaudis;* kinkajou, *Potos flavus;* spectacled bear, *Tremarctos ornatus;* and harbor seal, *Phoca vitulina.*

shown by the branch lengths assigned by the ACCTRAN and DELTRAN options (in parentheses on the figures, below the unambiguous character branch lengths), thereby dramatically influencing rate calculations. Similar optimization problems would apply if phenetic-distance analyses were used instead of parsimony techniques (see below). There is currently no empirical basis for rejecting the hypothesis that unambiguous character changes would be representative of the rate of change of all characters (both unambiguously distributed and ambiguously optimized characters). This method should yield reliable indicators of relative rates of molecular evolution across the phylogenetic tree, because it rejects the use of characters that cannot be determined unambiguously to have changed along a particular branch of the phylogeny. Clearly, because some variable characters are not included in the analysis, the branch lengths are not "total" changes that may have occurred along that branch, and rates of evolution are not "absolute" rates of change for this segment of the mitochondrial cytochrome *b* gene. Further, rates calculated using only the unambiguously optimized characters might differ from those including all characters, because ambiguously optimized characters might change more rapidly than others, and thus be more susceptible to homoplasious change.

The Method: Fossil Distributions and Estimates of the Range of Minimum Divergence Ages

An ideal calculation of "absolute" (temporally calibrated) rates of change, or among taxa rate comparisons, should incorporate an accurate calibration of the age of all divergence points for taxa within the group under study. Alternatively, one must assume a priori a rate of change calibrated either by experimental determination (rarely possible) or by data derived from other groups and assumed to be applicable across taxa. For example, the rates of change of molecules in primates or rodents are often used in rate studies of other mammalian groups. But all studies of rates of change must rely ultimately on calibration from the fossil record, because even extrapolated rates must be calibrated by fossil-based age estimates for at least one divergence event. The Carnivora provide an excellent study system for evaluating rates of change, because (1) the pattern of lineage divergences is well known from the base phylogeny, (2) the many higher-level lineage-divergence nodes have a broad range of divergence times (permitting evaluation of the effects of temporal averaging), and (3) the extensive fossil record for the group permits better temporal calibration for the minimum age of divergence than is possible in many other groups (e.g., primates have a much poorer fossil record). Figure 16.7 presents minimum lineage-divergence ages (12 nodes) for the study taxa. These minimum ages (estimates calibrated from the fossil record are always minima, because ranges can only become older with the discovery of new

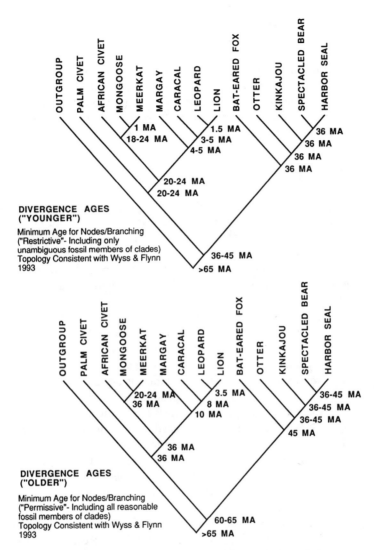

Figure 16.7. Minimum ages of clade divergence, estimated using the base phylogeny (see text) and the method of Norell (1992), for the taxa used in the cytochrome *b* study. The top figure uses a "Younger"/"Restrictive" definition of clade membership, which excludes fossil taxa that are not definitely members of the clade. The bottom figure uses an "Older"/"Permissive" definition, which includes all fossil taxa reasonably assigned as members of clades. Taxon names as in legend for Figure 16.6.

fossils) were determined using the approach of Norell (1992; see above) and the age of the earliest known taxon within each lineage, which is based both on the author's data and on stratigraphic occurrence information from M. C. McKenna's comprehensive mammalian classification (pers. comm.). A range of estimates is provided for each node, to provide limits on what the variance in branch temporal length might be in calculating rates of evolution. "Younger"/"Restrictive" refer to ages calculated using the ranges of only unambiguous members of clades, and "Older"/"Permissive" refer to ages calculated using all taxa that might reasonably be assigned as members of clades (see the more detailed discussion above). Figure 16.7 shows that the taxa analyzed in this study have a wide range of divergence ages, with minimum divergence ages for nodes ranging from 1–40 Ma (in the "Younger" estimates) or 3.5–62.5 Ma (in the "Older" estimates).

The Method: Calculating Rates of Evolution

The final step in this analysis consists of integrating the phylogenetic clade divergence age and molecular branch-length variation into a study of rates of evolution of the mitochondrial cytochrome *b* gene. The rates of evolution (Figures 16.8, 16.9) are calculated by:

1. Using the branch lengths (unambiguously assigned characters only) for "transversions only," "amino acids," and "all [nucleotide] data."

2. Dividing branch lengths by the amount of time between bracketing nodes. In cases where there are age ranges on the bracketing nodes, the average of the range is used for the node. In some cases (Arctoidea, *Nandinia*/rest of Feloidea), sequential nodes have the same minimum divergence ages; in these cases, an arbitrary difference of 1 m.y. (beginning at the minimum [or averaged minimum] age and extending sequentially older) is assigned to the intervening branches. Given that recognizable cladogenesis (e.g., cladogenesis supported by the acquisition of apomorphies among these branches) had occurred between these "collapsed" nodes, it seems reasonable to assume that 1 m.y. does not radically overestimate the branch duration, and that the nodes have the same apparent minimum age, because we have not yet discovered fossils that permit us to tease apart the precise ages of cladogenesis. If the 1 m.y. duration is an overestimate (e.g., if these proximate nodes represented an extremely rapid lineage radiation), then the rates of evolution will be underestimates. Alternatively, if significantly more time fell between these nodes (e.g., the fossil record is severely inadequate), then the rates will be overestimates. I know of no reasonable method for compensating for these possible minimum-age-estimate problems.

3. Determining "absolute" rates of evolution (branch lengths [total change] /time [time between the minimum age estimates for nodes bracketing the branches]) on the basis of both the "Younger"/"Restrictive" and the "Older"/"Permissive" temporal constraints.

The Results: Branch Lengths

Pairwise comparisons of the branch lengths, or rates of change (branch length/time since divergence) along branches, from the last common ancestor

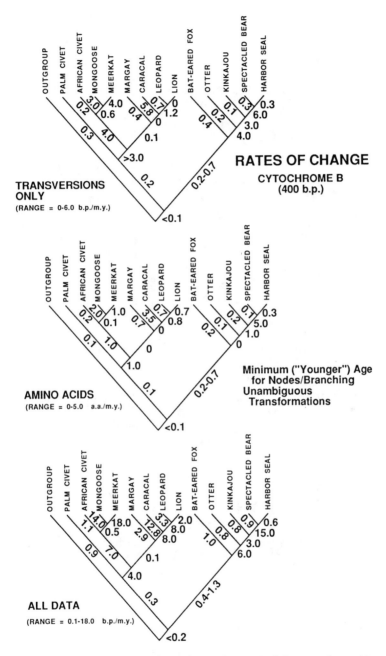

Figure 16.8. Rates of change (number of unambiguously optimized character changes/time, for all branches) in 399 base pairs of the mitochondrial cytochrome *b* gene sequence, using the "Younger" minimum divergence-age estimates of Figure 16.7. The rates are partitioned to include all base-pair changes (*bottom*), amino acid changes (*middle*), and transversions only (*top*). Taxon names as in legend for Figure 16.6.

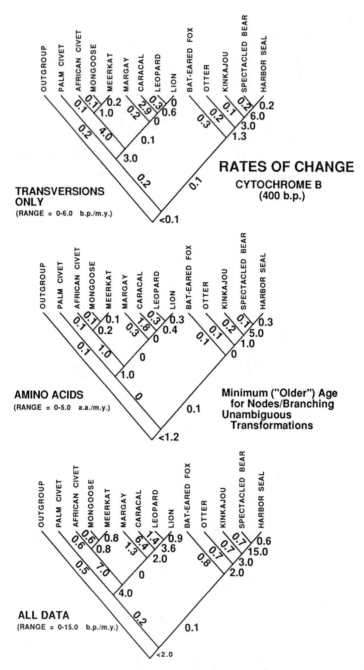

Figure 16.9. Rates of change (see legend for Figure 16.8), using the "Older" minimum divergence-age estimates of Figure 16.7. Taxon names as in legend for Figure 16.6.

of terminal taxa, provide relative-rate tests for constancy of molecular change. Most of these reciprocal comparisons for the taxa in Figure 16.6 indicate that rate of change in the unambiguously assignable (and even in reciprocal comparisons of total changes, when optimizing ambiguous characters using either ACCTRAN or DELTRAN options) variable mitochondrial cytochrome *b* characters is not constant, and is quite variable. To illustrate, comparisons of branch lengths in Figure 16.6 for pairwise comparisons of the number of changes (along the lineages from the last common ancestor of the pair of taxa) indicate up to 1.2–6 times more changes in the harbor seal lineage than in the lineages for most other species. Other pairwise comparisons are similarly, but not as extremely, skewed (e.g., spectacled bear to other carnivorans in all measures, kinkajou for transversions, palm civet for amino acids and all data, and most felids [except the caracal] relative to nonfelids for transversions, all data, and possibly amino acids; Figure 16.6). The caracal lineage has branch lengths, for all measures of molecular change, that are longer than those for any other taxon in the matrix (as well as relatively high absolute rates of change; see below). Either the amount and rates of change in the caracal have been exceptionally high (clearly above any average value for the rest of the lineages of Carnivora), or this branch accumulated changes over a long time, or there has been some kind of DNA contamination (although this sequence clearly differs from any of the other Carnivora analyzed in this study, and from outgroups, including human) or taxic misidentification of the tissue sample. Some reciprocal comparisons do, however, indicate relatively constant ("clocklike") changes between lineage pairs, such as leopard/lion (with a young divergence age) in all comparisons, or meerkat/mongoose (with an equivalent, or possibly older, divergence age) in all comparisons.

Preliminary analyses among several three-taxon comparisons, using Tajima's (1993) conservative, probabilistic, relative-rate test of phenetic differences in nucleotide sequences, provide similar results and permit rejection of the null hypothesis of a molecular clock in some cases. For example, in comparisons among the harbor seal, kinkajou (caniform ingroup), and palm civet (feliform outgroup), there was a significant variation in transversions (but not total substitutions) leading to rejection of the molecular-clock hypothesis (at $P < 0.01$ [$\chi^2 = 12.85$] using the 2D test of Tajima 1993). Therefore, the small number of nucleotides analyzed in this data set do not appear to create insurmountable problems in evaluating rates of molecular evolution.

The Results: "Absolute" Rates of Change

Calculations of "absolute" rates of change (Figures 16.8, 16.9), calibrated by the minimum divergence ages, show patterns of high variability along branch segments, in all three subsets of the data (transversions, amino acids, all data; Figures 16.8, 16.9) that are similar to those present in the reciprocal comparisons of branch lengths only. Because these cytochrome *b* sequences do not appear to be saturated (see Appendix 16.1), it is likely that "rates derived from parsimony analysis closely approximate absolute rates (Saitou 1989; Hillis et al. 1992)" (Smith and Littlewood 1994). It is important, however, to bear in mind that even moderate differences in "absolute" rates may be less significant statistically than would differences observed in simple reciprocal comparisons, because of the small number of unambiguously optimized changes in some of the data partitions (Figure 16.6), and variation in the possible minimum divergence ages for some nodes (Figure 16.7). The observed rate differences suggest that it is important to partition the molecular data, even within a single gene sequence, so as to analyze potentially variable rates of change (under the influence of selection, mechanistic constraints on molecule function, or varying probabilities of particular kinds of chemical-molecular changes such as transitions vs. transversions at a single base-pair site or synonymous vs. nonsynonymous changes at individual codon positions, etc.).

The lowest rates of change tend to lie along the longest branches (longest in absolute temporal duration), as would be expected given potential problems with "long branch attraction" and higher probability of unresolved/unobservable multiple hits. This is not exclusively true, however, for there are very low rates of change along branches that represent moderate to very short periods of time between divergence events (e.g., the lion in transversions, or the Caniformia/common ancestor of the Carnivora branch for all three data partitions). The relatively short branch lengths in most felids (except the caracal) seem to be due to shorter periods of separation rather than to unusually low rates of change, since rates of change are lower in many other lineages (most Caniformia branches, and most nonfelid Feliformia branches). The very slow rates of molecular evolution in the Ursidae apparent from some previous distance studies (see Wayne et al. 1991) are not duplicated in the cytochrome *b* data (using either unambiguously optimized changes only, or all changes optimized with either ACCTRAN or DELTRAN options), which suggests that phenetic distance measures may not adequately reflect rates of change in phylogenetically informative characters and/or that different molecules evolve at different rates. The ursid in this data set, the spectacled bear, has rates of evolution comparable to those of most other Carnivora, as well as all Caniformia (except possibly the kinkajou, in rate of transversion changes).

The highest rates of change tend to be concentrated along branches associated with recent divergences, or with radiation of major groups (inferred to have approximately equivalent node-divergence ages). These may simply reflect stochastic variation in rates of change, and the artifact of concentration of apparent slower rates along branches with long temporal durations. Alternatively, this pattern might reflect truly faster rates of molecular evolution concentrated at times of speciation, including those leading to what appear later to be "radiations" of major clades. This alternative retains the possibilities (1) that recent speciation events (e.g., lion/leopard) have had too little time to average the initial high amount of change over a longer time period to reach some "average" rate of evolution and/or (2) that the ages of branch points associated with apparently rapid "radiations" actually reflect an inadequate fossil record that has not yet closely approximated the true time of divergence for various clades (and thus the branch points for the different lineages within the "radiation" actually were spread over a longer temporal interval than the 1 m.y. separation assigned to them, with more time between each two nodes, than is reflected in the information currently available from the fossil record).

Implications of the Case Study: Testing Variability of Rates of Molecular Evolution

Any attribute that changes in a clocklike fashion should plot linearly in molecular distance versus divergence-age plots. Plots (Figure 16.10) of total numbers of unambiguous changes (transversions, amino acids, all data) versus both the shorter and longer possible minimum divergence ages for the cytochrome *b* data set indicate a very poor linear fit for any of the indicators of molecular change (*r*-squared values of 0.33–0.36 for transversions, 0.20–0.27 for amino acids, 0.39–0.41 for all data). If it is present at all, there is only a weak relationship between distance and age. These preliminary results provide little or no support for a strong relationship between divergence time and amount of molecular change, as would be necessary for the application of a molecular clock. Similar results have been obtained by others (e.g., Smith et al. 1992; Smith and Littlewood 1994). There are, however, some statistical problems associated with direct plotting of all pairwise distance comparisons, because the *r*-squared value can be skewed by outlier points. These outliers, and their possibly aberrant measurement of amount of change, will affect every pairwise comparison in which they are involved, thus presumably reducing the goodness of fit of any measure of relationship between distance and age. Phenetic-distance techniques typically average out distance across the entire reciprocal matrix, and thus avoid this statistical problem. But these approaches may create other problems not associated with direct pairwise comparisons. Most distance techniques average out real irregularity in change, especially in partitioning variance equally along branches, even when some branches may really be exceptionally long or short. Plotting total variation will partition ambiguously optimized character change in ways that differ dramatically among optimizing algorithms, and in ways that may not reflect the true pattern of change along branches. Many analyses of nucleotide data "correct" the distances to accommodate estimated substitutions at synonymous and nonsynonymous sites (e.g., models of Kimura 1980; Wu and Li 1985; Li et al. 1990), thereby making a priori assumptions about rates of change.

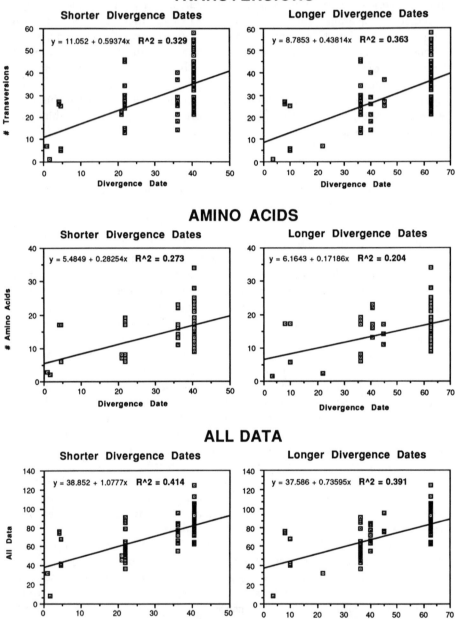

Figure 16.10. Molecular distance/divergence age plots for the Carnivora. Plots use all pairwise comparisons of unambiguously optimized changes in the 399 base pairs of the mitochondrial cytochrome *b* gene sequence, using both the shorter ("Younger") and longer ("Older") divergence-age estimates, and all three data partitions of Figures 16.7–16.9.

Additional testing of the variability in rates of molecular evolution observed both in reciprocal comparisons of branch lengths and in measures of "absolute" rates of change of mitochondrial cytochrome *b* gene sequences will be provided by more complete sampling of the gene, more genes (both mitochondrial and nuclear), and/or more taxa. This is, however, one of the first investigations of constancy of change in molecular data that includes discrete molecular character data, mapping of molecular change onto a phylogeny (using a phylogeny generated by an independent data set), a large taxic sample with a wide variety of divergence ages, and good, independent age calibration for all the branching nodes.

Summary

We continue to make significant advances in understanding the patterns and rates of evolution of the Carnivora. Aspects of the higher-level phylogeny of the Carnivora are supported by both morphological and molecular evidence, such as the basic Caniformia/Feliformia dichotomy, the position of Canidae as sister-group to all other Caniformia, and the probable monophyly of Pinnipedia and their placement somewhere within a monophyletic Arctoidea. Yet major questions remain regarding the higher-level phylogenetic placement of many groups, including the red panda and other taxa within arctoids, the various feloid families, and many fossil taxa. Determining the phylogenetic relationships of species within families, and including the broadest spectrum of taxic sampling possible, pose additional challenges. These will be among the major areas of inquiry for future systematic research, and efforts must continue in incorporating and reconciling diverse data (e.g., molecular, morphological, behavioral) in phylogenetic studies. Past studies, and those outlined in this chapter, illustrate the fact that patterns and rates of taxic evolution and diversity change calculated directly from the fossil record may differ dramatically from those determined through the integration of both phylogenetic and temporal information. Is the notion of "molecular clocks" valid, or are there highly variable rates of molecular evolution? The few available studies suggest that there may be great variability in rates of evolution among taxa and across different measures of genetic change (e.g., phenetic distance vs. discrete characters, within molecules, between different molecules, etc.). Much work remains to be done to resolve these questions. Analyses of the Carnivora promise to provide insights into the nature and rate of molecular change, as well as into broader issues in evolution, ecology, conservation biology, biogeography, and earth history. Yet the most valuable future studies will be those that integrate phylogeny, the temporal constraints on divergence provided by the fossil record, and a broad sampling both of taxa and of measures of evolutionary change.

Appendix 16.1.
Molecular Techniques, Cytochrome *b*

DNA Extraction

A small (50–100 mg) tissue sample from each taxon is selected from the Field Museum frozen-tissue collections, minced with a sterile blade, and digested overnight at 37°C in extraction buffer (10 mM tris-acetate, pH 8.0, 2 mM Na$_2$EDTA, 10 mM NaCl, 1% sodium dodecyl sulfate, 10 mg/ml dithiothreitol, and 0.5 mg/ml Proteinase K). Whole genomic DNA is extracted twice with phenol/chloroform solution (1:1), and twice with chloroform solution (Sambrook et al. 1989). The extracted DNA is ethanol-precipitated, dried, and resuspended in a small volume of distilled water or TE buffer.

DNA Amplification and Sequencing

Oligonucleotide primers L14724 (5'-CCAATGATATAAAAACCATCGTTG-3') and H15915 (5'-TTCATCTCCGGTTTACAAGAC-3')—where the numbers refer to the corresponding location of the 3' end of the primer in the human mtDNA genome (Anderson et al. 1981; see also Kocher et al. 1989)—are used to amplify the entire cytochrome *b* gene. Double-stranded amplifications (100 µl) are conducted in 0.5 ml microcentrifuge tubes, containing 10 mM tris, pH 8.3, 50 mM KCl, 3 mM MgCl$_2$, 150 µM dNTP's, 0.5 µM each amplification primer and 2.5 units *Taq* polymerase. Genomic DNA (10–1000 ng) is added after overlaying the reaction mixture with mineral oil. Reaction conditions consist of 35 cycles, each of 1 min at 93°C, 1 min at 51°C, and 2 min at 72°C, for the denaturation, annealing, and extension steps, respectively. The first cycle is preceded by a 3-min denaturation step, and the last cycle is followed by a further 3-min extension.

Following amplification, 5 µl of the double-stranded PCR (dsPCR) product is then applied to a 3% NuSieve agarose minigel and separated from free primers and nucleotides by electrophoresis at 60 V in 40 mM tris-acetate buffer, pH 8.0. The gel is stained with ethidium bromide, and the dsPCR product is excised from the gel and melted in 300 µl water. A small volume of a dilution of this solution is then used as template in a 100 µl asymmetric PCR (ssPCR) reaction for the purpose of generating single-stranded DNA template for direct sequencing (Gyllensten and Erlich 1988). The amplification conditions are similar to those for dsPCR except for a 20- to 100-fold dilution of one of the amplification primers and the raising of the annealing temperature to 55°C. Following 30–35 cycles of amplification, excess primers, salts, and free nucleotides are removed by three cycles of centrifugal dialysis (Centricon-30, Amicon). An aliquot of the washed and concentrated PCR product is then sequenced by the dideoxy chain-termination method (Sanger et al. 1977), using ^{35}S-a-dATP and Sequenase (U.S. Biochemical) with the oligonucleotide primer that had been limiting in the second-stage PCR.

Alternatively, the dsPCR product is freed of remaining nucleotides and primers, typically using a glass-powder suspension (GeneClean, Bio101, Inc.) and sequenced directly by the protocol of Thein (1989), which has been modified by the addition of 10% dimethyl sulfoxide to the annealing, labeling, and termination reactions.

Aliquots of the sequencing reactions are typically loaded on a 60 × 20-cm field gradient (0.4–1.2 mm) 6% polyacrylamide gel and subjected to electrophoresis for 3 hr at 55 W. After fixing in 5% methanol and 5% acetic acid, the gel is dried and exposed to x-ray film for one to three days. Typical autoradiographs from gels run in this manner yield 300 to 350 readable bases.

Sequencing Results

Sequences were easily aligned visually, and were corroborated by alignment against published sequences for seal, cow, mouse, and human (Anderson et al. 1981, 1982), using the computer program ESEE. For the pilot analyses, we sequenced at least 399 base pairs of the mitochondrial cytochrome *b* gene (positions 15488–15887; all positions numbered according to human sequence: Anderson et al. 1981). The pattern and amount of variation in these cytochrome *b* data are comparable to those in other studies of mammalian mitochondrial genes (e.g., Brown et al. 1982; Irwin et al. 1991; Allard and Miyamoto 1992; Allard and Honeycutt 1992; Arnason and Johnsson 1992). Mean base compositions at first, second, and third codon positions were as follows: third positions were low in G, second positions were relatively rich in T, and first positions were relatively equal in composition (but with somewhat higher A), all of which is similar to the patterns reported by Irwin et al. (1991; table 3) for cytochrome *b* of artiodactyls. Again as noted by Irwin et al. (1991), second positions showed the least interspecific variability (because substitutions here can result in more radical amino acid changes than can those in first positions), and third positions showed the greatest interspecific variability (because of more silent substitutions). Base composition bias is similar between all the species compared in the pilot analyses.

Synonymous transition substitutions occur much more frequently in mitochondrial DNA than do silent transversions or amino acid substitutions (fig. 7 of Li et al. 1985; Irwin et al. 1991). As appears to be the case in other vertebrates (e.g., Moritz et al. 1987), substitutions were heavily biased, with transitions (A-G, C-T) observed at many more sites than were shared implied transversions (purine to pyrimidine, or vice versa); the transition/transversion ratio was about 2:1 in the cytochrome *b* data set, suggesting increasing, but not complete, saturation.

Acknowledgments

I thank Paul Vrana for early access to his then unpublished manuscript; Malcolm McKenna for making his as yet unpublished mammalian classification available for this study; John Hall for collaboration in generating the mitochondrial cytochrome *b* pilot data; Lars Werdelin and an anonymous reviewer for valuable manuscript critiques; and the editor, John Gittleman, for the invitation to contribute to this volume and for his patience and his editorial review and assistance.

References

Allard, M., and Honeycutt, R. 1992. Nucleotide sequence variation in the mitochondrial 12S rRNA gene and the phylogeny of African mole-rats (Rodentia: Bathyergidae). *Mol. Biol. Evol.* 9:27–40.

Allard, M., and Miyamoto, M. 1992. Testing phylogenetic approaches with empirical data, as illustrated with the parsimony method. *Mol. Biol. Evol.* 9:778–786.

Anderson, S., Bankier, A., Barrell, B., de Bruijn, M., Coulson, A., Drouin, J., Eperon, I., Nierlich, D., Roe, B., Sanger, F., Schreier, P., Smith, A., Staden, R., and Young, I. 1981. Sequence and organization of the human mitochondrial genome. *Nature* 290:457–465.

Anderson, S., de Bruijn, M., Coulson, A., Eperon, I., Sanger, F., and Young, I. 1982. Complete sequence of bovine mitochondrial DNA: Conserved features of the mammalian mitochondrial genome. *J. Mol. Biol.* 156:683–717.

Arnason, U. 1974. Comparative chromosome studies in Pinnipedia. *Hereditas* 76:179–226.

Arnason, U., and Johnsson, E. 1992. The complete mitochondrial DNA sequence of the Harbor Seal, *Phoca vitulina. J. Mol. Evol.* 34:493–505.

Arnason, U., and Ledje, C. 1993. The use of highly repetitive DNA for resolving cetacean and pinniped phylogeny. In: F. Szalay, M. Novacek, and M. McKenna, eds. *Mammal Phylogeny: Placentals,* pp. 74–80. New York: Springer-Verlag.

Arnason, U., and Widegren, B. 1986. Pinniped phylogeny enlightened by molecular hybridizations using highly repetitive DNA. *Mol. Biol. Evol.* 3:356–365.

Berta, A., Ray, C., and Wyss, A. 1989. Skeleton of the oldest known pinniped, *Enaliarctos mealsi. Science* 244:60–62.

Brown, W., Prager, E., Wang, A., and Wilson, A. 1982. Mitochondrial DNA sequences of primates: Tempo and mode of evolution. *J. Mol. Evol.* 18:225–239.

Bryant, H. 1988. The anatomy, phylogenetic relationships and systematics of the Nimravidae (Mammalia: Carnivora). Ph.D. dissert., University of Toronto.

Bryant, H. 1989. An evaluation of cladistic and character analyses as hypothetico-deductive procedures, and the consequences for character weighting. *Syst. Zool.* 38(3):214–227.

Cope, E. 1880. On the genera of Creodonta. *Proc. Amer. Phil. Soc.* 19:76–82.

Couturier, J., and Dutrillaux, B. 1986. Evolution chromosomique chez les carnivores. *Mammalia* 50:128–162.

Czelusniak, J., Goodman, M., Koop, B., Tagle, D., Shoshani, J., Braunitzer, G., Kleinschmidt, T., deJong, W., and Matsuda, G. 1990. Perspectives from amino acid and nucleotide sequences on cladistic relationships among higher taxa of Eutheria. In: H. Genoways, ed. *Current Mammalogy,* 2:545–572. New York: Plenum Press.

de Jong, W. 1986. Protein sequence evidence for monophyly of the carnivore families Procyonidae and Mustelidae. *Mol. Biol. Evol.* 3:276–281.

de Jong, W., and Goodman, M. 1982. Mammalian phylogeny studied by sequence analysis of the eye lens protein alpha-crystallin. *Zeit. Saug.* 47:257–276.

Donoghue, M., Doyle, J., Gauthier, J., Kluge, A., and Rowe, T. 1989. The importance of fossils in phylogeny reconstruction. *Ann. Rev. Ecol. Syst.* 20:431–460.

Ewer, R. 1973. *The Carnivores.* Ithaca, N.Y.: Cornell Univ. Press.

Faith, D., and Cranston, P. 1991. Could a cladogram this short have arisen by chance alone? On permutation tests for cladistic structure. *Cladistics* 7:1–28.

Farris, J. 1972. Estimating phylogenetic trees from distance matrices. *Amer. Nat.* 106:645–668.

Fay, F., Rausch, V., and Feltz, E. 1967. Cytogenetic comparisons of some pinnipeds (Mammalia: Eutheria). *Canadian J. Zool.* 45:773–778.

Feng, W., Luo, C., Ye, Z., Zhang, A., and He, G. 1985. The electrophoresis comparison of serum protein and LDH isoenzyme in 5 Carnivora animals: giant panda, red panda, asiatic black bear, cat, and dog. *Acta Ther. Sinica* 5:151–155.

Fisher, D. 1992. Stratigraphic parsimony. In: W. Maddison and D. Maddison, eds. *MacClade: Analysis of Phylogeny and Character Evolution, Version 3,* pp. 124–129. Sunderland, Mass.: Sinauer.

Flower, W. 1869. On the value of the characters of the base of the cranium in the classification of the order Carnivora, and on the systematic position of *Bassaris* and other disputed forms. *Proc. Zool. Soc. London* 1869:4–37.

Flower, W., and Lydekker, R. 1891. *An Introduction to the Study of Mammals Living and Extinct.* London: Adam and Charles Black.

Flynn, J. 1992. Rates of evolution in the Carnivora (Mammalia): The importance of phylogeny and fossils. *Fifth North American Paleontological Convention, The Paleontological Society, Spec. Publ.* 6:100.

Flynn, J. In press. "Miacidae" (Carnivora). In: C. Janis, L. Jacobs, and K. Scott, eds. *The Tertiary Mammals of North America.* Cambridge: Cambridge Univ. Press.

Flynn, J., and Galiano, H. 1982. Phylogeny of Early Tertiary Carnivora, with a description of a new species of *Protictis* from the Middle Eocene of Northwestern Wyoming. *Amer. Mus. Novitates* 2725:1–64.

Flynn, J., Neff, N., and Tedford, R. 1988. Phylogeny of the Carnivora. In: M. Benton, ed. *The Phylogeny and Classification of the Tetrapods, Vol. 2: Mammals,* pp. 73–116. The Systematics Association Spec. Vol. No. 35B. Oxford: Clarendon Press.

Gauthier, J., Kluge, A., and Rowe, T. 1988. Amniote phylogeny and the importance of fossils. *Cladistics* 4:105–209.

Geffen, E., Mercure, A., Girman, D., Macdonald, D., and Wayne, R. 1992. Phylogenetic relationships of the fox-like canids: Mitochondrial DNA restriction fragment, site and cytochrome *b* sequence analyses. *J. Zool. (London)* 228:27–39.

Gittleman, J. L., ed. 1989. *Carnivore Behavior, Ecology, and Evolution.* Ithaca, N.Y.: Cornell Univ. Press.

Goodman, M., Czelusniak, J., and Beeber, J. 1985. Phylogeny of primates and other eutherian orders: A cladistic analysis using amino acid and nucleotide sequence data. *Cladistics* 1:171–185.

Goodman, M., Romero-Herrera, A., Dene, H., Czelusniak, J., and Tashian, R. 1982. Amino acid sequence evidence on the phylogeny of the primates and other eutherians. In: M. Goodman, ed. *Macromolecular Sequences in Systematic and Evolutionary Biology,* pp. 3–51. New York: Plenum Press.

Gyllensten, U., and Erlich, H. 1988. Generation of single-stranded DNA by the polymerase chain reaction and its application to direct sequencing of the HLA-DQA locus. *Proc. Natl. Acad. Sci.* 85:7652–7656.

Hennig, W. 1965. Phylogenetic systematics. *Ann. Rev. Entomol.* 10:97–116.

Hillis, D., Bull, J., White, M., Badgett, M., and Molineux, I. 1992. Experimental phylogenetics: Generation of a known phylogeny. *Science* 255:589–592.

Hunt, R. 1987. Evolution of the aeluroid Carnivora: Significance of auditory structure in the nimravid cat *Dinictis*. *Amer. Mus. Novitates* 2886:1–74.

Hunt, R. 1989. Evolution of the aeluroid Carnivora: Significance of the ventral promontorial process of the petrosal, and the origin of basicranial patterns in the living families. *Amer. Mus. Novitates* 2930:1–32.

Hunt, R. 1991. Evolution of the aeluroid Carnivora: Viverrid affinities of the Miocene carnivoran *Herpestides*. *Amer. Mus. Novitates* 3023:1–34.

Hunt, R., and Tedford, R. 1993. Phylogenetic relationships within the aeluroid Carnivora and implications of their temporal and geographic distribution. In: F. Szalay, M. Novacek, and M. McKenna, eds. *Mammal Phylogeny: Placentals*, pp. 53–73. New York: Springer-Verlag.

Irwin, D., Kocher, T., and Wilson, A. 1991. Evolution of the cytochrome *b* gene of mammals. *J. Mol. Evol.* 32:128–144.

Kimura, M. 1980. A simple method for estimating evolutionary rate of base substitutions through comparative studies of nucleotide sequences. *J. Mol. Evol.* 16:111–120.

Kluge, A. 1989. A concern for evidence and a phylogenetic hypothesis of relationships among *Epicrates* (Boidae, Serpentes). *Syst. Zool.* 38:7–25.

Kocher, T., Thomas, W., Meyer, A., Edwards, S., Paabo, S., Villablanca, F., and Wilson, A. 1989. Dynamics of mitochondrial DNA evolution in mammals: Amplification and sequencing with conserved primers. *Proc. Natl. Acad. Sci.* 86:6196–6200.

Krajewski, C. 1989. Phylogenetic relationships amongst cranes (Gruiformes: Gruidae) based on DNA hybridization. *The Auk* 106:603–618.

Leone, C., and Wiens, A. 1956. Comparative serology of carnivores. *J. Mammalogy* 37:11–23.

Li, W.-H., Gouy, M., Sharp, P., O'hUigin, C., and Yang, Y.-W. 1990. Molecular phylogeny of Rodentia, Lagomorpha, Primates, Artiodactyla, and Carnivora and molecular clocks. *Proc. Natl. Acad. Sci.* 87:6703–6707.

Li, W.-H., Luo, C.-C., and Wu, C.-I. 1985. Evolution of DNA sequences. In: R. MacIntyre, ed. *Molecular Evolutionary Genetics*, pp. 1–94. New York: Plenum Press.

Marshall, C. 1990. The fossil record and estimating divergence times between lineages: Maximum divergence times and the importance of reliable phylogenies. *J. Mol. Evol.* 30:400–408.

Marshall, C., and Swift, H. 1992. DNA-DNA hybridization phylogeny of sand dollars and highly reproducible extent of hybridization values. *J. Mol. Evol.* 34:31–44.

Matthew, W. 1909. The Carnivora and Insectivora of the Bridger Basin, middle Eocene. *Mem. Amer. Mus. Nat. Hist.* 9:289–567.

McKenna, M. 1975. Toward a phylogenetic classification of the Mammalia. In: W. Luckett and F. Szalay, eds. *Phylogeny of the Primates: An Interdisciplinary Approach*, pp. 21–46. New York: Plenum Press.

McKenna, M. 1987. Molecular and morphological analysis of high-level mammalian interrelationships. In: C. Patterson, ed. *Molecules and Morphology*, pp. 55–93. Cambridge: Cambridge Univ. Press.

Mivart, St.-G. 1882a. On the classification and distribution of the Aeluroidea. *Proc. Zool. Soc. (London)* 1882:135–208.

Mivart, St.-G. 1882b. Notes on some points in the anatomy of the Aeluroidea. *Proc. Zool. Soc. (London)* 1882:459–520.

Miyamoto, M., and Goodman, M. 1986. Biomolecular systematics of eutherian mammals: Phylogenetic patterns and classification. *Syst. Zool.* 35(2):230–240.

Moritz, C., Dowling, T., and Brown, W. 1987. Evolution of animal mitochondrial DNA: Relevance for population biology and systematics. *Ann. Rev. Ecol. Syst.* 18:269–292.

Neff, N. 1983. The basicranial anatomy of the Nimravidae (Mammalia: Carnivora): Character analyses and phylogenetic inferences. Unpublished Ph.D. dissert., City University, New York.

Neff, N. 1986. A rational basis for a priori character weighting. *Syst. Zool.* 35:110–123.

Norell, M. 1988. Cladistic approaches to evolution and paleobiology as applied to the phylogeny of the alligatorids. Unpublished Ph.D. dissert., Yale University, New Haven.

Norell, M. 1992. Taxic origin and temporal diversity: The effect of phylogeny. In: M. Novacek and Q. Wheeler, eds. *Extinction and Phylogeny*, pp. 89–118. New York: Columbia Univ. Press.

Norell, M., and Novacek, M. 1992. The fossil record and evolution: Comparing cladistic and paleontological evidence for vertebrate history. *Science* 255:1690–1693.

Novacek, M. 1989. Higher mammal phylogeny: The morphological-molecular synthesis. In: B. Fernholm, K. Bremer, and H. Jornvall, eds. *The Hierarchy of Life*, pp. 421–435. Amsterdam: Elsevier.

Novacek, M. 1990. Morphology, paleontology, and the higher clades of mammals. In: H. Genoways, ed. *Current Mammalogy*, vol. 2, pp. 507–543. New York: Plenum Press.

Novacek, M. 1991. "All tree histograms" and the evaluation of cladistic evidence: Some ambiguities. *Cladistics* 7:345–349.

Novacek, M. 1992. Fossils as critical data for phylogeny. In: M. Novacek and Q. Wheeler, eds. *Extinction and Phylogeny*, pp. 46–88. New York: Columbia Univ. Press.

Novacek, M., and Wyss, A. 1986. Higher-level relationships of the Recent eutherian orders: Morphological evidence. *Cladistics* 2:257–287.

Novacek, M., Wyss, A., and McKenna, M. 1988. The major groups of eutherian mammals. In: M. Benton, ed. *The Phylogeny and Classification of the Tetrapods, Vol. 2: Mammals*, pp. 31–71. The Systematics Association Special Volume no. 35B. Oxford: Clarendon Press.

O'Brien, S., Nash, W., Wildt, D., Bush, M., and Benveniste, R. 1985. A molecular solution to the riddle of the giant panda's phylogeny. *Nature* 317:140–144.

Pauly, L., and Wolfe, H. 1957. Serological relationships among members of the Order Carnivora. *Zoologica* 42:159–166.

Romero-Herrera, A., Lehmann, H., Joysey, K., and Friday, A. 1978. On the evolution of myoglobin. *Phil. Trans. Roy. Soc. London* B283:61–163.

Saitou, N. 1989. A theoretical study of the underestimation of branch lengths by the maximum parsimony principle. *Syst. Zool.* 38:1–6.

Sambrook, J., Fritsch, E., and Maniatis, T. 1989. *Molecular Cloning: A Laboratory Manual*. New York: Cold Spring Harbor Laboratory.

Sanger, F., Nicklen, S., and Coulson, A. 1977. DNA sequencing chain-terminating inhibitors. *Proc. Natl. Acad. Sci.* 74:5463–5467.

Sarich, V. 1969a. Pinniped origins and the rate of evolution of carnivore albumins. *Syst. Zool.* 18:286–295.

Sarich, V. 1969b. Pinniped phylogeny. *Syst. Zool.* 18:416–422.

Seal, U. 1969. Carnivora systematics: A study of hemoglobins. *Comp. Biochem. Physiol.* 31:799–811.

Sheldon, F. 1987. Rates of single-copy DNA evolution in herons. *Mol. Biol. Evol.* 4:56–69.

Shoshani, J. 1986. Mammalian phylogeny: Comparison of morphological and molecular results. *Mol. Biol. Evol.* 3(3):222–242.

Simpson, G. 1944. *Tempo and Mode in Evolution.* New York: Columbia Univ. Press.

Simpson, G. 1945. The principles of classification and a classification of mammals. *Bull. Amer. Mus. Nat. Hist.* 85:1–350.

Simpson, G. 1953. *The Major Features of Evolution.* New York: Columbia Univ. Press.

Simpson, G. 1971. *The Meaning of Evolution.* New York: Bantam Books.

Smith, A., Lafay, B., and Christen, R. 1992. Comparison of morphological and molecular rates of evolution: 28S ribosomal RNA in echinoids. *Phil. Trans. Roy. Soc. London* B338:365–382.

Smith, A., and Littlewood, D. 1994. Paleontological data and molecular phylogenetic analysis. *Paleobiology* 20:259–273.

Swofford, D. L. 1990. *PAUP: Phylogenetic Analysis Using Parsimony, Version 3.0.* Champaign: Illinois Natural History Survey. (Computer program.)

Tajima, F. 1993. Simple methods for testing the molecular evolutionary clock hypothesis. *Genetics* 135:599–607.

Tedford, R. 1976. Relationship of pinnipeds to other carnivores (Mammalia). *Syst. Zool.* 25:363–374.

Teilhard de Chardin, P. 1915. Les carnassiers des phosphorites de Quercy. *Annales Paleontologie (Paris)* 9:103–192.

Thein, S. 1989. A simplified method of direct sequencing of PCR amplified DNA with Sequenase T7 DNA polymerase. *U.S.B. Comments* 16:8–9.

Trouessart, E. 1885. Catalogue des mammifères vivants et fossiles. Fasc. IV: Carnivores (Carnivora). *Bull. Soc. Études Sci. Angers* 15:1–108.

Vrana, P., Milinkovitch, M., Powell, J., and Wheeler, W. 1994. Higher level relationships of the arctoid Carnivora based on sequence data and "total evidence." *Mol. Phylogenetics Evol.* 3:47–58.

Wayne, R., Benveniste, R., Janczewski, D., and O'Brien, S. 1989. Molecular and biochemical evolution of the Carnivora. In: J. Gittleman, ed. *Carnivore Behavior, Ecology, and Evolution,* pp. 495–535. Ithaca, N.Y.: Cornell Univ. Press.

Wayne, R., Van Valkenburgh, B., and O'Brien, S. 1991. Molecular distance and divergence time in carnivores and primates. *Mol. Biol. Evol.* 8(3):297–319.

Werdelin, L., and Solounias, N. 1990. Studies of fossil hyaenids: The genus *Adcrocuta* and the interrelationships of some hyaenid taxa. *Zool. Jour. Linnean Soc.* 98:363–386.

Werdelin, L., and Solounias, N. 1991. The Hyaenidae: Taxonomy, systematics and evolution. *Fossils and Strata* 30:1–104.

Wortman, J., and Matthew, W. 1899. The ancestry of certain members of the Canidae, the Viverridae, and Procyonidae. *Bull. Amer. Mus. Nat. Hist.* 12:109–139.

Wozencraft, W. 1989. The phylogeny of the Recent Carnivora. In: J. Gittleman, ed. *Carnivore Behavior, Ecology, and Evolution,* pp. 495–535. Ithaca, N.Y.: Cornell Univ. Press.

Wu, C.-I., and Li, W.-H. 1985. Evidence for higher rates of nucleotide substitution in rodents than in man. *Proc. Natl. Acad. Sci.* 82:1741–1745.

Wurster, D., and Benirschke, K. 1968. Comparative cytogenetic studies in the Order Carnivora. *Chromosoma (Berlin)* 24:336–382.

Wurster-Hill, D., and Bush, M. 1980. The interrelationships of chromosome banding patterns in the giant panda (*Ailuropoda melanoleuca*), hybrid bear (*Ursus middendorffi × Thalarctos maritimus*), and other carnivores. *Cytogen. Cell. Genet.* 27:147–154.

Wurster-Hill, D., and Centerwall, W. 1982. The interrelationships of chromosome banding patterns in canids, mustelids, hyaenids, and felids. *Cytogen. Cell. Genet.* 34:178–192.

Wurster-Hill, D., and Gray, C. 1975. The interrelationships of chromosome banding patterns in procyonids, viverrids and felids. *Cytogen. Cell. Genet.* 15:306–331.

Wurster-Hill, D., Ward, O., Davis, B., Park, J., Moyzis, R., and Meyne, J. 1988. Fragile sites, telomeric DNA sequences, B chromosomes, and DNA content in raccoon dogs, *Nyctereutes procyonoides,* with comparative notes on foxes, coyote, wolf, and raccoon. *Cytogen. Cell. Genet.* 49:278–281.

Wyss, A. 1987. The walrus auditory region and the monophyly of pinnipeds. *Amer. Mus. Novitates* 2871:1–31.

Wyss, A., and Flynn, J. 1993. A phylogenetic analysis and definition of the Carnivora. In: F. Szalay, M. Novacek, and M. McKenna, eds. *Mammal Phylogeny: Placentals,* pp. 32–52. New York: Springer-Verlag.

Wyss, A., Novacek, M., and McKenna, M. 1987. Amino acid sequence versus morphological data and the interordinal relationships of mammals. *Mol. Biol. Evol.* 4(2):99–116.

CHAPTER 17

Carnivoran Ecomorphology:
A Phylogenetic Perspective

LARS WERDELIN

When we speak of carnivores we mean one of two things. The more common usage is for animals of all kinds whose diet includes a large proportion of meat. The more restricted usage is reserved for the monophyletic order Carnivora, diagnosed on the basis of features such as the presence and position of the carnassial shearing teeth. Since this is a book on the order Carnivora I shall be concerned mainly with the second usage of the term, which here, as in many other publications, is modified to carnivoran to avoid confusion. Other mammalian carnivores will also be considered, however, since these have had substantial impact on carnivoran ecomorphology through time.

I see the evolution of carnivores over the past 55 million years or so as an interplay of the two major forces that shape the morphology of any organism: ecology and phylogeny. Only by studying both the proximal (ecological, behavioral, etc.) and historical (phylogenetic) forces that have shaped an animal's morphology can we fully understand its adaptations to its environment. Though this is not a new idea, it does deserve some attention, not least because textbooks and popular articles abound with examples of purported "convergence," such as that between the marsupial thylacine (*Thylacinus cynocephalus*) and the placental wolf (*Canis lupus*), that are in fact little more than superficial resemblance when one looks beneath the skin of these animals.

The terms *ecomorphology* and *constraint,* which I shall use, have been applied in many contexts, and it is not my purpose here to argue for one or another interpretation of their utility. Still, some operational definitions are required to ensure clarity in the discussions to follow. I will view ecomorphology as a "taxon-free" concept describing how animals function in their environment at a more detailed level than simply as "omnivore" or "carnivore." Martin (1989) uses terms such as "doglike" and "catlike" ecomorph. I shall try to avoid these as descriptive terms, for I feel that they raise difficulties on two counts. First, they tend to steer thoughts of ecomorphology along taxonomic lines, and taxonomy is but one aspect of the whole. Second, they

582

circumscribe the notion of what ecomorphology a cat or a dog should have, which seems to me unwarranted and unnecessary. Here, I shall be using terms like "stalk-and-ambush" or "ambush-and-slash" instead. Carnivores can, of course, be subdivided in many ways along these lines, and if the system is to lead to testable hypotheses it must be quantified and formalized. That program is beyond the scope of this contribution, however. A brief list of groupings I shall mention, and some identifying characteristics for each, are given in Table 17.1.

"Constraint" has been used in even more ways than has "ecomorphology" (Antonovics and van Tienderen 1991). In a recent review, McKitrick (1993) has discussed this subject, suggesting (p. 309) that phylogenetic constraint is "any result or component of the phylogenetic history of a lineage that prevents an anticipated course of evolution." I find several problems with this definition. First, it allows *relative* constraints, in the sense that a trait can be relatively less frequent or show more reduced variability in one lineage than in another, owing to some constraint's operating in the one group but not in the

Table 17.1. A list of some ecomorphological groupings as used herein, some of their characteristics, and some typical taxa. There is some overlap between categories.

Grouping	Characteristics	Extant taxa	Fossil taxa
Scavenger/omnivore	Limited specific adaptations to carnivory; moderately developed crushing molars or premolars	Striped and brown hyenas	Some mesonychids, enteledontids
Bone-crushing	Surface-to-surface grinding and comminuting of bone; flat, robust molars	Larger canids	Larger canids, some amphicyonids
Bone-cracking	Point-to-point cracking open of bone; pyramidal premolars	Hyenas, especially *Crocuta*	Hyenas, percrocutids, *Borophagus*
Stalk-and-pounce (or stalk-and-ambush)	Killing of prey by neck bite or shake after short chase and/or rapid pounce; oval, pointed canines, limited premolar function, noncursorial postcranially	Felids, viverrids, some canids, some mustelids	Felids, some mustelids, some hyaenids
Ambush-and-slash	Killing of prey by slashing, shallow wound; long, compressed canines	None	Nimravids, some felids, creodonts, *Thylacosmilus*
Pursuit carnivore	Killing of prey after extended chase, either in packs or solitary; strong cursorial adaptations in postcranium, no specific killing adaptations in dentition	Many canids	Some canids, some hyaenids

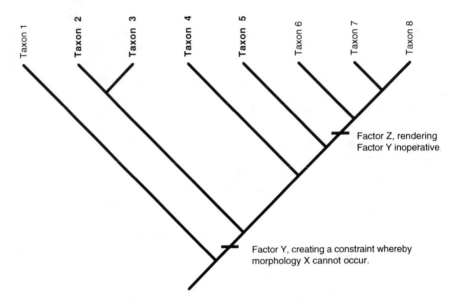

Figure 17.1. Schematic illustration of the concept of constraint, as used herein. A factor Y comes into being at a certain time in the phylogeny of a group of taxa, leading to a constraint rendering the presence of morphology X impossible. At a later time, factor Z evolves, eliminating the effects of factor Y. The taxa shown in bold are those that exhibit the constraint. If any of these taxa can be shown to have morphology X, the constraint would be considered invalid in the present view.

other. Second, it begs the question of how an "anticipated course of evolution" can be identified. Therefore, I have here and elsewhere preferred a more restrictive viewpoint, in which a constraint is seen as an absolute phenomenon as long as it is in effect. In other words: morphology (or behavior, etc.) X is entirely absent from group A as long as factor Y is in force. At a time after the origination of factor Y, the evolution of factor Z can render Y nonoperational, thereby releasing its constraint on morphology and allowing the presence of morphology X (see Figure 17.1). As thus defined, constraints are possible even in paraphyletic groups. Factors Y and Z must have a logical existence independent of their effects, something that would not always be the case under a relative view of constraint.

There are many examples of the interplay between proximal and historical forces. Marsupial and placental carnivores exhibit some degree of convergence, owing to the fact that both hunt prey and eat meat. Their dental morphologies and growth patterns, however, are quite different, because of their long, separate phylogenetic histories, with dramatic consequences for their evolution (Werdelin 1987). The same is true for the bone-cracking forms of the carnivoran families Canidae and Hyaenidae (Werdelin 1989). This effect is seen in a study of iterative hypercarnivory in canids by Van Valkenburgh (1991). Her study shows that under certain conditions a number of canids have approached a hypercarnivorous dental morphology by reducing the crushing portion of the dentition and increasing the sectorial portion. The sectorial ascendancy is seen particularly in the unicuspid m1 talonid of some hypercarnivorous canid taxa. The different canids showing this suite of features converge on extremely similar solutions owing to their phylogenetic relatedness, but none gives any appearance of approaching the extreme of hypercarnivory seen in more distantly related groups such as the Hyaenidae and Felidae, in which the postcarnassial dentition is fully reduced, the m1 talonid is minimal or entirely absent, and crushing emphasis has shifted entirely to the premolars, as is seen particularly well in hyaenid evolution (Werdelin and Solounias 1991).

In my review of ecomorphology I will try to include phylogenetic aspects where possible, and in some cases suggest questions that would be fruitfully pursued from a phylogenetic point of view. I will make no attempt to explain observed patterns of taxonomic or other change, or the presence or absence of specific ecomorphologies in terms of changing climates or environments; that is, there will be no discussion of proximal causes. Though clearly important, this type of treatment lies beyond the scope of this paper. See Janis (1993) for a recent review of mammalian evolution along these lines.

The chapter will be organized as follows. After a brief introduction to carnivorous mammals that are not carnivorans, I shall present some notes on family-level interrelationships within Carnivora, in order to provide a framework for some of the subsequent discussions. The major carnivoran radiations will be presented in a more or less chronological fashion, starting with early Tertiary carnivorans, and followed by sections on the caniform and feliform radiations of the Oligocene, Miocene, and Plio-Pleistocene. Though this material will mostly be a survey of the literature, it will draw heavily on my own recent work on carnivorans of the Old World. The data will then be synthesized chronologically for North America, Eurasia, and Africa, and I shall conclude with an attempt to draw some general conclusions from the observed patterns.

Carnivores That Are Not Carnivorans

Wholly or partly carnivorous species have appeared in many mammal groups. In four groups, however, there have evolved sets of taxa that can be considered to show some explicit adaptations to carnivory. Two of these, the mesonychids (Acreodi *sensu lato,* here taken to include taxa such as *Andrewsarchus* that may belong to the Arctocyonia) and the entelodontids (Artiodactyla), show limited adaptations to carnivory in their dentition, and may best be considered scavenger/omnivores. The Mesonychidae (Figure 17.2) appear to represent the first attempt by a mammal group to adapt to the large-carnivore niche (Szalay and Gould 1966), but they were rapidly replaced by

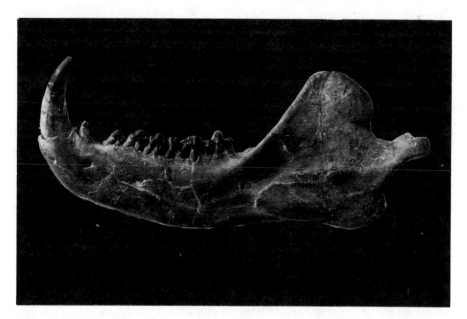

Figure 17.2. Left mandibular ramus of *Dissacus saurognathus*, a Torrejonian mesonychid from New Mexico, USA. Swedish Museum of National History, Palaeozoology (RM PZ) M5640 (cast of AMNH 2454). ×¹/₃.

the creodonts. The mesonychids died out in Europe in the Lutetian and in North America in the Uintan (both Eocene), though in Asia they appear to have survived into the Oligocene (Qi 1975).

The Entelodontidae are a strictly North American group of piglike artio-dactyls showing some dental adaptations to both shearing and durophagy (Peterson 1909; Scott 1941). It has been suggested (e.g., Van Valkenburgh 1988) that in the absence of hyenas in North America they filled the role of bone-crushing scavengers, a function they may have had partly by virtue of their size and partly because of their relatively high-crowned and broad-based premolars.

There are, however, two major noncarnivoran mammalian groups that in-clude significant radiations of carnivorous forms. The first of these is the Marsupialia. The marsupial carnivores comprise two distinct radiations, the Australian Dasyurida and the South American Borhyaenida. Both of these groups have been studied intensively in recent decades (Marshall 1978; Archer 1982), particularly from a phylogenetic point of view.

Despite their separate origins, borhyaenids and dasyurids are very similar in their craniodental anatomy, so much so that the thylacine, *Thylacinus cyno-cephalus,* has at times been suggested to be a borhyaenid, most recently by Archer (1976). After renewed study of both molecular and morphological data, however, it now seems clear that it belongs to the Australian marsupial radiation (Kirsch and Archer 1982; Sarich et al. 1982). Both borhyaenids and dasyurids have the full marsupial complement of four molars, all of which exhibit the basic carnassial pattern of a long shearing blade and an almost entirely reduced crushing basin. This stereotypic pattern has been hypothesized by Werdelin (1987) to be due to a phylogenetically based developmental constraint on the growth of the molar row in polyprotodont marsupials. The reduced morphological diversity (relative to that of placental carnivores) that this constraint has led to has had as a consequence that marsupial carnivores also have a relatively low diversity of species-level taxa compared with the placental carnivores (Werdelin 1987). This distinction shows the interplay between adaptation and phylogeny in the evolution of mammalian groups. In their dental anatomy these taxa show strong adaptations to a carnivorous mode of life, but their phylogenetic background places powerful limits on how that anatomy is integrated into a complete dental apparatus.

Two diprotodont marsupial groups include members that are partly or wholly carnivorous. One is the propleopine kangaroos, some of which show moderate adaptations to carnivory in their sectorial premolar dentition (Archer and Flannery 1985). The other is the Thylacoleonidae, or marsupial lions. Within the latter group there are two main evolutionary lineages, represented by the genera *Wakaleo* and *Thylacoleo* (Archer and Dawson 1982). The former includes smaller, carnivorous or frugivorous species emphasizing the posterior molars. The latter genus includes large, strictly carnivorous species with enormous pseudocarnassials and reduced molars. These adaptations were possible because the evolution of diprotodont marsupials involved changes to the craniodental apparatus that removed the primitive polyprotodont constraint on the evolution of the molar dentition (Werdelin 1988a).

The second group of highly carnivorous, noncarnivoran mammalian taxa is the Creodonta, an order consisting of two groups generally viewed as families, the Oxyaenidae and the Hyaenodontidae. Although the evidence is limited, these families are probably sister-groups and together form the sister-group to the order Carnivora (Wyss and Flynn 1993).

The Oxyaenidae and Hyaenodontidae were distinct from each other at their earliest known occurrence, with the first major radiation occurring in the Clarkforkian (Rose 1981), although both families are known from older deposits in North America. In Europe, creodonts did not appear until the Lower Eocene, although this finding may be due simply to a limited knowledge of Paleocene mammal faunas in Europe. Oxyaenids, which had their main radiation in North America, were represented in Eurasia by only a few taxa, such as *Oxyaena menui* (Rich 1971) and the largest of all creodonts, *Sarkastodon mongoliensis* (Granger 1938). In North America their always limited radiation was made up principally of the genera *Oxyaena* and *Patriofelis*, and they became extinct before the end of the Eocene.

The Hyaenodontidae have a quite different history. Although their origin may have been in North America they spread rapidly into Eurasia and were present in Africa by the late Eocene (Savage 1978). Their subsequent history on the three continents is very curious. After a brief radiation in the middle to late Eocene (Bridgerian-Duchesnean), North American hyaenodontid diversity became limited to a number of species of the genus *Hyaenodon,* the North American members of which have been analyzed by Mellett (1977). The youngest North American *Hyaenodon* appears to be *H. brevirostris* from the early Arikareean (late Oligocene). In Eurasia the same pattern is evident: a brief radiation of different genera in the middle to late Eocene (Robiacian and Headonian), after which *Hyaenodon* remained the only survivor for some time (Lange-Badré 1979). The subsequent Eurasian history, however, was quite different. Eurasian hyaenodontids displayed a resurgence during the early and middle Miocene, with the appearance of massive animals in such genera as *Hyainailouros* (Ginsburg 1980) and *Dissopsalis* (Barry 1988). The origin of these forms is obscure, although if Ginsburg (1980) is correct in placing both Eurasian and African material in the genus *Hyainailouros,* they may have been immigrants from the latter continent, which has an even more remarkable hyaenodontid history. Few Oligocene localities producing creodonts are known from Africa, but there must have been a hyaenodontid radiation on the continent at that time, because by the middle Miocene (Rusingan), hyaenodontid diversity was apparently greater in Africa than at any other time or place in the history of the group (Savage 1965, 1973, 1978). This diversity was short-lived, however, and by the beginning of the late Miocene the creodonts were extinct, except for a few lingering forms on the Indian subcontinent (Barry 1988).

Several comparative taxonomic analyses of creodonts have been published (e.g., Denison 1938; Springhorn 1980; Gebo and Rose 1993), but, unfortunately, very few phylogenetic hypotheses of creodont interrelationships have appeared, that of Barry (1988) being an exception. Still, the analysis of the postcranial anatomy of some early Eocene hyaenodontids and oxyaenids by Gebo and Rose (1993) provides some hints about the possible overall pattern of creodont evolution. These authors show that the postcranial anatomy of the limnocyonine hyaenodontid *Prolimnocyon* has adaptations to arboreality, a feature seen to a lesser degree in *Thinocyon* and *Limnocyon*. The oxyaenid *Oxyaena* differs from these forms in being, probably, confined largely to the ground. One might generalize from this difference to suggest that oxyaenids from the start were basically terrestrial and hyaenodontids were more or less arboreal. When the oxyaenids became extinct, for reasons possibly related to their generally limited adaptations to carnivory, the hyaenodontids developed true terrestrial forms (*Hyaenodon* spp.). At the same time, arboreally adapted hyaenodontids became extinct.

One other general pattern is apparent in the history of creodonts. Both oxyaenids and hyaenodontids had a strong tendency to develop into larger and larger forms, and in both groups the last representatives (*Patriofelis* and *Sarkastodon* among the Oxyaenidae and *Hyainailouros* and *Megistotherium* among the Hyaenodontidae) showed clear adaptations to scavenging and du-

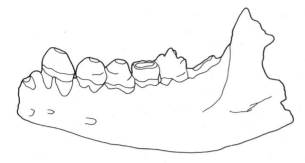

Figure 17.3. Partial left mandibular ramus of *Quercytherium tenebrosum,* a late Eocene hyaenodontid creodont from the Quercy beds, France. Note the bulbous premolars, an adaptation to durophagy. Natural size. (After Filhol 1882.)

rophagy, in both their large size and their pyramidal premolars, which in analogy with hyenas were presumably used for cracking bones.

In order that students of carnivore paleontology not become complacent about their ability to pigeonhole every taxon into a functional category, it should be noted that hyaenodontid evolution provided the remarkable genus *Quercytherium,* with two species, *Q. tenebrosum* and *Q. simplicidens* (Filhol 1880a, b; Lange-Badré 1979). In both these species, and in particular the more derived *Q. tenebrosum* (Figure 17.3), all premolars are broad-based and adapted to durophagy, the most extreme such adaptation seen in any carnivore. It has been suggested by many authors since Filhol (e.g., Savage 1977) that this adaptation is analogous to that of hyenas. I agree with Lange-Badré (1979), however, that the hyaenoid aspect of the dentition has been overemphasized at the expense of other interpretations. *Quercytherium* is quite small (all of its lower teeth are considerably less than 10 mm in length), thereby limiting the size of bones that could be cracked to a range where such an extreme specialization would hardly have been necessary. In addition, the premolars of *Quercytherium* are situated far forward in the jaw, in a position that is mechanically disadvantageous for maximum bite force (Greaves 1983). On the other hand, both the shape of the premolars and the wear pattern exhibited in *Q. tenebrosum* indicate that the animal ate some hard foodstuff. What that may have been must at present be considered a mystery.

Here, we have briefly surveyed major radiations of mammalian carnivores that do not belong to the Carnivora. We may note that these radiations differ from those of the Carnivora in a number of respects, such as the number and ecomorphological diversity of species, that seem to have their roots in the phylogenetic background of these groups. There are also present in these radiations several forms that apparently have no modern analogue in terms of ecomorphology.

Carnivoran Phylogeny

It is clear that understanding the evolution of carnivoran ecomorphology requires a good grasp of carnivoran phylogeny. In the past 20 years, a series of important, morphologically based papers on carnivoran phylogeny have appeared, in the wake of a dramatically increased interest in and understanding of phylogenetic reconstruction. Papers such as those of Hunt (1974a, 1987, this volume), Tedford (1976), Schmidt-Kittler (1981), Flynn and Galiano (1982), Neff (1983), and Flynn et al. (1988) have served to further our knowledge of the interrelationships of the higher taxa within the Carnivora. Much of this work has recently been summarized by Wyss and Flynn (1993; cf. Flynn, this volume), and we need not go into detail here regarding these matters. Molecular phylogenies of carnivorans have also appeared during this time (e.g., Collier and O'Brien 1985; Wayne et al. 1989). The majority of these have been based on distance criteria, and it has accordingly been difficult to evaluate them alongside the morphological studies under the same criteria. But although few sequence-based studies have as yet appeared, that situation is about to change (see Flynn, this volume), and we may expect significant interplay between morphological and molecular phylogenetic studies of Carnivora in the near future, along the lines of the studies of Wayne et al. (1991) and Flynn (this volume).

There are still, however, a number of points of disagreement regarding higher carnivoran phylogeny, as is evidenced by the paper by Hunt and Tedford (1993), published in the same volume as that of Wyss and Flynn but espousing an entirely different set of interrelationships within the Feloidea.

Beyond these disagreements, which are due mainly to a difference of opinion regarding the weight to be given to different character complexes, there is the problem of the place of fossils in the reconstruction of carnivoran phylogeny. Two issues are involved here (Novacek 1992; Werdelin 1993). One is the inclusion or exclusion of exclusively fossil taxa, such as the Nimravidae. For the most part, this issue has been satisfactorily resolved (Gauthier et al. 1988; Donoghue et al. 1989), and fossil groups are indeed incorporated in the phylogeny of Wyss and Flynn (1993). The second, less well resolved, issue concerns the use of fossils of groups with extant representatives in determining character polarities and primitive states for the groups.

That second problem is especially clear in the phylogeny of the Feloidea, where the topology of Wyss and Flynn (1993) hinges to a great extent on the characters posited as uniting Felidae and Hyaenidae (cf. their figure 4.4). They are (numbered as in Wyss and Flynn): alisphenoid canal absent (#12), major a2 arterial shunt large, intracranial rete (#28), p1 absent (#40), m1 talonid absent (#41), M1 reduced or absent (#45), M2 absent (#46), m2 absent (#50), and hallux greatly reduced or absent (#56). With the exception of #28, which is a soft-anatomy character not determinable on fossil material, the data from fossil taxa unambiguously assigned to the Hyaenidae (Werdelin and Solounias 1991) show that all of the characters thus posited are parallel developments in Felidae and Hyaenidae. I have recoded the data accordingly and reanalyzed the matrix presented by Wyss and Flynn (1993) using PAUP version 3.1.1 Swofford (1990). This results in a single tree of 123 steps (with multiple states within taxa considered as uncertainty, as in the analysis of Wyss and Flynn). The resulting tree is shown in Figure 17.4.

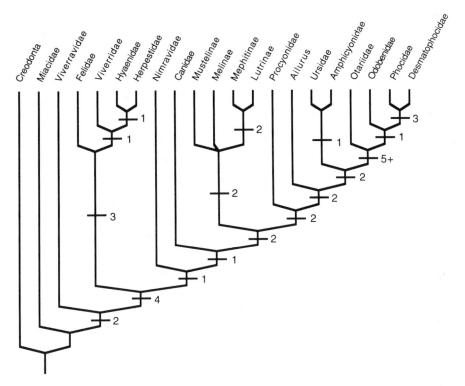

Figure 17.4. Phylogenetic hypothesis regarding the interrelationships of carnivoran families. The numbers at the nodes represent the Bremer support values for those nodes (see Källersjö et al. 1992). See also discussion in text and Table 17.2.

This tree differs in some respects from that presented by Wyss and Flynn (1993). First, the branching topology within the Feloidea is, as expected (given the changes to the data matrix), altered. Hyaenidae is the sister-group to Herpestidae (two unambiguous synapomorphies), and the sister-group to these is Viverridae (two unambiguous synapomorphies), while Felidae is the sister-group to all other Feloidea (seven unambiguous synapomorphies). The synapomorphy structure supporting this topology is given in Table 17.2. The other change relative to the topology of Wyss and Flynn (1993) is the position of the Nimravidae, which have gone from being the sister-group to Feloidea to being the sister-group of the Caniformia of Wyss and Flynn. The latter position accords with the earlier viewpoint of Flynn and Galiano (1982), but the support for this node is weak. The only unambiguous supporting character is #38 (reduced P4 protocone), which has a complex distribution, but is here (and in Flynn and Galiano 1982) seen as a synapomorphy linking Nimravidae + Caniformia, with various reversals and parallelisms throughout the Carnivora. A second character, showing support for the exclusion of the Nimravidae from the Feliformia in some optimizations, is the development of the P4 parastyle. I

Table 17.2. Synapomorphy scheme (unambiguously optimized characters only) for the Feloidea, as shown in Figure 17.4. Character formulations follow Wyss and Flynn (1993).

Groups united by specified characters	Characters uniting indicated groups
Hyaenidae and Herpestidae	Ectotympanic contributes to external auditory meatus
	Enlarged anal glands and anal sac
Hyaenidae, Herpestidae, and Viverridae	Anterior palatine foramina located anterior of palatine/maxilla suture
	Medio-ventral surface of petrosal with distinct ventral petrosal process
Feloidea	Postglenoid foramen absent or reduced
	Hypoglossal and posterior lacerate foramen closely associated
	Posterior entrance of carotide artery into auditory capsule anterior and not enclosed in tube
	Ectotympanic septum present
	Paroccipital process cupped anteriorly around posterior surface of bulla
	Cowper's gland absent
	Prostate gland large, ampulla bilobed

conclude—as have Flynn and Galiano (1982), Bryant (1991), and Wyss and Flynn (1993)—that the evidence for the phylogenetic position of Nimravidae is weak at best, though there is some evidence for placing them at the base of the Caniformia, and this topology will be used in the following discussion.

The Early Tertiary Carnivorans

Relationships of the "Miacoidea"

For a long time, all carnivorans that could not be conveniently assigned to either the Feloidea or the Caniformia were lumped together as "miacids," or "Miacoidea" (Simpson 1931, 1945), and this obviously paraphyletic group was understood to include the ancestors of later taxa in both main lineages. Even in early analyses (e.g., Matthew 1909) it was clear that there was a division within this group into, on the one hand, the Viverravinae, as typified by *Viverravus* (Figure 17.5) and, on the other, the Miacinae, as typified by *Vulpavus* (also Figure 17.5). This division was noted by Tedford (1976) in his analysis of relationships within the Carnivora, and he considered the relationships of Miacoidea to modern carnivorans "one of the largest gaps in our knowledge of the phylogeny of the Carnivora" (Tedford 1976:364).

The question of this gap was addressed by Flynn and Galiano (1982), who presented a phylogeny in which the Miacinae were related to the Caniformia, whereas the Viverravinae were included in the Feliformia. In a recent analysis of Carnivora (Wyss and Flynn 1993) this disposition was changed again, with the most parsimonious phylogeny indicating that the "miacids" were the sister-group to "viverravids" plus all remaining carnivorans (the latter being the

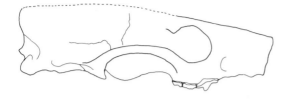

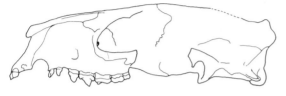

Figure 17.5. Skulls of *Viverravus minutus* (*above*, ×³/₄) and *Vulpavus profectus* (*below*, ×1.4), Bridgerian carnivorans from North America. (After Matthew 1909.)

Carnivora of Wyss and Flynn). In the same volume, however, Hunt and Ted-ford (1993) suggest the possibility that Caniformia and Feliformia have evolved from separate ancestors within the *"Cimolestes"* group, that is, that Carnivora as traditionally conceived is not a monophyletic group.

The matter of "miacoid" relationships is far from settled, at present hinging on the polarity of dental characters (loss of posterior molars, development of shearing locus). This situation can be resolved only with more detailed study of "miacoids" and a better understanding of the relationships within the *"Cimolestes"* group, and also leaves unanswered the question of the position of taxa such as *Quercygale* (Figure 17.6; also Werdelin, in prep.), which has basicranial characters indicating affinity with Caniformia, but dental characters (such as loss of posterior molars) suggesting affinity with Feliformia (although this character is paralleled in many derived caniforms).

Paleoecology of the "Miacoidea"

Because of the problems attendant upon early Tertiary carnivoran interrelationships, our understanding of the ecomorphological evolution of these taxa is limited. Presenting a causal explanation for the development of viverravids from a miacid-like ancestor, as implied by most schemes, is quite a different matter from positing one based on a parallel development of these groups from a *"Cimolestes"*-like ancestor as suggested by Hunt and Tedford (1993). The phylogeny shown in Figure 17.4 indicates, for example, that the posteriormost lower molar seen in some caniforms (m3) is a neomorph, whereas if caniforms and feliforms are parallel derivations from *"Cimolestes,"* this tooth is plesiomorphic to the former group.

From the beginning of carnivoran evolution there have been two basic

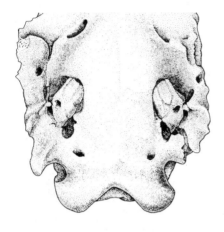

Figure 17.6. Basicranium of *Quercygale angustidens,* a probable late Eocene miacid from the Quercy beds, France. RM PZ M2329. ×⁴/₃.

morphotypes, one with a tendency to reduction of posterior molars and one in which these molars tended to be retained. The difference between these types can be seen already in *Viverravus* and *Vulpavus* (Figure 17.5). In the former, the tooth row including the canines and incisors makes up somewhat over 40% of the length of the skull, whereas in the latter the tooth row is 45% of skull length. It is not clear which of these is the primitive condition, although the long tooth row of creodonts and the correlation between a short tooth row and derived features such as the loss of posterior molars both argue for a long tooth row being primitive. More important, the tooth row is situated quite differently in the two forms. In *Viverravus* the anterior rim of the orbit is situated approximately above the anterior end of P4, and the infraorbital foramen is above P3. In *Vulpavus,* by contrast, though the infraorbital foramen is still placed above P3, the anterior rim of the orbit lies above the front of M1. This difference reflects in an incipient form a basic dichotomy seen in much later forms belonging to the Caniformia and Feliformia (Werdelin 1989). The division is not a firm one, but there is a tendency for forms with the *Vulpavus* pattern to belong to the Caniformia, and forms with the *Viverravus* pattern to belong to the Feliformia. Radinsky (1982:192), found no evidence of "family differences in ecology that might have been involved in their differentiation from each other," but the taxa he studied (*Vulpavus, Oodectes, Uintacyon,* and *Procynodictis*) all belong to the Miacidae and all have the *Vulpavus* skull pattern. Conjoint analysis of *Viverravus* might have yielded a different interpretation. The two structural types seem to be two functionally and morphogenetically persistent, apparently equally successful, approaches to carnivory within the Carnivora.

Among early Tertiary carnivorans, the postcranial skeleton is known for some genera of Miacidae and for the viverravid (as defined here and in Wyss and Flynn 1993) *Didymictis.* The majority of the former show adaptations to an arboreal mode of life, while the latter, although retaining a residue of such a lifestyle, shows adaptations to terrestrial locomotion as well (Rose 1990; MacLeod and Rose 1993). Unfortunately, very few postcranial elements of other Viverravidae and of many genera of Miacidae are known, and it is therefore not known whether they were arboreally, terrestrially, or otherwise adapted.

Craniodentally, the *Vulpavus* pattern seems, as suggested by Matthew (1909), more indicative of an omnivore/insectivore niche, whereas the *Viverravus* pattern indicates more highly carnivorous pursuits. The following three ecomorphological types were thus present in the Eocene: an arboreal omnivore/insectivore group (*Vulpavus* group, i.e., Miacidae, some Caniformia), a group moderately adapted to carnivory but having unknown locomotory habitus (*Viverravus* group, i.e., Viverravidae, some Caniformia), and a terrestrial or semiterrestrial group with apparently both omnivore/insectivore and carnivore adaptations in the dentition (Didymictida [Feliformia] of Flynn and Galiano 1982). The third of these groups also shows a tendency toward larger size in derived taxa. This range of ecomorphologies in the early Tertiary Carnivora echoes that seen within certain extant carnivoran families, for example Viverridae, while spanning several clades among the early Tertiary genera. Apparently absent in the early Tertiary are true adaptations to large size and to the aquatic habitat.

These patterns seem equivalent on both sides of the Atlantic. As has long been known, Europe and North America shared a number of carnivoran taxa in the early Eocene, including *Viverravus, Miacis,* and possibly *Vulpavus* (e.g., Quinet 1966). Later in the Eocene, isolation led to some endemicity, and genera such as the probable miacid *Quercygale* (Figure 17.6) were confined to Europe. The basic ecomorphological patterns are the same, however. Despite the loss of M3 (and in some individuals m3 as well), the upper tooth row of *Quercygale* is 46% of skull length, a number that is close to that seen in *Vulpavus,* and one that represents the presumed primitive condition for Carnivora.

In summary, there are clear indications of two different, stable patterns of skull morphology in early Tertiary carnivorans, patterns that remain visible in skulls of extant carnivorans. Further study of the phylogeny of early Tertiary Carnivora will show whether these morphologies have had a substantial impact on subsequent evolution within Caniformia and Feliformia, in terms of biasing morphological evolution in certain directions.

First Attempts: Late Eocene and Oligocene Carnivoran Radiations

Phylogeny and Paleoecology of the Nimravidae

It is convenient to separate the Nimravidae from the rest of the Late Eocene-Oligocene carnivoran radiation, for two reasons. First, their phylogenetic position is unclear, and, second, their ecomorphological characteristics are quite distinct.

A number of papers discussing the relationships of nimravids to other carnivorans have been published in recent years (Martin 1980, 1989; Flynn and Galiano 1982; Neff 1983; Hunt 1987, this volume; Flynn et al. 1988; Bryant 1991; Wyss and Flynn 1993). These studies have variously hypothesized a sister-group relationship between Nimravidae, on the one hand, and either Felidae, Feliformia, Caniformia, or the remaining Carnivora, on the other, or have concluded that the issue is unresolved at present. The lability of the situation is clear from the discussion above, in which changing the primitive character state in just a few characters of Hyaenidae moved the Nimravidae from the feloid clade to the caniform clade.

Not only are the relationships of Nimravidae debated, but the content of the family, as well, is not clear. There are two groups that may be monophyletic, although even this is disputed (Martin 1980). One is the Nimravinae (Figure 17.7), a late Eocene-Oligocene North American and Eurasian radiation that includes the genera *Nimravus, Dinaelurus, Dinictis, Eusmilus, Hoplophoneus,* and *Pogonodon* (e.g., Scott and Jepsen 1936; Ginsburg 1979; Martin 1980; Bryant 1991). The second group is the Barbourofelinae, a middle to late Miocene Eurasian, African, and North American radiation that includes the genera *Barbourofelis, Prosansanosmilus, Sansanosmilus, Syrtosmilus,* and *Vampyrictis* (Ginsburg 1961; Schultz et al. 1970; Bryant 1991; Werdelin, in press, a). Unfortunately, the phylogenetic relationship between these two groups is not clear. Most studies (e.g., Bryant 1991) have hypothesized a sister-group rela-

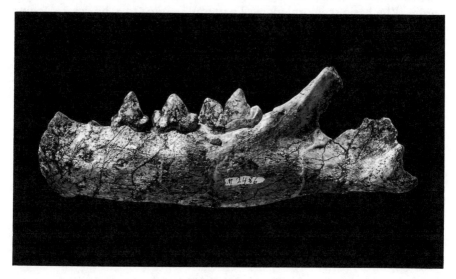

Figure 17.7. Partial left mandibular ramus of *Nimravus intermedius,* an early Oligocene nimravid from the Quercy beds, France. RM PZ M2486. ×³/₄.

tionship between them, but Neff (1983) found clear differences in basicranial characters between Nimravinae and Barbourofelinae and suggested that they might be more distantly related than previously supposed. To compound these difficulties, the interrelationships of genera and species within these two groups are also unclear.

As obscure as is the phylogenetic status of nimravids, however, their ecomorphological status is eminently obvious. For many years they were viewed as an early radiation within the Felidae, under the guise of "paleo-felids," and even now, in the view of some (Martin 1980), they represent the sister-group of Felidae. This phylogenetic position, which is disputed on the basis of basicranial and bullar characters (Flynn and Galiano 1982; Neff 1983; Hunt 1987), was established by using presumed functionally significant characters of the dentition, such as the compressed canines, the reduced premolar and molar counts, and the structure of the carnassials, which are reduced to shearing blades with only a vestige of the talonid remaining. These and other characters of the skull and postcranial skeleton show that nimravids in all probability occupied much the same niches in the Oligocene and early Miocene as felids have occupied since the middle Miocene. With the rise of the Felidae in the Miocene, the nimravids gradually became extinct, although this sequence probably does not represent cause and effect (Werdelin, in press, a).

Unlike the Felidae, all nimravids were essentially sabertoothed, that is, they had laterally compressed canines, the length of which could vary from moderate (e.g., *Nimravus*) to extreme (e.g., *Barbourofelis*). Indeed, nimravids in many ways surpassed the sabertoothed felids in this adaptation, not least in the existence of small sabertoothed nimravids of about the size of a modern domestic cat (Martin 1991). The sabertooth morphology is often viewed as an adaptation to killing large prey (relative to the size of the carnivore doing the killing). If this theory is correct, nimravids as a group would have to have killed relatively larger prey than do modern felids. This corollary would in turn suggest that nimravids, even the smallest of them, did not pursue and kill mouse- or rat-sized prey, as do most small and medium-sized felids of today. Why this should be so is not clear, but the postcranial skeleton of nimravids may not have been sufficiently modified to allow for that sort of stalk-and-pounce hunting technique. Alternatively, the nimravids may have been actively excluded from this niche by the hesperocyonine canids, which were probably faster and more agile than the smaller nimravids. Hesperocyonines did not exist in Eurasia, where the conical-toothed cats (Felidae) may have originated.

Thus, the relationships of Nimravidae are obscure, as regards both the phylogenetic position of the group as a whole and the interrelationships of its member taxa. Its ecomorphological position is nonetheless clear. During the Oligocene and part of the Miocene the Nimravidae occupied a portion of the niche subsequently held by Felidae: that portion which we can term ambush-slash predation on large prey (relative to carnivore size). The other part of the modern felid niche (leaving aside the cooperative hunting of lions and the pursuit hunting of cheetahs), the stalk-and-ambush killing of small or large prey, was never encroached upon by nimravids.

Origins of the Families of Caniformia

The European faunal turnover known as "La Grande Coupure" (Stehlin 1909) has long been known to be one of the most important in the history of the terrestrial fauna. Traditionally, it has been correlated with the Eocene-Oligocene boundary. However, recently (Berggren and Prothero 1992; Hooker 1992), the turnover has been shown to have occurred some few million years into the lower Oligocene, and is thus not correlative with the (more limited) faunal turnover seen at the Chadronian-Orellan boundary in North America, which now is correlated with the Eocene-Oligocene boundary (Prothero and Swisher 1992). Nevertheless, the importance of this "event" or series of events naturally extends to the Carnivora. The early Tertiary groups of Carnivora, the Miacidae and Viverravidae, did not pass this boundary, whereas the caniform families all originated just before or just after "La Grande Coupure." The compilation of Stucky (1992) shows at least Canidae and Amphicyonidae (Arctoidea) from the late Duchesnean (latest Eocene). The Ursidae must therefore also go back to the Eocene, since they are the sister-group to the Amphicyonidae, and the Mustelidae and Procyonidae are surely not much younger.

Whatever the reasons for the faunal turnover at the Eocene-Oligocene boundary and "La Grande Coupure" (discussed extensively in Prothero and Berggren 1992), it is clear that they had a profound impact on the eco-morphological evolution of carnivorans. Previously, the Carnivora had been mainly small, arboreal, or, in a few cases, terrestrial carnivores, insectivores, and omnivores. After the Eocene, carnivorans radiated into all of the niches occupied by them throughout the rest of the Cenozoic.

Phylogeny and Paleoecology of the Oligocene Canids

The first of three great radiations of canids occurred in the Oligocene of North America. This early radiation has generally been gathered under the heading of Hesperocyoninae (Tedford 1978), but recently X. Wang (1992, 1994) has shown that that heading probably represents a paraphyletic assemblage, with different species of *Hesperocyon* giving rise to the later subfamilies Borophaginae and Caninae. The "Hesperocyoninae" radiated from their early origins in small, probably omnivorous forms such as *H. gregarius* (Figure 17.8) into a number of niches. These include niches that are associated with canids today, such as frugivory/insectivory (e.g., the early Miocene *Phlaocyon* [Mat-

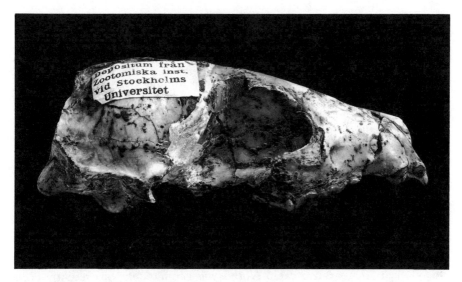

Figure 17.8. Skull of *Hesperocyon gregarius,* an early canid, Lower Oligocene, White River Formation, South Dakota, USA. RM PZ M2324. Natural size.

thew 1899; Dahr 1949] and omnivory/carnivory (e.g., *Mesocyon* and *Cynodesmus* [Matthew 1907]).

Most interesting from an ecomorphological point of view, as well as most derived relative to the basic canid craniodental pattern, is the genus *Enhydrocyon* (Cope 1879; Matthew 1907). This form has a very short face and a dentition reduced beyond the state seen in any other canid, in its lack of P1 as well as p1 and m3. It thus forms a counterexample to the constraint proposed for canids by me (cf. Werdelin 1989; Van Valkenburgh 1991). The short face would seem to indicate a powerful bite, despite the relatively small size of the animal, but aside from the robust carnassials there are no indications in the cheek dentition of *Enhydrocyon* of any adaptations to bone-cracking or scavenging. The enormous enlargement of I3 might, however, suggest some such specialization, in view of the similarly enlarged I3 in some derived scavengers, such as the hyaenid *Pachycrocuta brevirostris.* Perhaps *Enhydrocyon* was a stalk-and-ambush predator and/or scavenger of medium-sized prey, a niche occupied today by felids and in some areas by the mustelid *Gulo gulo,* the wolverine. The closest analogy to *Enhydrocyon* in the Eurasian fossil record is the late Miocene *Simocyon* (Roth and Wagner 1854; Gaudry 1862–1867), which also had a short face and reduced cheek dentition emphasizing the carnassial. Unfortunately, neither the phylogenetic position nor the ecomorphology of *Simocyon* is at present clear.

The work of Wang (1994) will greatly assist us in deciphering the functional morphology of the early canid ("hesperocyonine") radiation, allowing it to become one of the best known of all carnivoran radiations, in terms of eco-morphology. For now, we can merely note that in ecomorphological terms it was an extensive radiation. One aspect, however, is missing: there are no known "hesperocyonines" larger than the modern coyote, and the radiation therefore lacks analogues to the larger forms among the Borophaginae and Caninae. This lack may be due to the fact that the large, generalized carnivore niche had already been occupied by amphicyonids, a family that may be said to be the dominant carnivore (and carnivoran) group of the Oligocene and early Miocene.

The Early Radiation of the Arctoids

The canids were not the only group to radiate rapidly in the late Eocene-Oligocene. The same is true of the other, later caniforms, the arctoids, al-though the distinctions between the various arctoids are not always clear in the early faunas. Neither are the distinctions between the canids and arctoids always clear: the putative canid *Nothocyon geismarianus* has recently been shown to be an arctoid (Wang and Tedford 1992).

Ursids seem to be descended from generalized forms such as *Cephalogale* (Figure 17.9), which is characterized by a long, low carnassial with a wide, flat talonid basin. Their fossil record is limited during the late Eocene-Oligocene, and they do not radiate in earnest until the early Miocene of the Old World, when hemicyonines, including genera such as *Hemicyon* and *Plithocyon*, appear.

Mustelids also originate at this time. Early mustelid phylogeny, as well as the distinction between mustelids and procyonids, has recently been the focus of

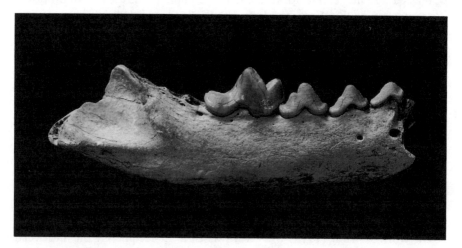

Figure 17.9. Partial right mandibular ramus of *Cephalogale* sp., an Oligocene ursid from the Quercy beds, France. RM PZ M2330. ×1½.

Figure 17.10. Partial skull of *Pseudobassaris* sp., an Oligocene mustelid from the Quercy beds, France. RM PZ M2349. Natural size.

several studies (Schmidt-Kittler 1981; Baskin 1982, 1989; Cirot and de Bonis 1993; Wolsan 1993). These authors differ somewhat among themselves, but they agree that the origins of both mustelids and procyonids must be sought among Oligocene genera such as *Amphictis, Mustelictis,* and *Bavarictis,* and that the generalized genus *Plesictis* may be the most primitive known musteline (de Beaumont and Weidmann 1981). Among the Procyonidae the same position is held by *Pseudobassaris* (Figure 17.10) and *Broiliana.*

During the Oligocene and early Miocene these groups radiated into a number of niches, including aquatic, semiaquatic (e.g., *Potamotherium*), and fossorial (e.g., *Zodiolestes*) ones. There is one group, however, that is of particular interest from an ecomorphological point of view, but has hitherto been little considered. I will therefore emphasize the evolution of large-sized mustelids into a stalk-and-ambush niche, with adaptations closely paralleling those of modern pantherine felids. This "radiation" really consists of only two rare genera, *Aelurocyon* and *Megalictis* (Peterson 1906; Matthew 1907), but it is unlikely, in any case, that any radiation into such a niche would be extensive. Even today there are only a few species of pantherines, and these are, aside from the "cave lion" of the Pleistocene, not common in the fossil record. The significance of this mustelid radiation lies not only in its existence in the early Miocene of North America, but also in its having been (almost certainly)

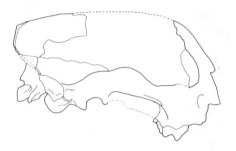

Figure 17.11. Reconstructed skulls of *Aelurocyon* (*above*) and *Megalictis* (*below*), early Miocene mustelids from North America. In *Aelurocyon* the snout is broken, but the high nasals can clearly be seen. Most of the face of *Megalictis* is missing, but enough exists to show the vertical aspect of the frontals. Both ×³/₁₀. *Aelurocyon* is a composite after Peterson (1906) and Riggs (1945); *Megalictis* is after Matthew (1907).

independently paralleled by other mustelids in the late Miocene (see below). It thus appears that mustelids, perhaps because of their short faces, a phylogenetic legacy, were "preadapted" to step into this niche when (for unclear reasons) felids did not occupy it.

Of the two genera, *Aelurocyon* (Figure 17.11) most resembles a large pantherine with its oval (in cross-section) canines, short and high skull, raised nasals, and short but straight face. *Megalictis* (Figure 17.11), by contrast, resembles the cheetah in having a high frontal region and nearly vertical face. There are, however, no other indications of a functional similarity to the cheetah in the known material of *Megalictis,* and the two skull morphologies may be seen as two ways of achieving the same end: a short face and jaw and a concomitantly powerful bite at the carnassials and canines, as also seen in felids and nimravids. In both forms the posterior cheek dentition is reduced, and at least in *Megalictis* the anterior premolars are reduced as well. The lower carnassial has a long, relatively low trigonid and a short talonid with trenchant cusp, while the upper carnassial has a long blade and distinct protocone. All these features are reminiscent of pantherines, and these two genera may represent the best example of ecomorphological convergence on felids in the "cat gap," a time in the early Miocene when (evidently) neither felids nor nimravids were present in North America (Hunt and Joeckel 1988; Van Valkenburgh 1991).

The dominant arctoid group of the late Eocene to early Miocene was, however, the Amphicyonidae. In North America this radiation, like that of the Canidae, occurred in three waves—the Daphoeninae, Temnocyoninae, and

Amphicyoninae—that had only partial temporal overlap (Hunt and Tedford 1993). The situation is little different in Europe, where both Daphoeninae and Amphicyoninae are found as far back as the early Oligocene (Ginsburg 1966; Springhorn 1977). Unfortunately, no hypotheses of interrelationships within the Amphicyonidae are currently available, and studies of ecomorphological pathways are accordingly hampered.

When a phylogenetic hypothesis does emerge, the Amphicyonidae should, in view of their morphological diversity, provide fertile ground for studies of the interplay between phylogeny and ecology in shaping morphology. Both the North American and the European radiations are extensive in ecomorphological terms, and a number of basic "adaptive" types can be discerned in the family. These include generalized carnivore/omnivores, such as *Daphoenus* (Scott and Jepsen 1936), bone-crackers with bulbous premolars, such as *Sarcocyon* (Ginsburg 1966; Springhorn 1977), forms specialized for meat-eating, such as *Daphoenictis* (Hunt 1974b), and short-faced forms, such as *Brachycyon* (Ginsburg 1966), that may have pursued a stalk-and-ambush carnivore lifestyle.

Approaching the Modern: Miocene Carnivoran Radiations

The Miocene radiation of carnivorans did not really begin until the end of the early Miocene, MN 4–5 in the European biostratigraphic scheme (Steininger et al. 1989). Before that time the taxa and ecomorphological patterns seen are essentially those of the Oligocene, as represented by many amphicyonids of various sizes, along with bears, mustelids, and nimravids, and, in North America, canids. The Oligocene radiation was essentially a caniform radiation. The only feliforms present at that time were the putative felid *Proailurus* and a few morphologically primitive forms such as *Stenoplesictis*, the phylogenetic positions of which are uncertain. Certain specimens of the former have been shown to be felids on the basis of bullar and basicranial characters (Hunt 1989), but it does not follow that *all* are (i.e., that *Proailurus* as currently conceived is monophyletic), and the age of the genus *qua* felid is therefore not clear. The phylogenetic position of the stenoplesictines is disputed (Hunt 1991; Werdelin and Solounias 1991), but they perhaps may best be considered stem feloids. Aside from *Proailurus,* the earliest feloid to be definitely referred to one of the modern families is the early Miocene viverrid *Herpestides* (MN 2) (de Beaumont 1967; Hunt 1991). At about the same time, the felid *Pseudaelurus* appears.

The Radiation of the Extant Feloid Families

It is not until the end of the early and the beginning of the middle Miocene (MN 4–6) that the feloid radiation in the Old World really accelerated. At that time we see the origin of hyaenids (Werdelin and Solounias 1991, in press) and the appearance of several primitive genera that are either viverrids or herpestids, such as *Semigenetta, Leptoplesictis, Viverrictis,* and others, as well as several new species of *Pseudaelurus.*

With the exception of the Hyaenidae (Werdelin and Solounias 1991), the interrelationships within these newly emerging feloid radiations are extremely poorly known. For the Herpestidae and Viverridae this lack is due mainly to the paucity of material. With the exception of *Herpestides,* the early and middle Miocene Eurasian members of these groups are known only from limited craniodental material (mostly lower dentitions) that does not include the characters used here and in Wyss and Flynn (1993) to define groups within the Feliformia. Further, the only thorough available study of viverrid interrelationships (Wozencraft 1984) deals mainly with characters not known in these fossils. Making the matter even more difficult is the fact that no comparable work on the Herpestidae is presently available. From what we currently know of African Miocene carnivorans (e.g., Savage 1965, 1978; Hendey 1974; Schmidt-Kittler 1987; M. G. Leakey et al., in prep.), indications are that these two families may always have found their greatest diversity on the African continent. At present they represent a difficult problem for phylogenetic studies, so far as the fossil record is concerned. There are, however, no indications of any special adaptations among the known Miocene representatives of viverrids/herpestids (undifferentiated) to make them exceptional in ecomorphological terms, and none appears to have been larger than the extant fossa (*Cryptoprocta ferox*) of Madagascar, which closely resembles early felids in skull and dental morphology.

During the later Orleanian and Astaracian (MN 4–8), more and more feliform taxa appeared in the Old World (Figure 17.12). The great majority of these forms were small, and none approached the sizes that feloids (hyaenids, percrocutids [Feloidea *incertae sedis*], and felids) were to attain in the late Miocene (MN 9–13) (Figure 17.13). The reasons for the lateness of this trend toward developing large-sized feloids are not clear, but the pattern parallels that of the Caniformia, which found large-sized forms appearing quite suddenly, relatively long after the origin of the Carnivora. Perhaps the reasons may be sought in the competition present at each time. The early carnivorans competed chiefly with creodonts, which apparently had something of a "head start" as far as size was concerned. For the Miocene feloids, there was competition from the earlier radiation of arctoid caniforms. In the early and middle Miocene a number of genera and species of amphicyonids and hemicyonine ursids were still present in the Old World, and, although not necessarily in direct competition with feloids, these may have retarded feloid diversification for a time. Both amphicyonids and hemicyonines declined in diversity during the early and middle Miocene and became extinct during the late Miocene, leaving only the agriotheriine ursids and the occasional large mustelid as competitors to large feloids.

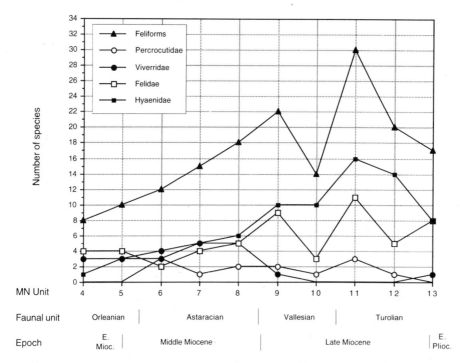

Figure 17.12. Diagram showing numbers of species in the feliform families, and total number of feliforms present in MN units from the late early to the terminal Miocene of western Eurasia. Note the general increase in numbers of species, especially of Hyaenidae and Felidae, until MN 11 and MN 12.

Whatever the reasons for the delay in their development, large-sized feloids appear in the early late Miocene (early Vallesian; MN 9). Among the Percrocutidae the gigantic *Dinocrocuta senyureki* appeared (Howell and Petter 1985), and in the Felidae we can trace the first records of several species referred to the genus *Machairodus,* which also embraced animals of large size. These developments were followed slightly later (late Vallesian; MN 10) by the appearance of the first bone-cracking hyaenid, *Adcrocuta eximia* (Werdelin and Solounias 1990). And as the late Miocene continued, the Percrocutidae, never common, became extinct, while the Hyaenidae and Felidae developed a multiplicity of forms. In the Hyaenidae these were mainly doglike, and some, such as *Lycyaena* and early *Chasmaporthetes,* even became somewhat felidlike. Among the Felidae, large sabertoothed forms such as *Machairodus, Paramachairodus,* and *Epimachairodus* dominated. Smaller, less extreme sabertoothed forms such as *Metailurus* (Figure 17.14) were also present, whereas true conical-toothed forms are extremely rare. If the true conical-toothed cats are defined as the monophyletic group composed of the common ancestor of all living Felidae and all of its descendants and diagnosed by the loss of m2 and

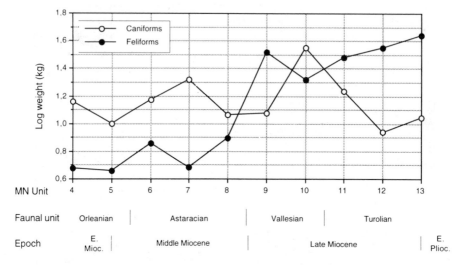

Figure 17.13. Diagram showing the increase in average mass of feliforms and caniforms during the time interval represented in Figure 17.11. Note that the average weight of feliforms increases dramatically between MN 8 and 9, whereas the average weight of caniforms fluctuates but shows no firm trends through the interval.

a round-to-oval upper canine cross-section, then the oldest of them is *"Felis" attica,* known from MN 10 onward (de Beaumont 1961).

In North America the pattern is analogous, although the taxa involved are slightly different. Felids made their first appearance in the Hemingfordian, but did not show increased diversity until the Barstovian, when several species of *Pseudaelurus* were present in the fauna (Ginsburg 1983). This is equivalent to the situation in Europe at the time, since the (middle and late) Barstovian is roughly equivalent to the Astaracian. No sabertooth forms seem to have been present in North America at this time, but in Europe, concurrently, the nimravid *Sansanosmilus* was occupying this niche. In the Clarendonian, as in the European equivalent Vallesian and lower Turolian, sabertooths were common, but in North America they were nimravids (Schultz et al. 1970), whereas in Europe this group was virtually extinct and had given place to felids (Werdelin, in press, a). Large felid sabertooths were not present in North America until the Hemphillian, a time that saw both long-sabered forms (*Machairodus*) and short-sabered forms analogous to (or possibly congeneric with) *Metailurus* (*Adelphailurus, Pratifelis:* Hibbard 1934). The development of large forms may have been hampered by competition with barbourofeline nimravids.

The Miocene feloid radiation can be summarized as follows. Viverrids and herpestids seem to have had the same ecomorphologies then as they do now, as small to medium-sized terrestrial and arboreal stalk-and-ambush predators incorporating a moderate amount of fruit and insects in their diet. Their radiation seems mainly to have been in Africa. The genus *Pseudaelurus,* the most

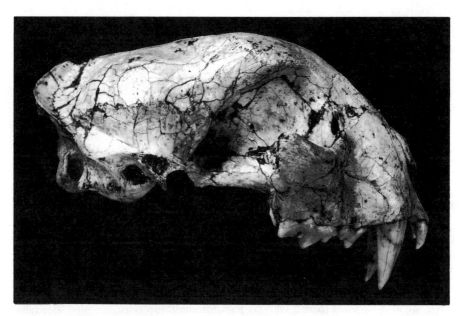

Figure 17.14. Skull of *Metailurus parvulus,* a late Miocene Eurasian felid with short sabers. Palaeontological Institute, University of Uppsala (PIU) M3895. ×²/₃.

common felid in the early and middle Miocene, includes wildcat- to lynx-sized forms that, ecomorphologically speaking, occupied the same niche as modern cats in the same size range, that is, terrestrial and arboreal stalk-and-ambush predators relying almost exclusively on meat. Just as today, there must have been considerable overlap between the lifestyles of viverrids and those of felids. Later in the Miocene, sabertoothed felids appeared and occupied the role of ambush-and-slash predator of large prey previously assumed by nimravids, while the smaller forms of *Pseudaelurus* were replaced by true conical-toothed cats directly related to the modern felid radiation. Unfortunately, the interrelationships of Miocene felids are not well enough known to allow us to say whether either the sabertoothed or the conical-toothed adaptations, or both, originated once or many times, although the suggestion has been made (de Beaumont 1978) that the genus *Pseudaelurus* is paraphyletic and gave rise to several independent radiations of sabertoothed forms.

Hyaenids are a different matter. One recent phylogenetic hypothesis (Werdelin and Solounias 1991) allows a distinction between adaptation and constraint in the ecomorphology of the group. This hypothesis is addressed in some detail elsewhere (Werdelin and Solounias, in press), and only a brief outline will be given here.

The earliest hyenas, from MN 4–5 (late Orleanian), were extremely similar in most respects to the coeval viverrids, suggesting similar adaptations. This interpretation is corroborated by the suggestion that at least some of the hyenas retained the retractile claws (Semenov 1989) that were subsequently lost in hyaenid evolution. In the Astaracian and Vallesian (MN 6–10), hyaenid diversity increased rapidly, reaching a peak in the Turolian (MN 11–13) (leaving aside the question of ghost lineages; cf. Flynn, this volume). The vast majority of these new taxa conformed to a doglike appearance, suggesting that their mode of life was similar to that of modern dogs (Werdelin and Solounias 1991; Werdelin, in press). Some may have been social, as are many canids today. This radiation into the terrestrial pursuit-carnivore/omnivore niche may have been made possible by the absence of canids from the Old World until the very end of the Miocene. Relatively late in this radiation, some forms evolved toward a more specialized pursuit-carnivore niche, but still with a mostly doglike morphology.

At the same time, the first scavenging hyaenids evolved. The first hyaenid with bone-cracking premolars known from the fossil record was the late Miocene (MN 10–13) *Adcrocuta eximia*. The development of P3/p3 into bone-cracking tools appears to have taken place only once in hyaenid evolution (*pace* Hunt and Tedford 1993), since the true bone-cracking/scavenging hyenas of the genera *Parahyaena, Hyaena, Pliocrocuta, Pachycrocuta, Adcrocuta,* and *Crocuta* form a well-corroborated monophyletic group (Werdelin and Solounias 1991).

The Miocene Canids

The canids of the Miocene belonged chiefly to the second of the major canid radiations, the Borophaginae (Munthe 1989). The earliest borophagines were small, relatively generalized forms such as *Leptocyon* and some small species of *Tomarctus* from the early Hemingfordian. Despite assertions by a number of early authors (e.g., Matthew 1930) that they were mainly scavenging bone-crushers (not bone-crackers), the Borophaginae radiated into a number of niches in the Miocene, including frugivory/omnivory (*Cynarctus, Carpocyon*), hunting large prey in open or closed terrain (*Aelurodon, Epicyon*), and true bone-cracking (*Osteoborus, Borophagus*) (Martin 1989; Munthe 1989). The bone-cracking adaptation was one of the last to appear in the group, and *Borophagus* was the only borophagine to survive into the Pliocene.

I have analyzed the bone-cracking adaptation of *Osteoborus* and *Borophagus* in relation to hyenas elsewhere (Werdelin 1989), and although the explanation for their morphology given there may or may not be successful, it is clear that there were drawbacks in the morphology of borophagine bone-crackers, insofar as the function of P4/p4 as bone-cracking teeth impinged on the function of the carnassials. This situation is unlike that in comparable hyaenids, where the bone-cracking function of P3/p3 leaves the carnassial function unimpeded. These drawbacks may explain why *Osteoborus* and *Borophagus,* unlike other bone-crackers, were relatively small forms within their family (cf. Munthe 1989, fig. 3). Other bone-crackers tended to be among the largest, if not the largest, species of their group (e.g., Hyaenidae, Percrocutidae, Oxyaenidae, Hyaenodontidae). *Osteoborus* and *Borophagus* nevertheless were North America's closest analogues to hyenas, and one of the most interesting natural experiments in convergent evolution.

The Arctoids of the Miocene

The arctoid radiation reached the peak of its diversity in the late Oligocene and early Miocene, when numerous species of mustelids, amphicyonids, and hemicyonine ursids appeared. Several of these groups became extinct or declined drastically in apparent diversity during the Miocene (Figure 17.15), and the caniforms of the late Miocene were taxonomically relatively impoverished, as compared with those groups present at the beginning of the period. The pattern of this decline is remarkably similar in the Old and New Worlds, despite the presence of competitors of separate ancestries—hyaenids (feliforms) in the Old World and canids (caniforms) in the New World.

As seen in Figure 17.15, observed amphicyonid diversity declined in a rela-

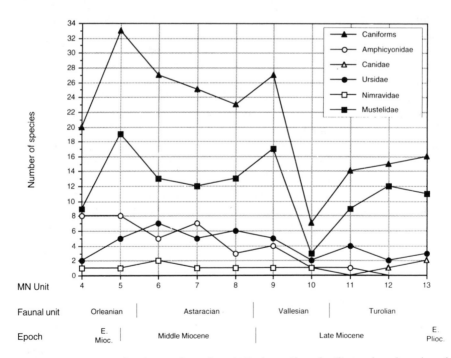

Figure 17.15. Diagram showing numbers of species in the caniform families, and total number of caniforms present in MN units from the late early to the terminal Miocene of western Eurasia. Note the decrease in numbers of species of amphicyonids through the time interval.

tively constant fashion through the middle and early late Miocene. In Europe, the first to go were archaic genera such as *Cynelos* and *Ysengrinia,* which enjoyed their greatest diversity in the Oligocene or early Miocene. More derived genera such as *Amphicyon, Thaumastocyon, Agnotherium,* and *Amphicyonopsis* persisted longer, but by the Turolian these were extinct as well. The youngest known European amphicyonid may be the apparently hypercarnivorous *Hubacyon pannonicus* (Kretzoi 1985), although the dating of this species is unclear. A pattern of marginalization is indicated, with later amphicyonids on the average having been larger and/or more ecologically specialized than were the earlier forms. Environmental changes in the Old World may have given other taxa a selective advantage in niches for smaller forms. Comparison of ecomorphological and phylogenetic patterns in the Amphicyonidae, integrated with patterns of environmental change, may help clarify this point in the future.

The same pattern seems to have occurred in North America (Savage and Russell 1983), although I know of no current taxonomic compilation that addresses the issue at the resolution of Figure 17.15. Amphicyonids are present until the Clarendonian, and are absent from the subsequent Hemphillian.

In terms of species diversity the arctoids of the Miocene are dominated, as today, by small mustelids. Unfortunately, there exist no good hypotheses concerning the interrelationships within this group. As far as can be determined, small mustelids of the Miocene conform to the basic ecomorphological types seen in modern faunas, as is reflected in the (probably erroneous) referral of certain species to the extant genera *Mustela* and *Martes.* Genera such as *Promeles* and *Parataxidea* are forerunners, ecomorphologically and perhaps phylogenetically, of modern badgers.

Just as in the mustelid radiation in the late Oligocene and early Miocene of North America (discussed above), the late Miocene of the Old World included forms with no modern counterpart in the Mustelidae. These were the large and very large specialized carnivores of the genera *Perunium* (Figure 17.16), *Eomellivora,* and *Hadrictis* (Zdansky 1924; Pia 1939; Orlov 1948; Zapfe 1948). The former two genera appear to be closely related, and the third may belong to the same group. The ecomorphological status of these forms has not been clear, but the find of a complete skeleton of a highly derived, leopard-sized

Figure 17.16. Skull of *Perunium ursogulo,* a late Miocene pantherine-like mustelid from the Ukraine. ×³/₁₀. (After Orlov 1948.)

species of *Perunium* from Kenya (M. G. Leakey et al., in prep) shows that at least this form, and by extension the others, would have occupied a large-body-size, stalk-and-ambush carnivore niche, much like that occupied today by pantherine cats. As we have seen, this was but one of many attempts by various caniform groups to evolve this ecomorphology. I shall return shortly to a consideration of the significance of this repeated pattern.

Brave New World: Pliocene and Pleistocene Carnivorans

Across the Miocene-Pliocene Boundary

In Europe, the events across the Turolian-Ruscinian boundary are the most momentous in carnivoran evolution since "La Grande Coupure" near the Eocene-Oligocene boundary (Werdelin and Turner, in press). Of 34 carnivoran species present in the mammalian fauna of MN13 (latest Turolian), only two survived into MN 14 (early Ruscinian): the hyaenid *Plioviverrops faventinus* and the canid *Nyctereutes donnezani* (Werdelin and Turner, in press). This is the greatest turnover between consecutive mammal zones seen in post-Eocene time. Not only did taxonomic composition change, but the eco-morphological makeup of Eurasian carnivoran faunas changed to what we see today. All hyenas occupying a cursorial-carnivore/omnivore niche became extinct (Werdelin and Solounias 1991) and were replaced by the descendants of canid immigrants from North America. The machairodont felid radiation of the late Miocene was greatly reduced, and the conical-toothed cats, both large and small, began a major radiation.

The reasons for this massive turnover are the subject of ongoing analysis (Werdelin and Turner, in press). It is tempting to relate the turnover to the Messinian salinity crisis at the end of the Miocene, but palynological analysis has shown that the effects of this event on the flora of the Mediterranean region were limited (Suc and Bessais 1990), leaving us no reason to believe that its effects on the fauna, and in particular the carnivoran fauna, would have been dramatic.

There was a similar turnover in North America, although its ecomorphological impact was a bit different, since no doglike hyenas were involved. Unfortunately, there is nothing in the data available at present that would allow us to determine whether this turnover was synchronous with that in Eurasia, whether it was earlier or later, or whether it was as rapid. Study of late Hemphillian and early Blancan faunas should resolve these questions.

New Beginnings: Reconstruction in the Plio-Pleistocene

Following the disappearances at the end of the Miocene, the carnivoran guild was slowly reconstructed. The new guild, however, differed in several respects from the old. The Old World hyenas were replaced by canids, the hyenas radiating instead into the scavenging/bone-cracking niche. The radiation of the third large subfamily of canids, the Caninae, is, in fact, one of the major features of the post-Miocene carnivoran fauna. The main characteristics of the Caninae that tend to set them apart from earlier canid radiations are the adaptations to cursoriality seen in many species. Some of the borophagines, such as *Epicyon,* may also have been cursorial to some degree (Munthe 1989), but the Caninae have taken this trait much further, to become open-terrain, small and medium-sized carnivore/omnivores. This is not to say that all Caninae are cursorial meat-eaters, for some are mainly insectivorous (e.g., *Otocyon*) or frugivorous (e.g., some species of *Vulpes*), and others show some convergence on felids in behavior and morphology (e.g., *Vulpes vulpes:* Macdonald 1987). Several Caninae are also pack hunters (e.g., *Canis lupus* and *Lycaon pictus*), a trait that may have been present in some Miocene hyaenids (Werdelin 1988b; Werdelin and Solounias 1991).

The second major feature of the post-Miocene carnivoran fauna was the radiation of pantherine cats. This radiation has been estimated on molecular grounds to have begun at approximately 2 Ma (Collier and O'Brien 1985), a date that may be too young, since fossils almost certainly referrable to *Panthera* are known from Laetoli, Tanzania (ca. 3.5 Ma: Barry 1987). The earliest pantherine taxa are very similar to modern pantherines, and the origins of the group are obscure. Perhaps they may be sought in forms very similar to "*Lynx*" *issiodorensis,* a species previously considered the ancestral *Lynx* (Kurtén 1978; Werdelin 1981), but one that may in fact be close to the common ancestor of *Lynx* and *Panthera.*

From its beginnings in Eurasia or Africa, *Panthera* spread rapidly to the rest of the world (except Australia), although numbering only a few, relatively eurytopic, species. For example, the lion at one time was present in most of Africa, Eurasia, and North America (Neff 1982), but records from South America previously thought to be of lion may pertain to large jaguars (Seymour, pers. comm. [1993]).

These two radiations were thus the most significant events in the carnivoran radiation after the Miocene. In terms of ecomorphology, the faunas of the Plio-Pleistocene were similar to those of today if we take into account the extinctions of the terminal Pleistocene, which included the last sabertooth felids.

The Geographical Perspective

Eurasia, North America, and Africa have been isolated from each other for much of the Tertiary and show interesting parallels and differences in the evolution of their carnivoran faunas. In my synthesis of Tertiary ecomorphological patterns within the Carnivora, I will present the three regions separately. Eurasia here implies mainly Europe, since the carnivoran fossil record for Asia is less well known, particularly for the Oligocene (B. Wang 1992).

Eurasia

The early Eocene European carnivore assemblages include a number of carnivorans (miacids and viverravids) occupying the small-carnivore niches in the trees and on the ground. Hyaenodontid creodonts such as *Proviverra* were slightly larger, but still unspecialized. A few oxyaenid creodonts, such as *Oxyaena*, were beginning to evolve into large-carnivore niches, which had been taken up by mesonychids such as *Dissacus* and *Pachyaena*—large to very large carnivore/scavengers with bone-crushing abilities.

As the Eocene progressed and Eurasia became separated from North America, endemic forms developed. Among the miacids the most notable is *Quercygale*, a highly carnivorous form, but the miacids remained small. Oxyaenids did not survive in the large-carnivore niche, and a radiation of hyaenodontids began to occupy this lifestyle, while the mesonychid *Dissacus* remained as a very large carnivore/scavenger. This state of affairs, with small miacids, medium- to large-sized hyaenodontids, and the very large mesonychid *Dissacus*, continued until the end of the Eocene. The exception was *Dissacus*, which became extinct and was replaced in the scavenger niche by hyaenodontids such as *Pterodon*.

In the late Eocene of Asia, which is well represented by Chinese and Mongolian finds, the situation was different. A few small miacids were present, along with some hyaenodontids, several of them, such as *Pterodon hyaenoides*, having adaptations indicative of scavenging. These forms were, however, accompanied by a number of other taxa of large to gigantic size, all with adaptations indicative of scavenging. These included the oxyaenid *Sarkastodon*, the mesonychids *Harpagolestes* and *Honanodon*, and the mesonychid (or arctocyonid) *Andrewsarchus*. What these forms could have eaten is not clear, although the radiation of brontotheres in Asia and North America (absent from Europe) may provide at least a partial answer. This was in any case the most extensive assemblage of scavenging carnivores seen in the Tertiary.

After "La Grande Coupure" the European carnivore fauna was quite different. The hyaenodontids were still present, but limited chiefly to taxa represented by large animals, especially *Hyaenodon* itself, a possible semicursorial large-prey specialist. The small-carnivore niches were now occupied by mustelids and stem-group feloids such as *Stenoplesictis* and *Paleoprionodon*. The ambush-and-slash niche was occupied for the first time, by nimravids; and some early ursids, such as *Cephalogale*, occupied an omnivore/carnivore niche. The early Oligocene was, however, notable as the beginning of the great diversity of amphicyonids, which radiated into a large number of niches, including omnivory, possible frugivory, durophagy, and pure carnivory.

This high diversity of amphicyonids continued into the early Miocene, although the number of taxa decreased somewhat. In the late early and middle Miocene, the amphicyonid radiation became more restricted in its taxonomic diversity, the majority of taxa belonging to the genus *Amphicyon*. This time also saw a radiation of hemicyonine ursids, some of relatively cursorial habits.

Mustelids were numerous in the small-carnivore niches, and the initial radiation of feloids had begun with the first viverrid-like hyaenas and the first felids to occupy the stalk-and-ambush niche (*Pseudaelurus*). Percrocutids occupied the scavenger/bone-cracker niche.

In the late Miocene, the amphicyonids and hemicyonines decreased rapidly in diversity and finally became extinct. This period was instead dominated by three groups of carnivorans: the mustelids as small carnivores, the hyaenas as omnivore/carnivore medium-sized forms (as well as the solitary bone-cracking scavenger, *Adcrocuta eximia*), and the felids as medium- to large-sized stalk-and-ambush carnivores. Most of the felids were sabertooths, but conical-toothed cats were beginning to make their appearance.

In the post-Miocene, the mustelids were still around, as were the felids. Canids moved into the omnivore/carnivore niche previously occupied by hyenas, while the latter moved into the scavenger/bone-cracker niche. This situation persisted, with some variations, until the megafaunal extinctions at or near the end of the Pleistocene.

North America

The carnivoran fauna of North America has been summarized recently by Hunt and Tedford (1993), and I refer the interested reader to their paper, and especially their fig. 5:15, for additional information.

The carnivores of the early Tertiary mammal faunas of North America were much the same as those of Europe, although the oxyaenid and mesonychid component was more diverse. By the late Eocene the oxyaenid element was gone, however, and the mesonychid one nearly so. Instead, there were radiations of miacids and viverravids in the small-carnivore niches, and hyaenodontids in the medium-sized ones. In the Bridgerian and Uintan the first true large-prey specialists to evolve among carnivores—the hyaenodont sabertooths *Machaeroides eothen* and *Apataelurus kayi*—appeared.

In the late Eocene and Oligocene the North American carnivore fauna is dominated by three elements: in the ambush-and-slash large-prey niche there were a number of genera of nimravids; the genus *Hyaenodon* was represented by a number of species of semicursorial, large-prey specialists; and the niches for generalized omnivore/carnivores were occupied by amphicyonids and hesperocyonine canids. For reasons that are not clear, this fauna appears impoverished relative to that of Europe at the same time (cf. Van Valkenburgh 1988), even if one considers the entelodont artiodactyls to have occupied the scavenger/bone-cracker niche. In part, this impoverishment may reflect taphonomic bias against small carnivores, but, even so, the amphicyonid radiation in North America during the Oligocene does not match the concurrent European radiation in terms of morphological diversity.

This situation changes drastically in the late Arikareean, when numerous amphicyonids, ursids, canids, and mustelids are found. The time between roughly 24 to 18 Ma was also the time of the "cat gap," when no felids or nimravids were present and a number of caniform taxa tried to occupy the stalk-and-ambush niche. These include mustelids such as *Megalictis* and *Aelurocyon*, as well as other taxa (Van Valkenburgh 1991).

In the later Miocene, mustelids occupied the small-carnivore niches, canids took the small-body and medium-sized-body omnivore/carnivore niches, and amphicyonids occupied niches similar to those of canids. The amphicyonids were probably more forest-dwelling (as they were in Europe), while canids were better adapted to open country. Felids and nimravids slowly migrated back into the ambush-and-slash niche, which in the late Miocene was dominated by barbourofeline nimravids.

Faunal change in the late Miocene and early Pliocene, though dramatic in taxonomic terms, did not have as profound an influence on ecomorphological patterns in North America as it did in Eurasia. Instead of the replacement of hyenas by canids in the omnivore/carnivore niches, we see the replacement of one canid group by another. We also see a relatively gradual appearance of modern taxa in other families, including the Felidae, which saw the conical-toothed-cat radiation build up steadily until the Pleistocene.

Africa

Africa has been notable by its absence from most of the presentation in this paper. This is so because the long isolation of Africa, from at least the late Eocene until the early Miocene, meant that carnivore evolution on the continent took a pathway entirely different from that of the Holarctic. During the Oligocene and most of the early Miocene, the carnivore niches were occupied by a wide array of hyaenodontid creodonts (Savage 1965, 1978), but in the early Miocene, carnivorans belonging to several families, including Amphicyonidae, Ursidae, Viverridae, and Felidae, appeared (Savage 1965; Schmidt-Kittler 1987). As noted by Schmidt-Kittler, the allocations of these taxa to particular niches were from the outset similar to those seen in Africa today, with one exception: the absence of canids. This group did not reach Africa until the early Pliocene. The later evolution of the larger carnivorans in Africa is described by Turner (1990), and requires no further comment here.

Conclusions

At the beginning of this chapter I emphasized the importance of integrating phylogeny and functional morphology in the study of ecomorphology. Phylogeny, however, has been relatively little in evidence throughout the text, which instead has been taken up mostly by narrative rather than testable hypotheses of ecomorphological patterns. What has come between the idea and the reality (T. S. Eliot 1925)? For the most part the lack of phylogenies hampers true understanding of ecomorphological patterns. The relationships among the higher groups of carnivorans have been much discussed in recent years (Flynn et al. 1988; Wayne et al. 1989; Wozencraft 1989; Hunt and Tedford 1993; Wyss and Flynn 1993), but there have been very few complete or even partial analyses of relationships of genera and species within families that incorporate fossil taxa (but see, for example, Baskin 1989; Werdelin and Solounias 1991; Cirot and de Bonis 1993; Wolsan 1993). This lack limits our understanding of the patterns that we know to exist but cannot fully analyze. This situation is improving, however, and with studies such as that of Wang (1994) on "Hesperocyoninae" the radiations of additional groups can be better understood. The dearth of within-family phylogenetic studies is the single most serious impediment to the study of carnivoran evolution in the Tertiary, but one that I hope will be removed within the next decade.

So what have we learned? I am in general agreement with the sentiments expressed by Martin (1989) regarding iterative evolution among carnivores, although I do not see this evolutionary pattern as a true impediment to understanding phylogeny. In a properly performed phylogenetic analysis, one that has been preceded by thorough character analysis, characters will "find their level of significance." This view implies that some suites of characters showing high homoplasy (e.g., dental characters) are generally useful at lower taxonomic levels, and others, with lower homoplasy (e.g. basicranial characters), are better used at higher levels, though this generalization does not always hold (Hunt and Tedford 1993). The most interesting aspect of iterative evolution, however, is that although ecomorphologies may evolve iteratively, they build on the constraints and morphology present in particular taxa. Thus, the stalk-and-ambush (or ambush-and-slash) niche has at one time or another been occupied by nimravids, felids, amphicyonids, and mustelids; the cursorial or semicursorial omnivore/carnivore niche has been occupied by amphicyonids, canids, and hyaenids; and the bone-cracker/scavenger niche has been occupied by oxyaenids, entelodontids, amphicyonids, percrocutids, canids, and hyaenids.

Certain elements of these niches can be further compared. The stalk-and-ambush niche is almost ubiquitously occupied throughout the Tertiary, but the taxa occupying it are few in number and rare in the fossil record. Some of the sabertooth taxa have long stratigraphic ranges, but the forms with normal canines tend to have very short ranges. This disparity suggests that the ecological circumstances favoring the latter ecomorphology may occur repeatedly, but are of short duration. The taxa occupying the cursorial or semicursorial omnivore/carnivore niche tend, by contrast, to be relatively common, in terms both of number of species and number of individuals. Studies of extant and fossil assemblages indicate that this is a highly competitive and highly structured niche (Dayan et al. 1992; Dayan and Simberloff, this volume; Werdelin, in press, b), and species longevity tends to be short. The scavenger/bone-cracker niche is generally, but not always, filled; some of the groups listed above have only limited adaptations for this niche, and seem to have survived in it mainly because of their large size. Except during the late Eocene of Asia, this niche supports few sympatric taxa, and these are generally not common at any given site. Many of them, particularly among the hyaenids, tend to be longer-lived and to have larger geographic ranges than do the taxa in the other niches mentioned above.

Many interesting questions remain, regarding not only the mechanics of iterative evolution, but also the constraints imposed on it by phylogeny. In order for research in this area to progress, it is, as mentioned above, necessary to have more and better phylogenies of carnivorans, especially within families. Without such phylogenies and a concomitant understanding of which features are primitive and which derived, in an organism, it will not be possible to place different ecomorphologies or functional traits in the adaptation/constraint, ecology/phylogeny spectrum. We also need some way of transforming narrative hypotheses, such as those posited here regarding the ecomorphological adaptations of certain taxa, into testable hypotheses. Still more complex is the prospect of generating predictions regarding evolutionary patterns within taxa on the basis of their ecomorphology. These are challenges that must be met in the next decade by more rigorous and structured analyses involving an interplay between phylogenies and quantitative morphological analysis. Though it is safe to say that we have a reasonable (though fragmented) idea about the whens and wheres of carnivoran ecomorphology, we still have only a tenuous grasp on the hows and whys.

Acknowledgments

This study has benefited greatly from discussions and cooperation with many scientists, both paleontologists and neontologists, over the years, only a few of whom can be mentioned here. Special thanks to J. J. Flynn, whose many constructive comments on the manuscript have significantly improved both content and presentation. I would also like to thank A. Biknevicius, M. Fortelius, V. Jaanusson, M. G. Leakey, J. S. Levinton, D. W. Macdonald, Yu. A. Semenov, N. Solounias, R. H. Tedford, A. Turner, B. Van Valkenburgh, M. Wolsan, W. C. Wozencraft, and many others for their assistance in various matters over the years. I particularly thank J. L. Gittleman for inviting me to contribute to this volume. My research is supported by the Swedish Natural Science Research Council.

References

Antonovics, J., and van Tienderen, P. H. 1991. Ontoecogenophyloconstraints? The chaos of constraint terminology. *Trends Ecol. Evol.* 6:166–168.

Archer, M. 1976. The dasyurid dentition and its relationships to that of didelphids, thylacinids, borhyaenids (Marsupicarnivora), and peramelids (Peramelina, Marsupialia). *Australian J. Zool., Supplement Series* 39:1–34.

Archer, M., ed. 1982. *Carnivorous Marsupials.* Sydney: Royal Society of New South Wales.

Archer, M., and Dawson, L. 1982. Revision of the marsupial lions of the genus *Thylacoleo* Gervais (Thylacoleonidae, Marsupialia) and thylacoleonid evolution in the late Cainozoic. In: M. Archer, ed. *Carnivorous Marsupials,* pp. 477–494. Sydney: Royal Zoological Society of New South Wales.

Archer, M., and Flannery, T. 1985. Revision of the extinct gigantic rat kangaroos (Potoroidae: Marsupialia), with description of a new Miocene genus and species and a Pleistocene species of *Propleopus. J. Paleontol.* 59:1331–1349.

Barry, J. C. 1987. The large carnivores from the Laetoli region of Tanzania. In: M. D. Leakey and J. M. Harris, eds. *Laetoli: A Pliocene Site in Northern Tanzania,* pp. 235–258. Oxford: Clarendon Press.

Barry, J. C. 1988. *Dissopsalis,* a middle and late Miocene proviverrine creodont (Mammalia) from Pakistan and Kenya. *J. Vert. Paleontol.* 8:25–45.

Baskin, J. A. 1982. Tertiary Procyoninae (Mammalia: Carnivora) of North America. *J. Vert. Paleontol.* 2:71–93.

Baskin, J. A. 1989. Comments on New World Tertiary Procyonidae (Mammalia: Carnivora). *J. Vert. Paleontol.* 9:110–117.

Beaumont, G. de. 1961. Recherches sur *Felis attica* Wagner du Pontien Eurasiatique avec quelques observations sur les genres *Pseudaelurus* Gervais et *Proailurus* Filhol. *Nouvelles Archives du Muséum d'Histoire Naturelle de Lyon* 6:17–45.

Beaumont, G. de. 1967. Observations sur les Herpestinae (Viverridae, Carnivora) de l'Oligocène supérieur avec quelques remarques sur des Hyaenidae du Néogène. *Arch. Sci., Genève* 20:79–108.

Beaumont, G. de. 1978. Notes complémentaires sur quelques félidés (carnivores). *Arch. Sci., Genève* 31:219–227.

Beaumont, G. de, and Weidmann, M. 1981. Un crâne de *Plesictis* (mammifère, carnivore) dans la molasse subalpine Oligocène fribourgeoise, Suisse. *Bull. Société Vaudoise des Sci. Nat.* 75:249–256.

Berggren, W. A., and Prothero, D. R. 1992. Eocene-Oligocene climatic and biotic evolution: An overview. In: D. R. Prothero and W. A. Berggren, eds. *Eocene-Oligocene Climatic and Biotic Evolution,* pp. 1–28. Princeton, N.J.: Princeton Univ. Press.

Bryant, H. N. 1991. Phylogenetic relationships and systematics of the Nimravidae (Carnivora). *J. Mammal.* 72:56–78.

Cirot, E., and Bonis, L. de. 1993. Le crâne d' *Amphictis ambiguus* (Carnivora, Mammalia): Son importance pour la compréhension de la phylogénie des mustéloïdes. *Comptes Rendus de l'Académie des Sciences, Paris* 316:1327–1333.

Collier, G. E., and O'Brien, S. J. 1985. A molecular phylogeny of the Felidae: Immunological distance. *Evolution* 39:473–487.

Cope, E. D. 1879. The relations of the horizons of extinct Vertebrata of Europe and America. *Bull. United States Geol. and Geogr. Survey of the Territories, F. V. Hayden, U.S. Geologist-in-Charge* 5:33–54.

Dahr, E. 1949. On the systematic position of *Phlaocyon leucosteus* Matthew and some related forms. *Arkiv för Zoologi* 41(11):1–15.

Dayan, T., Simberloff, D., Tchernov, E., and Yom-Tov, Y. 1992. Canine carnassials: Character displacement in the wolves, jackals, and foxes of Israel. *Biol. J. Linnean Soc.* 45:315–331.

Denison, R. H. 1938. The broad-skulled Pseudocreodi. *Annals New York Acad. Sci.* 37:163–257.

Donoghue, M. J., Doyle, J. A., Gauthier, J., Kluge, A., and Rowe, T. 1989. The importance of fossils in phylogeny reconstruction. *Ann. Rev. Ecol. and Syst.* 20:431–460.

Eliot, T. S. 1925. *The Hollow Men*. London: Faber & Faber.

Filhol, H. 1880a. Mémoire rélatif à quelques mammifères fossiles provenant des dépôts de phosphorites du Quercy. *Bull. Société des Sciences Phys. et Nat., Toulouse* 5:19–156.

Filhol, H. 1880b. Note sur les mammifères fossiles nouveaux provenant des phosphorites du Quercy. *Bull. Société Philomatique, Paris.* 7(3–4):120–126.

Filhol, H. 1882. *Mémoires sur quelques mammifères fossiles des Phosphorites du Quercy.* Toulouse: Imprimerie Vialette.

Flynn, J. J., and Galiano, H. 1982. Phylogeny of early Tertiary Carnivora, with a description of a new species of *Protictis* from the Middle Eocene of Northwestern Wyoming. *Amer. Mus. Nov.* 2725:1–64.

Flynn, J. J., Neff, N. A., and Tedford, R. H. 1988. Phylogeny of the Carnivora. In: M. J. Benton, ed. *The Phylogeny and Classification of the Tetrapods*, pp. 73–115. Oxford: Clarendon Press.

Gaudry, A. 1862–1867. *Animaux fossiles et géologie de l'Attique.* Paris.

Gauthier, J., Kluge, A. G., and Rowe, T. 1988. Amniote phylogeny and the importance of fossils. *Cladistics* 4:105–209.

Gebo, D. L., and Rose, K. D. 1993. Skeletal morphology and locomotor adaptation in *Prolimnocyon atavus*, an early Eocene hyaenodontid creodont. *J. Vert. Paleontol.* 13:125–144.

Ginsburg, L. 1961. La faune des carnivores Miocènes de Sansan (Gers). *Mémoires du Mus. Nat. d'Histoire Nat., Paris*, (n.s., sér. C) 9:1–190.

Ginsburg, L. 1966. Les amphicyons des phosphorites du Quercy. *Annales de Paléontologie (Vertébrés)* 52:23–64.

Ginsburg, L. 1979. Revision taxonomique des Nimravini (Carnivora, Felidae) de l'Oligocène des Phosphorites du Quercy. *Bull. Mus. Natl. d'Histoire Nat., Paris*, sér. 4 1:35–49.

Ginsburg, L. 1980. *Hyainailouros sulzeri*, mammifère créodonte du Miocène d'Europe. *Annales de Paléontologie (Vertébrés)* 66:19–73.

Ginsburg, L. 1983. Sur les modalités d'évolution du genre Néogène *Pseudaelurus* Gervais (Felidae, Carnivora, Mammalia). *Colloques Internationaux du CNRS.* 330:131–136.

Granger, W. 1938. A giant oxyaenid from the Upper Eocene of Mongolia. *Amer. Mus. Nov.* 969:1–5.

Greaves, W. S. 1983. A functional analysis of carnassial biting. *Biol. J. Linn. Soc., London* 20:353–363.

Hendey, Q. B. 1974. The late Cenozoic Carnivora of the south-western Cape province. *Annals South African Mus.* 63:1–369.

Hibbard, C. W. 1934. Two new genera of Felidae from the Middle Pliocene of Kansas. *Trans. Kansas Acad. Sci.* 37:239–255.

Hooker, J. J. 1992. British mammalian paleocommunities across the Eocene-Oligocene transition and their environmental implications. In: D. R. Prothero and W. A. Berggren, eds. *Eocene/Oligocene Climatic and Biotic Evolution*, pp. 494–515. Princeton, N.J.: Princeton Univ. Press.

Howell, F. C., and Petter, G. 1985. Comparative observations on some Middle and Upper Miocene hyaenids. Genera: *Percrocuta* Kretzoi, *Allohyaena* Kretzoi, *Adcrocuta* Kretzoi (Mammalia, Carnivora, Hyaenidae). *Geobios* 18:417–476.

Hunt, R. M., Jr. 1974a. The auditory bulla in Carnivora: An anatomical basis for reappraisal of carnivore evolution. *J. Morphology* 143:21–76.

Hunt, R. M., Jr. 1974b. *Daphoenictis*, a cat-like carnivore (Mammalia, Amphicyonidae) from the Oligocene of North America. *J. Paleontol.* 48:1031–1047.

Hunt, R. M., Jr. 1987. Evolution of the aeluroid Carnivora: Significance of the auditory structure in the nimravid cat *Dinictis*. *Amer. Mus. Nov.* 2886:1–74.

Hunt, R. M., Jr. 1989. Evolution of the aeluroid Carnivora: Significance of the ventral promontorial process of the petrosal, and the origin of basicranial patterns in the living families. *Amer. Mus. Nov.* 2930:1–32.

Hunt, R. M., Jr. 1991. Evolution of the aeluroid Carnivora: Viverrid affinities of the Miocene carnivoran *Herpestides*. *Amer. Mus. Nov.* 3023:1–34.

Hunt, R. M., Jr., and Joeckel, R. M. 1988. Mammalian biozones in nonmarine rocks of the North American continental interior: Biostratigraphic resolution within the "cat gap." *Geol. Soc. Amer. Abstracts with Programs, Rocky Mountain Section* 20:421.

Hunt, R. M., Jr., and Tedford, R. H. 1993. Phylogenetic relationships within the aeluroid Carnivora and implications of their temporal and geographic distribution. In: F. S. Szalay, M. J. Novacek, and M. C. McKenna, eds. *Mammal Phylogeny: Placentals*, pp. 53–73. New York: Springer-Verlag.

Janis, C. M. 1993. Tertiary mammal evolution in the context of changing climates, vegetation, and tectonic events. *Ann. Rev. Ecol. Syst.* 24:467–500.

Kirsch, J. A. W., and Archer, M. 1982. Polythetic cladistics, or, when parsimony's not enough: The relationships of carnivorous marsupials. In: M. Archer, ed. *Carnivorous Marsupials*, pp. 595–619. Sydney: Royal Zoological Society of New South Wales.

Kretzoi, M. 1985. Neuer Amphicyonide aus dem Altpannon von Pécs (Südungarn). *Annales Historico-Naturales, Musei Nationalis Hungarici* 77:65–68.

Kurtén, B. 1978. The lynx from Étouaires, *Lynx issiodorensis* (Croizet and Jobert), late Pliocene. *Annales Zoologici Fennici* 15:314–322.

Källersjö, M., Farris, J. S., Kluge, A. G., and Bult, C. 1992. Skewness and permutation. *Cladistics* 8:275–287.

Lange-Badré, B. 1979. Les créodontes (Mammalia) d'Europe occidentale de l'Éocène supérieur à l'Oligocène supérieur. *Mém. Mus. Nat. d'Histoire Nat.* (n.s., sér. C) 42:1–249.

Leakey, M. G., Feibel, C. S., Bernor, R. L., Cerling, T. E., Harris, J. M., Stewart, K. M., Storrs, G. W., Walker, A., Werdelin, L., and Winkler, A. J. Submitted. Lothagam: A record of faunal change in the late Miocene of East Africa. *J. Vert. Paleontol.*

Macdonald, D. 1987. *Running with the Fox*. London: Unwin Hyman.

MacLeod, N., and Rose, K. D. 1993. Inferring locomotor behavior in Paleogene mammals via eigenshape analysis. *Amer. J. Sci.* 293-A:300–355.

Marshall, L. G. 1978. Evolution of the Borhyaenidae, extinct South American predaceous marsupials. *Univ. California Publications in Geol. Sciences* 117:1–89.

Martin, L. D. 1980. Functional morphology and the evolution of cats. *Trans. Nebraska Acad. Sci.* 8:141–154.

Martin, L. D. 1989. Fossil history of the terrestrial Carnivora. In: J. L. Gittleman, ed. *Carnivore Behavior, Ecology, and Evolution*, pp. 536–568. Ithaca, N.Y.: Cornell Univ. Press.

Martin, L. D. 1991. A new miniature saber-toothed nimravid from the Oligocene of Nebraska. *Annales Zoologici Fennici* 28:341–348.

Matthew, W. D. 1899. A provisional classification of the fresh water Tertiary of the west: With lists of the mammals occurring in the formations. *Bull. Amer. Mus. Nat. Hist.* 12:19–75.

Matthew, W. D. 1907. A lower Miocene fauna from South Dakota. *Bull. Amer. Mus. Nat. Hist.* 23:169–219.

Matthew, W. D. 1909. The Carnivora and Insectivora of the Bridger Basin, Middle Eocene. *Mem. Amer. Mus. Nat. Hist.* 9:291–567.

Matthew, W. D. 1930. The phylogeny of dogs. *J. Mammal.* 11:117–138.

McKitrick, M. C. 1993. Phylogenetic constraint in evolutionary theory: Has it any explanatory power? *Ann. Rev. Ecol. Syst.* 24:307–330.

Mellett, J. S. 1977. Paleobiology of North American *Hyaenodon* (Mammalia, Creodonta). *Contrib. Vert. Evolution* 1:1–134.

Munthe, K. 1989. The skeleton of the Borophaginae (Carnivora, Canidae): Morphology and Function. *Univ. California Publ. Geol. Sciences* 133:1–115.

Neff, N. A. 1982. *The Big Cats: The Paintings of Guy Coheleach.* New York: Abrams.

Neff, N. A. 1983. The Basicranial Anatomy of the Nimravidae (Mammalia: Carnivora): Character Analyses and Phylogenetic Inferences. Ph.D. dissert., City University, New York.

Novacek, M. J. 1992. Fossils as critical data for phylogeny. In: M. J. Novacek and Q. D. Wheeler, eds. *Extinction and Phylogeny,* pp. 46–88. New York: Columbia Univ. Press.

Orlov, Y. A. 1948. *Perunium ursogulo* Orlov, a new gigantic extinct mustelid. (A contribution to the morphology of the skull and brain and to the phylogeny of Mustelidae). *Acta Zoologica* 29:63–105.

Peterson, O. A. 1906. The Miocene beds of western Nebraska and eastern Wyoming and the vertebrate faunae. *Ann. Carnegie Mus.* 4:21–72.

Peterson, O. A. 1909. A revision of the Entelodontidae. *Mem. Carnegie Mus.* 4:41–158.

Pia, J. 1939. Ein riesiger Honingsdachs (Mellivorine) aus dem Unterpliozän von Wien. *Annalen des Naturhistorischen Museums in Wien* 50:537–583.

Prothero, D. R., and Berggren, W. A., eds. 1992. *Eocene/Oligocene Climatic and Biotic Evolution.* Princeton, N.J.: Princeton Univ. Press.

Prothero, D. R., and Swisher, C. C., III. 1992. Magnetostratigraphy and geochronology of the terrestrial Eocene-Oligocene transition in North America. In: D. R. Prothero and W. A. Berggren, eds. *Eocene/Oligocene Climatic and Biotic Evolution,* pp. 46–73. Princeton, N.J.: Princeton Univ. Press.

Qi, T. 1975. An early Oligocene mammalian fauna of Ningxia. *Vert. Palasiatica* 13:217–224.

Quinet, G. E. 1966. Les mammifères du Landenien continental belge, Vol. 2. *Mém. de l'Institut Royale Belge des Sci. Nat.* 158:1–64.

Radinsky, L. 1982. Evolution of skull shape in carnivores, 3: The origin and early radiation of the modern carnivore families. *Paleobiology* 8:177–195.

Rich, T. H. V. 1971. Deltatheridia, Carnivora, and Condylarthra (Mammalia) of the Early Eocene, Paris Basin. *Univ. California Publ. Geol. Sci.* 88:1–72.

Riggs, E. S. 1945. Some early Miocene carnivores. *Field Mus. Nat. Hist., Geol. Series* 9:69–114.

Rose, K. D. 1981. The Clarkforkian land-mammal age and mammalian faunal composition across the Paleocene-Eocene boundary. *Papers on Paleontology* 26:1–189.

Rose, K. D. 1990. Postcranial skeletal remains and adaptations in early Eocene mammals from the Willwood Formation, Bighorn Basin, Wyoming. In: T. M. Bown and K. D. Rose, eds. *Dawn of the Age of Mammals in the Northern Part of the Rocky Mountain Interior, North America,* pp. 107–133. Boulder, Colo.: Geol. Soc. Amer., Special Paper 243.

Roth, J., and Wagner, A. 1854. Die fossilen Knochenüberreste von Pikermi in Griechenland. *Abhandl. bayerische Akad. von Wiss.* 7:371–464.

Sarich, V., Lowenstein, J. M., and Richardson, B. J. 1982. Phylogenetic relationships of *Thylacinus cynocephalus* (Marsupialia) as reflected in comparative serology. In: M. Archer, ed. *Carnivorous Marsupials,* pp. 707–709. Sydney: Royal Zoological Society of New South Wales.

Savage, D. E., and Russell, D. 1983. *Mammalian Paleofaunas of the World.* New York: Addison-Wesley.

Savage, R. J. G. 1965. Fossil mammals of Africa. 19: The Miocene Carnivora of East Africa. *Bull. Brit. Mus. (Nat. Hist.), Geol.* 10:241–316.

Savage, R. J. G. 1973. *Megistotherium*, gigantic hyaenodont from Miocene of Gebel Zelten, Libya. *Bull. Brit. Mus. (Nat. Hist.), Geol.* 22:485–511.

Savage, R. J. G. 1977. Evolution in carnivorous mammals. *Palaeontology* 20:237–271.

Savage, R. J. G. 1978. Carnivora. In: V. J. Maglio and H. B. S. Cooke, eds. *Evolution of African Mammals*, pp. 249–267. Cambridge: Harvard Univ. Press.

Schmidt-Kittler, N. 1981. Zur Stammesgeschichte des marderverwandten Raubtiergruppen (Musteloidea, Carnivora). *Eclogae Geologicae Helvetiae* 74:753–801.

Schmidt-Kittler, N. 1987. The Carnivora (Fissipedia) from the Lower Miocene of East Africa. *Palaeontographica* 197:85–126.

Schultz, C. B., Schultz, M. R., and Martin, L. D. 1970. A new tribe of saber-toothed cats (Barbourofelini) from the Pliocene of North America. *Bull. Univ. Nebraska State Mus.* 9:1–31.

Scott, W. B. 1941. The mammalian fauna of the White River Oligocene, 4: Artiodactyla. *Trans. Amer. Philos. Soc.* 28:363–746.

Scott, W. B., and Jepsen, G. L. 1936. The mammalian fauna of the White River Oligocene, 1: Insectivora and Carnivora. *Trans. Amer. Philos. Soc.* 28:1–153.

Semenov, Yu. A. 1989. *Iktiterii i morfologicheski skhodnye hieny neogena SSSR.* [Ictitheres and Morphologically Related Hyenas of the Neogene of the USSR.] Kiev: Naukova Dumka.

Simpson, G. G. 1931. A new classification of mammals. *Bull. Amer. Mus. Nat. Hist.* 59:259–293.

Simpson, G. G. 1945. The principles of classification and a classification of mammals. *Bull. Amer. Mus. Nat. Hist.* 85:1–350.

Springhorn, R. 1977. Revision der alttertiären europäischen Amphicyonidae (Carnivora, Mammalia). *Palaeontographica* 158:26–113.

Springhorn, R. 1980. Radiation der Hyaenodonta (Mammalia: Deltatheridia) unter gebisstrukturellen Gesichtspunkten und ein Vergleich mit fissipeden Carnivora. *Berichte der Naturforschenden Gesellschaft zu Freiburg im Breisgau* 70:97–109.

Stehlin, H. G. 1909. Remarques sur les faunules de mammifères des couches Éocènes et Oligocènes du Bassin de Paris. *Bull. Société Géologique de France* 9:488–520.

Steininger, F. F., Bernor, R. L., and Fahlbusch, V. 1989. European Neogene marine/continental chronologic correlations. In: E. H. Lindsay, V. Fahlbusch, and P. Mein, eds. *European Neogene Mammal Chronology*, pp. 15–46. New York: Plenum Press.

Stucky, R. K. 1992. Mammalian faunas in North America of Bridgerian to Early Arikareean "Ages" (Eocene and Oligocene). In: D. R. Prothero and W. A. Berggren, eds. *Eocene/Oligocene Climatic and Biotic Evolution*, pp. 464–493. Princeton, N.J.: Princeton Univ. Press.

Suc, J.-P., and Bessais, E. 1990. Pérennité d'un climat thermo-xerique en Sicile avant, pendant, après la crise de salinité messinienne. *C. R. Acad. Sci., Paris* 310:1701–1707.

Swofford, D. L. 1990. *PAUP: Phylogenetic Analysis Using Parsimony.* Champaign: Illinois Natural History Survey.

Szalay, F. S., and Gould, S. J. 1966. Asiatic Mesonychidae (Mammalia, Condylarthra). *Bull. Amer. Mus. Nat. Hist.* 132:127–174.

Tedford, R. H. 1976. Relationship of pinnipeds to other carnivores (Mammalia). *Syst. Zool.* 25:363–374.

Tedford, R. H. 1978. History of dogs and cats: A view from the fossil record. In: *Nutrition and Management of Dogs and Cats*, pp. 1–10. St. Louis: Ralston Purina Co.

Turner, A. 1990. The evolution of the guild of larger terrestrial carnivores during the Plio-Pleistocene in Africa. *Geobios* 23:349–368.

Van Valkenburgh, B. 1988. Trophic diversity in past and present guilds of large predatory mammals. *Paleobiology* 14:155–173.

Van Valkenburgh, B. 1991. Iterative evolution of hypercarnivory in canids (Mammalia: Carnivora): Evolutionary interactions among sympatric predators. *Paleobiology* 17:340–362.

Wang, B. 1992. The Chinese Oligocene: A preliminary review of mammalian localities and local faunas. In: D. R. Prothero and W. A. Berggren, eds. *Eocene-Oligocene Climatic and Biotic Evolution,* pp. 529–547. Princeton, N.J.: Princeton Univ.Press.

Wang, X. 1992. Phylogeny of Hesperocyoninae and origin of canids. *J. Vert. Paleontol.* 12(suppl.):58A.

Wang, X. 1994. Phylogenetic systematics of the Hesperocyoninae (Carnivora: Canidae). *Bull. Amer. Mus. Nat. Hist.* 221:1–207.

Wang, X., and Tedford, R. H. 1992. The status of genus *Nothocyon* Matthew, 1899 (Carnivora): An arctoid, not a canid. *J. Vert. Paleontol.* 12:223–229.

Wayne, R. K., Benveniste, R. E., Janczewski, D. N., and O'Brien, S. J. 1989. Molecular and biochemical evolution of the Carnivora. In: J. L. Gittleman, ed. *Carnivore Behavior, Ecology, and Evolution,* pp. 465–494. Ithaca, N.Y.: Cornell Univ. Press.

Wayne, R. K., Van Valkenburgh, B., and O'Brien, S. J. 1991. Molecular distance and divergence time in carnivores and primates. *Mol. Biol. Evol.* 8:297–319.

Werdelin, L. 1981. The evolution of lynxes. *Annales Zoologici Fennici* 18:37–71.

Werdelin, L. 1987. Jaw geometry and molar morphology in marsupial carnivores: Analysis of a constraint and its macroevolutionary consequences. *Paleobiology* 13:342–350.

Werdelin, L. 1988a. Circumventing a constraint: The case of *Thylacoleo* (Marsupialia: Thylacoleonidae). *Australian J. Zool.* 36:565–571.

Werdelin, L. 1988b. Studies of fossil hyaenids: The genera *Thalassictis* Gervais ex Nordmann, *Palhyaena* Gervais, *Hyaenictitherium* Kretzoi, *Lycyaena* Hensel, and *Palinhyaena* Qiu, Huang & Guo. *Zool. J. Linn. Soc., London* 92:211–265.

Werdelin, L. 1989. Constraint and adaptation in the bone-cracking canid *Osteoborus* (Mammalia: Canidae). *Paleobiology* 15:387–401.

Werdelin, L. 1993. Phylogenies, fossils, and evolutionary studies. *Quaternary Internatl.* 19:109–116.

Werdelin, L. In press, a. Carnivores, exclusive of Hyaenidae, from the later Miocene of Europe and western Asia. In: R. L. Bernor and V. Fahlbusch, eds. *Later Neogene European Biotic Evolution and Stratigraphic Correlation.* New York: Columbia Univ. Press.

Werdelin, L. In press, b. Community-wide character displacement in the lower carnassials of Late Miocene hyenas. *Lethaia.*

Werdelin, L., and Solounias, N. 1990. Studies of fossil hyaenids: The genus *Adcrocuta* and the interrelationships of some hyaenid taxa. *Zool. J. Linn. Soc., London* 98:363–386.

Werdelin, L., and Solounias, N. 1991. The Hyaenidae: Taxonomy, systematics, and evolution. *Fossils and Strata* 30:1–104.

Werdelin, L., and Solounias, N. In press. The evolutionary history of hyenas in Europe and western Asia during the Miocene. In: R. L. Bernor and V. Fahlbusch, eds. *Later Neogene European Biotic Evolution and Stratigraphic Correlation.* New York: Columbia Univ. Press.

Werdelin, L., and Turner, A. In press. Mio-Pliocene carnivore guilds of Eurasia. In: A. Nadachowski and L. Werdelin, eds. *Neogene and Quaternary Mammals of the Palaearctic: Papers in Honour of Professor Kazimierz Kowalski. Acta Zoologica Cracoviensia.*

Wolsan, M. 1993. Phylogeny and classification of early European Mustelida (Mammalia: Carnivora). *Acta Theriologica* 38:345–384.

Wozencraft, W. C. 1984. A phylogenetic appraisal of the Viverridae and its relationship to other Carnivora. Ph.D. dissert., University of Kansas, Lawrence.

Wozencraft, W. C. 1989. The phylogeny of the Recent Carnivora. In: J. L. Gittleman, ed. *Carnivore Behavior, Ecology, and Evolution,* pp. 495–535. Ithaca, N.Y.: Cornell Univ. Press.

Wyss, A. R., and Flynn, J. J. 1993. A phylogenetic analysis and definition of the Carnivora. In: F. S. Szalay, M. J. Novacek, and M. C. Mckenna, eds. *Mammal Phylogeny: Placentals,* pp. 32–52. New York: Springer-Verlag.

Zapfe, H. 1948. Neue Funde von Raubtieren aus dem Unterpliozän des Wiener Beckens. *Sitzungsberichte der österreichisches Akad. Wiss. mathematisch-naturwiss. klasse* 157:243–262.

Zdansky, O. 1924. Jungtertiäre Carnivoren Chinas. *Paleontologia sinica* (ser. C) 2(1):1–149.

Species and Subject Index

In the entries that follow, common and scientific names are indexed so as to equate each with the other. For example, "Kinkajou" refers the reader to "*Potos flavus,*" and "*Potos flavus*" is followed by "(kinkajou)." In each case, the complete citation is given under the species' scientific name. For species whose generic assignment is uncertain (e.g., *Puma concolor* or *Felis concolor*), all such assignments are listed, but just one referred to for complete index citations. Citations of material in figures and tables are designated by "f" and "t," respectively, following their page numbers. Text mentions of only passing interest are not indexed, and most fossil taxa (hundreds are mentioned in chapters 15 and 17) are not indexed.

The scientific and common names used both here and in the text agree, in the main, with Wozencraft, "Classification of the Recent Carnivora," an appendix in the 1989 predecessor to the present volume. As such, they represent both current usage and a consensus of the contributors to this volume.